Lecture Notes on Multidisciplinary Industrial Engineering

Series editor

J. Paulo Davim, Department of Mechanical Engineering, University of Aveiro, Aveiro, Portugal

"Lecture Notes on Multidisciplinary Industrial Engineering" publishes special volumes of conferences, workshops and symposia in interdisciplinary topics of interest. Disciplines such as materials science, nanosciences, sustainability science, management sciences, computational sciences, mechanical engineering, industrial engineering, manufacturing, mechatronics, electrical engineering, environmental and civil engineering, chemical engineering, systems engineering and biomedical engineering are covered. Selected and peer-reviewed papers from events in these fields can be considered for publication in this series.

More information about this series at http://www.springer.com/series/15734

Prasanta Sahoo · J. Paulo Davim
Editors

Advances in Materials, Mechanical and Industrial Engineering

Selected Contributions from the First International Conference on Mechanical Engineering, Jadavpur University, Kolkata, India

Springer

Editors
Prasanta Sahoo
Department of Mechanical Engineering
Jadavpur University
Kolkata, West Bengal, India

J. Paulo Davim
Department of Mechanical Engineering
University of Aveiro
Aveiro, Portugal

ISSN 2522-5022 ISSN 2522-5030 (electronic)
Lecture Notes on Multidisciplinary Industrial Engineering
ISBN 978-3-319-96967-1 ISBN 978-3-319-96968-8 (eBook)
https://doi.org/10.1007/978-3-319-96968-8

Library of Congress Control Number: 2018962772

This Springer imprint is published by the registered company Springer Nature Switzerland AG
The registered company address is: Gewerbestrasse 11, 6330 Cham, Switzerland

Preface

Mechanical engineering, possibly the oldest engineering branch, combines creativity and knowledge to carry out the difficult task of transforming an idea into reality. Mechanical engineering is a diverse subject that derives its breadth from the need to design and manufacture everything from small individual parts and devices to large systems. Since these skills are required for virtually everything that is built, mechanical engineering is perhaps the broadest and most diverse of engineering disciplines.

Mechanical engineers learn about materials, solid and fluid mechanics, thermodynamics, heat transfer, control, instrumentation, design, manufacturing and many other areas to understand mechanical systems. This diversity calls for a specialized knowledge, which divides mechanical engineering into various areas of specialization. The emerging trends in mechanical engineering involve interaction with other disciplines as well. The First International Conference on Mechanical Engineering (INCOM 2018) was held at Jadavpur University, Kolkata, India, during 4–6 January, 2018, which provided a platform for presenting original research work in diverse areas of mechanical engineering and interaction between mechanical engineers cutting across specializations. Apart from research articles, the conference provided platform for presenting innovative trends in mechanical engineering design, industrial practices and mechanical engineering education.

Mechanical Engineering Department of Jadavpur University, one of the oldest mechanical engineering departments in India dating back to 1909, organized INCOM 2018 which had sessions for contributed papers, plenary and keynote lectures, and panel discussion for preparing roadmaps for Mechanical Engineering Education and Research. INCOM 2018 was the first of a planned series of biennial conference in Mechanical Engineering to be developed into a major international event covering all aspects of Mechanical Engineering. This edition of the conference had 6 parallel sessions spread over 3 days consisting of 3 plenary lectures, 17 keynote lectures, 1 panel discussion and 202 contributed papers which have been selected from a pool of 258 submitted papers through peer review. The reviewers were chosen from acknowledged experts in their respective fields.

The book presents original, significant and visionary papers describing recent advances in materials, mechanical and industrial engineering. The book contains 34 chapters which are selected extended versions of peer-reviewed papers presented in the conference. The chapters contain a careful blend of fundamental and applied research in the frontiers of different areas of mechanical engineering including significant advances in materials engineering, design innovations and industrial engineering. This book will be useful to the students, teachers, research scholars, scientists and engineers working in the field of materials, mechanical and industrial engineering.

Chapter “Investigation and Optimization of Tribological Behavior of Electroless Ni–B Coating at Elevated Temperatures” presents friction and wear characteristics of heat-treated electroless Ni–B coating at different operating temperatures. Taguchi-based gray relational analysis is used to predict the optimal parametric setting to minimize wear loss and coefficient of friction of the coating sliding against hardened steel.

Chapter “Enhanced Thermal and Mechanical Performance of Functionalized Graphene Epoxy Nanocomposites: Effect Of Processing Conditions, Different Grades and Loading of Graphene” presents fabrication of uniform, homogenous and well-dispersed epoxy-based polyacroyl chloride functionalized grapheme nanocomposites. The role of incorporation of grapheme nanoplatelets at different weight percentage on the flexural properties of the nanocomposites is studied and improved mechanical properties are obtained.

Chapter “Behaviour Analysis and Comparison of Tribological Characteristics of Electroless Ni–B Coating under Dry and Lubricated Condition” presents deposition and tribological characterization of nickel–boron (Ni–B) coatings deposited by electroless technique on AISI 1040 steel specimens. Annealed coatings are tested on a pin-on-disc-type tribo-tester under dry and lubricated conditions at various loads, rotational speed and duration of sliding to evaluate friction and wear characteristics of the coatings

Chapter “Friction and Wear Characteristics of Heat Treated Electroless Ni–P–W Coatings Under Elevated Temperature” presents deposition of alkaline citrate-based Ni–P–W film on steel substrate. The role of elevated operating temperature on tribological behavior of heat-treated Ni–P–W coating is studied under different applied load and sliding velocity.

Chapter “Hardness, Friction and Wear Trends of Electroless Ni-W-P Coating Heat-Treated at Different Temperatures” presents wear performance of electroless Ni–W–P coatings undergoing different heat treatment temperatures. It evaluates the role of incorporation of tungsten to improve the tribological properties of Ni–P coatings when subjected to high-temperature heat treatment.

Chapter “Transition Reference Temperature for 20MnMoNi55 RPV Steel at Different Loading Rates” presents an adoption of Wallin’s methodology based on master curve and reference temperature to assess fracture toughness for 20MnMoNi55 low carbon steel. Effect of loading rate on tensile and fracture strength is studied in detail.

Chapter "Comparative Study of Cyclic Softening Modelling and Proposition of a Modification to 'MARQUIS' Approach" presents cyclic plastic response of SA333 steel subjected to uniaxial tension–compression cyclic loading. Predominant cyclic softening is observed with initially non-Masing plastic curvature. Three different softening models are approached with multi-segmented Ohno–Wang kinematic hardening rule in commercial finite element platform.

Chapter "Development and Finite Element Implementation of a Simple Constitutive Model to address Superelasticity and Hysteresis of Nitinol" presents a new constitutive model to replicate stress-induced phase transformation behavior of Nitinol. In addition to superelasticity, strain hardening and viscoplasticity are thoroughly explored and incorporated in the model.

Chapter "Abrasive Jet Machining: Drilling of Porcelain Tiles and Soda Lime Glass" presents machining performance of an indigenously built abrasive jet system on soda lime glass and porcelain tiles. The role of characteristics of nozzle, flow characteristics of the abrasive jet and workpiece material is investigated.

Chapter "Friction Stir Welding of Low-Carbon Steel" presents friction stir welding of 4-mm-thick low carbon steel at different welding conditions. Joints are characterized in respect of microstructure, tensile strength and hardness.

Chapter "Parametric Investigation of Various Electrolytes During Micro-electrochemical Texturing on Stainless Steel" presents maskless electrochemical micromachining method for generation of micro-surface textures. Friction tests are carried out to demonstrate the tribological effects of micro-circular patterned surfaces.

Chapter "Analysis of Casting Defects in Investment Casting by Simulation" presents the benefit of using casting simulation software in investment casting to analyze different defects like shrinkage porosity, cold shut, blow holes and hard zones. Different case studies are considered to evaluate the accuracy of the simulated models.

Chapter "Experimental Investigation of Fiber Laser Micro-Marking on Aluminum 6061 Alloy" presents laser marking on aluminum 6061 alloy using fiber laser. The role of influencing process parameters like power setting, duty cycle, pulse frequency, scanning speed and defocus height is studied.

Chapter "An Approach to Numerical Modeling of Temperature Field in Direct Metal Laser Sintering" presents the simulation of three-dimensional transient temperature field in the build part in direct metal laser sintering of AlSi10Mg alloy powder by using ANSYS platform. The simulations are carried out by considering radiation, convection and temperature-dependent thermo-physical properties of alloy powder.

Chapter "Parametric Optimization of TIG Welding Process on Mechanical Properties of 316L Stainless Steel Using RSM" presents tungsten inert gas (TIG) welding of 316L austenitic stainless steel material. The effects of process parameters such as welding current, welding speed and shielding gas flow rate on the weld joint quality are investigated.

Chapter "A Study on the Structural Behaviour of AFG Non-uniform Plates on Elastic Foundation: Static and Free Vibration Analysis" presents the effect of the stiffness value of the elastic foundation on large deflection and free vibration of axially functionally graded (AFG) non-uniform plate. The effects of the elastic foundations are presented through load vs. maximum deflection plot, deflected shape plot and backbone curves.

Chapter "Free Vibration of Rotating Twisted Composite Stiffened Plate" presents a first-order shear deformation-based finite element model to investigate the free vibration response of rotating twisted composite stiffened plate. Parametric studies are conducted to investigate the effect of angle of fibre orientation, pre-twist angle, stiffener depth to plate thickness ratio and rotational speeds on the fundamental frequency and mode shapes of the stiffened plate.

Chapter "Growth of Yield Front of Functionally Graded Non-uniform Bars under Thermal Load" presents the yield front propagation of functionally graded (FG) non-uniform bars subjected to thermal loads. Parametric study is attempted to reveal the nature of temperature field distribution, the effect of geometry parameters and material parameters on the elasto-plastic deformation of FG bars under thermal loading.

Chapter "Nonlinear Static Analysis and Superharmonic Influence on Nonlinear Forced Vibration of Timoshenko Beams" presents static analysis and subsequently superharmonic influence on the nonlinear dynamic behavior of externally excited thick beams. Effect of higher-order harmonics on the frequency–amplitude response is studied.

Chapter "Identification of Crack of Cantilever Beam Using Experimental Results and a Hybrid Neuro-GA Optimization Technique" presents dynamic behavior of free transverse vibration of an isotropic beam having single crack for cantilever boundary condition. It considers crack parameter prediction NDT technique using three optimization procedures to compare their suitability for structural health monitoring with practical significance.

Chapter "Experimental Investigation on Static and Free Vibration Behavior of Concentrically Stiffened Plates" presents experimental investigation and simulation study of different types of concentrically stiffened plates under static and dynamic conditions. Finite element simulation is performed through commercial software ANSYS Mechanical APDL 15.0 for independent validation and comparison of experimental data with an objective to establish the static deflection and natural frequency characteristics under loading.

Chapter "A Shaft Finite Element for Analysis of Viscoelastic Tapered and Hollow Tapered Rotors" presents the formulation of a finite element model for viscoelastic tapered rotor using tapered rotor elements. A comparative numerical study is performed for assessing rotor stability and frequency response using both tapered, hollow tapered rotor element and stepped cylindrical rotor element.

Chapter "Numerical Analysis of Heat Transfer and Entropy Generation for Natural Convection in a Quadrantal Cavity with Non-uniform Heating at the Bottom Wall" presents the buoyancy-induced fluid flow, thermal transport and irreversibility characteristics in a quadrantal cavity using numerical analysis. The irreversibility characteristics are presented in the form of distribution of local entropy generation due to heat transfer attributes and fluid friction attributes within the enclosure and the average Bejan number.

Chapter “Impact of Magnetic Field on Thermal Convection in a Linearly Heated Porous Cavity” presents combined effect of magnetic field and buoyancy on the thermo-fluid flow in a porous cavity considering top cold, bottom insulated and sidewalls linearly heated. The temperature and flow fields are analyzed using isotherms and streamlines, whereas the visualization of convective heat flow is presented using heatlines.

Chapter “Effect of Turbulence on Stability of Journal Bearing with Micropolar Lubrication: Linear and Non-linear Analysis” presents the effect of turbulence on the stability of finite hydrodynamic journal bearing lubricated with micropolar fluid. Both linear and nonlinear stability analyses are carried out.

Chapter “MHD Convection with Heat Generation in a Porous Cavity” presents the magnetic field affected heat generation/absorption undergoing natural convection in a differentially heated cavity packed with porous media. The flow through the porous medium is analyzed using Brinkman–Forchheimer–Darcy model (BFDM) and examined graphically using streamlines, isotherms and average Nusselt number plots.

Chapter “Studies on Fluid Flow Through an Elliptical Microchannel of Different Aspect Ratios” presents fluid flow analysis within a straight microchannel of elliptical cross section. Variational method with slip boundary conditions is adopted to determine velocity distribution. The velocity profile, Poiseuille number and slip coefficient are obtained as a function of aspect ratio and Knudsen number.

Chapter “Stresses and Deformation in Rotating Disk During Over-Speed” presents closed-form solution to predict permanent residual growth in rotating disk with variable thickness for linearly strain hardening material behavior using Tresca’s yield criteria and its associated flow rule. Results obtained using analytical solutions are compared with finite element method and experimental tests results for uniform thickness disks.

Chapter “Flow and Heat Transfer Characteristics of Surface-Mounted Cylinder in Presence of Rectangular Winglet Pair” presents combined effect of finite height circular cylinder and rectangular winglet pair (RWP) mounted on horizontal plate has been studied using three-dimensional computations. Effect of different locations of RWP corresponding to the cylinder center on heat and fluid flow characteristics is considered.

Chapter “GA Optimization of Cooling Rate of a Heated MS Plate in a Laboratory-Scale ROT” presents a laboratory-scale run-out table (ROT) experimentation to determine the initial set of data for genetic algorithm (GA) to maximize the cooling rate of a hot mild steel plate subject to variation in the air flow rate and the upper and lower nozzle bank distances.

Chapter “Experimental and Numerical Study of Velocity Profile of Air over an Aerofoil in a Free Wind Stream in Wind Tunnel” presents a numerical simulation to obtain the velocity profile for a non-standard aerofoil over five angles of attack including negative ones which are validated against the data obtained from wind tunnel experimentation for conditions compatible with the simulation.

Chapter "Numerical Simulation of Vortex Shedding from a Cylindrical Bluff-Body Flame Stabilizer" presents the qualitative aspects of the vortex shedding phenomena in two-dimensional, laminar transient flow past a cylindrical bluff body, with methane injection at cross-flow arrangement. The methane mass fraction and the injection velocity of methane injected from the slotted cylinder are altered simultaneously to investigate their effects on the combustion, flame characteristics and fluid mechanics.

Chapter "Entrepreneurial Culture-Driven Improvement of Technical Facets for Product Quality and Customer Satisfaction" presents an empirical study to accumulate primary data from different Indian manufacturing companies and identifies the role of entrepreneurial culture to promote the technical improvements to accelerate the new product development (NPD) success by developing quality products as per customer demand.

Chapter "Managerial Support in R&D Operations and Practices for Realizing Technological developments" presents an empirical study to identify the influential role of top management to control the R&D activities in optimal way to develop new products with technological advancement. Structural equation modeling (SEM) approach is used to develop the structural model using IBM SPSS AMOS 21.0 software package.

Kolkata, India — Prasanta Sahoo
Aveiro, Portugal — J. Paulo Davim

Acknowledgements

We are thankful to Prof. Dipankar Sanyal, the chairman of INCOM 2018, for allowing us to publish the book from extended versions of selected papers of the conference. We are also thankful to the organizing committee of INCOM 2018 for their motivation, encouragement, advice and support provided during organization of a very successful conference that led to the publication of this book. We sincerely thank all the authors for contributing their valuable research work as extended chapters for this book. The authors submitted the chapters and responded for the revisions on time which could make it possible to publish this book on time. We are thankful to all the reviewers who painstakingly reviewed all the papers and the revisions, wherever needed. We sincerely thank the organizers and the Advisory Committee members of INCOM 2018 for their guidance and support. We are grateful to Prof. Ishwar K Puri, McMaster University and Prof. Mustafizur Rahman, National University of Singapore for their encouragement and suggestions.

Last but not least, it is a nice experience to publish with Springer. We would like to express our sincere gratitude to the staff members of Springer, Dr. Mayra Castro, Ms. Petra Jantzen, Ms. Amudha Vijayarangan and all other Springer Editors and Staff, all of whom did an excellent job in getting this book published. This book would not be possible without their kind support from many aspects during the publication process.

Contents

About the Editors

Prasanta Sahoo Prasanta Sahoo received a Ph.D. degree in Mechanical Engineering in 2002 and M.Tech. degree in Mechanical Engineering (Machine Design) in 1994 both from Indian Institute of Technology, Kharagpur, Bachelor of Mechanical Engineering degree in 1991, from Jadavpur University, Kolkata. Currently, he is Professor at the Department of Mechanical Engineering of Jadavpur University, Kolkata, India. He has more than 25 years of teaching and research experience in Manufacturing, Materials and Mechanical Engineering, with special emphasis in Tribology of Materials. He has also interest in structural mechanics, contact mechanics, fractal modeling, composites and coatings. He has guided large numbers of Ph.D. and master's students as well as coordinated and participated in several financed research projects. He has worked as evaluator of projects for international research agencies as well as examiner of Ph.D. thesis for many universities and institutes. He is on the editorial board of several international journals and conferences. Presently, he acts as reviewer for more than 40 prestigious Web of Science journals. In addition, he has also authored 2 books, over 20 chapters and 400 articles in peer-reviewed journals and conference proceedings (h-index 25+/ 3000+ citations).

J. Paulo Davim J. Paulo Davim received a Ph.D. degree in Mechanical Engineering in 1997, M.Sc. degree in Mechanical Engineering (materials and manufacturing processes) in 1991, Mechanical Engineering degree (5 years) in 1986, from the University of Porto (FEUP), the Aggregate title (Full Habilitation) from the University of Coimbra in 2005 and the D.Sc. from London Metropolitan University in 2013. He is Eur Ing by FEANI-Brussels and Senior Chartered Engineer by the Portuguese Institution of Engineers with an MBA and Specialist title in Engineering and Industrial Management. Currently, he is Professor at the Department of Mechanical Engineering of the University of Aveiro, Portugal. He has more than 30 years of teaching and research experience in manufacturing, materials, mechanical and industrial engineering, with special emphasis in machining and tribology. He has also interest in management, engineering education and higher education for sustainability. He has guided large numbers of postdoc, Ph.D. and master's students as well as coordinated and participated in several financed research projects. He has received several scientific awards. He has worked as evaluator of projects for international research agencies as well as examiner of Ph.D. thesis for many universities. He is Editor in Chief of several international journals, Guest Editor of journals, books Editor, book Series Editor and Scientific Advisory for many international journals and conferences. Presently, he is Editorial Board member of 25 international journals and acts as reviewer for more than 80 prestigious Web of Science journals. In addition, he has also published as editor (and co-editor) more than 100 books and as author (and co-author) more than 10 books, 80 chapters and 400 articles in journals and conferences (more than 200 articles in journals indexed in Web of Science core collection/h-index 45+/6000+ citations and SCOPUS/h-index 53+/8500+ citations).

Chapter 1
Investigation and Optimization of Tribological Behavior of Electroless Ni–B Coating at Elevated Temperatures

Arkadeb Mukhopadhyay, Tapan Kumar Barman and Prasanta Sahoo

1.1 Introduction

Electroless nickel boron alloys have high hardness and excellent tribological characteristics though the corrosion resistance is moderate [1]. Due to this, the deposits are being widely investigated in tribological contact applications working in demanding environments. The inherent self-lubricating morphology of the coatings results in a reduction in the actual contact area and consequently a low COF and wear [2–4]. In electroless deposition, there is no requirement of electricity. Hence, they have the advantage of being deposited over different varieties of substrates with suitable pretreatments and activation [5–8].

The microhardness, tribological behavior, and corrosion resistance of the coatings may be suitably controlled by tailoring the bath properties. The important bath parameters include content of nickel chloride, sodium borohydride, and bath temperature. Optimum electroless bath composition for enhanced wear depth and COF has been reported by Das and Sahoo [9] using grey relational analysis (GRA). In the same study, 85 °C bath temperature, 0.6 g/L sodium borohydride, and 20 g/L nickel chloride have been suggested to improve the tribological characteristics of Ni–B coatings [9]. An improvement of 19% in the microhardness of Ni–B coatings could be achieved by controlling the bath parameters according to Taguchi's experimental design and optimization technique [10]. Heat treatment also leads to an increase in hardness of Ni–B coatings and a resultant improvement in friction and wear characteristics [11, 12]. Heat treating Ni–B coatings lead to precipitation of crystalline Ni and its hard boride phases [13]. Therefore, the hard coatings present a virtually incompatible surface to the counterface material which results in

A. Mukhopadhyay (✉) · T. K. Barman · P. Sahoo
Department of Mechanical Engineering, Jadavpur University, Kolkata 700032, India
e-mail: arkadebjume@gmail.com

P. Sahoo and J. P. Davim (eds.), *Advances in Materials, Mechanical and Industrial Engineering*, Lecture Notes on Multidisciplinary Industrial Engineering,
https://doi.org/10.1007/978-3-319-96968-8_1

high wear resistance. In fact, vacuum heat treatment results in a chromium equivalent high hardness [14]. Therefore, electroless Ni–B coatings may prove to be an environmentally friendly replacement to chromium.

The tribological test parameters also influence the wear behavior and COF of Ni–B coatings [11, 15]. Krishnaveni et al. [11] observed an increase in specific wear rate and COF with an increase in applied normal load, while Madah et al. [15] reported a decrease in specific wear rate of as-deposited Ni–B coatings with an increase in normal load. Furthermore, an increase in applied normal load also leads to a decrease in the 'endurance life' (effective sliding distance before failure of coatings) [15]. A change in dominant wear mechanism also takes place at different applied normal loads and sliding distances [15]. In general, as-deposited Ni–B coatings exhibit adhesive wear phenomenon [16] due to high mutual solubility of iron and nickel. Heat-treated Ni–B coatings exhibit abrasive wear with micro-grooves and scratches along the direction of sliding [11]. The deposition of composite electroless Ni–B–TiO_2 [17], Ni–B–Al_2O_3 [18], Ni–B–Si_3N_4 [19], Ni–B–B_4C [20], etc., coatings with high hardness and wear resistance has been carried out keeping in mind the demanding industrial challenges and working conditions.

Apart from the deposition of nanoparticles along with Ni–B coatings, the coating characteristics may be also enhanced by alloying with W or Mo [21, 22]. Electroless Ni–B–W coatings exhibit higher hardness compared to the binary Ni–B variant [23]. The oxidation resistance also improves by the inclusion of W in Ni–B coatings [24]. Ni–B–Mo films exhibit higher thermal stability [22, 25]. Due to the high thermal stability and oxidation resistance achieved by inclusion of Mo or W, recent investigations are being carried out to evaluate tribological characteristics of the deposits at elevated temperatures [23, 25–27]. High-temperature tribological behavior of Ni–B-based coatings is characterized by the formation of lubricious oxide patches and microstructural changes due to the in situ heat treatment effect [23, 25–27].

From the ongoing discussion, it is noticed that the friction and wear characteristics of Ni–B-based coatings have received widespread attention from researchers. Recent studies have also reported the high-temperature tribological performance of Ni–B-based coatings. The investigation and optimization of Ni–B coatings at high temperatures have been scarcely investigated. The present work therefore aims to carry out a systematic study and optimize the tribological behavior of Ni–B coatings considering operating temperature as a process parameter along with normal load and sliding speed. Wear (measured from mass loss) and COF are the responses under investigation. Taguchi's experimental design technique is coupled with grey relational analysis for the optimization of wear and COF of the coatings simultaneously. Grey relational analysis has been used in the present work due to its proven capabilities in optimizing multivariate and multi-objective optimization problems. Analysis of variance is carried out to find the significance of process parameters in influencing the wear and COF of the coatings. Wear mechanism at elevated temperatures is also studied.

1.2 Grey Relational Analysis

The use of grey relational analysis (GRA) has been suggested by various researchers for the optimization of a system with multiple responses due to its simplicity. In fact, a significant improvement in tribological characteristics and microhardness could be achieved by the use of GRA coupled with Taguchi's experimental design technique [9, 10]. Taguchi proposed the use of orthogonal arrays (OA) to reduce the number of experiments without compromising on the quality of a system which would ultimately lead to minimization of cost. The GRA or the grey system theory was first introduced by Deng [28]. According to the grey theory, a system with no information is black while a system with complete information is white. A system with incomplete information is termed as grey. It would be interesting to note that the relationship between the tribological test parameters and the tribological characteristics is vague or uncertain. Therefore, this problem may be efficiently dealt by using GRA. The steps followed for the optimization using Taguchi's experimental design and GRA may be summarized as follows:

(a) Selection of OA based on degrees of freedom (DOF) of the experiment.
(b) Performing experiments based on the combinations laid down in OA for process parameters.
(c) Normalizing the responses based on higher-the-better, lower-the-better, and nominal-the-best characteristics. In the present work, since wear and COF are to be minimized, they are normalized based on lower-the-better characteristics which is given as follows:

$$x_i^*(k) = \frac{\max x_i(k) - x_i(k)}{\max x_i(k) - \min x_i(k)} \tag{1.1}$$

where $x_i^*(k)$ is the value of response after normalization, $\max x_i(k)$ and $\min x_i$ are the largest and smallest values of $x_i(k)$ for the *kth* response.

(d) After normalization of responses, grey relational coefficient (GRC) is calculated. The GRC gives a comparison between the ideal (best = 1) and the sequences obtained experimentally, i.e., deviation from the ideal. It may be computed as:

$$\xi_i(k) = \frac{\Delta_{\min} + r\Delta_{\max}}{\Delta_{0i}(k) + r\Delta_{\max}} \tag{1.2}$$

where $\Delta_{0i} = \left\| x_0(k) - x_i^*(k) \right\|$ = difference of the absolute value between $x_0(k)$ and $x_i^*(k)$. From the absolute value of differences, $\Delta_{\max}$ and $\Delta_{\min}$ are the corresponding maximum and minimum values, respectively. Furthermore, the term '*r*' is the distinguishing coefficient. This controls the effect of $\Delta_{\max}$ when it becomes too large. Generally, the value of '*r*' used is 0.5 since it gives stable outcomes and has moderate distinguishing ability [9].

(e) The next step is the calculation of the multiple performance index, i.e., grey relational grade (GRG). Here, a single response is obtained from the multiple responses, i.e., the GRG. The grade may be evaluated by computing average of GRCs of each sequence. After the calculation of GRG, optimal process parameters are obtained from main effects plot.

(f) Finally, a confirmation or validation test is carried out. This is done to find out if the application of GRA has resulted in improved performance at the optimum parametric combination compared to any initial trial run. If the system exhibits improvement in GRG compared to the initial test run, then the optimization may prove to be beneficial. Moreover, the GRG obtained at optimal condition by conducting experiments should be also in accordance with the predicted grade. The predicted grade may be calculated as follows:

$$\hat{\gamma} = \gamma_m + \sum_{i=1}^{n} (\bar{\gamma}_i - \gamma_m) \tag{1.3}$$

where γ_m is the mean GRG, $\bar{\gamma}_i$ is the mean grade at optimal level, and '*n*' is the number of process parameters considered to affect the system performance.

1.3 Experimental Details

Ni–B films are deposited on AISI 1040 steel. The dimensions are 6 mm diameter and 30 mm length, referred to as 'pin' specimens. Prior to coating deposition, they are cleaned thoroughly using deionized water. Degreasing is carried out in acetone. The specimens are dipped into HCl and given a pickling treatment to remove any oxide products and again rinsed in deionized water. Finally, the specimens are dipped into the electroless bath. A schematic diagram of electroless deposition setup is shown in Fig. 1.1. The electroless Ni–B bath is a sodium borohydride-reduced alkaline EN bath with nickel chloride as the source of Ni ions. The detailed description of coating deposition may be found in the work carried out by Mukhopadhyay et al. [12]. Chemical composition of the coating bath is given in Table 1.1. Post-deposition, the specimens are rinsed, dried in air, and heat treated at 350 °C for 1 h in a muffle furnace. After heat treatment, the coated specimens are allowed to cool down to ambient temperature in the furnace itself. On heat treating at 350 °C for 1 h, electroless Ni–B coatings exhibit enhanced tribological behavior at high temperatures [12], while heat treating at 400 or 450 °C along with high test temperature may lead to an aging effect and grain coarsening [12]. Therefore, from the literature review 350 °C heat treatment temperature is selected.

Composition of the coatings is determined using energy-dispersive X-ray (EDX) analysis (Make: EDAX CORPORATION). Surface morphology of the coatings is observed using scanning electron microscope (SEM) (Make: FEI QUANTA FEG 250). Phase transformation analysis is carried out using X-ray

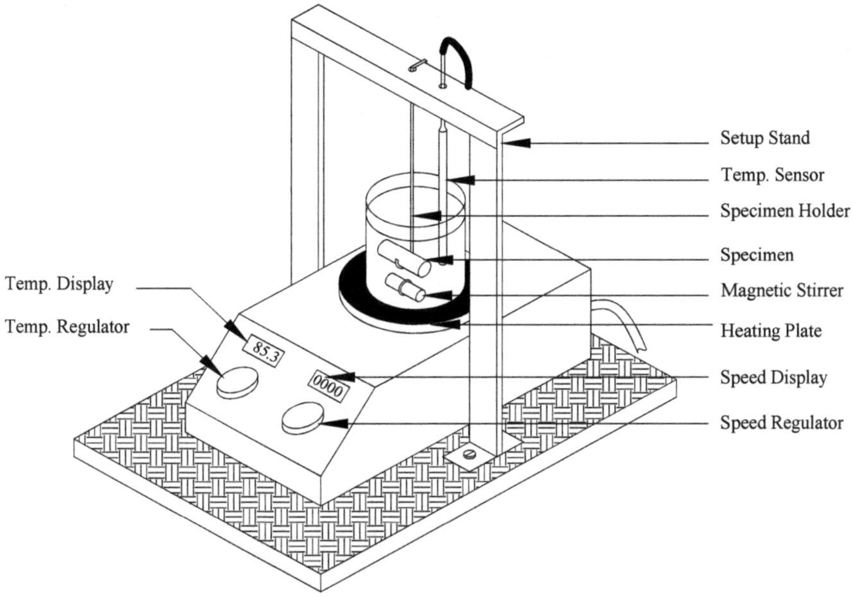

Fig. 1.1 Schematic diagram of coating deposition setup

diffraction (XRD) technique (Make: Rigaku Ultima III). Vicker's microhardness is determined at 100 gf load, 25 μm/s approach velocity, and 15 s dwell time (Make: UHL). Microhardness measurement setup is shown in Fig. 1.2.

Tribological tests are carried out on a pin-on-disk-type tribometer (Make: Ducom India, TR-20-M56). Figure 1.3 illustrates the high-temperature tribological test setup used in the present work. The counterface disk is of hardened EN 31 specification. Heat-treated Ni–B-coated specimens are held stationary against the rotating disk. Process parameters under consideration are operating temperature, applied normal load, and rotational speed of the counterface disk. The process parameters with their levels are enlisted in Table 1.2. Track diameter and sliding duration are kept fixed at 80 mm and 5 min, respectively. Since the present case considers three factors at three levels, experiments are carried out based on Taguchi's L_{27} orthogonal array (OA) so that individual and interaction effects may be determined. Mid-level combination of test parameters is the initial test condition. The COF of the deposits is displayed in real time on a computer attached to the test setup, while wear is obtained from mass loss after each experiment using a precision weighing balance. COF and wear are the responses considered in the present investigation. Wear mechanism of the coatings is ascertained from SEM micrographs of worn specimens.

Table 1.1 Electroless bath composition and deposition conditions [12]

Parameter	Value
Bath composition	
Nickel chloride	20 g/L
Sodium borohydride	0.8 g/L
Ethylenediamine	59 g/L
Lead nitrate	0.0145 g/L
Sodium hydroxide	40 g/L
Operating conditions	
Deposition time	4 h
Bath temperature	90 ± 2 °C
Bath pH	12.5
Bath volume	200 ml

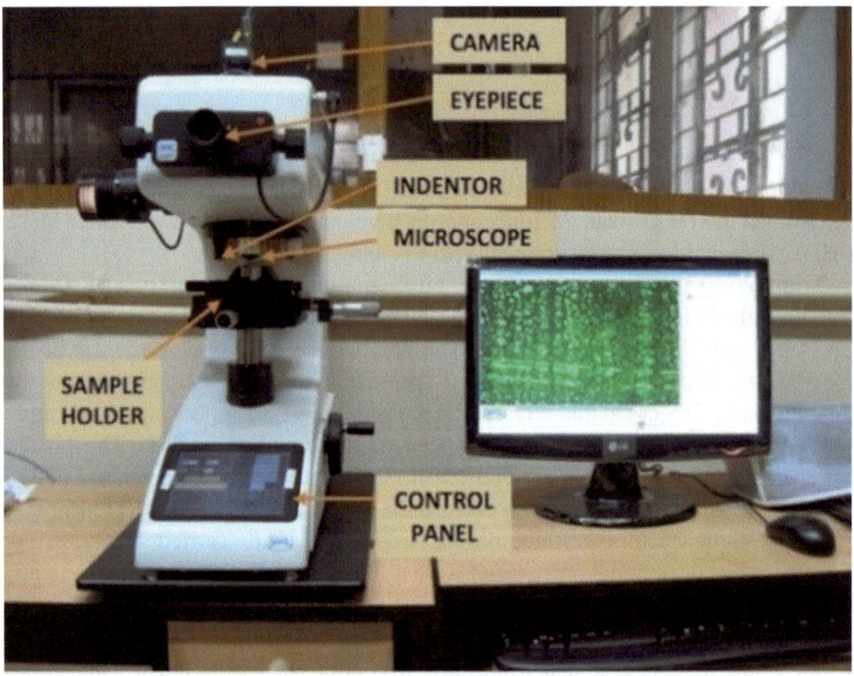

Fig. 1.2 Microhardness measurement instrument

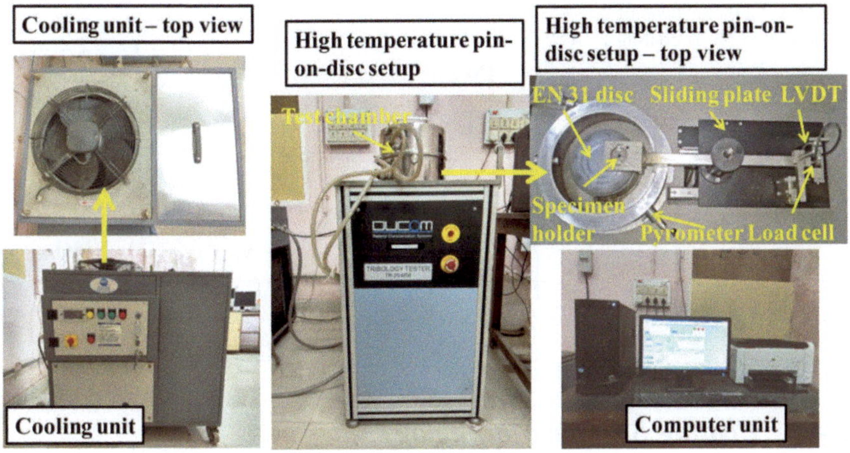

Fig. 1.3 High-temperature tribological test setup (pin-on-disk)

Table 1.2 Process parameters and their levels

Parameters	Unit	Levels		
		1	2	3
Temperature (A)	°C	100	300[a]	500
Load (B)	N	10	30[a]	50
Speed (C)	rpm	60	80[a]	100

[a]initial test run condition

1.4 Results and Discussions

1.4.1 Coating Composition

EDX analysis reveals that the percentage by weight of Ni and B in the coatings (as-plated condition) is 94.5 and 5.5%, respectively. Peaks of Ni and B may be observed in Fig. 1.4, where the EDX spectrum is presented. In general, coatings with B content of ~5–6.5% by weight are mid-B coatings [29]. Electroless Ni–B coatings considered in this study are therefore concluded to lie in the mid-B range. From the B content, it is expected that the coatings may be amorphous or a mixture of amorphous and nanocrystalline phases in as-deposited condition.

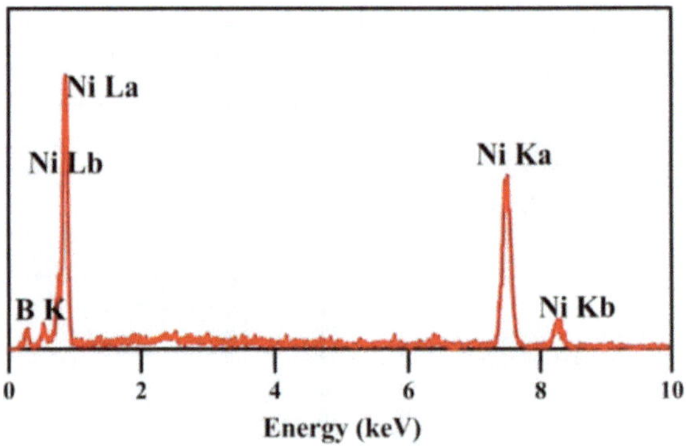

Fig. 1.4 EDX analysis of as-deposited Ni–B coating

1.4.2 Coating Surface Morphology

The surface morphology of the coatings observed under SEM is presented in Fig. 1.5. As-deposited Ni–B coatings resemble densely nodulated surface (Fig. 1.5 (a)). EN coatings in as-deposited condition have nodulated structures similar to that observed in the present work [3]. The coatings heat treated at 350 °C appear like a cauliflower with the formation of specific boundaries around the nodules (Fig. 1.5 (b)). The enlargement of nodules and appearance of cellular boundaries indicate changes in microstructure due to heat treatment. Similar feature has been observed by other researchers for electroless Ni–B coatings and its variants [7–12]. Low COF of Ni–B coatings has been attributed to the nodulated morphology of the coatings. SEM micrograph of cross section of Ni–B coating in as-plated state is shown in Fig. 1.5(c). The cross-cut section indicates columnar growths. These growths are efficient retainers of lubricants [7, 8]. Furthermore, the actual contact area is reduced due to the columnar structures. This enhances the tribological performance of the binary Ni–B alloy.

1.4.3 Phase Transformations

To understand the structural aspects, XRD of the as-deposited and heat-treated coatings is carried out and shown in Fig. 1.6. The XRD spectrum of as-deposited Ni–B coating may be observed in Fig. 1.6(a) which indicates a broad hump superimposed by small peaks. Amorphous nature of as-deposited coatings is manifested in the XRD spectrum as a broad peak. Higher segregation of B prevents the nucleation of Ni phases resulting in an amorphous structure [11]. Post-heat treatment at 350 °C, crystalline peaks of Ni, Ni_3B, and Ni_2B appear in Fig. 1.6(b). Heat treatment results in the formation of crystalline Ni and its boride phases. Therefore, it is concluded that the inflation in nodules and appearance of specific cellular boundaries (Fig. 1.5(b)) are due to the crystallization of the coatings.

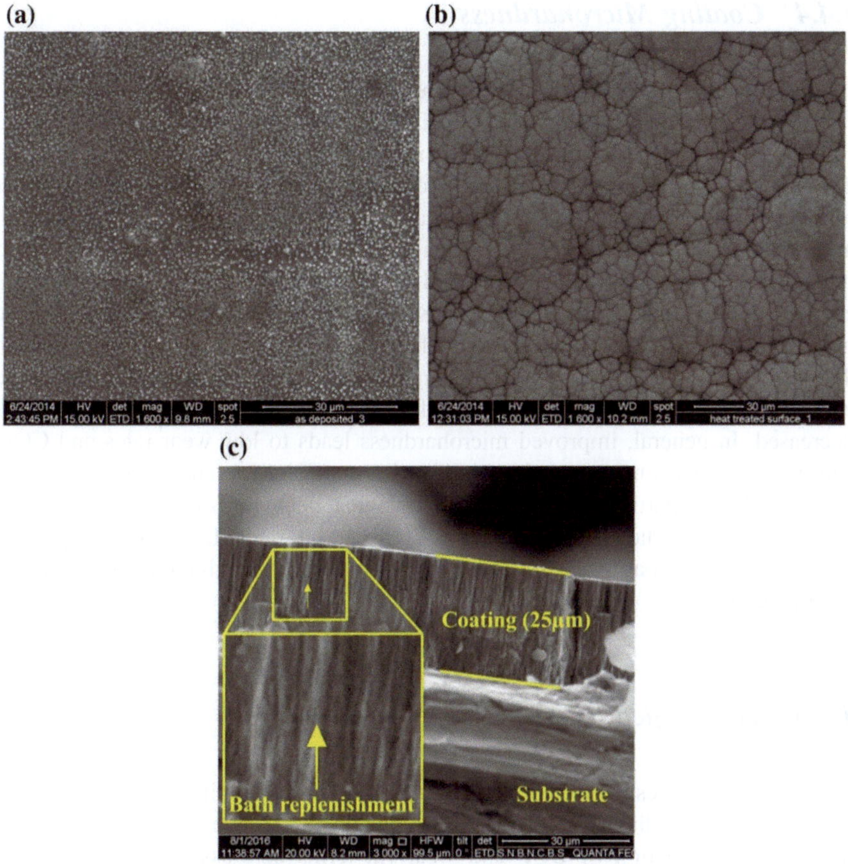

Fig. 1.5 SEM of electroless Ni–B coating: **a** as-deposited, **b** heat treated, and **c** cross section

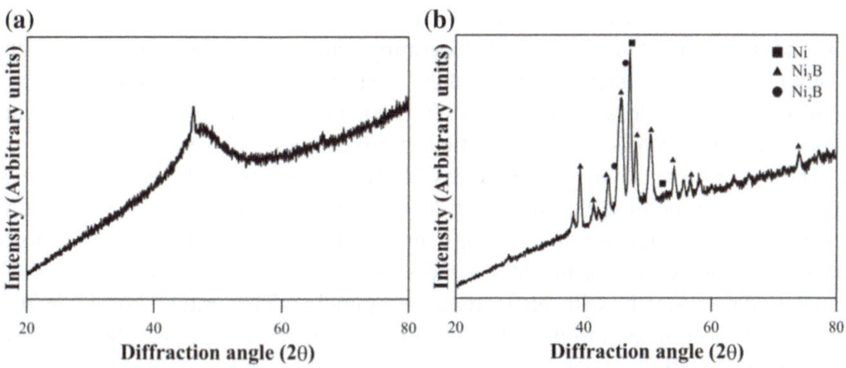

Fig. 1.6 XRD results of Ni–B coatings: **a** as-deposited and **b** heat treated

1.4.4 Coating Microhardness

As-deposited Ni–B coating exhibits microhardness of 670 HV_{100}. Krishnaveni et al. [11] observed a microhardness of 570 HV_{100} in the case of as-deposited Ni–B coating (6.5 wt% B). Madah et al. [15] reported a microhardness of 825 $HV_{0.05}$. The microhardness of as-deposited Ni–B coating is therefore comparable with other research works. On heat treatment, the hardness increases to 1060 HV_{100}. A high hardness of Ni–B coatings is observed on heat treatment. This may be attributed to precipitation hardening phenomenon and presence of nickel borides. Krishnaveni et al. [11] observed a high hardness of 908 HV_{100} when the deposits underwent heat treatment at 450 °C. On heat treating the coatings at temperatures greater than 450 °C, degradation in microhardness may take place. This happens due to the conglomeration of borides [11]. Therefore, the number of hardening sites is decreased. In general, improved microhardness leads to low wear rates and COF. But it is not always necessary that high hardness will lead to enhanced tribological behavior. This is particularly true when tribological behavior is observed at elevated temperatures. Mukhopadhyay et al. [12] observed degradation in tribological behavior of Ni–B coatings heat treated at 400 or 450 °C at high temperatures due to such phenomenon.

1.4.5 Tribological Behavior of Ni–B Coatings

The experimental results following Taguchi's L_{27} OA (combination of test parameters and corresponding COF/mass loss) are laid down in Table 1.3. From Table 1.3, the trends in variation of COF and wear (in terms of mass loss) with varying operating temperature, load, and speed are observed and the same is illustrated in Fig. 1.7. For all values of load and speed, mass loss of the coatings decreases with an increase in operating temperature as can be seen in Fig. 1.7(a). The mass loss increases with an increase in load at 100 and 500 °C (at all values of speed, i.e., 60, 80, and 100 rpm). Only at 300 °C operating temperature, the mass loss of the coatings at 50 N is intermediate to that observed at 10 and 30 N (Fig. 1.7(a)). A decrease in mass loss of electroless Ni–B coatings with an increase in operating temperature may be due to the formation of protective tribolayers [30, 31].

The variation of COF of electroless Ni–B coatings with the tribological test parameters may be observed in Fig. 1.7(b). At 60 rpm, the COF of Ni–B coating increases with an increase in operating temperature. Compared to 10 or 30 N load, the COF at 50 N load is lower. As the speed is increased to 80 rpm, the COF remains almost steady with an increase in operating temperature but again it is lower at 50 N compared to that observed at 10 or 30 N. At highest value of speed, i.e., 100 rpm, COF decreases with an increase in operating temperature. With load, a significant change is not observed at 100 rpm. It is well known that the formation

Table 1.3 Combination of tribological test parameters along with the obtained values of responses

Exp. No.	Temperature (°C)	Load (N)	Speed (rpm)	Experimental data	
				COF	Wear (gm)
1	100	10	60	0.288	0.00321
2	100	10	80	0.456	0.00401
3	100	10	100	0.675	0.00477
4	100	30	60	0.521	0.00514
5	100	30	80	0.681	0.00535
6	100	30	100	0.845	0.00681
7	100	50	60	0.321	0.0063
8	100	50	80	0.553	0.0081
9	100	50	100	0.766	0.0102
10	300	10	60	0.511	0.00224
11	300	10	80	0.523	0.00289
12	300	10	100	0.477	0.00238
13	300	30	60	0.615	0.00595
14	300	30	80	0.569	0.00673
15	300	30	100	0.488	0.00631
16	300	50	60	0.412	0.00411
17	300	50	80	0.468	0.00503
18	300	50	100	0.499	0.00428
19	500	10	60	0.572	0.00079
20	500	10	80	0.56	0.00179
21	500	10	100	0.47	0.00254
22	500	30	60	0.537	0.00157
23	500	30	80	0.532	0.00219
24	500	30	100	0.417	0.00352
25	500	50	60	0.475	0.00265
26	500	50	80	0.441	0.00325
27	500	50	100	0.354	0.00315

of oxide layers is expedited at high operating temperatures [30, 31]. At higher speeds, the replenishment of the lubricious oxides is higher [31]. Hence, the COF at 100 rpm (highest value of speed) decreases with increase in temperature.

1.4.6 Wear Mechanism at High Temperatures

The wear mechanism of Ni–B coatings at highest value of load and speed (50 N, 100 rpm) at 100, 300, and 500 °C may be observed in Fig. 1.8. At 100 °C, deep grooves along the direction of sliding may be observed in Fig. 1.8(a). It seems that

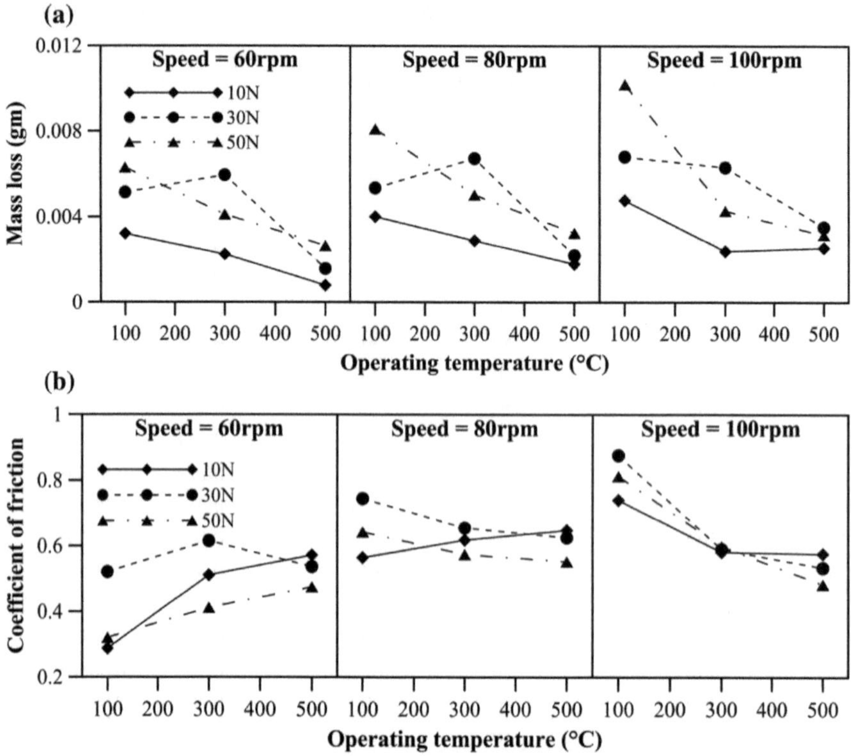

Fig. 1.7 Variation of **a** wear and **b** COF of electroless Ni–B coating

strain-hardened debris particles cut through the coatings to produce deep grooves. Consequently, the mass loss and COF increase at 100 °C as can be observed in Fig. 1.7. Similar behavior has been also observed for electroless Ni–P coatings [30]. At 300 °C, the coatings appear flattened in Fig. 1.8(b). These flattened patches contribute to the formation of load bearing areas. Moreover, compared to 100 °C (Fig. 1.8(a)) fewer deep scratches are observed at 300 °C. In fact, the tough coating resists wear due to microstructural changes occurring at high temperatures [31]. Such microstructural changes occur due to synergistic effect of high operating temperature and load. The wear mechanism at 500 °C may be observed in Fig. 1.8 (c). The worn surface at 500 °C is mainly characterized by tribo-oxide glazes. Tribo-oxide reactions get enhanced especially at high temperatures [15]. These oxide glazes improve the wear behavior and COF of EN coatings.

1.4.7 Optimization of Tribological Behavior

The initial step in GRA is normalization of the responses. COF and mass loss of the coatings (given in Table 1.3) are normalized using lower-the-better criteria.

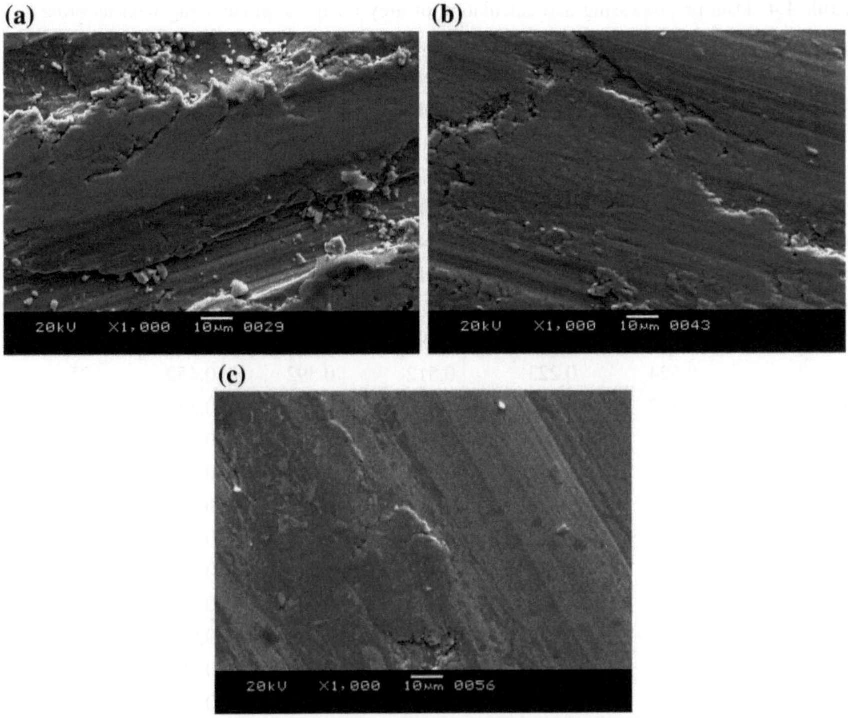

Fig. 1.8 Worn specimen at 50 N load and 100 rpm speed at **a** 100 °C, **b** 300 °C, and **c** 500 °C

The normalization is carried out according to Eq. 1.1 and laid down in Table 1.4. The GRC for COF and mass loss is calculated using Eq. 1.2 from the normalized values. The GRC of both the responses is also given in Table 1.4. Finally, the GRG is found out and presented in Table 1.4 along with its order. The GRG is indicative of multi-performance index.

After initial preprocessing and calculation of GRG, the response table for means is found out and the results are listed in Table 1.5. In Table 1.5, for each process parameter/factor, the mean GRG at each level is laid down. For example, if factor A (temperature) is considered, the mean grade at levels 1, 2, and 3 is computed by calculating the average grade of experiments 1–9, 11–18, and 19–27, respectively. Similarly, the average grades of factors B (load) and C (speed) are calculated. In the response table for means, ranks have been assigned for each parameter based on delta value. The delta value is the difference between highest and lowest mean in a column. Rank 1 is assigned to the factor with highest delta value. In Table 1.5, Rank 1 is assigned to temperature while Ranks 2 and 3 have been assigned to load and speed, respectively. From this, it is preliminarily deduced that highest significance in controlling the tribological behavior is of operating temperature followed by load and speed. The main effects plot corresponding to Table 1.5 is given in Fig. 1.9. The combination of parameters as predicted by the main effects plot is

Table 1.4 Data preprocessing and calculation of grey relational grade along with its order

Exp. No.	Normalized data		Grey relational coefficient		Grade	Order
	COF	Wear	COF	Wear		
1	1.000	0.743	1.000	0.660	0.830	1
2	0.698	0.658	0.624	0.594	0.609	16
3	0.305	0.577	0.418	0.542	0.480	21
4	0.582	0.538	0.544	0.520	0.532	19
5	0.294	0.515	0.415	0.508	0.461	24
6	0.000	0.360	0.333	0.439	0.386	26
7	0.941	0.414	0.894	0.461	0.677	5
8	0.524	0.223	0.512	0.392	0.452	25
9	0.142	0.000	0.368	0.333	0.351	27
10	0.600	0.846	0.555	0.764	0.660	9
11	0.578	0.777	0.542	0.691	0.617	15
12	0.661	0.831	0.596	0.747	0.672	6
13	0.413	0.452	0.460	0.477	0.468	23
14	0.496	0.369	0.498	0.442	0.470	22
15	0.641	0.413	0.582	0.460	0.521	20
16	0.777	0.647	0.692	0.586	0.639	14
17	0.677	0.549	0.607	0.526	0.567	18
18	0.621	0.629	0.569	0.574	0.572	17
19	0.490	1.000	0.495	1.000	0.748	2
20	0.512	0.894	0.506	0.825	0.665	8
21	0.673	0.814	0.605	0.729	0.667	7
22	0.553	0.917	0.528	0.858	0.693	4
23	0.562	0.851	0.533	0.771	0.652	12
24	0.768	0.710	0.683	0.633	0.658	10
25	0.664	0.802	0.598	0.717	0.657	11
26	0.725	0.739	0.645	0.657	0.651	13
27	0.882	0.749	0.808	0.666	0.737	3

A3B1C1, i.e., 500 °C operating temperature, 10 N load, and 60 rpm speed. The slope of main effects plot also indicates that operating temperature has the highest significance followed by applied normal load and speed in controlling the friction and wear performance of the coatings.

After optimization, a confirmation test is carried out where the optimal condition predicted by GRA is compared with an initial test condition (mid-level combination of parameters). Moreover, from Eq. 1.3, the predicted value of GRG is obtained. The results of confirmation test are presented in Table 1.6. An improvement of 59.15% in grade is obtained on application of GRA which is quite significant. Therefore, the optimization of process parameters results in a noteworthy improvement in tribological behavior of the coatings. But a slight increase in COF

Table 1.5 Response table for means of grey relational grade

Level	A	B	C
1	0.5309	0.6608	0.6561
2	0.5761	0.538	0.5715
3	0.6809	0.5892	0.5604
Delta	0.15	0.1228	0.0957
Rank	1	2	3
Mean grey relational grade = 0.596			

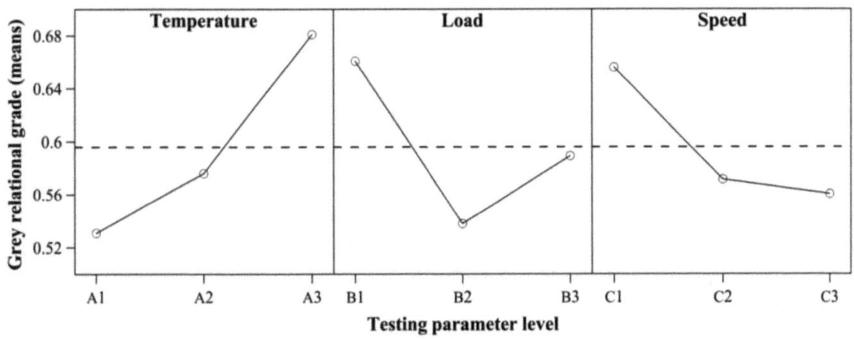

Fig. 1.9 Main effects plot for means of grey relational grade

Table 1.6 Results of confirmation test

	Initial	Optimal	
		Predicted	Experiment
Level	A2B2C2	A3B1C1	A3B1C1
COF	0.569		0.572
Wear	0.00673		0.00079
Grade	0.470	0.806	0.748
Improvement of Grade = 0.278 (59.15%)			

is observed compared to initial test run. This is attributed to the fact that some compromise needs to be done for achieving multi-characteristics optimization.

1.4.8 Analysis of Variance

The response Table 1.5 and main effects plot (Fig. 1.9) give a preliminary idea on the significance of process variables in influencing the tribological characteristics. But the statistical significance may be ascertained from analysis of variance (ANOVA). The significance of a parameter is indicated by the F-ratio,

Table 1.7 ANOVA results of grey relational grade

Source	DOF	Seq SS	Adj SS	Adj MS	F-ratio	Contribution (%)
A	2	0.106558	0.106558	0.053279	26.15[a]	30.49
B	2	0.068503	0.068503	0.034252	16.81[a]	19.60
C	2	0.049335	0.049335	0.024667	12.11[a]	14.12
A × B	4	0.028419	0.028419	0.007105	3.49	8.13
A × C	4	0.071771	0.071771	0.017943	8.81[a]	20.54
B × C	4	0.008609	0.008609	0.002152	1.06	2.46
Error	8	0.016299	0.016299	0.002037		4.66
Total	26	0.349495				100.00

[a]Significant at 95% confidence level

i.e., variance ratio. The F-ratio may be defined as the ratio of regression mean square and the mean square error. A parameter may be considered significant at any confidence level if the calculated F-ratio is higher than the tabulated value. ANOVA results of GRG for the present work are given in Table 1.7. The highest significance is concluded for operating temperature and then for load and speed. The interaction of operating temperature and speed also has some contribution in influencing the friction and wear characteristics.

1.5 Conclusions

In the present work, friction and wear characteristics of heat-treated crystalline Ni–B coatings are investigated following Taguchi's experimental design. Mass loss of the coatings (signifying wear) and COF are examined by carrying out experiments in an orderly manner following combinations laid down in L_{27} OA. Operating temperature, load, and speed are the design variables. Finally, GRA is used to find out the optimal parametric combination of tribological test parameters. The important conclusions that can be deduced from the study are:

(a) The composition of as-deposited Ni–B coating in terms of weight percentage of Ni and B is 94.5 and 5.5%, respectively.
(b) SEM micrographs of coatings in as-deposited condition show densely nodulated structures. The heat-treated coatings appear like a cauliflower.
(c) The deposits in as-deposited condition are amorphous. Precipitation of crystalline Ni, Ni_3B, and Ni_2B phases are observed post-heat treatment.
(d) As-deposited Ni–B coatings have a high microhardness of the order of 670 HV_{100}. Heat treatment causes further improvement in microhardness to ~1060 HV_{100} due to the precipitation of borides.
(e) The optimal combination of tribological test parameters predicted by GRA is A3B1C1 which is 500 °C operating temperature, 10 N load, and 60 rpm speed.

(f) Furthermore, the predicted and experimentally obtained optimal grades are in agreement. In comparison with initial test condition, an improvement of 59.15% in GRG is achieved at optimal parametric condition. Hence, it is concluded that the application of GRA results in a significant enhancement in tribological characteristics of the coatings.
(g) Finally, ANOVA results indicate the highest contribution from operating temperature followed by load and speed in influencing the friction and wear characteristics of the coatings.

References

1. Anik, M., Körpe, E., Şen, E.: Effect of coating bath composition on the properties of electroless nickel–boron films. Surf. Coat. Technol. **202**, 1718–1727 (2008)
2. Sudagar, J., Lian, J., Sha, W.: Electroless nickel, alloy, composite and nano coatings–A critical review. J. Alloy. Compd. **571**, 183–204 (2013)
3. Sahoo, P., Das, S.K.: Tribology of electroless nickel coatings–a review. Mater. Des. **32**(4), 1760–1775 (2011)
4. Riddle, Y.W., Bailerare, T.O.: Friction and wear reduction via an Ni-B electroless bath coating for metal alloys. JOM **57**(4), 40–45 (2005)
5. Correa, E., Zuleta, A.A., Guerra, L., Gómez, M.A., Castaño, J.G., Echeverría, F., Liu, H., Skeldon, P., Thompson, G.E.: Tribological behavior of electroless Ni–B coatings on magnesium and AZ91D alloy. Wear **305**(1–2), 115–123 (2013)
6. Zhang, W.X., Jiang, Z.H., Li, G.Y., Jiang, Q., Lian, J.S.: Electroless Ni-P/Ni-B duplex coatings for improving the hardness and the corrosion resistance of AZ91D magnesium alloy. Appl. Surf. Sci. **254**(16), 4949–4955 (2008)
7. Bülbül, F., Altun, H., Ezirmik, V., Küçük, Ö.: Investigation of structural, tribological and corrosion properties of electroless Ni–B coating deposited on 316L stainless steel. Proc. Inst. Mech. Eng., Part J **227**(6), 629–639 (2013)
8. Çelik, İ., Karakan, M., Bülbül, F.: Investigation of structural and tribological properties of electroless Ni–B coated pure titanium. Proc. Inst. Mech. Eng., Part J **230**(1), 57–63 (2016)
9. Das, S.K., Sahoo, P.: Tribological characteristics of electroless Ni–B coating and optimization of coating parameters using Taguchi based grey relational analysis. Mater. Des. **32**(4), 2228–2238 (2011)
10. Das, S. K., Sahoo, P.: Influence of process parameters on microhardness of electroless Ni-B coatings. Adv. Mech. Eng. **4**, 703168 (2012)
11. Krishnaveni, K., Sankara Narayanan, T.S.N., Seshadri, S.K.: Electroless Ni–B coatings: preparation and evaluation of hardness and wear resistance. Surf. Coat. Technol. **190**(1), 115–121 (2005)
12. Mukhopadhyay, A., Barman, T.K., Sahoo, P.: Effects of heat treatment on tribological behavior of electroless Ni–B coating at elevated temperatures. Surf. Rev. Lett. **24**(Supp01), 1850014 (2017)
13. Oraon, B., Majumdar, G., Ghosh, B.: Improving hardness of electroless Ni–B coatings using optimized deposition conditions and annealing. Mater. Des. **29**(7), 1412–1418 (2008)
14. Dervos, C.T., Novakovic, J., Vassiliou, P.: Vacuum heat treatment of electroless Ni–B coatings. Mater. Lett. **58**(5), 619–623 (2004)
15. Madah, F., Dehghanian, C., Amadeh, A.A.: Investigations on the wear mechanisms of electroless Ni–B coating during dry sliding and endurance life of the worn surfaces. Surf. Coat. Technol. **282**, 6–15 (2015)

16. Balaraju, J.N., Priyadarshi, A., Kumar, V., Manikandanath, N.T., Kumar, P.P., Ravisankar, B.: Hardness and wear behaviour of electroless Ni–B coatings. Mater. Sci. Technol. **32**(16), 1654–1665 (2016)
17. Niksefat, V., Ghorbani, M.: Mechanical and electrochemical properties of ultrasonic-assisted electroless deposition of Ni–B–TiO_2 composite coatings. J. Alloy. Compd. **633**, 127–136 (2015)
18. Ghaderi, M., Rezagholizadeh, M., Heidary, A., Monirvaghefi, S.M.: The effect of Al_2O_3 nanoparticles on tribological and corrosion behavior of electroless Ni–B–Al_2O_3 composite coating. Prot. Met. Phys. Chem. Surf. **52**(5), 854–858 (2016)
19. Krishnaveni, K., Sankara Narayanan, T.S.N., Seshadri, S.K.: Electroless Ni-B-Si_3N_4 composite coating: deposition and evaluation of its characteristic properties. Synth. React. Inorg., Met.-Org., Nano-Met. Chem. **42**(7), 920–927 (2012)
20. Rezagholizadeh, M., Ghaderi, M., Heidary, A., Monirvaghefi, S.M.: The effect of B_4C nanoparticles on the corrosion and tribological behavior of electroless Ni-B-B_4C composite coatings. Surf. Eng. Appl. Electrochem. **51**(1), 18–24 (2015)
21. Aydeniz, A.I., Göksenli, A., Dil, G., Muhaffel, F., Calli, C., Yüksel, B.: Electroless Ni-B-W coatings for improving hardness, wear and corrosion resistance. Materiali in Tehnologije **47** (6), 803–806 (2013)
22. Serin, I.G., Göksenli, A.: Effect of annealing temperature on hardness and wear resistance of electroless Ni–B–Mo coatings. Surf. Rev. Lett. **22**(05), 1550058 (2015)
23. Mukhopadhyay, A., Barman, T.K., Sahoo, P.: Tribological behavior of electroless Ni–B–W coating at room and elevated temperatures. Proc. Inst. Mech. Eng., Part J **232**(11), 1450–1466 (2018)
24. Eraslan, S., Ürgen, M.: Oxidation behavior of electroless Ni–P, Ni–B and Ni–W–B coatings deposited on steel substrates. Surf. Coat. Technol. **265**, 46–52 (2015)
25. Mukhopadhyay, A., Barman, T.K., Sahoo, P.: Effect of heat treatment on tribological behavior of electroless Ni-B-Mo coatings at different operating temperatures. Silicon **10**, 1203–1215 (2018)
26. Mukhopadhyay, A., Barman, T.K., Sahoo, P.: Effect of operating temperature on tribological behavior of as-plated Ni-B coating deposited by electroless method. Tribol. Trans. **61**(1), 41–52 (2018)
27. Mukhopadhyay, A., Barman, T.K., Sahoo, P.: Tribological behavior of sodium borohydride reduced electroless nickel alloy coatings at room and elevated temperatures. Surf. Coat. Technol. **321**, 464–476 (2017)
28. Deng, J.: Introduction to grey system theory. J. Grey Syst. **1**(1), 1–24 (1989)
29. Vitry, V., Bonin, L.: Increase of boron content in electroless nickel-boron coating by modification of plating conditions. Surf. Coat. Technol. **311**, 164–171 (2017)
30. Kundu, S., Das, S.K., Sahoo, P.: Tribological behaviour of electroless Ni-P deposits under elevated temperature. Silicon **10**(2), 329–342 (2018)
31. Mukhopadhyay, A., Barman, T.K., Sahoo, P.: Friction and wear performance of electroless Ni-B coatings at different operating temperatures. Silicon, https://doi.org/10.1007/s12633-017-9678-y (2018)

Chapter 2
Enhanced Thermal and Mechanical Performance of Functionalized Graphene Epoxy Nanocomposites: Effect of Processing Conditions, Different Grades and Loading of Graphene

Saswata Bose, Arit Das and Anirban Ghosh

2.1 Introduction

Graphene is a monolayer sheet of carbon having exceptional mechanical strength as well as high electrical and thermal conductivities due to the π-conjugated network of sp^2-hybridized carbons [1, 2]. Moreover, researchers have found that the material exhibits high surface area and optical transmittance besides being chemically stable [3, 4]. These unique properties of graphene have led to its application in a variety of different branches such as batteries and supercapacitors [5, 6], electronic devices [7], tribology [8] and fillers for nanocomposites [9, 10]. The fabrication of graphene-based polymer nanocomposites holds much promise since both the graphitized and layered structure can be retained in the hybrid nanocomposite [11]. Epoxy resin, a thermoset polymer, possessing decent mechanical, thermal and tribological properties has received much attention as a matrix for high-performance materials as evident from previous studies [10, 12, 13]. Epoxy-based composites have been widely used in aerospace, adhesive and coatings industries [14, 15]. In order to use these composites in a variety of multifunctional applications, the mechanical, thermal, and barrier properties of the said materials must be improved. This can be achieved via the incorporation of filler materials like graphene within the epoxy matrix. The properties of the nanocomposite depend on the available surface area of fillers which can be achieved by homogenous dispersion [16]. The presence of functional groups in GnP facilitates its dispersion in organic solvents, thereby enhancing the robust physico-chemical interaction between the functionalized GnP and epoxy polymer [17]. However, it must be kept in mind that the

S. Bose (✉) · A. Das · A. Ghosh
Department of Chemical Engineering, Jadavpur University, Kolkata, India
e-mail: saswataboseiit@gmail.com

P. Sahoo and J. P. Davim (eds.), *Advances in Materials, Mechanical and Industrial Engineering*, Lecture Notes on Multidisciplinary Industrial Engineering,
https://doi.org/10.1007/978-3-319-96968-8_2

functionalization of graphene involving strong acids can lead to disruption of the basic graphitic structure [18, 19]. A viable route for overcoming this problem is to treat the GnPs with mild acid solutions and then functionalizing the same. The mild acid treatment can help to preserve the integrity of the carbon backbone of graphene retaining the C-sp^2 configuration. In addition to keeping the structural stability of graphene, the mild acid treatment also generates sufficient functional sites at the edges of the GnPs while keeping the basal plane of the GnPs intact, thereby promoting the interaction between the polymer and GnP. Therefore, to ensure that the GnPs are uniformly dispersed throughout the polymer matrix, functionalization of GnPs must be carried out which in turn plays a pivotal role in reinforcing the thermal and mechanical properties of the GnP-based polymer composites. Researchers have looked into possibility of using different carbon-based nanofillers (CNT, graphene) for the performance improvement of the nanocomposites [20–31]. The dispersion and adhesion of the GnP particles within the epoxy matrix was addressed by carrying out functionalization. Recently, various efforts have been directed to functionalize graphene by the reaction in the presence of various different organic molecules depending upon the final applications [32–39]. Thermally reduced GnPs have been covalently functionalized employing the Bingel reaction to improve their dispersibility and interfacial bonding within the epoxy matrix which resulted in 22% improvement in the flexural strength [40].

In the current study, we have functionalized acid-treated GnP by employing PACl as reported by Bose et al. [10] and then fabricated the epoxy/functionalized GnP nanocomposites via the solution-processing route to investigate its effect on the material properties. In order to gauge the effect functionalization on the thermal and mechanical properties, we also prepared epoxy/GnP composites without functionalization. Control experiments were also carried out where the composites were prepared in the absence of any solvents. All the samples were prepared using two different grades of GnPs, namely M-15 and C-750 for studying the influence of nanofiller surface area on the nanocomposite properties. Moreover, the effect of functionalized GnP loading on the thermal and flexural properties was also investigated under the current work.

2.2 Methods

GnP powders of two different grades (M-15 grade having an average particle diameter of ~5–50 μm and C-750 grade having an average particle diameter of ~0.5–2 μm) were purchased from XG Sciences, USA. Reagent grade nitric acid (HNO_3) and Polyacroyl chloride (25% solution in dioxane) were obtained from Merck and Polysciences Inc., USA, respectively. Tetrahydrofuran (THF) and 1,4-dioxane were obtained from Sigma Aldrich. The treatment of GnP with nitric acid and subsequent functionalization was carried out as reported in a recent work by Bose et al. [10].

2.2.1 Preparation of Acid-Treated GnP (AGNP)

GnPs were treated using 3 (M) aqueous nitric acid solution. From our previous work, it was found that GnPs treated with 3 (M) HNO_3 retained the majority of their original structure with a high enough O/C ratio required for further functionalization [10]. The dispersion of GnPs in nitric acid was then refluxed for 4 h at 120 °C while being constantly stirred magnetically. The obtained product was washed with RO water repeatedly using a 0.22 μm PTFE membrane to neutralize the excess acid present. The collected AGNP was vacuum dried overnight.

2.2.2 Reaction of AGNPs with PACl

Weighed amount of AGNP was dispersed in 1,4 dioxane under an inert atmosphere. The dispersion was slowly added to PACl solution (weight ratio of GnP to PACl was kept at 1:4), and the mixture was magnetically stirred for two days at 60 °C. The resultant mixture was washed with THF to remove unreacted PACl and collected using a 0.22-μm PTFE membrane by vacuum filtration. The product (AGNP–PACl) was dried under vacuum for 6 h at 65 °C.

2.2.3 Preparation of Epoxy/AGNP–PACl Nanocomposite by Solution Processing

Neat epoxy (Epon 828) was dispersed in acetone by magnetic stirring for 30 min. Simultaneously, AGNP–PACl was also dispersed in acetone and the mixture of epoxy and AGNP–PACl was ultrasonicated for 30 min at 45 W. The reaction was continued for 2 h at 60 °C with continuous stirring. The solvent was then removed completely by degassing under vacuum for an hour at 60 °C. The hardener (meta-phenylene diamine, MPDA) was melted at 75 °C and added to the mixture and was further mixed for 2 min at 2000 rpmin a Flack-tek mixer followed by degassing under vacuum for 5 min at 70 °C to remove any remaining air bubbles. The obtained mix was cured in two stages: first at 75 °C for 2 h and then at 125 °C for a further couple of hours. Finally, the samples were polished (StruersAbramin Polishing Machine) to get the desired composite. The nanocomposites of pure epoxy and epoxy-unmodified GnP were also prepared following the above procedure. Figure 1 illustrates the schematic of the solvent-processing process for the preparation of the epoxy/AGNP–PACl nanocomposite. Control samples were prepared in the absence of acetone as solvent.

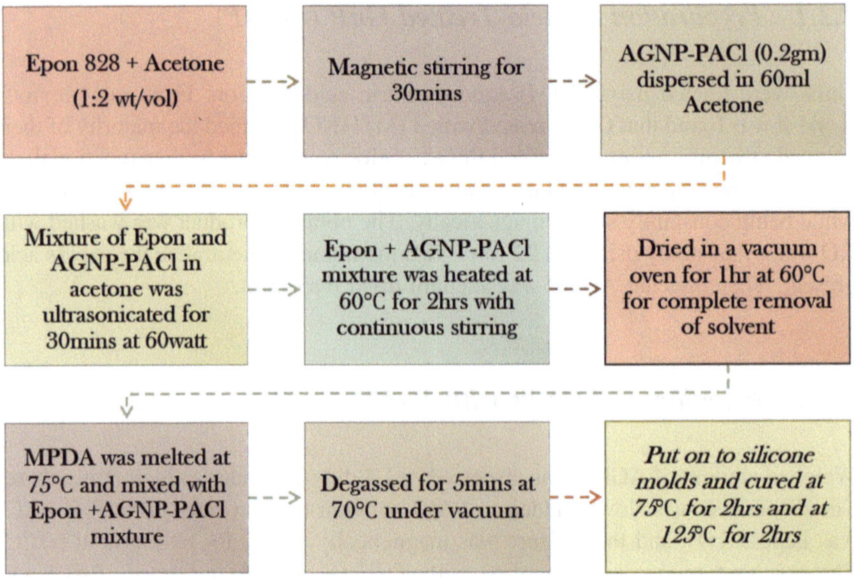

Fig. 1 Flow chart outlining the solution-processing steps for the nanocomposite preparation

2.2.4 *Characterizations*

A Perkin-Elmer 2000 spectrometer was used to carry out the Fourier transfer infrared (FTIR) spectroscopy of the epoxy and epoxy/GnP composite. The Raman spectrum of GnP and AGNP–PACl was recorded in a LabRAM ARAMIS confocal Raman Microscope. The morphology of fracture surfaces of the different nanocomposites was observed by using a scanning electron microscopy (Carl Zeiss Variable Pressure SEM EVO LS25). Thermogravimetric analyses (TGA) of the prepared materials were carried out using a TGA 500 (TA Instruments). Dynamic mechanical analyses (DMA 800; TA Instruments) of the prepared samples were implemented in single cantilever mode at a frequency of 1 Hz. Flexural tests were performed in accordance to the ASTM D790 in a United Testing System (UTS, SFM 20 load frame).

2.3 Results and Discussion

2.3.1 *Reaction Mechanism*

The mechanism of the functionalization reaction between AGNP and PACl has been proposed previously [10]. Briefly, the acid treatment generates –OH on the surface or edges of the GnPs following which the PACl chains attached onto the AGNP since the

–COCl groups of PACl and the –OH groups of AGNP reacted with each other. The epoxide groups present in the Epon 828 reacted with the remaining –COCl groups and created a strong covalent bond with GnPs which could be attributed as the reason behind the improved thermal and mechanical properties of the nanocomposite as discussed later. The details of reaction mechanism is shown in Scheme 1.

2.3.2 FTIR Spectra

Figure 2 shows the FTIR spectra of neat epoxy and epoxy-unmodified GnP composite. The peak at 3387 cm^{-1} due to the –OH group became less intense after incorporation of GnP suggesting that the hydroxyl groups were the sites of reaction. The epoxy (–CH–(O–CH_2)) stretching at 3060 cm^{-1} was less intense in epoxy GnP composite while the band at ~1100 cm^{-1}can probably be attributed to the formation of secondary alcohols after the composite preparation. The sharp peaks at 2927 and 2866 cm^{-1} were due to the methylene (–CH_2, –CH_3) asymmetric stretching and vibrations of the epoxy backbone. The oxirane functional groups of Epon 828 and the aromatic 1, 4-substituted ring were the reason behind the peak at 825 cm^{-1}. The aromatic –C=C–H stretching gave rise to the peaks at 1602 and 1510 cm^{-1}. The stretching of –C–C–O–C and ether bonds gave rise to the peaks at 1239, 1178 and 1030 cm^{-1}.

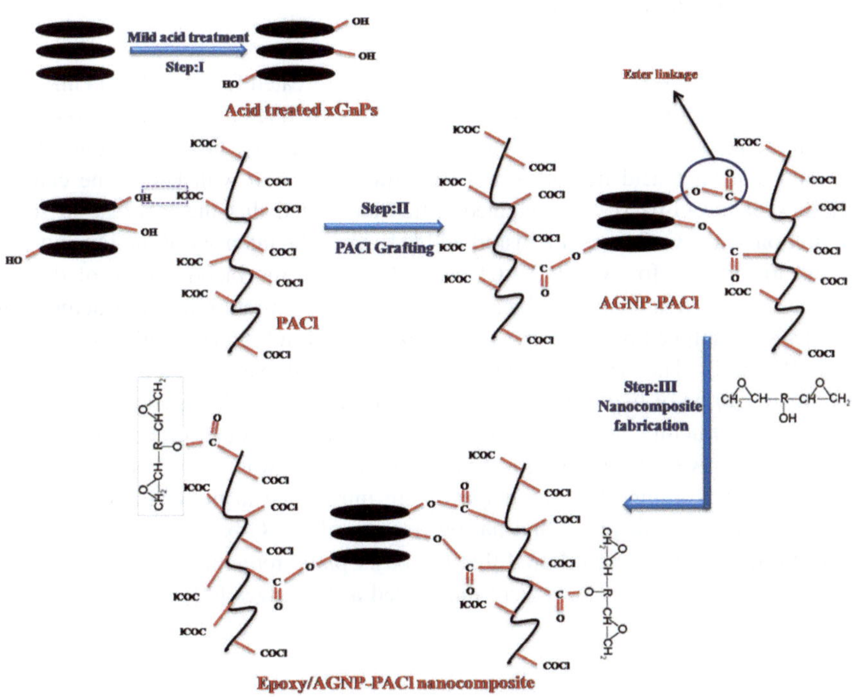

Scheme 1 Proposed reaction mechanism [10]

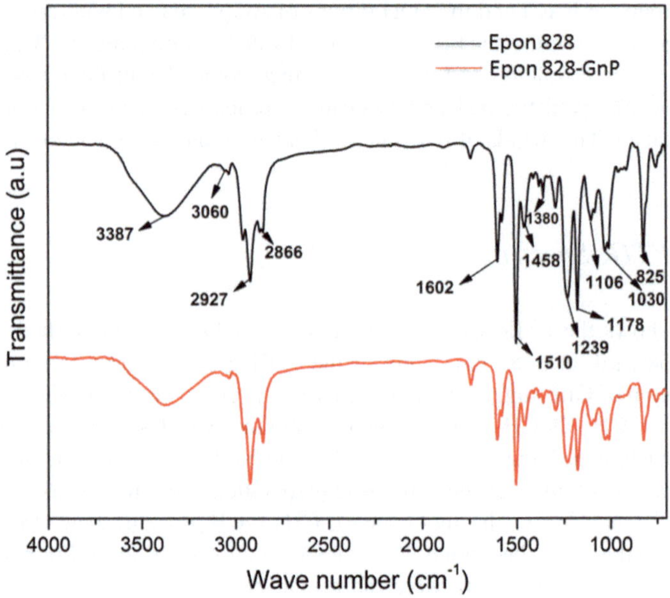

Fig. 2 FTIR spectra of neat epoxy and epoxy/GnP nanocomposites

2.3.3 Raman Spectroscopy

Figure 3 shows the Raman spectra of GnP and acid-treated functionalized GnP. The pristine GnP exhibited the characteristic Raman signatures at 1331 and 1564 cm^{-1} which are associated with the D- and G-band, respectively. From the figure, it can be said that after acid treatment and subsequent reaction with PACl, the characteristics of the G-band were retained which suggested that the inherent sp^2 configuration of the GnPs was not disturbed [41]. Also, the intensity of the D-band was found to increase for AGNP–PACl probably due to the incorporation of defects during the acid treatment. The increase in defect concentration in the structure was further substantiated by the increase in I_D:I_G ratio from 0.56 (for GnP) to 0.94 (for AGNP–PACl). The defects introduced into the GnP structure due to the acid treatment can be of two different types, namely defects on the basal plane that can damage the structure and the edge defects which can retain the original structure of GnP to a relatively greater extent. Since the oxidation was carried out under mild conditions, we assume that the defects were introduced at the edges which were also supported by the sharp and narrow feature of the G-band was retained after oxidation; so it could be inferred that the basal plane remained intact to a greater extent and mostly the defects were introduced at the edges [42].

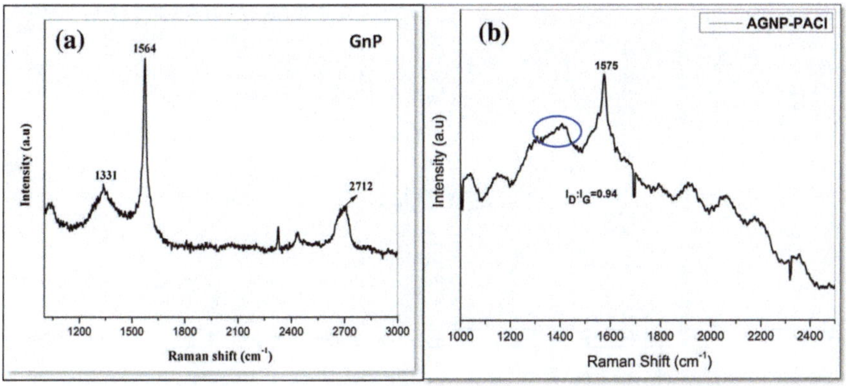

Fig. 3 Raman spectra of **a** GnP and **b** AGNP–PACl

2.3.4 SEM Observations: Morphology of the Nanocomposite

The SEM micrographs of the GnPs at different magnifications are shown in Fig. 4. The tensile-fractured surfaces of GnPs, acid-treated functionalized GnPs (AGNP–PACl) and Epon 828/GnP composites are shown in Fig. 5. The pure Epon 828 had smooth and even surface with ultra-fine rivulets indicative of brittle fracture similar to previously reported work [43]. Epoxy/GnP nanocomposites contained aggregates of GnP in the matrix. Moreover, a smooth surface was observed at higher magnification due to void formation which was caused due to removal of GnP from the Epon 828 (polymer) matrix. It can be concluded that the poor interfacial adhesion created areas that were prone to stress concentration, thereby aiding the propagation of cracks which finally resulted in failure of the material [44]. However, the SEM images of Epoxy/AGNP–PACl showed the homogenous distribution of GnP within the nanocomposite. Moreover, the GnPs were strongly attached to the epoxy matrix which led to better adhesion at the interfaces due to reduction in the void density and resulted in superior mechanical performance of the functionalized nanocomposites as confirmed by our following results.

2.3.5 Thermogravimetric Analysis

The onset degradation temperature (ODT) of epoxy/AGNP nanocomposite was found to be higher than that for epoxy/unfunctionalized GnPs (both M-15 and C-750) but was lower than onset degradation temperature of neat epoxy. This observation can be attributed to the incorporation of oxygenated moieties into the graphitic structure of GnPs due to the mild oxidation of GnP by nitric acid. Figure 6 shows the weight loss curves of pristine Epon 828 as well as the nanocomposites prepared with both unfunctionalized and functionalized GnP

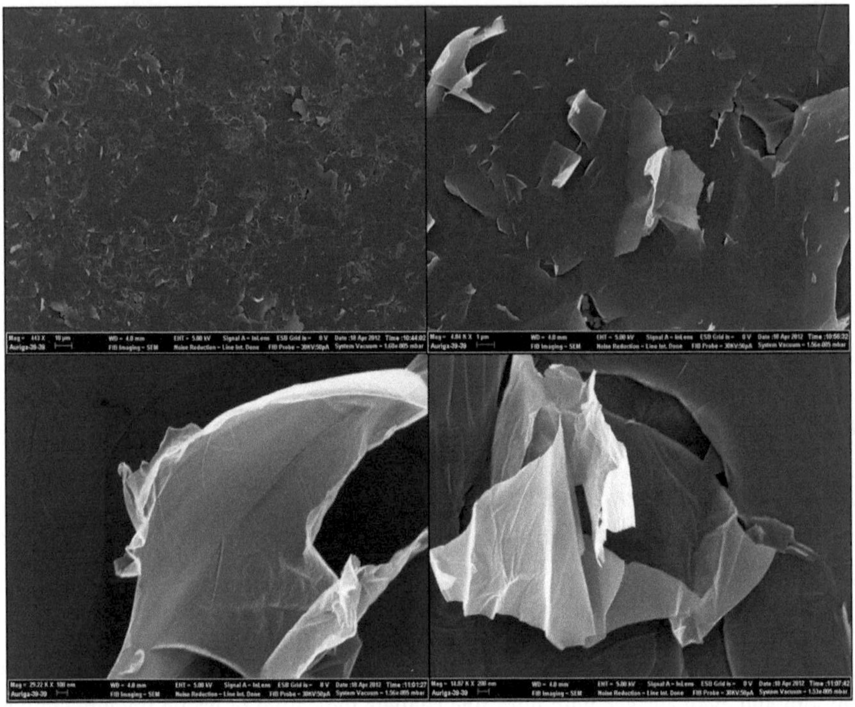

Fig. 4 SEM images of GnPs at different magnifications

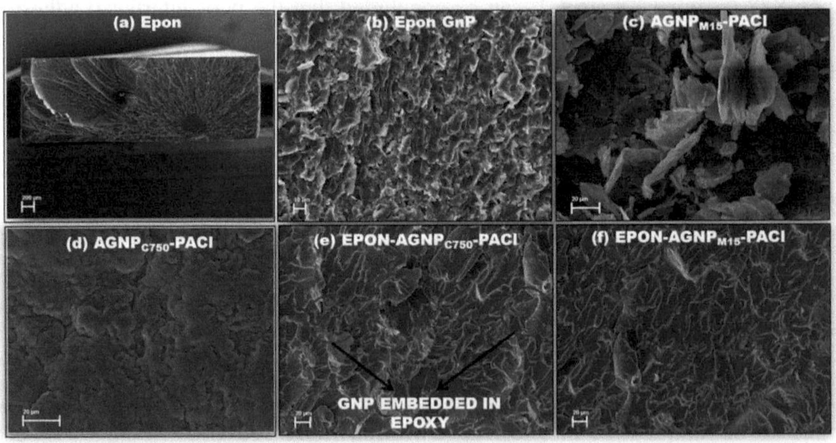

Fig. 5 SEM images of **a** neat Epoxy **b** Epoxy GnP **c** AGNP$_{M15}$–PACl **d** AGNP$_{C750}$–PACl **e** Epoxy/AGNP$_{M15}$–PACl and **f** Epoxy/AGNP$_{C750}$–PACl

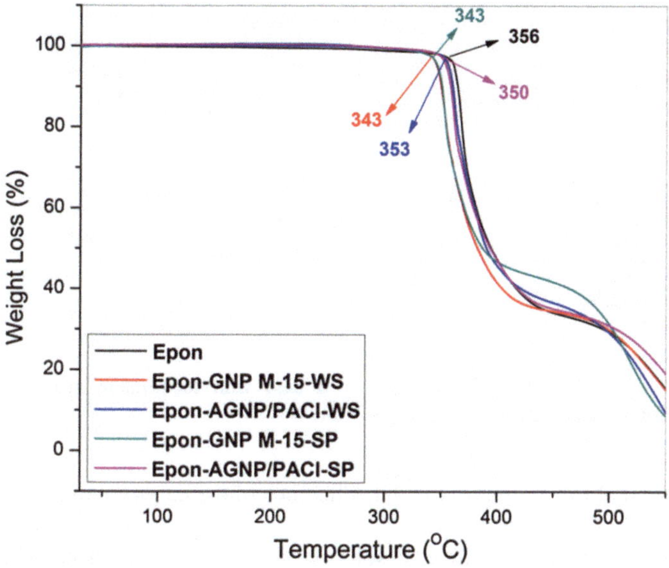

Fig. 6 TGA of neat epoxy and its nanocomposites under different processing conditions

prepared using SP and WS. The loss in weight observed in the range of 350–400 °C was assigned to the removal of the oxygenated groups. The subsequent loss in weight was observed around 450–550 °C which was due to the breakdown of the carbon backbone of the nanocomposites. An inherent property of the GnP particles is that they have a propensity towards aggregation which can lead to the formation of an interface between the polymer matrix and filler material with poor adhesion. The ODT of neat epoxy was 356 °C while that of the Epoxy/GnP was found to reduce to 343 °C (for both WS and SP) due to poor interfacial adhesion. The ODTs of the functionalized composites were 350 and 353 °C for SP and WS, respectively. The increase in ODT after functionalization was attributed to the formation of covalent linkages.

2.3.6 Dynamic Mechanical Analysis (DMA)

The effect of temperature on the viscoelasticity (storage modulus and tan δ) of the prepared nanocomposites is shown in Fig. 7. The T_g of neat epoxy was at 154 °C as evident from the tanδ maxima peak. The epoxy/unmodified GnP prepared by solution processing (SP) and without solution (WS) showed a decrease in the T_g values (141 and 145 °C, respectively). The drop in T_g values was attributed to the inadequate dispersion of GnP which inherently tends to form aggregates and poor filler-matrix adhesion. But the T_g values of Epoxy/AGNP–PACl for SP and WS were 153 and 157 °C, respectively, which suggested that the formation of a covalent linkage between the respective functional groups of Epon 828 and PACl

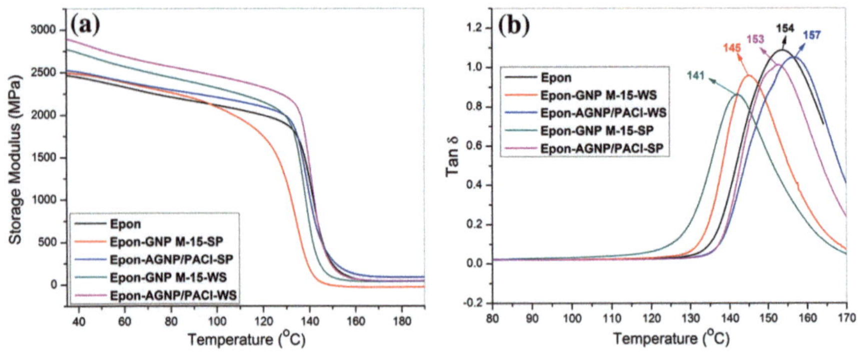

Fig. 7 Variation of **a** storage modulus and **b** tanδ with temperature neat epoxy and its nanocomposites under different processing conditions

played a pivotal role in improving the adherence of the nanocarbon filler material to the Epon 828 (polymer) matrix.

Figure 8 illustrates the effect of solution processing of the nanocomposites on their flexural modulus and strength. From the analysis, it was observed that the flexural strength of the Epon 828/unmodified GnP nanocomposites decreased irrespective of the SP or WS while the corresponding enhancement in flexural modulus was only 2.5% compared to pure Epon 828. However, upon the incorporation of AGNPM15–PACl within the epoxy nanocomposites resulted in an increase in flexural modulus by 16.4 and 10.8% when the composites were processed in WS and SP, respectively. Hence, it was concluded from the viscoelastic, thermal and flexural studies that the WS processing yielded better thermal and mechanical properties and it was followed to prepare the nanocomposites using GnPC-750. Table 1 lists the T_g, ODT and flexural properties of pure epoxy and nanocomposites prepared in presence of acetone as well as without any solvents.

The influence of AGNP–PACl functionalization on the flexural properties of epoxy/GnP nanocomposites using C-750 grade GnP prepared in the absence of solvent has been represented in Fig. 9. It was observed that the incorporation of functionalized C-750 GnP resulted in improvement of the flexural strength as well as flexural modulus of the prepared composites. The flexural strength of Epon 828/ $AGNP_{C750}$–PACl was found to increase by 14.6 and 17% in comparison with pure Epon 828 when the filler material ($AGNP_{C750}$–PACl) was present in the nanocomposite at 1 and 2.5 wt%, respectively. The flexural modulus of Epon 828/ $AGNP_{C750}$–PACl increased by 18.7 and 22% in comparison with pure Epon 828 when the nanocomposites were prepared with 1 and 2.5 wt% of $AGNP_{C750}$–PACl, respectively. However, in both the cases the epoxy-unmodified GnP composites showed negligible improvement in flexural properties. The $GnP_{C\text{-}750}$ showed enhanced performance as a filler which was attributed to increased number of reactive sites present due to its higher surface area. Table 2 lists the flextural properties of pure epoxy and the nanocomposites prepared with different functionalized GnP loading processed without solvent.

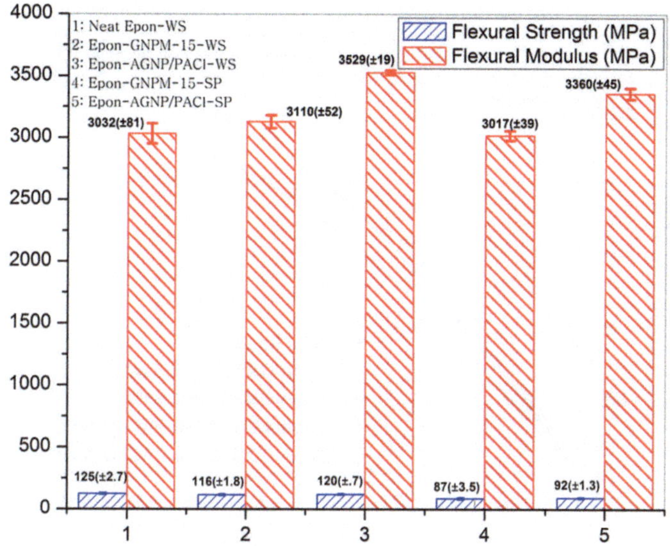

Fig. 8 Variation of flexural properties of the epoxy nanocomposites prepared using $GnP_{M\text{-}15}$ under different processing conditions

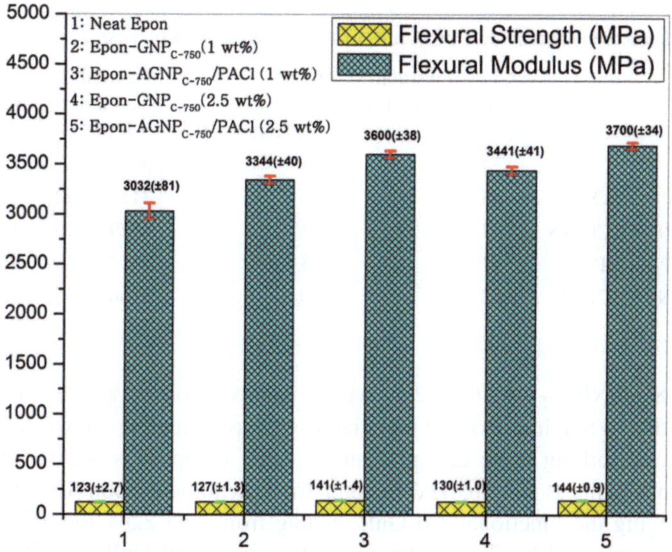

Fig. 9 Flexural properties of the epoxy nanocomposites prepared using $GnP_{C\text{-}750}$ (WS)

Figure 10 shows the load versus extension curve for neat pure Epon 828, Epon 828/unmodified GnP and Epon 828/PACl-functionalized GnP nanocomposites. The area under the stress–strain curve was highest (15.54 lb_f-inch/1.76 N-m) for the

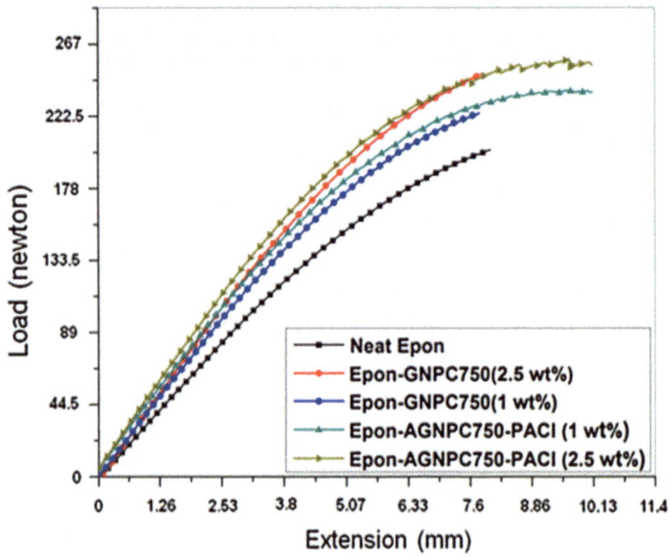

Fig. 10 Load versus extension curve for neat epoxy, epoxy/GnP and epoxy/AGNP–PACl nanocomposites

Table 1 Thermal and mechanical properties of the prepared nanocomposites

Material	Glass transition temperature (T_g) (°C)	Onset degradation temperature (°C)	Flexural modulus (MPa)	Flexural strength (MPa)
Epon	154	356	3032 ± 81	125 ± 2.7
Epon–GNP_{M15}–WS	145	343	3110 ± 52	116 ± 1.8
Epon–AGNP/PACl–WS	157	353	3529 ± 19	120 ± 0.7
Epon–GNP_{M15}–SP	141	343	3017 ± 39	87 ± 3.5
Epon–AGNP/PACl–SP	153	350	3360 ± 45	92 ± 1.3

Epon 828/$AGNP_{C\text{-}750}$–PACl (2.5 wt%) nanocomposite which corroborated the strength and toughness of the material as well as validating the capability of the material to absorbing more energy than the other composites before failure. The increase in toughness was marked in comparison with neat Epon 828 (87%) while upon changing the functionalized GnP loading from 1 to 2.5% the area under the curve only increased by 7.6%. The epoxy-functionalized GnP was 55.56% more tough than its unfunctionalized counterpart. The average particle diameter for C-750 grade GnP is ~0.5–2 μm which contributed to its homogenous diffusion within the polymer (Epon 828) matrix, thereby resulting in an appreciable improvement in the mechanical properties of the prepared nanocomposites.

Table 2 Effect of GNP loading on mechanical properties of the nanocomposites prepared without solution processing

Material	Flexural modulus (MPa)	Flexural strength (MPa)
Epon	3032 ± 81	125 ± 2.7
Epon–GNP_{C750}–1wt%	3344 ± 40	127 ± 1.3
Epon–AGNP/PACl–1wt%	3600 ± 38	141 ± 1.4
Epon–GNP_{C750}–2.5wt%	3441 ± 41	130 ± 1.0
Epon–AGNP/PACl–2.5wt%	3700 ± 34	144 ± 0.9

2.4 Conclusion

In summary, we have fabricated uniform, homogenous and well-dispersed epoxy-based PACl-functionalized graphene nanocomposites. The robust interfacial bonding in the nanocomposites evidenced by SEM observations resulted from the covalent ester linkage formed due to the reaction between PACl and the epoxy matrix. The nanocomposites prepared in the absence of solvent (acetone) yielded superior thermal and mechanical performance than the solution-processed ones. Also, the $GnP_{C\text{-}750}$ performed as a superior nanofiller than $GnP_{M\text{-}15}$ due to their increased surface area and high density of functional sites available which were responsible for the increment in the toughness of the material. The consequence of incorporation of GnP at different weight percentage on the flexural properties of the nanocomposites was also studied and improved mechanical properties were obtained at 1 wt% $AGNP_{C\text{-}750}$ loading which increased marginally upon changing the loading to 2.5 wt%. Overall, an enhancement of 17 and 22% in flexural strength and modulus was observed at 2.5 wt% loading of functionalized GnPs.

References

1. Geim, A.K., MacDonald, A.H.: Graphene: exploring carbon flatland. Phys. Today **60**, 35–41 (2007)
2. Stoller, M.D., Park, S.J., Zhu, Y.W., An, J.H., Ruoff, R.S.: Graphene-based ultracapacitors. Nano Lett. **8**, 3498–3502 (2008)
3. Chabot, V., Higgins, D., Yu, A., Xiao, X., Chen, Z., Zhang, J.: A review of graphene and graphene oxide sponge: material synthesis and applications to energy and the environment. Energy Environ. Sci. **7**(5), 1564–1596 (2014)
4. Cai, W., Zhu, Y., Li, X., Piner, R.D., Ruoff, R.S.: Large area few-layer graphene/graphite films as transparent thin conducting electrodes. Appl. Phys. Lett. **95**(12), 123115 (2009)
5. Wang, G., Shen, X., Yao, J., Park, J.: Graphene nanosheets for enhanced lithium storage in lithium ion batteries. Carbon **47** (8), 2049–2053 (2009)
6. Bose, S., Basu, S., Das, A., Rahman, M., Drzal, L.T.: Fabrication of a sulfonated aramid-graphene nanoplatelet composite paper and its performance as a supercapacitor electrode. J. Appl. Polym. Sci. **134**(29), 45099 (2017)
7. Guo, C.X., Yang, H.B., Sheng, Z.M., Lu, Z.S., Song, Q.L., Li, C.M.: Layered graphene/quantum dots for photovoltaic devices. Angew. Chem., Int. Ed. Engl. **49**(17), 3014–3017 (2010)

8. Eswaraiah, V., Sankaranarayanan, V., Ramaprabhu, S.: Graphene-based engine oil nanofluids for tribological applications. ACS Appl. Mater. Interfaces. **3**(11), 4221–4227 (2011)
9. Bose, S., Das, A., Basu, S., Drzal, L.T.: Edge stitching of graphene nanoplatelets (GnPs) and their effectiveness as a filler for epoxy nanocomposites. ChemistrySelect **2**(20), 5769–5774 (2017)
10. Bose, S., Das, A., Basu, S., Drzal, L.T.: Covalent functionalization of graphene using polyacryloyl chloride and performance of functionalized graphene-epoxy nanocomposite. Polym. Compos. **298**, 339 (2017)
11. Liu, Z., Shen, D., Yu, J., Dai, W., Li, C., Du, S., Jiang, N., Li, H., Lin, C.T.: Exceptionally high thermal and electrical conductivity of three-dimensional graphene-foam-based polymer composites. RSC Advances **6**, 22364–22369 (2016)
12. Zeng, C., Lu, S., Song, L., Xiao, X., Gao, J., Pan, L., He, Z., Yu, J.: Enhanced thermal properties in a hybrid graphene-alumina filler for epoxy composites. RSC Advances **5**, 35773–35782 (2015)
13. Yao, Y., Wang, J., Lu, H., Xu, B., Fu, Y., Liu, Y., Leng, J.: Thermosetting epoxy resin/thermoplastic system with combined shape memory and self-healing properties. Smart Mater. Struct. **25**, 015021 (2016)
14. Wajid, A.S., Ahmed, H.S., Das, S., Irin, F., Jankowski, A.F., Green, M.J.: High-performance pristine graphene/epoxy composites with enhanced mechanical and electrical properties. Macromol. Mater. Eng. **298**, 339–347 (2013)
15. Balakrishnan, S., Start, P.R., Raghavan, D., Hudson, S.D.: The influence of clay and elastomer concentration on the morphology and fracture energy of preformed acrylic rubber dispersed clay filled epoxy nanocomposites. Polymer **46**, 11255–11262 (2005)
16. Zhang, Y., Wang, Y., Yu, J., Chen, L., Zhu, J., Hu, Z.: Tuning the interface of graphene platelets/epoxy composites by the covalent grafting of polybenzimidazole. Polymer **55**, 4990–5000 (2014)
17. Mittal, V.: Functional polymer nanocomposites with graphene: a review. Macromol. Mater. Eng. **299**, 906–931 (2014)
18. Sofer, Z., Simek, P., Pumera, M.: Complex organic molecules are released during thermal reduction of graphite oxides. Phys. Chem. Chem. Phys. **15**, 9257–9264 (2013)
19. Cheng, M., Yang, R., Zhang, L., Shi, Z., Yang, W., WangD, Xie G., Shi, D., Zhang, G.: Restoration of graphene from graphene oxide by defect repair. Carbon **50**, 2581–2587 (2012)
20. Yang, K., Gu, M., Guo, Y., Pan, X., Mu, G.: Effects of carbon nanotube functionalization on the mechanical and thermal properties of epoxy composites. Carbon **47**, 1723–1737 (2009)
21. Kim, M.G., Moon, J.B., Kim, C.G.: Effect of CNT functionalization on crack resistance of a carbon/epoxy composite at a cryogenic temperature. Compos. A Appl. Sci. Manuf. **43**, 1620–1627 (2012)
22. Rahman, M.M., Hosur, M., Zainuddin, S., Jajam, K.C., Tippur, H.V., Jeelani, S.: Mechanical characterization of epoxy composites modified with reactive polyol diluent and randomly-oriented amino-functionalized MWCNTs. Polym. Testing **31**, 1083–1093 (2012)
23. Damian, C.M., Garea, S.A., Vasile, E., Iovu, H.: Covalent and non-covalent functionalized MWCNTs for improved thermo-mechanical properties of epoxy composites. Compos. B Eng. **43**, 3507–3515 (2012)
24. Zou, W., Du, Z.J., Liu, Y.X., Yang, X., Li, H.Q., Zhang, C.: Functionalization of MWNTs using polyacryloyl chloride and the properties of CNT–epoxy matrix nanocomposites. Compos. Sci. Technol. **68**, 3259–3264 (2008)
25. Ho, S.S., Park, K.H., Kim, B.H., Choi, Y.W., Jun, G.H., Lee, D.J., Kong, B.S., Paik, K.W., Jeon,S.: Enhanced thermal conductivity of epoxy–graphene composites by using non-oxidized graphene flakes with non-covalent functionalization. Adv. Mater. **25**, 732–737 (2013)
26. Lee, J.K., Song, S., Kim, B.: Functionalized graphene sheets-epoxy based nanocomposite for cryotank composite application. Polym. Compos. **33**, 1263–1273 (2012)
27. Jin, F.L., Ma, C.J., Park, S.J.: Thermal and mechanical interfacial properties of epoxy composites based on functionalized carbon nanotubes. Mater. Sci. Eng., A **528**, 8517–8522 (2011)

28. Nadler, M., Werner, J., Mahrholz, T., Riedel, U., Hufenbach, W.: Effect of CNT surface functionalisation on the mechanical properties of multi-walled carbon nanotube/ epoxy-composites. Compos. A Appl. Sci. Manuf. **40**, 932–937 (2009)
29. Geng, Y., Liu, M.Y., Li, J., Shi, X.M., Kim, J.K.: Effects of surfactant treatment on mechanical and electrical properties of CNT/epoxy nanocomposites. Compos. A Appl. Sci. Manuf. **39**, 1876–1883 (2008)
30. Shen, J., Huang, W., Wu, L., Hu, Y., Ye, M.: Thermo-physical properties of epoxy nanocomposites reinforced with amino-functionalized multi-walled carbon nanotubes. Compos. A Appl. Sci. Manuf. **38**(5), 1331–1336 (2007)
31. Kim, J.A., Seong, D.G., Kang, T.J., Youn, J.R.: Effects of surface modification on rheological and mechanical properties of CNT/epoxy composites. Carbon **44**, 1898–1905 (2006)
32. Zhang, Y., Ren, L., Wang, S., Marathe, A., Chaudhuri, J., Li, G.: Functionalization of graphene sheets through fullerene attachment. J. Mater. Chem. **21**, 5386–5391 (2011)
33. Niyogi, S., Bekyarova, E., Itkis, M.E., McWilliams, J.L., Hamon, M.A., Haddon, R.C.: Solution properties of graphite and graphene. J. Am. Chem. Soc. **128**, 7720–7721 (2006)
34. Worsley, K.A., Ramesh, P., Mandal, S.K., Niyogi, S., Itkis, M.E., Haddon, R.C.: Soluble graphene derived from graphite fluoride. Chem. Phys. Lett. **445**, 51–56 (2007)
35. Zhang, Y., Ren, L., Wang, S., Marathe, A., Chaudhuri, J., Li, G.: Functionalization of graphene sheets through fullerene attachment. J. Mater. Chem. **21**, 5386–5391 (2011)
36. Salavagione, H.J., Gomez, M.A., Martinez, G.: Polymeric modification of graphene through esterification of graphite oxide and poly (vinyl alcohol). Macromolecules **42**, 6331–6334 (2009)
37. MohammadiA, Peighambardoust S.J., Entezami, A.A., Arsalani, N.: High performance of covalently grafted poly (o-methoxyaniline) nanocomposite in the presence of amine-functionalized graphene oxide sheets (POMA/f-GO) for supercapacitor applications. J. Mater. Sci.: Mater. Electron. **28**, 5776–5787 (2017)
38. Sulleiro, M.V., Quiroga, S., Peña, D., Pérez, D., Guitian, E., Criado, A., Prato, M.: Microwave-induced covalent functionalization of few-layer graphene with arynes under solvent-free conditions. Chem. Commun. **54**, 2086–2089 (2018)
39. Criado, A., Melchionna, M., Marchesan, S., Prato, M.: The covalent functionalization of graphene on substrates. Angew.Chem. Int. Ed. **54**, 10734–10750 (2015)
40. Naebe, M., Wang, J., Amini, A., Khayyam, H., Hammed, N., Li, L.H., Chen, Y., Fox, B.: Mechanical property and structure of covalent functionalised graphene/epoxy nanocomposites. Sci. Rep. **4**, 4375 (2014)
41. Bose, S., Kuila, T., Mishra, A.K., Kim, N.H., Lee, J.H.: Preparation of non-covalently functionalized graphene using 9-anthracene carboxylic acid. Nanotechnology **22**, 405603 (2011)
42. Bose, S., Kuila, T., Mishra, A.K., Kim, N.H., Lee, J.H.: Dual role of glycine as a chemical functionalizer and a reducing agent in the preparation of graphene: an environmentally friendly method. J. Mater. Chem. **22**, 9696–9703 (2012)
43. Park, S., Lee, K.S., Bozoklu, G., Cai, W., Nguyen, S.T., Ruoff, R.S.: Graphene oxide papers modified by divalent ions-enhancing mechanical properties via chemical cross-linking. ACS Nano **2**, 572–578 (2008)
44. Ahmadi-Moghadam, B., Sharafimasooleh, M., Shadlou, S., Taheri, F.: Effect of functionalization of graphene nanoplatelets on the mechanical response of graphene/epoxy composites. Mater. Des. **66**, 142–149 (2015)

Chapter 3
Behaviour Analysis and Comparison of Tribological Characteristics of Electroless Ni–B Coating under Dry and Lubricated Condition

Santanu Duari, Arkadeb Mukhopadhyay, Tapan Kumar Barman and Prasanta Sahoo

3.1 Introduction

A successful design of machine element depends on the understanding of tribological principles. The word 'tribology' has been derived from Greek word 'tribos' which means rubbing. Tribology may be defined as the science and technology of contacting surfaces under relative motion. Tribological behaviour depends on rubbing surface of material with or without the effect of environment. Due to advancements in technology, nowadays various methods have been developed for surface modification [1]. In tribology, friction and wear are necessary evils which cannot be eliminated but needs to be controlled or minimized for the effective working of a machine element as well as improved working life. Thus, the study of tribology is an important and essential part. It applies operational analysis for attributes of great economic importance such as reliability, maintainability and wear of technical equipment ranging from a simple household appliance to a complex spacecraft.

In the present era, the purpose of deposition of metallic coatings is to serve as a protective layer to the base material against wear and corrosion. Friction and wear are two decaying marvels which are the wellsprings of significant misfortune in mechanical components. They do not just decrease the life of the modern machine components but also result in increase in cost and maintenance. Since friction, corrosion and wear are surface phenomenon, they can be controlled by surface treatment. Electroless deposition is by far a much better way to overcome such type of problems rather than any other coatings. Electroless deposition of metals was first demonstrated in 1844 [2]. However, the first practical system for electroless coating was developed by Brenner and Riddell after a century [3].

S. Duari (✉) · A. Mukhopadhyay · T. K. Barman · P. Sahoo
Department of Mechanical Engineering, Jadavpur University, Kolkata, India
e-mail: santanu_duari@yahoo.co.in

P. Sahoo and J. P. Davim (eds.), *Advances in Materials, Mechanical and Industrial Engineering*, Lecture Notes on Multidisciplinary Industrial Engineering,
https://doi.org/10.1007/978-3-319-96968-8_3

Electroless plating is an autocatalytic procedure where the substrate builds up a potential when it is immersed in an electroless setup called bath that contains a source of metallic ions, complexing agent, reducing agent, buffers, stabilizer, wetting agents and different components. Nickel sulphate and nickel chloride salts are generally used as the source of nickel ion. However, nickel hypophosphite [Ni $(H_2PO_2)_2$] has been considered the ideal source of nickel ion [4–7]. Sodium borohydride ($NaBH_4$) is a strong reducing agent for the electroless nickel (EN) plating process [8–11]. Investigations have been also carried out using hydrazine as a reducing agent for deposition of pure nickel, palladium and gold [12]. But the use of hydrazine has been discontinued since it is very difficult to control and its associated hazards [13]. In order to prevent the decomposition of the solution and to make the bath stable, the complexing agents are added to the electroless bath [14]. Buffering agents are used to control the pH of the bath. Effect of pH on the plating rate of electroless nickel deposits has been also carried out [15]. Stabilizer content even in significantly low amounts may slow down the reaction rate [14]. Surfactants tend to lower the surface tension and hence permit lowering the interfacial tension between two liquids or a liquid and solid surface. Sodium dodecylsulphate (SDS) and cetyltrimethyl ammonium bromide (CTAB) are used as a surfactant for electroless Ni–P coating [16].

A number of advantages of using electroless nickel coating method over other coating method are as follows [5]:

- Resistance against wear and corrosion;
- Increase in microhardness and thermal resistance;
- Smooth and uniform deposit regardless of the geometry of the substrates;
- Good solderability, bondability, weldability, chemical stability and non-magnetic properties;
- Greater lubricity and high thermal conductivity.

The physical and chemical properties of electroless nickel coatings depend upon bath formulation and conditions of operation. The microstructures and properties of EN coatings are dependent on phosphorous content alloyed in the deposit [17–19]. Annealing affects the properties of electroless nickel coatings by changing its microstructure [6, 20, 21]. Electroless Ni–B coatings are found to exhibit an extremely smooth surface [22]. Heat treatments of the coating cause a further degradation of normal and maximal roughness [22, 23].

From the literature, it may be comprehended that several studies have been focused on the investigation of tribological properties of electroless nickel coatings. Electroless nickel–phosphorus is the most popular variant due to its excellent tribological and anti-corrosion characteristics [24–26]. Ni–B coating is comparatively newer and is a hard variant of the EN family. But most of the literatures concerning the tribological characteristics of Ni–B coatings have been carried out under dry sliding condition [27–32]. But it has also been observed that the microstructure of Ni–B coatings is such that they are naturally/inherently self-lubricious. Furthermore, the nodulated structures and columnar growths also have the capability of retaining lubricants. Therefore, it is necessary to investigate the tribological behaviour of Ni–B coatings under lubrication. The suitability of electroless Ni–B as a tribo-coating under lubricated condition is assessed in the present work. A comparison with the tribological behaviour under dry condition is also carried out. Flow diagram of the present work is shown in Fig. 3.1.

3.2 Materials and Methods

3.2.1 Preparation of Substrate and Coating Deposition

Coating deposition is carried out on cylindrical steel (AISI 1040) substrates of dimension Ø6 × 30 mm. This specific measurement of the specimen is selected in accordance with its counterpart in the tribotesting setup where the sample must be attached for tribological testing. Electroless coatings follow the profile of the substrate on which it is deposited. Hence, wear and frictional force may be dependent on the surface roughness of the coating, which in turn depends on surface roughness of the substrate. In the present work, the centre line average roughness (R_a) of the coatings is 0.4 μm (corresponding to N5 roughness grade). Before coating, the substrates are rinsed in deionized water. Then, acetone is used for degreasing and removal of corrosive or any remaining organic products. Finally, before deposition, the substrates are subjected to pickling treatment in HCl solution for the removal of oxide layers which may have formed. Finally, the samples are rinsed in deionized water and dipped into the electroless bath.

The electroless bath is prepared by mixing the chemicals as shown in Table 3.1. The chemicals are measured on an electronic balance of high precision. Two beakers with 250 ml volume are properly cleaned and then rinsed using deionized water for preparation of the electroless bath. In one of the beakers (A), 100 ml deionized water is taken and the same volume of water is taken in another beaker (B). In beaker A, the chemicals (nickel chloride, ethylenediamine, sodium hydroxide and lead nitrate) are taken in proper quantity along with sequence given and mixed on a magnetic heater cum stirrer. In beaker B, the chemicals (sodium hydroxide and sodium borohydride) are taken sequentially and mixed by stirrer. The solutions in the two beakers are mixed to form a total bath volume of 200 ml. The pH of the solution is measured to be around 12.5 using pH strips and it was maintained constant throughout the deposition process. Upon heating the solution, the substrate initiates the autocatalytic chemical reduction process. The samples are dipped into the chemical bath keeping temperature constant at 90 ± 2 °C for deposition time of about 2 h. It is ensured that a uniform value of coating thickness is maintained in all samples by controlling chemical bath composition and deposition condition. The coated specimen is taken out from bath and washed with deionized water. The schematic diagram of the electroless coating deposition setup is presented in Fig. 3.2.

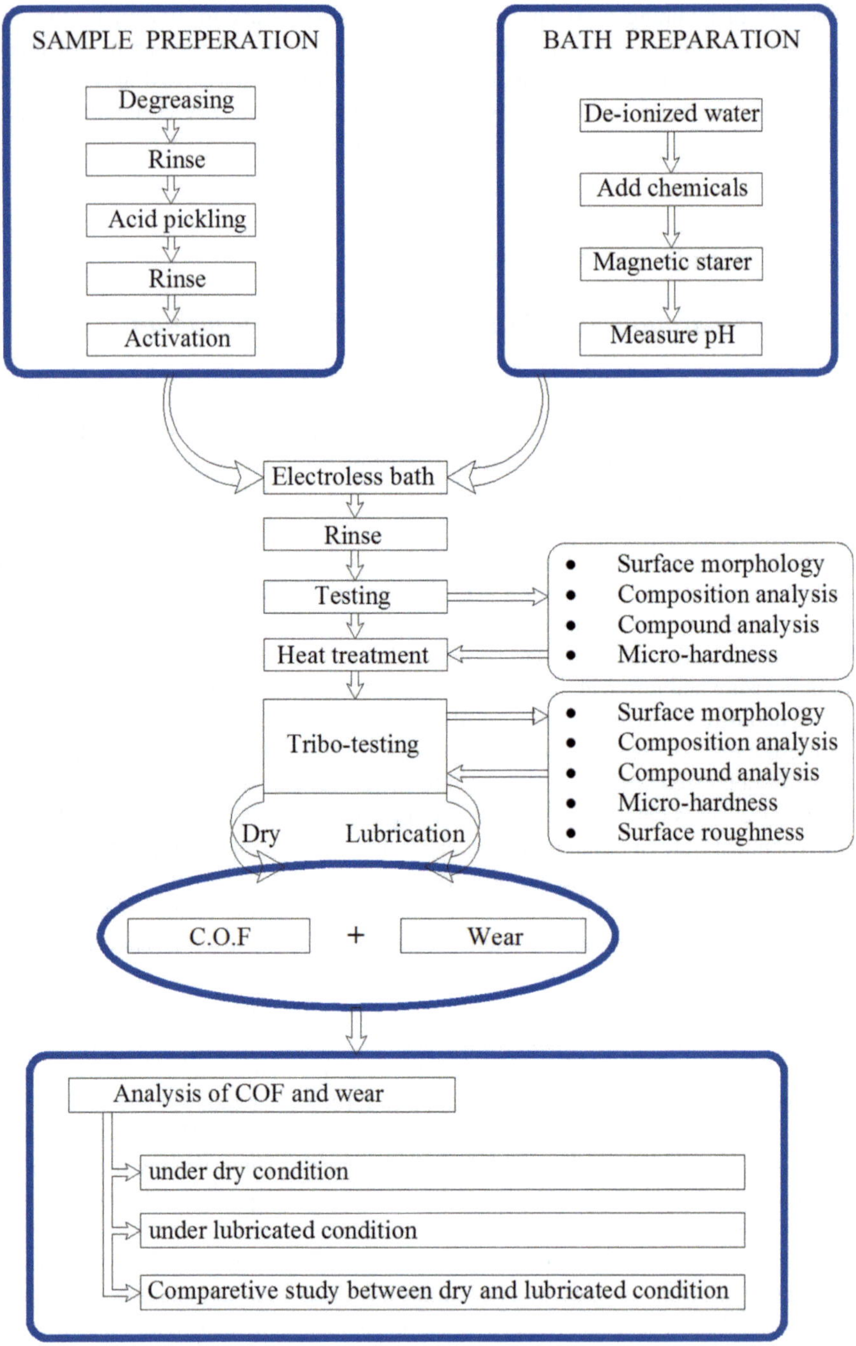

Fig. 3.1 Flow diagram of experiment and analysis

Table 3.1 Composition and deposition condition of chemical bath

Bath constituents	Role of chemicals	Values	Unit
Nickel chloride	Source of nickel ion	20	g/l
Sodium borohydride	Reducing agent	0.8	g/l
Ethylenediamine	Complexing agent	59	g/l
Lead nitrate	Stabilizer	0.0145	g/l
Sodium hydroxide	Buffer	40	g/l
Bath temperature		90 ± 2	°C
pH of solution		12.5	
Deposition time		2	hr
Bath volume		200	ml

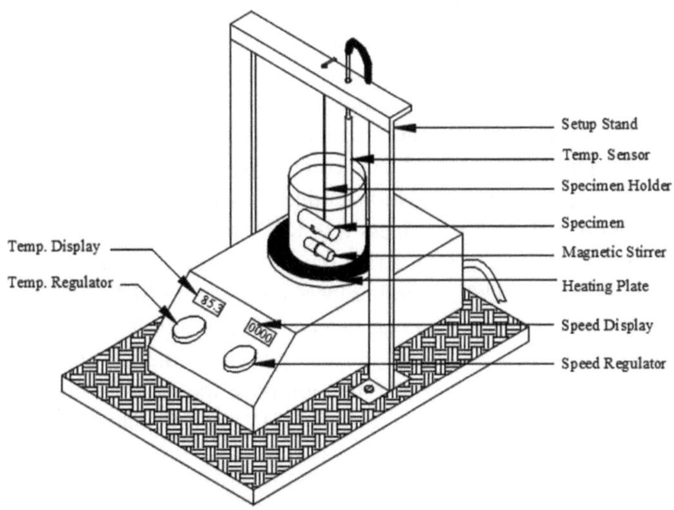

Fig. 3.2 Schematic diagram of the electroless coating deposition setup

The coated specimens are annealed in a muffle furnace with a size of 100 × 100 × 220 mm^3. Annealing temperature is kept constant to maintain almost same hardness of all the specimens. Heat treatment is done at 350 °C for one hour. After heating, the samples are held inside the chamber and allowed to cool down to ambient temperature without the application of any artificial cooling.

3.2.2 Coating Characterization

The composition of EN depositions significantly determines the phase structure and tribological behaviour. Heat treatment results in an improved hardness, wear resistance and low friction coefficient. Due to this, it is important to study the coating characteristics in detail and to understand their effect on wear behaviour. The surface morphology, bath composition and phase transformation characteristics of the coatings have been examined by scanning electron microscope (SEM), energy dispersive X-ray analysis (EDX) and X-ray diffraction analysis (XRD), respectively. SEM and EDX analysis are done on a FEI Quanta 250 equipment, whereas XRD is carried out on a Rigaku Ultima III equipment.

3.2.3 Hardness Study of the Coating

Hardness is defined as the reluctance of a material to permanently deform, or indent. Hardness indicates the ability of a material to resist permanent deformation by another harder material and its resistance to plastic deformation. There are several hardness testing mechanisms such as scratch hardness, indentation hardness and rebound hardness. The hardness of a material depends on the ductility, elastic stiffness, plasticity, strain, strength and toughness of the material. Scratch hardness denotes the resistance of a material to fracture or plastic deformation due to friction from a sharp object. Various indentation hardness tests include Vickers hardness test, Knoop hardness test, Brinell's hardness test, Rockwell's hardness test, etc.

Here, Vickers hardness test is carried out to measure the microhardness of thin coatings with the help of UHL microhardness apparatus (VMHT MOT, Technische Mikroskopie) with a Vickers diamond indenter (Fig. 3.3a). It is generally based on the indentation formed by a diamond indenter onto the material. The basic principle of Vickers hardness test is to observe the ability of the material under question to resist plastic deformation. Vickers hardness (HV) is calculated according to the following formula:

$$\mathrm{HV} = 0.1891\frac{F}{d^2} \tag{3.1}$$

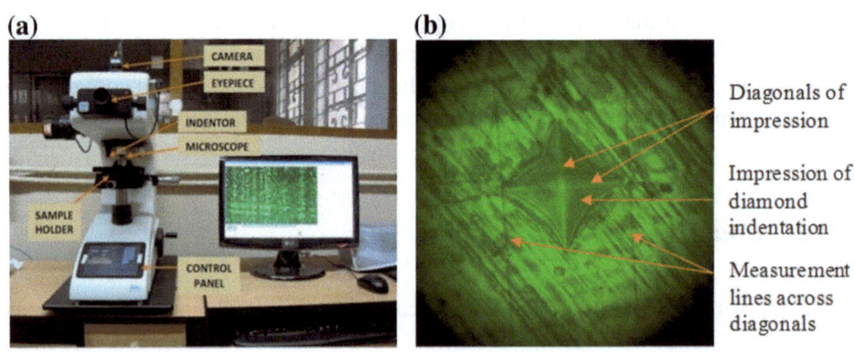

Fig. 3.3 **a** Photograph of microhardness testing machine **b** Microscopic image of impression

where F is the load in kilogram-force and d is the average length of the two diagonals (shown in Fig. 3.3b) created by the pyramidal diamond indenter onto the surface of the material in millimetres. The indentation is done with Vickers diamond indenter with load of 500 g, dwell time of 15 s and approach velocity of 25 μm/s. A mean of at least six hardness data for each sample is taken.

3.2.4 Roughness Study of the Coating

Surface roughness describes the morphological features of a surface. Any surface is not plane or smooth as it appears but covered with microscopic hills, valleys and even scratches. Surface roughness is a bulk measure of the average size of the hills and valleys. Surface roughness is critical to the performance of wear-resistant and self-lubricating coatings, stress corrosion and fatigue. Wear and friction are influenced by surface roughness and in many instances; an optimum roughness can be found which provides a minimum of wear and friction. For coatings proposed to present high corrosion resistance, excessive roughness can be an indication of low-quality coatings.

Centre line average or R_a is the arithmetic mean value of the departure of the profile from the centre line along the length given below:

$$R_a = \frac{1}{L}\int_0^L |Z(x)|\mathrm{d}x \tag{3.2}$$

where $Z(x)$ is the height of the surface above mean line at a distance x from the origin and L is the measured length of the profile as shown in Fig. 3.4. The surface roughness parameter R_a on the coated surfaces is estimated with the Talysurf (Surtronic 3+, Make—Taylor Hobson, UK). It is equipped with a diamond stylus with a tip radius of 5 μm.

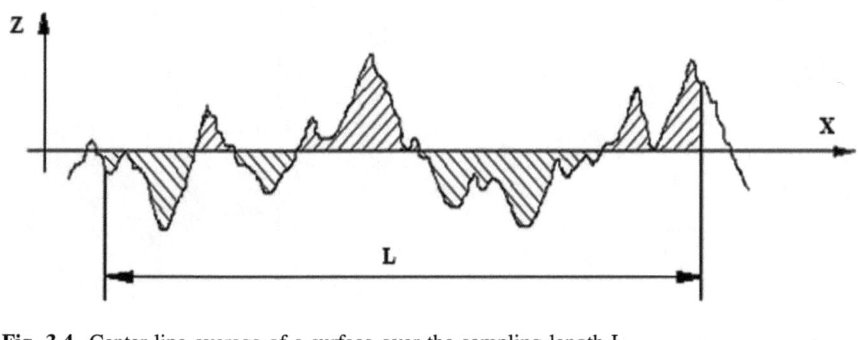

Fig. 3.4 Center line average of a surface over the sampling length L

In this study, the measurements are taken at 0.8 mm sampling length and 4 mm traverse length. Roughness measurements on the coated samples have been repeated six times. The measured profile of Talysurf is digitized by the advanced surface finish analysis software Talyprofile for evaluation of the roughness parameter R_a.

3.2.5 Tribological Tests

The friction and wear tests of EN coatings are carried out on a pin-on-disc type setup with a data acquisition system (TR-20LE-CHM-400, Ducom, India). The applied normal load, rotational speed of counterface disc and sliding duration are varied. The track diameter is set at 80 mm. A photograph of the tribotester is shown in Fig. 3.5. The coated specimens press against a rotating counterface disc of dimensions Ø165 × 8 mm (material: hardened EN31 steel). The speed and time for which the disc rotates can be controlled using the controller device provided with the tribotester. The normal load is applied by placing dead weights on a loading pan attached to a loading lever that transmits the normal load to the specimen at a ratio of 1:1. A load cell measures the frictional force. A linear variable differential transformer (LVDT) measures the wear depth (μm) of the specimens. Friction and wear tests are conducted under both dry and lubricated condition. For dry condition, testing is done under normal ambient condition without providing any lubricant. For the sliding wear tests under lubricated condition, Servo PRIDE 40,

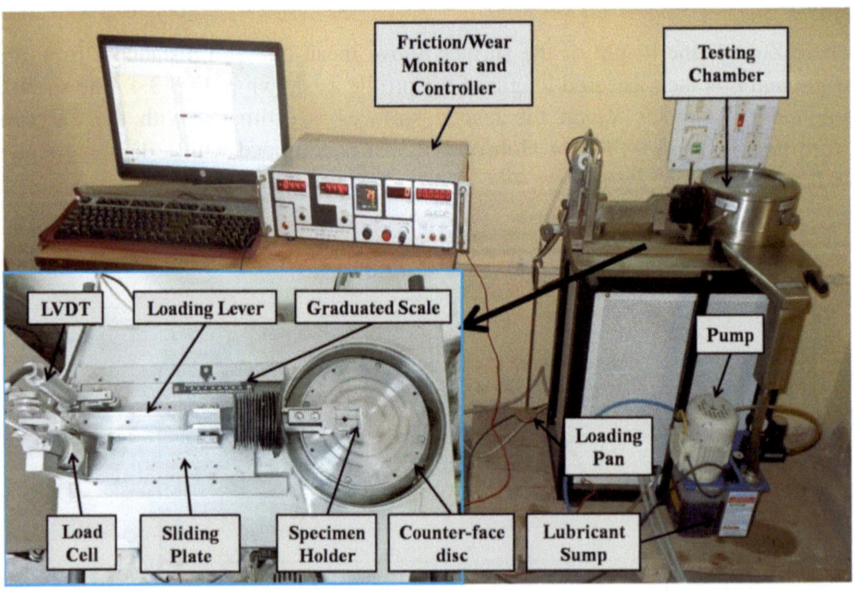

Fig. 3.5 Pin-on-disc type tribotester (TR-20LE-CHM-400)

Table 3.2 Characteristics of lubricant used

SAE Grade	PRIDE 40
Kinematic viscosity, cSt @ 100 °C	13–15
Viscosity index, Min.	90
Flash point (Cleveland open-cup method), °C Min.	220
Pour point, °C Max.	(−) 6
Total base number (TBN) mg KOH/gm	9.5–12.5

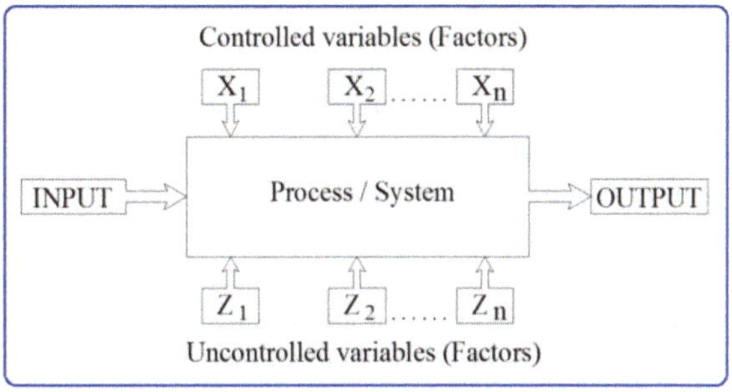

Fig. 3.6 General model of a particular process or system

Table 3.3 Test parameters along with their levels for behaviour analysis

Design factors	Unit	Levels				
		1	2	3	4	5
Load (L)	N	10	15	20	25	30
Speed (S)	RPM	60	70	80	90	100
Time (T)	Min.	3	6	9	12	15

manufactured by Indian Oil, is used which is an engine oil. The properties of the lubricating oil are shown in Table 3.2.

Experiments are carried out following full factorial design [33, 34]; that is, it includes all and every individual combination of levels and parameters. A general model of experimental design technique for investigation of a system is shown in Fig. 3.6. Here, three most important parameters, viz. applied load, rotating speed of counterface disc and duration of sliding, have been selected as tribological testing parameters. Their values at different levels are laid down in Table 3.3. The design parameters have been divided into equally spaced levels.

3.3 Results and Discussion

Electroless Ni–B is deposited on AISI 1040 specimen. Coating characterization is done using SEM, EDX and XRD to ensure and analyse the uniformity of the deposits. The microhardness and surface roughness of electroless Ni–B deposits are also reported. Variations of friction and wear with process parameters are described for dry and lubricated conditions. An attempt has been also made to compare the coefficient of friction (COF) and wear depth under dry and lubricated sliding condition.

3.3.1 Coating Characterization

3.3.1.1 Surface Morphology

SEM micrographs of electroless Ni–B in as-deposited condition and post-annealing (350 °C for 1 h) condition have been presented in Fig. 3.7. The coated surface seems to have a cauliflower-like appearance which makes it self-lubricating in nature. Also, no surface damage can be observed. The surface appears to be grey in colour. Compared to the as-deposited SEM micrograph, the heat-treated deposit is more inflated. The nodules of the coating increase in size with heat treatment, whereas in as-deposited condition they are more flat and deflated.

A cross section of the deposit has been shown in Fig. 3.8 that reveals uniform deposit thickness and columnar structures. The coating thickness is ~30 μm and seems to be well connected to the substrate.

3.3.1.2 Analysis of Composition

To ensure the deposition of nickel and boron on the mild steel substrate, EDX is used. The EDX spectrum of the coated samples in as-deposited condition is displayed in Fig. 3.9. The peaks of nickel and boron are quite specific. Boron content is seen to be around 7.51% by weight while the remaining is mostly nickel.

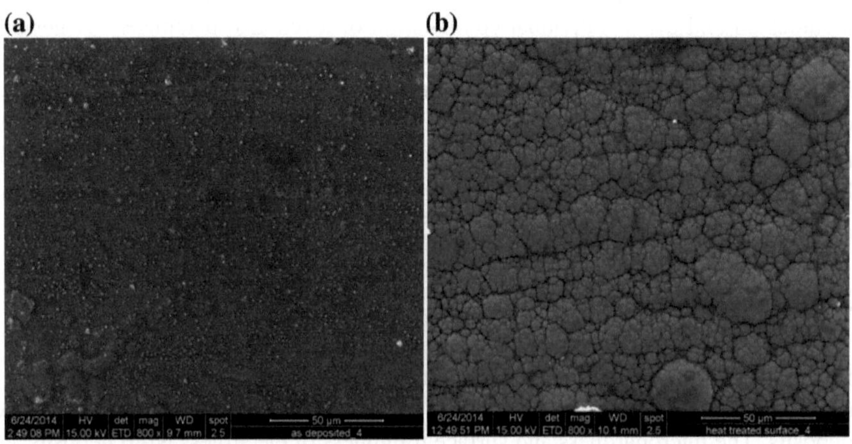

Fig. 3.7 SEM micrograph of Ni–B coating **a** as-deposited **b** annealed at 350 °C

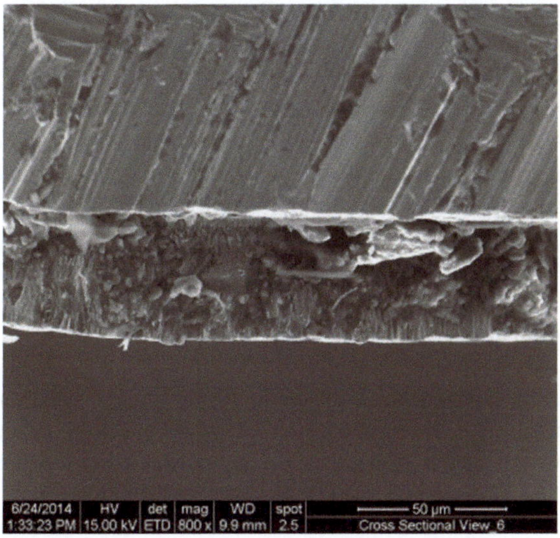

Fig. 3.8 SEM micrograph of cross cut of as-deposited Ni–B coating

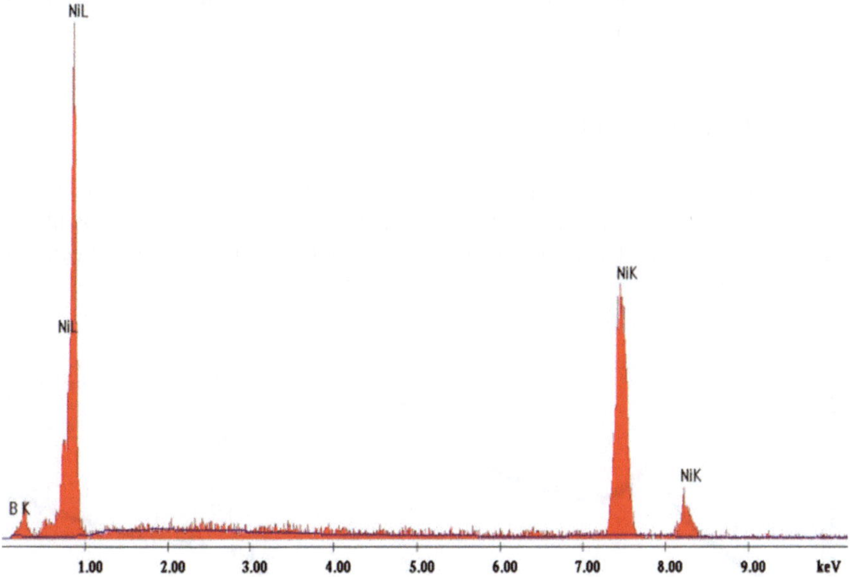

Fig. 3.9 EDX spectrum of Ni–B coating

3.3.1.3 Analysis of Phase Transformation

The XRD plots for as-deposited as well as heat-treated deposits are shown in Fig. 3.10a, b, respectively. It is observed from the plots that in as-plated condition, electroless Ni–B coating exhibits a single broad peak which indicates that it is amorphous in nature. Transformation of amorphous state to crystalline takes place due to heat treatment. The XRD pattern of electroless Ni–B coating post-annealing confirms the formation of Ni_3B and Ni_2B crystalline phases.

3.3.1.4 Microhardness Study

The microhardness values of the deposits have been measured in as-deposited and heat-treated condition to mark the effect of heat treatment on electroless Ni–B coating. The average values of hardness in as-deposited and heat-treated condition are seen to be 855 $HV_{0.5}$ and 1160 $HV_{0.5}$, respectively. Thus, there is a significant rise in the value of hardness. The values of hardness for all the 125 samples are presented in Fig. 3.11. The formation of hard nickel borides results in the increased hardness of electroless Ni–B coatings in heat-treated condition resulting from precipitation hardening phenomenon. Nickel borides may be impeding the movement of dislocations (defects in crystal lattice) and in this manner contributes to the strength of the deposition.

3.3.1.5 Roughness Study

Surface roughness plays an important role in controlling friction and wear. Before subjecting the Ni–B coatings to tribological tests, surface roughness of all the samples is measured. To ensure that results of tribological tests are not affected by

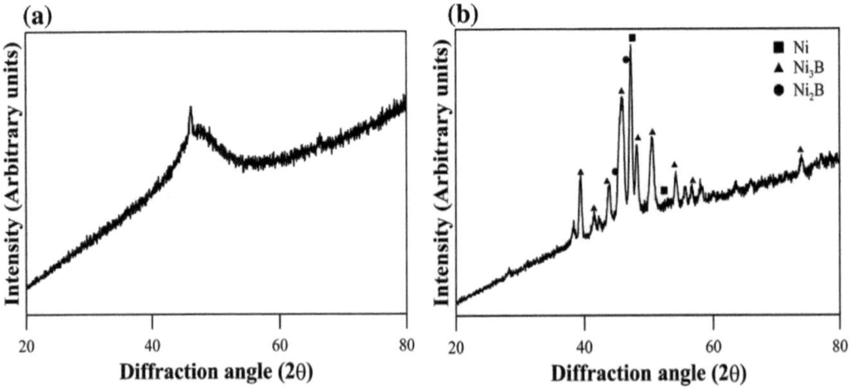

Fig. 3.10 XRD pattern of Ni–B coating **a** as deposited and **b** heat treated at 350 °C

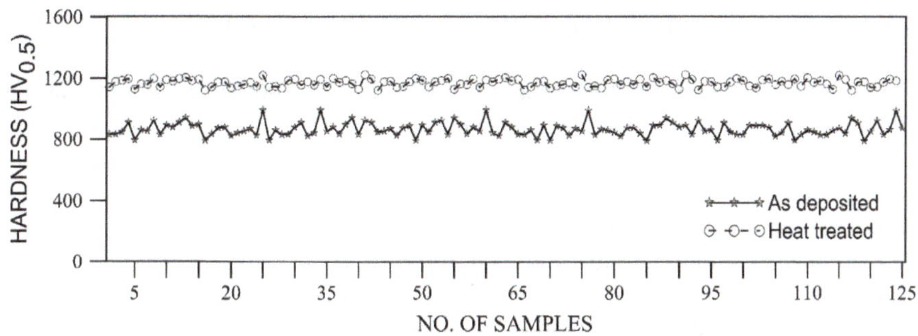

Fig. 3.11 Microhardness of Ni–B coating

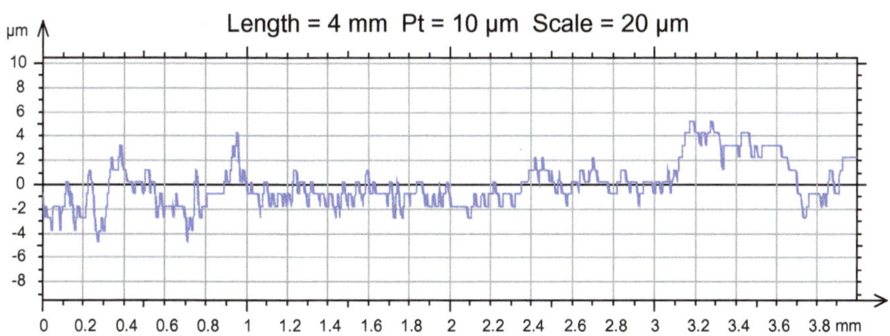

Fig. 3.12 Surface roughness of Ni–B coating

the surface roughness, the coatings having R_a value ~0.7–0.8 μm are selected for the experimental runs. Surface profile of one of the samples is shown in Fig. 3.12.

3.3.2 *Friction and Wear Behaviour*

Experiments for investigation of friction and wear are conducted on a pin-on-disc type tribotester as shown in Fig. 3.5 following the full factorial design. Tests are completed in two different environments, i.e. dry and lubricated condition. Properties of the lubricant are also presented in Table 3.2. Data for friction and wear depth are collected from the tribotester for all possible combination of design parameters, i.e. applied load, speed and sliding duration. Variation of friction and wear with process parameters under dry and lubricated conditions are studied.

3.3.2.1 Friction and Wear Behaviour Under Dry Condition

Figure 3.13a, b represents the variation of COF versus sliding duration at various sliding speeds when the load is kept constant at 10 and 30 N, respectively. It is noticed that with the increase of sliding speed, COF increases. So it can be concluded that COF varies inversely with load and directly with speed. Figure 3.14a, b shows the variations of COF versus sliding time with varying applied normal loads and at constant sliding speeds of 60 and 100 rpm, respectively. It is found that for any constant sliding speed, friction coefficient gradually increases with sliding duration. The COF decreases with an increment in applied load at 60 rpm as well as 100 rpm. But the COF seems to be quite low and comparable with that reported by other researchers for EN coatings [21, 23]. The low COF of electroless Ni–B coating can be mainly attributed to its microstructure and self-lubricating property. The rise in frictional force is not significant enough to cause an increase in the value of COF of the deposits.

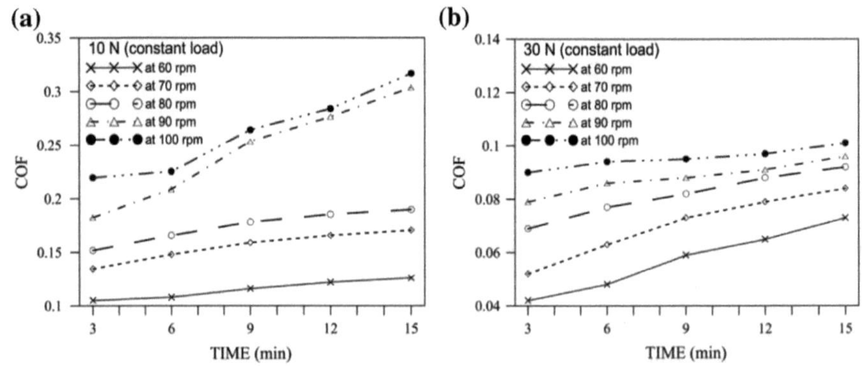

Fig. 3.13 COF versus sliding duration of Ni–B coating with different sliding speed at constant load of **a** 10 N and **b** 30 N under dry condition

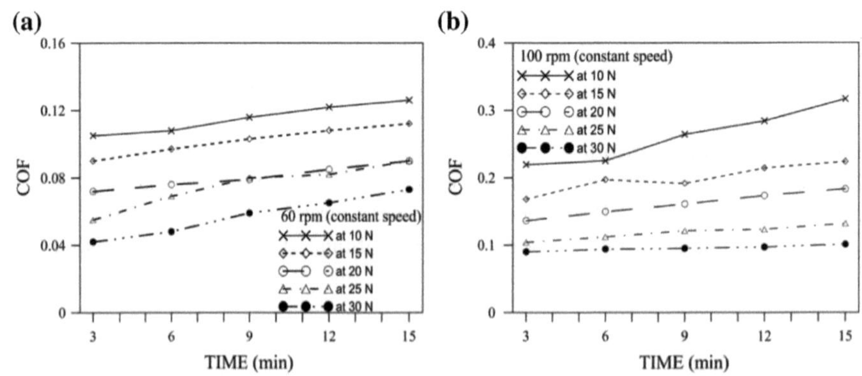

Fig. 3.14 COF versus sliding duration of Ni–B coating with different load at constant speed of **a** 60 rpm and **b**100 rpm under dry condition

Figure 3.15a, b shows the variation of wear depth with sliding duration at various speeds keeping the load constant at 10 and 30 N, respectively. In Fig. 3.15a, b, wear depth increases with speed at 10 and 30 N, respectively. In fact, the wear depth also increases significantly with duration of sliding. But the increase in wear depth with sliding speed is gradual and not drastic within 60–100 rpm at 10 N as well as 30 N.

Figure 3.16 shows the variation of wear depth with sliding time at constant speed and at different normal loads. Figure 3.16a represents the variation of wear depth at a constant speed of 60 rpm (at the lowest sliding speed). From this plot, the following observations may be made: (i) with sliding time, wear depth increases and (ii) as the normal load increases, the wear depth increases. Figure 3.16b represents the same trend at a sliding speed of 100 rpm (at the highest sliding speed). From these two graphs, it is also seen that in general at higher speed the wear depth

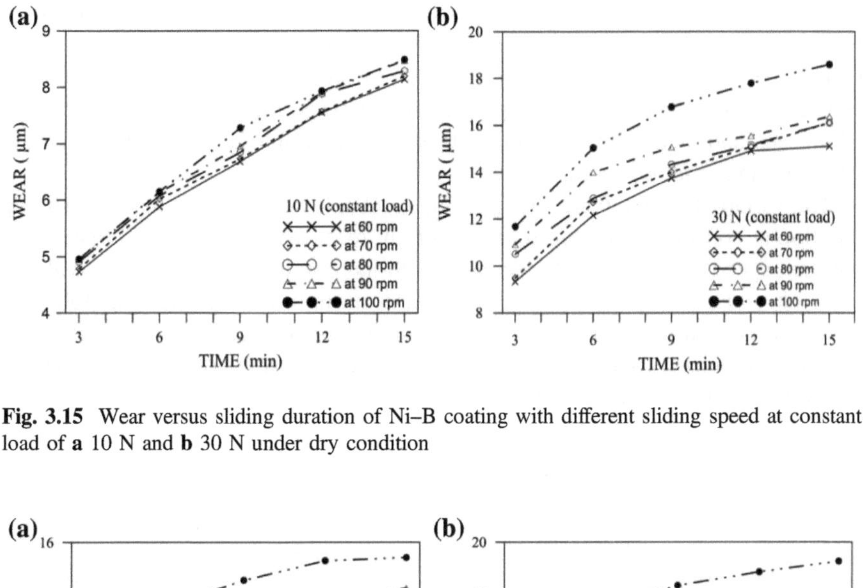

Fig. 3.15 Wear versus sliding duration of Ni–B coating with different sliding speed at constant load of **a** 10 N and **b** 30 N under dry condition

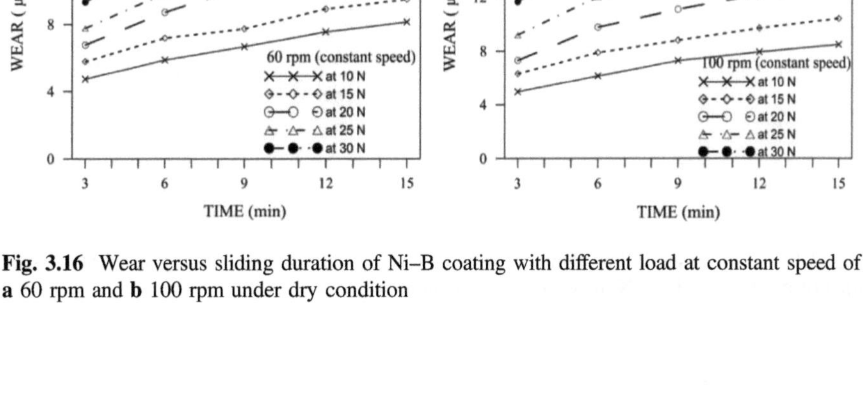

Fig. 3.16 Wear versus sliding duration of Ni–B coating with different load at constant speed of **a** 60 rpm and **b** 100 rpm under dry condition

is higher. In fact at 100 rpm (Fig. 3.16b), the increase in wear depth is quite high. Compared to 10 N load, wear depth increases by almost three times when the load is increased to 30 N. Finally, it is concluded that wear depth increases with applied load as well as sliding speed.

When normal load increases, the interacting surfaces advance more towards each other. A greater number of asperities come in contact and also the contact radius of the individual asperities increases which leads to an increase in contact area of the asperities. This results in an increase in the volume of material that is being shared by the rotating counterface material of the tribotester. Hence, the wear depth increases. So the minimum wear may be found at low levels of load, speed and sliding time.

3.3.2.2 Friction and Wear Behaviour Under Lubricated Condition

In this section, COF and wear characteristics are studied under lubricated condition. Figure 3.17 represents the effect of normal load on COF at different sliding speed with respect to duration of sliding. At 10 N as well as 30 N, the COF is almost constant with an increase in sliding duration. At 10 N, the COF initially decreases with speed and then increases. But at 30 N, the variation of COF at 70–100 rpm is very less and is almost comparable.

The variations of COF are shown in Fig. 3.18a, b with different load and sliding duration at constant speed of 60 and 100 rpm, respectively. It is found that at any constant speed, COF gradually increases with increasing load. The COF for all the cases observed in Figs. 3.17 and 3.18 exhibits an initial abrupt increase, followed by a slow rise during the first few seconds of sliding. The decrease in friction with the increase in load could be due to the self-lubricating nature of the deposition and the ability of the nodules to retain lubricants.

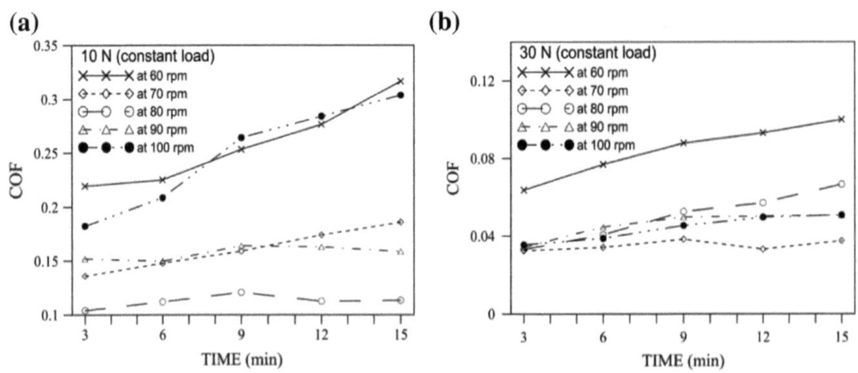

Fig. 3.17 COF versus sliding duration of Ni–B coating with different sliding speed at constant load of **a** 10 N and **b** 30 N under lubricated condition

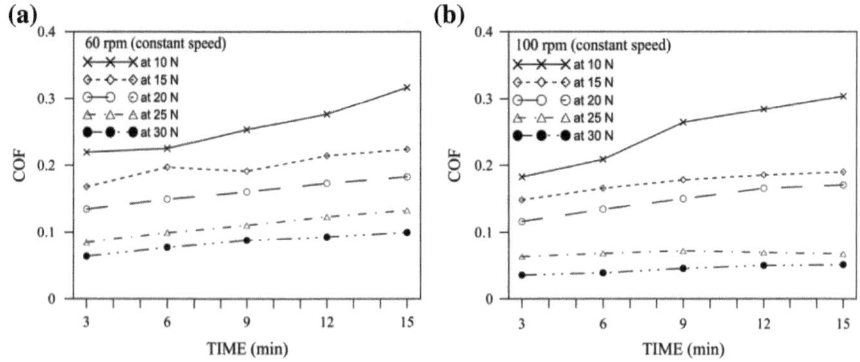

Fig. 3.18 COF versus sliding duration of Ni–B coating with different load at constant speed of **a** 60 rpm and **b** 100 rpm under lubricated condition

Figure 3.19 shows the effect of normal load on wear depth at different sliding speed regarding sliding duration. Figure 3.19a, b represents the variation of the depth of wear with sliding speed and sliding duration at constant load of 10 and 30 N, respectively. It is found that any constant load wear is gradually increasing with increase in rotating speed and sliding duration.

Figure 3.20a, b represents the wear depth with the variations of normal load and sliding time at constant speeds of 60 and 100 rpm, respectively. It is found that at any constant speed, wear gradually increases with load as well as sliding time. At 60 rpm as well as 100 rpm, the wear depth at 30 N is quite higher compared to that observed at 10 N ($\sim$3 times) especially after a sliding duration of 15 min.

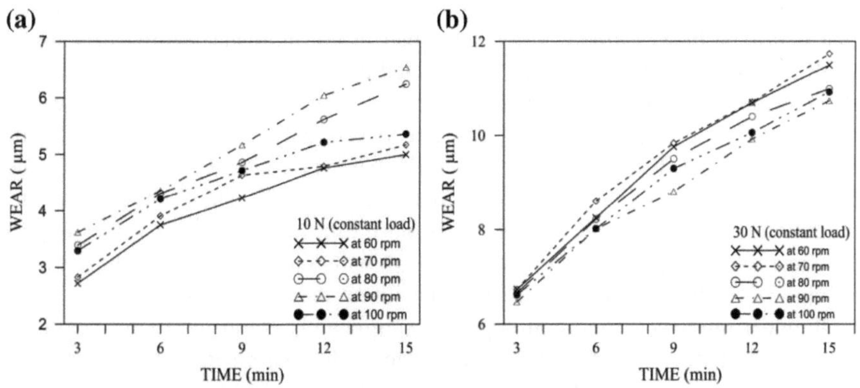

Fig. 3.19 Wear versus sliding duration of Ni–B coating with different sliding speed at constant load of **a** 10 N and **b** 30 N under lubricated condition

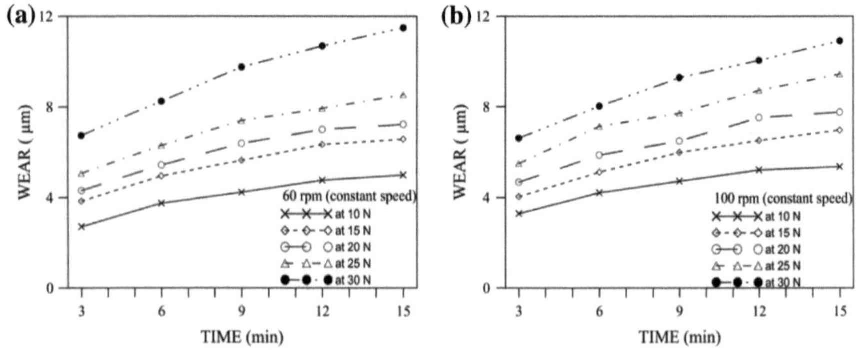

Fig. 3.20 Wear versus sliding duration of Ni–B coating with different load at constant speed of **a** 60 rpm and **b** 100 rpm under lubricated condition

3.3.2.3 Comparison of Friction and Wear Characteristics Under Dry and Lubricated Conditions

Comparative study of tribological characteristics of electroless Ni–B deposition has been carried out by generating 3D surface plots and shown in Figs. 3.21 and 3.22. In each of the 3D surface plots, two design parameters are varied while the third variable is kept fixed at its mid-level value.

The variation of COF under dry and lubricated conditions with load and speed is shown in Fig. 3.21a. It is noticed that COF under lubricated condition is lower at higher values of normal load and sliding speed compared to the dry condition. But at lower loads and speeds, the COF under dry condition is lower compared to lubricated condition. This may be due to inadequate film formation at lower values of applied load and rotating speed. Moreover, at a lower speed, there are chances that debris particles formed may get trapped in the lubricant leading to a higher COF [35]. But at higher speeds, it is expected that adequate lubrication occurs and the film provides lubricating effect. Moreover, at higher speeds, due to the centrifugal action of the counterface disc the debris may get carried away by the flowing lubricant. It has been also observed that tribo-oxidative layers form under lubricated condition [36]. They tend to lower the COF. Thus, under lubricated condition, an improvement in COF takes place especially at higher values applied load and rotating speed. The variation of COF with load and time may be observed in Fig. 3.21b. In this plot, it is seen that with an increase in sliding duration, the COF under lubricated condition improves significantly compared to dry condition. This could be due to the fact that a stable tribo-patch is formed with an increase in time. The variation of COF with sliding speed and duration may be observed in Fig. 3.21c. A curvature in the plot may be seen and the COF is quite lower under lubricated condition compared to dry condition especially at higher values of speed.

The wear depth of the deposits under dry and lubricated condition with different load and sliding speed is shown in Fig. 3.22a. In fact, the trends observed for the

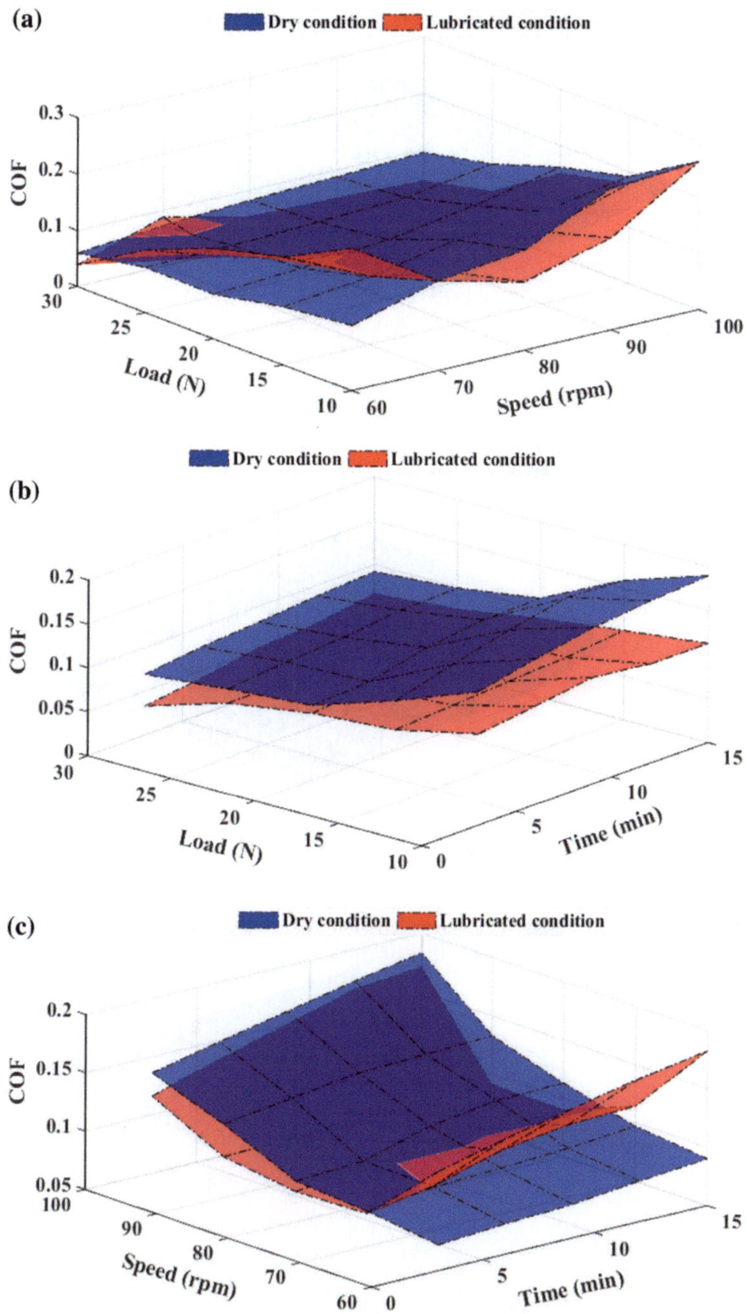

Fig. 3.21 Comparative study of friction behaviour of Ni–B coating under dry and lubricated condition **a** load versus speed, **b** load versus time and **c** speed versus time

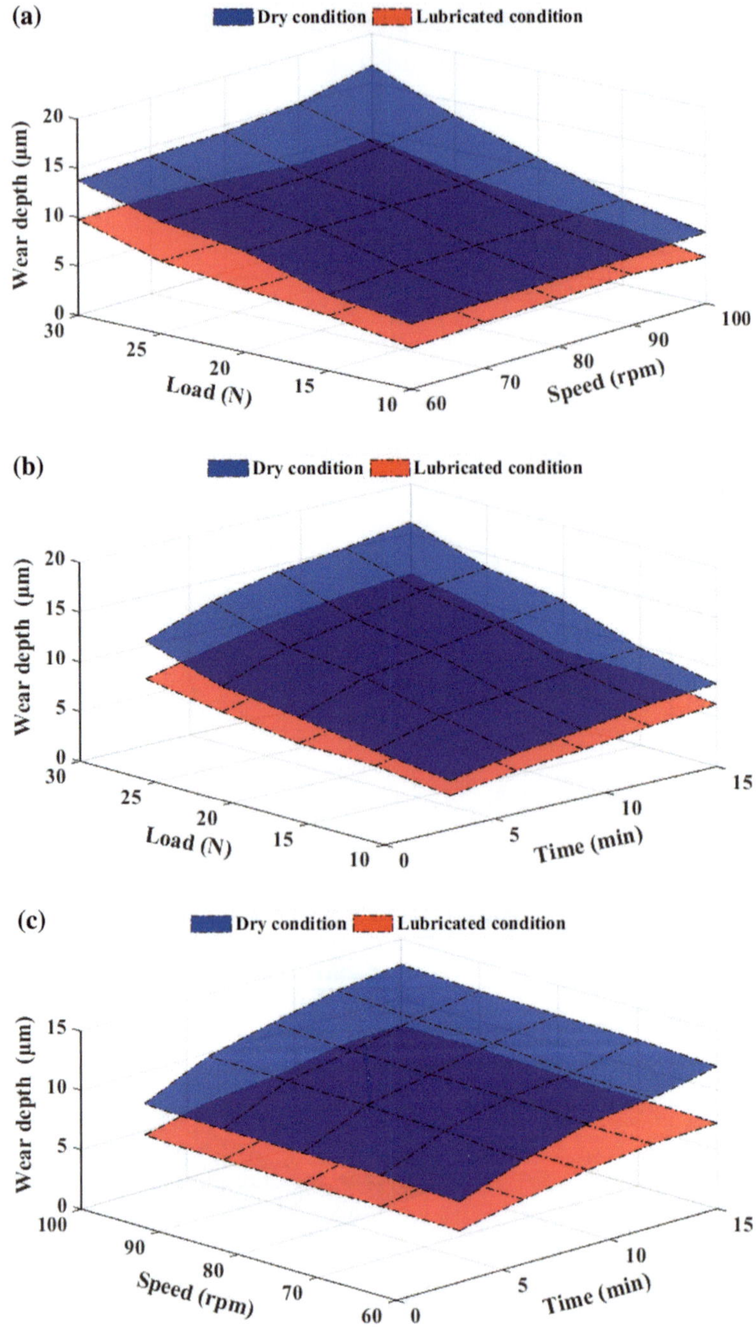

Fig. 3.22 Comparative study of wear behaviour of Ni–B coating under dry and lubricated condition **a** load versus speed, **b** load versus time and **c** speed versus time

variation of wear depth with applied load and sliding duration (Fig. 3.22b) as well as speed and time (Fig. 3.22c) are almost same. In all the three cases, wear depth increases with an increase in values of the tribological test parameters. The variation of wear depth under dry condition is also noticed to be in accordance with Archard's law. Furthermore, the wear depth under lubricated condition is significantly lower compared to dry condition. Therefore, tribological characteristics of the coatings improve under lubrication. This may be attributed to the self-lubricating nature of Ni–B deposits, i.e. the cauliflower-like microstructure, the ability of the columnar growths to retain lubricants and the flowing lubricant carrying away heat as well as wear debris generated along with it.

3.3.2.4 Microstructure Study of the Worn-Out Surfaces

The SEM image of the worn surface of a heat-treated Ni–B coating under dry condition is shown in Fig. 3.23a. The coatings are seen to undergo very less wear. Grooves may be observed along the sliding direction. Coating delamination or adhesion patches are also not visible. This is indicative of mild abrasive wear being the predominant mechanism under dry sliding condition. Resistance to plastic deformation increases due to an increase in hardness [37] preventing adhesion between the coated pins and the counterface disc. Figure 3.23b presents the worn surface of electroless Ni–B coating on which wear tests have been performed under lubricated condition. Fine scratches along sliding direction can be seen. No ploughs or furrows can be seen. Furthermore, the flowing lubricant has a cooling effect, thus reducing the possibility of adhesive wear. The wear mechanism is fundamentally mild abrasive in nature. Since the columnar structure of electroless Ni–B coating helps in retaining lubricant, a very low COF is indicated. A similar phenomenon is confirmed from the SEM micrograph of the worn surface (Fig. 3.23). Blackish patches may be may be seen due to the retained lubricant as well as formation of oxide layers.

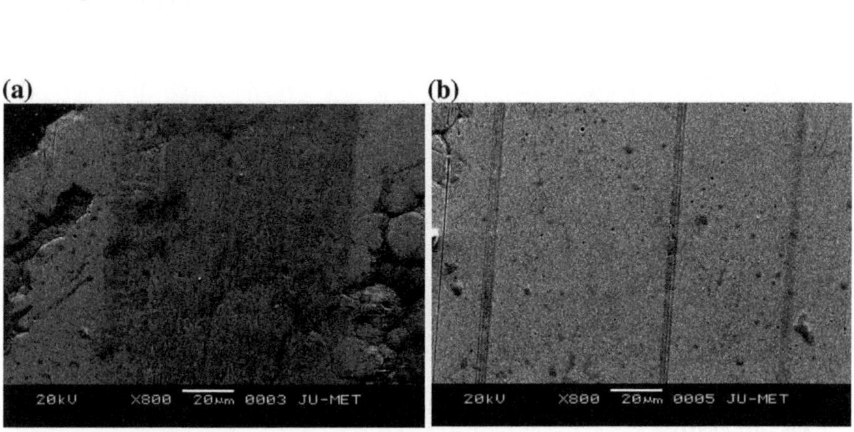

Fig. 3.23 SEM micrograph of wear track of Ni–B coating tested at **a** dry condition and **b** lubricated condition

3.4 Conclusion

In this investigation, electroless Ni–B coating has been deposited on mild steel substrate. Characterization of the deposited Ni–B coating has been done utilizing SEM, EDX and XRD technique. Microhardness and surface roughness are studied utilizing Vickers micro-indentation method and surface profilometry, respectively. Variations of friction and wear are studied under dry and lubricated conditions. Wear mechanism of coatings has been also reported. Following conclusions are drawn from the present work:

- The surface of the coating appears grey in as-deposited condition and heat-treated condition. The structure of the heat-treated coated samples resembles the surface of a cauliflower which makes it self-lubricating in nature.
- EDX affirms the presence of Ni and B in the coating. The B content is almost 7.51% by weight.
- The XRD plots reveal that the deposits are amorphous in as-deposited condition while it becomes crystalline on heat treatment due to the precipitation of nickel borides (Ni_2B and Ni_3B).
- Average values of hardness in as-deposited and heat-treated (1 h at 350 °C) condition are seen to be 855 $HV_{0.5}$ and 1160 $HV_{0.5}$, respectively.
- R_a of the coatings is found to be ~0.7–0.8 μm.
- Investigation of the tribological characteristics of the coatings under dry and lubricated condition indicates that Ni–B is suitable as a tribo-coating under lubricated condition. This is attributed to its inherent self-lubricating capabilities and columnar structures that act as retainer of lubricant.
- The wear mechanism under dry and lubricated condition is seen to be mild abrasive.

From the present study, it may be concluded that electroless Ni–B coatings have good friction and wear reduction capabilities under the influence of a lubricant. The coatings may find use in applications where tribological contact takes place under lubricated condition.

References

1. Mellor, B.G.: Surface Coatings for Protection Against Wear. Woodhead Publishing Series, Cambridge (2006)
2. Wurtz, A.: C. R. Hebd. Seances Acad. Sci. **18**, 702–705 (1844)
3. Brenner, A., Riddell, G.E.: Nickel coating on steel by chemical reduction. J. Res. Natl Bur. Stand. **37**(1), 31–34 (1946)
4. Mallory, G.O., Hadju, J.B.: Electroless Plating: Fundamentals and Applications. AESF, Orlando (1990)
5. Agarwala, R.C., Agarwala, V.: Electroless alloy/composite coatings: a review. Sadhana **28**(3–4), 475–493 (2003)

6. Hur, K.H., Jeong, J.H., Lee, D.N.: Microstructure and crystallization of electroless Ni–P deposits. J. Mater. Sci. **25**(5), 2573–2584 (1990)
7. Sahoo, P.: Wear behaviour of electroless Ni–P coatings and optimization of process parameters using Taguchi method. Mater. Des. **30**(4), 1341–1349 (2009)
8. Anik, M., Körpe, E., Sen, E.: Effect of coating bath composition on the properties of electroless nickel–boron films. Surf. Coat. Technol. **202**(9), 1718–1727 (2008)
9. Das, S.K., Sahoo, P.: Tribological characteristics of electroless Ni–B coating and optimization of coating parameters using Taguchi based grey relational analysis. Mater. Des. **32**, 2228–2238 (2011)
10. Das, S.K., Sahoo, P.: Optimization of electroless Ni-B coatings based on multiple roughness characteristics. J. Tribol. Surf. Eng. **2**(1/2), 85–106 (2011)
11. Mukhopadhyay, A., Duari, S., Barman, T.K., Sahoo, P.: Tribological performance optimization of electroless Ni–B coating under lubricated condition using hybrid grey fuzzy logic. J. Inst. Eng. (India): Ser. D **97**(2), pp. 215–231 (2016)
12. Steinmetz, P., Alperine, S., Friant-costantini, A., Josso, P.: Electroless deposition of pure nickel, palladium and platinum. Surf. Coat. Technol. **43–44**, 500–510 (1990)
13. Riedel, W.: Electroless Nickel Plating. ASM International, Metals Park, OH (1991)
14. Sahoo, P., Das, S.K.: Tribology of electroless nickel coatings–A review. Mater. Des. **32**, 1760–1775 (2011)
15. Lowenheim, F.A.: Electroplating: Fundamentals of Surface Finishing. McGraw- Hill Book Co, New York (1978)
16. Elansezhian, R., Ramamoorthy, B., Nair, P.K.: The influence of SDS and CTAB surfactants on the surface morphology and surface topography of electroless Ni–P deposits. J. Mater. Process. Technol. **209**(1), 233–240 (2009)
17. Allen, R.M., VanderSande, J.B.: The structure of electroless Ni-P films as function of composition. Scripta Metall. **16**(10), 1161–1164 (1982)
18. Berkh, O., Eskin, S., Zahavi, J.: Properties of electrodeposited Ni-P-SiC composite coatings. Met. Finish. **94**(3), 35–40 (1996)
19. Baudrand, D., Bengston, J.: Electroless plating processes: developing technologies for electroless nickel, palladium and gold. Met. Finish. **93**(9), 55–57 (1995)
20. Apachitei, I., Duszczyk, J., Katgerman, L., Overkamp, P.J.B.: Electroless Ni-P composite coatings: the effect of heat treatment on the microhardness of substrate coating. Scripta Mater. **38**(9), 1347–1353 (1998)
21. Apachitei, I., Tichelaar, F.D., Duszczyk, J., Katgerman, L.: The effect of heat treatment on the structure and abrasive wear resistance of autocatalytic NiP and NiP–SiC coatings. Surf. Coat. Technol. **149**(2–3), 263–278 (2002)
22. Vitry, V., Delaunois, F., Dumortier, C.: Mechanical properties and scratch test resistance of nickel–boron coated aluminium alloy after heat treatments. Surf. Coat. Technol. **202**(14), 3316–3324 (2008)
23. Palaniappa, M., Seshadri, S.K.: Friction and wear behavior of electroless Ni–P and Ni–W–P alloy coatings. Wear **265**(5–6), 735–740 (2008)
24. Staia, M.H., Castillo, E.J., Puchi, E.S., Lewis, B., Hintermann, H.E.: Wear performance and mechanism of electroless Ni-P coating. Surf. Coat. Technol. **86–87**, 598–602 (1996)
25. Ramalho, A., Miranda, J.C.: Friction and wear of electroless NiP and NiP + PTFE coatings. Wear **259**(7–12), 828–834 (2005)
26. Sahoo, P., Pal, S.K.: Tribological performance optimization of electroless Ni–P coatings using the Taguchi method and Grey relational analysis. Tribol. Lett. **28**(2), 191–201 (2007)
27. Srinivasan, K.N., Meenakshi, R., Santhi, A., Thangavelu, P.R., John, S.: Studies on development of electroless Ni–B bath for corrosion resistance and wear resistance applications. Surf. Eng. **26**(3), 153–158 (2010)
28. Balaraju, J.N., Anandan, C., Rajam, K.S.: Morphological study of ternary Ni–Cu–P alloys by atomic force microscopy. Appl. Surf. Sci. **250**(1–4), 88–97 (2005)
29. Balaraju, J.N., Anandan, C., Rajam, K.S.: Electroless deposition of ternary Ni–W–P alloys from sulphate and chloride based baths. Surf. Eng. **21**(3), 215–220 (2005)

30. Das, S.K., Sahoo, P.: Wear performance optimization of electroless Ni-B coating using Taguchi design of experiments. Tribol. Ind. **32**(4), 17–27 (2010)
31. Roy, S., Sahoo, P.: Optimization of electroless Ni-P-W coating for minimum friction and wear using grey Taguchi method, Article ID: 608140. J. Coat **2013**, 1–13 (2013)
32. Roy, S., Sahoo, P.: Friction performance optimization of chemically deposited Ni-P-W coating using Taguchi method, Article ID: 136740. ISRN Tribol. **2013**, 1–9 (2013)
33. Fisher, R.A.: Design of Experiments. Oliver & Boyd, Edinburgh (1951)
34. Montgomery, D.C.: Design and Analysis of Experiments. Wiley, New York (2001)
35. Correa, E., Mejía, J.F., Castaño, J.G., Echeverría, F., Gómez, M.A.: Tribological characterization of electroless Ni–B coatings formed on commercial purity Magnesium. J. Tribol. **139** (5), 051302 (2017)
36. Xu, Y., Zheng, X., Hu, X., Yin, Y., Lei, T.: Preparation of the electroless Ni–P and Ni–Cu–P coatings on engine cylinder and their tribological behaviors under bio-oil lubricated conditions. Surf. Coat. Technol. **258**, 790–796 (2014)
37. Krishnaveni, K., Sankara Narayanan, T.S.N., Seshadri, S.K.: Electroless Ni–B coatings: preparation and evaluation of hardness and wear resistance. Surf. Coat. Technol. **190**(1), 115–121 (2005)

Chapter 4
Friction and Wear Characteristics of Heat Treated Electroless Ni–P–W Coatings Under Elevated Temperature

Sanjib Kundu, Suman Kalyan Das and Prasanta Sahoo

4.1 Introduction

Brenner and Riddell [1] proposed a process of coating without any external current, known as electroless deposition in which electroless nickel (EN) became very popular. Since invention, it is widely used because it provides uniformity concerning deposit, low porosity and requires low process temperature [2]. The as-deposited Ni–P perhaps is invigorated through growth of nickel phosphide crystallites under satisfactory heating conditions. Yet, higher heat treatment period or heating [3] temperatures impair the solidity of Ni–P covering because of grain coarsening [4].

A typical electroless solution consists of supplying agent of nickel, reducing agent, complexing agent, stabilizer, surfactant, buffering agent, etc. The function of nickel source is to supply the nickel ions for deposition on the substrate while reducing agent supplied the requisite electrons intended for the reduction procedure of nickel [2]. Complexing agents are used to inhibit the breakdown of the solution as well as to make the bath stable. Stabilizers/inhibitors are also provided to avoid the random disintegration of the plating bath. Addition of surfactant results in better surface finish and increased hardness [5]. Moreover, existence of surfactant favours the film deposition reaction between the bath solution and the dipped substrate surface [5] which results in improved surface finish. Buffering agents are used to control the pH of the bath [2]. Primarily coating follows substrate roughness profile with marginal fluctuations. Surface roughness of the coating is hugely affected by the level of reducing agent [6]. It is seen to again decrease with heat treatment [7]. Chloride baths provide smoother surface than that developed from sulphate-based bath [8]. Several optimization methods [9] are employed to know the proper

S. Kundu (✉) · S. K. Das · P. Sahoo
Mechanical Engineering Department, Jadavpur University, Kolkata 700032, India
e-mail: skundu_me@rediffmail.com

P. Sahoo and J. P. Davim (eds.), *Advances in Materials, Mechanical and Industrial Engineering*, Lecture Notes on Multidisciplinary Industrial Engineering,
https://doi.org/10.1007/978-3-319-96968-8_4

deposition parameter combination for increasing the efficiency of electroless nickel coatings. Addition of third component [10, 11] which forms a ternary compound has been recommended to enhance the properties of chemically developed Ni–P coatings. These include Ni–Cu–P, Ni–W–P and Ni–Zn–P coatings. Moreover, elements, viz. Co, Cu, Mo, W, Zn, are also incorporated to give rise to various other ternary coatings. Higher phosphorous content (above 7 wt%.) leads to amorphous structure in case of Ni–Cu–P but the same under 3.9 wt% results in a crystalline Ni–Cu part which is visible [12]. Structure of the coating is noticeably influenced by tungsten content in the coating [13]. At a concentration of WO_4^{2-} ions under 0.35 mol dm^{-3}, an unstructured arrangement is attained. Crystalline structure of the nickel is unaltered until the zinc intensification up to 23 wt% for Ni–Zn–P coatings [14] is obtained. Several ternary and quaternary coatings are being developed to meet the industrial requirements such as Ni–P–W [15–18], Ni–P–Cu [19–22], Ni–P–B [23, 24], Ni–P–W–Cu [25, 26]. Co-deposition of materials, viz. diamond, silicon carbide (SiC), aluminium oxide (Al_2O_3), polytetrafluoroethylene (PTFE) particles, with electroless coatings, gives rise to composite coating. Wear-resistant composite coatings such as Ni–P–TiO_2 [27, 28], Ni–P–Al_2O_3 [29–32] can be formed by the co-deposition of fine hard particulate substance and lubricating composite coatings like Ni–P–PTFE [33–35], Ni–P–MoS_2 [36, 37] could be produced by solid lubricants PTFE and MoS_2. Dispersion of nanoparticles into the EN matrix gives rise to electroless nano-coatings. Among these coatings, carbon nanotubes (CNTs) [38, 39]-based coatings and nano-TiO_2 [40, 41]-based coatings have received considerable attention.

Mass loss during wear test is observed to be lowermost when the coating is exposed to heat treatment at 400 °C. It may be because of extreme solidity of the coatings attained by heating in that temperature [6]. In general, electroless nickel covering is amorphous or microcrystalline in as-deposited condition and goes crystalline through proper heating by formation of phosphides and borides. Annealing is done at 400 °C to Ni–P film yield Ni_5P_2, Ni_2P and NiP_2 as chief constituent compounds [42]. The crystal arrangement of electroless Ni–P film significantly alters through heating. A structure comprising of dual crystalline stages nickel and nickel phosphide (primarily Ni_3P) articulates as outcome of heat treatment (meant for both fine crystalline and amorphous coverings). Ni_3P forms around a temperature beyond 200 °C and at a temperature greater than 400 °C, and the agglomeration of the same takes place [43]. Proper heat treatment strengthens the Ni–P coating by crystallization of nickel and nickel phosphide though the solidity of the coating is spoiled in case of too much heating at higher temperature. Momentous change in the wear resistance of the coatings has been recounted, once hard elements are amalgamated. Selections of the elements are influenced by the properties desired from the coating. It is reported that the integration of a typical transformation metal, for instance, W, Co, Mn, Re and Mo in the binary Ni–P compound might lead to grander properties.

The third constituent in the coating governs the crystallization temperature of ternary coatings. Ni–Cu–P coatings that processed through heat treatment undergo crystallization and growth in grain dimensions. The existence of tungsten slows down the crystallization process of coatings during the course of heat treatment. Crystallographic changes take place at a temperature around 350 and 410 °C, respectively, for chemically deposited Ni–W–P and Ni–P film [44], respectively. Further, researchers detected the beginning of crystallographic alteration of Ni–W–P at 375 °C [45]. In controlling wear characteristics, heat treatment as well as bath temperature has remarkable impacts on electroless Ni–P coating, whereas source of nickel along with heat treatment temperature is the two main factors for Ni–P–W coatings [42, 46]. The life of the coated component gets shortened [3] because of material softening at high temperature.

Ni–W–P ternary coating was primarily proposed by Pearlstein and Weightman [47] in the year 1963, after which it received overwhelming response from researchers around the globe. Tungsten (W) is a non-compliant metal and cannot be deposited from any aqueous bath though tungsten as compounds with iron group transition metals can be deposited of solutions containing tungsten ions [48]. The unification of tungsten in the nickel matrix leads to the solute strengthening and enriched the coating hardness. In addition, the coating shows outstanding properties like mechanical properties, good thermal stability, tribological behaviour and corrosion resistance. Metal and amalgams/ceramic coatings have been in use for a very long period for protection of components presented to high temperatures in power plant, gas turbines and oil refineries with extraordinary results. A thorough literature survey reveals that research work in the field of electroless nickel and its alloy coatings has been focused mostly on the study of tribological behaviours under ambient test conditions. But there is very limited number of examinations that address application of EN coatings under elevated temperature. Ni–P (as-deposited) is found to be suitable at 550 °C because of creation of oxide layers and microstructural modifications during wear test [49]. Chemically deposited Ni–P–MoS_2 coatings display low friction coefficient and mass loss during high temperatures of 300–500 °C which is attributed to the development of oxide coats of molybdenum [50]. Ag-based coatings are called chameleon coatings as they have self-lubricating characteristics at raised temperatures [51]. Ni–P–SiC also provides satisfactory results up to an assessment temperature of 200 °C but again increment occurs through heat treatment [52]. Electroless Ni–P coating shows excellent wear resistance at high temperature which degrades after 500 °C due to severe oxidation and softening of the deposits [53]. Since addition of W to EN coatings enhances its thermal dependability and oxidation opposition [54], the present article attempts to research the tribological characteristics of chemically developed Ni–P–W coatings at raised temperatures and contrast the same with room temperature tests. Attention towards high-temperature self-lubricating coatings development is given to increase the lifespan of the components that are exposed to tribological interactions at raised temperatures, as experienced in case of hot metal working (hot shaping, hot illustration, and so on), aerospace industry, heat exchangers, extruders, plungers, rolls for rolling mills, gas turbine blades, crushers, etc. [55]. EN coatings have the potential to act as a thermal barrier to this adverse situation as nickel has a high-melting point [56]. This paper presents the tribological performance of nickel–phosphorus–tungsten units developed over the AISI 1040 steel specimens by electroless deposition technique. Deposition is followed by heating the sample in proper environment to enhance mechanical properties of the coatings. The coatings are kept in box furnace at 400 °C for 1 h. Portrayals of the Ni–P–W coverings are basically centred on microhardness, microstructure, phase distribution and wear mechanism. Field emission scanning electron microscope (FESEM) in combination with energy-dispersive X-ray (EDAX) and X-ray diffraction (XRD) techniques is used to examine the coating microstructure and phase transformation. Microhardness is deliberated via Vicker's technique, and surface roughness is measured by stylus-based profilometer because both have effect on the tribological behaviour of the coating. The developed heat treated films are exposed to tribological investigations at room and raised temperatures (100–500 °C) on a pin-on-disc arrangement, and the friction wear characteristics are assessed. The current investigation does not try to mimic the actual high-temperature applications mentioned above but attempts to test Ni–P–W coating as a thermal barrier coating under high temperature.

4.2 Experimental Details

4.2.1 Substrate Preparation and Film Deposition

Cylindrical mild steel (AISI 1040 steel) specimens that have 30 mm length and 6 mm diameter are used as substrate so that it fits into the attachment of the tribotester. Substrates conforming to N5 (R_a = 0.40 μm) roughness grade were chosen for the deposition of ternary coating (Ni–P–W) as EN coatings follow substrate profile. It is also important as equally friction and wear are external properties, and their values depend very much on the roughness Thus, specimens having a roughness variation around 0.1% are selected for coating deposition. Properly cleaned samples are etched by 50% hydrochloric acid for 60 s. Afterwards, they are washed in distilled water and subjected to methanol cleaning before going to be submerged into the bath.

The deposition conditions for Ni–P–W coating are fixed through numerous trials, and suitable ranges of the parameters are selected consequently. Table 4.1 specifies the deposition conditions. Sodium tungstate is used as a source of W ions, and its concentration is kept at 20 g/L because W content in the coating decreases beyond that concentration [25]. A diminishing trend in W inclusion as well as deposition speed was observed by Palaniappa et al. [15]. They found that on addition of excess sodium tungstate (>20 g/l) led to impulsive disintegration of the plating solution. This is typically indicated beforehand by increased hydrogen evolution and the production of finely separated dark precipitates of nickel and nickel phosphide all through the solution. The precipitates for the most part comprised of nickel or nickel phosphide. The pH of the arrangement is kept up at around 7–8 by continuous observation through a pH metre. The washed specimens

Table 4.1 Bath composition details

Bath constituents (g/l)	Values	Operating conditions	Values
Nickel sulphate	20	pH	7–8
Sodium hypophosphite	20	Temperature (°C)	90 ± 2
Sodium citrate	35	Duration of coating (hour)	3
Ammonium sulphate	30	Bath volume (mL)	200
Lactic acid	5		
Sodium tungstate	20		

are initiated in palladium chloride arrangement at 55 °C. This is necessary to activate the samples for autocatalytic reaction. This activation step is included intentionally to achieve good adhesion to the substrate, better deposition rate and expressive deposit thickness [57]. Activated specimens are then immersed in the solution which is kept up at a temperature in the vicinity of 90 and 92 °C through the assistance of a hot plate cum stirrer, joined by a temperature measuring device which is additionally submerged into the arrangement. Schematic chart showing steps engaged with coating deposition is shown in Fig. 4.1. The development process runs for a period of 3 h. Experimental set-up of electroless deposition is given in Fig. 4.2. Coating development period and bath volume are retained equal for all samples with the goal that the covering thickness and bath load remain roughly steady, and the approximate film depth is observed to be about 35 μm. After the coating, the samples are removed from the chemical solution and cleaned by distilled water. After that specimens are heated (at 400 °C for 1 h) in a muffle furnace and allowed to cool in air.

4.2.2 Coating Characterization

Vicker's hardness of the coating is studied at an indentation load of 100 g, dwell period of 15 s and a speed of 25 μm/s. Microhardness (Vicker's) of the Ni–P–W coatings (both before and after heating) are taken according to ASTM standard E384-16 utilizing a hardness analyser (VMHTOT, Technische Mikroskopie) equipped with a pyramidal diamond indenter. Three impressions are done on every sample of the reverse surfaces of the specimen, and their mean is reported. The length of the corner to corner of the indents formed on the surface of the coatings is obtained by means of a low-amplification optical magnifying instrument. Surface roughness of the Ni–P–W coverings (both condition as before and after heating) has been determined by means of Talysurf profilometer (Surtronic 3+), that has a diamond stylus with a tip radius of 5 μm. Sampling length of 0.8 mm with a traverse length of 4 mm is used for roughness measurement which is further processed through Talyprofile software to observe the surface profile of the coatings along with centreline average roughness (R_a) value. Each sample is tested four

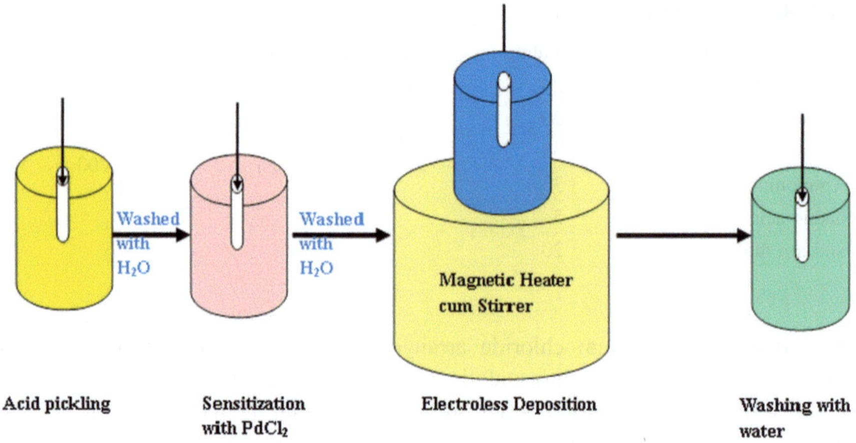

Fig. 4.1 Schematic representation of steps for electroless coating on substrate surface

times, and average value of roughness is reported. Characterization tests are conducted after deposition to guarantee about the proper development of the coating and also after the tribological examination to have idea about the underlying mechanisms. Elemental composition of the deposits with respect to weight proportions of Ni, P and W is found out by means of energy-dispersive X-ray (EDX) analyser using a Si (Li) superultrathin window (SUTW) intended for the transmission of low energy X-rays. Field emission scanning electron microscope (FESEM) with an Everhart Thornley secondary electron detector (FEI, Quanta FEG 250) is used to observe surface structure of Ni–P–W coatings (as-deposited and heat treated). The phase structure of the coatings is determined before and after heat treatment with the help of X-ray diffraction (XRD) method through a Cu Kα source (Rigaku, Ultima III) for 2θ values in between 20° and 80° with an scanning rate of 1°/min which is utilized for phase structure examination. The worn-out coating surfaces are watched again utilizing FESEM to investigate the wear components. EDX examination of the damaged/deformed areas is also conducted to watch the post-wear changes in the composition.

4.2.3 Tribological Characterization

Pin-on-disc-based tribotester (Ducom, TR-20-M56) is applied to assess friction and wear behaviour of the EN-coated samples at room ($\sim$30 °C) and high temperatures (100, 300 and 500 °C). The tests are carried out in line with ASTM standard G99-05 (reapproved 2010). Schematic diagram for high-temperature friction wear test is shown in Fig. 4.3. The impacts of sliding speed and applied load under dry, non-lubricated conditions on tribological behaviour under various testing

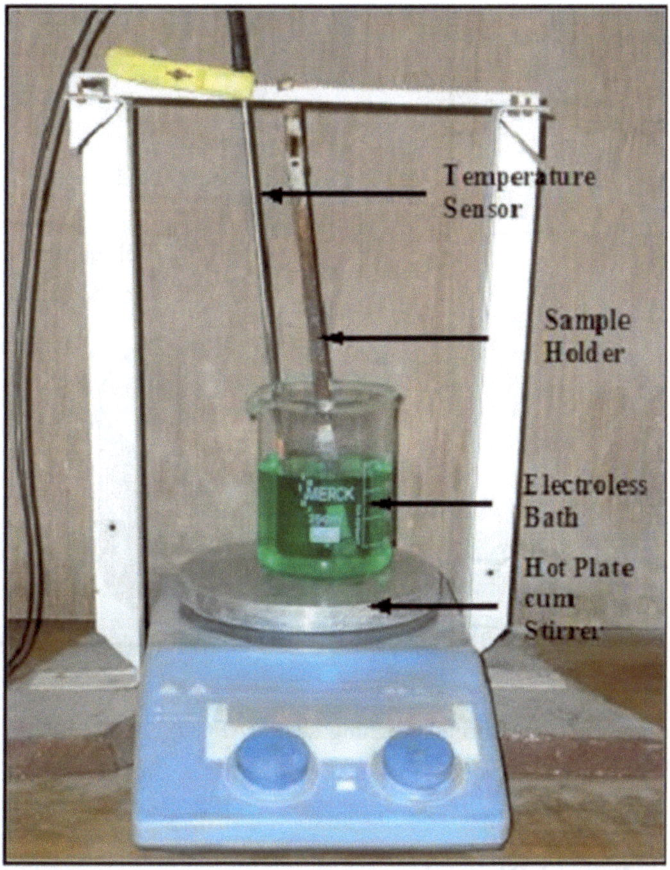

Fig. 4.2 Experimental set-up of electroless deposition

temperatures are contemplated. Heat treated coatings are slide at a track radius measuring 60 mm on a counterface prepared of EN-31 steel hardened to 58–62 HRC. The loading of the pin is done precisely through a stepper motor and load cell pair. The pivoting counter face disc is heated inductively to the desired temperature. A 15-kVA high-recurrence induction heating machine with coils prepared of beryllium copper is used to heat the counterface disc. The temperature of the disc is uninterruptedly recorded by a pyrometer having a smallest count of 1 °C and correctness of $1 \pm 1\%$ of the measured temperature. Wear is measured by means of the standard weight loss technique and described as the weight loss per unit travelling distance. The frictional force is estimated through a load cell of 100 N limits and an exactness of $0.1 \pm 1\%$ of estimated constrain. For the present investigation, considering the low thickness of the coating, duration of the tests is set at 300 s with varying load, sliding velocity and temperature. The mass of the films is taken

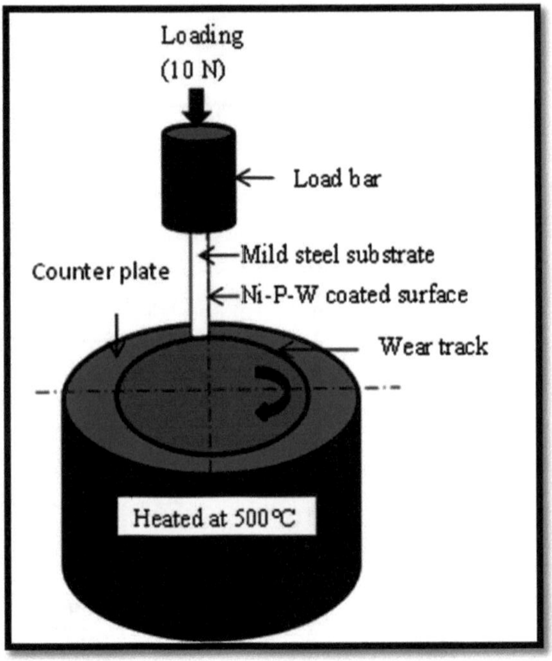

Fig. 4.3 Schematic diagram for high-temperature friction wear test

utilizing a precision measuring weighing scale (AFCOSET, ER 182A) having a resolution of about 0.01 mg.

4.3 Results and Discussion

4.3.1 Coating Characterization, Phase Transformation and Microhardness

Optical micrograph of both before and after heat treated Ni–P–W coating is shown in Fig. 4.4 proclaiming surface morphology qualitatively. The detailed analysis is carried out by SEM micrograph of the samples revealing its surface morphology and cross-cut section as shown in Fig. 4.5. The thickly concentrated nodular arrangement is observed for Ni–P–W coatings in as-deposited form in the high-magnification SEM micrograph (Fig. 4.5). Nodular surface characteristics with low porosity are very common in EN coatings [8, 15, 25, 26]. The quick growth of nuclei and subsequent development of the EN deposits endorse the nodular structure. A few pores in the coatings may come because of the evolution of hydrogen amid the electroless process. Pores emerge as dark regions, but these are extremely tiny in size and do not penetrate to the substrate face. Such structures augment the self-lubricating ability of the coatings by providing pockets for holding

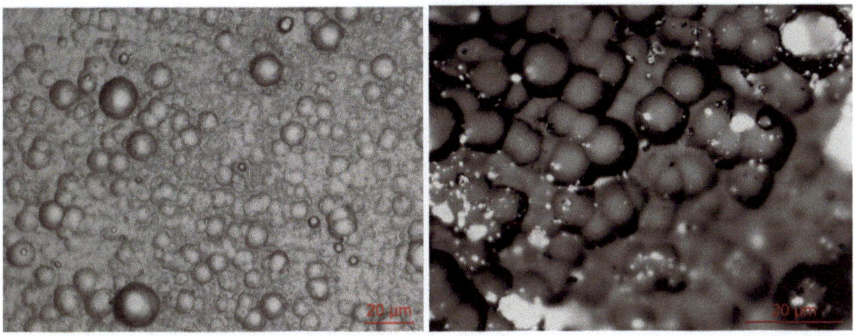

Fig. 4.4 Optical micrographs of Ni–P–W coating **a** as-deposited and **b** heat treated at 400 °C

oil. A sectional area of the as-deposited Ni–P–W coating is shown in Fig. 4.5c. It appears that the coating is associated closely with the substrate exhibiting good adhesion. Depth of the film supposedly lies for the most part around 30–35 µm. After heating at 400 °C over 60 min, the nodules get expanded resulting in a coarse-grained structure (Fig. 4.5b). With heat treatment, coating looks to be coarser and also turns bluish. Nickel oxide formation during heat treatment may be a reason behind this bluish colour of the coating. EDX study (Fig. 4.6) shows that co-deposition of tungsten in Ni–P matrix indeed led to a ternary Ni–W–P compound with a tungsten and phosphorous content of 4–4.5 and 8–8.5 wt%, respectively, while the rest is nickel. For Ni–W–P coating, decrease in phosphorus content was noted because of inclusion of tungsten in Ni–P deposit [26]. EDX of Ni–P–W coating heat treated at 400 °C for 1 h is shown in Fig. 4.6b which also shows the presence of oxygen (around 2%). Hence, the formation of oxide scales due to heat treatment could be reaffirmed from EDX results. Rests of the elements do not exhibit much variation in their composition. The oxide scale might be available as oxides of Ni and W.

Structural forms of chemically deposited Ni–P–W coating (before and after heat treatment) are ascertained by conducting XRD and related graphs are shown in Fig. 4.7. Crystalline of the electroless Ni–P–W coatings is dependent on phosphorus content in them. If percentage of P is more than 8 wt%, coating shows an amorphous configuration. It is because, lattice strain of Ni (111) arranged particles builds, that results in amorphous deposits in X-ray diffraction [58]. A blend of indistinct and nanocrystalline structure is accomplished when P lies in between 4 and 7 wt%. It exhibits a microcrystalline structure with P value below 3 wt%. In the current study, electroless ternary Ni–P–W layer greater than 8 wt% P displays an amorphous/nanocrystalline arrangement as expected. The existence of tungsten decelerates the crystallization of coatings during the course of heat treatment. Subsequent to heat treatment, a few of the peaks widen compared to as-deposited condition and generate crystalline phases. In the wake of heating at 400 °C for 60 min, the real peaks of Ni–W and Ni_3P are obtained. Higher crystallization

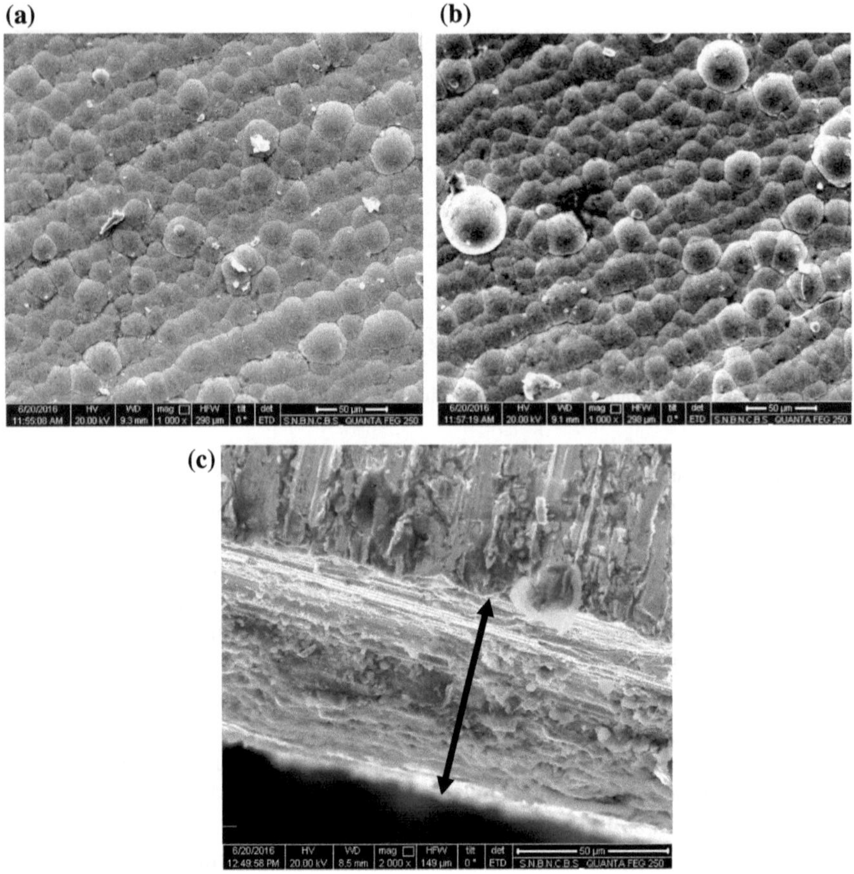

Fig. 4.5 SEM micrographs of Ni–P–W coating **a** as-deposited, **b** heat treated at 400 °C and **c** cross-cut section indicating thickness

temperatures are acquired because of W co-deposition in Ni–P module. Occurrence of stable Ni and Ni_3P is identified as similar to Balaraju et al. [59] for annealed specimen (400 °C/1 h).

Microhardness of the as-deposited together with heat treated ternary Ni–W–P films is studied. The typical hardness assessment of the samples is originated to be nearly 670 $HV_{0.1}$ in as-deposited condition and 1090 $HV_{0.1}$ for heat treated (400 °C/1 h) case. The results obtained are very close to that by Balaraju et al. [59]. The increase in hardness is because of the double impact of phosphide precipitation and solid solution reinforcing. The hardness of the crystallized Ni–W–P compound is more compared to Ni–P composite because of the development of the Ni–W solid solution [60, 61] separately from nickel phosphide precipitation. Along these lines, presentation of tungsten component in the Ni–P layer enhances the thermal stability as well as the microhardness of the electroless coatings. In fact, hardness of the

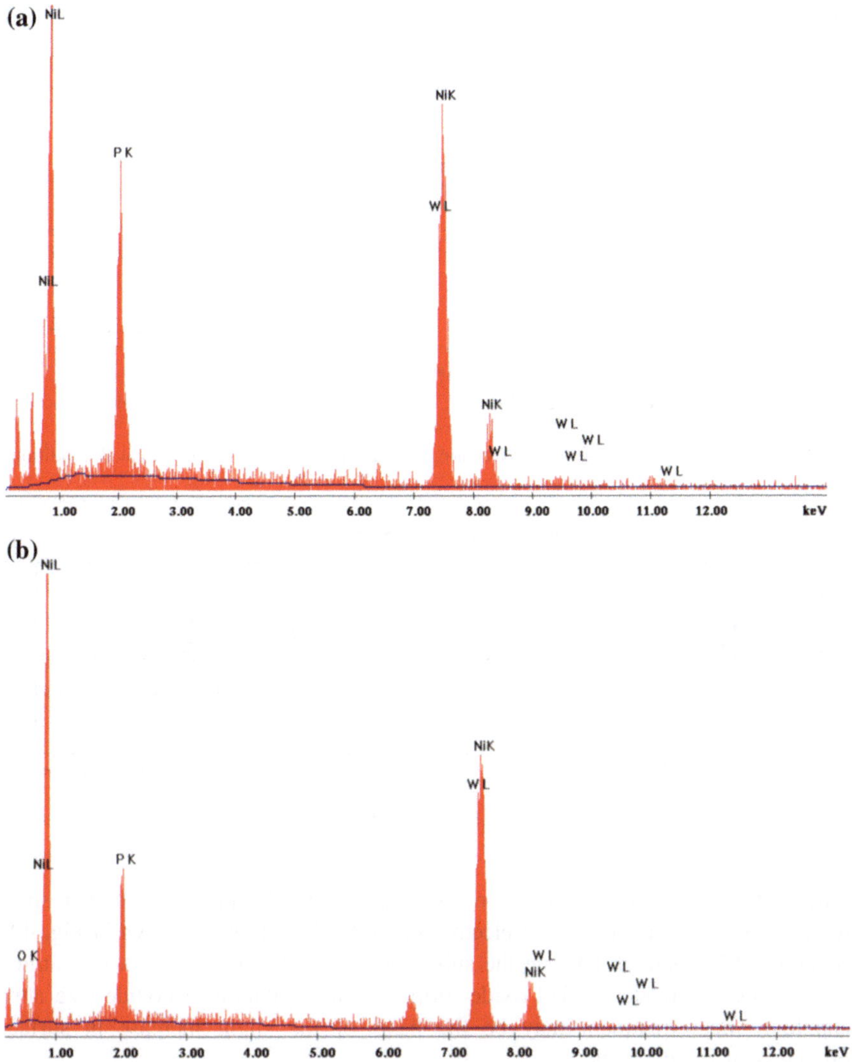

Fig. 4.6 EDX spectrum of Ni–P–W coating **a** as-deposited and **b** heat treated at 400 °C, 1 h

coating increases with tungsten in the coating. This is because of the solid solution strengthening of the nickel matrix by the dissolved tungsten. Proper heating and cooling arrangement can result in better hardness, however, the coating when heated at 600 °C demonstrates diminish in the hardness results for both binary and ternary coatings [15].

Tribological behaviour is influenced by surface roughness, and hence, it is desired to have smooth surface for better lifespan. In this regard, EN coatings play a vital role by providing smooth surface roughness. Centreline average roughness for

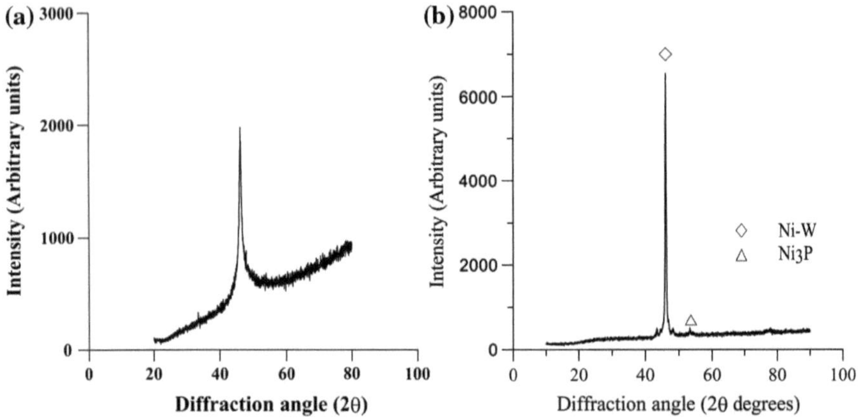

Fig. 4.7 X-Ray diffractions of Ni–P–W coating in **a** as-deposited and **b** heat treated condition

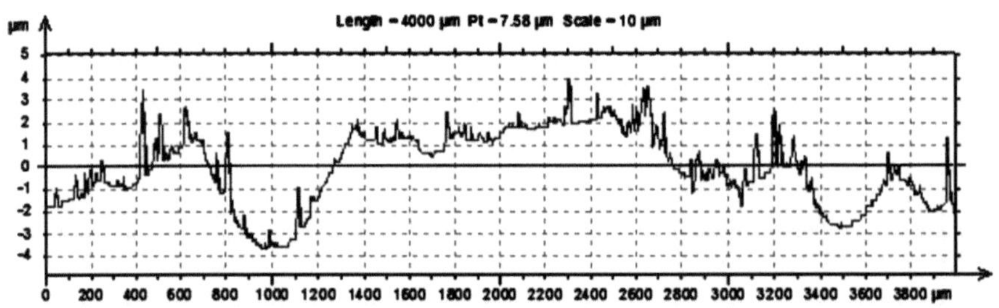

Fig. 4.8 Typical surface profile of electroless Ni–P–W coating

as-deposited and heat treated coverings is around 0.447 and 0.451 mm, respectively. Typical surface profile of electroless Ni–P–W coating is shown in Fig. 4.8. Generally, EN coatings take on the substrate profile. In the current work, similar observations are made. Nickel oxide formation and inflation in nodule size attributable to heat treatment may be the basis of the slight increase in surface roughness of the deposits.

4.3.2 Wear Loss and Friction Coefficient at Room Along and Elevated Temperature Test

Friction coefficient (COF) and mass loss in terms of wear of heat treated Ni–P–W coating at room and high temperature are shown in Fig. 4.9. It is found that the test conducted at 100 °C exhibits the highest COF and that conducted at room temperature shows the lowermost COF. But similar trend is not observed in case of wear behaviour of the coating (Fig. 4.9b). The heat treated coating, tried at room

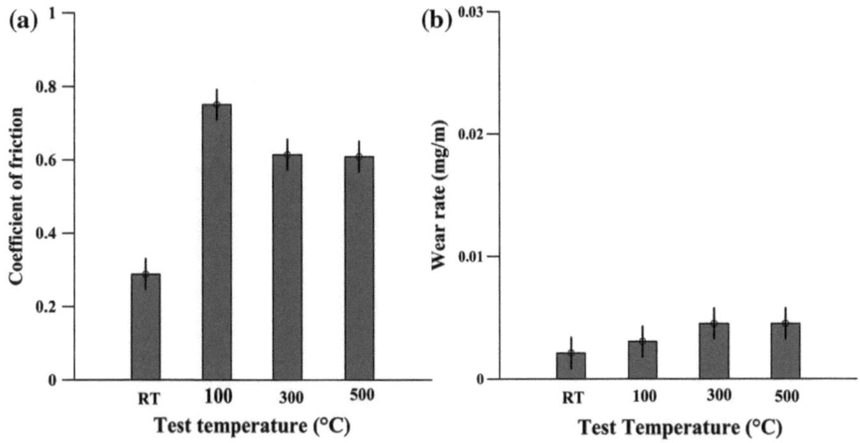

Fig. 4.9 Effect of test temperature on **a** COF and **b** wear rate of heat treated electroless Ni–P–W coating

temperature, has the most reduced wear rate yet after that the wear rate increases with the increase in the experiment temperature. Since the microstructure of as-deposited coating is amorphous (Fig. 4.5a) and the same is tough, the wear resistance of these structures is significantly controlled by the hardness [2, 15, 33]. Wear resistance increases as the hardness increases. Heat treatments result higher hardness and less adhesive (wear) amount for hypophosphite-based coating, compared with those in the as-deposited one which shows up as improved wear resistance of the heat treated samples at room temperature test [49]. Increase in test temperature causes massive scale deformation of the surface that leads to an applied stress value crossing the yield strength of the material. This acts result in severe adhesion and causes lower wear resistance. Similar kind of phenomenon is also observed with an increase in the heat treating temperature higher than 400 °C which contributes to the increase of the matrix ductility and, in this manner, to its softening and monstrous wear. Figure 4.10a shows the relationship between the coefficient of friction and the sliding distance (or the time) at various test temperatures for the tribological pair of EN 31 disc with Ni–P–W coating heat treated at 400 °C. As could be seen, more often than not, the increase in the test temperature increases the friction coefficient because of the number of interactions among the nodules and asperities of the mating surfaces as the contact region between the mating surfaces expands exponentially with temperature [62]. The advancement of COF at high temperature (500 °C) regarding sliding time at various speeds appears in Fig. 4.10b. The effect of temperature is pronounced over the sliding velocity. High fluctuations in COF are seen for the heat treated samples at room temperature and tests done at 100 °C (Fig. 4.10a). These changes are related to the occasional formation and removal of oxide patches. Raised test temperatures of 300 and 500 °C provide mostly stable values of COF. Friction coefficient and wear rate of heat treated Ni–P–W coating as a function of elevated temperature under diverse

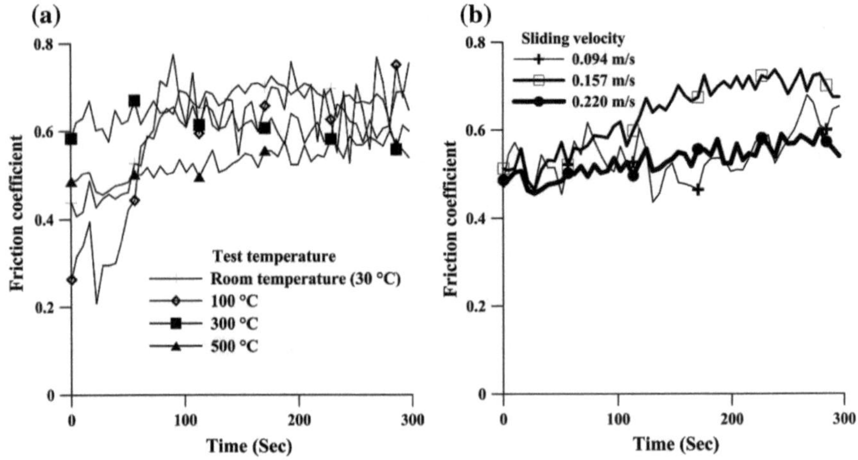

Fig. 4.10 Evaluation of friction coefficient for heat treated sample with time at typical load of 10 N at **a** different test temperature with a sliding velocity of 0.220 m/s and **b** different sliding velocity with a test temperature of 500 °C

load and sliding velocity are shown in Figs. 4.11 and 4.12, respectively. The effect of load under various operational temperatures over friction and wear role of heat treated Ni–P–W coating is shown in Fig. 4.11a, b, correspondingly. Both COF and wear show a positive response with load. Overall, COF increases with increase in contact load which is credited to the expansion in contact area and resulting higher force required to break them. It is observed that except room temperature starting from 100 °C tests, the friction coefficient mostly decreases with the increase in temperature running with any of the chosen velocities. An inverse relationship exists in between friction coefficient and elevated temperature, and it may be because of the softening nature of the coating at elevated temperature. Similar kind of phenomenon has been noticed by Kundu et al. [53] for Ni–P coatings. Wear rate is low at room temperature for heat treated sample as it provides higher amount of hardness but with the increase in temperature wear rate mostly increases in all cases as is shown in Fig. 4.9b. Interestingly, sliding velocity does not influence the friction behaviour of the coating especially when tested at temperatures greater than room temperature (Fig. 4.12a) but the same has great impact on the wear rate of the coating (Fig. 4.12b). Further it is noticed that with enhancement in sliding velocity, the wear rate decreases. This may be due to the low interaction between the mating surfaces due to insufficient time in establishing contact with considerable area. Moreover, due to inertia at higher sliding speed, some of the potential contact points may be missed due to vertical motion of the workpiece.

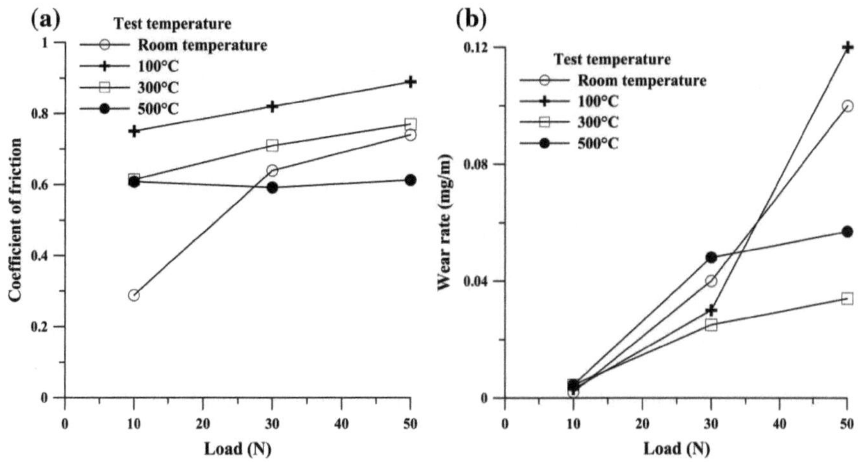

Fig. 4.11 **a** Variation of friction coefficient and **b** wear rate of heat treated electroless Ni–P–W coating against sliding velocity of 0.22 m/s with increase in temperature at different applied load

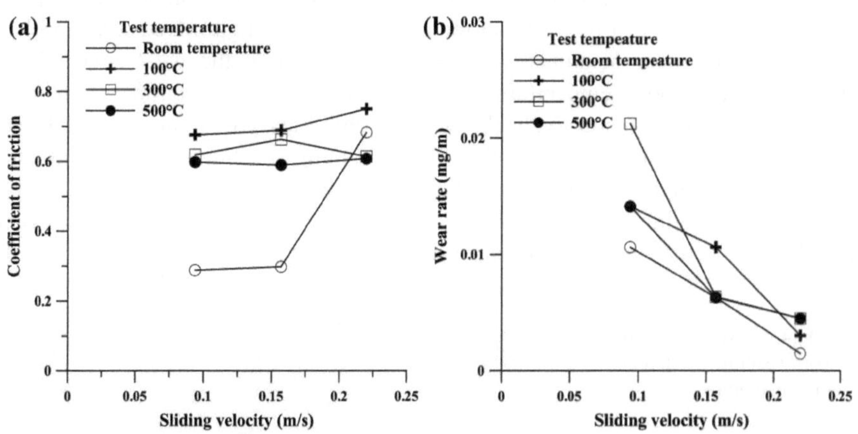

Fig. 4.12 **a** Variation of friction coefficient and **b** wear rate of heat treated electroless Ni–P–W coating against applied load of 10 N with increase in temperature at different sliding velocity

4.3.3 Investigation of Wear Mechanism

The wear mechanism of heat treated Ni–P–W films at room temperature (30 °C) and at high temperatures of 100–500 °C is investigated through optical microscope (OM), SEM micrographs and EDX of the worn surfaces. The trends of tribological performance of the ternary coatings exhibited by Figs. 4.9, 4.10, 4.11 and 4.12 are further consolidated by the SEM images of the wear scars and its corresponding EDX results. A preliminary idea of the wear mechanism is gained by taking OM

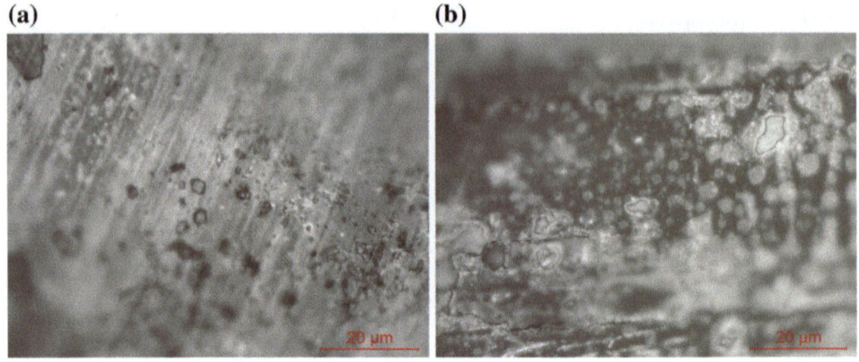

Fig. 4.13 Optical micrographs of wear tracks at **a** room temperature and **b** 500 °C

images of the wear track, which are shown in Fig. 4.13. The presence of pits and prows over the wear tracks are quite easily noticed for the sample tested at room temperature (Fig. 4.13a). Pits and prows are characteristic of adhesive wear mechanism, whereas wear tracks denote the occurrence of abrasive wear phenomenon. Hence, it can be said that the coating encounters a mixed wear mechanism under room temperature tests. SEM micrograph of the wear zone of the sample (tested at room temperature) and corresponding EDX spectrum is shown in Figs. 4.14 and 4.15, respectively. EDX result indicates post-wear elemental distribution which consists of mostly the original coating constituents. Ploughing and micro-cutting together with torn patches in the direction of sliding can also be observed for coating tested at 100 °C (Fig. 4.16) which again point towards a mixed wear mechanism. The uneven wear zone also supports towards the high COF

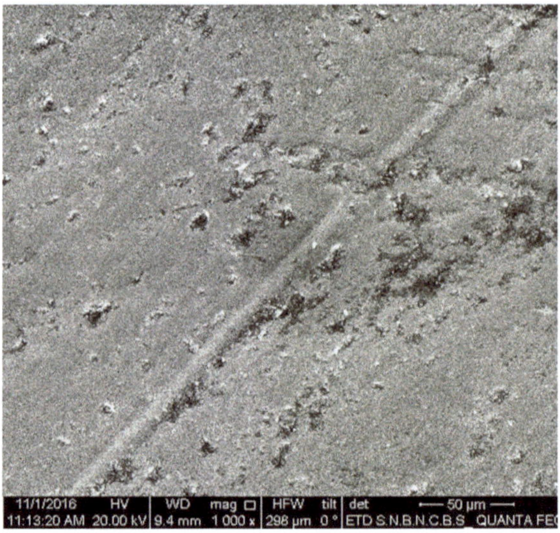

Fig. 4.14 SEM micrographs of worn specimen tested at room temperature

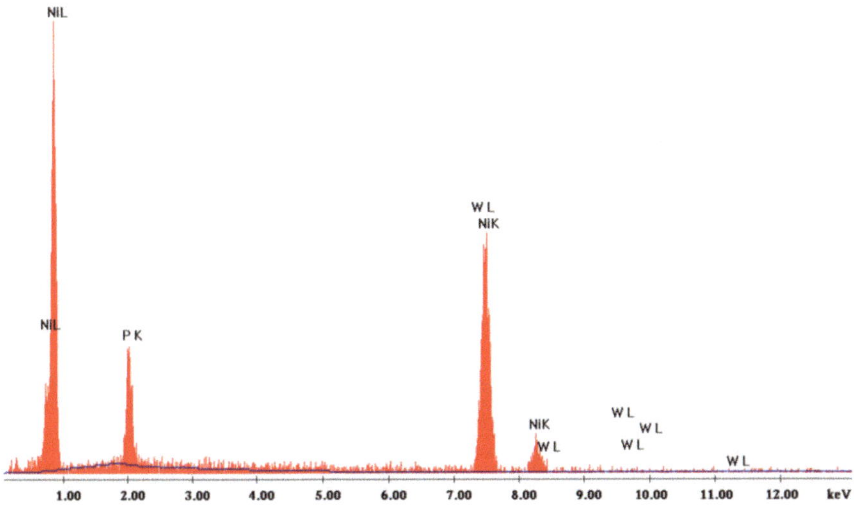

Fig. 4.15 EDX results of worn specimens of heat treated Ni–P–W coatings at room temperature

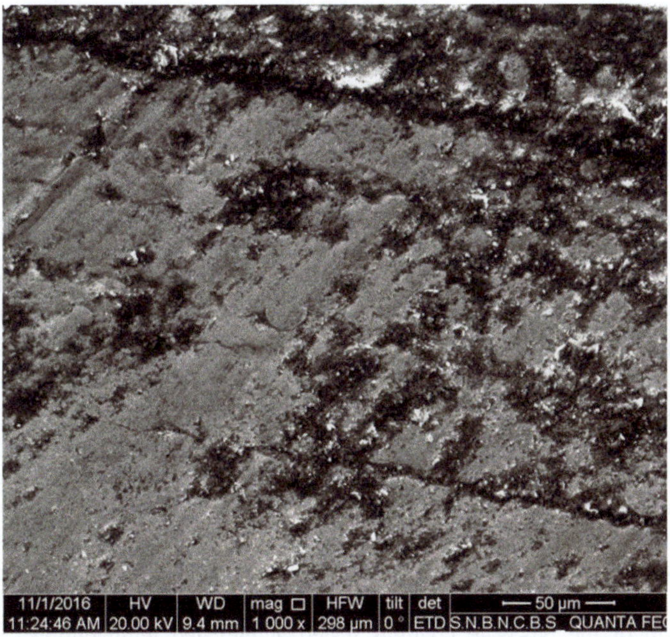

Fig. 4.16 SEM micrographs of worn specimen tested at 100 °C

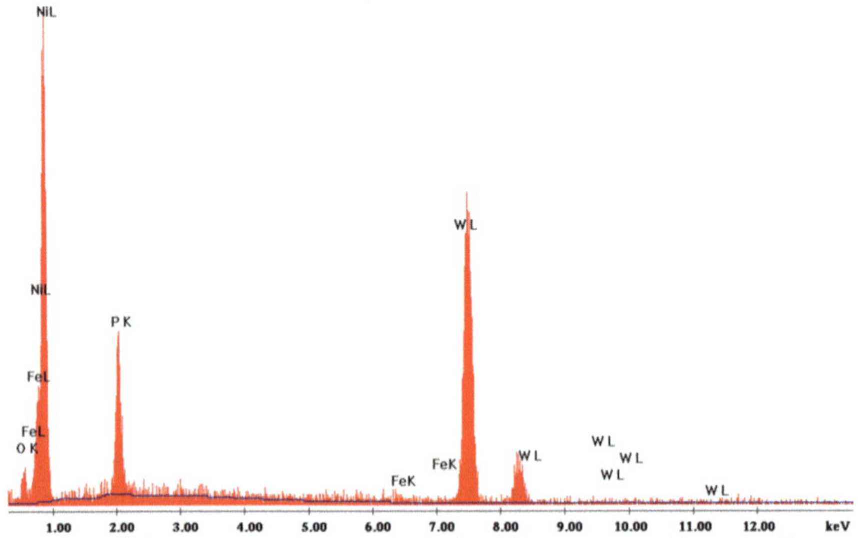

Fig. 4.17 EDX results of worn specimens of heat treated Ni–P–W coatings at 100 °C

observed at this condition (Fig. 4.9a). The EDX of the wear zone (Fig. 4.17) shows the presence iron which supports the occurrence of abrasive wear as the iron must have come from the counterface (steel part) owing to high mutual solubility among nickel and iron. Thus, adhesive wear seems to be the dominant wear mechanism at this test condition. Moreover, the presence of oxygen in EDX results point towards the formation of oxides in the wear zone. At 300 °C, test temperature the wear surface again exhibits micro-cutting and micro-ploughing together with torn patches (Fig. 4.18). However, overall coating surface appears to suffer ductile failure with a high level of plasticity in the path of sliding. The EDX spectrum (Fig. 4.19) again exhibits the presence of oxygen and iron. The coating tested at 500 °C again exhibits parallel grooves along with micro-ploughing (Fig. 4.20) which is again an indication of an abrasive wear phenomenon. However, there is a distribution of materials in the form of small spots on the entire coating surface which may be the wear debris being welded to the coating surface due to the pressure at the contact surface and under immense heat coming from the heated counterface. There are also some blackish regions which indicate towards heavy oxidation under 500 °C (corresponds mainly to NiO), as the same observed by several researchers [49–52]. The presence of oxygen in the EDX plot (Fig. 4.21) also supports this claim. Overall, it can be said that both adhesive plus abrasive wear ruled the mass loss mechanism of the Ni–W–P coatings tested at elevated temperatures. High temperature is generated at the coating-counterface contact region due to the heated counterface which is augmented by friction heating due to sliding of the workpiece. Thus, the genreated wear debris which undergo local hardening and get trapped between the coating and the counterface, are responsible for creating the grooves on the coating surface. At the highest temperature (500 °C), however, the debris tends

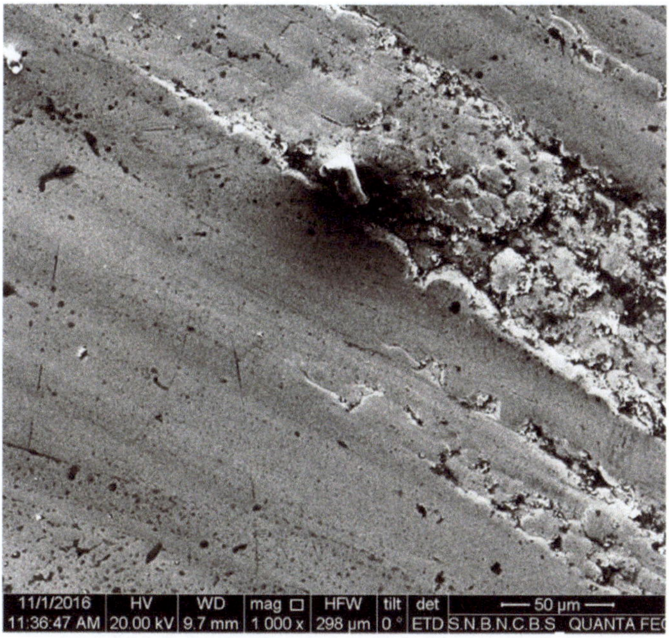

Fig. 4.18 SEM micrographs of worn specimen tested at 300 °C

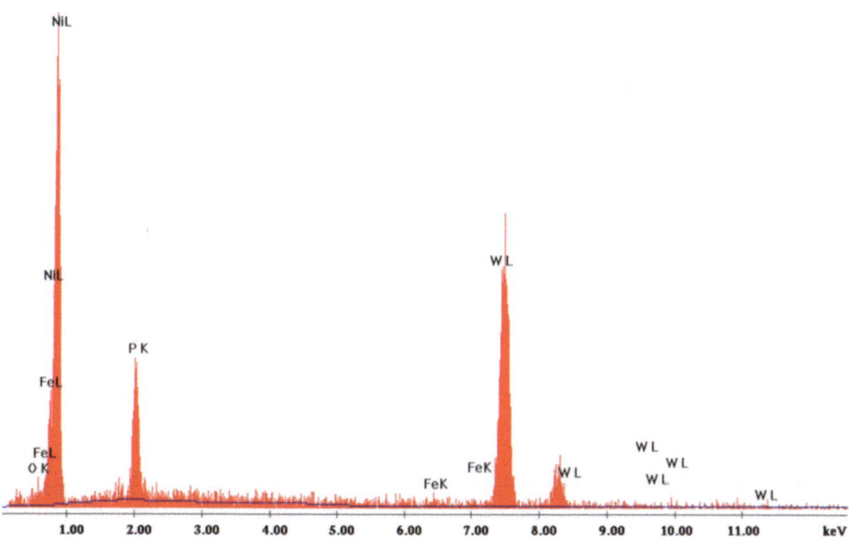

Fig. 4.19 EDX results of worn specimens of heat treated Ni–P–W coatings at 300 °C

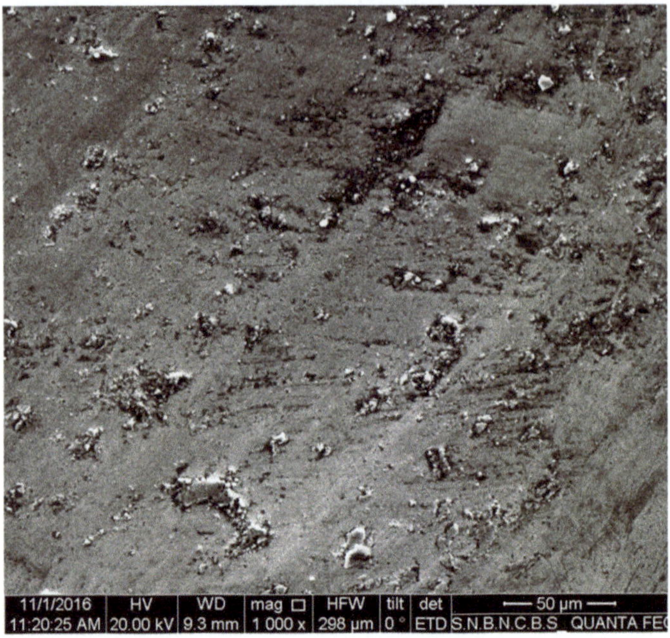

Fig. 4.20 SEM micrographs of worn specimen tested at 500 °C

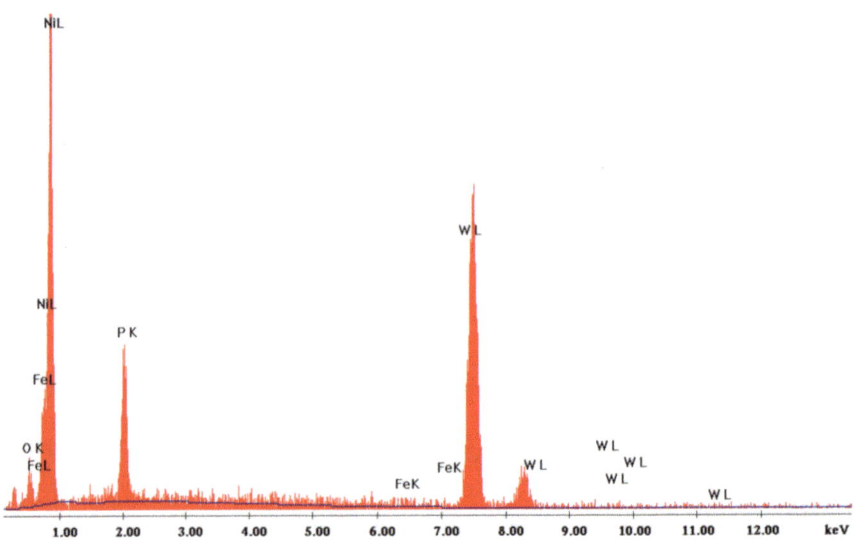

Fig. 4.21 EDX results of worn specimens of heat treated Ni–P–W coatings at 500 °C

to get attached to the coating face as observed from Fig. 4.21. The presence of oxygen in EDX spectrums (Figs. 4.17, 4.19 and 4.21) confirms oxidation of the coats and the construction of a tribo-oxide layer during sliding wear tests at 100, 300 and 500 °C, respectively. From the examination of the wear mechanism of heat treated Ni–P–W films, it can be understood that the mechanism changes from adhesive toward a mixture of adhesive and abrasive when temperature changes from RT to 500 °C.

4.4 Conclusion

Alkaline citrate-based Ni–P–W film is developed on AISI 1040 steel pin substrate. The consequence of elevated operating temperature over tribological behaviour of heat treated chemically deposited Ni–P–W covering is studied under different applied load and sliding velocity. The Ni–P–W layer displays an unstructured arrangement in the as-deposited case and presented a moderately small hardness of roughly 670 $HV_{0.1}$. As temperature was increased to 400 °C, the hardness increased to 1090 $HV_{0.1}$ because of the evolution of Ni and Ni_3P in the amorphous medium. Presence of steady Ni and Ni_3P was recognized for the heat treated deposits (400 °C, 1 h) comprising lesser amount of W and greater amount of P. Higher crystallization temperatures were due to of W incorporation in Ni–P matrix. Under elevated temperature tests, it is seen that wear rate increases with the increases in test temperature. The raise in applied load is found to negatively affect the wear resistance of the coating. However, wear rate is inversely related with sliding velocity for a fixed value of load. COF does not seem to be affected very much under elevated temperature tests. The SEM micrograph of the corroded and worn outer layer of the coating reveals that both abrasive and adhesive wear mechanism exist behind wear under high temperature. Oxide formation is another interesting phenomenon for the coatings tested under elevated temperatures.

REFERENCES

1. Brenner, A., Riddell, G.E.: Nickel plating on steel by chemical reduction. J. Res. Natl Bur. Stan. **37**(1), 31–34 (1946)
2. Sahoo, P., Das, S.K.: Tribology of electroless nickel coatings—a review. Mater. Des. **32**(4), 1760–1775 (2011)
3. Kundu, S., Das, S.K., Sahoo, P.: Properties of electroless nickel at elevated temperature-a review. In: Procedia Engineering, **97**, 1698–1706 (2014)
4. Loto, C.A.: Electroless nickel plating—a review. Silicon **8**(2), 177–186 (2016)
5. Elansezhian, R., Ramamoorthy, B., Nair, P.K.: Effect of surfactants on the mechanical properties of electroless (Ni–P) coating. Surf. Coat. Technol. **203**(5–7), 709–712 (2008)
6. Das, S.K., Sahoo, P.: Optimization of electroless Ni-B coatings based on multiple roughness characteristics. J. Technol. Sci. Educ. **2**(1/2), 85–106 (2011)

7. Vitry, V., Delaunois, F., Dumortier, C.: Mechanical properties and scratch test resistance of nickel–boron coated aluminium alloy after heat treatments. Surf. Coat. Technol. **202**(14), 3316–3324 (2008)
8. Balaraju, J.N., Rajam, K.S.: Electroless deposition of Ni–Cu–P, Ni–W–P and Ni–W–Cu–P alloys. Surf. Coat. Technol. **195**(2–3), 154–161(2005)
9. Kundu, S., Das, S.K., Sahoo, P.: Optimization studies on electroless nickel coatings: a review. Int. J. Manuf, Mater. Mech. Eng. **4**(4), 1–25 (2014)
10. Agarwala, R.C., Agarwala, V.: Electroless alloy/composite coatings: a review. Sadhana **28**(3–4), 475–493 (2003)
11. Sudagar, J., Lian, J., Sha, W.: Electroless nickel, alloy, composite and nano coatings—a critical review. J. Alloy. Compd. **571**, 183–204 (2013)
12. Yu, H.S., Luo, S.F., Wang, Y.R.: A comparative study on the crystallization behavior of electroless Ni–P and Ni–Cu–P deposits. Surf. Coat. Technol. **148**, 143–148 (2001)
13. Bangwei, Z., Wangyu, H., Qinglong, Z., Xuanynan, Q.: Properties of electroless Ni-W-P amorphous alloys. Mater. Charact. **37**, 119–122 (1996)
14. Bouanani, M., Cherkaoui, F., Fratesi, R., Roventi, G., Barucca, G.: Microstructural characterization and corrosion resistance of Ni-Zn-P alloys electrolessly deposited from a sulphate bath. J. Appl. Electrochem. **29**, 637–645 (1999)
15. Palaniappa, M., Seshadri, S.K.: Friction and wear behavior of electroless Ni–P and Ni–W–P alloy coatings. Wear **265**(5–6), 735–740 (2008)
16. Lu, G., Zangari, G.: Corrosion resistance of ternary Ni-P based alloys in sulfuric acid solutions. Electrochim. Acta **47**, 2969–2979 (2002)
17. Huang, Y., Huang, F., Zhao, W., Shi, K., Zhao, L., Wang, Y.: The study of electroless Ni-W-P alloy plating on glass fibre. Rare Met. **26**(4), 365–371 (2007)
18. Roy, S., Sahoo, P.: Optimization of electroless Ni-P-W coating for minimum friction and wear using grey Taguchi method, Article ID 608140. J. Coat **2013**, 1–13 (2013)
19. Ashassi-Sorkhabi, H., Dolati, H., Parvini-Ahmadi, N., Manzoori, J.: Electroless deposition of Ni-Cu-P alloy and study of the influences of some parameters on the properties of deposits. Appl. Surf. Sci. **185**, 155–160 (2002)
20. Chen, C.J., Lin, K.L.: Internal stress and adhesion of amorphous Ni–Cu–P alloy on aluminum. Thin Solid Films **370**, 106–113 (2000)
21. He, S.Z., Huang, X.M., Zheng, H.M., Li, P., Lin, Z.P., Shan, C.L.: Research on wear behavior of Ni-Cu-P co acting in different mediums. MocaxueXuebao/Tribol. **29**(4), 362–367 (2009)
22. Roy, S., Sahoo, P.: Optimization of wear of electroless Ni-P-Cu coating using artificial bee colony algorithm. In: Procedia Technology, **14**, 320–327 (2014)
23. Stremsdoerfer, G., Omidvar, H., Roux, P., Meas, Y., Ortega-Borges, R.: Deposition of thin films of Ni–P and Ni–B–P by dynamic chemical plating. J. Alloy. Compd. **466**, 391–397 (2008)
24. Srinivasan, K.N., Meenakshi, R., Santhi, A., Thangavelu, P.R., John, S.: Studies on development of electroless Ni–B bath for corrosion resistance and wear resistance applications. Surf. Eng. **26**(3), 153–158 (2010)
25. Balaraju, J.N., Kalavati. and Rajam, K.S. Electroless deposition of Ni–Cu–P, Ni–W–P and Ni–W-Cu–P alloy. J. Alloys Compd. **486**, 468–473 (2009)
26. Balaraju, J.N.Millath, Jahan, S., Anandan, C., Rajam, K.S.: Studies on electroless Ni–W–P and Ni–W–Cu–P alloy coatings using chloride-based bath. Surf. Coat. Technol. **200**, 4885–4890 (2006)
27. John, S.S., Srinivasan, K.N., Kavimani, P.M., Krishnan, K.H., Praveen, J., Ganesan, M.: Technical article: Electroless Ni-P-TiO_2 based composite coatings for surface engineering applications. Plat. Surf. Finish. **92**(5), 62–66 (2005)
28. Gadhari, P., Sahoo, P.: Effect of TiO_2 particles on micro-hardness, corrosion, wear and friction of Ni–P–TiO_2 composite coatings at different annealing temperatures. Surf. Rev. Lett. **22**(1550082), 1–15 (2015). https://doi.org/10.1142/S0218625X15500821

29. Novák, M., Vojtěch, D., Vlťů, T.: Influence of heat treatment on tribological properties of electroless Ni–P and Ni–P–Al_2O_3 coatings on Al–Si casting alloy. Appl. Surf. Sci. **256**(9), 2956–2960 (2010)
30. Ozimina, D., Madej, M., Kaldonski, M.T., Senatorski, J.: Production and tribological properties of anti wear Ni-P-Al_2O_3 composite coatings. Tribol. Schmierungstech. **55**(2), 37–41 (2008)
31. Winowlin Jappes, J.T., Ramamoorthy, B., Kesavan Nair, P.: Novel approaches on the study of wear performance of electroless Ni–P/diamond composite deposits. J. Mater. Process. Technol. **209**(2), 1004–1010 (2009)
32. Gadhari, P., Sahoo, P.: Effect of annealing temperature and alumina particles on mechanical and tribological properties of Ni-P-Al_2O_3 composite coatings. Silicon **9**(5), 761–774 (2017)
33. Ramalho, A., Miranda, J.C.: Friction and wear of electroless NiP and NiP + PTFE coatings. Wear **259**(7–12), 828–834 (2005)
34. Srinivasan, K.N., John, S.: Studies on electroless nickel-PTFE composite coatings. Surf. Eng. **21**(2), 156–160 (2005)
35. Lu, X.R., Liu, Y., Li, S.: Tribological behaviors of electroless composite plating with nano-particle PTFE. MocaxueXuebao/Tribol. **29**(2), 116–122 (2009)
36. Zou, T.Z., Tu, J.P., Zhang, S.C., Chen, L.M., Wang, Q., Zhang, L.L., He, D.N.: Friction and wear properties of electroless Ni-P-(IF-MoS_2) composite coatings in humid air and vacuum. Mat. Sci. Eng. A **426**(1–2), 162–168 (2006)
37. Hu, X., Jiang, P., Wan, J., Xu, Y., Sun, X.: Study of corrosion and friction reduction of electroless Ni–P coating with molybdenum disulfide nanoparticles. J. Coat. Technol. Res. **6** (2), 275–281 (2009)
38. Zhao, G., Deng, F.: Electroless plating of Ni-P-CNTs composite coating. Key Eng. Mater. **280–283**, 1445–1448 (2005)
39. Zhao, G., Zhang, H., Wang, Z.: Friction and wear resistance of the electroless Ni-P-CNTs composite coatings. In: Proceedings of SPIE, 6423, 64235S. (2007)
40. Li, Z.L., Wang, B., Guan, H.Y.: Preparation and properties of nano-TiO_2-Ni-P composite coating. Corros. Prot. **27**(8), 326–328 (2006)
41. Makkar, P., Agarwala, R.C., Agarwala, V.: Chemical synthesis of TiO_2 nano particles and their inclusion in Ni–P electroless coatings. Ceram. Int. **39**(8), 2741–2745 (2013)
42. Sahoo, P.: Wear behaviour of electroless Ni–P coatings and optimization of process parameters using Taguchi method. Mater. Des. **30**, 1341–1349 (2009)
43. Duncan, R.N.: The metallurgical structure of electroless nickel deposits: effect on coating properties. Plat. Surf. Finish. **83**(11), 65–69 (1990)
44. Tsai, Y.Y., Wu, F.B., Chen, Y.I., Peng, P.J., Duh, J.G., Tsai, S.Y.: Thermal stability and mechanical properties of Ni–W–P electroless deposits. Surf. Coat. Technol. **146–147**, 502–507 (2001)
45. Tien, S.K., Duh, J.G., Chen, Y.I.: The influence of thermal treatment on the microstructure and hardness in electroless Ni–P–W deposit. Thin Solid Films **469**, 333–338 (2004)
46. Roy, S., Sahoo, P.: Wear performance optimization of electroless Ni-P-W coating parameters using the Taguchi method. Paper ID: TSI 812541. In: ICIT-2012, The westin Pune Koregaon Park, India, 7–9th December 2012
47. Pearlstein, F., Weightman, R.F., Wick, R.: Nickel-Tungsten electroless deposition. Met. Finish. **61**, 77–81 (1963)
48. Brenner,A.: Electro Deposition of Alloys, vol. 2, Academic press, New York (1963)
49. Masoumi, F., Ghasemi, H.R., Ziaei, A.A., Shahriari, D.: Tribological characterization of electroless Ni–10% P coatings at elevated test temperature under dry conditions. Int. J. Adv. Manuf. Technol. **62**, 1063–1070 (2012)
50. Li, Z., Wang, J., Lu, J., Meng, J.: Tribological characteristics of electroless Ni–P–MoS_2 composite coatings at elevated temperatures. Appl. Surf. Sci. **264**, 516–521 (2013)
51. Alirezaei, S., Monirvaghefi, S.M., Saatchi, A. €Urgen, M., Motallebzadeh, A.: High temperature friction and wear behaviour of Ni-P-Ag-Al_2O_3 hybrid nanocomposite coating. Trans. Inst. Met. Finish. **91**, 207–213 (2013)

52. Franco, M., Sha, W., Aldic, G., Malinov, S., Cimeno_glu, H.: Effect of reinforcement and heat treatment on elevated temperature sliding of electroless Ni–P/SiC composite coatings. Tribol. Int. **97**, 265–271 (2016)
53. Kundu, S., Das, S.K., Sahoo, P.: Tribological behaviour of electroless Ni-P deposits under elevated temperature. Silicon **10**(2), 329–342 (2018)
54. Wu, F.B., Tien, S.K., Chen, W.Y., Duh, J.G.: Microstructure evaluation and strengthening mechanism of Ni–P–W alloy coatings. Surf. Coat. Technol. **177**, 312–316 (2004)
55. He, X.M., Liu, X.B., Wang, M.D., Yang, M.S., Shi, S.H., Fu, G.Y., Chen, S.F.: Elevated temperature dry sliding wear behavior of nickel-based composite coating on austenitic stainless steel deposited by a novel central hollow laser cladding. Appl. Surf. Sci. **258**(1), 535–541 (2011)
56. Mukhopadhyay, A., Barman, T.K., Sahoo, P.: Tribological behavior of electroless Ni–B–W coating at room and elevated temperatures. In: Proceedings of the Institution of Mechanical Engineers, Part J: Journal of Engineering Tribology, 1–17, (2018) https://doi.org/10.1177/1350650118755781
57. Panja, B. Das, S.K., Sahoo, P.: Tribological Behavior of Electroless Ni–P Coatings in Various Corrosive Environments. Surf. Rev. Lett. **23**(5), 1650040, 1–18 (2016)
58. Lewis, D.B., Marshall, G.W.: Investigation into the structure of electrodeposited nickel-phosphorus alloy deposits. Surf. Coat. Technol. **78**, 150–156 (1996)
59. Balaraju, J.N., Kalavati, Manikandanath, N.T., William Grip, V.K.: Phase transformation behavior of nanocrystalline Ni–W–P alloys containing various W and P contents. Surf. Coat. Technol. **206**, 2682–2689 (2012)
60. Osaka, T., Sawai, H., Otoi, F., Nihei, K.: Autocatalytic (electroless) nickel-tungsten-phosphorus plating. Met. Finish. **80**(8), 31–35 (1982)
61. Keong, K.G., Sha, W., Malinov, S.: Artificial neural network modelling of crystallization temperatures of the Ni–P based amorphous alloys. Mater. Sci. Eng. A, **365**, 212–218 (2004)
62. Maugis, D. In: Georges, J.M. (ed) Microscopic Aspects of Adhesion and Lubrication, Tribology Series 7, p. 221, Elsevier, Amsterdam (1982)

Chapter 5
Hardness, Friction and Wear Trends of Electroless Ni-W-P Coating Heat-Treated at Different Temperatures

Abhijit Biswas, Suman Kalyan Das and Prasanta Sahoo

5.1 Introduction

The term 'electroless coating', rather famous by the name 'electroless nickel' or EN coating, is a deposition technique which is in much demand in modern hard surface industries. This method has got overwhelming response from every field because of its high hardness and corrosion resistance properties and also praised for its splendid tribological characteristics [1]. The EN coating deposition process was first invented by Brenner et al. [2] in nineteenth century and then onwards it is employed considerably as a substitute for hard chrome coating on different substrates. However, EN coatings are widely used in aerospace, aviation, automotive, oil and gas processing, food processing, microelectronics, radio, electrical, electronics, computer engineering, chemical processing, textiles, machinery, mining and metallization of plastics, etc. [3]. The EN coating is deposited by an autocatalytic method and has the ability to deposit in a quite uniform manner on complex geometries. Besides, EN coating possesses other exclusive properties which are listed in Table 5.1.

In an aqueous solution, the alloy of nickel and phosphorus is chemically deposited onto the substrate without using any kind of external current source. The aqueous solution of electroless bath includes the metal ions, complexing agents, reducing agents and stabilizer. To achieve a successful coating deposition, the operating condition of electroless bath should be maintained by an exact concentration of metal ions, temperature and pH ranges. Moreover, both deposition rate and microstructural behavior of the coatings are very much depended on the aforesaid bath conditions [4]. The electrochemical reaction of EN coating is as follows [1]:

A. Biswas (✉) · S. K. Das · P. Sahoo
Department of Mechanical Engineering, Jadavpur University, Kolkata, India
e-mail: abhijitbiswas.mechanical@gmail.com

P. Sahoo and J. P. Davim (eds.), *Advances in Materials, Mechanical and Industrial Engineering*, Lecture Notes on Multidisciplinary Industrial Engineering,
https://doi.org/10.1007/978-3-319-96968-8_5

Table 5.1 Distinct characteristics of EN coatings [1]

Properties	Benefits
High hardness	Low wear characteristics
Deposit uniformity	Eliminates post plate finish
Good chemical resistance	Acts as protective coating
Excellent corrosion resistant	Good coating durability
Low friction co-deposits available	Self-lubricating coating
Fast plating rate	High production output
Solderability/weldability/brazeability	Functional in many applications
Non-magnetic/magnetic	Magnetic property selectivity
Good brightness	Attractive finish

$$H_2PO_2^- + H_2O \rightarrow H_2PO_3^- + 2H + 2e^- \tag{5.1}$$

$$Ni^{++} + 2e^- \rightarrow Ni \tag{5.2}$$

$$2H^+ + 2e^- \rightarrow H_2 \tag{5.3}$$

$$H_2PO_2^- + 2H+ \ + e^- \rightarrow P + 2H_2O \tag{5.4}$$

Some amount of hydrogen is produced during catalytic dehydrogenation of hypophosphite and adsorbed onto the substrate which eventually leads to the reduction of nickel on the substrate surface [1]. This adsorbed hydrogen reduces hypophosphite in small amount to water, hydroxyl ion and phosphorous. The atomic hydrogen mechanism is as follows [1]:

$$H_2PO_2^- + H_2O \rightarrow H_2PO_3^- + H^+ + 2H_{ads} \tag{5.5}$$

$$2H_{ads} + Ni^{++} \rightarrow Ni + 2H^+ \tag{5.6}$$

$$H_2PO_2 + H_{ads} \rightarrow H_2O + OH^- + P \tag{5.7}$$

$$(H_2PO_2)^{2^-} + H_2O \rightarrow H^+ + (HPO_3)^{2^-} + H_2 \tag{5.8}$$

Properties of EN coatings are greatly influenced by its bath composition and with appropriate heat treatment, the same enhances further to a great extent. Generally, both microstructure and crystallization performance of EN coatings are dependent on its phosphorus content. It has been reviewed that low phosphorus content (<3wt% P) coating generally exhibits both microcrystalline and crystalline structure in as-deposited condition. But coating structure with high phosphorus content (>9wt% P) exhibited either amorphous or a compound of both amorphous and microcrystalline [5, 6]. According to Yoon et al. [7], in high P content EN coating, as the phosphorus dissociation is very high, which dominates the FCC (face centered cubic) nickel nucleation and ultimately leads to an amorphous structure in as-deposited EN coating. A sharp reflection of nano-crystalline Ni (111) along with a board amorphous peak is seen in as-deposited condition, as reported by Liu et al. [8] and with appropriate heat treatment, this amorphous nickel is turned into crystalline nickel and its phosphides. In continuation, this nickel phase dominates the entire crystallization process from start to end and again the volume of nickel phosphides (Ni_3P) is also determined by nickel. Moreover, these nickel peaks are estimated by the phosphorus content at the time of deposition. However, properties (viz. higher hardness, wear resistance, smoothness) of EN coating can be further enhanced by heat treatment. Generally, in EN coating, amorphous phases turn into crystalline phases by precipitation of harder nickel phosphide phases (mainly Ni_3P) and strong orientation toward Nickel (111) phase is observed. In some studies, the phase transformation of EN coating mainly occurs beyond the heat treatment temperature of 300 °C. Moreover, a few researchers have also mentioned about some intermediate phases (i.e., Ni_5P_2 and $Ni_{12}P_5$) before obtaining the most stable Ni_3P phases [9–11]. Hardness of coating is found to follow an increasing trend with heating temperature, but beyond 400 °C, hardness falls, as mentioned by Hou et al. [12]. They also attributed the highest hardness value, i.e., above 1000 HV at 400 °C to the synergistic effect of both precipitations of harder Ni_3P phase and smaller grain size (<5 nm). However, hardness of the EN coating is reduced beyond heat treatment temperature of 500 °C. Increase of hardness ultimately influences the wear resistance properties of EN coating [4]. Instead of high melting point of nickel (1455 °C), a few researchers have reported about oxide formation over the coating morphology at higher heat treatment temperature and coatings begins to degrade due to same [13]. However, heat-treated at 400 °C for 1 h specimen shows a better tribological performance than 260 °C for 25 h heat-treated sample [6]. But the other research group [4] has reported that the EN coating exhibited more effectiveness when heat-treated at 400 °C but no such attribution on friction coefficient is mentioned.

Usefulness of EN coating is widely accepted, especially regarding its mechanical and tribological aspects. Besides, the incorporation of third element (i.e., copper, cobalt, tungsten) into binary Ni-P coating can further improve its properties. According to Balaraju and Rajam [14], incorporation of tungsten in EN coating enhances its thermal stability along with its wear, corrosion and electrical resistances. Co-deposition of tungsten into EN coating improves the atomic binding force against regional plastic deformation which leads to higher hardness but the lattice disorder on crystalline phase decreases with increasing tungsten content in EN deposit. Increment in tungsten percentage of coating may also be directing to the reduction of phosphorus percentage and hence leading to a crystalline structure in as-deposited Ni-W-P coating. Also the addition of tungsten enhances the crystallization temperature that indicates more thermal stability compared to binary EN coatings [15, 16].

Tungsten is chosen as surface protection material by many of the researchers because of its higher wear resistance and corrosion resistance properties. Moreover, nickel-tungsten electrodeposition not only yields a proper replacement of hard chromium coating but also protects against environmental hazards [17]. Now, tungsten being a refractory metal cannot itself be deposited onto the substrate from an aqueous solution. Thus, tungsten is deposited onto a substrate in combination with iron group metal [18]. However, according to Hu et al. [19], electroless Ni-W-P coating is well developed on Al alloy matrix and a clear image of intermetallic Ni–Al compound appears on crosscut after heat treatment. As-deposited Ni-W-P coatings exhibit an amorphous or nano-crystalline structure and turns into crystalline by precipitation of stable Ni_3P and solid solution of Ni-W matrix during heat treatment, which leads to increase in the hardness of the coating [14]. Compared to the nodular structure of Ni-P coatings, Ni-W-P exhibits a more flat surface [20]. Wear behavior of EN coating can be significantly enhanced by addition of hard particles as studied by Tsai et al. [21]. The abrasive wear mechanism is prevalent for both dry and lubricated test condition of Ni-W-P coatings during tribological performance and the latter yields better wear resistance [22]. At higher heat treatment temperature, the oxide layer is formed on the coating as reported by Eraslan and Ürgen [23].

From the literature review, it can be seen that the investigation of the result of heat treatment temperature on tribological characteristics of Ni-W-P coatings has almost remained unaddressed. Hence, the current study is motivated in this direction. Ni-W-P coating is prepared in the laboratory by electroless method. Then those newly developed samples are heat-treated which is followed by micro-hardness and tribological tests (friction and wear). Therefore, the effect of heat treatment temperature on Ni-W-P coating performance is investigated. Moreover, X-ray diffraction analyzer (XRD) for phase structure, scanning electron microscope (SEM) for coating surface and energy-dispersive X-ray analysis (EDX) for elemental analysis are utilized for the present Ni-W-P coating. Finally, the wear mechanisms of worn specimens of current coatings are studied by SEM micrographs.

5.2 Methods

5.2.1 Development of Ni-W-P Coating

Both square and pin-shaped mild steel (AISI1040) substrates are used for electroless Ni-W-P deposition with dimension of 15 × 15 × 2 mm^3 and Ø6 × 30 mm, respectively. Square and pin samples are considered for microstructural and tribological characterization, respectively. All specimens are produced with high precision and grinded to a smooth surface finish to achieve N5 (R_a = 0.4 μm) roughness grade. Specimens are subjected to acid pickling (35% hydrochloric acid) for 60 s to get rid of the foreign matter and organic particles from the same. After that they are washed in deionized water followed by methanol cleaning and dried in air. Prior to deposition, all cleaned specimens are dipped in palladium chloride ($PdCl_2$) solution at temperature of 55 °C for 8–10 s and finally dipped into the

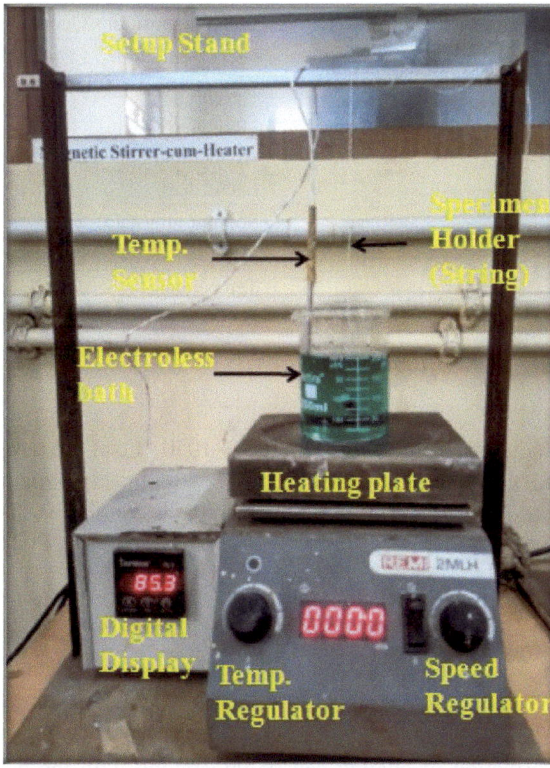

Fig. 5.1 Electroless Ni-W-P deposition setup

electroless solution for a period of 3 h. Several trials are conducted to get proper bath composition [18]. Coating deposition rate is found to be around 10–12 μm/h. Coating deposition setup of Ni-W-P is shown in Fig. 5.1. Bath composition and deposition conditions for electroless Ni-W-P coatings are presented in Table 5.2. The deposition conditions are kept constant for all specimens so that the thickness of each specimen stays nearly the same. After the coating deposition is over, specimens are taken out from the solution and thoroughly washed by the deionized water followed by acetone cleaning.

5.2.2 Heat Treatment of Electroless Ni-W-P Coatings

Heat treatment temperature and also heat treatment period has great influence upon both phase transformation and microstructural behavior of EN coating [4]. Here, only the effect of heat treatment on Ni-W-P coating is evaluated. The coated specimens (both square and pin) are heat-treated in box furnace (Make SUNBIM,

Table 5.2 Bath composition and operating conditions of electroless Ni-W-P coating

Bath compositions/Operating conditions	Values
Nickel sulfate	20 g/L
Sodium hypophosphite	20 g/L
Sodium citrate	35 g/L
Ammonium sulfate	30 g/L
Lactic acid	5 g/L
Sodium tungstate	20 g/L
pH	7–8
Temperature	85 ± 2 °C
Duration of coating	3 h
Bath volume	250 mL

India and Temperature Range 0–1050 °C) at the temperatures 200, 400 and 800 °C for constant time duration of 1 h. After heat treatment, specimens are removed from the furnace and cooled under ambient condition.

5.2.3 Coating Characterization

Coating characterization is introduced into the study of microstructure and its elemental composition of the developed Ni-W-P coatings. Moreover, phase transformation of coating can be understood with the help of phase structure analysis. Microstructural studies of the coated samples are conducted using SEM and EDX (make FEI, FEG Quanta 250) to look into the morphology and find the actual composition of the present coating, respectively. Generally, in SEM electrons in a higher energy state hits the substrate that produces different signals at the sample surface. Information about the sample (viz. surface morphology, chemical composition, and crystalline structure and orientation of materials making up the sample) is obtained from these signals which are generated due to electron-sample interactions. The microstructure of the present coatings is analyzed by using SEM on the both as-deposited and heat-treated specimens.

EDX is a technique employed to define the elemental composition of materials. EDX systems are attached to electron microscopy instruments, where the imaging capability of the microscope identifies the specimen of interest. The data produced by EDX analysis consist of spectra showing peaks corresponding to the elements making up the true composition of the sample being analyzed. The surface morphology of worn-out specimens is also studied by SEM to find the predominating wear mechanism existing both in case of as-deposited and heat-treated coating under dry condition.

Different phases of both as-deposited and heat-treated samples are identified by using XRD analysis (make Rigaku, Ultima). Basically, XRD is a non-destructive technique to determine crystalline phases and orientation of crystal structure.

This technique is based on diffraction which is the elastic scattering (change of direction) of the electromagnetic waves without any energy loss. The use of diffraction method is of great importance in the analysis of crystalline solids. Not only can they reveal the main features of the structure, i.e., the lattice parameter and type of structure, but also other details such as the arrangement of different kinds of atoms in crystals, the presence of imperfections, the orientation, sub-grain and grain size, the size and density of precipitates. In the present study, XRD analysis is carried out for 2θ values ranging from 30° to 90° and a scanning speed of 1°/min (based on literature) [6].

5.2.4 Micro-Hardness Study

Hardness is the property of a material that enables to resist the penetration, scratching, abrasion or cutting. The hardness of material is measured with the help of a specific device that usually measures the indentation depth made by an indenter (made of hard material like diamond) of a definite shape, with definite amount of load and for a certain time duration. If the microstructure of the material is coarse and heterogeneous then lager indentation depth is required than homogeneous material. On the basis of applied compressive load, the indentation of hardness can be serially termed as macro (and micro)-indentation or nano-indentation [24]. According to Biswas et al. [4], wear resistance of EN coating (specifically Ni-(high) P) is found to possess a good correlation with the hardness of the same. Hence, for the present study, Vicker's micro-hardness tester (make VMHTMOT, Technische Mikroskopie) is used to measure the hardness of both as-deposited and heat-treated Ni-W-P coating. A pyramid shaped square base indentation is left by a diamond indenter with an apex angle of 136° (see Fig. 5.2). The equation of Vicker's hardness is as follows [24]:

$$\mathrm{HV} = 1854.4\frac{\mathrm{F}}{\mathrm{d}^2} \tag{5.9}$$

where 'F' is the applied force (in gf) and 'd' is the mean of the diagonal lengths. The present coating being thin, lower value of applied load (100 gf) is considered such that indenter may not penetrate the coating and substrate hardness does not influence the hardness calculation. Dwell time and indentation speed are taken as 15 s and 25 μm/s, respectively. Average of ten hardness measurements for a particular sample is reported.

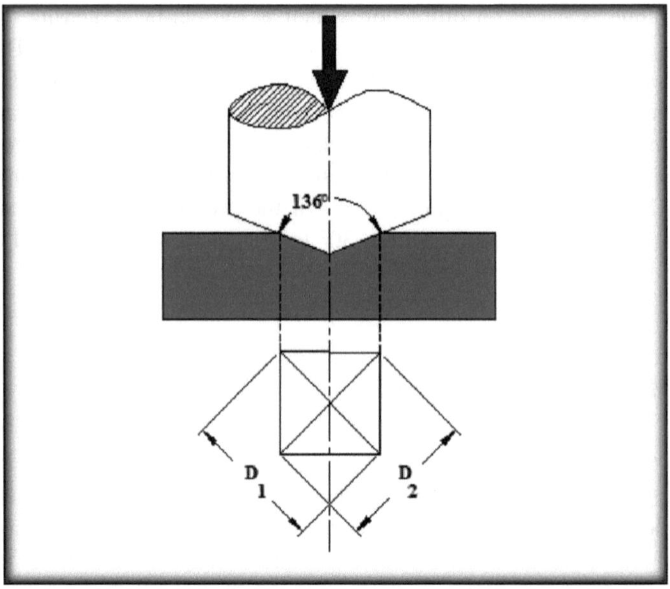

Fig. 5.2 Principle of Vicker's hardness measurement

5.2.5 *Tribological Study*

As-deposited and heat-treated pin specimens are analyzed on pin-on-disk type tribo-tester setup (TR-20LE-CHM-400, Ducom, India) (see Fig. 5.3) to study the tribological behavior (friction coefficient and wear) of the present coating under dry condition and at ambient temperature (32 °C) with a relative humidity of 45%. The developed Ni-W-P coated specimens are held firmly against a rotating EN31 (hardness ~62HRc) counter face disk. The diameter and width of the counterface disk are 165 and 8 mm, respectively. Normal load is applied with 1:1 ratio of loading lever by placing dead weights on a loading pan which is attached to the lever. The frictional force and wear depth are measured by LVDT and load cell with accuracies of 0.1 ± 2% and 1 ± 1%, respectively, of measured value. The wear depth and frictional force are directly recoded on a computer in real time during the tests [24]. Each test is performed for 600 s based on present coating thickness with a fixed track diameter of 60 mm and speed of 50 RPM. Moreover, the sliding speed (157 mm/s) and sliding distance (94,200 mm) are kept constant during test [4]. As the counterface is made of hardened steel (EN31), it is assumed that only wear of coatings takes place. However, wear is calculated in terms of wear rate by using the following equation [4]:

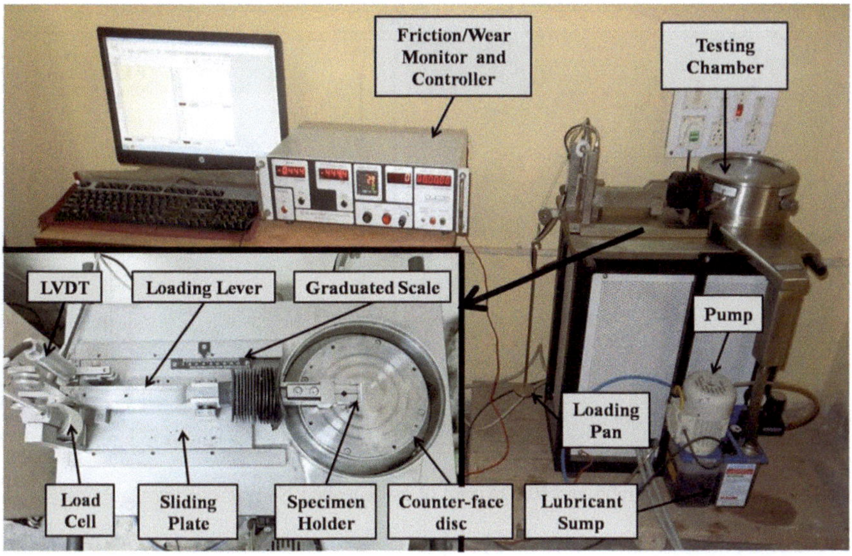

Fig. 5.3 Pin-on-disk type tribo-tester setup

$$K = \frac{V}{S \times F} \tag{5.10}$$

where, V, S and F are wear volume (mm^3), sliding distance (m) and normal load (N), respectively. For the present study, density of electroless Ni-W-P coating is taken around 8.6 gm/cm^3 as calculated from the percentage composition of the individual elements obtained from EDX results.

5.3 Result and Discussion

5.3.1 SEM and EDX Analysis of as-Deposited Ni-W-P Coating

To understand whether the coating is developed properly, microstructural analysis of deposition is useful. However, SEM micrograph and EDX plot of the coatings are presented in Fig. 5.4. From the microscopic images, as-deposited coating is found to be in nodular structure without any pores over the coating morphology which leads to a higher corrosion resistant [25] and the nodule size is around 30–40 μm (Fig. 5.4a). The nodular structure of the coating indicates about its self-lubricating nature and ultimately results in lower friction coefficient [6]. A nodular structure of nickel-based alloy coating has been reported by many researchers [4, 6, 18, 25–28]. This kind of structure is also acts as lubricant pockets

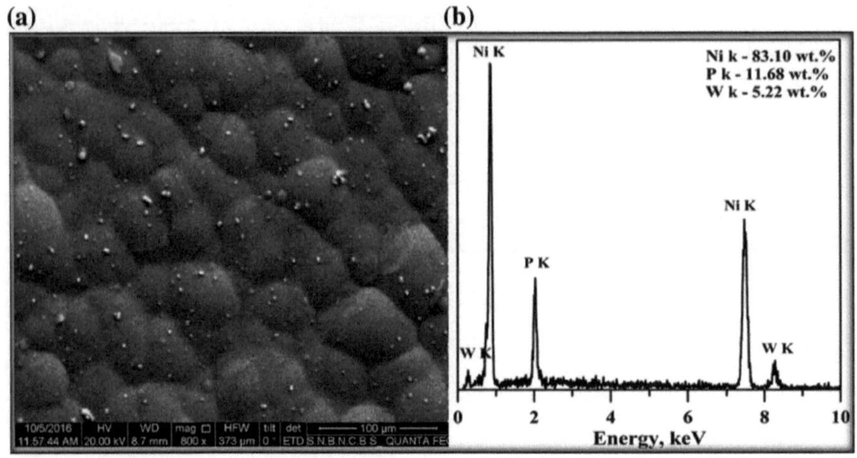

Fig. 5.4 **a** SEM micrograph and **b** EDX plot of as-deposited Ni-W-P coating

[29]. In comparison with some EN coatings [4], morphology of present coating is found to be more compact. Hence, inclusion of tungsten may be lucrative for enhancing corrosion resistance of binary EN coating [29]. A few bright particles are scattered over the coating surface due to excess growth of nickel [20].

It is an established fact that phosphorous percentage in the coating decides whether the structure of the corresponding EN coating would be amorphous or crystalline. A research group has reported about the presence of both amorphous and nano-crystalline structure in as-deposited Ni-W-P coating. They attributed this to the presence of tungsten (around 4 wt%) [27]. However, in the present case, tungsten is detected around 5.22% by weight, whereas nickel and phosphorus are obtained as 83.10 weight % and 11.68 weight %, respectively, (Fig. 5.4b). Therefore, it is believed that the present coating would be amorphous in nature as phosphorus content [1, 27] is in the higher range (>10 wt% P). Moreover, coating appears homogeneous throughout without any apparent damage.

5.3.2 *Phase Structure Study*

To get an idea about the phase structure of electroless Ni-W-P coating, XRD analysis of the coating is conducted for both as-deposited and heat-treated samples. XRD analysis of as-deposited Ni-W-P coating is also carried out at various heat treatment temperatures. From XRD plot (Fig. 5.5), the broad peak of as-deposited Ni-W-P coating is clearly an evidence of amorphous structure [26]. According to Duari et al. [24], a broad peak is generally found due to chaos in atomic arrangement in structure. Now, the dissociation rate of nickel and phosphorus atoms in coating determines its crystallinity [30]. As nickel diffusion rate is lower than

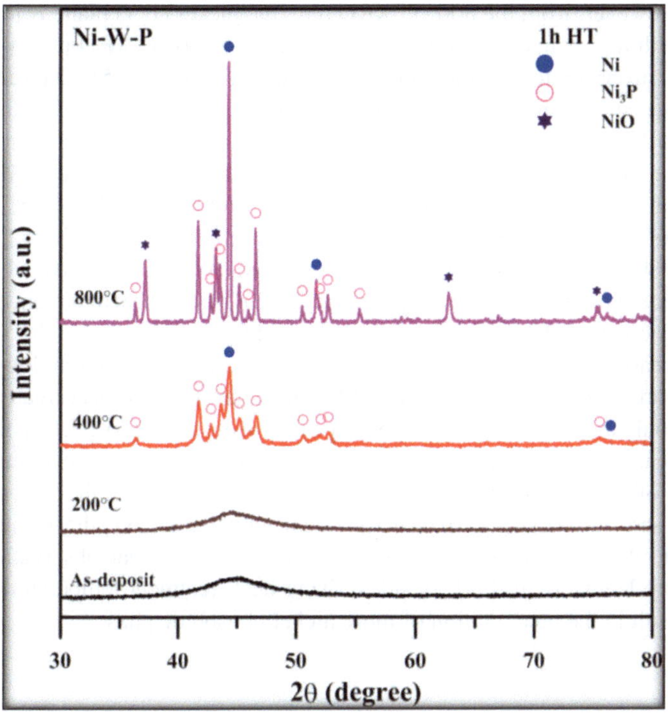

Fig. 5.5 XRD patterns of Ni-W-P coating in its as-deposited and heat-treated condition

phosphorus atom, higher amount of phosphorus is required to segregate the nickel as reported by Palaniappa and Seshadri [30]. This is why the coating with higher phosphorus concentration exhibits amorphous structure [24].

However, XRD plot of 200 °C for 1 h heat-treated specimen displays the same broad peak as obtained from as-deposited coating (see Fig. 5.5). This signifies that no such transformation occurs at that temperature, and hence, the coating still retains its amorphous structure. This behavior is also quite similar to EN coatings [6]. But, in case of 400 °C for 1 h heat-treated specimen, sharp peak of nickel and nickel phosphide are observed in XRD plot. The presence of those sharp peaks in XRD plot indicates that amorphous phase turns into crystalline. Differential scanning calorimetry (DSC) analysis may reveal the actual crystallization temperature of the present electroless Ni-W-P coating.

The coating appears crystalline based on the presence of sharp nickel (111) peak and a few small peaks of Ni_3P (321), (301), (420), etc. at 400 °C plot (Fig. 5.5). These peaks are formed by the precipitation of nickel and its phosphides into the coating without any intermediate phosphide phases [27]. In ternary deposits, the amount of tungsten appears as a form of solid solution and the solubility limit of tungsten in nickel is around 39 wt% [31]. A peak of tungsten nitride oxide is seen in 400 °C XRD plot, which reveals the occurrence of oxide layer on the coating surface. However, from the 800 °C XRD plot, the nickel (111) plane is found more sharper with higher intensity than the nickel peak of 400 °C XRD plot, which may indicate a strong preferred orientation in that direction (Fig. 5.5). The reduction of phosphorus content in coating may be the reason behind this higher nickel peak, as mentioned by other researchers [24]. Similar observation has also been made by Balaraju et al. [32], due to increment of tungsten content (or decrement in P content) in coating which results in sharper nickel peaks. Moreover, increased intensity of nickel phosphide (Ni_3P) peak is also observed from Fig. 5.5.

From 800 °C XRD plot (Fig. 5.5), Bunsenite (NiO) peak is detected, which reveal the presence of oxides on the coating surface [13, 23]. As the heat treatment of Ni-W-P coatings is conducted in box furnace and allowed to cool in open air, this could be a probable reason of the formation of oxide layer into the coating morphology. But it is quite important to note that no metastable phases are seen for heat-treated Ni-P-W coatings; only major peaks of Ni and Ni_3P peaks are observed [30, 33, 34]. But in case of heat-treated EN coating, a few metastable phases of nickel are found prior to formation of stable Ni_3P phase [6]. It is worth noting that though the peak intensities are increased due to rising of heat treatment temperature but the position of Ni (111) and other phosphide peaks (Ni_3P) remain same. Similar observation is also reported by Koiwa et al. [35]. Besides, rising heat treatment temperature may not always result in the precipitation of more nickel phosphide phase.

5.3.3 Surface Morphology of Heat-Treated Ni-W-P Coating

SEM micrograph of electroless Ni-W-P coating heat-treated at different temperature for constant period of time (i.e.,1 h) are illustrated in Fig. 5.6. From the SEM images, it is seen that these heat-treated Ni-W-P coatings appear to be quite denser and nodular in structure and also maintain its homogenity throughout the coating surface. The heat-treated surface of present coating is found to be apparently defect free and without any porosity up to 200 °C heat treatment temperature. A few pores (marked by arrow) are noticed on further heating but it is believed that their sizes are too tiny to be able to penetrate up to the substrate surface. With heat treatment, coatings seem to undergo volume contraction [36]. The typical cauliflower-like structure is seen to be retained even after heat treatment [29]. This type of structure has also been reported by Mukhopadhyay et al. [25]. But interestingly, the coatings gradually turned into flat morphology with increasing heat treatment temperature. Nodular morphology can be perceived up to 400 °C heat treatment temperature (see Fig. 5.6a, c) beyond which it turns into flat morphology. Similar observation is also made by other research group for EN coating with high phosphorus content [6]. However, increase of nodule size may be attributed to grain coarsening phenomenon. Krishnaveni et al. [37] have observed the reduction of several hardening sites due to higher heat treatment temperature.

Moreover, these nodular structures obtained from SEM (Fig. 5.6) appear as columnar ridges [21]. This columnar growth in nickel-based alloy coating directs to

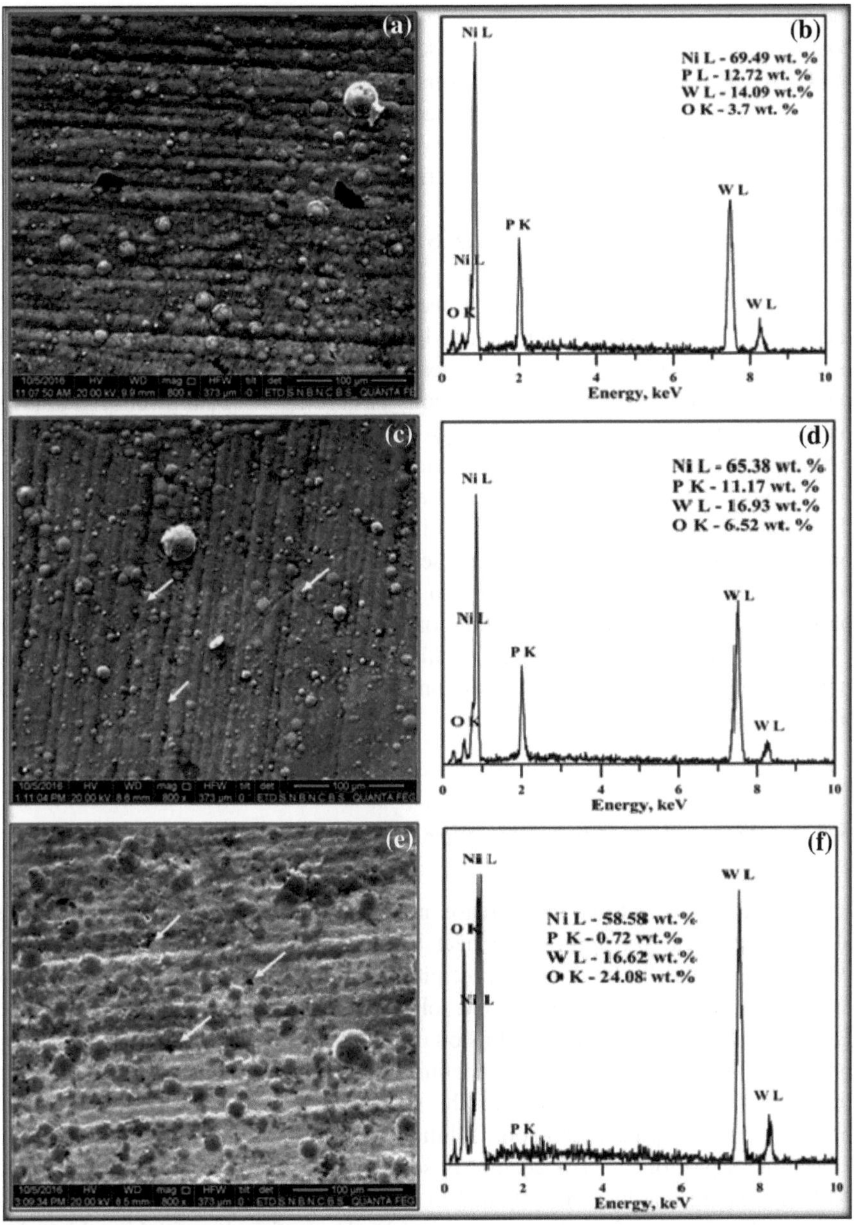

Fig. 5.6 SEM micrograph and EDX spectra of heat-treated (1 h) Ni-W-P coating: **a**, **b** 200 °C. **c**, **d** 400 °C and **e**, **f** 800 °C

the reduction of original contact area leading to wear resistant property of the coating [29]. From Fig. 5.6e, a few blackish spots are seen in coatings heat-treated at 800 °C. This could be due to the formation of oxide scale on the coating surface [23]. From Fig. 5.6a, i.e., for 200 °C heat-treated specimen, there is no such dissimilarity in the appearance with as-deposited coating, but the formation of oxide layer is present as found in EDX result (Fig. 5.6b). However, there is no evidence of oxide phase in XRD analysis (Fig. 5.5) for the sample. A Glow Discharge Optical Emission Spectroscopy (GDOES) of the crosscut specimen may unfold the definite reason behind this observation.

However, it is quite clear from the EDX spectra (Fig. 5.6d, f) that with increasing heat treatment temperature, large amount of oxygen incorporation is observed in the coating [6]. The presence of higher oxidation may be due to some other mechanism of inter layer diffusion. According to Tomlinson et al. [13], oxygen enters into the coating through micro cracks by intermetallic diffusion which may result in the formation of oxide layer of the coating, which is apparent for the present case. From Fig. 5.6f, it is also observed that phosphorous content of the coating drastically decreases with raising the heat treatment temperature. Similar observation is also reported by other researchers [6, 38]. Massive incorporation of oxygen at elevated temperature may also result in obvious decrement of nickel and phosphorus weight percentage in the coating. According to Biswas et al. [6], in electroless Ni-(High) P coating, formation of oxide starts beyond 600 °C heat treatment temperature and the aberration of phosphides and exhalation of phosphorus occur above 800 °C.

5.3.4 Micro-Hardness Behavior

It is an established fact that the increase in hardness of nickel-based alloy coating is always dependent on appropriate heat treatment. Hardness of heat-treated EN coating is mainly the function of precipitation of harder phase into nickel matrix and strengthening mechanism of solid solution due to heat treatment. According to Biswas et al. [6], the elemental diffusion of phosphorus and nickel in solid solution produced higher strength in EN coating. Generally, with increase in the heat treatment temperature, hardness of the coating is increased by precipitation of harder nickel phosphides and solid solution strengthening mechanism [4, 14].

Hardness plot of Ni-W-P coating is seen in Fig. 5.7a. In case of as-deposited Ni-W-P coating, hardness is obtained in the range 600–620 HV0.1 at room temperature. But in case of Ni-(High) P coating, hardness of as-deposited is achieved in the range 582–600 HV0.1 (Fig. 5.7b). Therefore, the inclusion of W definitely improves the as-deposited hardness of Ni-W-P coatings.

At 200 and 400 °C for 1 h post heat-treated specimens, hardness is drastically changed from 693 $HV_{0.1}$ to 1113 $HV_{0.1}$. Similar hardness results are also obtained by Balaraju and Rajam [14]. By comparing with Ni-(High) P, hardness value is almost same when heat-treated at 200 and 400 °C. This highest value of hardness

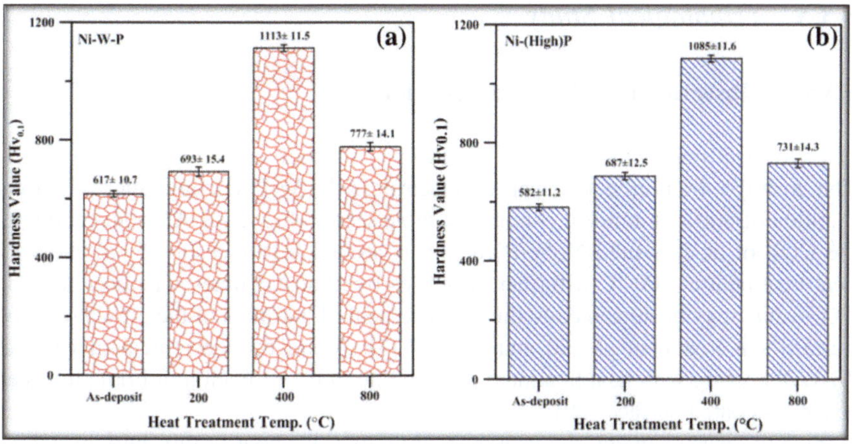

Fig. 5.7 Hardness plot of electroless **a** Ni-W-P coating and **b** Ni-(High) P coating

achieved may be due to the synergistic effect of strong orientation toward nickel (111), precipitation of Ni_3P phases and the solid solution strengthening mechanism of Ni-W matrix. All these contribute to the enhancement in the wear resistance properties of coating [26]. Moreover, this precipitated harder stable phase may obstruct any dislocation within the crystal lattice [4]. Generally, the crystalline structure of coating exhibits a higher strength as the dislocation within the material structure may be obstructed before it reaches the grain boundary. In continuity, if the grain size is reduced, means exaggeration of grain boundaries which ultimately results in higher hardness of coating [6].

But at higher heat treatment temperature (800 °C), hardness is obtained as 777 $HV_{0.1}$ and 731 $HV_{0.1}$ for Ni-W-P and Ni-(high) P coating, respectively, (Fig. 5.7). This lower value of hardness may be attributed to the grain coarsening and intermetallic formation of iron oxide into the coating [13, 23]. By increasing grain size, the chance of obstruction by grain boundary lessens which may show up as lower hardness of the coating [4]. The hardness value does not increase significantly after heat treating at higher temperatures. This signifies the thermal stability of the coatings. However, the increase in micro-hardness of Ni-W-P coating is mainly attributed to the precipitation of crystalline Ni and its hard phosphide phases [29].

However, the micro-hardness feature is very much dependent upon microstructural characteristics of material and heat treatment temperature and heating period as well. Moreover, crystallization behavior of EN deposit is dependent on phosphorus content in the coating. If grain size is raised, then grain boundary area is also decreased. Therefore, the barrier to the motion of dislocation could be dominated during plastic deformation which directs toward lower hardness value of the coating [4].

5.3.5 Tribological Study

5.3.5.1 Observation on Friction Behavior

Effect of heat treatment on friction behavior of nickel-based alloy coating is vivid. The average friction coefficient of electroless Ni-W-P coating after a sliding duration of 600 s at ambient condition can be seen in Fig. 5.8a. At room condition, the COF of as-deposited Ni-W-P coating is around 0.18 (see Fig. 5.8a) and with heat treatment at 200 °C it further decreases by almost 3 times. The decrement in COF value may be attributed to the enhancement of hardness of the present coating during heat treatment at 200 and 400 °C [4]. Again this decrement of COF of Ni-W-P coating with heat treatment may be due to the formation of brittle tungsten oxide on coated surface [29].

Due to heat treatment, the coating morphology is about to turn into crystalline structure (i.e., for 200 °C) or fully converted into crystalline structure (i.e., for 400 °C) and along with the nodular boundaries as seen from the XRD and SEM images (see Figs. 5.5 and 5.6). These nodular boundaries acted as diffusion channels for tungsten and through which the tungsten is able to move into the surface of the coating and form oxides [29]. Hence, the heat-treated coatings, having increased grain boundaries, display higher COF. The friction results of Ni-(High) P coatings [6] at different heat-treated condition are also provided in Fig. 5.8b for comparison purpose. For Ni-(High) P as-deposited condition, COF is obtained around 0.43 which is significantly higher than the current Ni-W-P coatings. COF of Ni-(High) P coating is greatly reduced when heat-treated at 200 °C for 1 h. The overall trend of the friction behavior of both the coatings seems to be similar. Oxide formation at high temperature is also observed in case of Ni-(High) P coating [6]. However, in terms of high-temperature performance, Ni-W-P coating exhibits more consistent

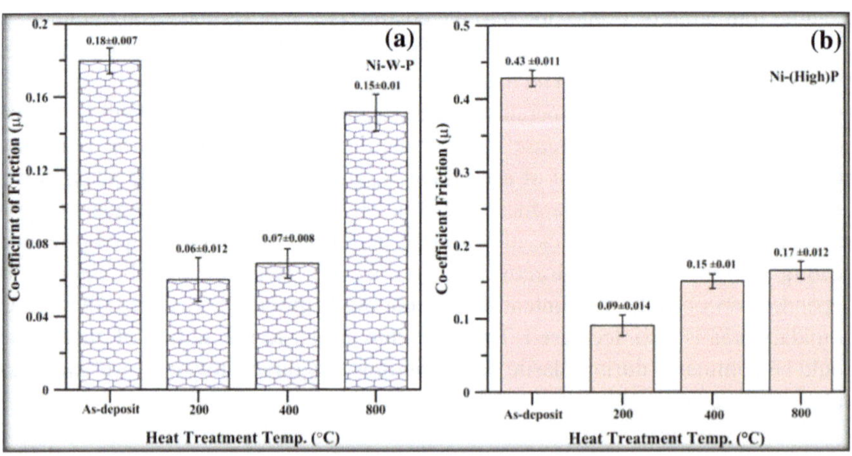

Fig. 5.8 COF plot for electroless **a** Ni-W-P coating and **b** Ni-(High) P coating

friction behavior compared to Ni-(High) P coating. At 800 °C, the COF value of both the coatings is very nearby which may be due to the oxide layer formed at the interface whose properties are nearly similar for both the coatings. The oxides which are formed may have poor adhesion strength to the interlayer of the coating are relatively easy to shear.

Generally, plastic deformation of harder material is low. Thus, a chance of increased contact area between the coating and the counter face is significantly less [4]. There are so many factors including coating microstructure, grain size, surface roughness, etc. which may influence the friction performance of coating. Comparing between electroless ternary (Ni-W-P) and binary (Ni-(high) P) coating, former shows lower friction coefficient than latter. Thus, incorporation of tungsten in binary EN coating is found to be beneficial in terms of friction reduction. Moreover, from the present study, it is observed that friction coefficient of coating is reduced for a particular heat treatment temperature.

Figure 5.9 shows the COF trend with testing time for both the coatings. The testing period is kept lower due to low coating thickness as previously discussed. Comparing both Fig. 5.9a, b, it can be said that the coatings display more or less consistent COF value over the entire test duration. The most stable COF plot is obtained for Ni-(High) P coatings heat-treated at 200 °C (Fig. 5.9b). However, the value of COF is found to be highly undulating w. r. t. sliding duration particularly for 800 °C heat-treated sample which may again be attributed to the formation and breakage of intermediate oxide layers as well as changes in the microstructural mechanisms during sliding [25, 29]. However, further exploration is required in this facet.

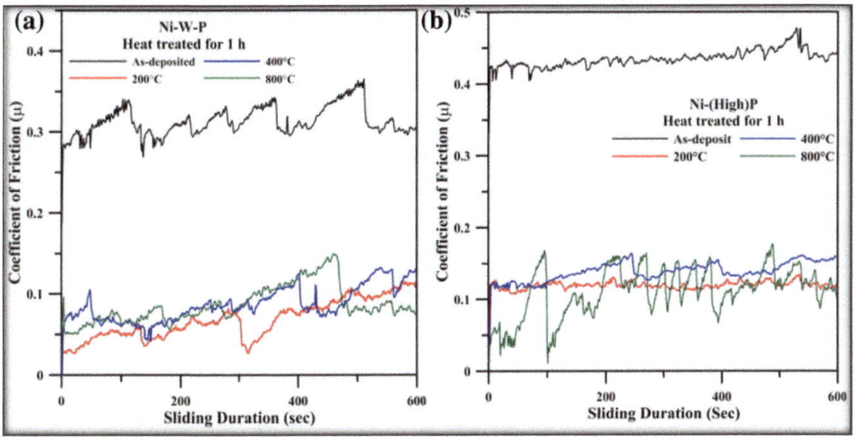

Fig. 5.9 COF versus sliding duration plot for electroless **a** Ni-W-P coating and **b** Ni-(High) P coating

5.3.5.2 Observation of Wear Behavior

Wear rate of both electroless Ni-W-P and Ni-(high) P coatings are exhibited in Fig. 5.10a, b, respectively. For the present case, strong correlation is observed between hardness and wear resistance. The wear rate of Ni-W-P coating is achieved around 5.86 mm^3/N-m for as-deposited and 0.93 mm^3/N-m for heat-treated at 400 °C (reduced by 6 times compared to as-deposited coating) as shown Fig. 5.10a. For the present study, the coatings exhibit higher wear resistance due to solid solution strengthening mechanism of Ni-W matrix [25].

According to Biswas et al. [6], as-deposited Ni-(High) P coating displays a wear rate of around 8.6 mm^3/N-m which is somewhat higher compared to Ni-W-P coatings. Thus, the improvement in the wear resistance of EN coating by introducing tungsten as the ternary element is directly evident. From Fig. 5.10, it is evident that wear resistant is achieved maximum when both coatings are heat-treated at 400 °C for 1 h with Ni-W-P displaying a very low wear rate. Thus, 400 °C heat treatment temperature is again found to very optimum for EN coatings when hardness and wear resistance is the main concern. Now, the wear rate of amorphous structure is mainly depended on their hardness. The coatings also invariably display higher hardness at this particular temperature (see Fig. 5.7). In fact if Figs. 5.10 and 5.7 are compared, a good correlation between hardness and wear resistance of EN coatings is observed.

However, hardness is not only the key factor to influence the wear resistance property of EN coatings. Other factors, such as formation of various phases, oxides, diffusion mechanisms, grain coarsening, etc. determine a significant influence over the wear characteristics of the coating [6]. However, heat treatment temperature has a profound effect on wear behavior of EN coating [39]. Increase of wear resistance may be accredited to the precipitation of hardened phase into coating, as previously

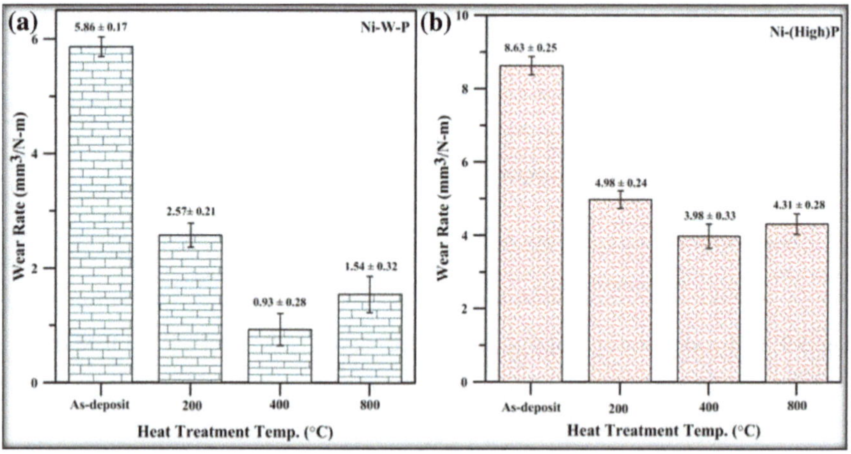

Fig. 5.10 Wear rate of electroless **a** Ni-W-P coating and **b** Ni-(High) P coating

discussed. At 800 °C, wear rate of electroless Ni-W-P coating is found to be around 1.54 mm^3/N-m, which is higher than 400 °C. This higher wear rate is caused by the presence of higher amount of oxide layer over the coating morphology which is evident from XRD result (Fig. 5.5). Similar observation is also made by other research group [6]. They reported that at higher heat treatment temperature wear resistant properties may degrade due to the presence of higher amount of oxide layer over the coating morphology [6]. However, more interpretation on this wear behavior can be obtained by observing SEM images of worn specimens.

5.3.5.3 Wear Mechanism

SEM micrographs of worn surfaces are presented in Fig. 5.11 by which the wear mechanism can be predicted. By observing the fine longitudinal grooves along with the sliding direction (marked by arrow), it can be concluded that abrasive wear one of the wear mechanisms encountered for Ni-W-P coatings. These fine grooves may have appeared due to synergistic effect of both the micro-cutting and micro-plowing effect of the hardened counterface (i.e., EN 31), as observed by others. [6]. Apart from these fine grooves, a few irregular pits, micro-cracks and prows as well as patch delamination are seen (marked by ellipse) which are characteristic features for adhesive wear mechanism [40]. According to Biswas et al. [6], adhesive wear mechanism occurred due to the experimental conditions that induce a substantial attractive force between the coatings, and thus, the counterface surface is leading to a high mutual solubility of nickel and iron. This is also substantiated in the present case as iron is detected in EDX plot (Fig. 5.12) of the coating post wear test. This iron must have definitely come from the steel counterface, and hence, it is confirmed that adhesive wear also occurs in case of Ni-W-P coatings. Hence, the wear mechanism for the present coatings is governed by both abrasive and adhesive phenomena. Moreover, from worn surface morphology, it is also observed that the wear debris plays an important role in influencing the tribological behavior of the coatings. This debris is consisted of fine particles and several coarse ones [29]. The wear debris generated from the coating gets trapped at the contact interface and gets strain hardened. Now, when the two surfaces slide on each other, the debris affects the friction and wear by modifying the contact interface.

Hence, it can be concluded that the present Ni-W-P coating shows a mixed mechanism of both the wear mechanisms (abrasive and adhesive). Figure 5.11d exhibits delamination of some portion of the coating after wear test (marked by rectangle). This may be caused by the loss of adhesion between the top layer (mostly containing oxides) and the coating under layer [6, 41, 42]. Besides, the delamination effect upon coated surface may be seen due to severe oxidation and high wear rate of the present coatings during heat treatment at high temperature [29].

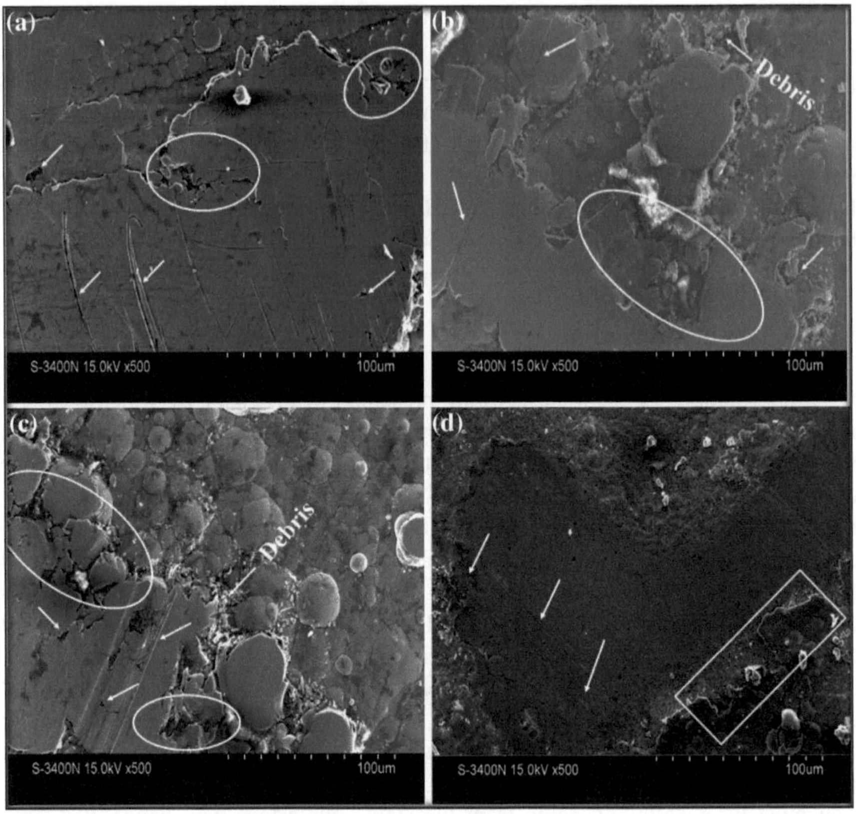

Fig. 5.11 SEM micrographs of worn surface **a** as-deposited, **b** 200 °C, **c** 400 °C and **d** 800 °C for 1 h

5.4 Conclusions

Wear performance of present electroless Ni-W-P coating is investigated under different heat treatment temperature (200, 400 and 800 °C for 1 h). However, the hardness of coating increases with rising heat treatment temperature except for 800 °C. Maximum hardness is obtained around 1113 $HV_{0.1}$, while coating is heat-treated at 400 °C for 1 h. Increase of hardness is due to precipitation of hard phosphide and Ni-W phases into the coating and decrease of hardness may because of grain coarsening effect. Moreover, the wear resistance is seen to have correlation with hardness of coating, and it is increased by 5 times when heat-treated at 400 °C. Besides, the present Ni-W-P coating also displays a low and stable COF value with corresponding lower (i.e., for 200 and 400 °C) heat treatment temperature. Both as-deposited and heat-treated of Ni-W-P coatings exhibit a nodular in structure. As-deposited Ni-W-P coatings are found to possess both amorphous and nano-crystalline structure and gradually turn into crystalline by suitable heat

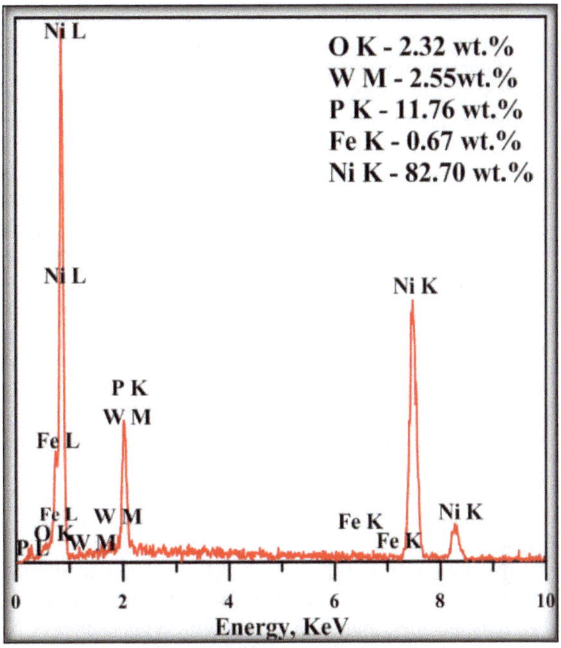

Fig. 5.12 EDX plot of post wear test

treatment. The formation of oxide layer is also revealed by XRD, while heat-treated at high temperature (800 °C). The wear mechanism of present coating is governed by both abrasive and adhesive phenomena under dry sliding condition. It is finally concluded that incorporation of tungsten improves the tribological properties of Ni-P coatings even when subjected to high temperature heat treatment.

Acknowledgements The first author would like to acknowledge the financial assistance and facilities provided by University Grants Commission (UGC), Govt. of India through Rajiv Gandhi National Fellowship.

References

1. Sahoo, P., Das, S.K.: Tribology of electroless nickel coatings—a review. Mater. Des. **32**(4), 1760–1775 (2011)
2. Brenner, A., Riddell, G.E.: Nickel plating on steel by chemical reduction. J. Res. Natl. Bur. Stand. **37**(1), 31–34 (1946)
3. Agarwala, R.C., Agarwala, V.: Electroless alloy/composite coatings: a review. Sadhana **28**(3), 475–493 (2003)
4. Biswas, A., Das, S.K., Sahoo, P.: Effect of heat treatment duration on tribological behavior of electroless Ni-(high)P coatings. IOP Conf. Ser. Mater. Sci. Eng. **149**(1), 012045 (2016)
5. Keong, K.G., Sha, W., Malinov, S.: Hardness evolution of electroless nickel–phosphorus deposits with thermal processing. Surf. Coat. Technol. **168**(2), 263–274 (2003)

6. Biswas, A., Das, S.K., Sahoo, P.: Correlating tribological performance with phase transformation behavior for electroless Ni-(high)P coating. Surf. Coat. Technol. **328** (Supplement C), 102–114 (2017)
7. Yoon, J.W., Park, J.H., Shur, C.C., Jung, S.B.: Characteristic evaluation of electroless nickel–phosphorus deposits with different phosphorus contents. Microelectron. Eng. **84**(11), 2552–2557 (2007)
8. Liu, H., Guo, R.X., Liu, Z.: Effects of laser nanocrystallisation on the wear behaviour of electroless Ni–W–P coatings. Surf. Coat. Technol. **219**(Supplement C), 31–41 (2013)
9. Guo, Z., Keong, K.G., Sha, W.: Crystallisation and phase transformation behaviour of electroless nickel phosphorus platings during continuous heating. J. Alloy. Compd. **358**(1), 112–119 (2003)
10. Erming, M., Shoufu, L., Pengxing, L.: A transmission electron microscopy study on the crystallization of amorphous Ni-P electroless deposited coatings, Metallurgical Coatings. In: Proceedings of the 15th International Conference on Metallurgical Coatings, pp. 273–280 (1988)
11. Keong, K.G., Sha, W., Malinov, S.: Crystallisation kinetics and phase transformation behaviour of electroless nickel–phosphorus deposits with high phosphorus content. J. Alloy. Compd. **334**(1), 192–199 (2002)
12. Hou, K.H., Jeng, M.C., Ger, M.D.: The heat treatment effects on the structure and wear behavior of pulse electroforming Ni–P alloy coatings. J. Alloy. Compd. **437**(1), 289–297 (2007)
13. Tomlinson, W.J., Wilson, G.R.: The oxidation of electroless Ni-B and Ni-P coatings in air at 800 to 1000°C. J. Mater. Sci. **21**(1), 97–102 (1986)
14. Balaraju, J.N., Rajam, K.S.: Electroless deposition of Ni–Cu–P, Ni–W–P and Ni–W–Cu–P alloys. Surf. Coat. Technol. **195**(2), 154–161 (2005)
15. Palaniappa, M., Seshadri, S.K.: Structural and phase transformation behaviour of electroless Ni–P and Ni–W–P deposits. Mater. Sci. Eng. A **460–461**(Supplement C), 638–644 (2007)
16. Hu, Y., Wang, T., Meng, J., Rao, Q.: Structure and phase transformation behaviour of electroless Ni–W–P on aluminium alloy. Surf. Coat. Technol. **201**(3), 988–992 (2006)
17. Allahyarzadeh, M.H., Aliofkhazraei, M., Rezvanian, A.R., Torabinejad, V., Rouhaghdam, A. R.S.: Ni-W electrodeposited coatings: characterization, properties and applications. Surf. Coat. Technol. **307**(Part A), 978–1010 (2016)
18. Roy, S., Sahoo, P.: Optimization of multiple roughness characteristics of chemically deposited Ni-PW coating using weighted principal component analysis. ISRN Mechanical Engineering, Article ID 495857, pp. 7 pages (2012)
19. Hu, Y.J., Xiong, L., Meng, J.: Electron microscopic study on interfacial characterization of electroless Ni–W–P plating on aluminium alloy. Appl. Surf. Sci. **253**(11), 5029–5034 (2007)
20. Selvi, V.E., Chatterji, P., Subramanian, S., Balaraju, J.N.: Autocatalytic duplex Ni–P/Ni–W–P coatings on AZ31B magnesium alloy. Surf. Coat. Technol. **240**(Supplement C), 103–109 (2014)
21. Tsai, Y.Y., Wu, F.B., Chen, Y.I., Peng, P.J., Duh, J.G., Tsai, S.Y.: Thermal stability and mechanical properties of Ni–W–P electroless deposits. Surf. Coat. Technol. **146–147** (Supplement C), 502–507 (2001)
22. Mukhopadhyay, A., Duari, S., Barman, T.K., Sahoo, P.: Investigation of wear behavior of electroless Ni-PW coating under dry and lubricated conditions using RSM and fuzzy logic. Portugaliae Electrochimica Acta **34**(4), 231–255 (2016)
23. Eraslan, S., Ürgen, M.: Oxidation behavior of electroless Ni–P, Ni–B and Ni–W–B coatings deposited on steel substrates. Surf. Coat. Technol. **265**(Supplement C), 46–52 (2015)
24. Duari, S., Mukhopadhyay, A., Barman, T.K., Sahoo, P.: Tribological performance optimization of electroless nickel coatings under lubricated condition. In: Design and Optimization of Mechanical Engineering Products IGI Global, pp. 250–280 (2018)
25. Mukhopadhyay, A., Barman, T.K., Sahoo, P.: Tribological behavior of sodium borohydride reduced electroless nickel alloy coatings at room and elevated temperatures. Surf. Coat. Technol. **321**, 464–476 (2017)

26. Mukhopadhyay, A., Duari, S., Barman, T.K., Sahoo, P.: Wear behavior of electroless Ni-P-W coating under lubricated condition—a Taguchi based approach. IOP Conf. Ser. Mater. Sci. Eng. **149**(1), 012004 (2016)
27. Duari, S., Mukhopadhyay, A., Barman, T.K., Sahoo, P.: Study of wear and friction of chemically deposited Ni-P-W coating under dry and lubricated condition. Surf. Interfaces **6**, 177–189 (2017)
28. Sahoo, P., Roy, S.: Tribological behavior of electroless Ni-P, Ni-P-W and Ni-P-Cu coatings: a comparison. Int. J. Surf. Eng. Interdisc. Mater. Sci. (IJSEIMS) **5**(1), 1–15 (2017)
29. Mukhopadhyay, A., Barman, T.K., Sahoo, P.: Tribological behavior of electroless Ni–B–W coating at room and elevated temperatures. In: Proceedings of the Institution of Mechanical Engineers, Part J: Journal of Engineering Tribology, pp. 1350650118755781 (2018)
30. Palaniappa, M., Seshadri, S.K.: Friction and wear behavior of electroless Ni–P and Ni–W–P alloy coatings. Wear **265**(5), 735–740 (2008)
31. Balaraju, J.N., Raman, N., Manikandanath, N.T.: Nanocrystalline electroless nickel poly-alloy deposition: incorporation of W and Mo. Trans. IMF **92**(3), 169–176 (2014)
32. Balaraju, J.N., Kalavati, Rajam, K.S.: Surface morphology and structure of electroless ternary NiWP deposits with various W and P contents. J. Alloy. Compd. **486**(1), 468–473 (2009)
33. Wu, F.B., Tien, S.K., Chen, W.Y., Duh, J.G.: Microstructure evaluation and strengthening mechanism of Ni–P–W alloy coatings. Surf. Coat. Technol. **177–178**(Supplement C), 312–316 (2004)
34. Tien, S.K., Duh, J.G., Chen, Y.I.: The influence of thermal treatment on the microstructure and hardness in electroless Ni–P–W deposit. Thin Solid Films **469–470**(Supplement C), 333–338 (2004)
35. Koiwa, I., Usuda, M., Osaka, T.: Effect of heat-treatment on the structure and resistivity of electroless Ni-W-P alloy films. J. Electrochem. Soc. **135**(5), 1222–1228 (1988)
36. Mukhopadhyay, A., Duari, S., Barman, T.K., Sahoo, P.: Optimization of wear behavior of electroless Ni-PW coating under dry and lubricated conditions using genetic algorithm (GA). Jurnal Tribologi **11**, 24–48 (2016)
37. Krishnaveni, K., Sankara Narayanan, T.S.N., Seshadri, S.K.: Electroless Ni–B coatings: preparation and evaluation of hardness and wear resistance. Surf. Coat. Technol. **190**(1), 115–121 (2005)
38. Karthikeyan, S., Vijayaraghavan, L.: Investigation of the surface properties of heat treated electroless Ni–P coating. Trans. IMF **94**(5), 265–273 (2016)
39. Sahoo, P.: Wear behaviour of electroless Ni–P coatings and optimization of process parameters using Taguchi method. Mater. Des. **30**(4), 1341–1349 (2009)
40. Sahoo, P., Pal, S.K.: Tribological performance optimization of electroless Ni–P coatings using the Taguchi method and grey relational analysis. Tribol. Lett. **28**(2), 191–201 (2007)
41. Gawne, D.T., Ma, U.: Wear mechanisms in electroless nickel coatings. Wear **120**(2), 125–149 (1987)
42. Kundu, S., Das, S.K., Sahoo, P.: Tribological behaviour of electroless Ni-P deposits under elevated temperature. Silicon, 1–14 (2016)

Chapter 6
Transition Reference Temperature for 20MnMoNi55 RPV Steel at Different Loading Rates

Swagatam Paul, Sanjib Kumar Acharyya, Prasanta Sahoo and Jayanta Chattopadhyay

Nomenclature

a	Length of crack (mm)
$a0$	Initial notch length (mm)
$b0$	Length of uncracked ligament (mm)
$B0$	Thickness of specimen (mm)
E	Young's modulus (GPa)
K	Stress intensity factor (MPa$\sqrt{m}$)
K_{JC}	Plain strain fracture toughness (MPa$\sqrt{m}$)
T	Test temperature (°C)
Pf	Probability of failure
W	Width of specimen (mm)
YS	Yield strength (MPa)
UTS	Ultimate tensile strength (MPa)
N	Poisson's ratio
σ_{ys}	Yield stress (MPa)

6.1 Introduction

In nuclear power plant, a nuclear reactor produces energy from splitting of atoms by fission reactions. This energy is used as a heat to produce steam which is further used in steam turbine to generate electricity. 16% of the world's electricity is produced from nuclear energy. In the time of fission reaction, high neutron irradiation takes place at the core of reactor. Components in or near to core are exposed

S. Paul (✉) · S. K. Acharyya · P. Sahoo
Department of Mechanical Engineering, Jadavpur University, Kolkata, India
e-mail: kgec.swagy@gmail.com

J. Chattopadhyay
Reactor Safety Division, Bhaba Atomic Research Centre, Mumbai, India

P. Sahoo and J. P. Davim (eds.), *Advances in Materials, Mechanical and Industrial Engineering*, Lecture Notes on Multidisciplinary Industrial Engineering,
https://doi.org/10.1007/978-3-319-96968-8_6

to fluxes of neutrons with energies ranging from several MeV to 0.025 eV and 10 keV. This high-energy neutron irradiation can degrade the materials of critical components. Material loses its ductility and becomes brittle. The degree of embrittlement increases with neutron fluency (neutron flux x time), and thus, the service life of the components become shorter rapidly and high maintenance is required. For the maintenance and design of nuclear power plant components, it is necessary to study the embrittlement nature of reactor pressure vessel (RPV) steel. RPV steels are used generally to make critical components like shells, nozzles, flanges, studs, turbine blades. The choice of RPV materials is also a critical concern. The material should have sufficient ductility with high fracture toughness in adverse conditions like high temperature, high corrosive environment. To characterize the embrittlement phenomenon of RPV steel, several experiments are needed on the degraded material. But in general laboratory, it is very risky to handle neutron-irradiated materials because of severe health hazards. So, it was a big challenge to scientists and engineers for maintaining the structural integrity of nuclear power plant.

At the time of World War II, Liberty ships were built with welding in spite of traditional riveting techniques because of high demands of cargo ships. Some of them were deployed to Atlantic Ocean. Some Liberty ships experienced structural damages when cracks initiated, and within no time of maintenance, they broke into two pieces catastrophically without any traceable prominent reasons. Scientists suspected that the temperature of North Atlantic Ocean might be playing some dominant roles on this failure of Liberty ships. The temperature of North Atlantic Ocean was far below of fridging temperature those days. The ships were built by steels which were experiencing ductile-to-brittle transition upon cooling through a critical range of temperature. Steel losses its ductility at decreasing of temperature and fails with cleavage failure. This type of phenomenon is quite similar to neutron-irradiated embrittlement. And in general, laboratory temperature controlling is easy and hazard-free compared to handling radiated materials.

In early days, Charpy v-notch (CVN) tests were done to capture the tendency of material behaviour with the changing temperature [1]. CVN energy was plotted with temperature scale to find out the temperature at which ductile-to-brittle transition took place. This curve is shown in Fig. 6.1. It looks like an S-type curve where above a certain temperature (T_1) CVN energy is constant and the failure is ductile in nature. This region is called upper self-energy region. Similarly at a low-temperature zone below a certain point (T_5), the CVN energy is constant and failure is completely cleavage fracture. This point is known as nil ductility temperature (NDT). Primarily, NDT was considered as the measure of critical transition temperature [2]. Later, another reference temperature was introduced named RT_{NDT} at which the CVN energy was 41 J [3]. The shift of RT_{NDT} is the measure of loss of ductility. According to ASME code, K_{IC} defines that static crack initiation and $K_{ID}/K_{IA}/K_{IR}$ are for dynamic crack arrest based upon an approach that utilizes a material indexing parameter [4]. The determination of K_{IC} is adjusted by finding RT_{NDT} of CVN curves of irradiated material. These K_{IC} and K_{ID} are lower bound of CVN curves. The main drawback of CVN approach is it requires large number of experiments

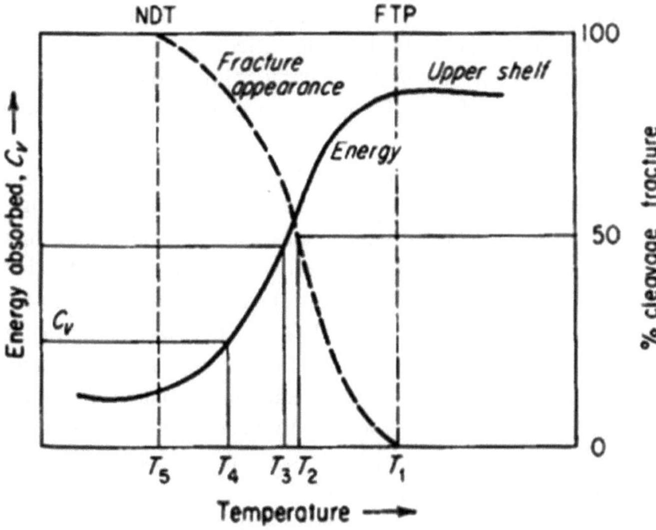

Fig. 6.1 CVN energy versus temperature

which is very costly and time-consuming. CVN energy gives only lower bound of fracture toughness curves. And in between ductile-to-brittle zone, the CVN energy is scatter and probabilistic in nature. Most importantly, CVN approach cannot address this statistical scatter of fracture toughness values in ductile-to-brittle transition region. To overcome this problem, Wallin [5, 6] proposed a methodology based on three-parameter Weibull statistical distribution which can capture the scatter of ductile-to-brittle transition zone. This methodology is called master curve methodology which can be characterized by a single parameter named reference temperature (T_0). T_0 signifies the position of the master curve on temperature scale. T_0 is defined as the temperature at which the median fracture toughness for 1T-CT fracture toughness specimen equals 100 MPa$\sqrt{\text{m}}$ [7]. One of a feature of master curve method is it defines a bounding curve on fracture toughness values. Lower bound of fracture toughness is 95% degree of confidence of master curve which means 95% of all fracture toughness values should fall into the tolerance bound. Kim et al. [8] proposed an alternative approach to estimate the reference temperature in master curve method using Charpy impact test data. The uniqueness of master curve is that it is applicable for all ferritic steels. For different materials, the only thing which governs the master curve position is T_0 but the shape of master curve is same for all ferritic steels. As master curve is formed with the J-integral-based cleavage fracture toughness K_{Jc}, materials are considered homogenous at macroscopic level. This is one of an advantage of master curve that it requires only six valid fracture toughness data which can be obtained with simple fracture testing. There are some factors which can influence the T_0 and master curve. Geometric factors like crack-to-width ratio (a/W), thickness have considerable effects over T_0. It is found in research [9] that T_0 obtained from shallow flaw ($a/W = 0.1$) and deep flaw ($a/W = 0.5$) is different. A shifting of transition temperature is found due to constraint loss. Similarly, some researches [10, 11] show that effect of specimen size on cleavage fracture toughness is not only due to constraint loss but also due to some local stress distribution near crack tip. Basically, fracture toughness is highly dependent on triaxiality factors at crack tip.

There is another factor named loading rate which has influencing effects on fracture toughness and T_0. Here loading rate is defined as a change of stress intensity factors to time ($\mathrm{d}K/\mathrm{d}t$). Loading rates can be divided into three different ranges such as (1) quasi-static, (2) dynamic and (3) impact range. Quasi-static range is ranging from 0.1 to 2 $\mathrm{MPa}\sqrt{\mathrm{m}}$, whereas dynamic range is defined as above 2 $\mathrm{MPa}\sqrt{\mathrm{m}}$ and impact range is above 10^5 $\mathrm{MPa}\sqrt{\mathrm{m}}$. Effect of loading rate is basically a combination of three effects such as rate effect on material parameters, effect of inertia and effect of reflected stress wave. Material parameters like yield strength, ultimate strength and flow behaviour are highly dependent on strain rate. For some materials, it is seen that yield and ultimate strength are increased with increasing strain rate. Some renowned material models are available to characterize the strain rate effects on viscoplastic behaviour of materials [12–14]. Effect of inertia is also a key factor which plays crucial role in impact loading rate. Large errors can be found in calculating the bending stress at specimens if the effect of inertia forces is not taken into account [15]. International Atomic Energy Agency has conducted series of coordinated research projects focusing on RPV structural integrity application to capture size effects and effect of loading rates [16]. Similar projects were going on in USA and Europe to establish a standard method of dynamic fracture testing and computational technique. In late 1970, a US-based research group (Electric Power Research Institute) published a report on testing and analysing of dynamic fracture toughness from pre-cracked Charpy impact tests [17]. Later, ASTM technical committee and European Structural Integrity Society's technical committee separately published some standard test and computing procedure of dynamic fracture toughness from pre-cracked Charpy impact tests. Equivalent energy approach [18], J-integral approach [19], impact response curve methods [20, 21] are some analytical approaches to find out dynamic fracture toughness from pre-cracked Charpy impact tests. Lucon [22] has described all the above methods briefly. In this work, the effect of loading rate on $\mathrm{T_0}$ is studied within dynamic range where effect of inertia is neglected. The master curve approach was standardized at ASTM E1921-05 for the quasi-static loading rate. But master curve for dynamic and impact toughness was developed later. Wallin [23] has done several dynamic and impact tests on different materials and proposed an empirical relationship between loading rate and increase of reference temperature. This relationship was later adopted in ASTM E1921-09a [24]. Wallin proposed the empirical relationship using test data sets for different materials to describe the effect of loading rate on $\mathrm{T_0}$ for a wide range of loading rates ranging from dynamic to impact. Thus, it is generalized one with respect to loading rate and materials. In the present work, verification of Wallin's prediction for 20MnMoNi55 steel has been explored with the experimental T_0 in dynamic loading range (above 2 $\mathrm{MPa}\sqrt{\mathrm{m}}$/s) using 1T-CT pre-cracked fracture testing. Sreenivasan [25] proposed an empirical relation to estimate quasi-static T_0 from T_0 calculated in dynamic range. The relationship is useful when dynamic T_0 is available and quasi-static T_0 is required to be estimated without doing any experiments. The same empirical relation is used here to estimate quasi-static T_0 at different higher loading rates ranging between 4.18 and 417.3 MPa$\sqrt{m}$/s.

6.2 Material Description

RPV steels are exposed to high neutron irradiation, enormous heat and corrosive environments. So, the material should have some property to withstand that adverse condition. Mostly alloying elements are added in such a way that the material can resist corrosion, thermal shock reasonably without losing its strength and ductile nature. For example, due to high thermal conductivity and low thermal expansion low-alloyed Ni-Mo-Mn ferritic steels can resist thermal shock [26]. Some RPV steels have Cr as an alloying element. Cr is good corrosive-resistant material (e.g. 9Cr-1Mo, 22NiMoCr37). In the present study, 20MnMoNi55 low-carbon reactor pressure vessel steel is used as a material. Chemical composition of this material is given in Table 6.1.

6.3 Experimental Details

In the present work, three types of experiments have been carried out: (1) tensile tests, (2) fatigue pre-cracking and (3) fracture toughness tests. Tensile tests, fatigue pre-cracking and fracture experiments are conducted according to ASTM E8, ASTM E647 and ASTM E1820 standards, respectively.

For the tensile testing, a dynamic extensometer is used at different deformation rates at 20 °C. The tensile test specimen dimensions are shown in Fig. 6.2. Extensometer is fitted at the parallel portion of specimen which is called specimen gauge length. Tensile extension of specimen gauge length is calculated from the strain at extensometer. Tensile extension values are extrapolated from extension of extensometer. This extrapolation from extension of extensometer to tensile extension is valid up to ultimate point, before necking starts. Tensile extension is calculated from the extensometer value at 20 °C temperature. But at subzero temperature, extensometer could not be attached. So, tensile extension at subzero temperature is calculated from ram extension. The correlation of the difference

Table 6.1 Chemical composition of 20MnMoNi55 RPV steel

Name of element	C	Si	Mn	P	S	Al	Ni	Mo	Cr	Nb
Percentage composition (in weight)	0.20	0.24	1.38	0.011	0.005	0.068	0.52	0.30	0.06	0.032

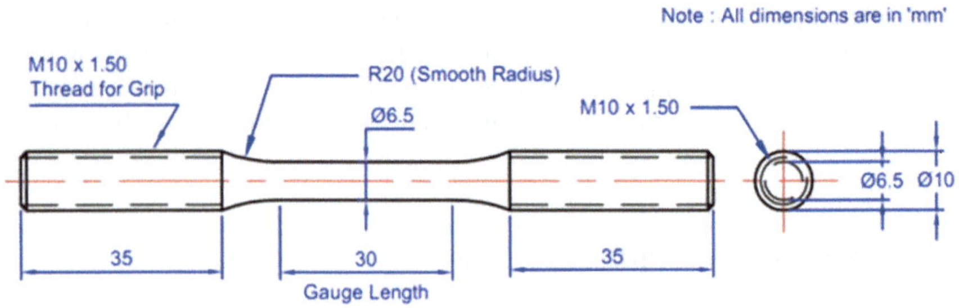

Fig. 6.2 Round tensile specimen dimension

between ram extension and tensile extension with load response at 20 °C temperature is established for respective deformation rate. Ram extension is the sum of specimen extension (uniform section + non-uniform section shown in Fig. 6.2) and other machine parts. Extension, calculated from extensometer values (strain × specimen gauge length), is specimen extension at uniform cross section only. So, to get the contribution of other parts (extension − strain × Gauge length), verses load curves are plotted. This plot consists of two separate sections: one is linear section and another is nonlinear section. Linear section is up to yield point and nonlinear section is where the material gets plastically deformed. At the linear portion, linear curve fitting is done. The slope of the fitting curve is a constant. And for the nonlinear portion, a second-order polynomial curve fitting is done. This polynomial fit is valid up to ultimate point. Using these fitting curves both linear and nonlinear, undesired extensions are calculated and subtracted from ram extension. Thus, both linear and nonlinear fitting parameters are used in subzero temperature for the respective deformation rates to obtain tensile extension. From tensile testing, true strain and true stress data are computed and further used as an input of fatigue pre-cracking and fracture toughness test.

Fatigue pre-cracking is necessary before fracture testing because in fracture toughness test sharp crack is needed without any prior yielding at crack tip. If the plastic zone size at the crack tip can satisfy the small-scale yielding condition, the stress state is uniquely defined as stress intensity factor (K), and the crack growth rate is characterized by K_{min} and K_{max}. The functional relationship for crack growth can be expressed in the following form

$$\frac{\mathrm{d}a}{\mathrm{d}N} = f(\Delta K, R)$$

where $\Delta K = (K_{max} - K_{min})$ and $R = K_{max}/K_{min}$ $\frac{\mathrm{d}a}{\mathrm{d}N}$ = crack growth per cycle and A number of expressions for f have been proposed, most of which are empirical.

According to ASTM E1820, fracture toughness test is done at −100 °C on 1T-CT specimens. Specimen dimensions are shown in Fig. 6.3. COD gauge is used to obtain the crack opening displacement. All the tests are performed in INSTRON 8801 UTM machine using liquid nitrogen environment. The experimental set-up is shown in Fig. 6.4. The test matrices are shown in Tables 6.2 and 6.3.

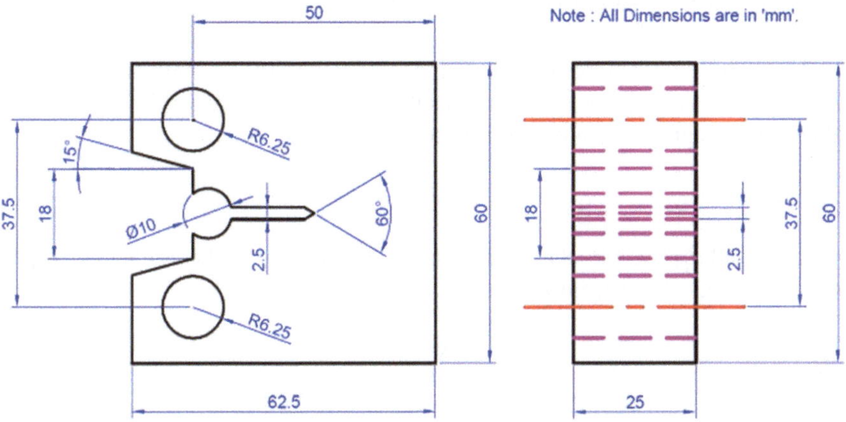

Fig. 6.3 1T-CT specimen dimension

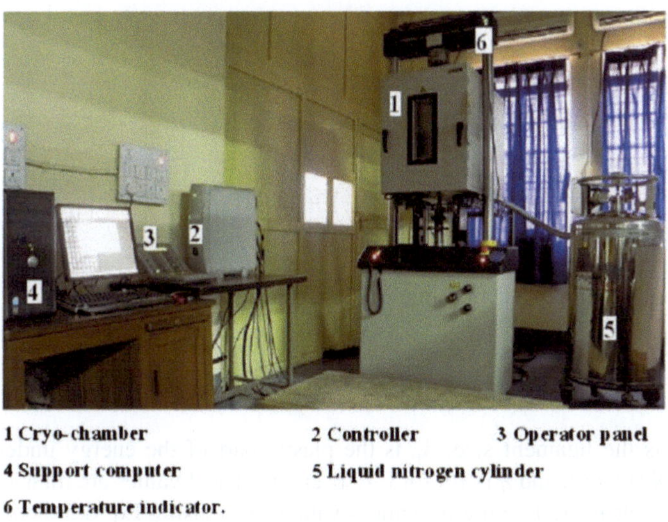

Fig. 6.4 INSTRON 8801 with cryogenic chamber

Table 6.2 Test matrix for tensile tests

No. of samples	30
Test temp (°C)	20, −60, −80, −100, −120, −140
Deformation rates (mm/min)	1, 20, 50, 100, 150

Table 6.3 Test matrix of 1T-CT fracture testing at $a/W = 0.5$ at −100 °C

No. of samples	35
Loading rate ($\mathrm{d}K/\mathrm{d}t$)	0.625, 4.18, 80.25, 215.5, 417.3

6.4 Fracture Toughness Evaluation

The J-integral is determined according to ASTM E1921-09a. J-integral is calculated from the load versus crack opening displacement (COD) curve. Total J-integral value can be segregated into two parts as shown in Eq. 6.1:

$$J = J_{\mathrm{el}} + J_{\mathrm{pl}} \tag{6.1}$$

where J_{el} and J_{pl} are the elastic and plastic component of the J-integral, respectively. J_{el} is calculated from Eq. 6.2

$$J_{\mathrm{el}} = \frac{K^2(1-\nu^2)}{E} \tag{6.2}$$

where K is stress intensity factor which is computed from load versus COD curve by Eq. 6.3

$$K = P\left(\frac{1}{B\sqrt{w}}\right)f\left(\frac{a}{W}\right) \tag{6.3}$$

where $f\left(\frac{a}{W}\right)$ is an empirical function. For CT specimen,

$$f\left(\frac{a}{W}\right) = \frac{(2+a/W)}{(1-a/W)^{3/2}}\left[0.886+4.64\left(\frac{a}{W}\right) - 13.32\left(\frac{a}{W}\right)^2 + 14.72\left(\frac{a}{W}\right)^3 - 5.6\left(\frac{a}{W}\right)^4\right]$$

The plastic component of J-integral is calculated from Eq. 6.4

$$J_{\mathrm{pl}} = \frac{\eta A_{\mathrm{pl}}}{Bb_0} \tag{6.4}$$

where b_0 is the ligament size, A_{p} is the plastic part of the energy under the load versus COD record, and η is 1.9 for CT specimen. The J values are transformed into plain strain cleavage fracture toughness values, K_{JC}, using Eq. 6.5

$$K_{\mathrm{JC}} = \sqrt{J\left(\frac{E}{1-\nu^2}\right)} \tag{6.5}$$

where E is Young's modulus and v is Poisson's ratio. The K_{JC} limit is calculated according to the ASTM E1921-02 [12] standard and is expressed in the form of Eq. 6.6

$$K_{JC(limit)} = \sqrt{\frac{Eb_0\sigma_{YS}}{30(1-v^2)}} \tag{6.6}$$

where

E Modulus of elasticity of the material
b_0 Initial specimen ligament length
v Poisson's ratio (equal to 0.33)
σ_{YS} Material yield strength at the test temperature.

In addition to size requirement, there is the maximum ductile crack growth criterion of 0.05 (W $-$ a_0) or 1 mm, whichever is smaller. The K_{JC} values above the validity criteria shall be censored to the validity limit. The following weighting system from ASTM E1921-02 specifies the required minimum number of valid K_{JC} data points.

$$\sum_{i=1}^{3} r_i n_i \geq 1 \tag{6.7}$$

Here,

r_i Number of valid K_{JC} tests within $(T - T_0)$ range
n_i Specimen weight factor.

The value of T_0 is calculated with all valid and censored values according to the single- or multi-temperature methods.

6.5 Master Curve Analysis and Determination of T_0

Wallin showed that the brittle fracture probability which is defined as P_f for a given temperature in the transition region can be described by a three-parameter Weibull model in Eq. 6.8 as

$$P_f = 1 - \exp\left[-\left(\frac{K_{JC} - K_{min}}{K_0 - K_{min}}\right)^4\right] \tag{6.8}$$

where

P_f Probability of fracture
K_0 Scale parameter

K_{min} Minimum possible fracture toughness
K_{JC} Fracture toughness value obtained from J_C according to Eq. 6.5.

According to Wallin and International Atomic Energy Agency, K_{min} should be assumed to be equal to 20 MPa$\sqrt{m}$. According to weakest link theory, effect of specimen thickness is important; larger thickness is needed for the higher temperature to get a valid K_{IC}. Therefore, the fracture toughness values obtained have been normalized to those of the IT size using Eq. 6.9

$$K_{JC(1T)} = K_{min} + [K_{JC} - K_{min}]\left(\frac{B_0}{B_{1T}}\right)^4 \tag{6.9}$$

where

K_{min} Lower bound fracture toughness (20 MPa$\sqrt{m}$)
B_0 Thickness of the tested specimen (without side grooves)
B Thickness of 1T-CT specimen.

Evaluation of K_0 is calculated according to Eq. 6.10 and the fracture toughness for a median (50%) cumulative probability of fracture, K_{JC} (med), according to Eq. 6.10 of a data set at the applied test temperature:

$$K_0 = \left[\sum_{i=1}^{N}\frac{\left(K_{JC(i)} - K_{min}\right)}{N}\right]^{1/4} + K_{min} \tag{6.10}$$

where

K_{JC} (i) = Individual K_{JC} (1T) value
N = Number of K_{JC} values.
K_{JC} (med) is then calculated with the help of Eq. 6.11.

$$K_{JC(med)} = K_{min} + (K_0 - K_{min})(\ln 2)^{1/4} \tag{6.11}$$

The K_{JC} (med) value thus determined from Eq. 6.7 for the data set at test temperature (T) is used to calculate T_0 using Eq. 6.12

$$T_0 = T - \left(\frac{1}{0.019}\right)\ln\left(\frac{K_{JC(med)} - 30}{70}\right) \tag{6.12}$$

The upper bound (95%) and lower bound (5%) tolerance levels are then calculated using Eq. 6.13. Any data failing to fall within this tolerance bound are considered to be invalid.

$$K_{JC(0.xx)} = 20 + \left[\ln\left(\frac{1}{1-0.xx}\right)\right]^{\frac{1}{4}}[11 + 77\exp\{0.019(T - T_0)\}] \tag{6.13}$$

Here, 0.*xx* represents the cumulative probability.

Multi-temperature evaluation:
The multi-temperature option of ASTM E1921-02 represents a tool for the determination of T_0 with K_{JC} values distributed over a restricted temperature range, namely, $T_0 \pm 50$ °C. The value T_0 can be evaluated by an iterative solution of Eq. 6.14:

$$\sum_{i=1}^{N} \frac{\delta_i \exp[0.019(T_i - T_0)]}{11 + 77\exp[0.019(T_i - T_0)]} - \sum_{i=1}^{N} \frac{(K_{JC(i)} - K_{min})^4 \exp[0.019(T_i - T_0)]}{\{11 + 77\exp[0.019(T_i - T_0)]\}^5} = 0 \tag{6.14}$$

where

T_i Test temperature corresponding to $K_{JC(i)}$
δi Censoring parameter.

if the $K_{JC(i)}$ datum is valid $\delta i = 1$, or if the $K_{JC(i)}$ datum is not valid and censored $\delta i = 0$.

Establishment of the Master Curve and Tolerance Bounds
Values of K_{JC} tend to conform to a common toughness versus temperature curve. Both upper and lower tolerance bounds can be calculated using Eq. 6.15

$$K_{JC(0.xx)} = 20 + \left[\ln\left(\frac{1}{1-0.xx}\right)\right]^{1/4}\{11 + 77\exp[0.019(T_i - T_0)]\} \tag{6.15}$$

where $0._{xx}$ represents the cumulative probability level.

6.6 Result and Discussion

Tensile test results are used as input to fracture tests. Figure 6.5 shows the variation of yield strength and UTS with deformation rate. The strength of the material is mostly governed by the dislocation motion and therefore strongly influenced by its mobility. The mobility of dislocation depends on many factors those try to restrict the 'free glide'. The increased loading rate increases the diffusivity of atoms and the mobility of dislocations. Diffused atoms restrict the mobility of dislocations. As a consequence, high strength occurs for moving the arrested dislocations to continue deformation. Here it is seen that yield and ultimate strength both are increasing with deformation rate increase, so it can be stated that this particular material is positive

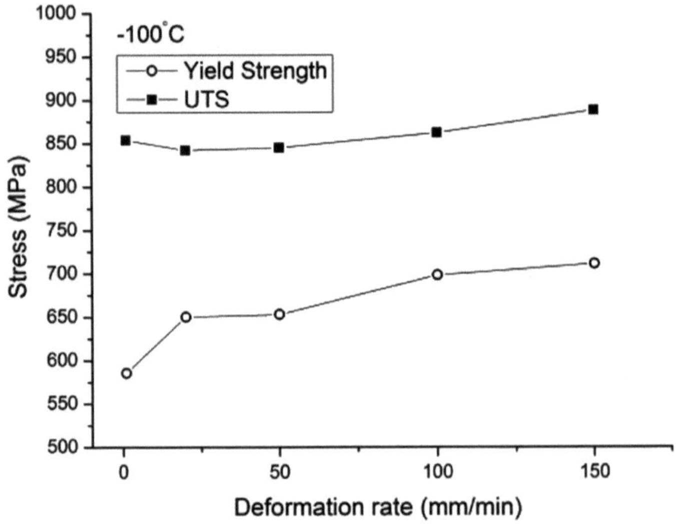

Fig. 6.5 Yield strength and UTS with deformation rate

strain rate sensitive. Figures 6.6 and 6.7 show the true stress versus true strain curves for 20 and −100 °C, respectively. Mechanical properties like Young's modulus, yield strength and ultimate strength are also given in Table 6.4 at 20 and −100 °C in different deformation rates. J-integral tests are performed according to ASTM E1820 standard on 1T CT at −100 °C and 0.5 *a*/*W* ratio at different deformation rate (mm/min). The stress intensity factor (*K*) is calculated using Eq. 6.16:

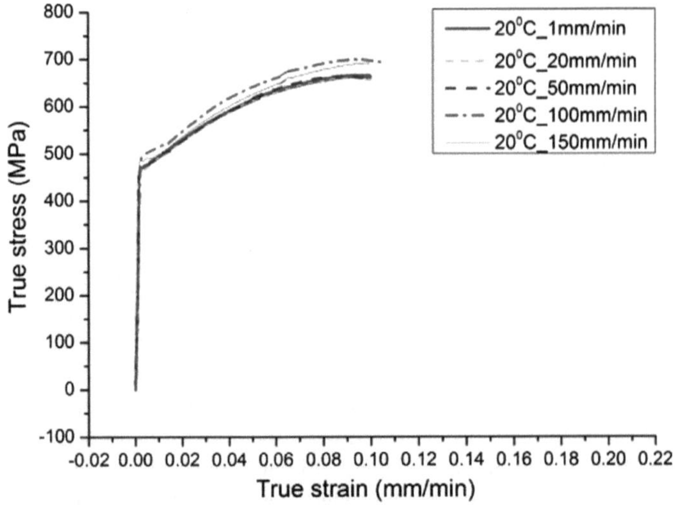

Fig. 6.6 True stress versus true strain curve at 20 °C in different deformation rate

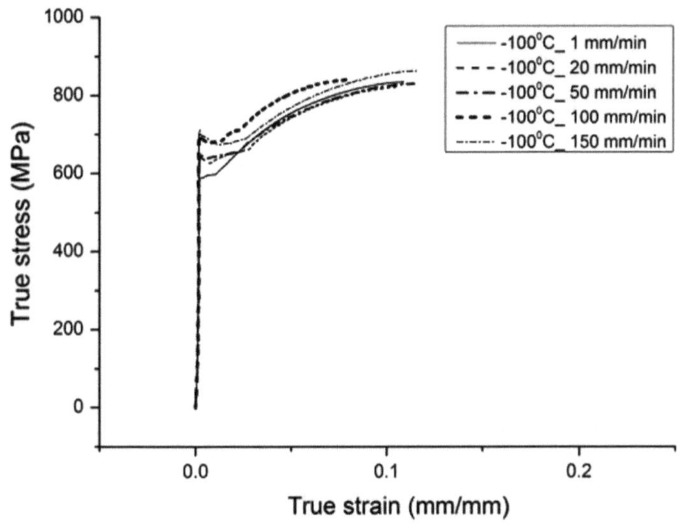

Fig. 6.7 True stress versus true strain curve at −100 °C in different deformation rate

Table 6.4 Mechanical properties at 20 and −100 °C in different deformation rates

Temperature (°C)	Mechanical properties	Deformation rate (mm/min)				
		1	20	50	100	150
20	Young's modulus (Gpa)	203.759	204.522	205.158	205.904	209.614
	Yield strength (Mpa)	464.127	469.108	461.854	492.483	485.122
	Ultimate strength (Mpa)	615.702	657.533	661.592	694.806	687.28
−100	Young's modulus (Gpa)	210.364	214.165	214.304	215.949	215.059
	Yield strength (Mpa)	585.758	643.482	647.086	691.743	700.946
	Ultimate strength (Mpa)	835.976	823.193	831.349	842.178	864.393

$$K = P\left(\frac{1}{B\sqrt{w}}\right)f\left(\frac{a}{W}\right) \tag{6.16}$$

where P is load, a is crack length, W is width, B is thickness of the specimen.

All tests are done in deformation rate (dΔ/dt) and then deformation rates are converted to loading rate (dK/dt) with the following procedure. The calculated K is plotted with deformation (Δ) as shown in Fig. 6.8. The value of the slope of the curve is dK/dΔ. The deformation rate (dΔ/dt) is kept constant for a specific loading

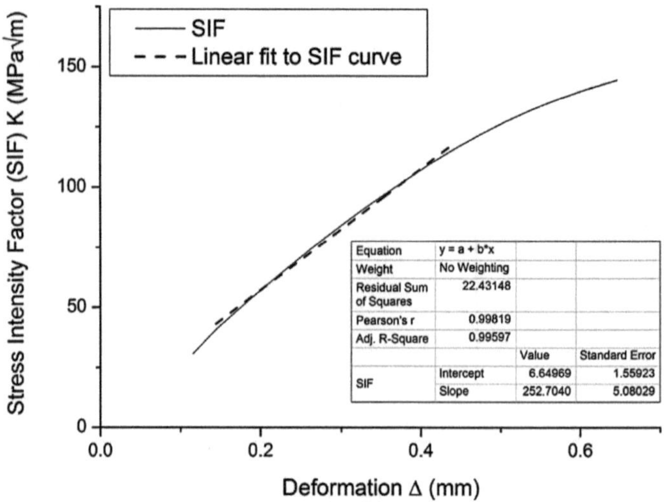

Fig. 6.8 Stress intensity factor with COD

rate. Hence, d*K*/d*t* is calculated with the relation $\frac{\mathrm{d}K}{\mathrm{d}t} = \frac{\mathrm{d}K}{\mathrm{d}\Delta} \cdot \frac{\mathrm{d}\Delta}{\mathrm{d}t}$. The fracture toughness test results at different loading rates are given in Table 6.5. From fracture test data, master curves are developed at each loading rate. Obtained master curves are shown in Figs. 6.9, 6.10, 6.11, 6.12 and 6.13; one multi-temperature master curve is also shown in Fig. 6.14 and master curves for two *a*/*w* ratio of 0.5 and 0.6 are shown in Fig. 6.15. From the results, it is observed that reference temperature and master curve are almost independent of the *a*/*w* ratio and method of computing T_0. Values of T_0 at different loading rates are determined using ASTM E1921 standard. The values of T_0 with corresponding loading rates are shown in Table 6.6. From Table 6.6, it is apparent that with increasing loading rate, reference temperature (T_0) increases. Wallin's relation is given as Eq. 6.17 to predict dynamic T_0.

$$T_{0,X}^{\text{est}} = \frac{(T_0 - 273.15) * \Gamma}{\Gamma - \ln(X)} - 273.15 \tag{6.17}$$

where $\Gamma = 9.9 * \exp\left[\left(\frac{T_0 + 273.15}{190}\right)^{1.66} + \left(\frac{\sigma_{\text{ys}}^{T_0}}{722}\right)^{1.09}\right]$ and $\sigma_{\text{ys}}^{T_0}$ is yield stress at T_0 and X = (d*K*/d*t*).

From the empirical relation, the values of T_0 are estimated and compared with experimental ones in Table 6.6. These values are also plotted in Fig. 6.16 where the experimental and estimated values of dynamic T_0 match well. The deviation of experimental T_0 from estimated values is insignificant at lower loading rate. At higher loading rate, it is found to be within ±5 °C. Hence, it can be concluded that Wallin's prediction is valid for the present material. However, large number of pre-cracked Charpy tests is required to validate the higher loading rates.

Table 6.5 Fracture toughness for 1T-CT specimen at −100 °C and $a/W = 0.5$ for varying loading rate

Sl. No.	Temperature (°C)	Loading rate (MPa√m/s)	Exp J-integral (kJ/m²)	K_{JC} (MPa√m)
1	−100	0.625	84.84076038	139.9237
2			179.5322687	203.5449
3			315.6068711	269.8747
4			323.0133184	273.023
5			58.6405308	116.329
6			96.14595518	148.9548
1		4.18	126.362	170.764
2			157.207	190.469
3			157.207	190.469
4			230.045	230.406
5			49.052	106.39
6			98.54	150.797
7			117.268	164.505
1		80.25	56.337	115.103
2			94.689	149.2237
3			91.432	146.6347
4			171.015	200.5412
5			98.121	151.9034
6			92.437	147.438
7			149.418	187.4515
1		215.5	52.87	110.4574
2			103.66	154.6659
3			158.977	191.5388
4			75.288	131.8109
5			79.697	135.6154
6			16.842	62.34281
7			160.661	192.5501
1		417.3	93.871	147.1823
2			84.836	139.9199
3			28.473	81.05999
4			181.249	204.5163
5			42.116	98.58563
6			144.482	182.5984
7			93.871	147.1823
1	−80	4.18	240.228	235.451
2	−120	4.18	47.784	105.011

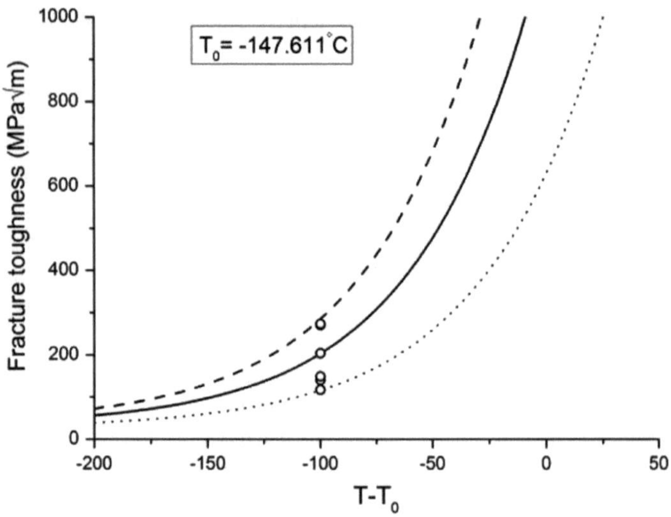

Fig. 6.9 Master curve at 0.625 MPa$\sqrt{\text{m}}$/s

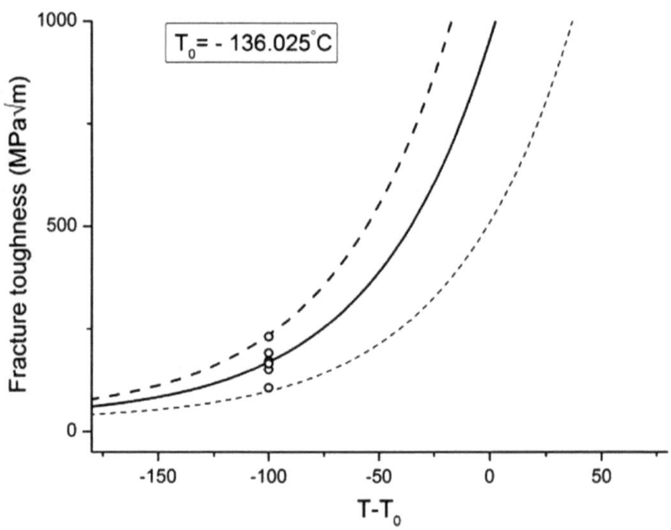

Fig. 6.10 Master curve at 4.18 MPa$\sqrt{\text{m}}$/s

Sreenivasan [25] described a method based on fitting of some reference temperature (T_0) data at different loading rates and for different materials to find quasi-static T_0 from dynamic T_0. The empirical relationship is given in Eq. 6.18.

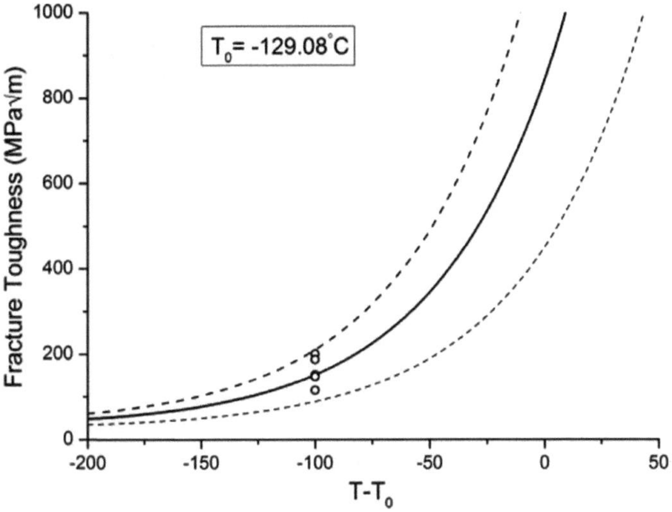

Fig. 6.11 Master curve at 80.25 MPa$\sqrt{m}$/s

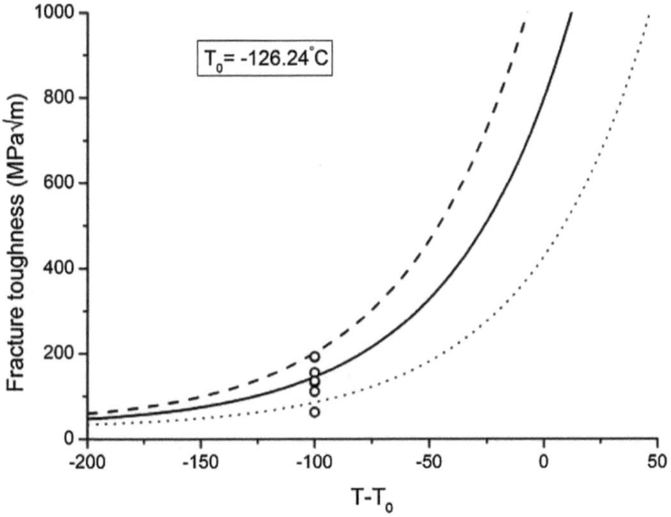

Fig. 6.12 Master curve at 215.5 MPa$\sqrt{m}$/s

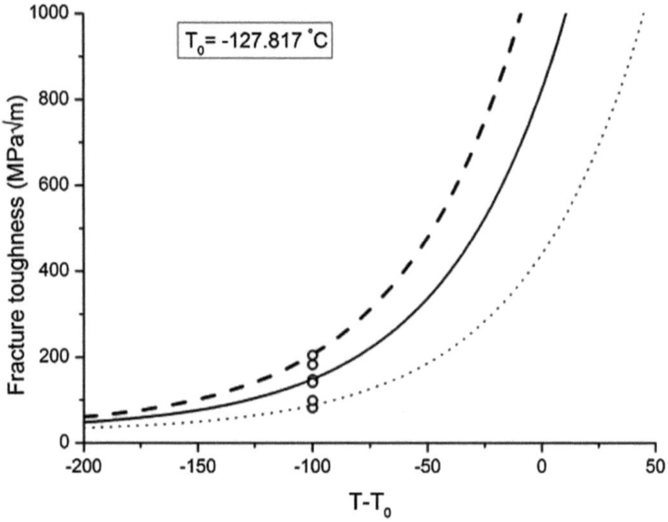

Fig. 6.13 Master curve at 417.3 MPa√m/s

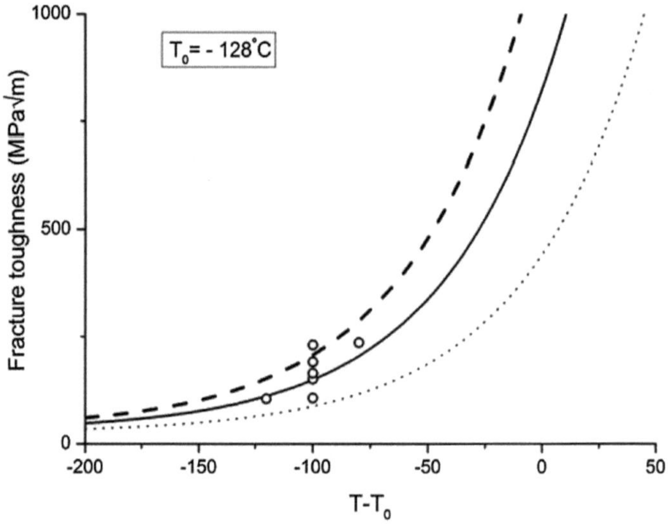

Fig. 6.14 Master curve from multi-temperature method at 4.18 MPa√m

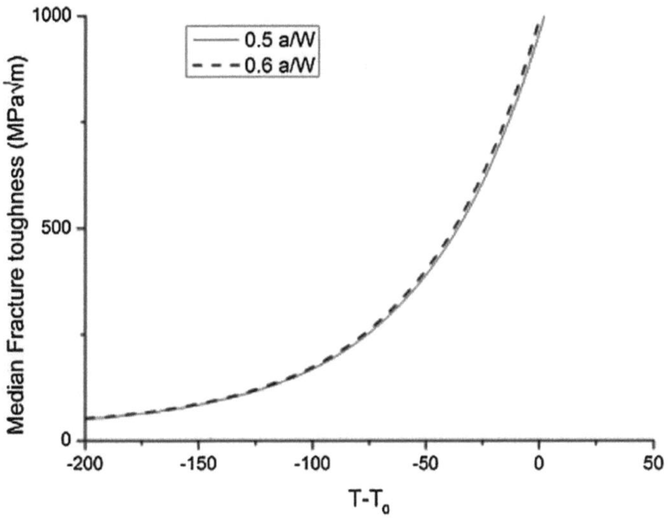

Fig. 6.15 Comparison of median fracture toughness at two different *a*/*W* ratios

Table 6.6 T_0 at different loading rates

No. of samples	Test temp (°C)	Loading rate (MPa$\sqrt{m}$/s)	Experimental T_0 (°C)	Estimated T_0 (°C)
6	−100	0.625	−147.612	−143
7		4.18	−136.026	−138
7		80.25	−129.08	−128
7		215.5	−126.248	−125
7		417.3	−127.817	−123
5 + 2	5(−100) & 2(−80, −120)	4.18	−128	–

$$T_{0,X}^{\mathrm{st}} = \frac{\left(T_0^{\mathrm{dy}} - 273.15\right) * \Gamma'}{\Gamma' - \ln(X)} - 273.15 \tag{6.18}$$

where

$$\Gamma' = -41.54 * \exp\left[\left\{\frac{\left(T_0^{\mathrm{dy}} + 273.15\right)}{72.86}\right\}^{-125.98} + \left(\frac{\sigma_{\mathrm{ys}}^{RT}}{0.142}\right)^{-0.4099}\right],$$

and $\sigma_{\mathrm{ys}}^{\mathrm{RT}}$ is yield stress at room temperature. Using this relation, quasi-static T_0 is estimated from T_0 at different higher loading rates and given in Table 6.7.

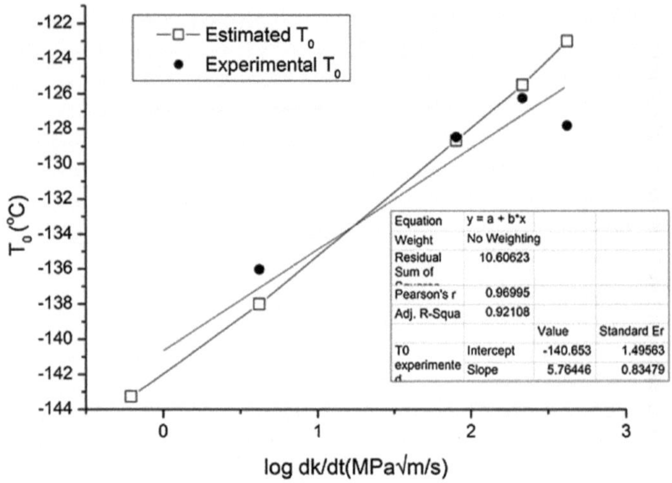

Fig. 6.16 Comparison between experimental and estimated T_0

Table 6.7 Estimation of quasi-static T_0 from dynamic T_0

Loading rate (MPa$\sqrt{m}$/s)	Experimental dynamic T_0 (°C)	Estimated T_0 (°C)
4.18	−136.026	−143.2
80.25	−129.08	−142.3
215.5	−126.248	−142.2
417.3	−127.817	−144.9

6.7 Conclusion

Estimation of degradation in fracture toughness values is important for nuclear components to assess residual life due to embrittlement by neutron irradiation. Wallin's methodology based on master curve and reference temperature to asses fracture toughness in DBTT is adopted because of its low material requirement for testing at specimen levels only. This methodology is applied for the material 20MnMoNi55 low-carbon steel and observed to follow the propositions in ASTM E1920. Effect of loading rate on tensile and fracture strength of the low-carbon steel 20MnMoNi55 has been studied in detail. For 20MnMoNi55 RPV steel, yield strength and ultimate strength increase with loading rates. The rate of variation depends on temperature also. Fracture toughness of the material decreases both with decrease in temperature and increase in loading rate which is expected. But the scatter in fracture toughness values is observed to match with Weibull distribution which is apparent from the master curve drawn at different temperatures. Reference temperature (T_0) increases with loading rates. Wallin developed a correlation between T_0 and loading rate and proposed an equation. It is observed that experimental T_0 matches well with the estimated T_0 derived from Wallin's empirical relation in the region of lower loading rates and departs slightly at higher loading rates. For evaluation at higher loading rates, Charpy test results are to be incorporated. It can be concluded from the current analysis that dependence of T_0 on loading rate can be predicted in dynamic range for 20MnMoNi55 steel. Again, quasi-static T_0 has been predicted with the help of dynamic T_0 as obtained from experimental results. It is found that all the predicted values of quasi-static T_0 closely match with each other.

References

1. Kim, S.H., Park, Y.W., Lee, J.H., Kang, S.S.: Estimation of fracture toughness in the transition region using CVN data. In: 15th International Conference on Structural Mechanics in Reactor Technology (SMiRT 15), Seoul, Korea, August, 15–20 (1999)
2. ASTM E208-06: Standard Test Method for Conducting Drop-Weight Test to Determine Nil-Ductility Transition Temperature of Ferritic Steels (2012)
3. Application of Master Curve Method Fracture Toughness Methodology for Ferritic Steel. EPRI Technical Report (1998)
4. Nanstad, R.K., Keeney, J.A., McCabe, D.E.: Prepared for the U.S. Nuclear Regulatory Commission Office of Nuclear Regulatory Research under Interagency Agreement No. 18986-80 1 1-93
5. Wallin, K.: Master curve analysis of the "Euro" fracture toughness dataset. Eng. Fract. Mech. **69**, 451–481 (2002)
6. Wallin, K.: Use of the Master Curve methodology for real three dimensional cracks. Nucl. Eng. Des. **237**, 1388–1394 (2007)
7. International Atomic Energy Agency: Guidelines for Application of the Master Curve Approach to Reactor Pressure Vessel Integrity in Nuclear Power Plants, Technical Reports Series No. 429, IAEA, Vienna (2005)
8. Kim, S.H., Park, Y.W., Kang, S.S., Chung, H.D.: Estimation of fracture toughness transition curves of RPV steels from Charpy impact test data. Nucl. Eng. Des. **212**, 49–57 (2002)
9. McAfee, W.J., Williams, P.T., Bass, B.R., McCabe, D.E.: An Investigation of Shallow-Flaw Effects on the Master Curve Indexing Parameter (*To*) in RPV Material. Division of Engineering Technology Office of Nuclear Regulatory Research U. S. Nuclear Regulatory Commission
10. Rathbun, H.J., Odette, G.R., Yamamoto, T., He, M.Y., Lucas, G.E.: Statistical and constraint loss size effects on cleavage fracture-implications to measuring toughness in the transition. J. Press. Vessel Technol. **128**(3), 305 (2006)
11. Henry, B.S., Luxmoore, A.R.: The stress triaxiality constraint and the Qvalue as a ductile fracture parameter. Eng. Fract. Mech. **57**(4), 375–390 (1997)
12. Johnson, G.R., Cook, W.H.: A constitutive model and data for metals subjected to large strains, high strain rates and high temperatures. In: The 7th International Symposium on Ballistic, p. 541 (1983)
13. Rusinek, A., Zaera, R.,. Klepaczko, J.R.: Constitutive relations in 3-D for a wide range of strain rates and temperatures—application to mild steels. Int. J. Solids Struct. **44**, 5611–5634 (2007)
14. Zerilli, F.J., Armstrong, R.W.: Dislocation-mechanics-based constitutive relations for material dynamics calculations. J. Appl. Phys. **61**(5), 1816–1825 (1987)

15. Venzi, S., Priest, A.H., May, M.J.: Influence of inertial load in instrumented impact tests, impact testing of metals, ASTM STP 466. American Society for Testing and Materials, pp. 165–180 (1970)
16. International Atomic Energy Agency: Master Curve Approach to Monitor Fracture toughness of Reactor Pressure Vessels in Nuclear Power Plants. IAEA-TECDOC-1631, Vienna (2009b)
17. Server, W.L., Scheckerd, J.W., Wullaert, R.A.: Fracture Toughness Data for Ferritic Nuclear Pressure Vessel Materials, EPRINP-119, Final Report (1976)
18. Witt, F.J.: Equivalent Energy Procedures for Predicting Gross Plastic Fracture, USAEC Report ORNL-TM 3172
19. Rice, J.R., Paris, P.C., Merkle, J.G.: Some Further Results of J-Integral Analysis and Estimates. In: Progress in Flaw Growth and Fracture Toughness Testing, ASTM STP 563, ASTM, pp. 231–245 (1973)
20. Kalthoff, J.F., Winkler, S., Bohme, W.: A Novel Procedure for Measuring the Impact Fracture Toughness Kld with Precracked Charpy Specimens. Journal de Physique, Colloque C5, Supplement au n°8, Tome 46, aout (1985)
21. Bohme, W.: Dynamic key-curves for brittle fracture impact tests and establishment of a transition time. Presented at the 21st ASTM National Symposium on Fracture Mechanics, Annapolis, Maryland, June 28–30 (1988)
22. Lucon, E.: Dynamic toughness testing of pre-cracked charpy V-notch specimens, convection ELECTRABEL – SCK.CEN, Mol, Belgium (1999)
23. Wallin, K.: Summary of the IAEA/CRP3 Fracture Mechanical Results. In: Proceedings IAEA Specialists Meeting, IAEA working material, Espoo (1999)
24. ASTM E1921-09a: https://www.astm.org/DATABASE.CART/HISTORICAL/E1921-09.htm
25. Sreenivasan, P.R.: Inverse of Wallin's relation for the effect of strain rate on the ASTM E-1921 reference temperature and its application to reference temperature estimation from Charpy tests. Nucl. Eng. Des. **241**, 67–81 (2011)
26. Blagoeva, D.T., Debarberis, L., Jong, M., ten Pierick, P.: Stability of ferritic steel to higher doses: survey of reactor pressure vessel steel data and comparison with candidate materials for future nuclear systems. Int. J. Press. Vessels Pip. **122**, 1–5 (2014)

Chapter 7
Comparative Study of Cyclic Softening Modelling and Proposition of a Modification to 'MARQUIS' Approach

Snehasish Bhattacharjee, Sankar Dhar, Sanjib Kumar Acharyya and Suneel Kumar Gupta

7.1 Introduction

If the applied cyclic load is high enough to deform the material irreversibly (plastic), it survives relatively low number of cycles before fatigue failure and accordingly termed as 'low cycle fatigue' (LCF). LCF has received much attention since the early work of Manson [1] and Coffin [2] where they explained fatigue crack growth in terms of plastic strain at the tip of the micro-cracks. Geometrical discontinuities, surface roughness, persistent slip bands (PSB), grain interfaces, inclusions, defects, etc., may cause localised plastic deformation even for loadings with small amplitude. Before the catastrophic failure or even crack initiation, unlike high cycle fatigue, different kind of stress–strain responses may be observed. The variety of material behaviours can be observed depending on the material and loading sequence and symmetry in the fatigue life. These material responses are of practical interest in design of power plant and other engineering components, as sometimes, those may indicate any underlying secondary damage mechanism that has been evolved and influencing the fatigue life. Thus, considering the safety aspect through structural integrity, comprehending the material behaviour in cyclic plastic loading is very important.

Extensive research work has been carried out in last a few decades by several researchers to understand the material behaviour for irreversible cyclic loading. However, a number of cyclic plasticity issues are yet to be fully understood [3, 4].

S. Bhattacharjee (✉) · S. Dhar · S. K. Acharyya
Department of Mechanical Engineering, Jadavpur University,
Kolkata 700032, India
e-mail: bhattacharjeesnehasish@gmail.com

S. K. Gupta
Reactor Safety Division, Bhabha Atomic Research Centre,
Mumbai 400085, India

P. Sahoo and J. P. Davim (eds.), *Advances in Materials, Mechanical and Industrial Engineering*, Lecture Notes on Multidisciplinary Industrial Engineering,
https://doi.org/10.1007/978-3-319-96968-8_7

The present work deals with experimentation of SA333 Grade-6 C-Mn steel subject to symmetric uniaxial cyclic plastic loading and to simulate the material behaviour by the finite element method. To simulate the material behaviour, a mathematical material model is required. In this study, depending on the experimental observations, the development of a material model is also attempted.

7.2 The Material

SA333 Grade-6 C-Mn steel is commonly used in primary heat transport (PHT) pipes at Indian pressurised heavy-water reactors (PHWR) where, during start-up and shutdown or, due to fluctuations in operating conditions or, seismic activities can trigger cyclic loading, which is most likely multi-axial and may be non-proportional in nature. Depending on the amplitude of loading and/or geometric or micro-structural discontinuities, the cyclic deformations may reach beyond the elastic range locally. From the safety point of view, the knowledge of different cyclic plastic behaviour of the material and their contribution to the failure mechanism must be accurately known.

The material, SA333, as reported in the literature, has shown a wide variety of cyclic plastic behaviour. Rigorous experiments by Paul and his co-workers [5–7], Khutia and his co-workers [8], Sivaprasad and his co-workers [9, 10] on monotonic, uniaxial and multi-axial behaviour of the material reports variations of its cyclic behaviours. Depending mainly on heat cycles and loading history, the hardening characteristics of the material have changed.

Typically, elastic modulus of the material is 210 GPa, and Poisson's ratio is 0.3. The material is of moderate strength (monotonic yield strength is 304 MPa and tensile strength is 495 MPa) and ductility (% elongation 24.5). There is about ±5% strength variation observed, changing the heat number alone.

Basic metallurgical investigations are also carried out to understand the experimentally observed phenomena. The steel mainly consists of Carbon (0.18% by wt.) and Manganese (0.9% by wt.), and the detailed chemical composition of the material is shown in Table 7.1.

The two major alloying elements Carbon and Manganese are austenite stabilisers. Carbon has a strong tendency to segregate at the defects (such as grain boundaries and dislocations) in steels. Elements like Fe, Mn may interact with Carbon and form carbides. The strengthening effect of Carbon in steel consists of interstitial solid solution strengthening and carbide dispersion strengthening. Carbon, simultaneously, decreases the ductility and weldability of the material [11, 12]. 0.18% Carbon in the steel generally deteriorates the surface quality [13].

Table 7.1 Chemical composition of SA333 C-Mn steel

Element	C	Mn	Si	P	Fe
Qty. by wt. (%)	0.18	0.90	0.02	0.02	Rest

Manganese is essentially used as a deoxidiser and a desulphuriser [14]. Manganese, unlike Carbon, has favourable effect on weldability and improves surface quality. Manganese forms stable carbides in steel but cannot form in competition with Iron (Cementite), or even alter the microstructures formed after transformation [15]. However, Manganese can infuse into solid solution Cementite at low as well as high concentrations. The presence of Manganese may enhance the segregation of Phosphorous to grain boundaries and induce temper embrittlement [12].

Silicon and Phosphorus do not form Carbide and normally found in the matrix. Silicon, which is the principal deoxidiser, dissolves completely in ferrite increasing its strength without greatly decreasing ductility when the amount is low [11]. If combined with Manganese, Silicon may produce greater hardenability. Silicon can eliminate stress corrosion and improves wear resistance of the material. Phosphorus segregates during solidification, but to a lesser extent than Carbon. Phosphorus dissolves in ferrite and increases the strength of steels. Phosphorus has a very strong tendency to segregate at the grain boundaries and causes the temper embrittlement of Mn-steels [13]. The effect of Silicon and Phosphorous is hardly observable macroscopically as their presence is negligible.

The typical microstructure of the material is given in Fig. 7.1 showing pearlite bands (volume fraction 30%) over ferrite matrix with average ferrite grain size of 22 μm. It is assumed that all the deformations are occurring mainly on the ferrite grains because ferrite is softer and susceptible to deformation than pearlite.

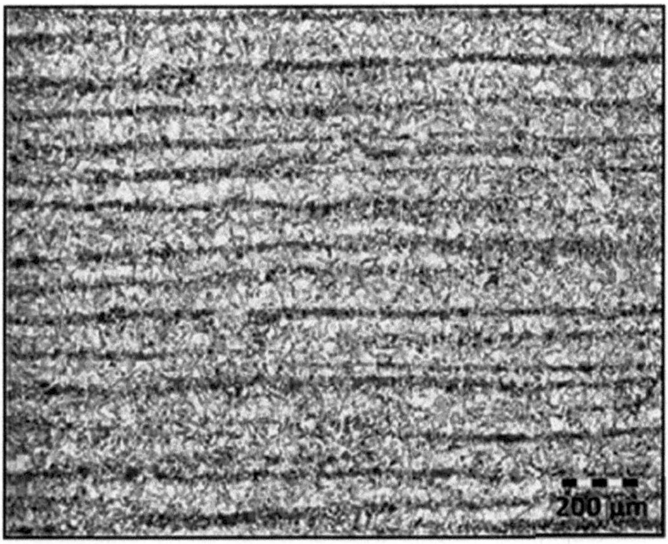

Fig. 7.1 Optical microstructure of SA333 (Scale: 200 μm)

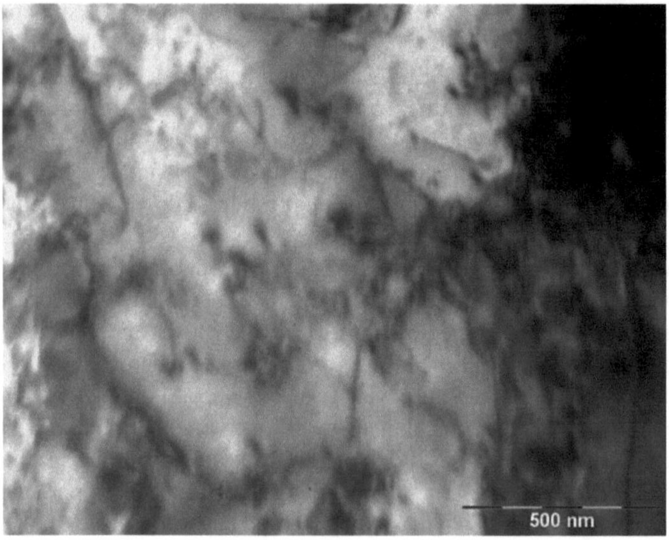

Fig. 7.2 TEM micrograph for SA333 C-Mn steel at 'as-received' condition

TEM imaging is done in Philips CM200 operated at 200 kV. TEM micrographs are obtained at 'as-received' condition (Fig. 7.2). The TEM micrograph shows no definitive dislocation structure at 'as-received' condition, though there are traces of obscure dislocation veins, possibly formed during previous extensive plastic deformation like rolling.

7.3 Experiment and Observation

It is already discussed that the field of application of the material SA333 Grade-6 C-Mn steel has critical safety issues and cyclic plastic deformation of the material may occur even for small amplitude loadings. Therefore, a comprehensive understanding of its cyclic plastic behaviour is necessary. The material behaviour, when subjected to uniaxial and symmetric loading, is discussed in this chapter. In comparison with the complex real-life loading condition of the material, the uniaxial and symmetric loading condition may sound idealistic. But, for modelling complex behaviour it is important to start from the basic loading situation.

According to ASTM E606, uniaxial (tension–compression) low cycle fatigue tests were conducted on the material, by Bhabha Atomic Research Centre, India, for various strain amplitudes ranging from ± 0.35 to $\pm 0.75\%$ at room temperature avoiding the rate effects, in standard environmental condition. Schematic representation of the specimen and actual experimental arrangement is shown in Fig. 7.3a, b, respectively.

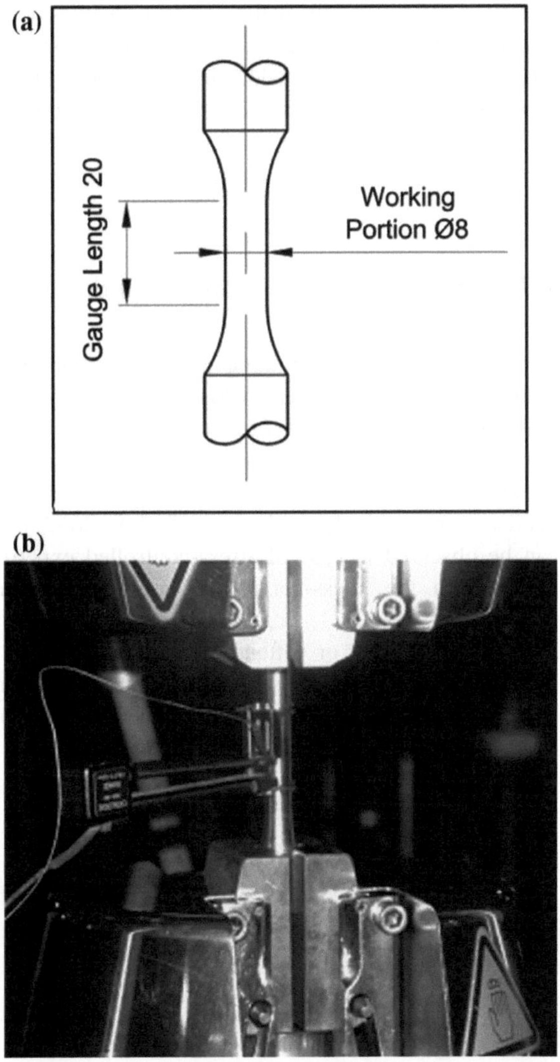

Fig. 7.3 **a** Schematic diagram of uniaxial LCF specimen. **b** Uniaxial testing system

In the experimental results, Bauschinger effect [16], cyclic softening [17, 18], non-Masing initially and later Masing behaviour [19] at saturation cycles are predominantly observed.

7.3.1 Bauschinger Effect

It is a fundamental and well-known cyclic plasticity behaviour observed in metals. According to Bauschinger [20], it is 'the phenomenon by which plastic deformation increases yield strength in the direction of plastic flow and decreases it in other direction'. It happens because of the dislocations with reversed Burger's vectors interact with the forward dislocations and get annihilated during loading in reversed slip direction. The Bauschinger effect gets saturated within first a few cycles.

The typical manifestation of the Bauschinger effect is observed in the cyclic plastic experiments of SA333 Grade-6 C-Mn steel, as shown in Fig. 7.4.

7.3.2 Cyclic Hardening and Softening

Cyclic hardening and softening refer to the hardening or softening response of a material subjected to repeated loading [21–25]. It is often reflected by the cyclic plastic experimental results of metals under fully reversed strain-controlled loading. For strain-controlled cyclic tests, the material response is observed to display cyclic hardening or softening (Fig. 7.5a, b) when the stress amplitude increases or decreases, respectively, with cycle [5–7, 9, 26–29]. Cyclic hardening or softening can be observed in case of stress-controlled experiments also. The plastic strain response in the stress-controlled case shows decreasing or increasing order, respectively.

Cyclic hardening or softening is influenced by a number of factors like: prior deformation history(microscopically, dislocation density and arrangement) [30], chemical composition of the material and undergone heat treatment (materials

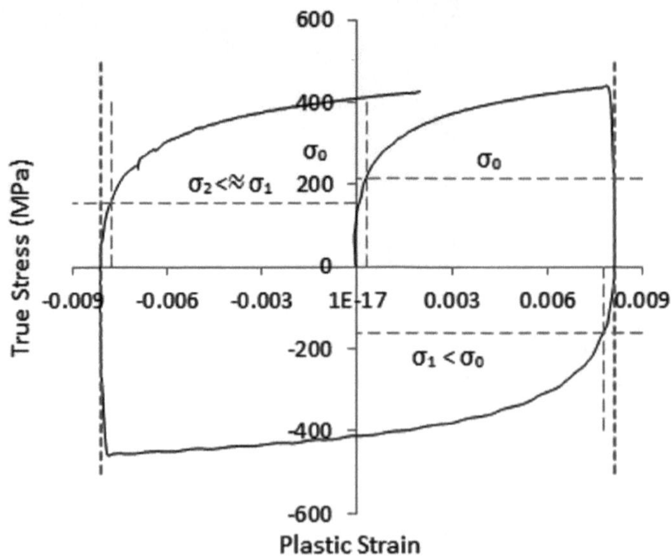

Fig. 7.4 Typical experimental results showing Bauschinger effect in SA333

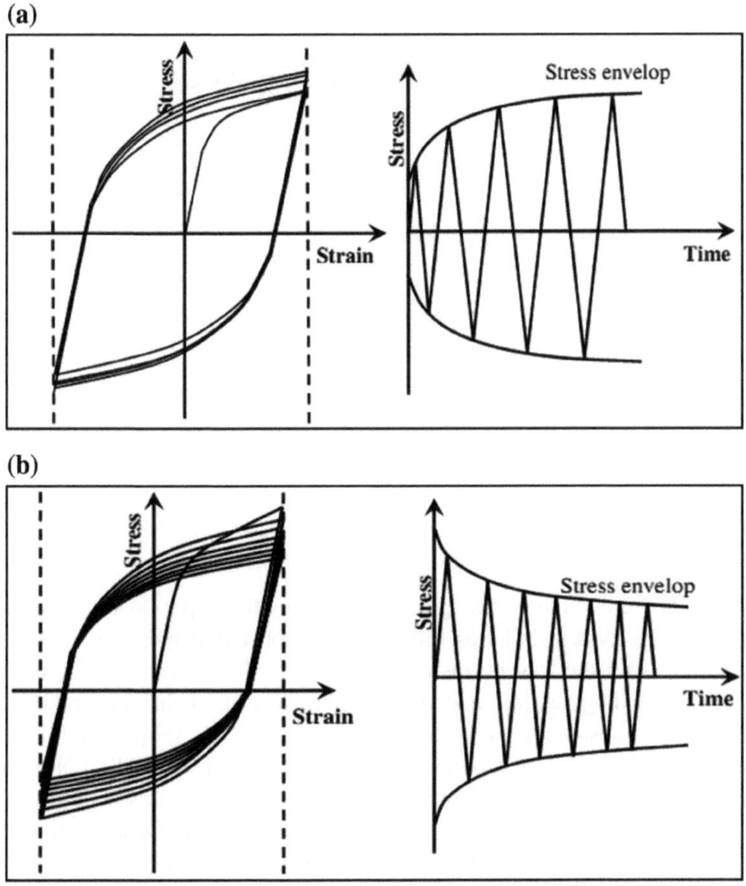

Fig. 7.5 **a** Schematic strain-controlled cyclic hardening. **b** Schematic strain-controlled cyclic softening

phase, i.e. planner slip or wavy slip), [31–33], applied stress state (mean stress and stress amplitude) [31, 32] and environment temperature [31]. Cyclic hardening or softening of materials depends on the initial dislocation population and structure. Previous loading alters the initial dislocation structure of the material and thereby changes its initial cyclic hardening/softening response [30]. As for example, for a cyclic hardening material, initial softening for first a few cycles and then usual cyclic hardening can be manifested in high–low (amplitude) loading sequence [29].

Usually it has been seen that, hard materials (i.e. cold deformed, quenched and tempered) cyclically soften and soft materials (i.e. annealed) cyclically harden [3, 31, 32]. The present material SA333 Grade-6 C-Mn steel shows cyclic softening behaviour in the initial transient (first to 30th) cycles, as shown in Fig. 7.6.

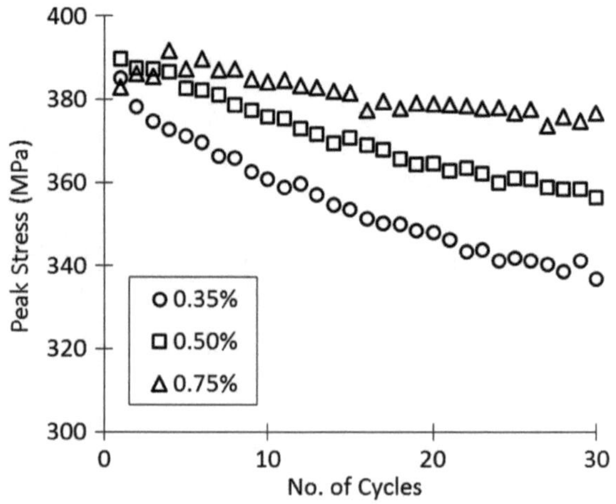

Fig. 7.6 Experimental results showing cyclic softening of SA333

7.3.3 Masing Behaviour

The material is said to have Masing behaviour when the upper (loading) branches of the stress–strain hysteresis loops follow a common curve which is known as the master curve, with the lowers tips are tied together for different strain amplitudes [34, 35]. Some engineering materials display Masing behaviour under certain conditions like pressure vessel steel at 300 and 420 °C [36], ultra-fine grained Copper [37], polycrystalline Copper under ramp loading [38], SS304LN when 30% cold worked prior testing [39]. Normally, for single-phase metals with low stacking fault energy and finely dispersed particles exhibit Masing behaviour [40]. For metals with high stacking fault energy, where the cyclic deformation is matrix controlled, the cyclic stress–strain response is non-Masing. However, Masing behaviour is generally observed below a threshold strain level, above which, non-Masing behaviour can be manifested. In this condition, the formation of a cellular dislocation microstructure [40] is observable.

The present material SA333 Grade-6 C-Mn steel shows atypical Masing behaviour. It non-Masing in the initial transient cycles, but near saturation it shows Masing behaviour up to 0.75% strain amplitude. Figure 7.7a, b shows the Masing behaviour of the present material.

7.4 Physical Explanation

Dislocation motion and lattice rotation are supposed to be the prime cause of plastic deformation. Through decades, several researchers were trying to find micro-mechanism-based explanations behind the cyclic plastic phenomena as well. The work of Gough [41] established that a metal deforms under cyclic loading as it

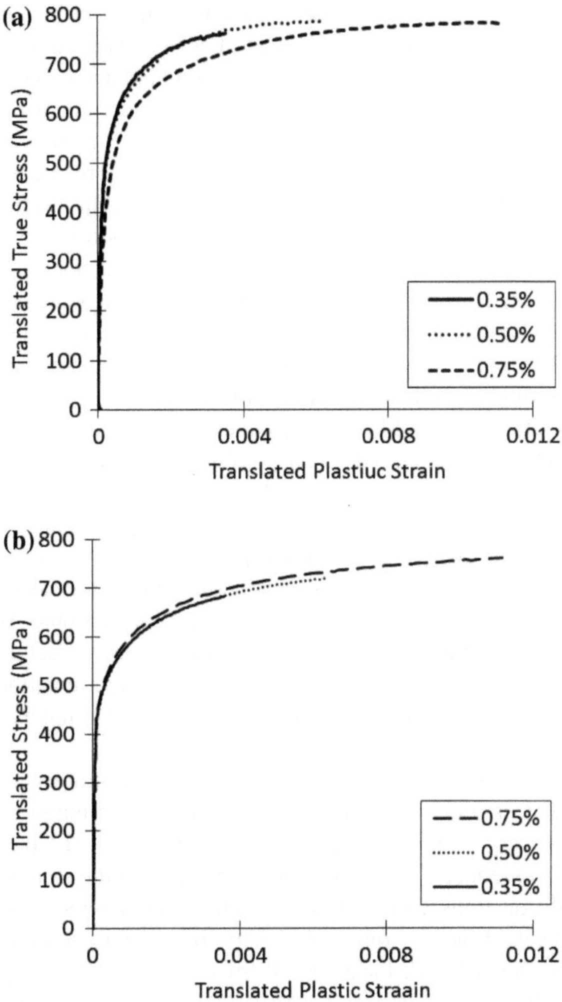

Fig. 7.7 **a** Experimental results showing non-Masing behaviour at initial cycle of SA333. **b** Experimental results showing Masing behaviour at 30th cycle of SA333

does in case of unidirectional or monotonic case. The deformation, in both the cases, is caused by slip of certain atomic planes in the certain crystallographic direction. Concurrent work of Orowan [42], recognises the relationship between the dislocation density and slip. Taylor [43] proposed the correlation between dislocation density and shear stress. There are detailed discussions found in the literature on the constants used in Taylor pile-up model. Taylor factor [44, 45] and parameter α [46] are important constants. Relation between slip and strain was established [47, 48]. Later, Rauch and Schmitt [49] have pointed out that plastic strain can evolve based on dislocation.

It is the nature of the dislocation interactions that differentiate macroscopic material behaviour. Years of meticulous study of the dislocation dynamics by several researchers in different polycrystals revealed more facts. In the work of

Kubin et al. [50], the possible interactions of dislocations are classified in six kinds. The Lomer–Cottrell lock, the Hirth lock and the glissile junction are mainly accounted for the forest interactions between non-coplanar slip systems. The self-interaction and the coplanar interaction are two non-contact interactions for dislocations gliding in parallel slip planes with same or different Burger's vectors, and the collinear interaction produces annihilations. The hardening responses are mainly the manifestation of dislocation entanglements while annihilation results in softening.

By the course of time, the work of Wood [51] suggested a mechanism of slip band extrusion and intrusion for the understanding of fatigue. Plumbridge and Ryder [52] studied the basic structural changes occur in the material subjected to cyclic loading. Ashby [53] reported deformation modes are complex in polycrystals than single crystal, and understanding of dislocation dynamics for polycrystals during large deformation became critical.

In BCC and FCC crystals, there are twelve such slip systems exist, which can be activated easily. The total possible interactions between mobile dislocations in a BCC crystal can be captured by a matrix of 144 (12×12) interaction coefficients [54]. This interaction matrix actually designates the generalised hardening character of the material. It is reported [50] that, there are only six independent coefficients can represent the matrix considering symmetry assumptions, which are associated with six different types of dislocation interactions. Apart from dislocation interactions, some reorganisations of the lattice structures continue to occur while deforming in the plastic regime. These reorientations can create and/or destroy many opportunities for further interactions, increasing the complexities in the imminent state of affairs.

During cyclic plastic deformation, continuous generation and annihilation of dislocations simultaneously occur and, the remaining dislocations at any instance, get aligned gradually assisted by the rotations of the neighbouring lattices [55] such as to reduce their individual contribution to the stored energy. The reorganisation of dislocations can resolve the rough entanglements quickly within a few cycles. These reorientations of the dislocations continue to occur to promote the formation of the stable dislocation substructures [56]. The fully developed substructures are cellular and surrounded by walls consist of rough tangles of dislocations. The interiors of the sub-cells are nearly dislocation free. During the initial stages of plastic deformation, the sub-cells are often disoriented due to the dissimilar lattice rotations. The simplest case of the 'misorientation' can be formed by the sub-grains of tilted symmetry [55, 57–59] with an array of edge dislocations, commonly known as low angle tilt boundary where the misorientation angle is less than 15°.

A moving dislocation, in the vicinity of such a sub-grain boundary, takes part either by pilling-up at sub-grain walls reinforcing the wall entanglements or, by getting annihilated with its compatible pair relieving some stored energy [60]. Obviously, entanglements resist the easy glide of remaining dislocations and thereby cumulatively raise the macroscopic hardness, and annihilations make free passages for the remaining dislocations thus reduce the hardness. The interactions of moving dislocations are direction dependent and may result in directional behaviour like ratcheting.

The overall manifestation of the annihilation is the recovery of energy (softening), but the dissimilar kinematics of the dislocations involved, has critical physical influences. As for example, when a mobile dislocation interacts with a fixed dislocation entangled in (forming) the sub-cell boundary wall and the pair gets annihilated, the reduction of boundary dislocation primarily resolve the boundary misorientation [60, 61] and eventually cause the dissolution of the boundary itself. Which is why, the effect of annihilations of boundary dislocations can be considered as a reasonable phenomenon to justify isotropic cyclic softening during cyclic plastic loadings. On the other hand, the annihilated mobile dislocation may result in the dynamic recovery and can be manifested as Bauschinger effect.

Mughrabi [62] researched the Masing behaviour in polycrystalline copper, and he proposed that the hysteresis loops can show Masing behaviour when the dislocation arrangements remain unchanged during cyclic deformation. The work of Li and Laird [63] reveals that the Masing behaviour is not a saturation phenomenon but, simply confirms Mughrabi's [62] work, investigations by Watanabe and his co-workers [64] reveal that dislocation structures are not always the same during Masing behaviour.

It can be concluded that from the experimental results and the TEM micrograph is that at the onset of plastic deformation, the probability of dislocation pile-ups is high. And similar reason can cause high magnitude softening when after stress reversal reorientation of crystals makes favourable conditions, which is clearly manifested in experimental results. Low strain amplitudes show more hardened loops with larger softening than high strain amplitudes, and non-Masing behaviour is believed to be the manifestation.

7.5 Material Modelling for Symmetric Uniaxial Cyclic Plastic Deformation of SA333

Finite element simulation is a useful tool for analysing the irregular geometry objects and large specimens, where physical experimentation is not quite possible. But before relying on the FE simulations, the whole method including constrains, load and/or boundary conditions and the material model must be validated with experimental observations for simple geometry and loading conditions. In the finite element simulation of a physical problem, the geometry is presented along with all constraints and loading and/or boundary conditions. The material behaviour is added to the geometry by the material model. Thus, accurately defined material model is required for good agreements between actual and simulated results.

7.5.1 General Plasticity Framework

The material is considered as homogeneous and isotropic, and linearly elastic. The material plasticity is based on infinitesimal strain theory or small deformation theory, where it is required to introduce a suitable yield criterion $\sigma_{\text{eq}} = \sigma_0$, which thresholds the plastic deformation. Here, σ_{eq} is the equivalent stress and σ_0 is the initial yield stress. Mathematically, $\sigma_{\text{eq}} - \sigma_0 = 0$ repeats the same condition with a function on the left-hand side, known as the yield function ϕ. In the elastic regime, $\sigma_{\text{eq}} < \sigma_0$ or in other words, $\phi < 0$. After yield function reaches certain threshold condition (=0), further loading causes plastic deformation of the material, and plastic strain increments $\left(\mathrm{d}\varepsilon_{ij}^{p}\right)$ are observed.

Beyond the elastic limit, the total strain can be additively decomposed into elastic and plastic strain components $\left(\mathrm{d}\varepsilon_{ij} = \mathrm{d}\varepsilon_{ij}^{e} + \mathrm{d}\varepsilon_{ij}^{p}\right)$. The present material (metal) is analysed based on continuum plasticity theories, i.e. the material is considered incompressible $\left(\varepsilon_{ij}^{p}\delta_{ij} = 0\right)$ in the plastic regime, or in other words, plastic deformation takes place without changing the volume of the material. Therefore, hydrostatic strain is purely elastic and plastic strain components are purely deviatoric. Incompressible plasticity comprises five-independent plastic strain components.

Incompressibility hypothesis endorses no effects of the hydrostatic stress component on yielding and the plastic deformation. Hence, the distortion energy criterion is considered suitable in this case. In its simplest form in case of perfectly plastic (no hardening) incompressible material, the von Mises yield function is read as:$\phi = \sigma_{\text{eq}} - \sigma_0 = \left[\frac{3}{2}S_{ij}S_{ij}\right]^{1/2} - \sigma_0$, where S_{ij} are the deviatoric stress components or reduced state of stress. The yield function is represented by an open cylinder of radius σ_0, on π-plane in the deviatoric stress space. For metals (isotropic and incompressible), the von Mises equivalent/effective stress is a function of the total plastic work as developed from the distortion energy principle. This is why the von Mises yield function, for most of the metallic material, is considered as the plastic potential surface.

In the plastic regime, the constitutive stress–plastic strain relation for elastic—perfectly plastic material was first proposed by Prnadtl [65] as an extension of Levy [66]—Mises [67] equations for plane strain deformation. The general form of the equations was given by Reuss [68] as $\mathrm{d}\varepsilon_{ij}^{p} = S_{ij}\mathrm{d}\lambda$. The total strain increment at any moment, therefore, according to Prandtl–Reuss equation is as follows: $\mathrm{d}\varepsilon_{ij} = \left[\frac{\mathrm{d}s_{ij}}{2G} + \frac{(1-2v)}{E}\delta_{ij}\mathrm{d}\sigma_m\right] + \left[S_{ij}\mathrm{d}\lambda\right]$.

Extensive multi-axial experimentation mainly on thin-walled tube by Lode [69], Taylor and Quinney [70], Hohenemser [71], Morrison and Shepherd [72] and others have shown approximately good agreement with the Prandtl–Reuss equation.

For the work-hardening material (metals), the complete quantification of the plastic strain is dependent on the hardening characteristics of the material. Hill [73] assumed that the resistance to the distortion is measured by the yield surface, expanding isotropically during plastic flow retaining its shape and position with respect to the hydrostatic state of stress. On the other hand, Prager [74] proposed a different hardening rule which considers the yield surface can move in the stress space in any direction of strain increment retaining its shape and size. Considering both the hardening mechanisms, the manifestation of the von Mises yield function becomes:

$$\phi = \left[\frac{3}{2}\left(S_{ij} - \alpha_{ij}\right)\left(S_{ij} - \alpha_{ij}\right)\right]^{1/2} - \sigma_c = 0 \tag{7.1}$$

where α_{ij} are deviatoric back stress components. Current yield stress σ_c at any instance can be additively decomposed into initial yield stress (σ_0) and isotropic hardening/softening (R). The hardening rules specifically describe the evolutions of isotropic hardening stress and/or back stress (kinematic hardening) components. For work-hardening materials, Hill [73] developed the complete stress–strain equation in a more general manner as: $\mathrm{d}\varepsilon_{ij} = \left[\frac{\mathrm{d}s_{ij}}{2G} + \frac{(1-2v)}{E}\delta_{ij}\,\mathrm{d}\sigma_m\right] + \left[\frac{3}{2}S_{ij}\frac{\mathrm{d}\varepsilon^p_{eq}}{\sigma_{eq}}\right]$.

Contemporarily, for plastic deformation of the perfectly plastic (neutrally stable) or strain-hardening (stable) material, Drucker [75–77] proposed a stability postulate, which is represented as $\mathrm{d}\sigma_{ij}\mathrm{d}\varepsilon^p_{ij} \geq 0$. The postulate practically restricts the possibility of material softening and instability. Drucker's postulate is broader than the statement summarises. It confirms the convexity of yield surface and suggests the maximum plastic dissipation [78–80]. Consequences of the maximum plastic dissipation postulate are of the highest importance in the plasticity theory. The normality of the plastic flow direction to the potential surface can be directed from the principle of maximum plastic dissipation.

The possibility and direction of plastic strain increment (plastic flow) for strain-hardening material are generally directed by the flow rule. The possibility and quantification of the plastic flow are ascertained by a non-negative scalar $(\mathrm{d}\lambda)$ similarly as Prandtl–Reuss equation. It can be shown that the plastic multiplier is actually the equivalent plastic strain increment, though the complete quantification of the plastic strain magnitudes is dependent on the specific hardening rules and consistency condition. From the normality rule, the direction of the plastic flow, called the flow vector $\left(n_{ij}\right)$ in plasticity analysis, is determined by the gradient of the potential surface in the stress direction. For von Mises plasticity for metals, the flow rule is called the associative or associated flow rule when the yield surface is identical to the potential surface. The equation of the associated plastic flow becomes:

$$\mathrm{d}\varepsilon^p_{ij} = \mathrm{d}\lambda\frac{\partial\phi}{\partial S_{ij}} \tag{7.2}$$

The consistency condition implies that the plastic deformation occurs for the state of stress is to be in such a way that their equivalent always lies on the revised yield surface. The hypothesis enables the quantification of the plastic multiplier $(\mathrm{d}\lambda)$ for associated flow. The condition is mathematically presented as: $\phi + \mathrm{d}\phi = 0$, in yielded condition , $\phi = 0$, therefore, $d\phi = 0$ and expressed as:

$$\mathrm{d}\phi = \frac{\partial \phi}{\partial S_{ij}} \mathrm{d}S_{ij} + \frac{\partial \phi}{\partial \alpha_{ij}} \mathrm{d}\alpha_{ij} + \frac{\partial \phi}{\partial \sigma_c} \mathrm{d}\sigma_c = 0 \tag{7.3}$$

The plastic modulus is important in this context when the equivalent plastic strain increment $\left(\mathrm{d}\varepsilon^p_{eq}\right)$ is computed using Newton-Raphson iterative method in numerical simulations. The plastic modulus is developed from the consistency condition $(\mathrm{d}\phi = 0)$ considering associative flow rule $\left(\mathrm{d}\varepsilon^p_{ij} = \mathrm{d}\lambda \frac{\partial \phi}{\partial S_{ij}}\right)$ and incorporating the specific hardening laws as:

$$h = n_{ij} \frac{\mathrm{d}S_{ij}}{\mathrm{d}\varepsilon^p_{\mathrm{eq}}} = n_{ij} \frac{\mathrm{d}\alpha_{ij}}{\mathrm{d}\varepsilon^p_{\mathrm{eq}}} + \frac{\mathrm{d}\sigma_c}{\mathrm{d}\varepsilon^p_{\mathrm{eq}}} \tag{7.4}$$

here, $n_{ij} = \frac{\partial \phi}{\partial S_{ij}} = \frac{3}{2} \frac{S_{ij}}{\sigma_{\mathrm{eq}}}$ and, $\mathrm{d}\varepsilon^p_{\mathrm{eq}} = \left[\frac{2}{3} \mathrm{d}\varepsilon^p_{ij} \mathrm{d}\varepsilon^p_{ij}\right]^{1/2}$. In the subsequent material modelling sections, in this chapter and later, the specific hardening laws suitable for the material, their quantification and proposed modifications are discussed.

7.5.2 The Kinematic Hardening Rule

Cyclic plasticity is still considered as one of the most critical structural problems to be simulated. Incidentally, the implementation of the kinematic hardening (KH) rules, which models the translation of the yield surface in the deviatoric stress space, enables the accurate simulations of basic cyclic plastic behaviour of metals. The development of kinematic hardening rule is an enormous benchmark in the history of cyclic plasticity modelling. The standard kinematic hardening rules include the forest pile-ups of dislocations as 'back stress' and the effect of annihilated mobile dislocations causing some recovery of the back stress is generally included in the 'dynamic recovery' term. But this was not like the first one formulated. The concept of KH rule is materialised first in the work of Prager [81]. After decades of study and research improved the formation of standard KH rules, some historical benchmarks are presented below chronologically:

- Prager [76] proposed the linear-type kinematic hardening rule suggesting a proportional stress raiser by dislocation pile-ups may be apprehended. When a material is loaded beyond its yield limit, microscopically, a large number of dislocations are activated by generation and multiplication at different sources, like grain and sub-grain boundaries, pinned dislocations, impurities, and other kind of defects. As plastic deformation continues, the probability of mobile dislocations being tangled with each other and caught up in the dislocation forest increases exponentially as more slip systems get activated. The entanglement of mobile dislocations at different sites resists the easy glide of dislocations, cumulatively resulting the macroscopic hardening.

- Mroz [82] improved the performance of the linear kinematic hardening model by a multi-surface model, where each surface represents a constant work-hardening modulus in the stress space. Besseling [83] introduced a multi-layer model without any notion of surfaces. Both the models functionally divide the flow curve into several linear segments. For a sufficient number of segments, the simulations of uniaxial hysteresis loops by these models yield very good results. But these rules, like the linear KH model, fail to simulate uniaxial stress cycle with mean stresses(uniaxial ratcheting).
- From the metallurgical point of view, Li [84, 85] proposed a model for local kinematic stress as a function of sub-grain size and low angle misorientation angle. In this model, no clear information was given about the source of the kinematic stress in the material.
- The linear hardening law was successful to simulate the Bauschinger effect but rudimentary to imitate the stress–strain hysteresis due to the linearity. The development of nonlinear KH rule consisting of a dislocation annihilation-based nonlinear 'Recovery' or 'Recall' term was first introduced by Armstrong and Frederick [86]. In support, transmission electron microscope (TEM) micrographs can show other than entanglement, when dislocations of opposite signs interact, they both get annihilated [87] reliving the stored energy and space for remaining dislocations to glide easily without barrier. Movable dislocations within a slip system (slip direction and plane) when interact with each other and dislocations of opposite signs annihilate, the process is designated as 'dynamic recovery'. Any kinematic hardening (KH) rule used in the material modelling, nowadays, is equipped with the hardening term that handles the dislocation generation and pile-ups and a recovery term mainly captures the annihilation of movable dislocations within the active slip systems.
- The Armstrong and Frederick (AF) [86] model incorporates a dynamic recovery term and was successful to simulate balanced nonlinear cyclic loading. The incorporation of the recovery term in KH rule is another landmark in the history of cyclic plasticity modelling. For unbalanced loading, the AF rule showed over-prediction might be due to lesser hardening and further modifications of KH rule are focussed mainly on dynamic recovery of the back stress. To get proper hardening effect, the recovery term needs to be understood and modelled properly because the annihilations of dislocations are 'not so controlled' physical phenomena.
- Chaboche and his co-workers [88, 89] proposed a 'segmented' nonlinear kinematic hardening rule, which can be considered as a superposition of several Armstrong–Frederick hardening rules implemented at different segments with different sets of KH parameters. Orthodox AF rule with single set of KH parameters often fails to simulate the typical nonlinear hysteresis curves, which can be simulated by the Chaboche model accurately, by virtually dividing it into three critical segments: initial high modulus at the onset of yielding, the transient nonlinear segment and the constant modulus segment at a higher strain range. Accordingly, the first rule (α^1) should start hardening with a very large modulus

and stabilizes very quickly. The second rule (α^2) should simulate the transient nonlinear portion of the hysteresis curve. Finally, the third rule (α^3) should be a linear hardening rule ($\gamma^3 = 0$) to represent the subsequent linear part of the hysteresis curve at a high strain range Chaboche model generally over-predict uniaxial ratcheting response in the initial cycles and highly over-estimates the biaxial cases. Chaboche [90] added a fourth hardening rule with a concept of 'threshold' in the modification. This kinematic hardening grows linearly unto the 'threshold' stress level and followed by Armstrong–Frederick rule. The incorporation of the fourth rule improves the symmetric hysteresis loop simulation, but does not improve much for the ratcheting (asymmetric) simulations for both uniaxial and biaxial cases.

- Guionnet [91] proposes a model that modifies the original AF rule by incorporating the effect of accumulated plastic strain in it. The Guionnet model uses parameters determined from the biaxial ratcheting experiments. The simulations of hysteresis loops and ratcheting response are not up to the mark, and the model has numerical divergence problem for higher uniaxial ratcheting rate responses as reported by Bari and Hassan [92]. For the bow-tie loading case, this model over predicts with constant rate of ratcheting, whereas for the reverse bow tie, it under predicts the ratcheting. Similar erratic behaviour is also observed in the simulations for bow tie and reverses bow-tie loading cycles obtained by the Dafalias–Popov [93] uncoupled model with Armstrong–Frederick kinematic hardening rule [94].
- One of the popular models, following the Besseling-type multi-linear formation, was proposed by Ohno and Wang [95]. The model, like other linear models, simulates closed hysteresis loops resulting no ratcheting response. Later, to rectify this error, they extended their model [96] to have nonlinearity and thus confirms the AF format, showing improvements in simulating nonlinearity in the stabilised hysteresis loops and uniaxial ratcheting curves than Chaboche model with 'threshold' term. Though there are further modifications to the OW model mainly targeting the ratcheting response. In 1994 [97], they extend the model further and replaced plastic strain increment with the accumulated plastic strain in the dynamic recovery term. This rule [97] becomes a direct modification of AF rule, where the contribution of the dynamic recovery term is modulated by the fitting parameter χ^k. For asymmetric cyclic loading simulations, over-predictions observed. For the present work, Ohno–Wang model is chosen for its considerable accuracy than the previous KH models and comparable simplicity than the following KH models.
- McDowell [98] and Jiang and Sehitoglu [99] modified the dynamic recovery power term χ as functions of non-proportionality parameter but the simulations were conducive to portray the experimental biaxial ratcheting curves.
- The kinematic hardening rule proposed by Voyiadjis and Sivakumar [100, 101] is a combination of the KH rules proposed by Phillips et al. [102, 103] and Tseng–Lee [104]. They reason that the Phillips rule stipulates the evolution of the back stress along the stress rate direction and therefore follows the

experimental trend better, but it does not ensure the tangential nesting of the yield and other surfaces in a multi-surface model. On the other hand, the Tseng–Lee rule invokes the desired nesting feature for both the proportional and non-proportional loading. Voyiadjis and Sivakumar [100] propose a rule by appropriately blending both the rules, to include the requirements of yield surface movement and the surface nesting. Although the model fails to simulate biaxial ratcheting responses,

- Voyiadjis–Basuroychowdhury [105] and Basuroychowdhury–Voyiadjis [106] attempted to improvise the Chaboche [90] KH rule by adding the direction of the stress rate, but could not achieve more than that of the modified Chaboche model.
- Abdel-Karim and Ohno [107] propose a new KH rule combining the initial multi-linear version of the Ohno–Wang [95, 96] and the AF [86] rule. This model produces a better biaxial ratcheting simulation, especially in reproducing the recorded steady rate of ratcheting, but under predicts the uniaxial ratcheting response.

The modifications in KH rule are focused on making the model compatible for both balanced and unbalanced, proportional and non-proportional, uniaxial and multi-axial loading conditions. Isotropic hardening rules [73, 108] also employed alongside the KH rule in the model to capture several hardenings or softenings manifested during irreversible cyclic loading. But these mathematical theories are mainly based on loading functions and justified over experimental observations.

The recent trend is directed towards constitutive modelling based on dislocation dynamics. Dislocation-based modelling delivers physically justified and more accurate hardening functions (both Kinematic and Isotropic), which can be effective tools for simulation of monotonic and cyclic loading conditions with or without considering the rate effects. Anisotropic material properties are also completely compatible with this type of formation. Furthermore, any orthodox numerical scheme can easily fit into incorporate the hardening function, though some complexity may be costly in terms of numerical computations.

7.5.3 Modelling for Saturated Cycles with Kinematic Hardening Rule

Accurate estimation of $d\varepsilon_{eq}^{p}$ requires incorporation of correctly characterised hardening laws. In cyclic plasticity modelling, the major role is played by the kinematic hardening rule. Its mathematical application is devised to generate stress–strain hysteresis loops and simulate the Bauschinger effect simultaneously.

Inspired by the Besseling [83]-type multi-layered linear formation and conforming nonlinear Armstrong–Frederick [86] philosophy in each segments, Ohno and Wang [95, 96] developed a multi-segmented nonlinear kinematic hardening rule. The first terms of each segments offer the hardening due to dislocation pile-up and are governed KH parameters C^k while the second term estimates the recovered energy due to annihilation of mobile dislocations. The recovery activities of each segments are controlled by a fitting parameter (χ). The activation of different segments is sequential and administrated by the back stress thresholds values $\left(r^k\right)$.

In this work, observing the critical stress–strain hysteresis curvature of the material, six-segmented nonlinear Ohno–Wang KH rule is employed to simulate basic stress–strain hysteresis loops with Bauschinger effect. According to the model architecture, the total deviatoric back stress is:

$$\alpha_{ij} = \sum_{k=1}^{m} \alpha_{ij}^{k} \tag{7.5}$$

Here, α_{ij}^k is the kth segment of the deviatoric back stress tensor. For each segment, back stress evolution law is given as:

$$\mathrm{d}\alpha_{ij} = C^k r^k \left[n_{ij} - \left(\frac{\left\| \alpha_{ij}^k \right\|}{r^k} \right)^{\chi^k + 1} l_{ij}^k \right] \mathrm{d}\varepsilon_{\mathrm{eq}}^p \tag{7.6}$$

C^k, r^k and χ^k are the material parameters for Ohno–Wang kionematic hardening rule, where C^k are the dimensionless hardening coefficients and r^k are the saturation values of α_{ij}^k for kth segment and also control the contribution of the recovery term. χ^k is the parameter that controls the degree of nonlinearity by modifying the contribution of the dynamic recovery part. In the above expression, l_{ij}^k is the direction vector to the corresponding segment of deviatoric back stress, and $l_{ij}^k = \frac{\alpha_{ij}^k}{\left\| \alpha_{ij}^k \right\|}$ where $\left\| \alpha_{ij}^k \right\| = \left[\frac{3}{2} \alpha_{ij}^k \alpha_{ij}^k \right]^{\frac{1}{2}}$.

The KH parameters are calculated from saturated loop of highest strain amplitude (0.75%). The saturated loops for all the strain amplitudes are 'Masing' in the plastic regime as observed in the experimental results.

For the simulation of saturated stress–strain hysteresis loops, the material model consists of only kinematic hardening rule with constant cyclic yield stress. The hardening modulus, for this case, is deduced to be:

$$h = \sum_{k=1}^{m} C^k r^k \left[\frac{3}{2} - \left(\frac{\left\| \alpha_{ij}^k \right\|}{r^k} \right)^{\chi^k + 1} l_{ij}^k n_{ij} \right] \tag{7.7}$$

7.5.4 Phenomenological Modelling for Cyclic Softening by Exponential Cyclic Softening

The modelling for cyclic hardening or softening can be done by incorporating suitable isotropic hardening rule. The experimental observations show equal amount of hardening or softening in forward and reversed loading conditions promote the simplicity in the modelling by assuming that, the cyclic hardening and softening are isotropic in nature [73].

An exponential evolution of isotropic hardening was proposed by Zaverl and Lee [108], which is mainly phenomenological mathematical fitting function. The isotropic hardening stress is a nonlinear exponential function of total plastic strain. The incremental form of the isotropic hardening rule is as follows:

$$\mathrm{d}R_p = b\left(R_{ps} - R_p\right)\mathrm{d}\varepsilon^p_{\mathrm{eq}} \tag{7.8}$$

where b and R_{ps} are the material parameters. R_{ps} is the magnitude of total softening and b is the softening rate. It is observed from experimental results that both R_{ps} and b depend on plastic strain amplitude for the present material.

For the combined model with isotropic hardening and kinematic hardening, the plastic modulus is recast into the following form:

$$h = \sum_{k=1}^{m} C^k r^k \left[\frac{3}{2} - \left(\frac{\left\|\alpha^k_{ij}\right\|}{r^k}\right)^{\chi^k+1} l^k_{ij} n_{ij}\right] + b\left(R_{\mathrm{ps}} - R_p\right) \tag{7.9}$$

7.5.5 Cyclic Softening by Original Marquis' Approach

To justify the ideology of the physical phenomena behind the cyclic hardening or softening, another popular way of modelling is by modifying the dynamic recovery term of the KH rule. Researchers like Marquis [109], Haupt et al. [110] and many others have proposed such modifications to calibrate hardening or softening effects. Their philosophies were same to modify the dynamic recovery contribution of KH rule, but they proposed different evolution equations. Haupt and his co-workers suggested a history-dependent equation where Marquis' approach was rather simple and instantaneous.

Marquis [109] proposed an accumulated plastic strain-based function that modifies the dynamic recovery contribution. The evolution equation of the KH rule with Marquis' modification is:

$$d\alpha_{ij}^k = C^k r^k \left[n_{ij} - \left[1 + \lambda e^{-\omega \varepsilon_{eq}^p}\right] \left(\frac{\left\|\alpha_{ij}^k\right\|}{r^k} \right)^{\chi^k + 1} l_{ij}^k \right] d\varepsilon_{eq}^p \qquad (7.10)$$

The function proposed by Marquis originally contains three parameters but for simplicity Hauptand his co-workers [110] used 1, λ and ω. The choice of the sign of λ in the above equation can produce hardening or softening effect. The cyclic yield stress is kept constant, and no separately constructed isotropic hardening rule is incorporated in the model. The hardening modulus, for this case, is recast to be:

$$h = \sum_{k=1}^{m} C^k r^k \left[\frac{3}{2} - \left[1 + \lambda e^{-\omega \varepsilon_{eq}^p}\right] \left(\frac{\left\|\alpha_{ij}^k\right\|}{r^k} \right)^{\chi^k + 1} l_{ij}^k n_{ij} \right] \qquad (7.11)$$

7.5.6 Cyclic Softening by Modified Marquis' Approach

The experimental results show that the cyclic hardening/softening has no directional influence in forward and reverse loading and therefore considered isotropic in nature. In view of the fact, the proposition by Marquis is used to modify the cyclic yield stress with the idea that the generation of dislocations due to plastic deformation can produce more dynamic recovery [109] at the present state or, it can contribute immediately after as isotropic softening.

Current yield stress (σ_c), for a particular strain amplitude, is not considered constant and additively decomposed into constant cyclic yield stress (σ_0) and isotropic softening stress (R). Initially, non-Masing behaviour is observed in the experimental results for the present material SA333 mainly in the elastic region and therefore, different initial cyclic yield stresses (σ_0) are to be used for different strain amplitudes. The equation for the current yield stress evolution is:

$$\sigma_c = \sigma_0 + R\left(\varepsilon_{eq}^p\right) = K\left[1 + \lambda e^{-\omega \varepsilon_{eq}^p}\right] \qquad (7.12)$$

where K is a constant, such that $K = \frac{\sigma_0}{1+\lambda}$. Parameter λ and ω control the softening. For the combined hardening plasticity, the plastic modulus is recast into the following form:

$$h = \sum_{k=1}^{m} C^k r^k \left[\frac{3}{2} - \left(\frac{\left\|\alpha_{ij}^k\right\|}{r^k} \right)^{\chi^k + 1} l_{ij}^k : n_{ij} \right] - K\left(\lambda \omega e^{-\omega \varepsilon_{eq}^p}\right) \qquad (7.13)$$

7.6 Material Parameters

The material properties are extracted from the monotonic (tensile) tests and the symmetric uniaxial cyclic plastic experiments and suitably tuned during FE simulation.

7.6.1 *Elastic Parameters*

Elastic parameters, such as Young's modulus, are extracted from the uniaxial tensile tests following standard procedures and found to be 210 GPa (rounded); Poisson's ratio is taken 0.3. The elastic constants are considered unchanged for all strain amplitudes as temperature and straining rate are kept the same for all tests and unaltered during experimentation.

7.6.2 *Cyclic Yield Stresses*

The non-Masing behaviour of the material has incisive impact on the modelling scheme. For non-Masing behaviour of the material, irregular strain-hardening character may be observed. For simulations of such behaviour, different KH parameters may be used for different strain amplitudes. It may sometimes be observed in the experimental results that there is insignificant difference in the plastic curvature and the dissimilar elastic portion is responsible for non-Masing behaviour. In those cases, one set of KH parameters is used for every strain amplitudes and modification is made to the cyclic yield stresses.

The initial cycles are non-Masing mainly in the elastic regime (Fig. 7.7a) and clearly reflected in the cyclic yield stress values calculated from the initial cycles. The plastic curvature is observed (Fig. 7.7b) same (Masing), and therefore, same kinematic hardening parameters are used. The suitable values of initial cyclic yield stresses are shown in Table 7.2.

7.6.3 *Kinematic Hardening Parameters*

It is realised while analysing the experimental results that the plastic curves have minor variations due to the non-Masing behaviour, while observed majorly in the

Table 7.2 Initial cyclic yield stresses (σ_0)

Strain amplitude (%)	Initial cyclic yield Stress (MPa)
0.35	248
0.50	234
0.75	222

elastic limits. To overcome the difficulty of non-Masing behaviour for the present material, variation of the cyclic yield stresses is considered, whereas the KH parameters are calculated from saturated loop of highest strain amplitude which are 'Masing' mostly in the plastic range with all other strain amplitudes.

For Ohno–Wang KH rule, the parameters (C^k and r^k) are determined from saturated (30th cycle) fully reversed loading branch of highest strain amplitude (0.75%) experimental data. A single set of KH parameters is sufficient to evaluate the kinematic hardening behaviour of the material due to the Masing behaviour of the plastic curves at saturation. The procedure is outlined in the work of Jiang and Kurath [111] and has been followed to determine the KH Parameters.

Parameters C^k and r^k regulate the curvature of the loading and unloading branch of the hysteresis loop. Therefore, from the fully reversed uniaxial loading branch of a saturated hysteresis curve, the values of C^k and r^k are re-extracted. Figure 7.8 schematically shows such a loading branch. Here $\Delta\varepsilon_p$ indicates plastic strain range, i.e. twice the plastic strain amplitude, and similarly $\Delta\sigma$ represents the stress range, again twice the stress amplitude. The above curve can be described by the following four-parameter equation:

$$\Delta\varepsilon^p = a\left(\frac{\Delta\sigma - \Delta\sigma_0}{\Delta\sigma_u - \Delta\sigma}\right)^M \tag{7.14}$$

where a and M are power-law coefficient and exponent, respectively, of the fitting function. $\Delta\sigma_0$ and $\Delta\sigma_u$ are the cyclic yield stress range and ultimate stress range, respectively. The objective of the fitting function is to get extrapolated data at high strain amplitude to cover maximum range in LCF tests. From the above equation, a set of seven pre-defined data points $\Delta\varepsilon_k^p$, and $\Delta\sigma_k$ are selected (Fig. 7.8) and the values of C^k and r^k are determined correspondingly as follows:

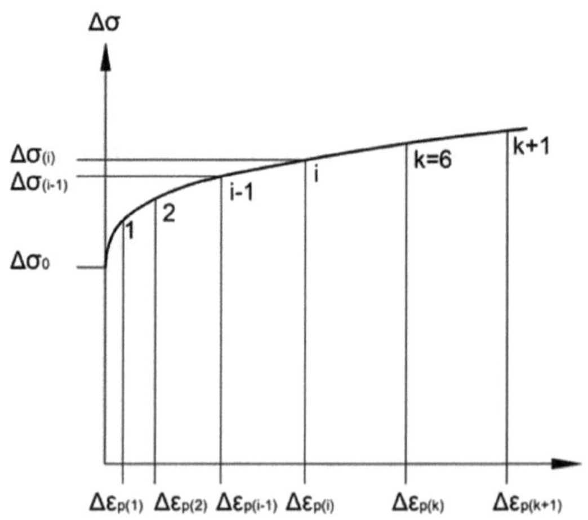

Fig. 7.8 Stress versus plastic strain curve of the loading branch of the first cycle

Table 7.3 C^k and r^k values

C^1 = 16,323	C^2 = 4082.5	C^3 = 1633	C^4 = 816.5	C^5 = 363	C^6 = 163
r^1 = 13.8 MPa	r^2 = 16.6 MPa	r^3 = 19.8 MPa	r^4 = 24.6 MPa	r^5 = 32.5 MPa	r^6 = 43.4 MPa

$$C^k = \sqrt{\frac{2}{3}}\frac{2}{\Delta\varepsilon_k^p} \tag{7.15}$$

where k = 1, 2, … 6. For calculating six r^k, it is necessary to calculate H^k first. The formula adopted for calculating is:

$$H^k = \frac{\Delta\sigma_k - \Delta\sigma_{k-1}}{\Delta\varepsilon_k^p - \Delta\varepsilon_{k-1}^p} \tag{7.16}$$

$$r^k = \frac{2}{3}\frac{H^k - H^{k+1}}{C^k} \tag{7.17}$$

The extracted values of C^k and r^k are shown in Table 7.3.

The mathematical fitting parameter χ^k (Eq. 7.6), which regulates the dynamic recovery contribution, is set to zero to ensure the fully active dynamic recovery process. Generally, for unbalanced or asymmetric loading, the parameter is set to allow partial recovery.

7.6.4 Exponential Isotropic Hardening Parameters

The cyclic softening parameters R_{ps} and b are determined from uniaxial cyclic softening data (peak stress vs. accumulated plastic strain) from initial cycle till the softening gets saturated. R_{ps}, the saturation value of the softening stress, is the algebraic difference between the maximum and the saturated peak stresses for a particular strain amplitude as shown in Fig. 7.9a. Figure 7.9b shows that the logarithmic values of the softening stresses normalised by saturated softening stress are plotted against the accumulated plastic strain. The slope of the linear fit to this curve indicates the softening rate.

It is observed from experimental data that both R_{ps} and b depend on plastic strain amplitude (q) for the present material and therefore should be calibrated with plastic strain amplitude as follows:

$$\begin{cases} R_{ps} = 6645.7q - 54.63 \\ b = 18.81\exp(-342q) \end{cases} \tag{7.18}$$

Figure 7.10a, b shows calibration curves of the above equations.

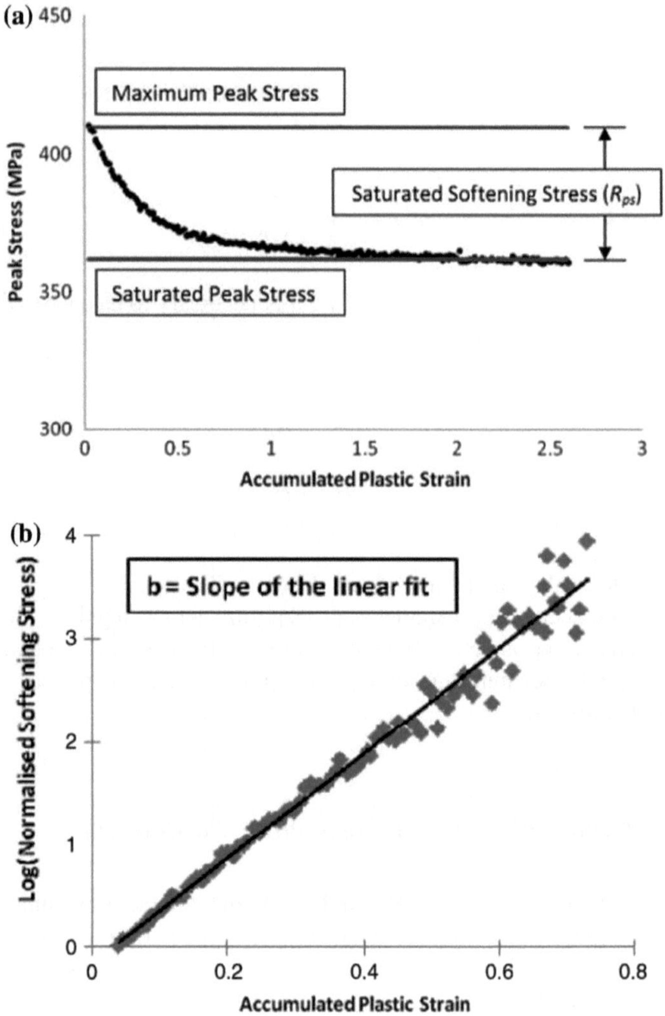

Fig. 7.9 **a** Schematic diagram for calculations of R_{ps}. **b** Schematic diagram for calculations of b

7.6.5 Marquis Parameters

λ and ω are calibrated from the softening characteristics of respective strain amplitudes and adjusted by trial and error method during simulations. The values of λ and ω are shown in Table 7.4.

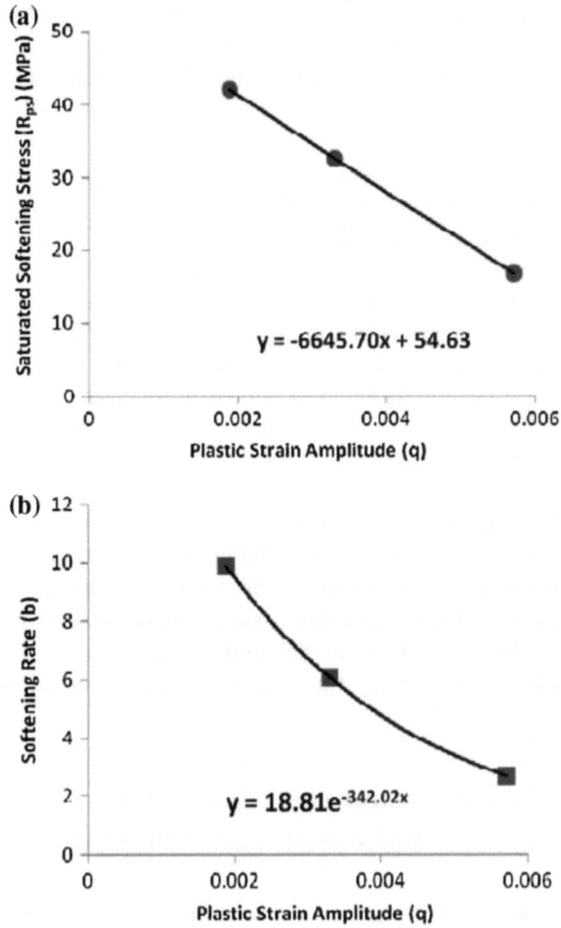

Fig. 7.10 **a** Dependence of R_{ps} with plastic strain amplitude. **b** Dependence of b with plastic strain amplitude

Table 7.4 Softening parameters (λ and ω)

Strain amplitude (%)	0.35	0.50	0.75
λ	1.6	0.5	0.16
ω	−2.2	−1.9	−1.8

7.7 Validation

Mathematical material model is validated through finite element analysis. Simulation results are compared to the experimental results to confirm the validity of the model. A commercial FEA package—SIMULIA ABAQUS CAE v6.8 [112]—is used for accountability of the validation process. The finite element method is widely used for structural analysis of nonlinear material. The finite element method employs constitutive modules to capture the material stress–strain response at each integration point. Structural response is the integrated output from all integration points. During structural analysis through finite element method, local and global level numerical integrations should be carried out simultaneously [113]. Local integration performs numerical calculations using constitutive equations for finite increments of loading at integration points in each element. Global integration solves for the nodal force equilibrium equations in an iterative manner. The accuracy of a structural solution is dependent on the accuracy of calculation at the integration points. The structural stiffness matrix in turn depends on the tangent modulus calculated for each integration point. Thus, efficient global solution is dependent on consistency between the local numerical scheme and tangent modulus at each integration point. In this study, constitutive models have been implemented through the user material subroutine (UMAT for static analysis) facility with radial return method as the local numerical scheme.

The working (gauge length) portion of the specimen, i.e. three-dimensional solid cylindrical geometry is analysed. The specimen geometry is as per the dimensions shown in Fig. 7.3a. The specimen is discretised adopting structural meshing. 8-noded linear hexahedral elements with 6-noded linear wedge elements at central zone are used to make a radial orientation. Displacement boundary conditions are given according to the strain amplitudes using balanced triangular amplitude function. Figure 7.11 shows such a model created in ABAQUS FE platform.

The major task, in the finite element analysis, is to devise the integration schemes for the constitutive mathematical material model.

7.7.1 Backward Euler Integration Scheme and the Constitutive Relations

Backward Euler integration scheme is same as Euler method, only it is implicit. A step forward in time takes the updated trial state of stress outside the yield surface

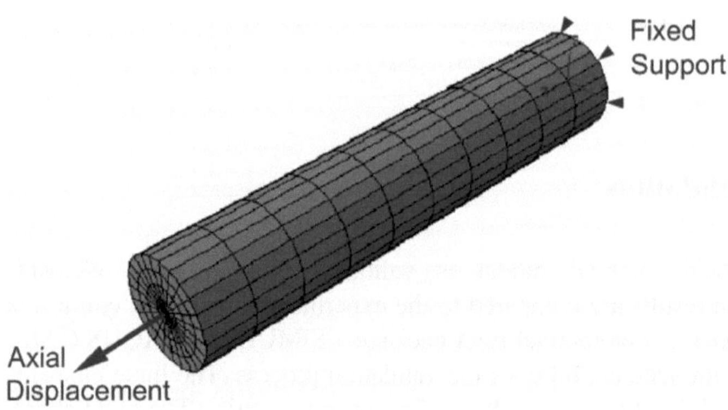

Fig. 7.11 Discretised specimen with loading arrangements

by elastic prediction, and then the state of stress is updated by plastic correction and brought back onto the yield surface in that particular time step. The function of the scheme named it 'Radial Return' or 'Return Mapping' algorithm, which consists of an elastic predictor and a plastic corrector [114–117]. A schematic flow chart of the integration algorithm is represented below:

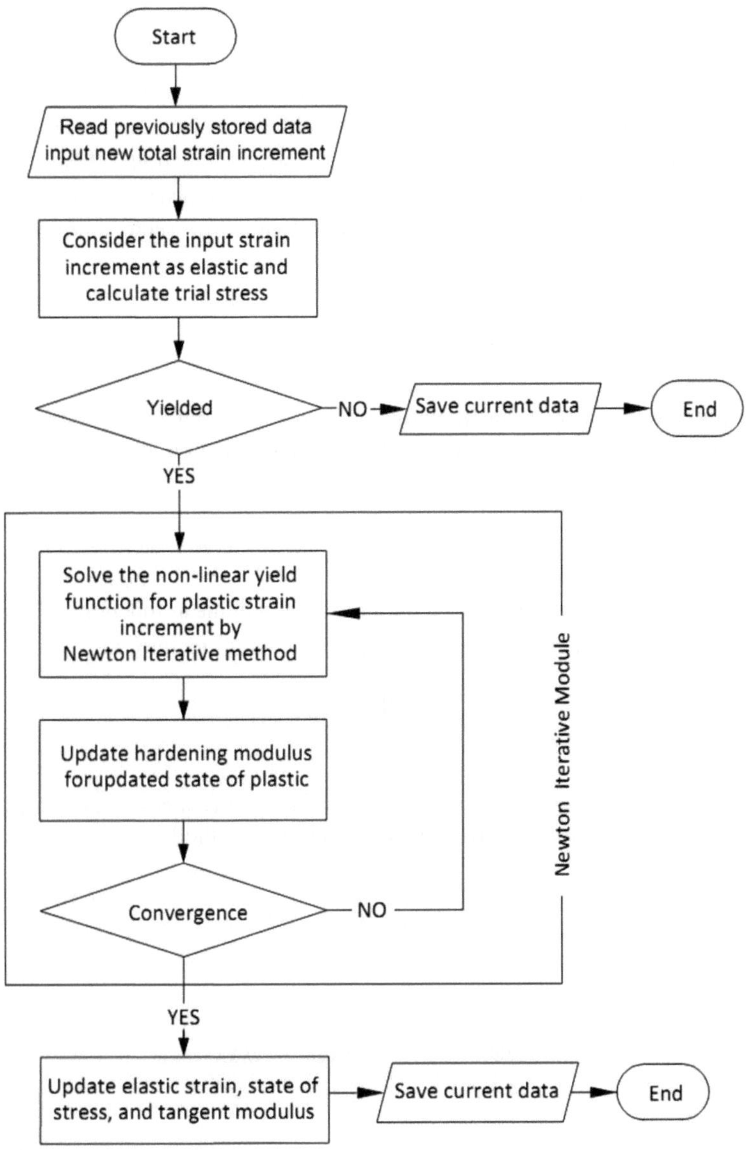

Initially, the components of the total strain increment $(\Delta\varepsilon_{ij})$ fed into 'UMAT' are assumed completely elastic and the elastic predictor stresses are calculated using generalised Hooke's law by elastic stiffness matrix. But the incremental plastic strain components within the total of strain increment modify the stress value to the correct ones [118].

$$^{1}\sigma_{ij} = {}^{0}\sigma_{ij} + \underbrace{\mu\Delta\varepsilon_{ij} + \lambda Tr(\Delta\varepsilon_{ij})}_{\text{Elastic Predictor}} - \underbrace{\mu\Delta\varepsilon^{p}_{ij}}_{\text{PlasticCorrector}} \tag{7.19}$$

where prefix 1 and 0 signify current and previous values, respectively. μ and λ are Lame's constants, required to construct elastic stiffness matrix. $\Delta\varepsilon_{ij}$ is the total strain increment in that particular time step $Tr()$ expresses the function $Tr(\Delta\varepsilon_{ij}) = \Delta\varepsilon_{ij}\delta_{ij}$ (δ_{ij} is the Kronecker delta) that represents the volumetric strain increment. Introducing equivalent plastic strain increment with flow vector current state of stress can be expressed as:

$$^{1}\sigma_{ij} = {}^{\text{trial}}\sigma_{ij} - 2G\Delta\varepsilon^{p}_{\text{eq}}\frac{3}{2}\frac{S_{ij} - \alpha'_{ij}}{S_{\text{mises}}} \tag{7.20}$$

where previous stress and elastic predictor stress are cumulatively known as trial stress, and G is the shear modulus (=μ/2). The yield function, using the trial stress, would be:

$$\Phi = S_{\text{mises}} - R - \sigma_0 = S^{\text{trial}}_{\text{mises}} - 3G\Delta\varepsilon^{p}_{\text{eq}} - R - \sigma_0 = 0 \tag{7.21}$$

The above expression can be identified as a nonlinear function of $\Delta\varepsilon^{p}_{\text{eq}}$. To maintain the consistency condition, dϕ should be equal to zero, i.e.

$$\mathrm{d}\Phi = \frac{\partial\Phi}{\partial\Delta\varepsilon^{p}_{\text{eq}}}\mathrm{d}\Delta\varepsilon^{p}_{\text{eq}} = 0 \tag{7.22}$$

And together, in conjunction with the yield function (i.e. ϕ = 0), the expression below can be solved for $\Delta\varepsilon^{p}_{\text{eq}}$ by Newton's iterative method.

$$\Phi + \frac{\partial\Phi}{\partial\Delta\varepsilon^{p}_{\text{eq}}}\mathrm{d}\Delta\varepsilon^{p}_{\text{eq}} = 0 \tag{7.23}$$

where

$$\frac{\partial\Phi}{\partial\Delta\varepsilon^{p}_{\text{eq}}} = h = n_{ij}\frac{\partial\Delta\alpha_{ij}}{\partial\Delta\varepsilon^{p}_{\text{eq}}} + \frac{\partial\Delta R}{\partial\Delta\varepsilon^{p}_{\text{eq}}} \tag{7.24}$$

Substituting above expressions into Newton-Raphson [119] formulation and rearranging after ith iteration, the value of current back stresses will be:

$$\alpha_{ij}^{\prime(i)} = \alpha_{ij}^{\prime(i-1)} + \Delta\alpha_{ij}^{\prime}\left(\Delta\varepsilon_{\mathrm{eq}}^{p(i)}\right) \tag{7.25}$$

As the value of back stresses changes depending on the equivalent plastic strain increment, the von Mises equivalent trial stress will also change. Similarly, the value of the current isotropic hardening stress will be:

$$R^{(i)} = R^{(i-1)} + \Delta R\left(\Delta\varepsilon_{\mathrm{eq}}^{p(i)}\right) \tag{7.26}$$

And, correspondingly the increment of plastic strain increment is expressed as:

$$d\Delta\varepsilon_{\mathrm{eq}}^{p} = \frac{S_{\mathrm{mises}}^{\mathrm{trial}(i)} - 3G\Delta\varepsilon_{\mathrm{eq}}^{p(i)} - R^{(i)} - \sigma_0}{3G + H^{(i*)}} \tag{7.27}$$

where * signifies that H is not completely modified within a time step for every iterations. The isotropic part of it remains unaltered and be updated after the time step. The integration scheme adopted in the present coding can be specifically named as semi-implicit integration scheme. Elastic prediction and radial return (plastic correction) make the scheme implicit. Incomplete update of the hardening modulus makes the process semi-implicit integration. For stability, this semi-implicit scheme is adopted.

Yield criterion is employed as the convergence criterion with some numerical tolerance. For convergence, if not attained, the equivalent plastic strain increment will be further adjusted by accumulating calculated increment $\left(d\Delta\varepsilon_{eq}^{p}\right)$ for the next iteration (i + 1th) in that time step as:

$$\Delta\varepsilon_{\mathrm{eq}}^{p(i+1)} = \Delta\varepsilon_{\mathrm{eq}}^{p(i)} + d\Delta\varepsilon_{\mathrm{eq}}^{p} \tag{7.28}$$

7.7.2 Newton Iterative Method

Newton's method or Newton-Raphson [119] method is an iterative technique for finding successively better approximations to the roots of a real-valued function. This is a very fast method involves the slope or the function derivative to search the root.

$$x_{n+1} = x_n - \frac{f(x_n)}{f'(x_n)} \tag{7.29}$$

Newton method is applied in the implicit plasticity algorithms to find plastic strain increments for a particular time step solving the nonlinear yield function of equivalent plastic strain increment.

7.7.3 Consistent Tangent Modulus

After convergence of the local iteration for the numerical scheme, it is important to calculate the consistent tangent modulus to preserve the quadratic convergence rate at global level of finite element calculation. Numerical method is said to be consistent with the tangent modulus when prescribed stress increment for given strain increment from tangent modulus is consistent with stress calculated by local numerical scheme. Such tangent matrix is termed as consistent tangent modulus for the numerical scheme adapted. The consistent tangent modulus is prescribed in a general form for the Armstrong and Frederick model with radial return algorithm by Kobayashi and Ohno [112], and Rahaman [108].

7.8 Results and Discussion

Uniaxial cyclic plastic deformations are simulated in ABAQUS CAE (v6.8) commercial finite element platform. The material model along with the parameters is incorporated through user material subroutine (UMAT). The simulation results are compared with the experimental results to validate the material model and the parameters.

7.8.1 Simulation of Saturated Stress–Strain Hysteresis Loops

The simulation of saturated loops is carried out with kinematic hardening rule alone. Masing behaviour of the material enables the use of single set of parameters. The comparison of simulated results with the corresponding experiments is shown in Fig. 7.12.

7.8.2 Discussion

It is observed in the comparison of results that the six-segmented kinematic hardening rule proposed by Ohno–Wang is suitable for the present material. The KH parameters calculated from the loading branch of the fully reversed saturated stress–strain hysteresis loop of highest (0.75%) strain amplitude which is appropriately captured the nonlinearity in hardening for the material for all the strain amplitudes.

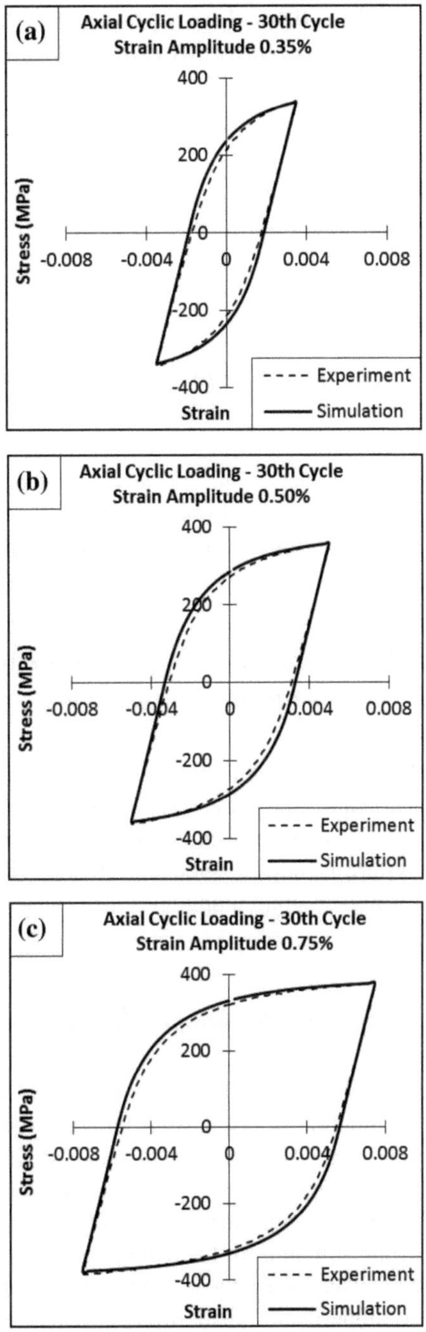

Fig. 7.12 **a** Comparison of (uniaxial) experimental and simulated stress–strain hysteresis loops for 0.35% strain amplitude at saturated (30th) cycle. **b** Comparison of (uniaxial) experimental and simulated stress–strain hysteresis loops for 0.50% strain amplitude at saturated (30th) cycle. **c** Comparison of (uniaxial) experimental and simulated stress–strain hysteresis loops for 0.75% strain amplitude at saturated (30th) cycle

7.8.3 Simulation Cyclic Softening with the Exponential Isotropic Softening Model

The cyclic softening of the material in the first a few transient cycles is attempted to model through the incorporation of exponential isotropic hardening rule as proposed by Zaverl and Lee [103]. The experimental and simulated saturated stress–strain hysteresis loops and peak stress softening plots with cycle are shown in Fig. 7.13 in a comparative manner.

7.8.4 Discussion for the Phenomenological Model

The cyclic softening is well simulated (Fig. 7.13) by the exponential isotropic hardening law. The related material parameters are found to be calibrated accurately. But the isotropic hardening law is purely empirical mathematical function and lacks sufficient physical justification.

7.8.5 Simulation Cyclic Softening with the Original 'Marquis' Approach

The original marquis approach is implemented with six-segmented OW kinematic hardening rule. The comparison of initial and 30th cycle of low (0.35%) strain amplitude experiment is shown in Fig. 7.14. The approach distorts the shape of the hysteresis loops to simulate softening (Fig. 7.14). The atypical hysteresis curvature for higher strain amplitudes also causes difficulty in numerical convergence.

7.8.6 Discussion for Original Approach

The Marquis model is based on the concept that, microscopically, dislocation annihilation is responsible for global softening and dislocation pile-up promotes hardening. The structure of standard KH rule constitutes of a hardening or pile-up term and a recovery or annihilation term. The philosophy of the Marquis model is to fortify or undermine the recovery contribution of the standard KH rules to endorse cyclic softening or hardening, respectively. The mechanism of the model, though, somewhat phenomenological uses a plastic strain-based exponential function. A single cyclic yield stress with no isotropic hardening rules is required in this approach.

The main difficulty of working with this model is that the mechanism distorts the hysteresis loop curvature as observed in Fig. 7.14b. At higher strain amplitudes

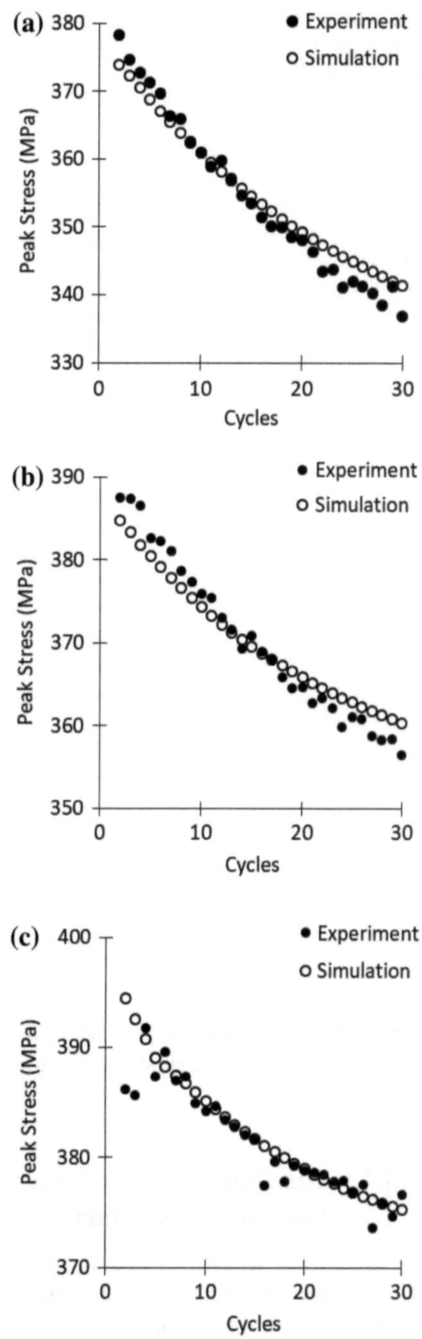

Fig. 7.13 **a** FE simulation with exponential isotropic softening rule compared with experimental peak stresses for 0.35% strain amplitude. **b** FE simulation with exponential isotropic softening rule compared with experimental peak stresses for 0.50% strain amplitude. **c** FE simulation with exponential isotropic softening rule compared with experimental peak stresses for 0.75% strain amplitude

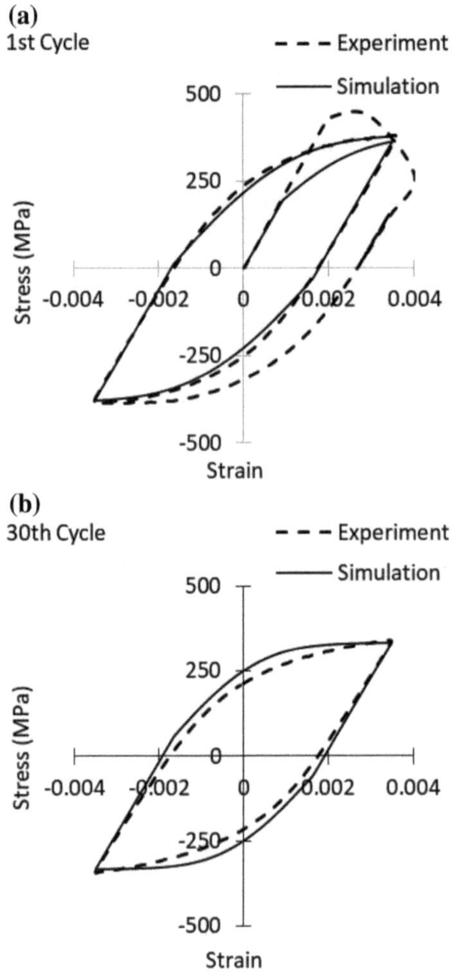

Fig. 7.14 **a** FE simulation with original Marquis' approach shows good agreement with experiment at initial cycles for low strain amplitude (0.35%). **b** FE simulation with original Marquis' approach shows mismatch with experiment in the plastic zone near saturation (30th cycle) for 0.35% strain amplitude

(like 0.5% or 0.75%), where hysteresis curvature shows low hardening, FE calculations lead to numerical divergence.

7.8.7 Simulation Cyclic Softening with the Modified 'Marquis' Approach

Though the original approach simulates peak stress softening curves in good agreement with the actual, it fails to simulate the shape of hysteresis loop and also face divergence in case of high (0.75%) strain amplitude. To overcome the difficulties, a new approach is devised and the simulation results for all strain amplitudes are shown in Fig. 7.15 in a comparative manner with the experimental results.

Fig. 7.15 **a** FE simulation with modified Marquis' approach shows good agreement with experimental hysteresis loop at saturated (30th) cycle for 0.35% strain amplitude. **b** FE simulation with modified Marquis' approach shows good agreement with experimental peak stress softening with cycles for 0.35% strain amplitude. **c** FE simulation with modified Marquis' approach shows good agreement with experimental hysteresis loop at saturated (30th) cycle for 0.50% strain amplitude. **d** FE simulation with modified Marquis' approach shows good agreement with experimental peak stress softening with cycles for 0.50% strain amplitude. **e** FE simulation with modified Marquis' approach shows good agreement with experimental hysteresis loop at saturated (30th) cycle for 0.75% strain amplitude. **f** FE simulation with modified Marquis' approach shows good agreement with experimental peak stress softening with cycles for 0.75% strain amplitude

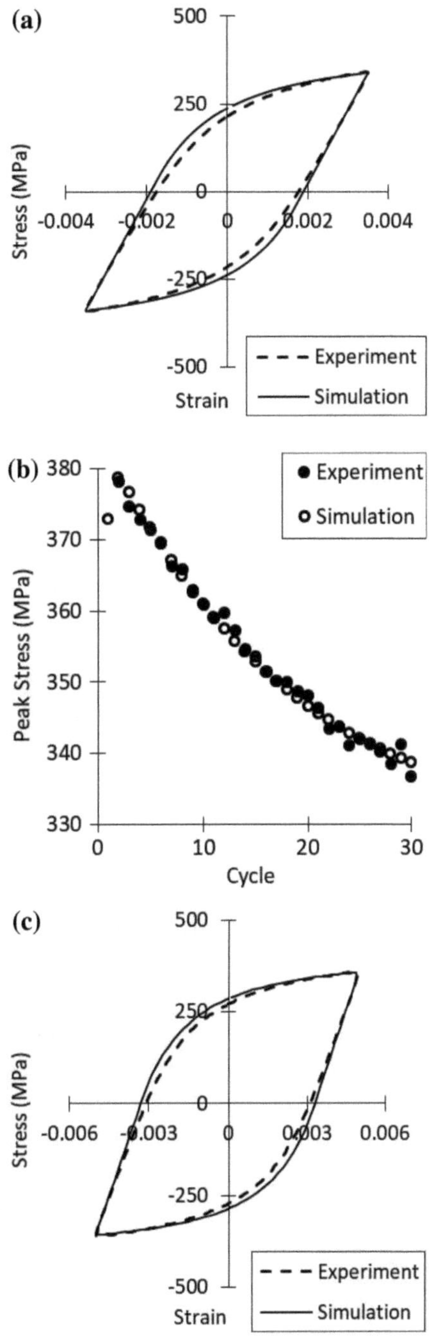

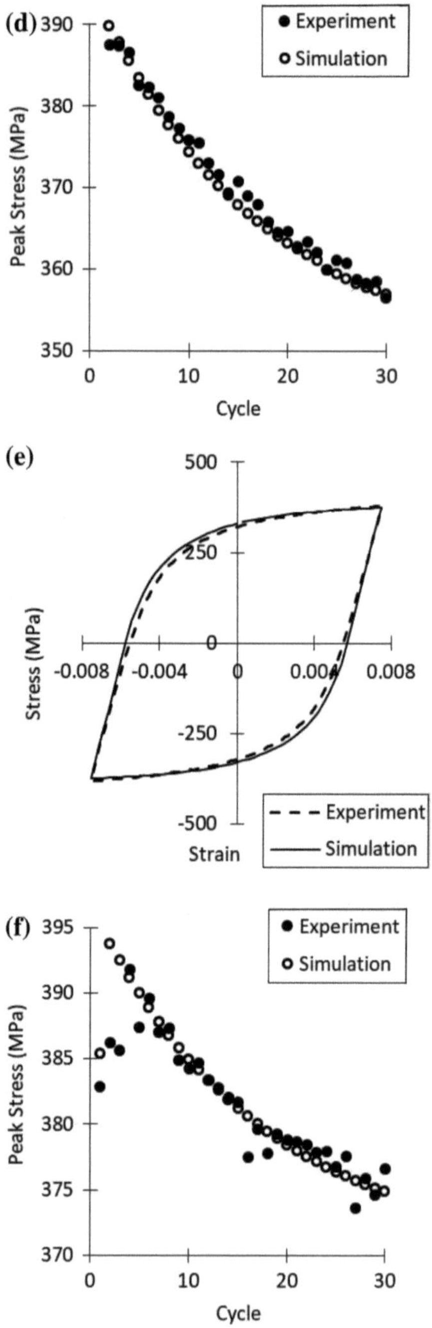

Fig. 7.15 (continued)

7.8.8 Discussion for Modified Approach

The original approach of Marquis, i.e. modification of dynamic recovery, is physically justifiable and reduction of the size of yield surface in deviatoric stress space without any damage mechanism is an absurd idea. Yet, the present work adopted such an idea for following reasons:

(1) The reduction of yield value is the manifestation of the increased recovery (dislocation annihilations) at higher plastic deformation. It is not to be treated as damage induced softening as per the conventional concept.
(2) In a segmented KH law, dynamic recover is a cumulative influence bounded by several thresholds. The Marquis approach can be incorporated in segmented law also, but not in the simplest form. The calibration of parameter would involve a meticulous procedure with no such improvements in the results than present model.

The non-Masing behaviour of the material is captured with a variation of initial yield stresses with strain amplitude. The values are extracted from experimentally observed initial cycles and finalised by trial and error.

The overall contribution of the 'Marquis' function for softening worked nicely. Cyclic softening is simulated (Fig. 7.15) with good agreements by the function when applied on cyclic yield stress.

7.9 Closing Remarks

Modelling of cyclic plastic deformation behaviour is quite complex, and a systematic approach is required. The objective of the work was to study the existing phenomenological modelling approaches for cyclic softening and apply the same for SA333 Grade-6 C-Mn steel. During the study, physical justification and numerical stability of the model motivated a modification and proposed accordingly. The comparison of results clearly shows that:

(1) The hysteresis loop and Bauschinger effect are well taken by six-segmented Ohno–Wang kinematic hardening law with fully active dynamic recovery.
(2) The set of parameters for Ohno–Wang law can be counted as appropriate for the material and validate the procedure of extraction from uniaxial LCF data.
(3) A single set of KH parameters was able to describe the plastic curvature for all strain amplitudes. The non-Masing behaviour of the material is limited in the elastic range and is captured by the variation of initial yield stresses with strain amplitudes.
(4) The exponential isotropic softening is successfully replaced by Marquis's approach; though for atypical hardening behaviour observed in the hysteresis loops for some material, the original Marquis approach may fail to converge in numerical simulations.

(5) The contribution of the 'Marquis' function as softening rule (modified approach) worked nicely. Cyclic softening is simulated with good agreements by the function when applied on cyclic yield stress.

Precise simulation of hysteresis loops incorporating accurate estimation of Bauschinger effect and the Masing behaviour was decisive for the proposed modification. The model can be used, and its transferability can be tested for in-phase multi-axial loading conditions, where similar dislocation interactions are involved.

Acknowledgements The authors acknowledge Bhabha Atomic Research Centre, Mumbai, for financial assistance through collaborative project and National Metallurgical Laboratory, Jamshedpur, for experimental support. The authors also acknowledge Dr. Surajit Kumar Paul, National Metallurgical Laboratory, Jamshedpur, for TEM micrographs.

References

1. Manson, S.S.: Behavior of materials under conditions of thermal stress. Heat Transf. Symp., 9–75 (1953)
2. Coffin, L.F.: A study of the effects of cyclic thermal stresses on a ductile metal. Trans. Am. Soc. Test. Mater. **76**, 931–950 (1954)
3. Suresh, S.: Fatigue of materials. Cambridge University Press, ISBN: 978-0-52-157847-9 (1998)
4. Mughrabi, H.: Fatigue, an everlasting materials problem—still en vogue. Proc. Eng. **2**(1), 3–26 (2010)
5. Kumar, P.S., Sivaprasad, S., Dhar, S., Tarafder, S.: Ratcheting and low cycle fatigue behavior of SA333 steel and their life prediction. J. Nucl. Mater. **401**(1–3), 17–24 (2010)
6. Kumar, P.S., Sivaprasad. S., Dhar. S., Tarafder. S.: Cyclic plastic deformation behavior in SA333 Gr. 6 C–Mn steel. Mater. Sci. Eng. A **528**(24), 7341–7349 (2011)
7. Kumar, P.S., Sivaprasad, S., Dhar, S., Tarafder, S.: Key issues in cyclic plastic deformation: experimentation. Mech. Mater. **43**(11), 705–720 (2011)
8. Khutia, N., Dey, P.P., Kumar, P.S., Tarafder, S.: Development of non masing characteristic model for LCF and ratcheting fatigue simulation of SA333 C-Mn steel. Mech. Mater. **65**, 88–102 (2013)
9. Sivaprasad, S., Kumar, P.S., Arpan, D., Narasaiah, N., Tarafder, S.: Cyclic plastic behaviour of primary heat transport piping materials: influence of loading schemes on hysteresis loop. Mater. Sci. Eng. A **527**(26), 6858–6869 (2010)
10. Sivaprasad, S., Bar, H.N., Kumar, G.S., Punit, A., Bhasin, V., Tarafder, S.: A comparative assessment of cyclic deformation behaviour in SA333 Gr.6 steel using solid, hollow specimens under axial and shear strain paths. Int. J. Fatigue **61**, 76–86 (2014)
11. Sinha, A.K.: Ferrous Physical Metallurgy. Butterworths, London (1989)
12. Zhang, S., Wu, C.: Ferrous Materials. Metallurgical Industry Press, Beijing (1992)
13. Totten, G.E.: Steel Heat Treatment—Metallurgy and Technologies, 2nd edn. CRC Press, Taylor and Francis Group, Boca Raton (2007)
14. Ross, R.B.: Metallic Materials Specification Handbook, 4th edn. Chapman & Hall, London (1992)
15. Mehran, M.: The Effects of Alloying Elements on Steels (I), Christian Doppler Laboratory for Early Stages of Precipitation, InstitutfürWerkstoffkunde, Schweißtechnik und SpanloseFormgebungsverfahren, TechnischeUniversität Graz (2007)

16. Bauschinger, J.: Über die Veranderung der Elasticitatsgrenze und elastcitatsmodulverschiedener, Metal Civiling N.F., **27**, 289–348 (1881)
17. Chai, H.-F., Laird, C.: Mechanisms of cyclic softening and cyclic creep in low carbon steel. Mater. Sci. Eng. **93**, 159–174 (1987)
18. Sivaprasad, S., Swaminathan, J., Tiwary, Y.N., Roy, P.K., Singh, R.: Remaining life assessment of service exposed reactor and distillation column materials of a petrochemical plant. Eng. Fail. Anal. **10**(3), 275–289 (2003)
19. Masing, G.: Eigenspannungen und verfestigungbeim messing. In: Proceedings of the Second International Congress for Applied Mechanics, Zurich, pp. 332–335 (1926)
20. Bauschinger, J.: Mitteilung XV ausdemMechanisch-technischenLaboratorium der KöniglichenTechnischenHochschule in München **13**, 1–115 (1886)
21. Boller, C., Seeger, T.: Materials Data for Cyclic Loading. Elsevier, ISBN: 978-0-44-442875-2 (1987)
22. Doong, S.H., Socie, D.F., Robertson, I.M.: Dislocation substructures and nonproportional hardening. ASME J. Eng. Mater. Technol. **112**(4), 456–465 (1990)
23. Jiang, Y., Kurath, P.: Nonproportional cyclic deformation: critical experiments and analytical modeling. Int. J. Plast. **13**, 743–763 (1997)
24. Jiang, Y.: An experimental study of inhomogeneous cyclic plastic deformation. J. Eng. Mater. Technol. **123**(3), 274–280 (2001)
25. Zhang, J., Jiang, Y.: A study of inhomogeneous plastic deformation of 1045 steel. J. Eng. Mater. Technol. **126**(2), 164–171 (2004)
26. Jiang, Y., Sehitoglu, H.: Cyclic ratchetting of 1070 steel under multiaxial stress states. Int. J. Plast. **10**(5), 579–608 (1994)
27. Jiang, Y., Sehitoglu, H.: Multiaxial cyclic ratchetting under multiple step loading. Int. J. Plast. **10**(8), 849–870 (1994)
28. Véronique, A., Philippe, Q., Suzanne, D.: Cyclic plasticity of a duplex stainless steel under non-proportional loading. Mater. Sci. Eng. A **346**(1–2), 208–215 (2003)
29. Jiang, Y., Zhang, J.: Benchmark experiments and characteristic cyclic plasticity deformation. Int. J. Plast. **24**(9), 1481–1515 (2008)
30. Estrin, Y., Vinogradov, A.: Extreme grain refinement by severe plastic deformation: a wealth of challenging science. Acta Materialla **61**(3), 782–817 (2013)
31. Mughrabi, H.: On the current understanding of strain gradient plasticity. Mater. Sci. Eng. A **387–389**, 209–213 (2004)
32. Lu, K., Lu, L., Suresh, S.: Strengthening materials by engineering coherent internal boundaries at the nanoscale. Science **324**(5925), 349–352 (2009)
33. Pan, Q.S., Lu, L.: Strain-controlled cyclic stability and properties of Cu with highly oriented nanoscale twins. Acta Mater. **81**, 248–257 (2014)
34. Masing, G.: ZurHeyn'schenTheorie der Verfestigung der Metalledurchverborgenelastische Spannungen. In: Harries, C.D. (ed.) WissenschaftlicheVeröffentlichungenausdem Siemens-Konzern, pp. 231–239. Springer, Heidelberg (1923)
35. Elline, F., Kujawaski, D.: Plastic strain energy in fatigue failure. J. Eng. Mater. Technol. Trans. **106**, 342–347 (1984)
36. Fan, Z., Jiang, J.: Investigation of low cycle fatigue behavior of 16MnR steel at elevated temperature, Zhejiang DaxueXuebao (Gongxue Ban)/Journal of Zhejiang University (Engineering Science) **38**, 1190–1195 (2004)
37. Maier, H.J., Gabor, P., Gupta, N., Karaman, I., Haouaoui, M.: Cyclic stress-strain response of ultrafine grained copper. Int. J. Fatigue **28**, 243–250 (2006)
38. Wang, Z., Laird, C.: Relationship between loading process and Masing behavior in cyclic deformation. Mater. Sci. Eng. A **101**, L1–L5 (1988)
39. Raman, S.G.S., Padmanabhan, K.A.: Effect of prior cold work on the room temperature low-cycle fatigue behavior of AISI 304LN stainless steel. Int. J. Fatigue **18**, 71–79 (1996)
40. Plumtree, A., Abdel-Raouf, H.A.: Cyclic stress-strain response and substructure. Int. J. Fatigue **23**, 799–805 (2001)

41. Gough, H.J.: Crystalline structure in relation to failure of metals—especially by fatigue, Edgar Marburg Lecture. In: Proceedings of the American Society for Testing and Materials, vol. 33, II, pp. 3–114 (1933)
42. Orowan, E.Z.: Zur Kristallplastizität. I, Tieftemperaturplastizität und Beckersche Formel, ZeitschriftfürPhysik **89**(9–10), 605–613 (1934)
43. Taylor, G.I.: Plastic strain in metals. J. Inst. Metals **62**, 307–324 (1938)
44. Estrin, Y.: Dislocation theory based constitutive modelling: foundations and applications. J. Mater. Process. Technol. **80–81**, 33–39 (1998)
45. Kassner, M.E., Kyle, K.: Taylor hardening in five power law creep of metals and class M alloys. Nano Microstructural Des. Adv. Mater. 255–271 (2003)
46. Mughrabi, H.: The α-factor in the Taylor flow-stress law in monotonic, cyclic and quasi-stationary deformations: dependence on slip mode, dislocation arrangement and density. Curr. Opin. Solid State Mater. Sci. **26**(6), 411–420 (2016)
47. Kocks, U.F.: On the temperature and stress dependence of the dislocation velocity stress exponent. ScriptaMetallurgica **4**(1), 29–31 (1970)
48. Mecking, H., Kocks, U.F.: Physics and phenomenology of strain hardening: the FCC case. In: Progress in Material Science, pp. 171–273 (2003)
49. Rauch, E.F., Schmitt, J.H.: Dislocation substructures in mild steel deformed in simple shear. Mater. Sci. Eng. A **113**, 441–448 (1989)
50. Kubin, L., Devincre, B., Hoc, T.: Modeling dislocation storage rates and mean free paths in face-centered cubic crystals. Acta Mater. **56**(20), 6040–6049 (2008)
51. Wood, W.A.: Bulletin/Institute of Metals **3**, 5–6 (1955)
52. Plumbridge, W.J., Ryder, D.A.: Metall. Rev. **14**(136), 119–142 (1969)
53. Ashby, M.F.: Philos. Mag. **21**, 399–424 (1970)
54. Franciosi, P.: The concepts of latent hardening and strain hardening in metallic single crystals. Acta Metall. **33**(9), 1601–1612 (1985)
55. Read, W.T., Shockley, W.: Dislocation models of grain boundaries. Phys. Rev. **78**(3), 275–289 (1950)
56. Sedláček, R., Blum, W., Kratochvíl, J., Forest, S.: Subgrain formation during deformation—physical origin and consequences. Metall. Mater. Trans. A **33A**, 319–327 (2002)
57. Bragg, W.L.: The structure of a cold worked metal. Proc. Phys. Soc. **52**(1), 105–109 (1940)
58. Burger, J.M.: Geometrical considerations concerning the structural irregularities to be assumed in a crystal. Proc. Phys. Soc. **52**(1), 23–33 (1940)
59. Read and Shockley: Quantitative predictions from dislocation models of crystal grain boundaries. Phys. Rev. **75**(4), 692 (1949)
60. Sauzay, M., Brillet, H., Monnet, I., Mottot, M., Barcelo, F., Fournier, B., Pineau, A.: Cyclically induced softening due to low-angle boundary annihilation in a martensitic steel. Mater. Sci. Eng. A 400–401, 241–244 (2005)
61. Fournier, B., Sauzay, M., Pineau, A.: Micromechanical model of the high temperature cyclic behavior of 9-12%Cr martensitic steels. Int. J. Plast. **27**(11), 1803–1816 (2011)
62. Mughrabi, H.: The long-range internal stress field in the dislocation wall structure of persistent slip bands. Physica Status Solidi (A) **104**(1), 107–120 (1987)
63. Li, Y., Laird, C.: Masing behavior observed in monocrystalline copper during cyclic deformation. Mater. Sci. Eng. A **161**(1), 23–29 (1993)
64. Watanabe, E., Asao, T., Toda, M., Yoshida, M., Horibe, S.: Relationship between Masing behavior and dislocation structure of AISI 1025 under different stress ratios in cyclic deformation. Mater. Sci. Eng. A **582**, 55–62 (2013)
65. Prnadtl, L.: Proceedings of the First International Congress of Applied Mechanics, Delft, 43 (1924)
66. Levy, M.: ComptesRendus. Académie des Sciences **70**, 1323 (1870)
67. Mises, R.: Mechanik der festenKörperimplastisch-deformablenZustand. Nachrichten von der Gesellschaft der WissenschaftenzuGöttingen, Mathematisch-PhysikalischeKlasse **4**, 582–593 (1913)

68. Reuss, A.: Berücksichtigung der elastischenFormänderung in der Plastizitätstheorie, ZeitschriftfürAngewandteMathematik und Mechanik (ZAMM—Journal of Applied Mathematics and Mechanics) **10**(3), 266–274 (1930)
69. Lode, W.: Versucheüber den Einfluss der mittlerenHauptspannung auf das Fliessen der MetalleEisen. Kupfer und Nickel, ZeitschriftfürPhysik **36**(11–12), 913–939 (1926)
70. Taylor, G.I., Quinney, H.: The Plastic Distortion of Metals. Philos. Trans. R. Soc. A **230**, 681–693 (1931)
71. Hohenemser, K.: Fließversuche an Rohrenaus Stahl beikombinierter Zug- und Torsionsbeanspruchung, ZeitschriftfürAngewandteMathematik und Mechanik (ZAMM—Journal of Applied Mathematics and Mechanics) **11**(1), 15–19 (1931)
72. Morrison, J.L.M., Shepherd, W.M.: An experimental investigation of plastic stress-strain relations. Proc. Inst. Mech. Eng. **163**(1), 1–17 (1950)
73. Hill, R.: Mathematical Theory of Plasticity. Oxford University Press, ISBN: 978-0-19-850367-5 (1950)
74. Prager, W.: The theory of plasticity: a survey of recent achievements. Proc. Inst. Mech. Eng. **169**(1), 41–57 (1955)
75. Drucker, D.C.: Stress-Strain Relations in the Plastic Range—A Survey of Theory and Experiment. O.N.R. Report, NR-041-032 (1950)
76. Drucker, D.C.: A more fundamental approach to plastic stress-strain relations. In: Proceedings of the First US National Congress of Applied Mechanics, ASME, pp. 487–491 (1951)
77. Drucker, D.C.: A definition of stable inelastic material. J. Appl. Mech. **26**, 101–106 (1959)
78. Mises, R.: Mechanik der plastischenFormänderung von Kristallen, Zeitschriftfür AngewandteMathematik und Mechanik (ZAMM—Journal of Applied Mathematics and Mechanics) **8**(3), 161–185 (1928)
79. Taylor, G.I.: A connexion between the criteria of yield and the strain ratio relationship in plastic solids. Proc. R. Soc. A **191**(1027), 441–446 (1947)
80. Hill, R.: A theory of the yielding and plastic flow of anisotropic metals. Proc. R. Soc. A **193** (1033), 281–297 (1948)
81. Prager, W.: A new method of analyzing stresses and strains in work hardening plastic solids. ASME J. Appl. Mech. **23**, 493–496 (1956)
82. Mróz, Z.: On the description of anisotropic work hardening. J. Mech. Phys. Solids **15**(3), 163–175 (1967)
83. Besseling, J.F.: A theory of elastic, plastic and creep deformations of an initially isotropic material. J. Appl. Mech. **25**, 529–536 (1958)
84. Li, J.C.M.: Some elastic properties of an edge dislocation wall. Acta Metall. **8**(8), 563–574 (1960)
85. Li, J.C.M.: Petch relation and grain boundary sources. Trans. Metall. Soc. AIME **227**, 239–247 (1963)
86. Armstrong, P.J., Frederick, C.O.: A Mathematical Representation of the Multiaxial Bauschinger Effect, CEGB Report No.-RD/B/N 731 (1967)
87. Essmann, U., Mughrabi, H.: Annihilation of dislocations during tensile and cyclic deformation and limits of dislocation densities. Philos. Mag. A **40**(6), 731–756 (1979)
88. Chaboche, J.L., DangVan, K., Cordier, G.: Modelization of the Strain Memory Effect on the Cyclic Hardening of 316 Stainless Steel, SMiRT-5, Berlin (1979)
89. Chaboche, J.L.: Time-independent constitutive theories for cyclic plasticity. Int. J. Plast **2**(2), 149–188 (1986)
90. Chaboche, J.L.: On some modifications of kinematic hardening to improve the description of ratchetting effects. Int. J. Plast. **7**(7), 661–678 (1991)
91. Guionnet, C.: Modeling of ratcheting in biaxial experiments. J. Eng. Mater. Technol. **114**, 56–62 (1992)
92. Bari, S., Hassan, T.: Anatomy of coupled constitutive models for ratcheting simulation. Int. J. Plast. **16**(3–4), 381–409 (2000)

93. Dafalias, Y.F., Popov, E.P.: Plastic internal variables formalism of cyclic plasticity. J. Appl. Mech. **43**, 645–650 (1976)
94. Corona, E., Hassan, T., Kyriakides, S.: On the performance of kinematic hardening rules in predicting a class of biaxial ratcheting histories. Int. J. Plast. **12**(1), 117–145 (1996)
95. Ohno, N., Wang, J.D.: Kinematic hardening rules with critical state of dynamic recovery, part I: formulation. Int. J. Plast. **9**(3), 375–391 (1993)
96. Ohno, N., Wang, J.D.: Kinematic hardening rules with critical state of dynamic recovery, part II: application. Int. J. Plast. **9**(3), 391–403 (1993)
97. Ohno, N., Wang, J.D.: Kinematic hardening rules for simulation of ratchetting behaviour. Eur. J Mech. A Solids **13**(4), 519–531 (1994)
98. McDowell, D.L.: Stress state dependence of cyclic ratcheting behavior of two rail steels. Int. J. Plast. **11**, 397–421 (1995)
99. Jiang, Y., Sehitoglu, H.: Modeling of cyclic ratchetting plasticity—Part I: development of constitutive relations. ASME J. Appl. Mech. **63**(3), 720 (1996)
100. Voyiadjis, G.Z., Sivakumar, S.M.: A robust kinematic hardening rule for cyclic plasticity with ratcheting effects, part I: theoretical formulation. Acta Mech. **90**, 105–123 (1991)
101. Voyiadjis, G.Z., Sivakumar, S.M.: Cyclic plasticity and ratchetting. Stud. Appl. Mech. **35**, 253–295 (1994)
102. Phillips, A., Tang, J.L.: The effect of loading paths on the yield surface at elevated temperatures. Int. J. Solids Struct. **8**(4), 463–474 (1972)
103. Phillips, A., Lee, C.W.: Yield surfaces and loading surfaces. Exp. Recommendations Int. J. Solids Struct. **15**, 715–729 (1979)
104. Tseng, N.T., Lee, G.C.: Simple plasticity model of two-surface type. ASCE J. Eng. Mech. **109**(3), 795–810 (1983)
105. Voyiadjis, G.Z., Basuroychowdhury, I.N.: A plasticity model for multiaxial cyclic loading and ratcheting. Acta Mech. **126**, 19–35 (1998)
106. Basuroychowdhury, I.N., Voyiadjis, G.Z.: A multiaxial cyclic plasticity model for nonproportional loading cases. Int. J. Plast. **14**(9), 855–870 (1998)
107. Abdel-Karim, M., Ohno, N.: Kinematic hardening model suitable for ratcheting with steady-state. Int. J. Plast. **16**(3–4), 225–240 (2000)
108. Zaverl Jr., F., Lee, D.: Constitutive relations for nuclear reactor core materials. J. Nucl. Mater. **75**, 14–19 (1978)
109. Marquis, D.: Modelisation et Identification de I'Ecrouisage Anisotropic des Metaux, These Paris VI (1979)
110. Haupt, P., Kamlah, M., TsakmakisCh, : Continuous representation of hardening properties in cyclic plasticity. Int. J. Plast. **8**, 803–817 (1992)
111. Jiang, Y., Kurath, P.: Characteristics of Armstrong-Frederick type plasticity models. Int. J. Plast. **12**(3), 387–415 (1996)
112. SIMULIA ABAQUS CAE v6.8 user manual
113. Rahaman, S.M.: Finite Element Analysis and Related Numerical Schemes for Ratcheting Simulation, Ph.D. thesis submitted in North Carolina State University, USA (2006)
114. Nagtegaal, J.C.: On the implementation of inelastic constitutive equations with special reference to large deformation problems. Comput. Methods Appl. Mech. Eng. **33**, 469–484 (1982)
115. Simo, J.C., Taylor, R.L.: Consistent tangent operators for rate-independent elastoplasticity. Comput. Methods Appl. Mech. Eng. **48**, 101–118 (1985)
116. Simo, J.C., Taylor, R.L.: A return mapping algorithm for plane stress elastoplasticity. Int. J. Numer. Meth. Eng. **22**, 649–670 (1986)
117. Kobayashi, M., Ohno, N.: Implementation of cyclic plasticity models based on a general form of kinematic hardening. Int. J. Numer. Meth. Eng. **53**, 2217–2238 (2002)
118. Dunne, F., Patrinik, N.: Introduction to Computational Plasticity. Oxford University Press, ISBN: 978-0-19-856826-1 (2005)
119. Ypma, T.J.: Historical development of Newton-Raphson method. Soc. Ind. Appl. Math. **37** (4), 531–551 (1995)

Chapter 8
Development and Finite Element Implementation of a Simple Constitutive Model to Address Superelasticity and Hysteresis of Nitinol

Siddhartha Patra, Sarmita Sinha and Abhijit Chanda

8.1 Introduction

In the modern era of engineering and technology, a new class of materials with exotic properties is coming in a big way. One such new material is Nitinol, an alloy of Nickel and Titanium, which has been developed by William Buhler in 1961. It shows unique mechanical properties like superelasticity, shape memory effect [1]. It also possesses special features like high fatigue resistance, kink resistance, high MR compatibility which make it a very important biomaterial. Already Nitinol has found wide application in biomedical engineering in the form of self-expandable stents, intramedullary pins, etc. [2–4].

In last few years, number of research groups worked on the unique mechanical behavior of this material and reported in details how these properties change with temperature, strain rate, type of loading, and other parameters [5–10]. Both superelastic behavior and shape memory effect are thoroughly addressed in these studies, and various aspects of austenite to martensite transformation have been reported in details. Some of the research groups have studied the typicality in its hysteresis loop and shown how it resembles the hysteresis of natural biomaterials (Fig. 8.1) [5, 11–13] like hair, bone.

Apart from the experimental studies, few attempts have been made to model the mechanical behavior of Nitinol for computational purpose. In one of the very first attempts a simple elasto-viscoplasticity-based model is presented [15]. In later studies, a number of attempts have been made to propose *phenomenological* models for superelastic Nitinol in both small [16, 17] and large deformation [18] formulations. All such formulations in essence use the theory of thermodynamics to

S. Patra · S. Sinha (✉) · A. Chanda
Department of Mechanical Engineering, Jadavpur University,
Kolkata 700032, India
e-mail: sinha.sarmita@gmail.com

P. Sahoo and J. P. Davim (eds.), *Advances in Materials, Mechanical and Industrial Engineering*, Lecture Notes on Multidisciplinary Industrial Engineering,
https://doi.org/10.1007/978-3-319-96968-8_8

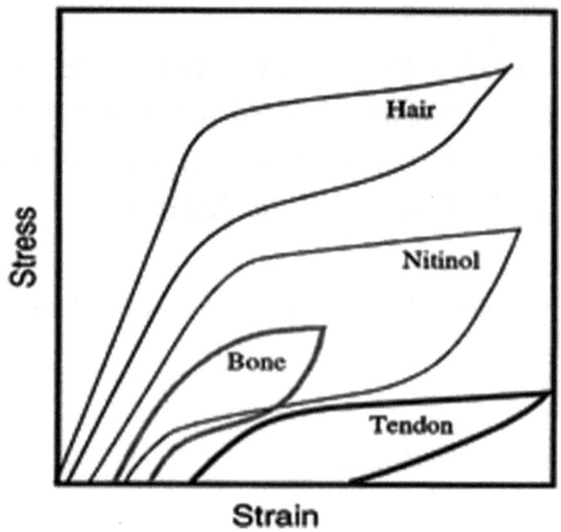

Fig. 8.1 Hysteresis loops of different materials [14]

incorporate the concept of a free energy ψ, which is a function of stress, temperature, and a quantity called transformation strain ϵ_{tr}, which is the manifestation of transformation from austenite to martensite phase. Such models try to explain macroscopic behavior from the microstructural aspects and involve parameters that cannot be directly fitted from the stress–strain diagram. Also, such models cannot be used for all the different crystalline and amorphous materials which exhibit superelasticity as shown in Fig. 8.1.

A critical overview of all these above-mentioned experimental studies and modeling suggests that a comprehensive study involving both experiments and modeling is not yet there.

The present paper reports a detailed experimental study on the superelastic behavior of Nitinol wire involving variation of strain rate and hysteresis. It also proposes a simple constitutive model based on a schematic assembly of basic mechanical components, which can replicate the superelastic behavior of Nitinol including the hysteresis loop. The detailed mathematical formulation is also presented here. The proposed model is based on the observed macroscopic behavior and unlike previous models does not try to correlate between the macroscopic behaviors and the microscopic phenomena.

8.2 Experimental Methodology

The material for the experiment was commercially available Nitinol alloy of 0.8 mm diameter. These high strength, super elastic, straight annealed, oxide surface Nitinol wires were of same composition 56%Ni 44%Ti (VRAS Traders,

India) and from a single batch. As per the information provided by the supplier, the composition conforms well with ASTM F-2063-05.

To study the hysteresis behavior, a set of controlled tests were done in tensile mode using a universal testing machine (INSTRON 4204 shown in Fig. 8.2) at a crosshead speed of 0.5 mm/min and at room temperature. Tests were done at least three times under identical conditions at room temperature (with ambient temperature around 22 °C) using standard load frame with 50 KN load cell. The load and displacement were measured by a load cell and crosshead movement, respectively.

All the tests including only loading as well as hysteresis tests were performed at a relatively low range of loading rates in order to minimize the self-heating effect caused by the transformation latent heat. Loading tests were done at three different loading rates so as to identify the effect (both quantitative and qualitative) of loading rate on deformation pattern.

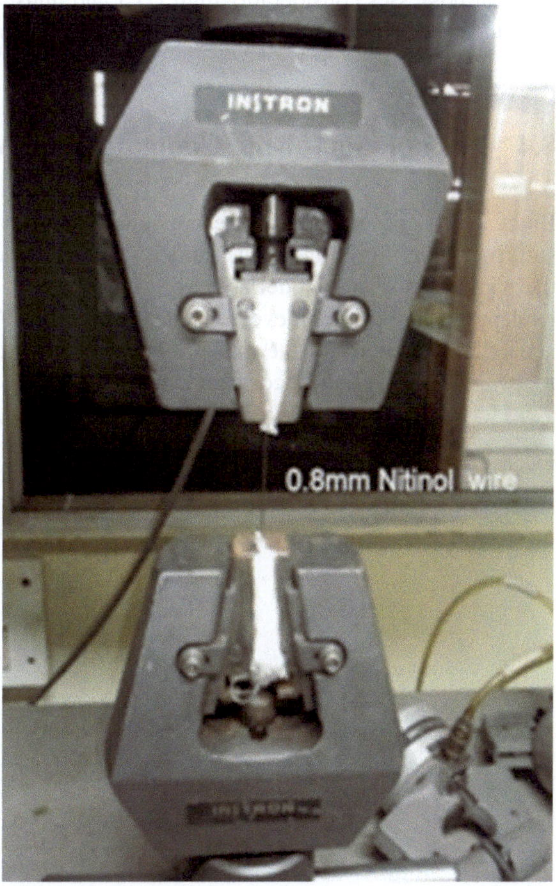

Fig. 8.2 Arrangement for tensile testing (arrow shows the padding used for holding Nitinol wire)

8.3 Results and Discussion

In the uniaxial loading–unloading test shown in Fig. 8.3, two things could be clearly noted. Firstly, it was found that even after crossing the horizontal plateau of loading curve, the specimen could be brought back to its original position, thus creating a hysteresis loop. Secondly, the loading and unloading curves did not follow identical paths; i.e., there is some sort of strain energy accumulation in the system.

Figure 8.4 shows how the load-deformation pattern varied with loading rates. All the curves revealed that almost identical superelastic nature was observed; however, there was minor variation in the horizontal part of the curve. The extent of horizontal part slightly varied showing a slight increase in the slope with an increase in loading rate. Table 8.1 presents the data available from the experimental studies which further revealed that low variation of loading rate did not have prominent effect on elastic strain, neither on ultimate tensile stress (UTS). It may be noted that the material possesses considerably high UTS and unusually high elastic strain, and in both the cases, the extent of variation was well within experimental scatter.

8.4 Formulation of the Computational Model

Based on the observation of the previous section, we have proposed a new constitutive model for Nitinol. The model we propose here is easy to implement, and the parameters of the model are easy to identify. In the following part, we would explain the step-by-step development of the model. At first, we would address the superelastic behavior of the material along with the development of the hysteresis loops. Later, we would incorporate strain hardening and rate dependency.

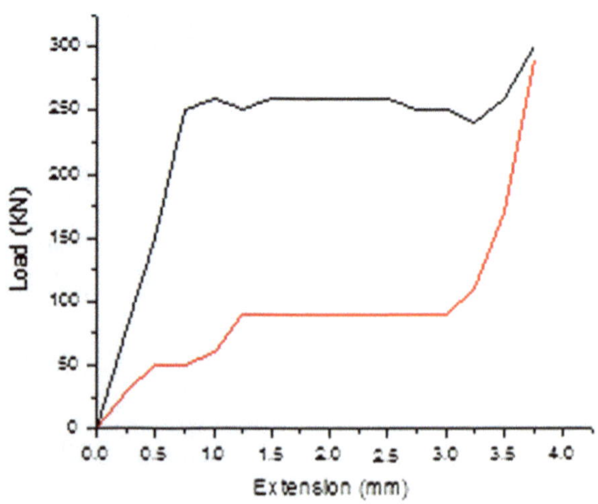

Fig. 8.3 Hysteresis at room temperature

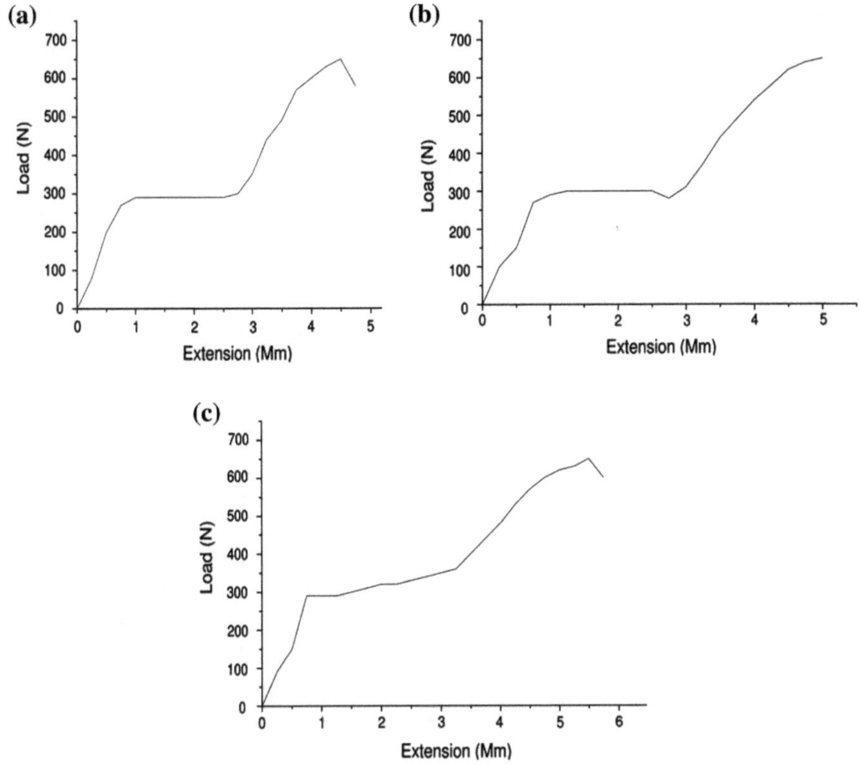

Fig. 8.4 Load-deformation diagrams of 0.8 mm diameter Nitinol wire at room temperature. **a** 0.5 mm/min speed. **b** 1 mm/min speed. **c** 5 mm/min speed

Table 8.1 Experimental data for Nitinol (0.8 mm) varying with strain rate at room temperature

S. No.	Parameters	Speed 0.5 mm/min	Speed 1 mm/min	Speed 5 mm/min
1	Gauge length (mm)	20.65	22.98	22.04
2	Peak load (KN)	0.6403	0.6510	0.6510
3	Peak extension (mm)	4.46	5.03	5.67
4	Break load (KN)	0.5503	0.5611	0.5691
5	Break extension (mm)	4.73	5.23	5.78
6	% elongation at peak	21.59	21.89	25.73
7	% elongation at break	22.90	22.76	26.23
8	Ultimate tensile strength (Mpa)	1273	1295	1295.1
9	Area under the curve (mm^2)	1725	1801.25	2240

Schematic representation of material behavior has inspired various material models including elasto-plasticity, viscoplasticity, viscoelasticity. The new material model has also been built upon such schematic representation. Table 8.2 depicts the schematic and stress strain behavior of various material models including the proposed new material model. Incorporation of rigid walls at the two sides of the moving block puts a limit on the extent of plastic deformation. In short, even with $|\sigma| > \sigma_y$ the stress continues to increase and the purely elastic deformation continues to happen if the moving block reaches one of the rigid walls and further pressed against it.

In order to model hysteresis loop of superelastic Nitinol, we additionally incorporate two parameters, namely the initial stress σ_{in} and the initial strain ϵ_{in} as shown in Fig. 8.5. The initial strain is merely the starting position of the block, and the initial stress can be thought of some sort of kinematic hardening stress (back stress) induced through a very slack spring (spring stiffness $\ll E$) which is heavily compressed.

In addition, we have kept the provision for linear isotropic and kinematic hardening.

The advantage of such a constitutive model is that the material parameters can be readily obtained from the stress–strain diagram and does not require an additional effort for fitting material parameters through an error minimization technique such as genetic algorithm. Also, the significance of each material parameter is readily understood. This helps in using the model for different materials which show similar macroscopic behavior (Fig. 8.1) irrespective of their microstructural behavior. Also since the concept of simple mechanics is used to develop such models, they automatically satisfy the laws of thermodynamics.

Here, we have presented the numerical scheme for implementing the multiaxial version of the said material model through user-defined routines available with any finite element tool (such as ABAQUS UMAT or ANSYS USERMAT). Such routines are called at each integration point at each global iteration. The main input to such routines is the total strain increment $\Delta\epsilon$ for the step, and the main outputs are the stress $\boldsymbol{\sigma}$, the plastic strain increment $\Delta\epsilon_{\text{p}}$, and the consistent tangent $\boldsymbol{C}_t : \frac{\text{d}\sigma}{\text{d}\Delta\epsilon}$. The superscript ^{0} is associated with quantities from the previous step. Von Mises yield criterion is used in this algorithm.

At first, we inspect the possibility of a completely elastic step.

$$\boldsymbol{\sigma}^* = \boldsymbol{\sigma}^0 + \boldsymbol{D}\Delta\epsilon \tag{8.1}$$

$\boldsymbol{D}$ in the above equation represents elastic stiffness tensor. Next, we check if the yield criterion is satisfied. For that, we need to define an effective stress as follows.

$$\boldsymbol{\sigma}_{\text{eff}} := \boldsymbol{\sigma}^* - \boldsymbol{\alpha}^0 - \boldsymbol{\sigma}_{\text{in}} \tag{8.2}$$

Table 8.2 Schematics of different material models

Material model	Schematic	Stress–strain diagram
Linear elastic (metals within elastic limits)	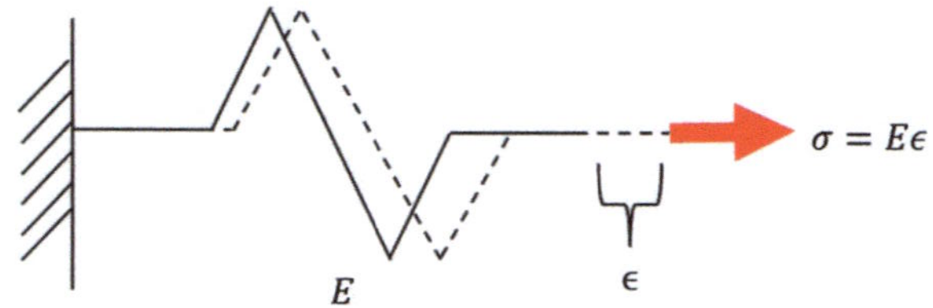	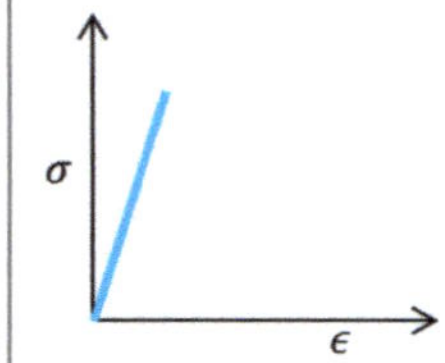
Elasto-plastic (metals)	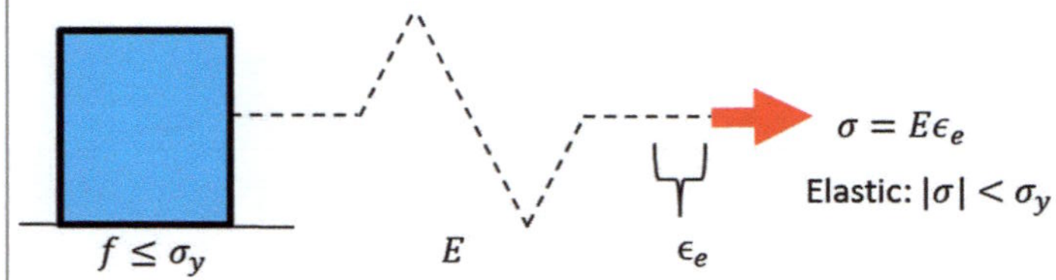	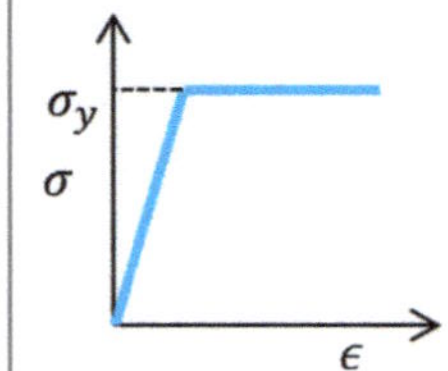
Viscoplastic (metals at high temperature)	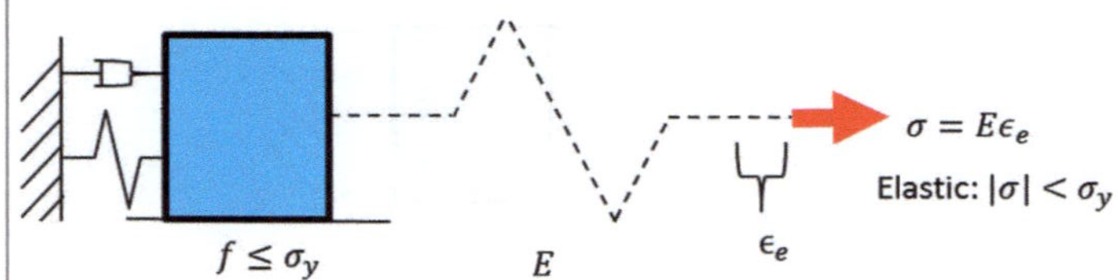	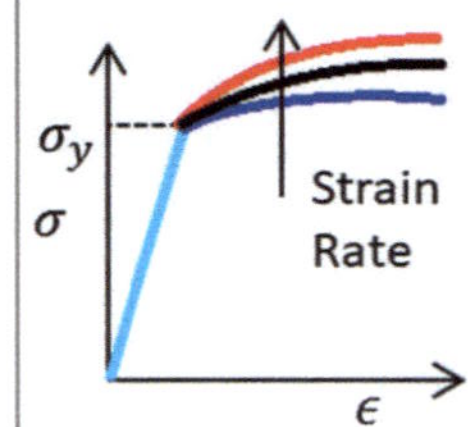

(continued)

Table 8.2 (continued)

Material model	Schematic	Stress–strain diagram
New model	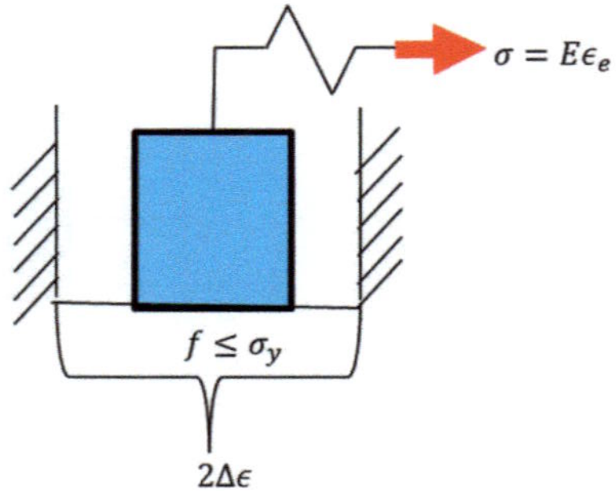	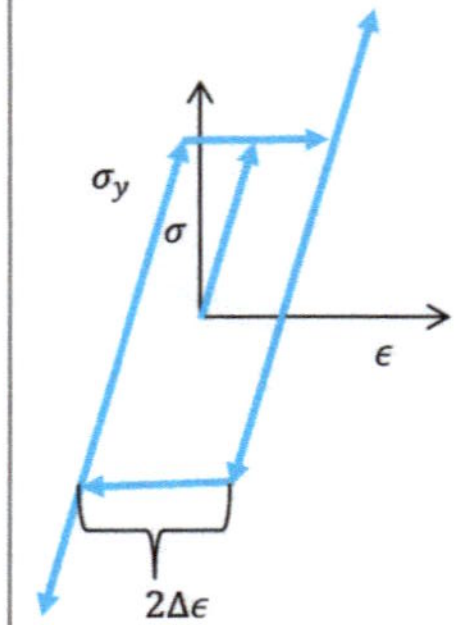

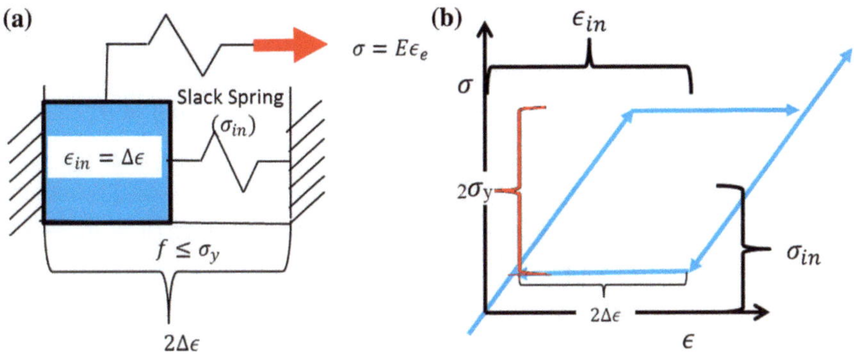

Fig. 8.5 Schematic model for superelasticity. **a** Schematic outline. **b** Stress–strain response

A new scalar variable x is introduced.

$$x : \frac{\sqrt{\frac{3}{2}\boldsymbol{\sigma}_{\text{eff}}^T \boldsymbol{P} \boldsymbol{\sigma}_{\text{eff}}}}{\sigma_y^0} \tag{8.3}$$

If $x \leq 1$, our assumption of completely elastic step holds true and thus $\boldsymbol{\sigma} = \boldsymbol{\sigma}^*$ and $\boldsymbol{C}_t = \boldsymbol{D}$ and the algorithm is exited. Else, one needs to proceed with the plastic deformation calculation. The isotropic hardening $\Delta\sigma_y = \lambda H$ and the kinematic hardening $\boldsymbol{\alpha} = \frac{2}{3} C \Delta\epsilon_p$, where λ is the equivalent plastic strain increment. Thus, we define a new parameter w_1 as follows.

$$w_1 = H + C + 3\mu \tag{8.4}$$

In Eq. (8.4), μ represents the shear modulus.

Using the radial return algorithm [19–20] depicted in Fig. 8.6, the yield surface can be represented by Eq. (8.5).

$$\begin{aligned} &\|\sigma - \boldsymbol{\alpha}\| := \left\|\sigma^* - 2\mu\Delta\epsilon_p - \boldsymbol{\alpha}^0 - \Delta\boldsymbol{\alpha}\right\| = \sigma_y^0 + \lambda H \\ &\Rightarrow (x - 1)\sigma_y^0 = w_1 \lambda \end{aligned} \tag{8.5}$$

The expression for the trial plastic strain is obtained as follows.

$$\Delta\epsilon_p^* := \frac{3}{2w_1} \frac{x - 1}{x} \boldsymbol{P} \boldsymbol{\sigma}_{\text{eff}} \tag{8.6}$$

Now, we would check if the strain limit criterion is satisfied. Thus, a new scalar variable y is introduced.

$$y := \frac{\sqrt{\frac{2}{3}\left(\Delta\epsilon_{p,\text{eff}}\right)^T \overline{\boldsymbol{P}} \Delta\epsilon_{p,\text{eff}}}}{\Delta\epsilon} \tag{8.7}$$

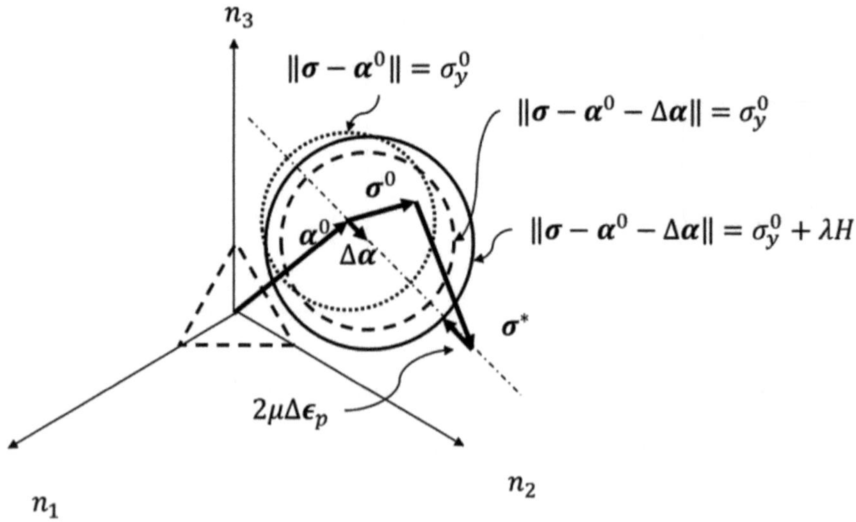

Fig. 8.6 Radial return on octahedral plane

The effective plastic strain increment, $\Delta\epsilon_{p,\text{eff}}$, in Eq. (8.6) is represented by Eq. (8.7).

$$\Delta\epsilon_{p,\text{eff}} := \epsilon_p^0 + \Delta\epsilon_p - \epsilon_{in} \tag{8.8}$$

If $y \leq 1$, the plastic strain increment $\Delta\epsilon_p = \Delta\epsilon_p^*$ and the consistent tangent is represented by Eq. (8.8).

$$\boldsymbol{C}_t = (\boldsymbol{I} - \boldsymbol{DA})\boldsymbol{D} \tag{8.9}$$

Else, the plastic strain increment

$$\Delta\epsilon_p = \frac{\Delta\epsilon_{p,\text{eff}}}{y} + \Delta\epsilon_p^* - \epsilon_p^0 \tag{8.10}$$

and the consistent tangent

$$\boldsymbol{C}_t = (\boldsymbol{I} - \boldsymbol{DBA})\boldsymbol{D} \tag{8.11}$$

Expressions for the matrices $\boldsymbol{A}, \boldsymbol{B}$ in Eqs. (8.9)–(8.11)are given in the Appendix. Stress for both the cases is given by Eq. (8.11).

$$\boldsymbol{\sigma} = \boldsymbol{\sigma}^* - \boldsymbol{D}\Delta\epsilon_p \tag{8.11}$$

The yield stress value must also be corrected for the next iteration. Thus,

$$\sigma_y^0 = \sigma_y^0 + \lambda H \tag{8.12}$$

Although we have presented a rate-independent plasticity-based model here, with a substantial number of experimental data suggesting rate dependency, viscoplasticity can be incorporated in the current model with minor modifications.

We implemented the above formulation using ABAQUS UMAT and conducted simulations on tensile test. The parameters are listed in Table 8.3. FE simulation results visually resemble the experimental data as could be verified from Fig. 8.7.

The rate dependence study was conducted with three different loading rates, i.e., 0.5, 1, and 2 mm/min. When converted to strain rates, they would be $\dot{\epsilon} = 0.000404/\mathrm{s}, 0.000725/\mathrm{s}$ and $0.003781/\mathrm{s}$, respectively. Figure 8.8 shows the

Table 8.3 Material parameters for superelastic behavior of Nitinol

E (MPa)	15,009
σ_y (MPa)	176.2
σ_{in}(MPa)	$[-118.9 \quad -118.9 \quad 237.9 \quad 0 \quad 0 \quad 0]^T$
ϵ_{in}	$[-0.029 \quad -0.029 \quad 0.058 \quad 0 \quad 0 \quad 0]^T$
$\Delta\epsilon$	0.058
C (MPa)	293.3
H (MPa)	0.0

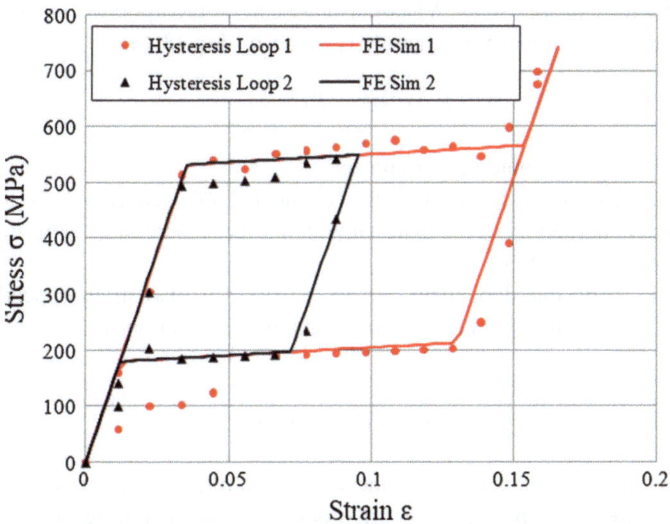

Fig. 8.7 Hysteresis in stress–strain plots, experiment versus simulation

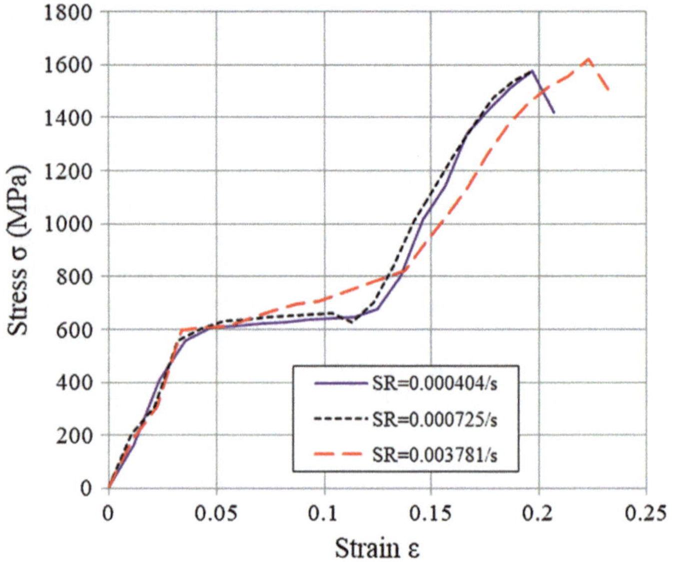

Fig. 8.8 True stress–true strain plots for different strain rates

true stress–true strain plots for three different strain rates. For each plot, the stress–strain curve can be divided into three linear segments, marked in Fig. 8.8 with numbers (1, 2, 3), with the last segment extending up to ultimate stress point.

Clearly, a rate dependency with positive strain rate sensitivity is observed. The difference of stresses between the highest and lowest strain rate (ratio $\approx$ 10) is found to be about 150 MPa. Thus, the material behavior in the segment 2 can be associated with viscoplasticity. Viscoplasticity can be incorporated in a constitutive model by using a simple power law [21] given by Eq. (8.13).

$$\sigma_v = K\dot{\epsilon}_p^n \tag{8.13}$$

Here, K and n are material parameters.

However, a closer inspection of the segment 2 of each curve revealed strain rate-dependent hardening of the material with hardness increasing with strain rate as could be seen in Fig. 8.9.

Thus, the strain rate-dependent part of the stress is also strain dependent. This behavior can be incorporated in the constitutive relation [22, 23] by using Eq. (8.15).

$$\sigma_d = \sigma_s + L(\epsilon_p)\dot{\epsilon}_p \tag{8.14}$$

where σ_d is the dynamic stress, i.e., the stress with a finite strain rate, and σ_s is the quasi-static stress. L represents the strain-dependent hardening term.

We defined the viscoplastic stress as

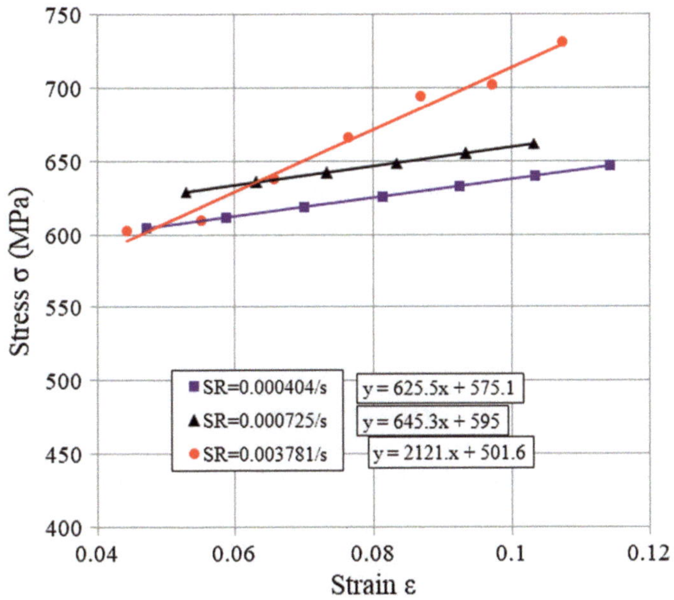

Fig. 8.9 Dependence of strain hardening on strain rate

$$\sigma_v = \overline{H}\dot{\epsilon}_p\epsilon_p \tag{8.15}$$

where $\overline{H}$ is a material parameter. While fitting $\overline{H}$ parameter, corrections are to be done to account for the conversion from total engineering strain rate to true plastic strain rate.

The incremental form of the above equation is given by Eq. (8.16).

$$\Delta\sigma_v = \overline{H}\left(\frac{\lambda}{\Delta t}\right)\lambda \tag{8.16}$$

where $\frac{\lambda}{\Delta t}$ represents the average strain rate during the step.

The viscoplastic stress, σ_v, is then incorporated in the model with a minor modification of Eq. (8.5) as given by Eq. (8.17).

$$(x-1)\sigma_y^0 = w_1\lambda + \Delta\sigma_v\left(\frac{\lambda}{\Delta t}, \lambda\right) \tag{8.17}$$

This is a nonlinear equation in λ, thus requiring a numerical technique for solution such as Pegasus method [24]. The expression for tentative plastic strain increment given by Eq. (8.6) should now be modified to more general expression given by Eq. (8.18).

$$\Delta\epsilon_p^* = \frac{3\lambda}{2x\sigma_y^0}\boldsymbol{P}\boldsymbol{\sigma}_{\text{eff}} \tag{8.18}$$

Then, from the expressions of the consistent tangent matrices given by Eq. (8.9) and Eq. (8.11), matrix $\boldsymbol{A}$ should now be replaced with matrix $\overline{\boldsymbol{A}}$, expression for which is given in the Appendix. Thus, the modified expressions for consistent tangents for $y \leq 1$ is

$$\boldsymbol{C}_t = \left(\boldsymbol{I} - \boldsymbol{D}\overline{\boldsymbol{A}}\right)\boldsymbol{D} \tag{8.19}$$

and for $y > 1$ is

$$\boldsymbol{C}_t = \left(\boldsymbol{I} - \boldsymbol{D}\boldsymbol{B}\overline{\boldsymbol{A}}\right)\boldsymbol{D} \tag{8.20}$$

Finally, the yield stress must be updated strain rate-dependent hardening. Thus,

$$\sigma_y^0 = \sigma_y^0 + \Delta\sigma_v + H\lambda \tag{8.21}$$

Table 8.4 lists the values of the material parameters for the rate-dependent model.

The comparison of the results from the finite element simulations with the experimental data is shown in Fig. 8.10. Figure 8.10a shows the comparison between experimental data and the simulation results with power law fit (Eq. 8.13). Simulation results show decent match with experimental data for two strain rates $\dot{\epsilon} = 0.000404/\text{s}$ and $\dot{\epsilon} = 0.000725/\text{s}$. However, for the highest strain rate $\dot{\epsilon} = 0.003781/\text{s}$ the viscoplastic stress matches the average value but the natures of the curves are different.

Figure 8.10b shows the comparison between experimental data and the simulation results using strain rate-dependent hardening (Eq. 8.15) and shows significant improvement over the previous fit. The simulation results show good agreement for each strain rate.

Table 8.4 Material parameters after incorporation of viscoplasticity

E (MPa)	16,590
σ_y (MPa)	193
$\boldsymbol{\sigma_{in}}$(MPa)	$[-135 \quad -135 \quad 270 \quad 0 \quad 0 \quad 0]^T$
$\boldsymbol{\epsilon_{in}}$	$[-0.022 \quad -0.022 \quad 0.044 \quad 0 \quad 0 \quad 0]^T$
$\Delta\epsilon$	0.0436
C (MPa)	381.8
H (MPa)	0.0
$\overline{H}$ (MPa-s)	630,158

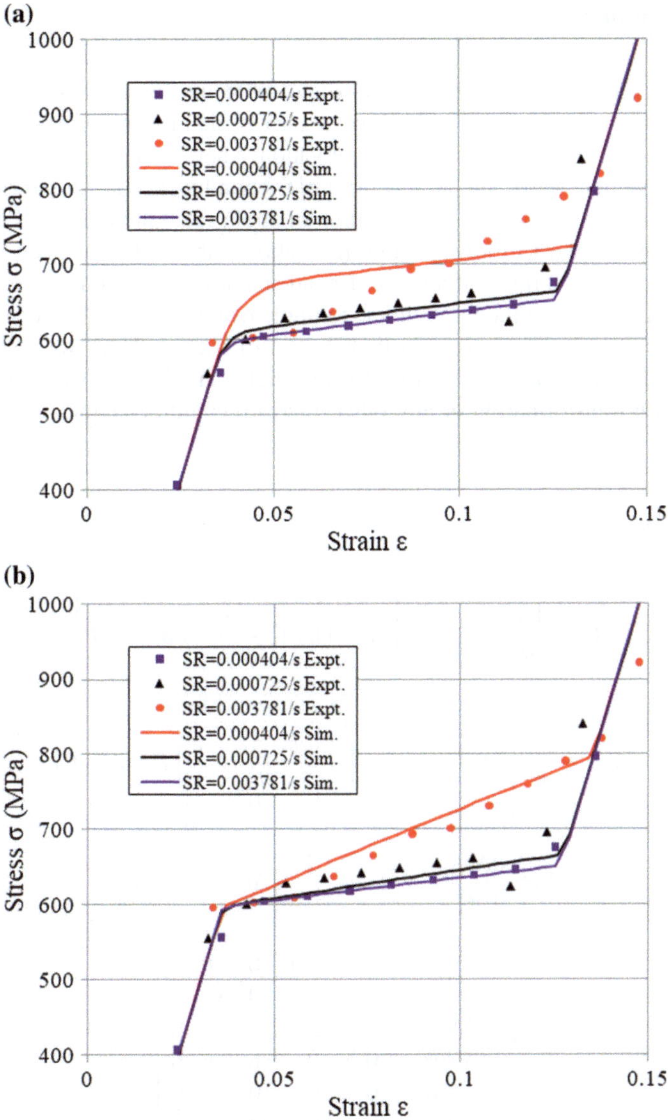

Fig. 8.10 Strain rate dependence: experiment versus simulation. **a** Using power law (Eq. 8.13). **b** Using strain rate-dependent hardening (Eq. 8.15)

8.5 Conclusion

The experimental results suggest that Nitinol wire undergoes superelastic deformation under uniaxial load and forms a clearly defined hysteresis loop while unloaded. Based on the experimental observation, a new material model is proposed using a convenient schematic arrangement of friction block, spring, and rigid walls. The variation of stress due to the variation of strain rate (about 150 MPa of stress variation for a highest to lowest strain rate ratio of about 10) has been accurately modeled by incorporating a strain rate-dependent hardening term in the constitutive relation. This proposed model is simple to implement yet efficient at replicating the superelastic as well the as viscoplastic behavior of Nitinol. Thus, we ascertain that the material model can also be used for materials that show superelasticity either due to phase transformation or any other microstructural behavior.

Appendix

$$\boldsymbol{A} := \frac{\mathrm{d}\Delta\epsilon_p^*}{\mathrm{d}\boldsymbol{\sigma}_{\text{eff}}} = \frac{3}{2w_1 x}\left[(x-1)\boldsymbol{P} + \frac{3}{2\boldsymbol{x}^2\boldsymbol{\sigma}_y^{0\,2}}\boldsymbol{P}\boldsymbol{\sigma}_{\text{eff}}\boldsymbol{\sigma}_{\text{eff}}^T\boldsymbol{P}\right];$$

$$\overline{\boldsymbol{A}} = \frac{3}{2x\sigma_y^0}\left[\lambda\boldsymbol{P} + \frac{1}{x\left(\sigma_y^0\right)^2}\left(\frac{\sigma_y}{w_1 + \frac{\mathrm{d}\Delta\sigma_y}{\mathrm{d}\lambda}} - \frac{\lambda}{x}\right)\boldsymbol{P}\boldsymbol{\sigma}_{\text{eff}}\boldsymbol{\sigma}_{\text{eff}}^T\boldsymbol{P}\right];$$

$$\boldsymbol{B} := \frac{\mathrm{d}\Delta\epsilon_p}{\mathrm{d}\Delta\epsilon_p^*} = \frac{1}{y}\left[I - \frac{2}{3y^2\Delta\epsilon^2}\Delta\epsilon_{p,\text{eff}}\Delta\epsilon_{p,\text{eff}}^T\overline{\boldsymbol{P}}\right];$$

$$\boldsymbol{P} = \frac{1}{3}\begin{bmatrix} 2 & -1 & -1 & 0 & 0 & 0 \\ -1 & 2 & -1 & 0 & 0 & 0 \\ -1 & -1 & 2 & 0 & 0 & 0 \\ 0 & 0 & 0 & 6 & 0 & 0 \\ 0 & 0 & 0 & 0 & 6 & 0 \\ 0 & 0 & 0 & 0 & 0 & 6 \end{bmatrix};$$

$$\overline{\boldsymbol{P}} = \frac{1}{3}\begin{bmatrix} 2 & -1 & -1 & 0 & 0 & 0 \\ -1 & 2 & -1 & 0 & 0 & 0 \\ -1 & -1 & 2 & 0 & 0 & 0 \\ 0 & 0 & 0 & 1.5 & 0 & 0 \\ 0 & 0 & 0 & 0 & 1.5 & 0 \\ 0 & 0 & 0 & 0 & 0 & 1.5 \end{bmatrix}.$$

References

1. Bellouard, Y.: Shape memory alloys for microsystems: a review from a material research perspective. Mater. Sci. Eng. A **481–482**, 582–589 (2008)
2. Mahtabi, M., Shamsaei, N., Mitchell, M.: Fatigue of Nitinol: the state-of-the-art and ongoing challenges. J. Mech. Behav. Biomed. Mater. **50**, 228–254 (2015)
3. Plotino, G., Grande, N.M., Cordaro, M., Testarelli, L., Gambarini, G.: A review of cyclic fatigue testing of nickel-titanium rotary instruments. J. Endod. **35**, 1469–1476 (2009)
4. Jani, J.M., Leary, M., Subic, A., Gibson, M.A.: A review of shape memory alloy research, applications and opportunities. Mater. Des. **56**, 1078–1113 (2014)
5. Nayan, N., Buravalla, V., Ramamurty, U.: Effect of mechanical cycling on the stress–strain response of a martensitic Nitinol shape memory alloy. Mater. Sci. Eng. A **525**, 60–67 (2009)
6. Adharapurapu, R.R., Jiang, F., Bingert, F.J., Vecchio, S.K.: Influence of cold work and texture on the high-strain-rate response of Nitinol. Mater. Sci. Eng. A **527**, 5255–5267 (2010)
7. Sadiq, H., Wong, B.M., Al-Mahaidi, R., Zhao, L.X.: The effects of heat treatment on the recovery stresses of shape memory alloys. Smart Mater. Struct. **19**, 1–7 (2010)
8. Schlun, M., Zipse, A., Dreher, G., Rebelo, N.: Effects of cyclic loading on the uniaxial behavior of Nitinol. J. Mater. Eng. Perform. **20**, 684–687 (2011)
9. Halani, R.P., Kaya, I., Shin, C.Y., Karaca, E.H.: Phase transformation characteristics and mechanical characterization of nitinol synthesized by laser direct deposition. Mater. Sci. Eng. A **559**, 836–843 (2013)
10. Pelton, A., Dicello, J., Miyazaki, S.: Optimisation of processing and properties of medical grade Nitinol wire. Minim. Invasive Ther. Allied Technol. **9**, 107–118 (2000)
11. Mckelvey, A., Ritchie, R.: Fatigue-crack growth behavior in the superelastic and shape-memory alloy Nitinol. Metall. Mater. Trans. A **32a**, 731–743 (2001)
12. McNaneyM, J., Imbeni, V., Jung, Y., Papadopoulos, P., Ritchie, R.O.: An experimental study of the superelastic effect in a shape-memory Nitinol alloy under biaxial loading. Mech. Mater. **35**, 969–986 (2003)
13. Shishkovsky, I.: Hysteresis modeling of the porous Nitinol delivery system, designed and fabricated by SLS method. Phys. Procedia **39**, 893–902 (2012)
14. Duerig, T., Pelton, A., Stockel, D.: An overview of Nitinol medical applications. Mater. Sci. Eng., A **273–275**, 149–160 (1999)
15. Whitcher, F.D.: Simulation of in vivo loading conditions of Nitinol vascular stent structures. Comput. Struct. **64**(5–6), 1005–1011 (1997)
16. Souza, A.C., Mamiya, E.N., Zouain, N.: Three-dimensional model for solids undergoing stress-induced phase transformations. Eur. J. Mech. A/Solids **17**, 789–806 (1998)
17. Auricchio, F., Coda, A., Reali, A., Urbano, M.: SMA numerical modeling versus experimental results: parameter identification and model prediction capabilities. J. Mater. Eng. Perform. **18**, 649–654 (2009)
18. Jung, Y., Papadopoulos, P., Ritchie, R.O.: Constitutive modelling and numerical simulation of multivariant phase transformation in superelastic shape-memory alloys. Int. J. Numer. Meth. Eng. **60**, 429–460 (2004)
19. Crisfield, M.A.: Nonlinear Finite Element Analysis for Solids and Structures, vol. 1, pp. 166–181. Wiley, Hoboken (2000)
20. Crisfield, M.A.: Nonlinear Finite Element Analysis for Solids and Structures, vol. 2, pp. 158–167. Wiley, Hoboken (2000)
21. Yaguchi, M., Takahashi, Y.: A viscoplastic constitutive model incorporating dynamic strain aging effect during cyclic deformation conditions. Int. J. Plast. **16**, 241–262 (2000)
22. Naghdi, P.M.: Constitutive restrictions for idealized elastic-viscoplastic materials. J. Appl. Mech. **51**, 93–101 (1984)
23. Kim, K.T., Cho, Y.H.: A temperature and strain rate dependent strain hardening law. Int. J. Press. Vessels Pip. **49**, 327–337 (1992)
24. Dowell, M., Jarratt, P.: The "Pegasus" method for computing the root of an equation. BIT **12**, 503–508 (1972)

References

Chapter 9
Abrasive Jet Machining: Drilling of Porcelain Tiles and Soda Lime Glass

Tina Ghara, G. Desta, Santanu Das and Barun Haldar

9.1 Introduction

The manufacturing industries often face problems while machining of hard-brittle materials. Sometimes, problem stands with dimensional accuracy and surface quality of the engineering components, and conventional machining processes often fail to fulfill these requirements. In such cases, special machining processes like some non-traditional machining processes may be undertaken to produce products as per specifications. In abrasive jet machining process, the micro-abrasive particles are being mixed into a high-velocity gas flow line to form an abrasive jet for removing materials by means of erosion. The abrasive particles beget a velocity near about 300 m/s [1] to target and hit the work surface. The mode of material removal for brittle material is micro-fracturing [2], and for ductile material, it is micro-cutting [3–5]. The AJM is advantageous due to

- it's flexible machine components.
- less heat effects on the workpiece as the growing micro-compressive residual stress produces micro-crack and removes material [6].
- no burr formation behind machined surfaces.

T. Ghara · S. Das
Department of Mechanical Engineering, Kalyani Government Engineering College, Kalyani 741235, India

G. Desta
School of Mechanical Engineering, JIT, Jimma University, 378, Jimma, Ethiopia

B. Haldar (✉)
Department of Mechanical Engineering, Global Institute of Management & Technology, Krishnagar, West Bengal 741102, India
e-mail: dr.barun.haldar@gmail.com

P. Sahoo and J. P. Davim (eds.), *Advances in Materials, Mechanical and Industrial Engineering*, Lecture Notes on Multidisciplinary Industrial Engineering,
https://doi.org/10.1007/978-3-319-96968-8_9

The abrasive jet system is utilized for various purposes like cutting [7], drilling [8], grooving [9], engraving [10] of different materials, viz. glass, ceramics, ferrous materials, and polymers [11] also. The abrasive jet has unique applications on the removal of bar [12] and rounding of sharp machined edges. The erosive property [13] testing of any material can be performed against abrasive jet. It is a promising technology for processing damage-free micro-part features like micro-slots, channels, micro-holes on ceramics and semiconductors for the microelectromechanical systems (MEMS) manufacturing, micro-fluidic devices, and optoelectronic components [14, 15] applications.

In AJM, the brittle and ductile both the materials are removed by means of erosion. The erosion rate of brittle material is high at the abrasive jet impact angle of 90°, whereas for ductile materials, erosion rate is maximum at the abrasive jet impingement angle of 20–30° [16]. However, the rate of material removal by abrasive jet increases with increasing nozzle diameter and jet pressure up to a particular level. Furthermore, the erosion rates of different materials tend to increase at elevated temperature. Balasubramaniam et al. [17] reported the influential effects of some machining parameters on the effectiveness of AJM. The MRR and penetration rates were found to increase with abrasive particle size. Also with increase in SOD, the entry side diameter of holes, MRR, and edge radius increase. The MRR and edge radius also increase with increasing velocity of the jet.

The different types of abrasive grit particles are utilized for abrasive jet material processing such as silicon dioxide, glass beads, sodium bicarbonate, dolomite, silicon carbides, aluminum oxides, boron carbides [18], etc. The abrasive micro-machining is one of the powerful methods for hard and brittle materials dimpling [19]. Fine abrasives are appropriate for micro-machining [20–22]. In conventional AJM, as the machining progresses, the taperness in machined hole can be controlled by changing standoff distance.

Some researchers focused on investigating the characteristics of abrasive jets and observed the effects of several machining parameters on the process efficiency. Some other investigations dealt with carrier gas, shape and size of nozzle, nozzle wear, abrasive jet velocity, jet pressure, stand off distance (SOD), etc. The performance of the process was explained in connection with MRR and surface finish of work pieces. A few works were performed on modeling the process. The carrier media, by which abrasive particles are mobilized, plays a significant role in improving the efficiency of the process. Few researchers [23, 24] employed hot air for carrying abrasives to explore the effect of that hot media on the process performance of the abrasive jet. Surprisingly, the material removal rate was found to increase due to that hot abrasive jet employed. Moreover, the surface roughness of the machined parts was observed to decrease.

The glass sheet drilling by abrasive jet was performed by El-Domiaty et al. [25]. Material removal rates obtained from this experiment were close to each other within an error of 20%. Kandpal et al. [26] carried out experiments to machine ceramic and glass using Al_2O_3 and SiC abrasives in AJM to analyze the effects of different process parameters. In another work, Rao et al. [27] investigated the effect of process parameters on the rate of material removal of abrasive jet machining of epoxy glass fiber composite by silicon carbide abrasive with size 60 μm using tungsten carbide nozzles having diameters 3, 4, and 5 mm. They used three pressure levels of 4, 6, 8 kg/cm^2 and three different nozzle tip distances of 6, 8, and 10 mm. The prime MRR value was investigated through RSM technique at 8 kg/cm^2 pressure and a 4-mm-diameter nozzle having 10 mm SOD.

The AJM performance of sea bed sand on glass was analyzed by Pawar et al. [28] using silicon carbide and mild steel nozzles. They noticed that with increasing work feed rate, the cut width increases. It is also found that taper cut becomes more at higher SOD and work feed rate. It was observed that silicon carbide nozzle has more life than mild steel nozzle. They also revealed that material removal takes place due to brittle fracturing with silicon carbide nozzle. In case of mild steel, nozzle material was eroded by means of ploughing action.

Verma and Lal [29] performed experiments in abrasive jet machining on a designed setup using aluminum oxide abrasives of grain sizes 25, 30, 38, and 48 μm, taking dry air as the carrier gas with pressure 9.81×10^4 to 29.43×10^4 N/m^2 using WC nozzle of diameter 0.7 mm. The effects of different parameters affecting material removal rate were observed. They observed that the maximum values of MRR and penetration rate are obtained at different SODs. They also suggested that deburring and finishing can be done effectively with larger SODs, whereas smaller SOD is good for micro-drilling.

Srikanth et al. [30] performed drilling by abrasive jet on glass sheets, by varying different parameters to assess the machinability of the material. The effects of machining parameters on MRR were explored by Taguchi technique. High MRR was observed at 8 kg/cm^2 pressure with 4 mm diameter nozzle at 10 mm SOD. Whereas air pressure of 6 kg/cm^2, 3 mm diameter nozzle with 9 mm SOD show in less kerf.

Some studies and experiments were conducted by Ray et al. [31] on abrasive jet drilling to study the effects of different machining parameters on MRR. They used SiC abrasive of grain size 60–120 μm and stainless steel nozzles having 1.83 and 1.63 mm diameters. It was observed that material removal factor (MRF) becomes maximum at a range of pressure 2–3 kgf/cm^2. They also found out that higher SOD is more preferable for higher MRR and higher pressure and lower SOD is preferable for precision work.

In another work, a model of AJM was designed and developed by Shriyan et al. [32] for drilling study on glass using silicon carbide abrasive having 50 μm and 90 μm grit sizes to investigate the effects of operating parameters on AJM process performance. They concluded from their work that abrasive size and jet pressure have a significant effect on material removal, whereas angle of taper depends only on the jet pressure.

Ghobeity et al. [33] worked on the improvement of process repeatability in AJM. A mixing device was introduced within the pressure reservoir to ensure that the abrasive powder remains loose and may flow freely.

The effects of AJM parameters on material removal rate and the diameter of drilled holes in glass plates using SiC were studied by Mahajan [34]. The upper and lower side diameters of drilled holes were observed to increase with SOD. Besides, the MRR was also found to increase with pressure. The machining of fiber-reinforced plastics (FRPs) was observed by abrasive jet. The parametric optimization of machining FRP was performed by Srikanth et al. [35] and Reddy et al. [36].

Sharma and Deol [37] also examined the effects of AJM parameters on MRR, overcut, taper cut on machining of glass fiber-reinforced plastic using tungsten carbide nozzle with diameters 1.2, 1.5, 2.3 mm by silicon carbide abrasive with grit size 70 μm. The MRR was found to increase with pressure and nozzle diameter. However, with increase in SOD, MRR decreases. It was also seen that taper cut and overcut reduce with increase in nozzle diameter and jet pressure and decrease with SOD.

Roughness of the machined surface is an area of concern in AJM. Gavaskar et al. [38] studied the process parameters which affects the surface roughness in AJM process. They identified the major process parameters affecting the surface roughness are abrasive flow rate, time and cutting speed, feed rate, abrasive particle size, etc.

Abrasive jet system was scarcely designed and fabricated indigenously for performing different activities. The objectives of the observations were to develop an abrasive jet machine setup for experimental investigations of various aspects of abrasive jet applications. Thus, the specific objectives of the investigation were:

- Development of a cost-effective experimental abrasive jet machine setup with abrasive flow rate controller.
- Experimental investigations on abrasive jet machining performance on cutting of various materials like glass, tiles, etc.
- To evaluate the influence of machining parameters like abrasive carrier gas pressure, flow rate, SOD on machined holes.

9.2 Experimental Investigations

9.2.1 Setup Development

An abrasive jet machine setup is designed indigenously to perform machining operations. The setup (Fig. 9.1) is manufactured and supplied by 'Asian Drilling Industries, Kolkata' as per the designed specification placed. The nozzle and nozzle-workpiece system are shown in Figs. 9.2 and 9.3 respectively.

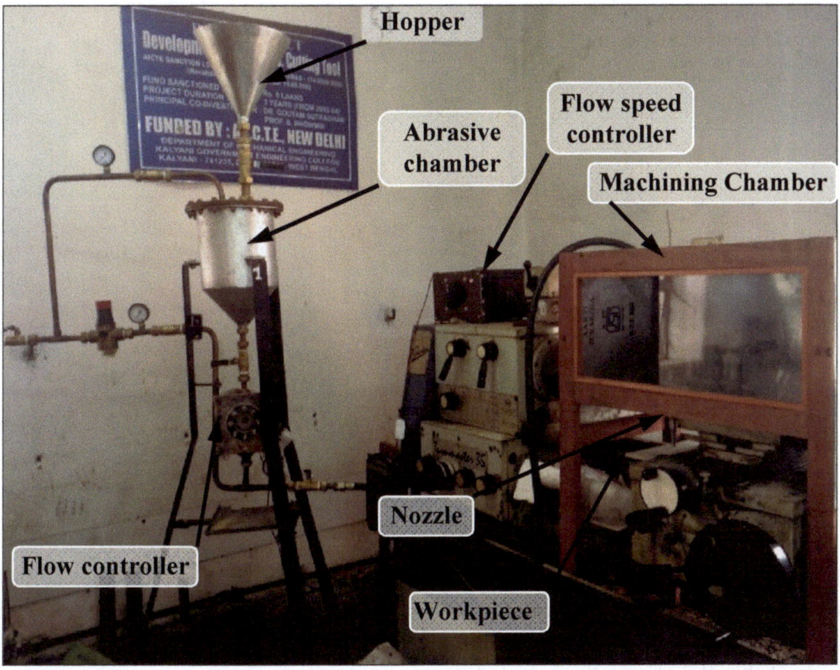

Fig. 9.1 Photograph of the developed abrasive jet processing setup at KGEC, Kalyani, India

Fig. 9.2 Photograph of the nozzle

Fig. 9.3 In set photograph of the nozzle-workpiece system

9.2.2 *Experimental Conditions and Process Parameters*

The whole experiment was performed in the newly developed setup under certain experimental conditions which are shown in Table 9.1. Three levels of two parameters (i.e., pressure and standoff distance) were chosen to conduct experiments. Table 9.2 shows the selected process parameters.

9.2.3 *Procedure of Experiment*

For this experimental work, riverbed sand of 425 μm grain size and 300 μm grain size was collected using a sieve shaker. It was observed that sand of 425 μm size

Table 9.1 Experimental conditions used in the investigation

Carrier medium	Air
Abrasive	Sand (300 μm), SiC (250 μm)
Flow rate of abrasive	700 g/min
Nozzle diameter	2 mm
Nozzle material	Stainless steel
Workpiece material	1. Porcelain tiles 2. Soda lime glass

Table 9.2 Process parameters used for the experiment

Parameter	Level 1	Level 2	Level 3
Pressure (kg/cm^2)	6	7	8
Standoff distance (mm)	2	4	6

was not suitable for 2-mm-diameter stainless steel nozzle as it was getting blocked the nozzle. So, sand of size 300 μm and SiC of size 250 μm were used for the experimental investigation.

For starting the experiments, first the compressor was set ON, and the particular pressure was set by pressure regulator. The nozzle was set at a particular SOD above the work piece. By using a particular type of abrasive, drilling operation was performed for 10 s that is kept constant for drilling all the holes. For different combinations of process parameters and all types of abrasives, the same procedure was followed. Then, the types of the holes generated were observed. The diameter, taper angle, and out of roundness of the holes were also measured.

9.3 Experimental Investigations

9.3.1 Results and Discussion on Experiment Set-I

In this set of experiment, drilling on 5.8 mm thick tiles was performed by river bed sand of 300 μm grain size. The results obtained from this set of the experiment are tabulated in Table 9.3. In experiment set-I, blind holes were generated (Fig. 9.4) for all combinations of process parameters. It may be due to high degree of friability of the sand used. The microscopic views of the holes are shown in Fig. 9.5, which show the edges and bottom surface of the holes clearly and may depict the quality of the holes.

Table 9.3 Results of drilling of tiles using sand as the abrasive (expt. set-I)

Sl. no.	Pressure (kg/cm^2)	SOD (mm)	Time (s)	Diameter of hole (mm)		Taper angle (°)	Out of roundness (mm)	Type of hole observed
				Upper surface	Lower surface			
1	6	2	10	4.136	–	–	0.552	blind hole of oval cross section
2	7	2	10	4.152	–	–	0.154	blind hole
3	8	2	10	4.086	–	–	0.179	blind hole
4	6	4	10	5.124	–	–	0.179	blind hole of larger diameter
5	7	4	10	5.076	–	–	0.178	blind hole with two step
6	8	4	10	5.434	–	–	0.178	blind hole
7	6	6	10	6.374	–	–	0.235	blind hole of larger diameter
8	7	6	10	5.92	–	–	0.284	blind hole
9	8	6	10	6.908	–	–	0.186	blind hole

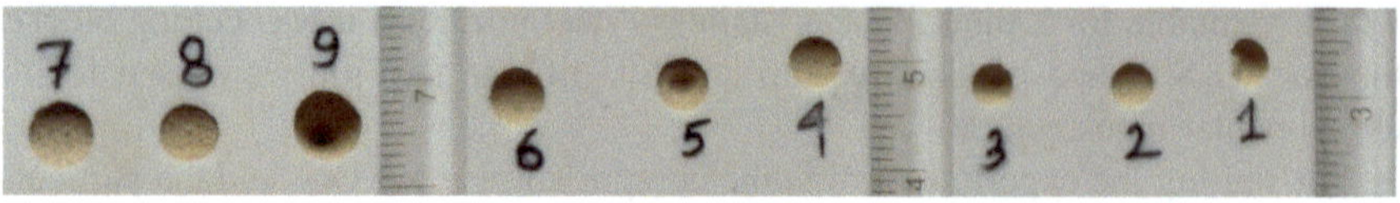

Fig. 9.4 Photograph of the machined holes (machined using sand) on tiles

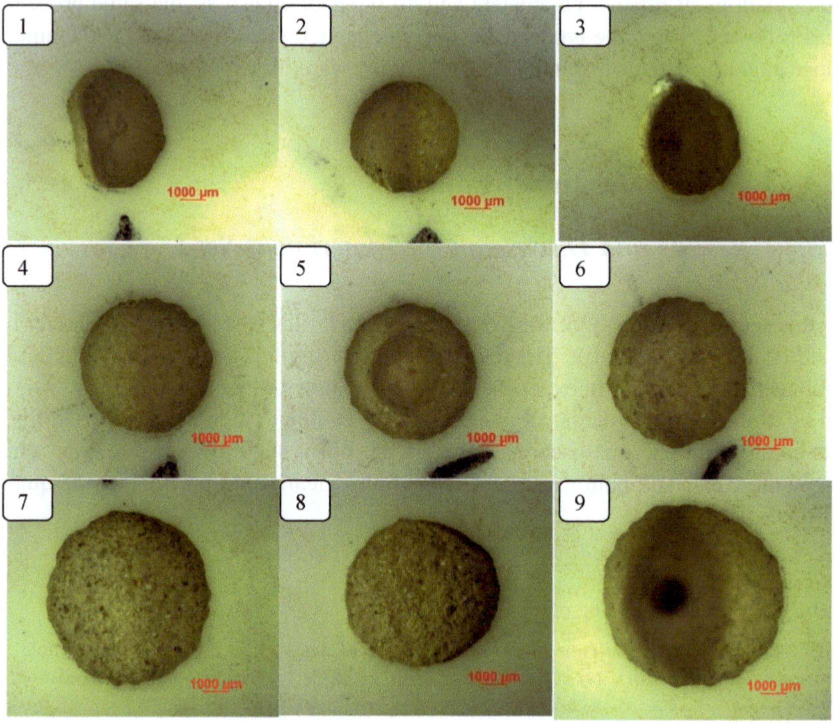

Fig. 9.5 Microscopic views of the machined hole using sand on tiles

The hole diameters are plotted against pressure as shown in Fig. 9.6. It is observed from Fig. 9.6 that the hole diameter increases with increasing SOD for a particular pressure, but it remains almost same with respect to pressure for a particular SOD. The reason behind this may be attributed to the diverging nature of the jet. With increase in stand off distance, the diversion of jet is more, so the diameter of cut becomes more.

As the generated holes are blind, taper angles are not measured due to the difficulty in measuring the lower surface diameter.

The out of roundness of the generated holes was measured and plotted against pressure at different SODs as shown in Fig. 9.7. From the plot, it can be said that the variation of out of roundness of the holes do not follow any direct relationship

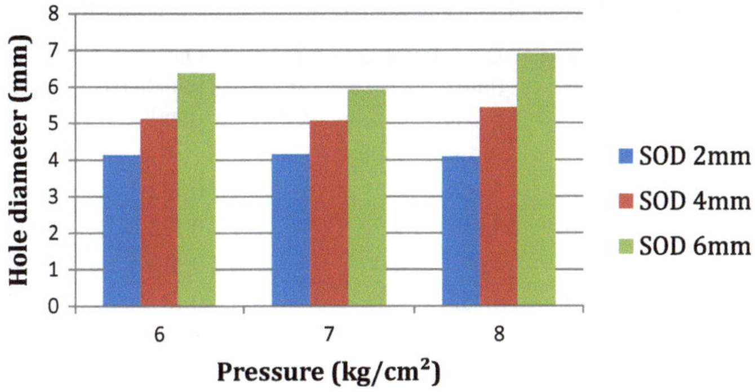

Fig. 9.6 Plot of variation of hole diameter with pressure at different SODs (expt. set-I)

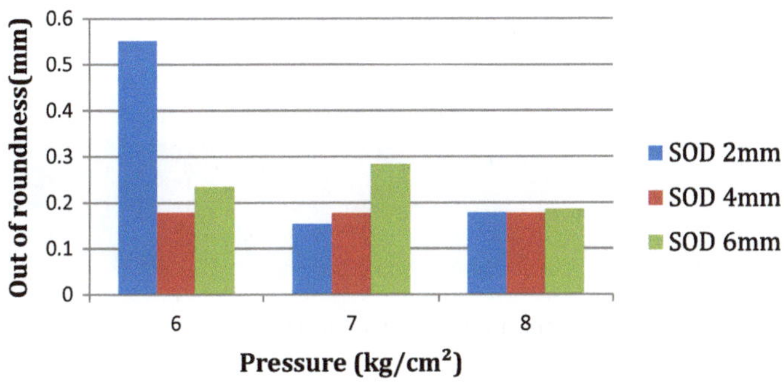

Fig. 9.7 Plot of variation of out of roundness with pressure at different SODs (expt. set-I)

with pressure and SOD. It may be due to the characteristics of the nozzle, flow characteristics of the jet, inhomogeneity in material characteristics, etc.

At the 6 kg/cm^2 pressure and 2 mm SOD, the machined hole took oval shape may due to some inhomogeneity nature of the machined material.

9.3.2 Results and Discussion on Experiment Set-II

In this set, drilling on 2.37 mm thick glass sheet was performed using sand having 300 μm grit size. The results obtained from this experiment set are shown in Table 9.4.

The blind holes were produced in this set of experiments on 2.37 mm thick glass plate which is shown in Fig. 9.8. The microscopic views of the holes are shown in Fig. 9.9 where the edges and bottom surface of the holes are clearly seen. In this case also, taper angles could not be measured due to blindness of the holes.

Table 9.4 Results of drilling of glass using sand as the abrasive (expt. set-II)

Sl. no.	Pressure (kg/cm^2)	SOD (mm)	Time (s)	Diameter of hole (mm)		Taper angle (°)	Out of roundness (mm)	Type of hole observed
				Upper surface	Lower surface			
1	6	2	10	4.168	–	–	0.122	blind hole
2	7	2	10	4.054	–	–	0.122	blind hole
3	8	2	10	4.07	–	–	0.113	blind hole
4	6	4	10	4.752	–	–	0.13	blind hole of larger diameter
5	7	4	10	5.012	–	–	0.138	blind hole
6	8	4	10	4.558	–	–	0.146	blind hole
7	6	6	10	5.482	–	–	0.122	blind hole of larger diameter
8	7	6	10	5.384	–	–	0.227	blind hole
9	8	6	10	5.434	–	–	0.13	blind hole

Fig. 9.8 Photograph of the machined holes on glass using sand

The results shown in Fig. 9.10 indicate that the diameter of hole increases with increasing SOD which is quite natural as observed for tiles also. But the values of hole diameter are nearer to each other with respect to pressure at a particular standoff distance.

Figure 9.11 shows the variation of out of roundness of holes with respect to pressure. The variation of out of roundness is observed to be random in nature without any direct relationship due to flow characteristics of jet and workpiece material characteristics.

9.3.3 Results and Discussion on Experiment Set-III

In this set of experiment, drillings on 5.8 mm thick tiles are performed by SiC having 250 μm grain size. The results obtained from this experiment set are shown in Table 9.5. It is observed that at a lower SOD, blind holes are generated. As SOD increased, holes generated become through as shown in Fig. 9.12a–c. It may be due to the fact that at lower SOD, a lot of congestion take place between

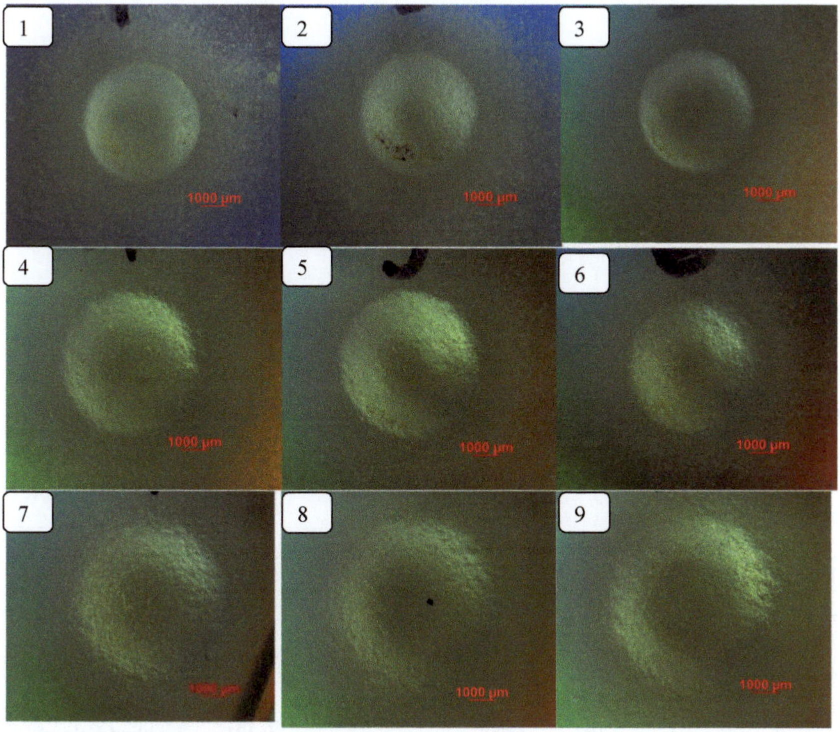

Fig. 9.9 Microscopic views of the machined holes on glass

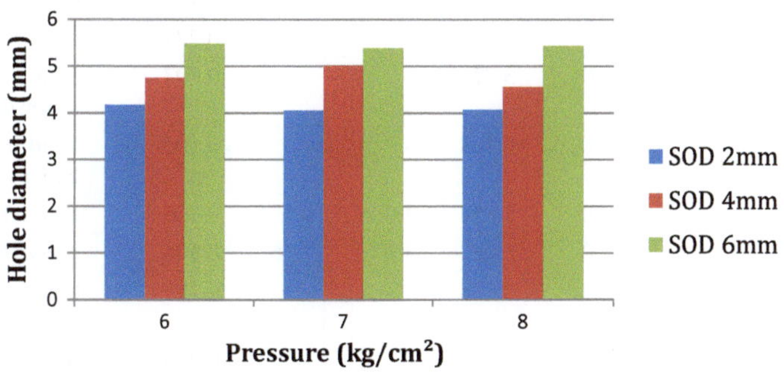

Fig. 9.10 Plot of variation of hole diameter with pressure at different SODs (expt. set-II)

abrasive and eroded particles. A lot of collision may take place between outgoing abrasive particles with rebounded abrasive particles and eroded particles. It is also observed that at further higher stand off distances, some blind holes are formed. It may be due to the reduction of pressure and jet velocity at a far-off position from the nozzle. The microscopic views of the holes are shown in Fig. 9.13.

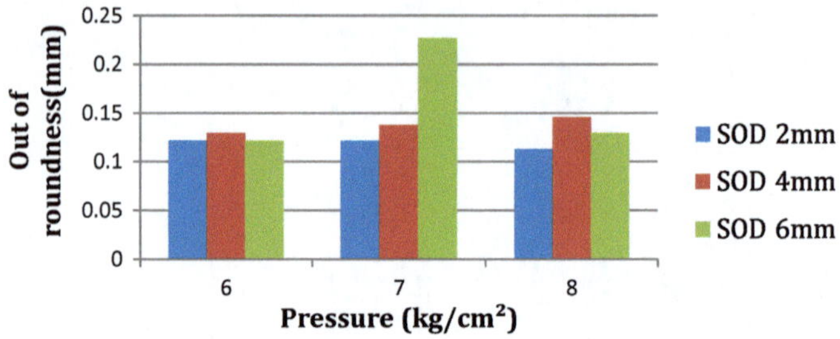

Fig. 9.11 Plot of variation of out of roundness with pressure at different SODs (expt. set-II)

Table 9.5 Results of drilling of tiles using SiC as the abrasive (expt. set-III)

Sl. no.	Pressure (kg/cm^2)	SOD (mm)	Time (s)	Diameter of hole (mm)		Taper angle (°)	Out of roundness (mm)	Type of hole observed
				Upper surface	Lower surface			
1	6	2	10	4.622	–	–	0.122	blind hole
2	6	2	10	4.346	–	–	0.105	blind hole
3	6	2	10	4.476	–	–	0.187	blind hole
4	7	2	10	3.908	–	–	0.121	blind hole of lesser diameter
5	7	2	10	3.958	–	–	0.121	blind hole of lesser diameter
6	7	2	10	4.412	–	–	0.170	blind hole
7	8	2	10	4.444	–	–	0.122	blind hole
8	8	2	10	4.444	–	–	0.227	blind hole
9	8	2	10	3.778	–	–	0.097	blind hole of lesser diameter
10	8	4	10	5.968	–	–	0.106	blind hole of larger diameter
11	8	4	10	5.854	4.006	9.036	0.105	bhrough hole
12	8	4	10	5.06	3.6	7.16	0.106	bhrough hole
13	7	4	10	4.638	–	–	0.187	blind tapered hole
14	7	4	10	4.978	–	–	0.219	blind tapered hole
15	7	4	10	4.996	3.358	8.023	0.171	through hole
16	6	4	10	5.092	–	–	0.138	blind hole
17	6	4	10	5.628	–	–	0.146	blind hole
18	6	4	12	5.416	3.958	7.15	0.170	through hole
19	6	6	10	7.054	–	–	0.113	blind 'U'-shaped hole of larger diameter

(continued)

Table 9.5 (continued)

Sl. no.	Pressure (kg/cm^2)	SOD (mm)	Time (s)	Diameter of hole (mm)		Taper angle (°)	Out of roundness (mm)	Type of hole observed
				Upper surface	Lower surface			
20	6	6	10	6.162	4.622	7.549	0.235	through hole
21	6	6	10	6.682	5.79	4.39	0.154	through hole
22	7	6	10	6.276	4.866	6.92	0.203	through hole
23	7	6	10	7.072	2.448	21.7	0.162	through hole, 'U'-shaped
24	7	6	10	6.228	4.996	6.052	0.146	through hole
25	8	6	10	5.904	3.632	11.06	0.121	through tapered hole
26	8	6	10	7.218	–	–	0.122	blind 'U'-shaped hole of larger diameter
27	8	6	10	6.276	4.558	8.41	0.146	through hole
28	8	6	10	6.13	4.022	10.28	0.138	through hole

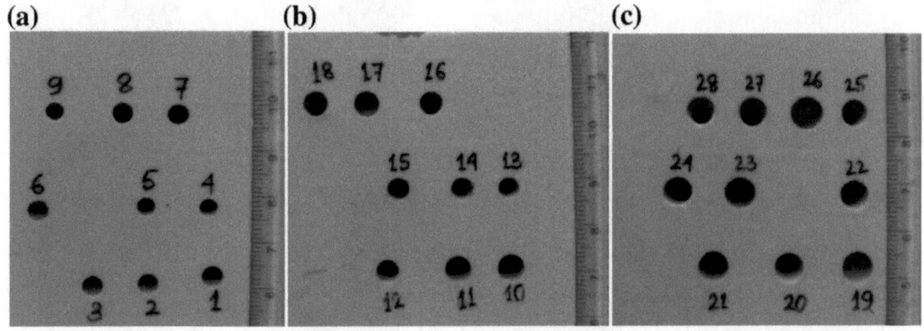

Fig. 9.12 (a–c) Photographs of the machined hole (using SiC) on tiles

The variation of hole diameter plotted against pressure is shown in Fig. 9.14. It is observed from the figure that hole diameter increases with SOD at a constant pressure. The reason may be due to the diverging nature of the jet as stated earlier.

The taper angles measured for the generated holes are plotted with respect to pressure shown in Fig. 9.15. The plot clearly shows that at 2 mm SOD, blind holes are produced by the abrasive jet. It is also observed that for a particular stand off distance, taper angle has an increasing trend with pressure.

The out of roundnesses of the generated holes are plotted against pressure for different fixed values of SOD as shown in Fig. 9.16. From the plot, it is observed that the variation of out of roundness of the holes also do not follow any direct relationship with pressure and SOD. Also, their spreads are random in nature.

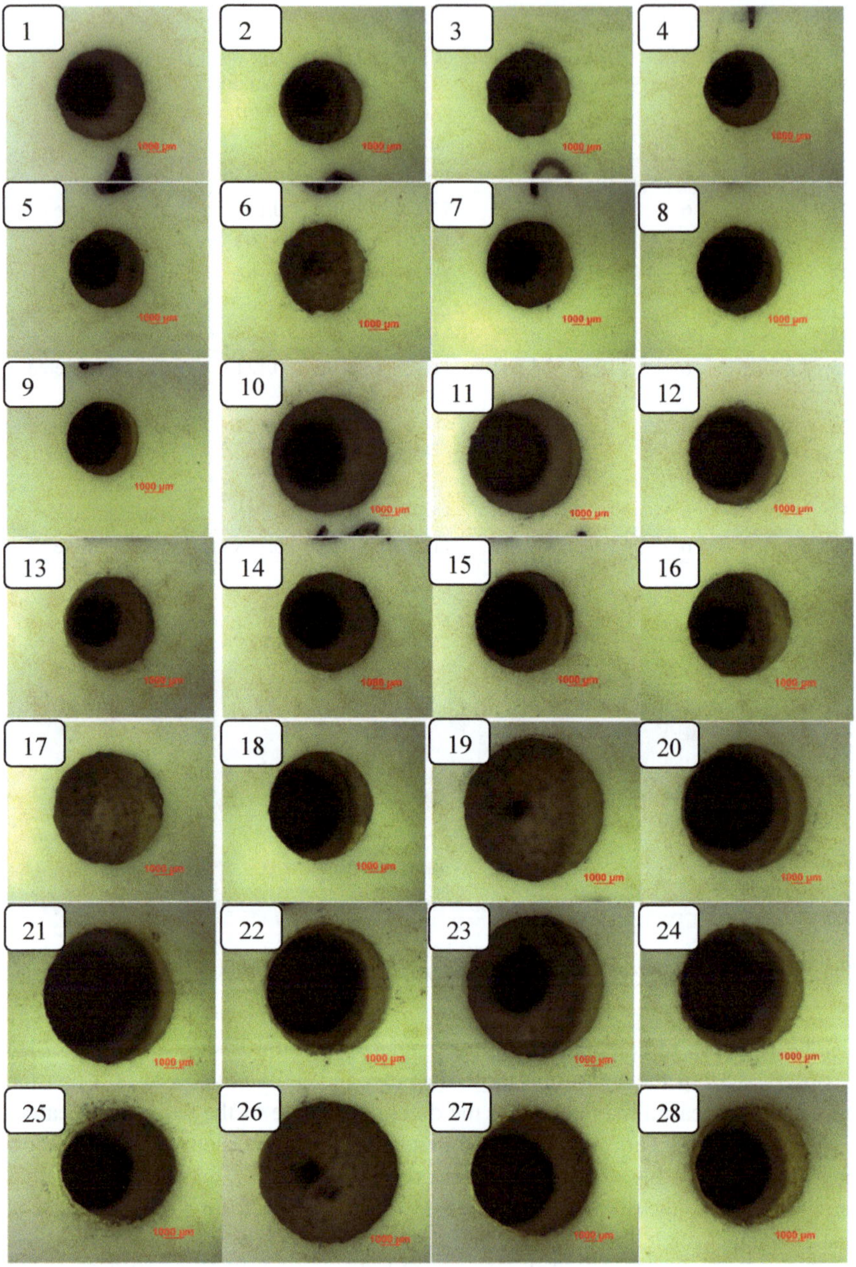

Fig. 9.13 Microscopic views of the machined holes (using SiC) on tiles

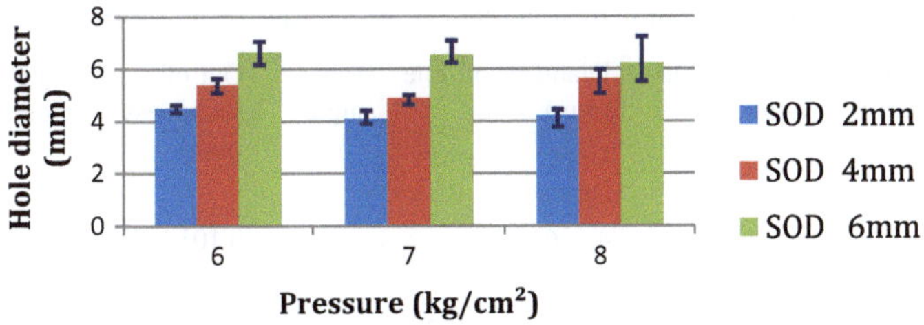

Fig. 9.14 Plot of variation of hole diameter with pressure at different SODs (expt. set-III)

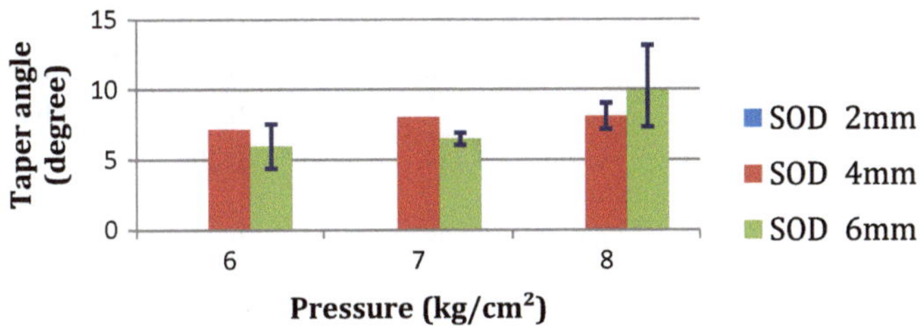

Fig. 9.15 Plot of variation of taper angle with pressure at different SODs (expt. set-III)

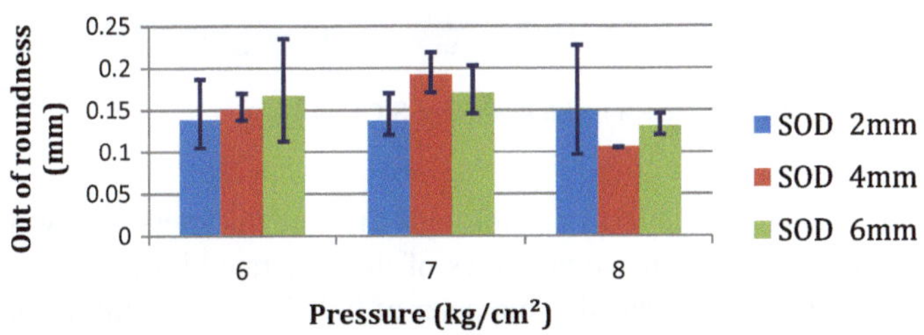

Fig. 9.16 Plot of variation of out of roundness with pressure at different SODs (expt. set-III)

It may be due to the characteristics of the nozzle, flow characteristics of jet, hardness and compactness of the workpiece material, etc.

9.3.4 *Results and Discussion on Experiment Set-IV*

In this experiment set, drilling of 2.37 mm thick glass is performed by SiC of 250 μm grain size. The results obtained are shown in Table 9.6.

In this set, blind holes are formed at lower pressure (Fig. 9.17) as hardness of the glass is high and the bonding strength is more, thereby possibly resulting in less

Table 9.6 Results of drilling of glass using SiC as the abrasive (expt. set-IV)

Sl. no.	Pressure (kg/cm^2)	SOD (mm)	Time (s)	Diameter of hole (mm)		Taper angle (°)	Out of roundness (mm)	Type of hole observed
				Upper surface	Lower surface			
1	6	2	10	4.136	–	–	0.105	blind hole
2	7	2	10	4.25	–	–	0.097	blind hole
3	8	2	10	4.168	1.102	32.89	0.219	through tapered hole
4	6	4	10	5.158	–	–	0.122	blind hole of larger diameter
5	7	4	10	5.254	3.86	16.38	0.219	through hole
6	8	4	10	4.768	3.032	20.11	0.235	through hole
7	6	6	10	5.724	–	–	0.137	blind hole
8	7	6	10	6.034	3.454	28.56	0.203	through tapered hole
9	8	6	10	6.584	4.688	21.80	0.098	through hole

Fig. 9.17 Photograph of the machined holes(using SiC) on glass

erosion. So, the jet is unable to cut through the full thickness at lower pressure. Figure 9.18 shows the microscopic views of the generated holes.

In Fig. 9.19, results indicate that diameter of hole increases with SOD. It is quite natural as observed during machining of tiles also. But at a particular stand off distance, values of hole diameter are nearer to each other with respect to pressure.

Taper angle of holes is also plotted against pressure as shown in Fig. 9.20. From the plot, it is visualized that at 6 kg/cm^2 pressure, no through hole is formed. It is also observed that at a pressure of 7 kg/cm^2, taper angle increases with SOD. But at 8 kg/cm^2 pressure, taper angle slightly decreases with increasing SOD.

In Fig. 9.21, variation of out of roundness of the holes with respect to pressure is shown. The variation of out of roundness is observed to be random in nature. There is also no direct relation found between out of roundness and pressure or SOD which may be due to the workpiece material characteristics, flow characteristics of jet, etc.

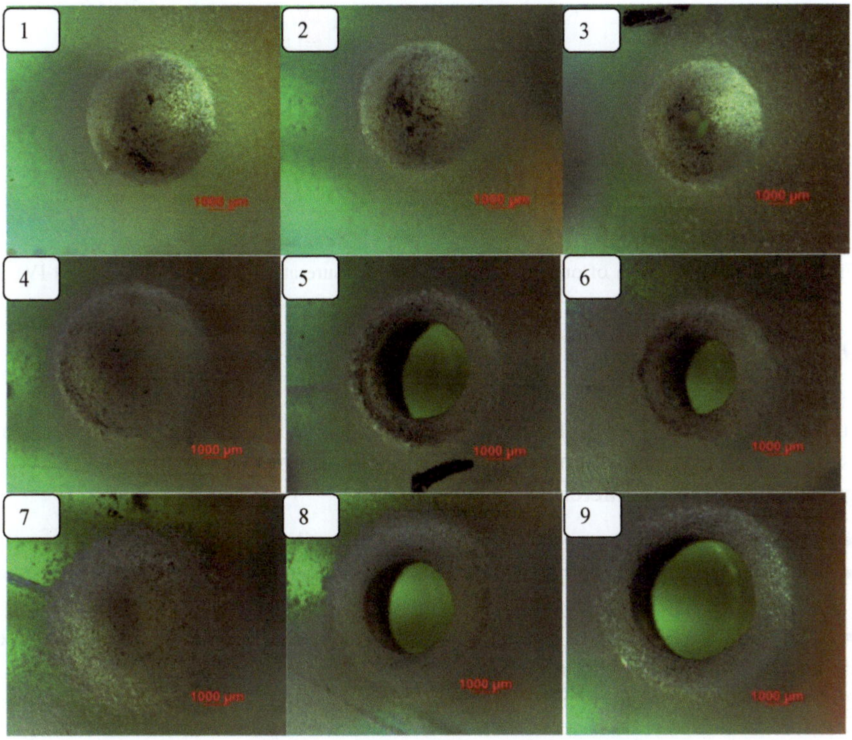

Fig. 9.18 Microscopic views of the machined holes (using SiC) on glass

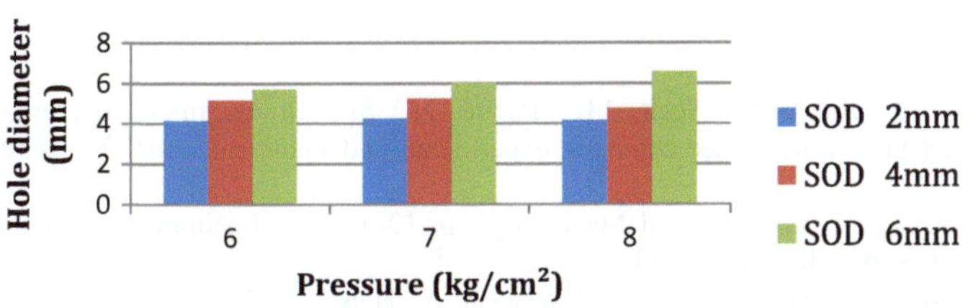

Fig. 9.19 Plot of variation of hole diameter with pressure at different SODs (expt. set-IV)

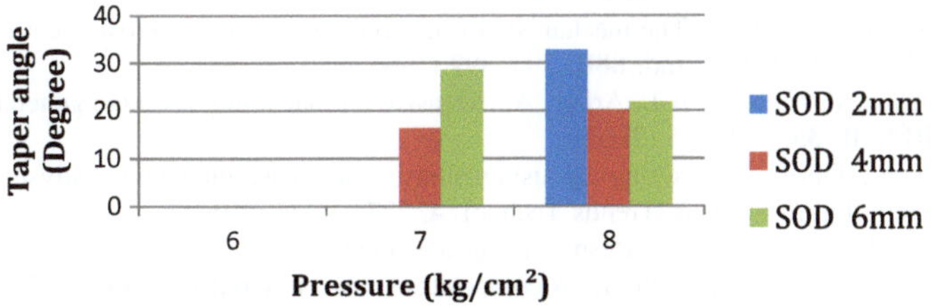

Fig. 9.20 Plot of variation of taper angle with pressure at different SODs (expt. set-IV)

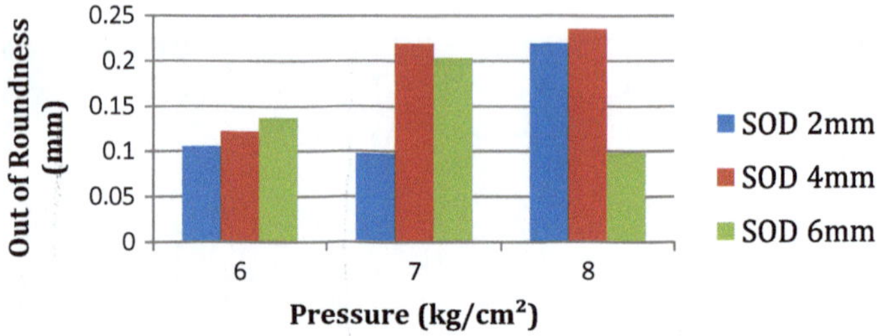

Fig. 9.21 Plot of variation of out of roundness with pressure at different SODs (expt. set-IV)

9.4 Conclusions

In this work, machining performance of an indigenously made abrasive jet system is investigated by performing drilling operations. From this experimental investigation, the following inferences may be drawn:

- The SiC abrasive has more cutting ability than silica sand.
- The machined hole diameters have an increasing trend with the increasing of SOD in a particular pressure.
- The variations of out of roundness in the machined holes are observed to be almost random which may due to the abrasive jet flow characteristics, and workpiece material inhomoginity.

References

1. Mohammad Jafar, R.H., Spelt, J.K., Papini, M.: Surface roughness and erosion rate of abrasive jet micro-machined channels: experiments and analytical model. Wear **303**, 138–145 (2013)
2. Mishra, P.K.: Nonconventional Machining, pp. 12–13. The Institution of Engineers (India) Textbook Series, Kolkata (2014)
3. Haldar, B., Adak, D.K., Ghosh, D., Karmakar, A., Habtamu, E., Ahmed, M., Das, S.: Present status and some critical issues of abrasive jet materials processing: a review. Proc. Manuf. **20**, 523–529 (2018)
4. Fan, J.M., Wang, C.Y., Wang, J.: Modeling the erosion rate in micro abrasive air jet machining of glasses. Wear **266**(9–10), 968–974 (2009)
5. Sheldon, G.L., Finnie, I.: The mechanism of material removal in the erosive cutting of brittle materials. ASME J. Eng. Ind. **88B**, 393–399 (1966)
6. Chastagner, M.W., Shih, A.J., Arbor, A.: Abrasive jet machining for edge generation. Trans. NAMRI/SME **35**, 359–366 (2007)
7. Raju, F.A.: Fibre glass cutting by using abrasive jet machining and analysis of process parameters. Int. J. Comput. Trends Technol. **4**, 2274–2278 (2013)
8. Srikanth, D.V., Rao, M.S.: Response surface methodology for optimization of process parameters in abrasive jet drilling of composites. J. Mech. Civil Eng. **11**, 20–26 (2014)

9. Park, D., Cho, M., Lee, H., Cho, W.: Micro-grooving of glass using micro-abrasive jet machining. J. Mater. Process. Technol. **146**, 234–240 (2004)
10. Oancea, A., Gherman, L., Braha, V.: Modified nozzle for abrasive jet engraving, Nonconv. Technol. Rev., 16–21 (2012)
11. Gradeen, A.G., Spelt, J.K., Papin, M.: Cryogenic abrasive jet machining of polydimethylsiloxane at different temperatures. Wear, 335–344 (2012)
12. Gillespie, L.K.: Deburring and edge finishing handbook. Society of Manufacturing Engineers, New York (1999)
13. Haldar, B., Ghara, T., Ansari, R., Das, S., Saha, P.: Abrasive jet system and its various applications in abrasive jet machining, erosion testing, shot-peening and fast cleaning. Mater. Today: Proc. **5**, 13061–13068 (2018)
14. Li, H.Z., Wang, J.Ã., Fan, J.M.: Analysis and modelling of particle velocities in micro-abrasive air jet. Int. J. Mach. Tools Manuf **49**, 850–858 (2009)
15. Getu, H., Spelt, J.K., Papini, M.: Conditions leading to the embedding of angular and spherical particles during the solid particle erosion of polymers. Wear, 292–293 (2012)
16. Malkin, T.W.H.S.: Grinding mechanisms for ceramics. CIRP Ann. **45**, 569–580 (1996)
17. Balasubramaniam, R., Krishnan, J., Ramakrishnan, N.: A study on the shape of the surface generated by abrasive jet machining. J. Mater. Process. Technol. **121**, 102–106 (2002)
18. Srikanth, D.V., Sreenivasa Rao, M.: Metal removal and kerf analysis in abrasive jet drilling of glass sheets. Proc. Mater. Sci. **6**, 1303–1311 (2014)
19. Verma, A.P., Lal, G.K.: An experimental study of abrasive jet machining. Int. J. Mach. Tool Des. Res. **2**, 19–29 (1983)
20. Finnie, I.: Some reflection on the past and future of erosion. Wear **186–187**, 1–10 (1995)
21. Abhishek, K., Hiremath, S.S.: machining of micro-holes on sodalime glass using developed micro-abrasive jet machine (μ-AJM). Proc. Technol. **25**, 1234–1241 (2016)
22. Suresh, R., Sohit, R.K., Shapur, K.: Abrasive jet machining for micro-hole drilling on glass and GFRP composites. Mater. Today Proc. **5**, 5757–5761 (2018)
23. Jagannatha, N., Hiremath, S.S., Sadashivappa, K.: Analysis and parametric optimization of abrasive hot air jet machining for glass using Taguchi method and utility concept. Int. J. Mech. Mater. Eng. **7**(1), 9–15 (2012)
24. Jagannatha, N., Somashekhar, S.H., Sadashivappa, K., Arun, K.V.: Machining of soda lime glass using abrasive hot air jet: an experimental study. Mach. Sci. Technol. **16**, 459–472 (2012)
25. El-Domiaty, A., Abd El-Hafez, H.M., Shaker, M.A.: Drilling of glass sheets by abrasive jet machining. Int. J. Mech. Aerosp. Ind. Mechatron. Manuf. Eng. **3**(8), 872–878 (2009)
26. Kandpal, B.C., Kumar, N., Kumar, R., Sharma, R., Deswal, S.: Machining of glass and ceramic with alumina and silicon carbide in abrasive jet machining. Int. J. Adv. Eng. Technol. **II**(IV), 251–256 (2011)
27. Rao, M.S., Srikanth, D.V.: Optimization of process parameters of abrasive jet machining on epoxy glass fibre composite. Int. J. Sci. Res. Educ. **3**(9), 4577–4587 (2015)
28. Pawar, N.S., Lakhe, R.R., Shrivastava, R.L.: A comparative experimental analysis of sea sand as an abrasive material using silicon carbide and mild steel nozzle in vibrating chamber of abrasive jet machining process. Int. J. Sci. Res. Publ. **3**(10), 1–4 (2013)
29. Verma, A.P., Lal, G.K.: An experimental study of abrasive jet machining. Int. J. Mach. Tool Des. Res. **24**(1), 19–29 (1984)
30. Srikanth, D.V., Rao, M.S.: Metal removal and Kerf analysis in abrasive jet drilling of glass sheets. In: Proceedings of the 3rd International Conference on Materials Processing and Characterization, Procedia Materials Science, vol. 6, pp. 1303–1311 (2014)
31. Ray, P.K., Paul, A.K.: Some studies on abrasive jet machining. J. Inst. Eng. (India) **68**(Part PE 2), 27–30 (1987)
32. Shriyan, G., Shinde, R., Ronge, H.: Study of effect of process parameters on the performance of abrasive jet machining. J. Eng. Res. Gen. Sci. **3**(3), 861–867 (2015)
33. Ghobeity, A., Getu, H., Krajac, T., Spelt, J.K., Papini, M.: Process repeatability in abrasive jet micro-machining. J. Mater. Process. Technol. **190**(1–3), 51–60 (2007)

34. Mahajan, G.: A study of effect of various process parameters on abrasive jet machining using silicon carbide as abrasive material. Int. J. Eng. Dev. Res. **3**(1), 25–31 (2014)
35. Srikanth, D.V., Rao, M.S.: Machining of FRP composites by abrasive jet machining optimization using Taguchi. Int. J. Mech. Aerosp. Ind. Mechatron. Manuf. Eng. **8**(3), 632–636 (2014)
36. Reddy, S.M., Hussain, S., Srikanth, D.V., Rao, M.S.: Experimental analysis and optimization of process parameters in machining of RCFRP by AJM. Int. J. Innovative Res. Sci. Eng. Technol. **4**(8), 7085–7092 (2015)
37. Sharma, P.K., Deol, G.S.: A comparative analysis of process parameters during machining of glass fibre reinforced plastic by abrasive jet machining. Int. J. Adv. Found. Res. Sci. Eng. **1**(3), 28–37 (2014)
38. Gavaskar, T., Devabalan, C., Revanth, M.A., Sakthiseelan, G., Rajivranjan, V., Kumar, A.K., Vishwanathan, H.: Analysing the factors affecting surface roughness in abrasive jet machining. Int. J. Innovative Res. Adv. Stud. **3**(3), 1–4 (2016)

Chapter 10
Friction Stir Welding of Low-Carbon Steel

Avinish Tiwari, Piyush Singh, Pankaj Biswas and Sachin D. Kore

10.1 Introduction

Joining of materials is an important part in manufacturing process. Welding is one of the joining techniques in which two similar and dissimilar materials are joined permanently with application of heat and with or without the application of pressure. Welding is broadly classified into mainly two categories (a) fusion welding and (b) solid-state welding techniques. Fusion welding is an important category of joining technique which requires direct application of heat with or without the requirement of pressure. Fusion welding is attained by melting the base materials with an intense heat source followed by solidification. Solid-state joining is another important category of welding process in which joining can be achieved at the temperature less than the melting point, i.e., around 80% melting point temperature of base material. Friction welding is one of the prominent solid-state joining techniques. Friction welding utilizes the frictional heat, and no external heat source is employed.

Joining of aluminum alloys by fusion welding is very difficult due to various critical issues like high thermal conductivity, oxide layer formation, and cracking. This difficulty in producing sound quality welds in aluminum and its various alloys leads to the development of a relatively new technology known as friction stir welding (FSW).

FSW was developed and patented by The Welding Institute (TWI), UK, in the year 1991. It is one of the derivatives of friction welding technique. It is a novel joining technique used to the join the material below its melting point. FSW has been proven a green technology since it does not emit any fumes and toxic gases during welding, unlike fusion welding techniques.

A. Tiwari (✉) · P. Singh · P. Biswas · S. D. Kore
Department of Mechanical Engineering, Indian Institute of Technology Guwahati, Guwahati, India
e-mail: avit252@gmail.com

P. Sahoo and J. P. Davim (eds.), *Advances in Materials, Mechanical and Industrial Engineering*, Lecture Notes on Multidisciplinary Industrial Engineering,
https://doi.org/10.1007/978-3-319-96968-8_10

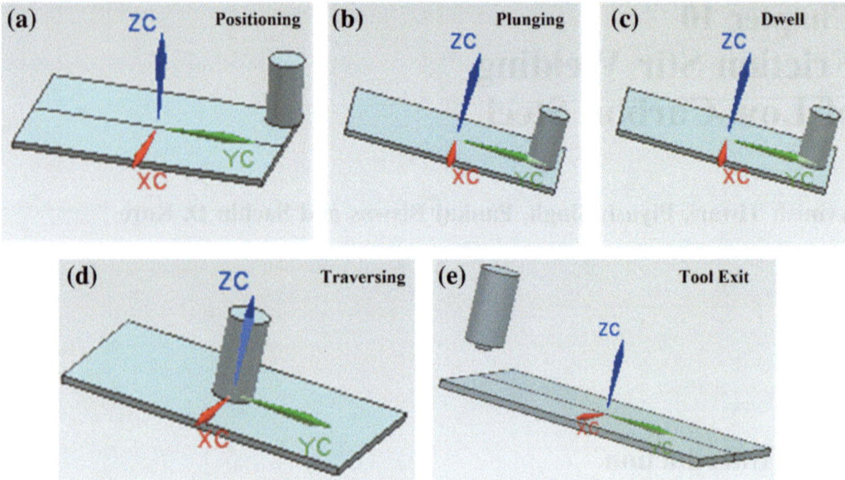

Fig. 10.1 Schematic representation of FSW process showing various stages **a** positioning, **b** plunging, **c** dwell, **d** traversing, **e** tool exit

FSW needs non-consumable tool having higher hardness and strength than the base material, which plunges into the workpiece material and produces the joint by the combined rotational and traverse movement of the tool. The amount of heat required to produce the joint is provided by both frictional heat and the heat generated during the plastic deformation of the workpiece. The various important stages during FSW are described by the schematic diagram as shown in Fig. 10.1.

(i) Positioning of tool
(ii) Plunging of tool
(iii) Dwell period
(iv) Traverse movement of tool
(v) Tool exit.

FSW technology has gained huge success in the joining of aluminum, magnesium, and other softer materials. FSW has various applications in various industries like aerospace, automobile, railway, and shipbuilding industries. FSW joints have superior mechanical properties than the base material. Due to its additional advantage of low heat input which results in minimization of distortion and residual stress, FSW is an attractive choice for joining of various materials.

Low-carbon steel is highly used structural material due to its excellent mechanical properties and good weldability. Steel has wide variety of alloys as compared to any other material. There are so many fusion welding processes in which steel can be joined on industrial scale depending upon their thickness and application. However, fusion joining of steel is associated with various defects like solidification cracking, liquification cracking, hydrogen cracking, alloying element segregation, porosity, high distortion and formation of dendritic structure [1].

Fusion joining of steel requires expensive preheat and post-heat treatments to control the weld joint properties. Additionally, the development of new steel alloys challenges the fusion welding process to join the material since it requires newly filler material.

To overcome these issues, FSW technique has gained huge attention in joining of ferrous alloys these days. FSW technology offers more advantageous over conventional welding of steels. It removes all welding defects associated with fusion welding. In addition to that, lower heat inputs of FSW are believed to reduce heat-affected zone (HAZ) section which is one of the critical issues in fusion welding of steel. Additionally, the hydrogen cracking associated with the fusion welding of steel is completely absent in this process. The additional benefits of being low heat input process are low distortion which results in minimizing residual stress in the joining of large structural steels parts. Minimization of distortion and residual stress is most important criteria in welding of thick plates, used in the shipbuilding industries, marine industries, and other heavy manufacturing industries. This process has also the potential to join some different grades of steel which are less prone to fusion weld, and steel with high carbon content and dissimilar alloy steels.

In FSW, process parameters play a vital role in producing good weld quality. Selection of appropriate process parameter is very important for good quality weld. Process parameters control the heat input and cooling rate which govern the microstructural evolution and mechanical properties. Few relevant studies regarding the selection of process parameters is highlighted in this section. Medium-carbon steel was studied at different traverse speeds (100, 67, and 150 mm/min) and at three rotational speeds (710, 900, and 1120 rpm) by Ghosh et al. [2]. Cooling rate was directly related to the welding speed and was increasing with increase in welding speed at a constant tool rotation. It was reported that the carbon diffusion time was reduced which resulted in higher concentration of carbon content in the austenite phase. This more carbon concentration in the austenite resulted in lowering the transformation temperature in steel. Interlamellar spacing of pearlite was decreased, and pearlite became finer. Higher cooling rate also resulted in more nucleation site [2]. Feasibility studies in FSW of carbon steel and stainless steel were studied up to 12-mm-thick plates [3]. However, no attempt was done to study the weld quality at different welding conditions, i.e., the rotational speed and traverse speeds. Although it is well known that the to produce successful welds [3]. Friction stir welding of DH 36 low-carbon steel was carried out at different traverse speeds of 3.4, 4.1, and 7.6 mm/s by Reynolds et al. [4]. Partial sticking was observed at the root of the weld due to higher heat input at low traverse speed. FSW of ultrahigh-carbon steel (1.02% C by wt.) was studied at different traverse speeds. Their study did not include the influence of traverse speed on the microstructural characteristics of the weld joint, although it is supposed that the microstructural characteristics are strongly related to the operating parameters [5]. FSW of L80 and X80 steel was carried out at different rotational speeds of 450 and 550 rpm [6]. Their results suggested that the mechanical properties like microhardness of the weldment can be tailored by controlling the FSW parameters like rotation speed, travel speed, and Z-load. FSW of M190 steel was carried out at different traverse speeds of 0.21, 0.85, and 1.69 mm/s [7]. Their study was mainly focused on the influence of heating and cooling rates on the weld quality, i.e., both mechanical and the microstructural properties of the joint. They reported that the peak temperature was around 1080–1090 °C. It was observed that the heat input to the weld was decreased and the cooling rate was increased with the increase in weld traverse speed [7]. Three types of steel of different carbon contents were investigated by Fujii et al. [8]. Microstructure and mechanical properties of S12C and S35C were significantly affected by welding process parameters which were absent in IF steel. It was observed on increasing welding speed, peak temperature decreased and cooling rate increased which was independent of the type of steel welded. Welding speed was studied on five types of steel of different carbon contents at rotational speed of 400 rpm. The maximum temperature attained was decreased with the increase in weld traverse speed. These values are independent of the types of steels [9]. When the peak temperature was controlled to be below A1, friction stir welding can be performed without any transformation [9]. FSW of DH 36 was carried out up to the welding speed of 500 mm/min [10]. Their work developed a wide range of process parameters envelope between rotational speed and traverse speed. They obtained welding speed which was similar to the conventional welding techniques of steel [10]. The effect of welding speed was studied during FSW of stainless steel [11]. They reported successful welds up to traverse speeds of 200 mm/min. It was observed that on increasing welding speed at a constant spindle improved the mean hardness and tensile strength due to larger reduction in the grain size [11]. Successful weld in FSW of high-carbon steel was obtained at both above and below A1 (eutectoid temperature) by controlling the rotational speed and traverse speed [1]. Friction stir welding was studied on carbon steels at different rotational speeds by Hussain et al. [12]. It was observed that either above or below the rotational speed of 1000 rpm, joint efficiency reduced [12]. FSW of 1018 mild steel was investigated at 1000 rpm and 50 mm/min as welding speed. The parametric studies and its effect on the microstructure and mechanical properties of the weld were not explored [13]. Influence of weld traverse speed and rotational speed was studied on the weld joint characteristics in FSW of S70C steel by Cui et al. [14]. In carbon steel, change in rotational speed also causes phase transformation, which significantly changes the mechanical properties of microstructure as well as of the weld joint.

The present research was aimed to establish the FSW process for low-carbon steel and to investigate the weld quality by microstructural and mechanical characterizations of the weld joint.

10.2 Experimental Procedures

In this work, 4-mm-thick low-carbon steel was used as the workpiece material. The mechanical properties and chemical composition of the workpiece material are given in Tables 10.1 and 10.2, respectively.

Table 10.1 Mechanical properties of workpiece material

Yield strength (MPa)	UTS (MPa)	Percentage elongation (%)	Hardness ($HV_{0.5}$)
255	320	13.7	89

Table 10.2 Details of chemical composition of low-carbon steel

Elements	C	Mn	P	S	Si	Fe
Weight (%)	0.08–0.1 max	0.25–0.40 max	≤ 0.3	≤ 0.05	≤ 0.2	Rest

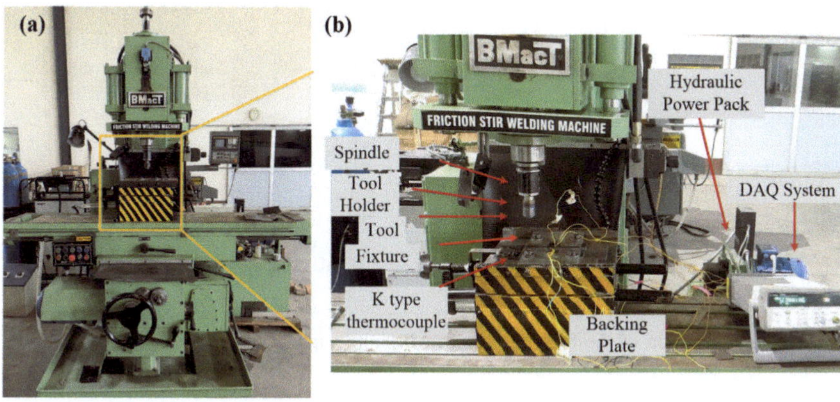

Fig. 10.2 **a** Friction stir welding machine, **b** Experimental setup

The friction stir welding machine with spindle motor power of 25 hp was used to perform the welds. Photograph of the welding machine is shown in Fig. 10.2.

A suitable tool holder was developed to carry out the experiments and to prevent any damage to the machine spindle from the red-heated FSW tool. Mild steel was used for the fabrication of tool holder. Figure 10.3a, b shows the illustration of tool holder and its photograph, respectively.

Non-consumable tungsten carbide (WC-10wt.% Co) tool was used to perform the welds in butt joint configuration. Tool shoulder had a diameter of 22 mm. Tapered pin of 9 mm base diameter and 7 mm at tip was used to perform the welding trials and experiment. Schematic diagram of FSW tool and tool photograph is shown in Fig. 10.4a, b.

A fixture was developed to clamp the workpiece material to prevent the separation of workpieces during the welding. Mild steel was chosen as the material for fixture. The plates were machined into rectangular samples with dimension of 160 mm × 55 mm × 4 mm. The weld plates were cleaned with acetone prior to welding. Single-pass butt joint procedure was performed along the rolling direction of the base plate.

Initially, trial runs were made to decide the working range of process parameters to produce sound welds. After performing the trials, welded samples were inspected

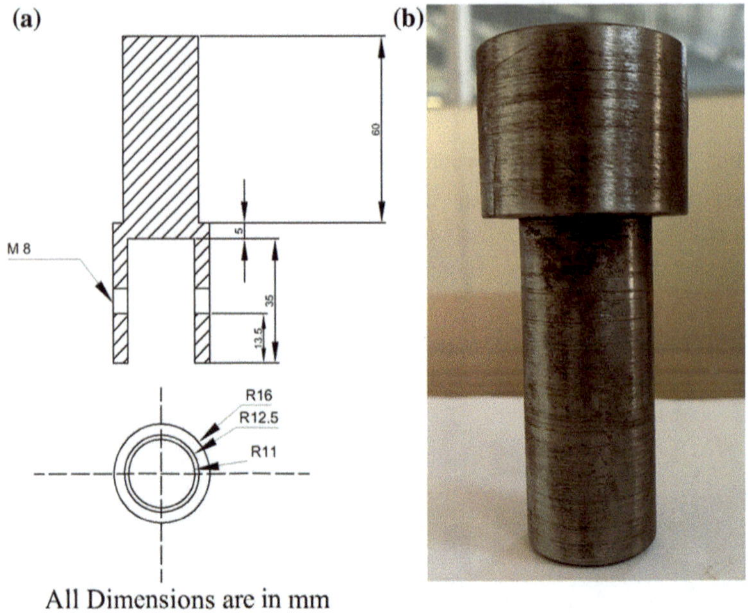

Fig. 10.3 **a** Illustration of the FSW tool holder, **b** photograph of the tool holder

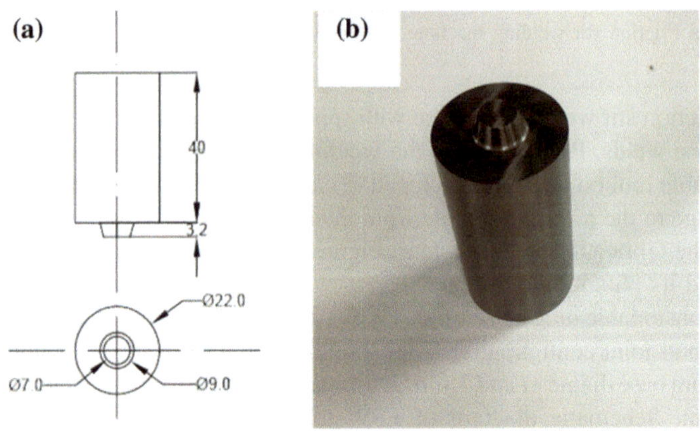

Fig. 10.4 **a** Illustration of the FSW tool, **b** FSW tool

visually from both the top and bottom sides to check whether the joints are free from any visual defects. After visual inspection, a section was cut along the transverse direction from the welded sample at both start and end positions of the welded joints. Samples were polished to different grades of emery papers and mirror finish was achieved by loth polishing on alumina solution.

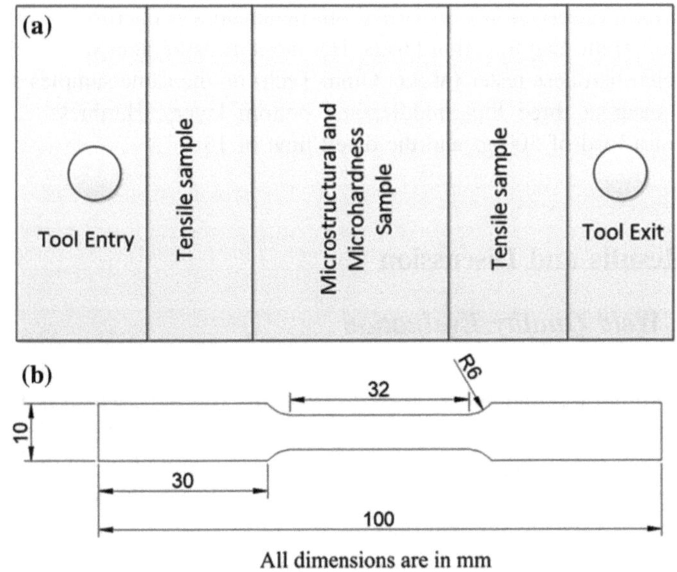

Fig. 10.5 **a** Location of sample extraction, **b** schematic diagram of tensile specimen as per ASTM E8 standards

Thermal history during the weld joint was recorded using K-type thermocouples, and thermal data was recorded with the help of data acquisition system. Samples extraction was carried normal to the weld direction for the metallographic evaluation and mechanical testing. The location of sample extraction from the welded plate is shown in Fig. 10.5a. Subsize tensile specimens were cut on CNC wire electric discharge machine (WEDM). Gauge length and gauge width of each specimen were 25 and 6 mm, respectively, as shown in Fig. 10.5b. Samples were cut to include entire weld zone and some parts of base material in the gauge length as shown in Fig. 10.5b.

After that machined tensile specimens' surface was polished using 1500 grit size emery paper. Tensile properties evaluation was carried out at the crosshead speed of 1 mm/min on universal tensile testing machine (Make BISS and Model MEDIAN 250). The yield stress, ultimate stress, and % elongation to fracture were recorded. Fractured tensile samples were investigated under FESEM (Model Sigma and Make Zeiss) to study the fracture behavior of tensile samples.

For microstructural and microhardness studies, the specimens were cut to the required dimensions which included weld zone and the base material. Initial polishing was carried out using different grades of emery paper from 320 to 2000. Final polishing was carried out using alumina solution on velvet cloth mounted on the rotating disk polishing machine. Etching was carried out on the polished samples by using Nital solution (95% ethanol and 5% nitric acid), and microstructural observation was carried out under optical microscope (Make: Carl Zeiss.) at different magnification levels. Hardness measurement was carried out on Vickers microhardness tester (Make: Omni Tech) on the same samples at intervals of 1 mm each at three top, middle, and bottom layers. Hardness values were recorded at a load of 500 gf and the dwell time of 15 s.

10.3 Results and Discussion

10.3.1 Weld Quality Evaluation

After performing welding, the weld characterization was done which is described in various sections. Initially, the weld joint was inspected by naked eyes from both top and the bottom sides. After that macro and microstructural observations were done to check the weld quality. Mechanical properties in this investigation are limited to the hardness and tensile strength only.

10.3.1.1 Visual Inspection

Figure 10.6a shows the red-hot photograph of the welding tool during welding. Figure 10.6b shows the top suface appearance of the welded sample. The visual inspection of the weld was carried out at both the top and bottom surfaces. Good quality weld joint was produced. Flashes were observed at the retreating side of the weld joint produced at the rotational speed of 600 rpm. From Fig. 10.6, it is clear that the weld was good and free from any visible defects.

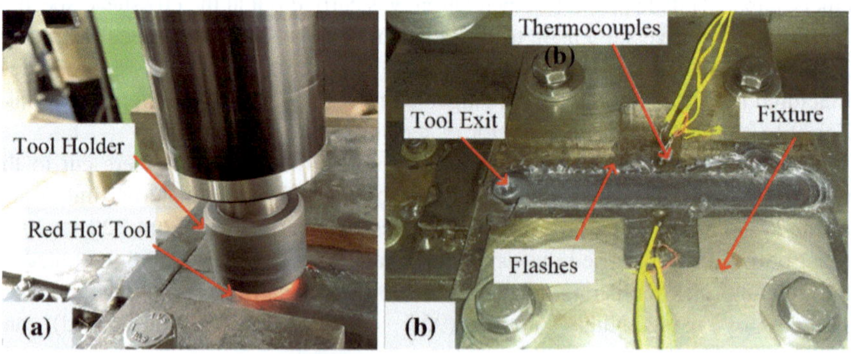

Fig. 10.6 **a** Photograph of the red-hot tool during welding, **b** successful weld joint

10.3.1.2 Thermal History

Figure 10.7a shows the schematic illustration of thermocouple location two K-type thermocouples were attached at the top surface of the workpiece using spot weld machine to record thermal history during welding on both sides of the weld joint. The thermocouples location was exactly in the middle length of the weld plates at different locations as shown in Fig. 10.7a.

Figure 10.7b represents the transverse transient thermal history that was recorded for weld 1(i.e., welding produced at high rotational speed). It was observed that the peak temperature was different on both sides of the weld joints. Higher peak temperature was observed on the advancing side relative to the retreating side of the joint.

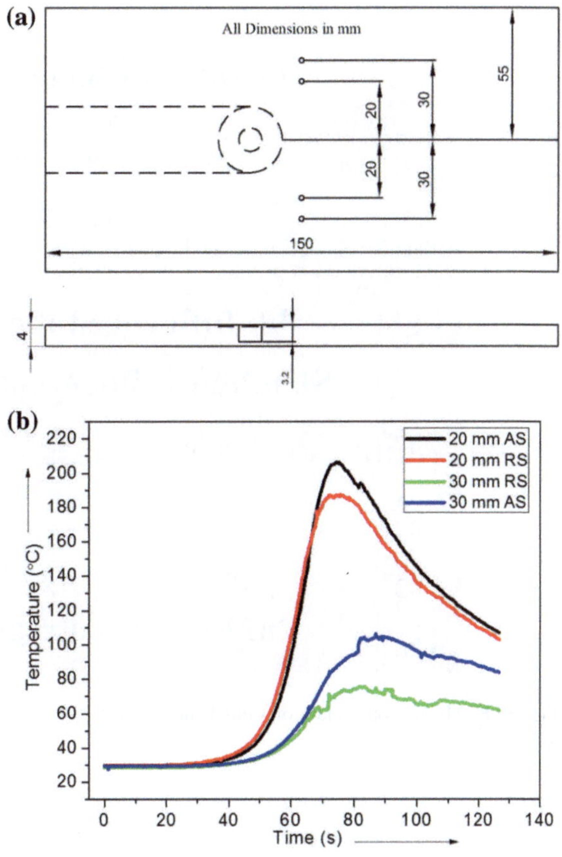

Fig. 10.7 **a** Illustration of location of K-type thermocouples in weld joints, **b** weld joint at 600 rpm and 132 mm/min

10.3.1.3 Macrostructure Evaluation

A section was cut in the transverse direction and was polished to reveal various FSW zones. Figure 10.8a, b shows the macrostructure of the weld 1 and weld 2, respectively. From Fig. 10.8, the boundary of the weld nugget is clearly seen. AS and RS represent the advancing side and the retreating side of the welded sample, respectively. The width of the weld zone was measured at top, middle, and bottom sides.

The boundary separating the TMAZ and HAZ portions are more clearly visible on the AS compared to the RS of the weld joint. The macrograph shows lack of symmetry along the centerline of the weld. The "basin shape" weld nugget was observed which broadens toward the crown surface demonstrating the dominance of material movement by tool shoulder. The thickness of the weld was decreasing on going from top to bottom which was also varying with the process parameters. Various characteristics zones of FSW were observed in the macrostructures like stir zone (SZ), thermal and mechanically influenced zone (TMAZ), heat-affected zones (HAZ).

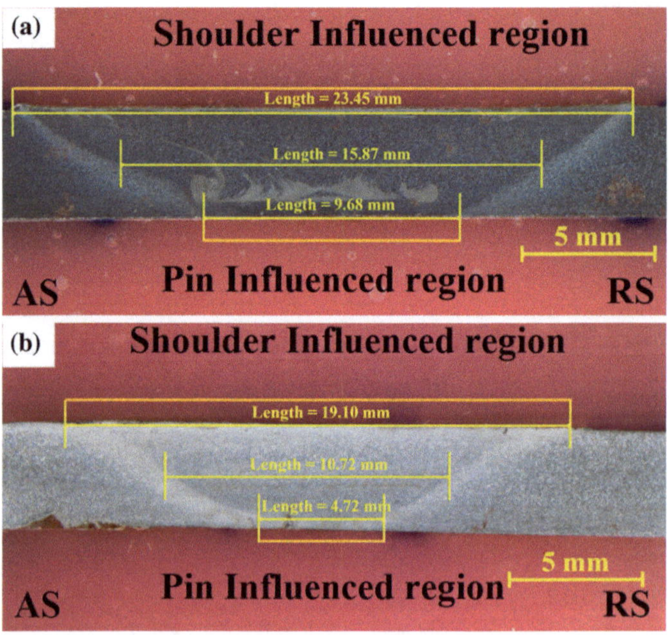

Fig. 10.8 Macroscopic view of weld 1 and weld 2

10.3.1.4 Microstructure

Figure 10.9 represents the microstructural features of the welded samples at the middle of the stir zone. Micrograph of the base material consists of ferrite (F) and pearlite (P) with ferrite as major constituent.

From Fig. 10.9a, b, different material movement patterns were observed in the weld joints produced at different process parameters. Figure 10.10a shows the formation of ring structure in the middle of stir zone. Figure 10.10b represents the layer-wise material movement in the weld microstructure. Different layers of material movement were observed separated by some distance in the weld 2. Significant grain size reduction was also observed in the weld zone (i.e., stir zone and the thermomechanically affected zone). As the HAZ is not subjected to any mechanical deformation that is why in HAZ, the grain structure is almost similar to

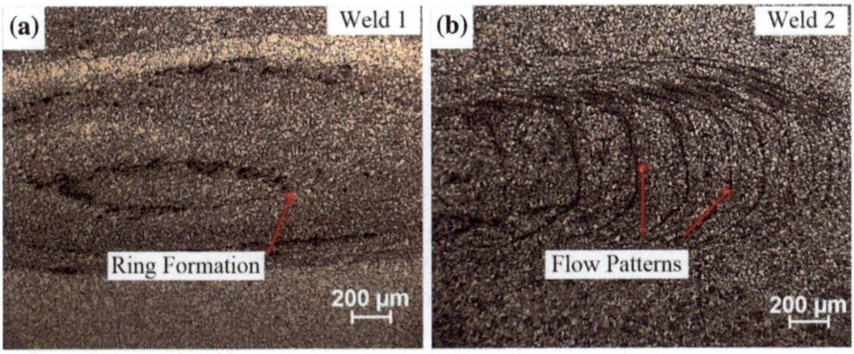

Fig. 10.9 Microstructure of the stir zone of the weld 1 (**a** and **c**) and weld 2 (**b** and **d**) and the (e) base material

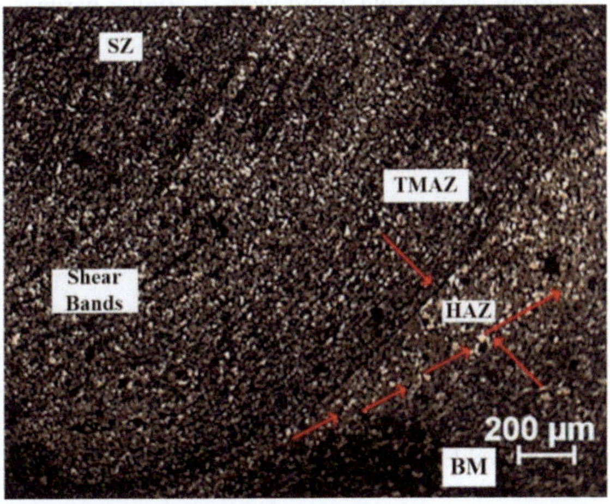

Fig. 10.10 Microstructure representing the shear bands in the tool pin-influenced region

that of the parent material. However, in this study, no attempt was made to measure the grain size of weld zones.

Figure 10.10 shows the welded samples cross-sectional micrograph at the bottom of the retreating side of the weld joint. This figure represents the microstructural feature of different weld zones and base material (BM).

Shear bands were clearly observed on both sides, i.e., RS and AS of the weld zone at the lower magnification. Shear bands were consistent with the material transportation in the pin-influenced region at the root of the weld. Because of the unique characterstic of material deformation and material movement, these shear bands occured in the weld zone. Material movement can be divided into two categories (i) shoulder-influenced region and (ii) pin-influenced region. Complex material movement developed the final microstructure and governed the mechanical properties of the weld joint.

10.3.1.5 Microhardness

The Vickers microhardness evaluation was carried at top line, middle line, and bottom line along thickness at an interval of 1 mm each. Figure 10.11 shows the obtained microhardness contours of the weld 1 and weld 2. The average microhardness value of the workpiece material is about 89 $HV_{0.5/15.}$

From Fig. 10.11a, b, it was observed that high hardness values were obtained in the weld 1 and weld 2 as compared to the base material. Peak hardness values were observed in the stir zone of both the weld 1 and weld 2. The hardness values were decreasing on moving from stir zone toward the base material. Another important observation from the hardness contour was the variation of hardness values along the thickness direction. Hardness values were more at the top of the weld. This can be explained as the intense deformation occurred in the shoulder-influenced region. It can also be concluded that the hardness values were asymmetric along both sides of the weld joint. Microhardness values were consistent with the microstructural evolution of the weld 1 and weld 2. Higher hardness values in the stir zone can be attributed to grain refinement microstructure in the weld 2 due to low rotational speed of 300 rpm. The variation in hardness values of weld 1 and weld 2 is the result of microstructural evolution due to the difference in the peak, temperature the material flow variation, strains and strain rate [15].

10.3.1.6 Tensile Properties

Tensile properties mainly the yield strength (YS), ultimate tensile strength (UTS), and percentage elongation of friction stir welded low-carbon steel joints were evaluated. The yield stress and ultimate stress obtained for the base material are 255 and 320 MPa respectively. Figure 10.12a, b shows the results of tensile tests of the weld 1, weld 2, and the base material.

Fig. 10.11 Microhardness profiles for the **a** weld 1 and **b** weld 2

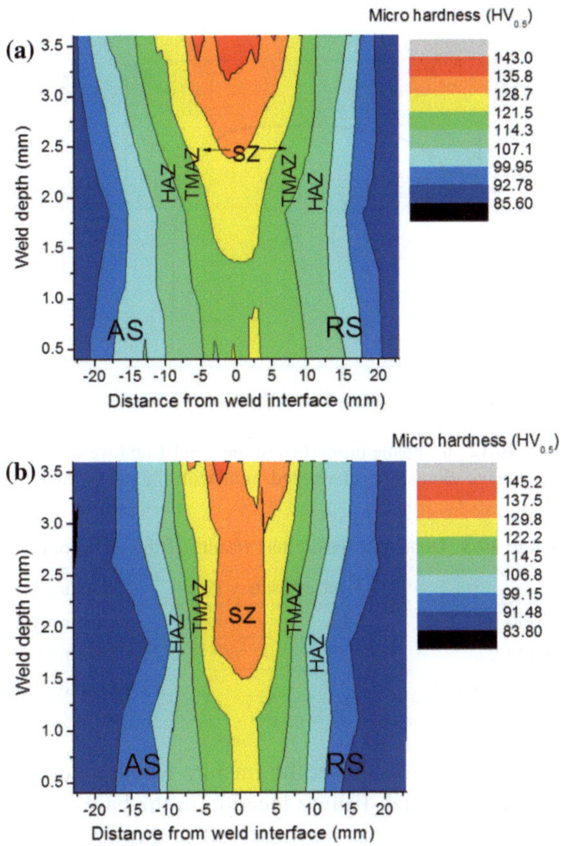

Both the yield strength and ultimate tensile strength of the weld 1 and weld 2 were higher than the base material. Tensile strength was more for weld 2 than the weld 1. The percentage increment in the yield strength (YS) was comparable for both welds. Joint efficiency was calculated as the ratio of UTS of the welded joint to the UTS of base material. The joint efficiency for the weld 1 and weld 2 were 102 and 107% respectively. The increased strength of the weld zone can be explained by happening of the various phenomena like grain size strengthening (Hall Petch effect), increase in dislocation density, and discontinuous dynamic recrystallization. Discontinues dynamic recrystallization is most likely to occur in materials having low stacking fault energy like mild steel.

Figure 10.12(b) shows the percentage elongation of the welded joints. The welded joints showed less ductility as compared to the base material. This decrease in ductility of the welded joints was caused by increase in yield strength as per the Hall Petch relationship and reduction in work hardening rate due to dynamic recovery. The percentage elongation up to failure decreased with reduction in rotational speed. The details of tensile test results is summarized in Table 10.3.

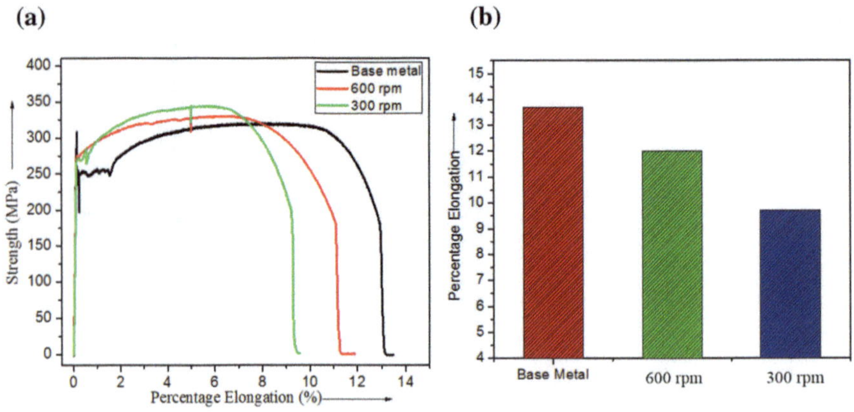

Fig. 10.12 **a** Tensile properties of the weld 1 and weld 2 with the base material. **b** Comparison of percentage elongation of welded joint with the parent material

Table 10.3 Details of tensile test results of the weld joint

	Joint efficiency (%)	Fracture location	Weld quality
Weld 1	102	Base material	Good
Weld 2	107	Base material	Good
Base material	–	Exactly in the middle	–

Figure 10.13a shows the fracture location comparison of the weld sample with the base material. Failure location in the base material was in the middle of gauge length. Fracture was not observed in the welds which were well outside the weld portion confirming the good quality of the weld. Failure occurred in the tensile sample near the boundary between heat-affected zone and base material. Figure 10.13b shows the fracture tensile surface of weld 1 and weld 2.

From the photograph of the fracture tensile sample, cup-and cone-type fractures can be seen clearly. From the fractured tensile specimens, samples were extracted for high magnification fractography analysis.

10.3.1.7 Fractography Analysis

Figure 10.14a, b shows the FESEM micrograph of the fractured surface of weld 1 and weld 2. From the fractography analysis, it was observed that dimples were present in abundance on both the fractured surfaces. It is well known, the presence of dimples confirms the ductile failure of the welded joint. Dimples of various shapes along with voids were distributed over the entire fracture surface. In dimple fracture mode, it was assumed that the overload is the major cause of fracture in the material and the failure is generally controlled by the growth and coalescence of these microvoids formed during the tensile strain.

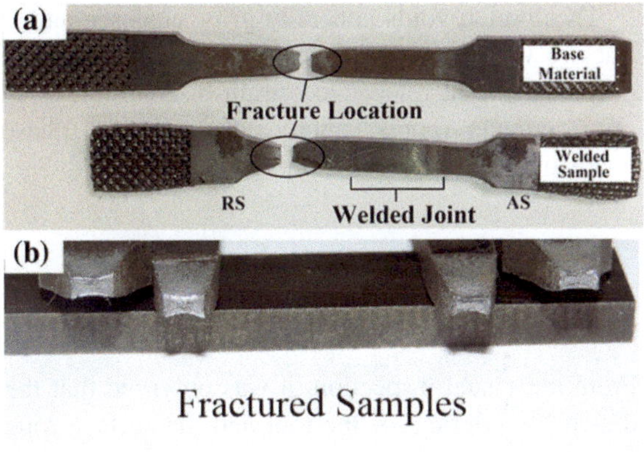

Fig. 10.13 **a** Fracture location comparison with the base material, **b** fracture tensile surface of weld 1 and weld 2

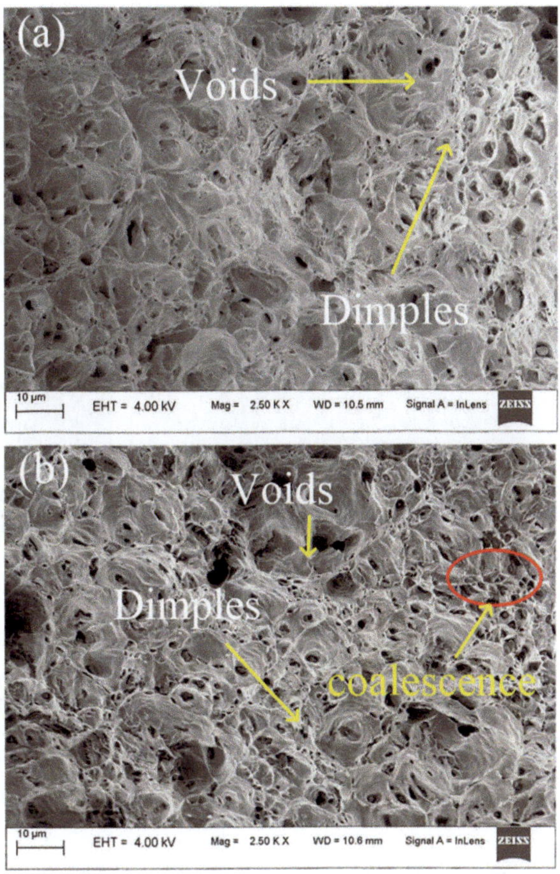

Fig. 10.14 FESEM images of fractured tensile samples of weld 1 and weld 2

These microvoids generally grow coalesce and then create the continuous-type fracture with time during tensile test in displace control mode. These microvoids may nucleate at relocalized strain discontinuities such as those associated with a priori defects (pores, microcracks), second phase particles, grain boundaries, inclusions, and dislocations pileup.

10.3.2 Tool Performance

From the visual inspection, it was observed that the tool undergone severe degradation. Photograph of the tool and the surface roughness was recorded after performing the welding.

10.3.2.1 Surface Roughness

Figure 10.15(a, b) shows the photograph of the tool after welding. Figure 10.15(b) show layer formation on the tool surface. Figure 10.15(c, d) shows the surface roughness values at the tool shoulder and tool pin respectvely. The surface roughness is the most important feature to characterize the surface integrity of the

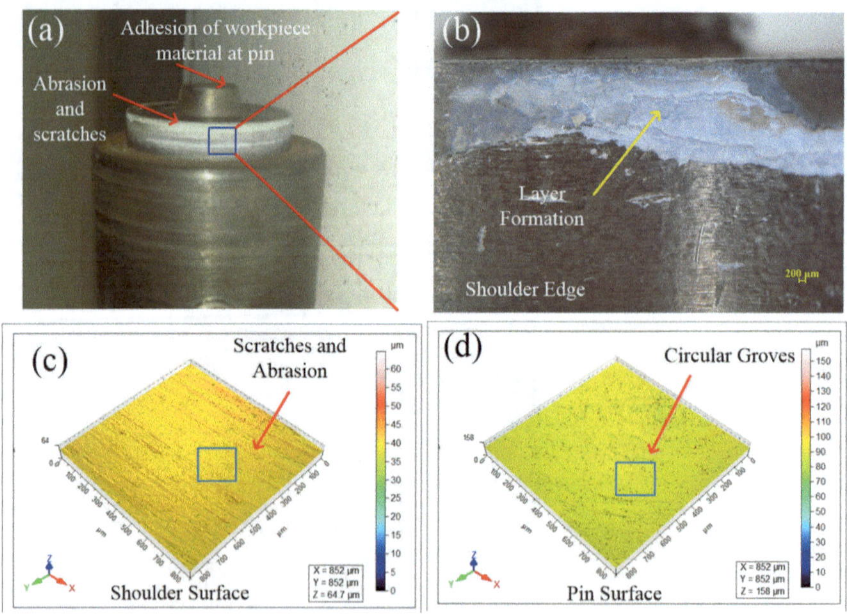

Fig. 10.15 **a** Tool photograph after performing welding, **b** surface roughness at the tool shoulder, and **c** tool pin

machined component. The surface quality is evaluating criteria for the productivity of machine tools as well as machined components. Hence, good quality surface finish is of great importance for less wear and higher tool life. The surface finish is characterized by average roughness (R_a) values. It was observed that the surface roughness values were higher at the tool pin than the tool shoulder. Adhesion of workpiece material was observed at the tool pin, and abrasion wear was observed at the tool shoulder. Adhesion of the workpiece on the tool pin may be due to the peak temperature generated at the tool pin. It was observed that abrasion and scratches were mainly responsible for the wear of the tool shoulder.

10.4 Conclusions

The present work indicated the capability of FSW to produce defect-free weld in low-carbon steel using tungsten carbide WC-10 wt.% Co-based tool. Successful joints were produced with superior mechanical properties. The major conclusions that can be drawn from the present investigations are as follows

- The hardness of the stir zone was higher than the base material. Higher hardness values of the stir zone can be due to grain-refined microstructure. Hardness values were decreasing on moving from stir zone toward the base material.
- Both the yield strength and the ultimate tensile strength of the welded samples were higher than the base material. Failure location was outside the weld zone confirming the good quality weld. The presence of dimples in the fractured surface of the tensile samples confirmed the ductile failure mode.
- Percentage elongation up to fracture was reduced in comparison to the base material. This reduction in percentage elongation is attributed to higher yield strength and reduction in work hardening.
- The tool wear was confirmed by visual inspection and surface roughness values measurement in FSW of steel. Surface roughness values were higher at the tool pin as compared to that of the tool shoulder.

Detailed systematic study will be conducted to study the effect of process parameters mainly the rotational speed and the traverse speeds on the weld joint quality. By controlling the process parameters, we can achieve sound weld quality and higher tool life to implement the process for industrial application.

Acknowledgements The authors thankfully acknowledge the funding and support provided by the Naval Research Board (NRB), Government of India. The authors are also grateful to the Management and Mechanical Engineering Department, Indian Institute of Technology Guwahati (IITG), Guwahati, India. The authors are also thankful to the Central Instruments Facility of IITG for providing the required research facilities.

References

1. Chung, Y.D., Fujii, H., Ueji, R., Tsuji, N.: Friction stir welding of high carbon steel with excellent toughness and ductility. Scr. Mater. **63**, 223–226 (2010). https://doi.org/10.1016/j.scriptamat.2010.03.060
2. Ghosh, M., Hussain, M., Gupta, R.K.: Effect of welding parameters on microstructure and mechanical properties of friction stir welded plain carbon steel. ISIJ Int. **52**, 477–482 (2012). https://doi.org/10.2355/isijinternational.52.477
3. Thomas, W.M., Threadgill, P.L., Nicholas, E.D.: Feasibility of friction stir welding steel. Sci. Technol. Weld. Join **4**, 365–372 (1999). https://doi.org/10.1179/136217199101538012
4. Reynolds, A.P., Tang, W., Posada, M., Deloach, J.: Friction stir welding of DH36 steel. Sci. Technol. Weld. Join **8**, 455–460 (2003). https://doi.org/10.1179/136217103225009125
5. Sato, Y.S., Yamanoi, H., Kokawa, H., Furuhara, T.: Microstructural evolution of ultrahigh carbon steel during friction stir welding. Scr. Mater. **57**, 557–560 (2007). https://doi.org/10.1016/j.scriptamat.2007.04.050
6. Ozekcin, A., Jin, H.W., Koo, J.Y., Bangaru, N.V., Ayer, R., Vaughn, G., et al.: A microstructural study of friction stir welded joints of carbon steels. Int. J. Offshore Polar Eng. **14**, 284–288 (2004)
7. Ghosh, M., Kumar, K., Mishra, R.S.: Friction stir lap welded advanced high strength steels: microstructure and mechanical properties. Mater. Sci. Eng., A **528**, 8111–8119 (2011). https://doi.org/10.1016/j.msea.2011.06.087
8. Fujii, H., Cui, L., Tsuji, N., Maeda, M., Nakata, K., Nogi, K.: Friction stir welding of carbon steels. Mater. Sci. Eng., A **429**, 50–57 (2006). https://doi.org/10.1016/j.msea.2006.04.118
9. Cui, L., Fujii, H., Tsuji, N., Nakata, K., Nogi, K., Ikeda, R., et al.: Transformation in stir zone of friction stir welded carbon steels with different carbon contents. ISIJ Int. **47**, 299–306 (2007). https://doi.org/10.2355/isijinternational.47.299
10. Toumpis, A., Galloway, A., Cater, S., McPherson, N.: Development of a process envelope for friction stir welding of DH36 steel—a step change. Mater. Des. **62**, 64–75 (2014). https://doi.org/10.1016/j.matdes.2014.04.066
11. Saeid, T., Abdollah-zadeh, A., Assadi, H., Malek, Ghaini F.: Effect of friction stir welding speed on the microstructure and mechanical properties of a duplex stainless steel. Mater. Sci. Eng., A **496**, 262–268 (2008). https://doi.org/10.1016/j.msea.2008.05.025
12. Husain, M.M., Sarkar, R., Pal, T.K., Prabhu, N., Ghosh, M.: Friction stir welding of steel: heat input, microstructure, and mechanical property co-relation. J. Mater. Eng. Perform. **24**, 3673–3683 (2015). https://doi.org/10.1007/s11665-015-1652-5
13. Lakshminarayanan, A.K., Balasubramanian, V., Salahuddin, M.: Microstructure, tensile and impact toughness properties of friction stir welded mild steel. J. Iron. Steel Res. Int. **17**, 68–74 (2010). https://doi.org/10.1016/S1006-706X(10)60186-0
14. Cui, L., Fujii, H., Tsuji, N., Nogi, K.: Friction stir welding of a high carbon steel. Scr. Mater. **56**, 637–640 (2007). https://doi.org/10.1016/j.scriptamat.2006.12.004
15. Arora, A., Zhang, Z., De, A., DebRoy, T.: Strains and strain rates during friction stir welding. Scr. Mater. **61**, 863–866 (2009). https://doi.org/10.1016/j.scriptamat.2009.07.015

Chapter 11
Parametric Investigation of Various Electrolytes During Micro-electrochemical Texturing on Stainless Steel

Sandip Kunar and Bijoy Bhattacharyya

11.1 Introduction

Micro-electrochemical texturing process plays a significant responsibility for fabrication of micro-circular patterns, which enhance the interfacial properties of mechanical components by reducing friction, wear, vibration, noise, power consumption, etc. This process has also been employed for various textured surface generations such as microgroove arrays, square micropatterns, rectangular micropatterns for enhancing of many advance fields, i.e., energy, electronics, optics, biology, information technology, etc [1]. Particular manufacturing technique is directly correlated with the generation of microsurface textures for definite functions, which are used as a part of realistic applications. Many researchers have engaged to provide interest in surface finish and modified surface features that influence the performance of mechanical components.

Maskless micro-electrochemical texturing process is used to produce the microsurface textures with controlled shape, size, and surface quality. Zhang et al. [2] have approached a process of sandwich-like EMM method to reduce the radial overcut of micro-circular pattern and to enhance the machining accuracy of micro-circular pattern using 10 wt% $NaNO_3$ electrolyte. Various types of textured patterns such as ellipses, squares, and hexagons have also been fabricated using $NaNO_3$ electrolyte at a concentration of 10 wt%. This method is time consuming because every workpiece needs individual masking before machining. Shin et al. [3] have used a method of electrochemical etching through laser masking for controlled anodic dissolution using 2 M sodium nitrite electrolyte. Laser masking process is used to mask the textured surface and acts as a temporary mask. The patterned surface is dissolved by electrochemical etching under a parametrical combination using 2 M sodium nitrite electrolyte. Laser masking process may

S. Kunar (✉) · B. Bhattacharyya
Department of Production Engineering, Jadavpur University, Kolkata 700032, India
e-mail: sandip.sandip.kunar@gmail.com

P. Sahoo and J. P. Davim (eds.), *Advances in Materials, Mechanical and Industrial Engineering*, Lecture Notes on Multidisciplinary Industrial Engineering,
https://doi.org/10.1007/978-3-319-96968-8_11

deteriorate the machining accuracy of patterned surface, and higher concentration of electrolyte is unsuitable for fabrication of micropatterned surfaces which may also reduce the dimensional uniformity. Hou et al. [4] have introduced the method of through-mask EMM (TMEMM) for fabrication of micro-circular pattern on large area of stainless steel workpiece using $NaNO_3$ solution having concentration of 100 g/L. WEDM produces mold with micropillar arrays on stainless steel for generation of microdimple. Through-mask EMM is a time-consuming process, and higher concentration of electrolyte is costly for fabrication of many machined samples. Mitchell-Smith et al. [5] have proposed the electrochemical jet processing technique to fabricate surface structuring on large area using higher molar concentrations of NaCl, $NaNO_3$, and NaI. Higher molar concentration is a costly process and may deteriorate the dimensional uniformity. Ming et al. [6] have used a modified TMEMM process with reusable mask for fabrication of microdimple array using 20 wt% $NaNO_3$ electrolyte and investigated the effect of voltage, mask thickness, external pressure, and duty ratio on dimple responses. Madore et al. [7] have employed TMEMM method to fabricate well-defined regular grid patterns and spiral-shaped channels on titanium using 3 M sulfuric acid. This process is not economical due to higher concentration of electrolyte which lowers the machining accuracy. Zhang et al. [8] have investigated the influences of TMEMM process parameters, i.e., machining time and voltage on machining accuracy of micro-circular pattern using 10 wt% $NaNO_3$ electrolyte. But, this process is costly and time consuming for fabrication of many machined samples. Wang et al. [9] have introduced combined technology of electro-discharge machining and electrochemical machining processes to generate microhole arrays using KOH electrolyte having concentration of 2 M/L. Firstly, rectangular columns are fabricated by the wire-EDM, and then EMM is utilized to generate arrays of cylindrical columns. These microelectrodes are used to fabricate micro-circular pattern using EMM method. Higher concentration of electrolyte is costly for fabrication of dimple arrays. Wang et al. [10] have used electrochemical micromachining method to fabricate the microholes and square micropatterns by disk-type microelectrodes using 5–20% $NaNO_3$ electrolyte. The influence of EMM parameters like feeding speed, pulse on time, and voltage on taper, localization, material removal rate, and hole diameter is investigated. This process is a time-consuming process for generation of disk-type electrodes via WEDM. Schuster [11] has described electrochemical microstructuring with short voltage pulses using $0.01mHClO_4$/ $0.1mCuSO_4$ electrolyte. Higher electrolyte concentration is costly for generation of microstructures using electrochemical reactions. Kawanaka et al. [12] have presented electrolyte jet machining for controlled microsurface texturing using 20wt% $NaNO_3$ and 20wt% NaCl electrolytes and investigated the effects of electrolyte and current density on the surface properties of stainless steel. It is a time-consuming process for generation of one-by-one hole. Costa et al. [13] have used maskless electrochemical texturing process for fabrication of textured surfaces using 250 g/l

$NaNO_3$ electrolyte. This method is utilized to describe the influence of electrolyte flushing conditions and current pulse history on material removal rate, feature definition, and current efficiency. This proposed technique is uneconomical for higher concentration of electrolyte which may reduce the dimensional uniformity of generated textured patterns. Parreira et al. [14] have improved the electrochemical dissolution with good reproducibility and high speed for industrial applications using 200 g/l NaCl solution. The effect of voltage, machining time, and inter-electrode gap is evaluated during fabrication of different texturing patterns such as trace dots, chevrons, and dots. EDM is utilized to produce the cathode tools. But, EDM may reduce the dimensional uniformity and higher concentration is costly for production of texturing patterns. Schonenberger et al. [15] have described an electrochemical process to transfer microscale patterns such as square and linear micropatterns on a fully exposed substrate using 0.1 M $CuSO_4$ and 0.5 M H_2SO_4 electrolytes. But, the etched depth of micropatterns is less. Kunar et al. [16] have investigated the influences of maskless EMM process variables, i.e., texturing time, frequency, duty cycle, voltage, and electrolyte concentration on microsurface textured responses including machining accuracy, current efficiency, and depth of micro-circular pattern using combined electrolytes of NaCl and $NaNO_3$. But, the size of circular pattern is large.

The textured patterns are widely used in many applications like automotive, aerospace, and defense. These micropatterns are also used for improving the tribological properties of interfacial components. Many fabrication techniques are used to manufacture the surface textures, such as abrasive jet machining, LBM, EDM, EMM, micromilling, chemical etching. But, maskless EMM process is a promising technique for fabrication of different surface textures due to its several benefits like higher material removal rate, lower surface roughness, higher surface quality, free from tool wear, cracks, and burr than other methods.

In this article, an alternative approach of maskless EMM is developed to produce microsurface textures containing micro-circular patterns by improving the textured characteristics. In this article, a maskless EMM setup with maskless EMM cell and vertical cross-flow electrolyte system has been indigenously developed for performing the experiments of micro-circular pattern generation. One masked patterned tool can fabricate numerous micro-circular patterns. The comparison of machined characteristics, i.e., overcut and surface roughness using three different types of electrolytes, i.e., $NaNO_3$, NaCl, and NaCl + $NaNO_3$, has been investigated for the influences of major influencing EMM process variables, i.e., inter-electrode gap(IEG), machining voltage, duty ratio, and pulse frequency. The analysis of machined characteristics for micro-circular patterns has been carried out to obtain the suitable machining conditions for three different electrolytes. The friction tests have also been conducted to show the tribological aspects on generated textured patterns.

11.2 Experimental Method

11.2.1 Experimental Setup and Planning

Figure 11.1 represents the indigenously developed maskless electrochemical micromachining setup consisting of various sub-systems, i.e., maskless electrochemical micromachining cell, electrode fixture devices, power supply unit, and electrolyte circulation arrangement for generation of microsurface textures. EMM cell consists of tool- and workpiece-holding devices, inlet and outlet arrangements for vertical cross-flow electrolyte system, and power supply arrangements. Maskless electrochemical micromachining cell with electrode fixtures is made of Perspex material, and outlet and inlet segments of EMM cell are made of stainless steel material for avoiding corrosion from environment. EMM cell is a well-built structure because all assembled components of cell can endure easily higher electrolyte flow rate. The pulsed current has been provided from pulsed DC power supply which has in-built function generator, protection functions, ultra-fast response, and compact size. The most significant feature of EMM cell is a vertical cross-flow electrolyte supply system which is suitable for generation of uniform micro-circular pattern generation, and it has capability to provide extra back pressure as it flows from downward to upward directions. This generated extra back pressure cleans electrolysis products from the micromachining zone. It can also be utilized for fabrication of intricate profiles with good surface quality compared to other flow systems. In this method, micro-circular patterned tool having 600-μm diameter is fabricated using SU-8 2150 mask having thickness of 230 μm by UV lithography process. The masked pattern is generated on stainless steel wafer. The gap between two consecutive holes across the margin is 1000 μm. This fabricated stainless steel sheet is attached to precision micrometer arrangements for inter-electrode gap settings.

Dimensional accuracy and good surface finish of machined circular patterns are directly related to masked patterned tool quality. Three types of electrolytes are used for these experiments, i.e., NaCl (0.25 M/L), $NaNO_3$ (0.17 M/L), and NaCl (0.25 M/L) + $NaNO_3$ (0.17 M/L). The selected ranges of machining voltage, inter-electrode gap (IEG), duty ratio, and pulse frequency are 10–16 V, 100–400 μm, 40–70%, and 2–5 kHz, respectively, with other supplementary constant process parameters like current cutoff of 2A, back pressure of 0.15 kg/cm^2, machining time of 3 min, and flow rate of 5.75 m^3/hr on the basis of many trial experiments. These process parameters are used to study the microsurface textured characteristics using three different electrolytes, and the suitable electrolyte is chosen from the experimental results. Each experiment has been repeated five times, and average results have been taken for each textured characteristic to eradicate the results' error. Below the ranges of above-mentioned process parameters, all micro-circular patterns do not produce appropriately. Above the ranges of above-mentioned process parameters, the shape and size of micro-circular patterns

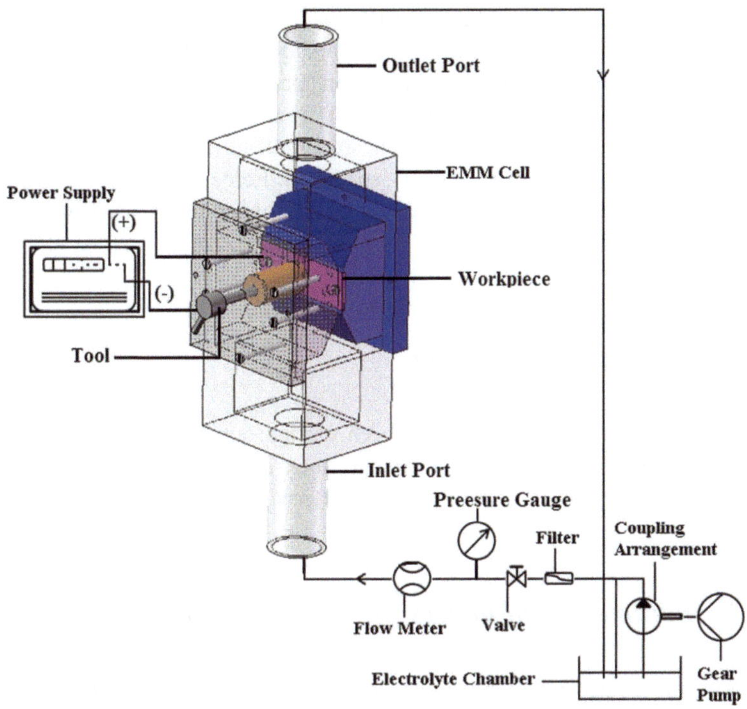

Fig. 11.1 Developed maskless EMM setup

are irregular and machining efficiency is very less. The microsurface textured responses are examined with 3D non-contact profilometer (CCI Sunstar, Taylor Hobson Ltd.) and optical microscope (Leica EZ4D, Germany).

11.2.2 Friction Test

The friction tests are conducted using a multi-tribotester block TR25 (Ducom). A rotating steel roller (material: EN8 steel and hardness: 55 HRC) is used as a counter-surface and stainless steel as a test specimen. All textured samples are kept fixed with the help of some attachments, and rotating steel roller is made to slide against the specimen. The tests are conducted under wet condition at ambient temperature of 28 °C with a relative humidity of 85% using specimen-on-roller geometry. Commercially available Castrol Gear Oil (SAE 90EP) is used as a lubricant (density of 897 kg/m^3 at 15 °C, kinematic viscosity of 178.1 mm^2/s at 40 °C, 17 mm^2/s at 100 °C, viscosity index of 95, and flash point of 202 °C), and the test specimen is supplied adequate volume of lubricant for full lubrication at the beginning of every test. The applied loads are 25, 50, and 75 N. The contact pressures are 1.081, 2.181, and 3.262 MPa for 25, 50, and 75 N, respectively. For each load, the test is carried out for 15 min at sliding speed of 300 rpm. Every test is repeated three times. Mean values of friction coefficients have been taken for each condition. The roller speed and interval of friction tests can be controlled by computer attached to the tribotester. The required normal load is applied to the loading lever on the upper part of every specimen. Force sensor is used to measure the frictional force. A computer-attached testing apparatus is used to record the frictional coefficient.

11.3 Results and Discussion

The experimental investigations have been carried out in the developed setup to explore the influence of machining voltage, inter-electrode gap (IEG), duty ratio, and pulse frequency on the micro-circular patterned characteristics.

11.3.1 Parametric Influences on Textured Characteristics

Influences of voltage on overcut and surface roughness for $NaNO_3$, NaCl, and $NaNO_3$ + NaCl electrolytes are studied with other constant parameters such as pulse frequency of 5 kHz, inter-electrode gap of 100 μm, and duty ratio of 40%.

Figure 11.2 represents that the overcut of micro-circular pattern increases with increasing machining voltage for different types of electrolytes. It occurs because machining current increases with increase in machining voltage. As a result, current density and stray current effect increase with increasing machining voltage. For higher stray current effect, material removal rate increases from the periphery of micro-circular pattern and radial overcut increases. For $NaNO_3$, the radial overcut occupies the middle zone between other two electrolytes without 12 V. It occurs because the machining localization is higher for lower electrical conductivity than that of NaCl. At 12 V, lower overcut is observed than that of NaCl and $NaNO_3$ + NaCl because controlled etching occurs. In case of NaCl, the overcut is higher than that of $NaNO_3$ and $NaNO_3$ + NaCl because it has higher electrical conductivity which is responsible for higher stray current effect and increases the machining area of micro-circular pattern. In case of $NaNO_3$ + NaCl electrolyte, lower radial overcut is obtained because the combined electrolyte performs controlled etching than that of NaCl and $NaNO_3$. Lower voltage with $NaNO_3$ + NaCl electrolyte is recommended for generation of uniform micro-circular pattern.

Influences of applied voltage on surface roughness are investigated for $NaNO_3$, NaCl, and $NaNO_3$ + NaCl electrolytes as shown in Fig. 11.3. The surface roughness increases with increasing applied voltage for non-uniform current distribution. For $NaNO_3$, surface roughness lies in between NaCl and $NaNO_3$ + NaCl due to controlled dissolution for lower electrical conductivity. In case of NaCl, higher

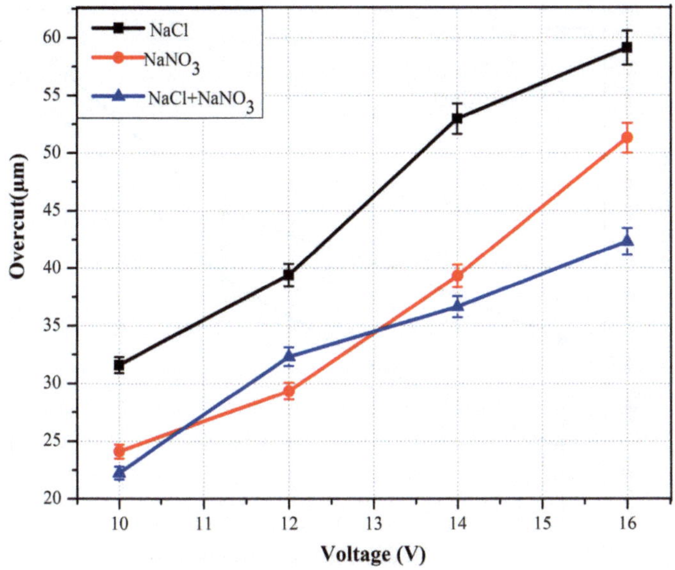

Fig. 11.2 Influence of machining voltage on radial overcut for $NaNO_3$, NaCl, and $NaNO_3$ + NaCl

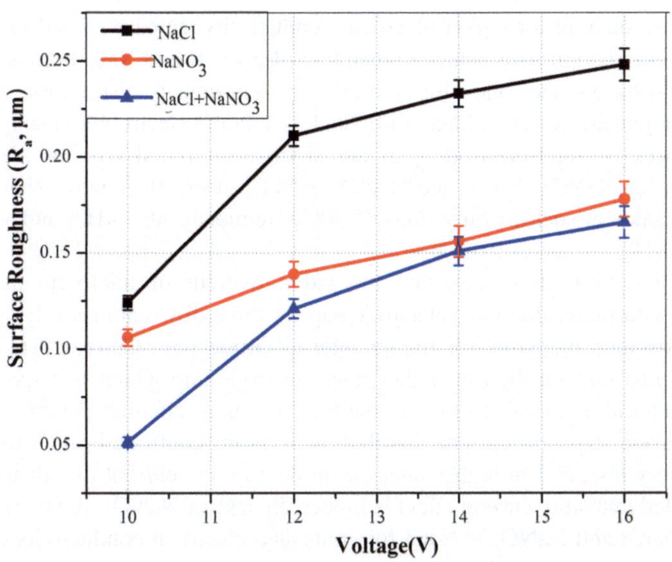

Fig. 11.3 Influence of machining voltage on surface roughness for $NaNO_3$, NaCl, and $NaNO_3$ + NaCl

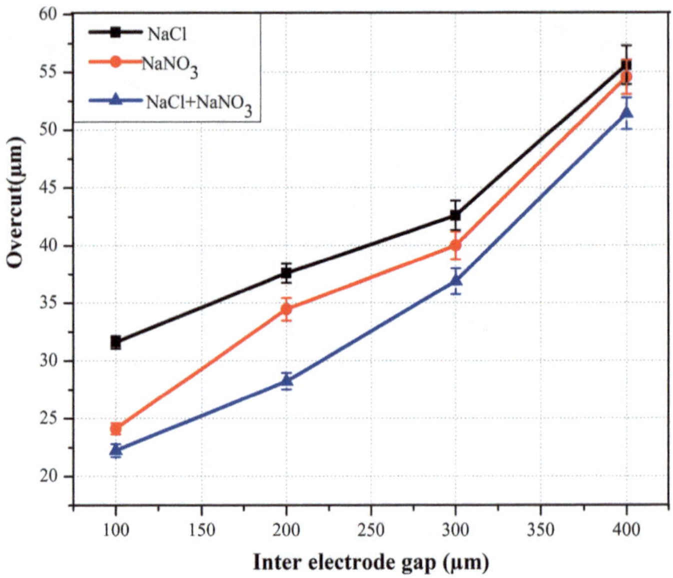

Fig. 11.4 Influence of inter-electrode gap on radial overcut for $NaNO_3$, NaCl, and $NaNO_3$ + NaCl

roughness is observed than that of $NaNO_3$ and $NaNO_3$ + NaCl. It occurs due to non-uniform etching for higher electrical conductivity. In case of $NaNO_3$ + NaCl, lower surface roughness is achieved than that of $NaNO_3$ and NaCl because uniform anodic dissolution takes place for controlled current density distribution. For lower surface roughness, lower voltage with $NaNO_3$ + NaCl electrolyte is suggested.

Influences of inter-electrode gap are studied on radial overcut and surface roughness for $NaNO_3$, NaCl, and $NaNO_3$ + NaCl electrolytes with other constant EMM parameters such as duty ratio of 40%, frequency of 5 kHz, and voltage of 10 V.

From Fig. 11.4, it is seen that the radial overcut of micro-circular pattern increases with increasing inter-electrode gap for three different electrolytes because ohmic resistance increases for higher inter-electrode gap. Current density distribution is also uncontrolled in higher inter-electrode gap which is responsible for higher material removal from the margin of micro-circular pattern. In lower inter-electrode gap, the current distribution is more controlled and stray current effect is very less, and in higher inter-electrode gap, the current flux distribution is uncontrolled and stray current effect is higher. In case of $NaNO_3$, the overcut lies in between NaCl and $NaNO_3$ + NaCl for controlled electrical conductivity compared to NaCl. In case of NaCl, higher overcut is obtained than that of $NaNO_3$ and $NaNO_3$ + NaCl for higher metallic ions. For $NaNO_3$ + NaCl, lower overcut is acquired than that of $NaNO_3$ and NaCl. It occurs due to controlled anodic dissolution. For good geometrical shape, lower inter-electrode gap with $NaNO_3$ + NaCl electrolyte is suggested.

Influences of inter-electrode gap on surface roughness (R_a) for $NaNO_3$, NaCl, and $NaNO_3$ + NaCl are investigated as shown in Fig. 11.5. The surface roughness increases with increase in inter-electrode gap. This is because non-uniform etching occurs for uncontrolled current flux distribution. In case of $NaNO_3$, the surface roughness occupies the middle zone in between $NaNO_3$ + NaCl and NaCl due to uniform anodic dissolution for controlled electrical conductivity compared to NaCl. In case of NaCl, higher surface roughness is achieved than that of $NaNO_3$ and $NaNO_3$ + NaCl due to non-uniform etching for higher electrical conductivity. In case of $NaNO_3$ + NaCl, lower surface roughness is acquired than that of $NaNO_3$ and NaCl due to controlled etching for uniform current flux distribution. For good surface quality, lower inter-electrode gap with NaCl + $NaNO_3$ electrolyte is suggested.

Influences of duty ratio are explored on the overcut and surface roughness for $NaNO_3$, NaCl, and $NaNO_3$ + NaCl electrolytes with other constant parameters such as inter-electrode gap of 100 μm, voltage of 10 V, and pulse frequency of 5 kHz.

The radial overcut of micro-circular pattern increases with increasing duty ratio for $NaNO_3$, NaCl, and $NaNO_3$ + NaCl electrolytes because pulse on time, i.e., machining time, increases and pulse off time decreases in higher duty ratio as shown in Fig. 11.6. The stray current effect is also higher for higher pulse on time. In case of $NaNO_3$ electrolyte, the radial overcut locates in between NaCl and $NaNO_3$ + NaCl electrolytes without duty ratio of 70% because the lower stray current effect is observed for lower electrical conductivity. At duty ratio of 70%, higher overcut is obtained than that of NaCl and $NaNO_3$ + NaCl due to

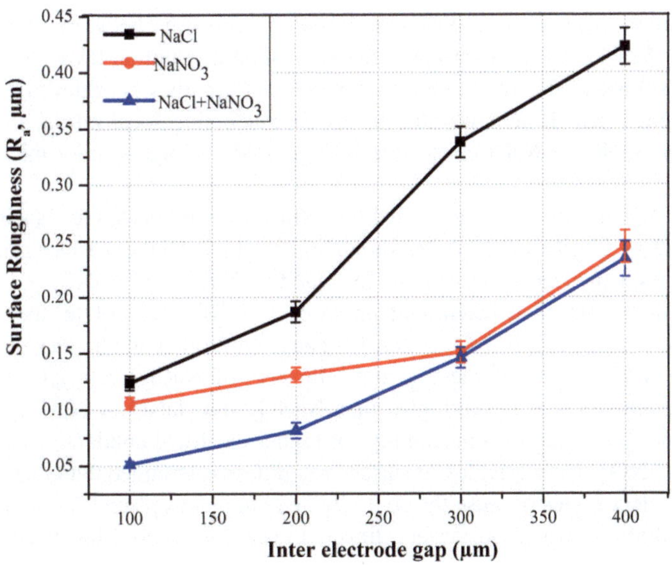

Fig. 11.5 Influence of inter-electrode gap on surface roughness for $NaNO_3$, NaCl, and $NaNO_3$ + NaCl

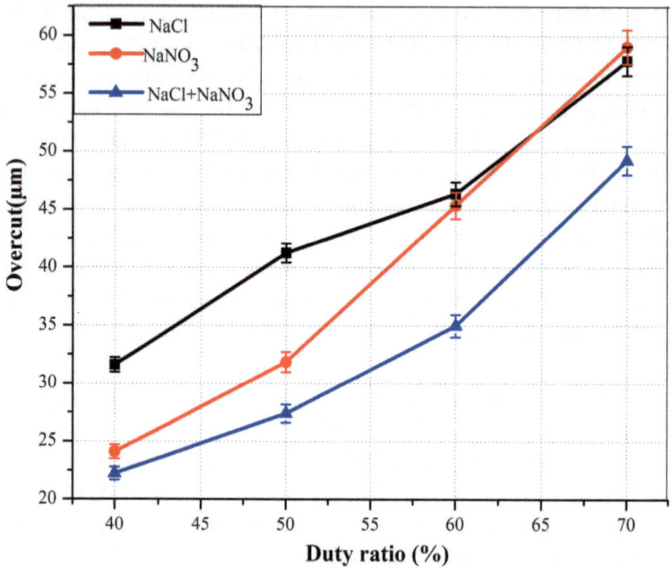

Fig. 11.6 Influence of duty ratio on overcut for $NaNO_3$, NaCl, and $NaNO_3$ + NaCl

uncontrolled machining for the availability of higher pulse on time in higher duty ratio. In case of NaCl electrolyte, higher overcut is achieved than that of $NaNO_3$ and $NaNO_3$ + NaCl electrolytes without duty ratio of 70% because higher electrical conductivity of NaCl is responsible for higher stray current effect. At duty ratio of 70%, radial overcut occupies the middle zone in between $NaNO_3$ and $NaNO_3$ + NaCl electrolytes because the controlled dissolution takes place. In case of NaCl + $NaNO_3$ electrolyte, higher-dimensional accuracy is achieved than that of $NaNO_3$ and NaCl. It achieves due to higher machining localization. Lower duty ratio with $NaNO_3$ + NaCl electrolyte is suggested for high-quality micro-circular pattern.

Influences of duty ratio are studied on surface roughness (R_a) for $NaNO_3$, NaCl, and $NaNO_3$ + NaCl electrolytes as shown in Fig. 11.7. Higher surface roughness is acquired with increase in duty ratio for $NaNO_3$, NaCl, and $NaNO_3$ + NaCl electrolytes for uncontrolled machining. In case of $NaNO_3$ electrolyte, the roughness lies between NaCl and $NaNO_3$ + NaCl because controlled anodic dissolution takes place for less number of metallic ions. In case of NaCl electrolyte, the higher surface roughness is achieved higher than $NaNO_3$ and $NaNO_3$ + NaCl electrolytes. This is due to non-uniform machining for higher electrical conductivity. In case of NaCl + $NaNO_3$ electrolyte, lower surface roughness is obtained than that of $NaNO_3$ and NaCl electrolytes because the controlled etching takes place for uniform current flux distribution. For good surface finish, lower duty ratio with $NaNO_3$ + NaCl electrolyte is recommended.

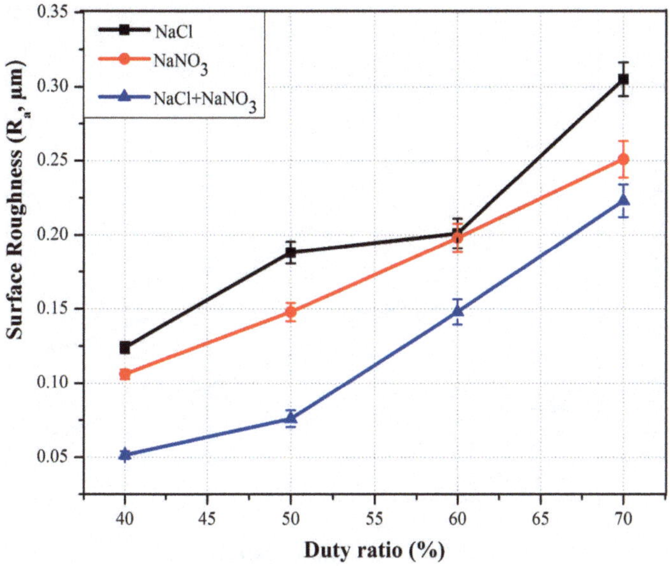

Fig. 11.7 Influence of duty ratio on surface roughness for $NaNO_3$, NaCl, and $NaNO_3$ + NaCl

Influences of pulse frequency are studied on overcut and surface roughness for three different $NaNO_3$, NaCl, and $NaNO_3$ + NaCl electrolytes with other constant parameters, i.e., inter-electrode gap of 100 µm, voltage of 10 V, and duty ratio of 40%.

Influence of pulse frequency on overcut for $NaNO_3$, NaCl, and $NaNO_3$ + NaCl electrolytes is investigated as shown in Fig. 11.8. The radial overcut decreases with increasing pulse frequency because the rate of repetition of pulse on time, i.e., machining time, decreases with increasing pulse frequency. So, the stray current effect is less in higher frequency range and higher in lower frequency range. In case of $NaNO_3$ electrolyte, the radial overcut occupies the middle zone in between $NaNO_3$ and NaCl. It occurs due to controlled electrical conductivity compared to NaCl. In case of NaCl electrolyte, higher overcut is observed than that of NaCl and $NaNO_3$ + NaCl because of higher electrical conductivity which is responsible for higher stray current effect. In case of $NaNO_3$ + NaCl, lower overcut is achieved compared to NaCl and $NaNO_3$ because of controlled etching with less stray current effect. Higher frequency with $NaNO_3$ + NaCl electrolyte is suggested for lower overcut.

Influences of pulse frequency on surface roughness for $NaNO_3$, NaCl, and $NaNO_3$ + NaCl are investigated as shown in Fig. 11.9. The surface roughness decreases with increasing pulse frequency for $NaNO_3$, NaCl, and $NaNO_3$ + NaCl due to uniform etching. For $NaNO_3$, the surface roughness possesses the middle zone in between $NaNO_3$ + NaCl and NaCl because controlled dissolution takes place for lower electrical conductivity. In case of NaCl, the surface roughness is higher than that of $NaNO_3$ and $NaNO_3$ + NaCl due to uncontrolled machining with

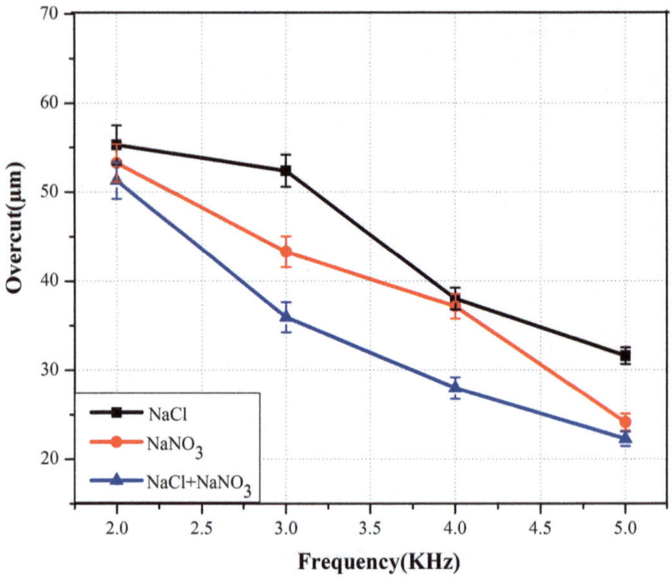

Fig. 11.8 Influence of pulse frequency on overcut for $NaNO_3$, NaCl, and $NaNO_3$ + NaCl

non-uniform current distribution for higher metallic ions. In case of NaCl +$NaNO_3$, lower roughness is achieved than that of $NaNO_3$ and NaCl because controlled etching takes place for uniform current flux distribution. Higher pulse frequency with $NaNO_3$ + NaCl electrolyte is recommended for good surface finish of micro-circular pattern.

11.3.2 Analysis of Fabricated Micro-Circular Patterns

SU-8 2150 mask is used on the patterned cathode tool for surface textured experiments. This mask has capability for fabricating more than twenty-five samples due to higher adhesiveness with stainless steel wafer and higher strength for enduring higher flow rate. It does not deform after generation of twenty-five samples. So, SU-8 2150 mask demonstrates the reusability of masked tool. Micro-circular pattern on SU-8 2150 mask is shown in Fig. 11.10a before machining. The patterned mask after use of twenty-five times is shown in Fig. 11.10b. So, the utility of SU-8 2150 mask has been experimentally verified for generating many high-quality micro-circular patterns.

The micrograph of improper micro-circular pattern is captured by the optical microscope as shown in Fig. 11.11, and it is fabricated at a particular parametric setting, i.e., inter-electrode gap of 100 μm, voltage of 10 V, duty ratio of 40%, and pulse frequency of 5 kHz using NaCl electrolyte. Some micro-circular impressions

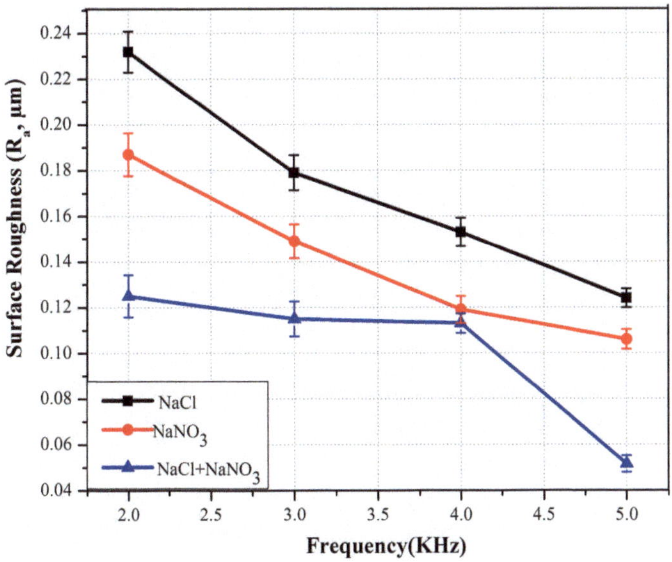

Fig. 11.9 Influence of pulse frequency on surface roughness for $NaNO_3$, NaCl, and $NaNO_3$ + NaCl

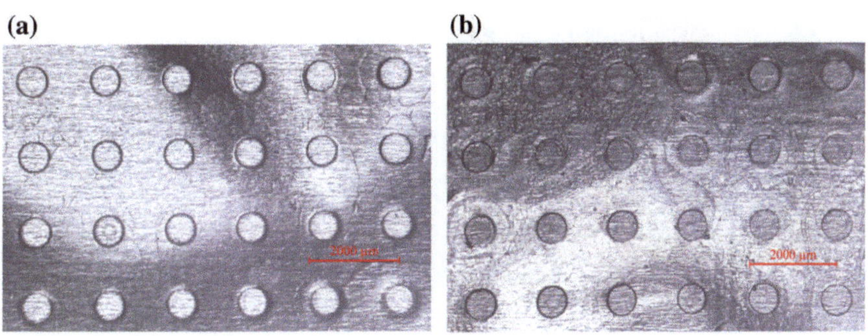

Fig. 11.10 Masked tool **a** before machining and **b** after machining of twenty-five times

of improper micro-circular pattern are irregular in terms of dimensional accuracy because of uncontrolled machining. This is because NaCl electrolyte has higher electrical conductivity which is responsible for higher stray current effect. Some impressions do not form properly due to accumulation of insoluble by-products. Machining depth of all impressions is non-uniform due to lower machining localization for the availability of higher metallic ions. The surface quality of micro-circular pattern also deteriorates due to uncontrolled etching.

A 3D image and the details of surface roughness (R_a) profile of micro-circular impression are shown in Fig. 11.12. The value of surface roughness of micro-circular impression is 0.220 μm.

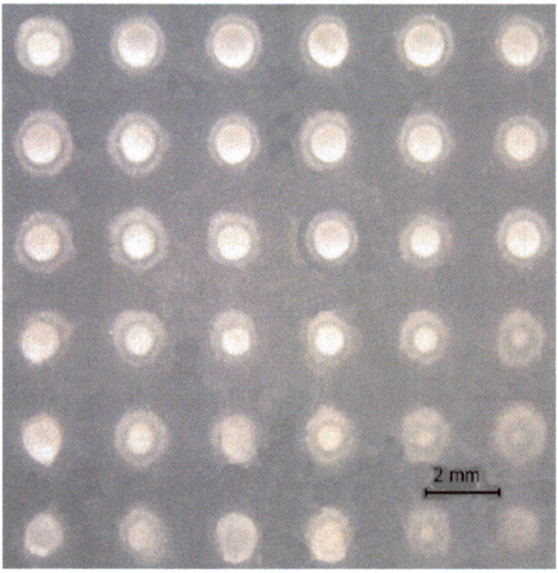

Fig. 11.11 Improper micro-circular pattern generated by NaCl

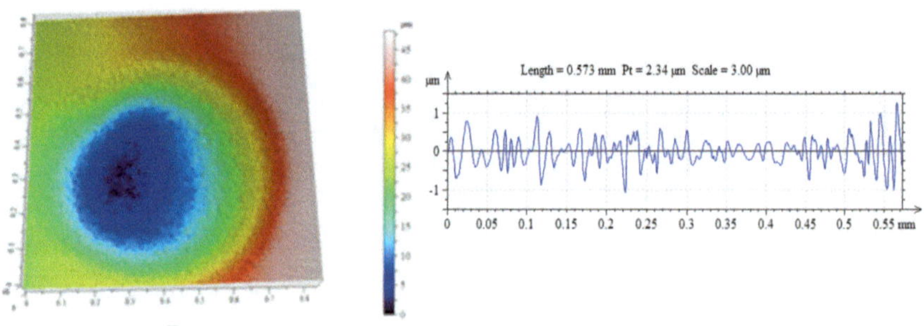

Fig. 11.12 Three-dimensional view with roughness profile of a micro-circular impression generated by NaCl

The optical microscopic image of irregular micro-circular pattern is produced by a particular parametric combination including pulse frequency of 5 kHz, voltage of 10 V, duty ratio of 40%, and inter-electrode gap of 100 µm using $NaNO_3$ electrolyte as shown in Fig. 11.13. Improper micro-circular pattern does not retain the dimensional uniformity of some micro-circular impressions. This is due to non-uniform etching for uncontrolled current density distribution. The generated machining depth of micro-circular pattern is non-uniform throughout the pattern due to uncontrolled machining localization. The surface roughness of micro-circular pattern is also higher due to uncontrolled etching.

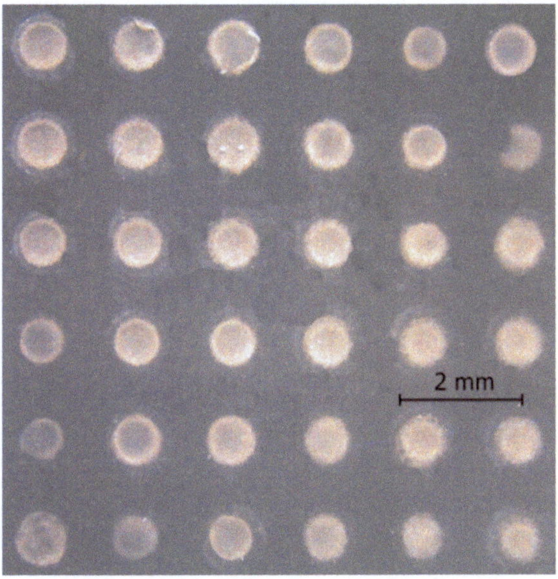

Fig. 11.13 Irregular micro-circular pattern generated by $NaNO_3$

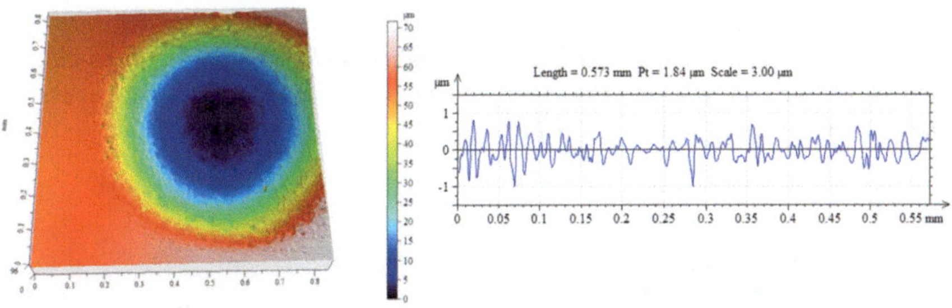

Fig. 11.14 Three-dimensional view with roughness profile of a micro-circular impression generated by $NaNO_3$

A 3D image and the details of surface roughness (R_a) profile of micro-circular impression are shown in Fig. 11.14. The value of surface roughness of micro-circular impression is 0.187 µm.

The micrograph of regular micro-circular pattern is generated at the particular machining conditions, i.e., voltage of 10 V, duty ratio of 30%, inter-electrode gap of 100 µm, and pulse frequency of 5 kHz using $NaNO_3$ + NaCl electrolyte as shown in Fig. 11.15. The dimensional uniformity of generated micro-circular pattern is homogeneous throughout the pattern. It occurs because of uniform anodic dissolution for mixed electrolyte than that of $NaNO_3$ and NaCl electrolytes. The machining depth of micro-circular pattern is also controlled across the pattern due

Fig. 11.15 Regular micro-circular pattern generated by NaNO3 + NaCl

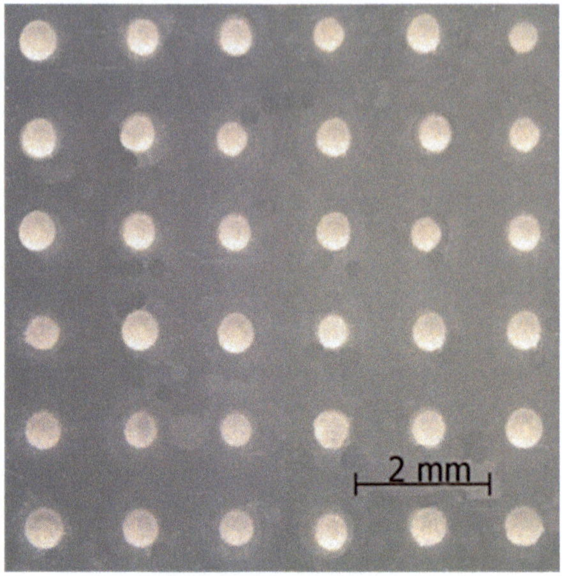

to higher anodic dissolution for uniform current flux distribution. The surface roughness of generated regular micro-circular pattern is also lower because of uniform machining.

A 3D image and the details of surface roughness (R_a) profile of micro-circular impression are shown in Fig. 11.16. The value of surface roughness of micro-circular impression is 0.123 μm.

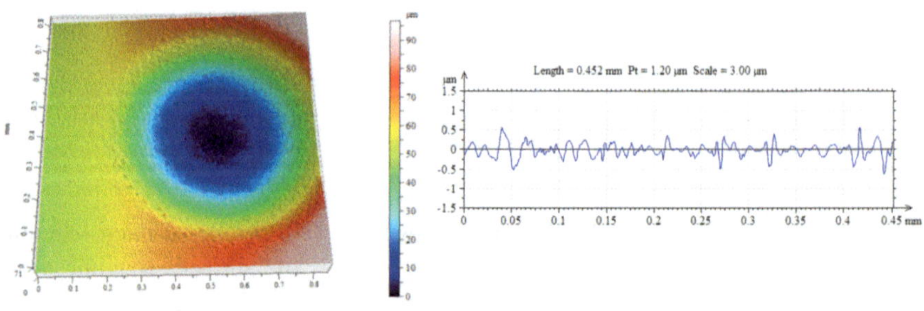

Fig. 11.16 Three-dimensional view with roughness profile of a regular micro-circular impression generated by $NaNO_3$ + NaCl

11.3.3 Analysis of Friction Test

To demonstrate the utility of textured surfaces, the friction tests are carried out on textured specimens containing micro-circular patterns with mean diameter of 630 μm. The friction tests are conducted with constant rotational speed of 300 rpm and time of 15 min, respectively. The tests are conducted at wet condition for different depths of 21.4, 48.5 μm and untextured surfaces. The area density of micro-circular pattern is defined as the ratio between textured area and untextured area. The area density of micro-circular pattern is approximately 12.63–18.52%.

The effect of normal load on coefficient of friction for micro-circular patterned and untextured surfaces at wet condition has been shown in Fig. 11.17. The friction coefficient of microsurface textures decreases with increasing normal loads compared to untextured surfaces. The micro-circular patterned surfaces have positive effect due to trapping of wear debris by reducing wear and tear. The most important surface effect is hydrodynamic effect under low load and high-speed conditions of conformal contact [17]. The aspect ratio of micro-circular pattern is the basic feature for determination of hydrodynamic effect between two sliding surfaces, which is experimentally and theoretically verified. The range of aspect ratio, i.e., 0.01 to 0.05, is responsible for hydrodynamic effect [18, 19]. The aspect ratios are 0.033 and 0.076 for the depths of 21.4 and 48.5 μm, respectively. So, the hydrodynamic effect is more responsible for depth of 21.4 μm. The hydrodynamic effect leads to a considerable increase in film thickness, which reduces the coefficient of friction. But, hydrodynamic effect is not significant for depth of 48.5 μm. Mixed lubrication

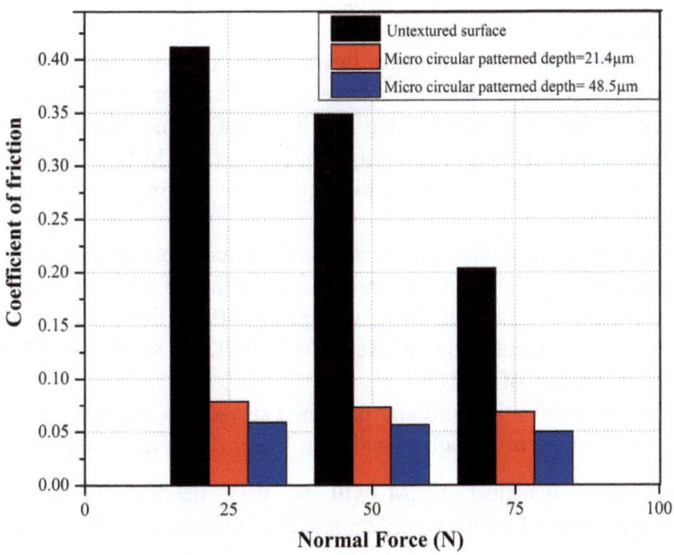

Fig. 11.17 Effect of normal force on coefficient of friction for untextured surface and different micro-circular patterned depths

is more effective for the depth of 48.5 μm. When the load increases, mixed lubrication is more significant and higher patterned depth can increase the amount of oil store, which improves the lubrication effect [20, 21]. The micro-circular patterns store the lubricant and reduce serious wear and tear between two sliding surfaces. Higher depth of patterned surfaces can hold the lubricant for longer time than lower depth and untextured surface, which assists to reduce the friction coefficient of moving surfaces for long time. The reduction of highest friction coefficient is observed at higher normal load and higher depth at wet condition.

11.4 Conclusions

In this study, maskless electrochemical micromachining is a promising technology for micro-circular pattern generation. Effects of process variables such as inter-electrode gap, machining voltage, duty ratio, and pulse frequency are investigated on overcut and surface roughness of micro-circular pattern. The following conclusions can be summarized:

(i) An alternative concept of maskless EMM is a distinctive proposed technique for micro-circular pattern generation economically. In this method, one patterned cathode tool with SU-8 2150 mask can produce more than twenty-five textured samples and avoids the masking of individual workpiece before machining.
(ii) In this experimental setup, EMM cell and vertical cross-flow electrolyte system have been developed indigenously for experimental investigations of microdimple pattern.
(iii) Machining with lower voltage, lower inter-electrode gap, higher pulse frequency, and lower duty ratio is recommended for minimum overcut and lower surface roughness using $NaNO_3$, NaCl, and $NaNO_3$ + NaCl electrolytes. Only the mixed electrolyte of NaCl (0.25 M/L) + $NaNO_3$ (0.17 M/L) generates the best array of high-quality microdimple patterns than other electrolytes.
(iv) From the experimental investigations of microdimple pattern generation, the best machining conditions are inter-electrode gap of 100 μm, machining time of 3 min, duty ratio of 30%, voltage of 10 V, flow rate of 5.75 m^3/hr, electrolyte concentration of NaCl (0.25 M/L) + $NaNO_3$ (0.17 M/L), and pulse frequency of 5 kHz.
(v) Higher depth of micro-circular patterned surfaces is suitable for reduction of friction coefficient at higher load than lower depth and untextured surfaces.

In this study, maskless EMM will be very needful for fabricating the microdimple patterns of various sizes on metallic surfaces for different applications. Hence, for generation of uniform surface textures, the design of a proper patterned tool, tool movement strategy, and controlled flushing conditions require to be improved further.

References

1. Bruzzone, A.A.G., Costa, H.L., Lonardo, P.M., Lucca, D.A.: Advances in engineered surfaces for functional performance. CIRP Ann. Manufact. Technol. **57**(2), 750–769 (2008)
2. Zhang, X., Qu, N., Chen, X., Zhang, X.: Sandwich-like electrochemical micromachining of micro-dimples. Surf. Coat. Technol. **302**, 438–447 (2016)
3. Shin, H.S., Chung, D.K., Park, M.S., Kim, B.H., Chu, C.N.: Electrochemical etching of stainless steel through laser masking. J. Micromech. Microeng. **20**(5), 055030, 1–10 (2010)
4. Hou, Z., Qu, N., Chen, X.: Electrochemical micromachining of large-area micro-dimple arrays with high machining accuracy. J. Eng. Manufact. 1–10 (2016)
5. Mitchell-Smith, J., Speidel, A., Clare, A.T.: Transitory electrochemical masking for precision jet processing techniques. J. Manufact. Process. **31**, 273–285 (2018)
6. Ming, P., Zhou, W., Zhao, C., Zhou, H., Ge, Q., Zhang, X.: Development of a modified through-mask electrochemical micromachining for micropatterning nonplanar surface. Int. J. Adv. Manufact. Technol. **93**(5–8), 2613–2623 (2017)
7. Madore, C., Piotrowski, O., Landolt, D.: Through-mask electrochemical micromachining of titanium. J. Electrochem. Soc. **146**(7), 2526–2532 (1999)
8. Zhang, X., Qu, N., Li, H., Xu, Z.: Investigation of machining accuracy of micro-dimples fabricated by modified microscale pattern transfer without photolithography of substrates. Int. J. Adv. Manuf. Technol. **81**(9–12), 1475–1485 (2015)
9. Wang, M.H., Zhu, D.: Fabrication of multiple electrodes and their application for micro-holes array in ECM. Int. J. Adv. Manuf. Technol. **41**(1–2), 42–47 (2009)
10. Wang, M., Bao, Z., Wang, X., Xu, X.: Fabrication of disk microelectrode arrays and their application to micro-hole drilling using electrochemical micromachining. Precision Eng. **46**, 184–192 (2016)
11. Schuster, R.: Electrochemical microstructuring with short voltage pulses. ChemPhysChem **8** (1), 34–39 (2007)
12. Kawanaka, T., Kato, S., Kunieda, M., Murray, J.W., Clare, A.T.: Selective surface texturing using electrolyte jet machining. Procedia CIRP **13**, 345–349 (2014)
13. Costa, H.L., Hutchings, I.M.: Development of a maskless electrochemical texturing method. J. Mater. Process. Technol. **209**(8), 3869–3878 (2009)
14. Parreira, J.G., Gallo, C.A., Costa, H.L.: New advances on maskless electrochemical texturing (MECT) for tribological purposes. Surf. Coat. Technol. **212**, 1–13 (2012)
15. Schonenberger, I., Roy, S.: Microscale pattern transfer without photolithography of substrates. Electrochim. Acta **51**(5), 809–819 (2005)
16. Kunar, S., Bhattacharyya, B.: Investigation on surface structuring generated by electrochemical micromachining. Adv. Manufact. **5**(3), 217–230 (2017)
17. Wang, X.L., Liu, W., Zhou, F., Zhu, D.: Preliminary investigation o f the effect of dimple size on friction in line contacts. Tribol. Int. **42**(7), 1118–1123 (2009)
18. Etsion, I., Kligerman, Y., Halperin, G.: Analytical and experimental investigation of laser-textured mechanical seal faces. Tribol. Trans. **42**(3), 511–516 (1999)
19. Chen, X., Qu, N., Hou, Z., Wang, X., Zhu, D.: Friction reduction of chrome-coated surface with micro-dimple arrays generated by electrochemical micromachining. J. Mater. Eng. Perform. **26**(2), 667–675 (2017)
20. Podgornik, B., Vilhena, L.M., Sedlacek, M., Rek, Z., Zun, I.: Effectiveness and design of surface texturing for different lubrication regimes. Meccanica **47**(7), 1613–1622 (2012)
21. Gachot, C., Rosenkranz, A., Hsu, S.M., Costa, H.L.: A critical assessment of surface texturing for friction and wear improvement. Wear **372–373**, 21–41 (2017)

Chapter 12
Analysis of Casting Defects in Investment Casting by Simulation

Sudip Banerjee and Goutam Sutradhar

12.1 Introduction

Investment casting is one of the oldest primary manufacturing processes as mankind has learnt to use liquid metal around ten millennia [1]. Since about 4000 years, products (low melting point cast products) of investment castings were produced by the traditional metal smiths, Dhokra tribes. They used to produce knives, jewelery, arrows, spears, etc., by using disposal pattern made of wax. Thus, investment casting is also called as lost-wax process [2]. In this process, they used to produce intricate shapes having complex geometry. Investment casting is considered as the most ancient metal casting process as well as most modern and versatile process of manufacturing components having complex geometry, higher-dimensional accuracy, and excellent surface finish [3]. The word 'investment casting' was originated from the concept of using mobile slurry or investment for making mold with extremely smooth surface and high-dimensional tolerance [4]. The dancing girl of bronze was found at Mohenjo-daro (3000 B.C) which was manufactured by investment casting. This is one of the finest examples of ancient investment casting process [5]. Nowadays, the versatility and integrity of investment casting have highly emphasized the foundry industry for using this process while producing near-net shape and thin products. Now, the advancement of technology has transformed it as the most modern and versatile metal casting process. Technologists reengineered the lost-wax process as investment casting process by applying different pattern materials, pattern-making mechanism, and metal feeding system. The wax replica of the desired pattern is made by injecting wax into the metal die. A casting cluster is formed by attaching a number of patterns to a central wax stick. Thus, the tree is formed which is then dipped again and again into refractory slurry

S. Banerjee (✉) · G. Sutradhar
Department of Mechanical Engineering, Jadavpur University, Kolkata, India
e-mail: banerjee.sudip71@gmail.com

P. Sahoo and J. P. Davim (eds.), *Advances in Materials, Mechanical and Industrial Engineering*, Lecture Notes on Multidisciplinary Industrial Engineering,
https://doi.org/10.1007/978-3-319-96968-8_12

made of grained silica, water, and binders to produce a fully hardened refractory shell. The wax pattern is then removed thermally or by steam autoclaving. Residual wax is removed by burning the shells into an oven. Now, the refractory shell is ready to hold molten metal. Molten metal is poured through pouring cup or drawn up by vacuum. The casting is then allowed to get solidified. Now, the ceramic mold is broken, and the parts are cut from the tree to have a number of pieces of desired product [3, 6]. Investment casting is one of the most versatile casting processes as it is able to cast intricate shapes with highest precision even shapes having undercuts in desirable numbers. Welding and assembling are not needed for finished product which means there will be no need of combining parts to get a complex shape which saves time and money as well as eliminates the chances of errors. Even larger sized product like turbine blade can be cast by this process. In spite of all these advantages, there are downsides also like initial cost, product size (heavy steel or Al-castings, door frame of aircraft, etc.), longer cycle time. [7].

India has a long-established custom of metal casting as the earliest castings were produced more than 5000 years ago [8]. The Chola bronze and Dhokra iron castings had made ancient Indian investment casting famous worldwide. At that time, major application areas of this casting were idols for worship, daggers, hooks, bowls, pillars, nails, frames, utensils, weapons, etc., of different materials like gold, copper, silver, lead, iron [2, 9]. The excavations at different parts of India (Delhi, Ropar, Nashik, Indus valley) have confirmed these facts. Many districts all over India (Tirupati, Salem, Cuttack, Mayurbhanj, Bankura, Kolhapur, Mandi) are still practicing this lost-wax process [10]. At ancient time, the mixture of natural resin, bee wax, and little amount of cooking oil were used as the material of wax pattern. This mixture is then heated and poured into sheets to produce the desired wax pattern. This pattern is then placed into the clay mold which is made of fine river clay, cow dung, natural resin, and gingelly oil. This mixture is followed by coarser clay. The models are then parched in sun for 2–4 weeks and then heated by cow dung cakes for melting out the wax. The mold is preheated. Metal charge is melted by crucible furnace and poured into the mold cavity through gates and risers. Then, the final product is removed by chiseling with iron nails [11].

The casting requirements of last decade have changed significantly due to the demand of market and technological push (CAD/CAM/CAE). Widely used casting processes like die casting and sand casting cannot meet all desired requirements (better appearance and competitive cost, zero or just finish machining, continuous design improvements with higher surface finish) of the present industry. The lost-wax process can meet these requirements. This process was reinvented in 1940 by west, and they have renamed it as investment casting process. Modern investment casting process can provide higher-dimensional accuracy (1 μm/mm) and best surface finish (1–2 μm) [12]. Here, the manual crafting of wax pattern of ancient process is replaced by wax injection die. Several advanced technologies have been used in this process to achieve better results. Wax additives have been used to increase strength and fluidity, whereas shell strength is improved by using fiber materials. Mold filling is improved by using new material and applying vacuum-assisted gravity casting. Computer-aided-technologies, i.e., intelligent methoding (feeders and gating design), creating solid model (for multiple casting), rapid prototyping (for fabrication of wax pattern), are used for speeding up the process of pattern making [13, 14].

Indian foundry industries generally cast components of auto, railways, sanitary, tractors, defense, earth moving, power machinery, pipe fittings, pumps, textile, etc. [15]. As per 51st world casting census, India has 4600 casting units and 90% of them are MSMEs. According to 2016–17 casting census, total production of castings of India is 11.35 million MT. Out of this total production, 70% is gray iron, 12% steel, 9% ductile iron, 8% non-ferrous, and 1% others [16]. Depending on locations and specific end-use markets, several foundry clusters are there (Kolhapur, Ludhiana, Pune, Rajkot, Ahmedabad, Coimbatore, Howrah, Kolkata, Belgaum, Indore, etc.) [17, 18]. India has only 250 investment casting foundries out of 4600 foundries. Most of the investment casting foundries are situated at Ahmedabad, Rajkot, Bhavnagar, Coimbatore, and Belgaum as cluster [16]. Recently, a few foundries are coming up at Howrah and Kolkata cluster. Particularly at Gujarat, the environmental conditions are in favor of Investment casting as the shells are getting dry in a shorter time compared to the foundries of eastern part of our country where de-humidification room is very essential [19]. Being the sixth largest producer of automotive vehicle, India has a huge possibility for investment casting. The Indian investment casting foundries used to produce pumps, machinery, and industrial valve for automotive industries and also for cookware industries which have extensive scope. The demands of railways and automobile industries for investment casting components are mainly fulfilled by Rajkot and Coimbatore clusters [20].

Literature review reveals that several researchers and industry personnel have concentrated on simulation-based study for predicting casting defects. Upadhya et al. have generated a computer-based model for calculating view factor and predicting defects [21]. Dabade et al. have tried to examine sand and mold-related casting defects by combing L_{18} orthogonal array-based design of experiment and computerized casting simulation [22]. Chalekar et al. have used CATIA V5 and casting simulation software to predict probable casting defects and perform necessary process development of 100NB pistons of piston valves which are cast by investment casting [23]. Jadhao et al. have reviewed simulation-based casting studies and also used casting simulation software to examine the feeder system of a cast iron block [24]. Dong et al. have used structural identification method to study the non-uniform shrinkage distribution of single-crystal hollow turbine blade which is manufactured by investment casting method [25]. Jabur et al. have predicted shrinkage porosity of Al–Si castings by choosing appropriate parameters and methoding with the help of casting simulation software. Numerical solution based on explicit 3-D finite difference method was used to describe the role of silicon percentage on defect morphology [26]. Shamasundar et al. have applied FEM-based casting simulation in several investment casting products to examine their defects and performed gating modification to minimize those defects [27]. Fu and Yong have applied CAE-based simulation to examine several process-related defects of two industrial component produced by four-cavity HPDC method [28]. Tiberto and Klotz have used two different casting simulation softwares (Flow-3D and PoligonSoft) to reduce micro-porosity-related defects in jewelery castings [29]. Beffel et al. have examined the effect of several parameters like grain size, alloy selection, part design in simulation of investment casting products [30]. Sabau et al. have performed numerical simulation on heat transfer criterion to study the alloy shrinkage factors of investment casting products [31]. Norouzi et al. have used MAGMA soft to study the hot tearing and residual stresses in investment casting products [32]. Hang et al. have used cellular automation-finite difference method to optimize and simulate directional solidification of a component [33]. Chattopadhyay has developed a model to estimate solidification time of investment casting [34]. Zhang et al. have found cast deformation by using coordinate measurement machine [35]. Guan et al. have formed a model to analyze the cooling behavior and cracking behavior of Ti–Al by investment casting process [36]. Sung et al. have generated a FEM model to predict gas porosity of a rectangular plate by using investment casting [37].

In most foundries, there is no scope for the foundry engineer to design the product at earlier stage of product development. The foundry engineer is bound to follow stringent instructions of the designer which put him under immense pressure to produce sound casting within given budget and stipulated time [29]. Customer demand in terms of quality consistency as well as decreasing availability of technically skilled and experienced manpower has led to computer-based approach to focus on pattern design and method optimization [38]. Also, defects are examined only after the prototype is made which is waste of resources. On the other hand, computer-based simulation uses 3D CAD model of the component to suggest the methoding solution, gating system, and predict probable defects. It helps the foundry engineer to design, model, and analyze a complex component within a shorter lead time. Thus, simulation improves the quality of castings as well as increases profit by enhancing net yield. In this process, several iterations can be done to predict the optimal methoding solution which provides defect-free castings with maximum possible yield [39]. The flowchart of simulation process is shown in Fig. 12.1, and the comparison between shop-floor trial and simulation process is shown in Fig. 12.2. Casting simulation softwares are regularly used by aluminum, copper, iron, and steel foundries using processes ranging from green sand, resin, and shell-bonded sand to investment and gravity die casting. Several casting simulation softwares like AutoCAST, MAGMASoft, ProCAST, SOLIDCast. are available in market. For the present study, AutoCAST and ZCast are used. Investment casting has got immense importance in modern foundry industry due to its versatility, integrity, and higher-dimensional accuracy. Therefore, this process is more suitable for turbine blades, aerospace industry, jewelery castings, and many more industrial and scientific components [2]. Modern technologies like rapid prototyping, rapid tooling, 3D printing, and casting simulation have helped to improve the product quality and also made it cost-effective [12]. Wax properties, autoclave de-waxing, blending effect, slurry preparation have taken the attention of researchers, but very few of them has shown the interest in simulation of investment

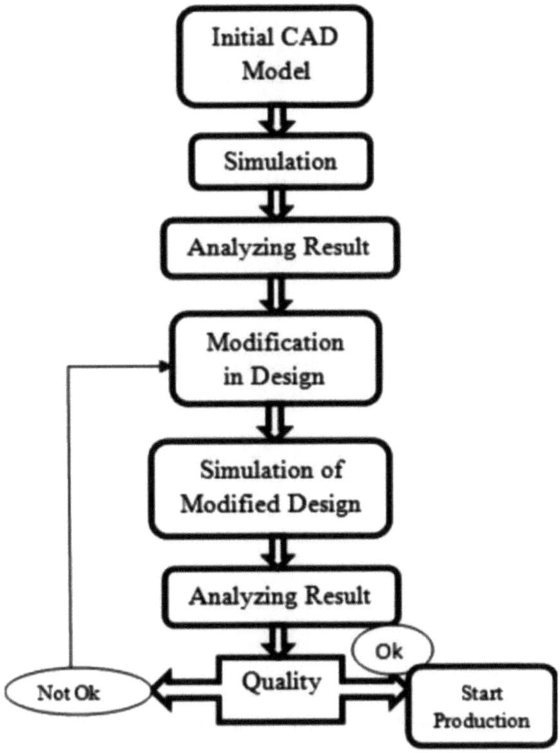

Fig. 12.1 Flowchart of simulation process

casting products though simulation can visualize mold filling, solidification and also can predict shrinkage porosity, blowholes, cold shut, and hard zones without shop-floor trials. Research on yield improvement also has not got much importance though yield of investment casting product is near about 50%.

Casting simulation is a powerful tool of imitating original process with the help of a set of mathematical equations in a program just to anticipate the mold filling and solidification [40]. Different defects such as porosity, cold shut, and shrinkage cavity could be forecasted provided proper shop-floor inputs are being given to computer. Thus, it is very helpful tool for developing the existing castings as well as new ones without or minimum shop-floor trials [34, 41]. Casting simulation in earlier step could figure out the defect-prone zones which will help to take protective measures [22, 42]. 3D CAD model of the casting part in STL format is the main input of the simulation, and the output is animated visualization of mold filling, cooling to room temperature, and solidification. Filling time, mold erosion, incomplete filling can be determined by mold filling option, whereas cooling rate, temperature, and gradients can be predicted from solidification simulation [43]. Some well-known casting simulation softwares are AutoCAST, Z-Cast, ProCAST, MAGMASoft, SOLIDCast, etc. [44]. For the present case study, AutoCAST and Z-Cast are used for simulation and solidification analysis along with the mold filling analysis.

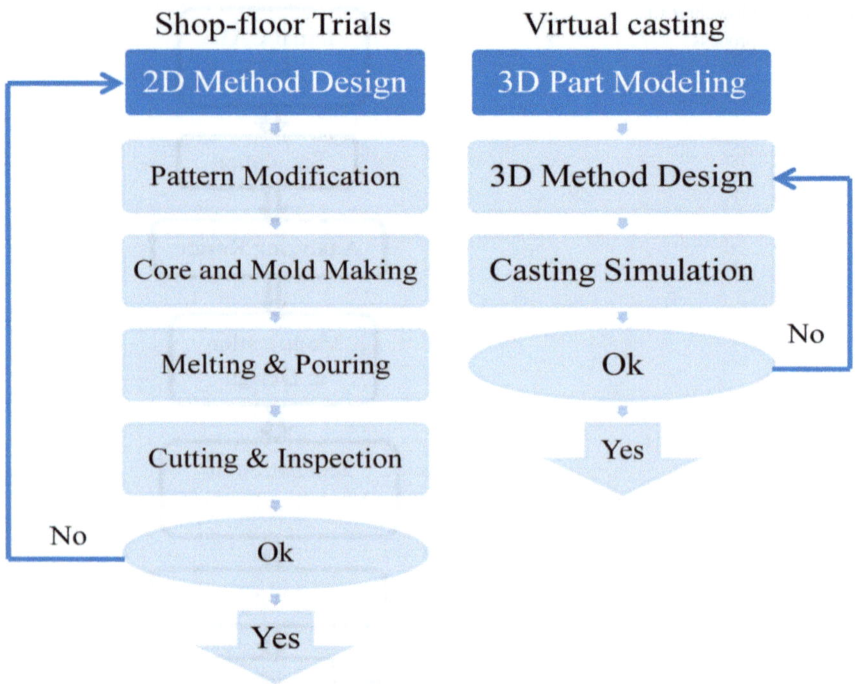

Fig. 12.2 Flowchart of comparison between simulation process and shop-floor trials

12.2 Results and Discussion

In casting simulation study, filling process and solidification process are of immense concern. Filling process is mainly affected by melt velocity, pressure, flow direction, venting conditions, while solidification depends on metallurgical, thermal, and physical phenomenon like location of in-gate, last solidification zone, solidification sequence, temperature distribution of the casting, rationality of cooling layout. Based on filling and solidification criterion, location of probable defects can be predicted. To demonstrate the role of filling and solidification criterion on casting defects, three industrial case studies of investment casting are discussed here.

12.2.1 Case Study I: 100NB Flange

In the present study, the foundry is producing 100NB flange of weight 6.85 kgs of cast steel-ss-austenitic by investment casting process. Effect of in-gate size and shape is studied here by taking three types of in-gate (i.e., flat, square, and taper)

Table 12.1 Composition of cast steel-ss-austenitic

Element	C	Si	Mg	Cr	Ni	Cu	Fe
wt%	0.019	0.27	1.40	18.46	10.20	0.276	Balance

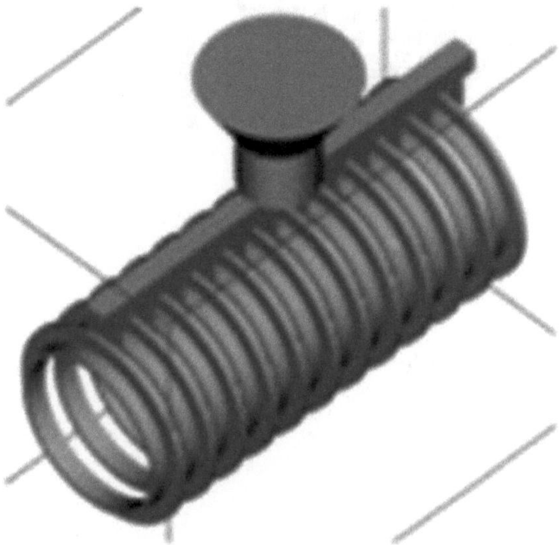

Fig. 12.3 3-D model of 100NB flange tree with flat in-gate

having different sizes (15 × 15 × 300 and 20 × 20 × 300) which are taken into consideration. Composition of casting material is shown in Table 12.1. 3-D model of 100NB flange tree with flat in-gate is shown in Fig. 12.3. This 3D model is generated in solid modeling software and converted to .STL format with the help of CAD model data exchange system. The .STL file is the direct input of AutoCAST-XI simulation software. Here, pouring temperature is taken as 1520 °C, and pouring time is 22 s. Molten metal is poured into the cavity at the flow velocity of 40–100 m/s.

In the first iteration, taper in-gate of size (15 × 15 × 300) is taken into consideration. The result of first iteration is shown in Fig. 12.4. Figure 12.4 shows that there is no probable shrinkage porosity, blowhole is there in the in-gate, but possibility of cold shut and hard zone is more. Figure 12.4a shows the post-solidification condition of the simulation where violet and blue colors indicate faster freezing zones, whereas white and yellowish white signify slower cooling zones. The whitish zone will remunerate the shrinkages. Here, the tapernature of the in-gate improves longitudinal temperature gradients; thus, no shrinkage is there, but the ends of in-gate solidify faster by improving internal stress of the casting which may cause blowholes at the end of the in-gate. Figure 12.4e shows that hard zones are also present in the casting. Hard zones are generally caused by impurities in the material or agglomeration of alloy elements which are added in small percentage.

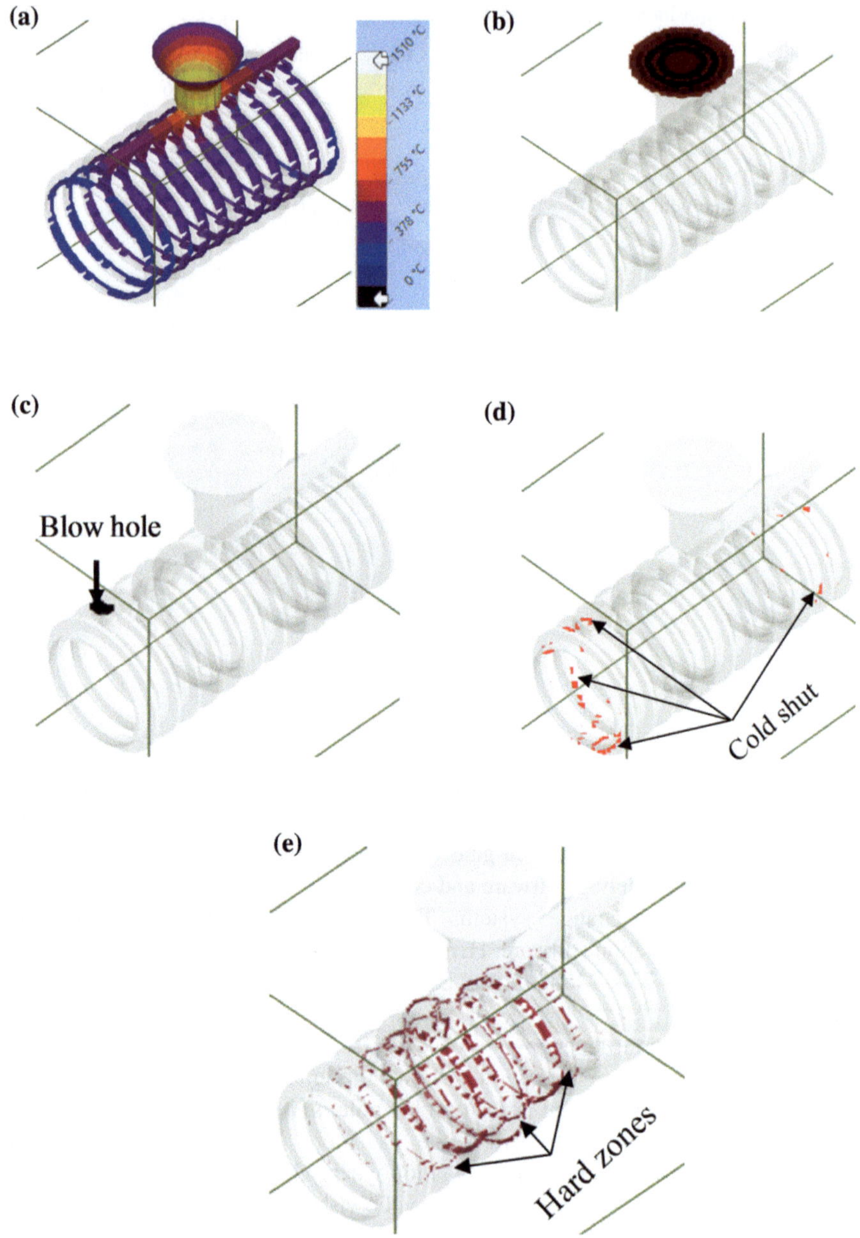

Fig. 12.4 Simulated casting defects for taper in-gate of size (15 × 15 × 300)

In next iteration, simulation of taper in-gate of size (20 × 20 × 300) is done for further improvement. Result of simulation is shown in Fig. 12.5. Result shows that some probable shrinkage porosities are there along with cold shut and hard zones,

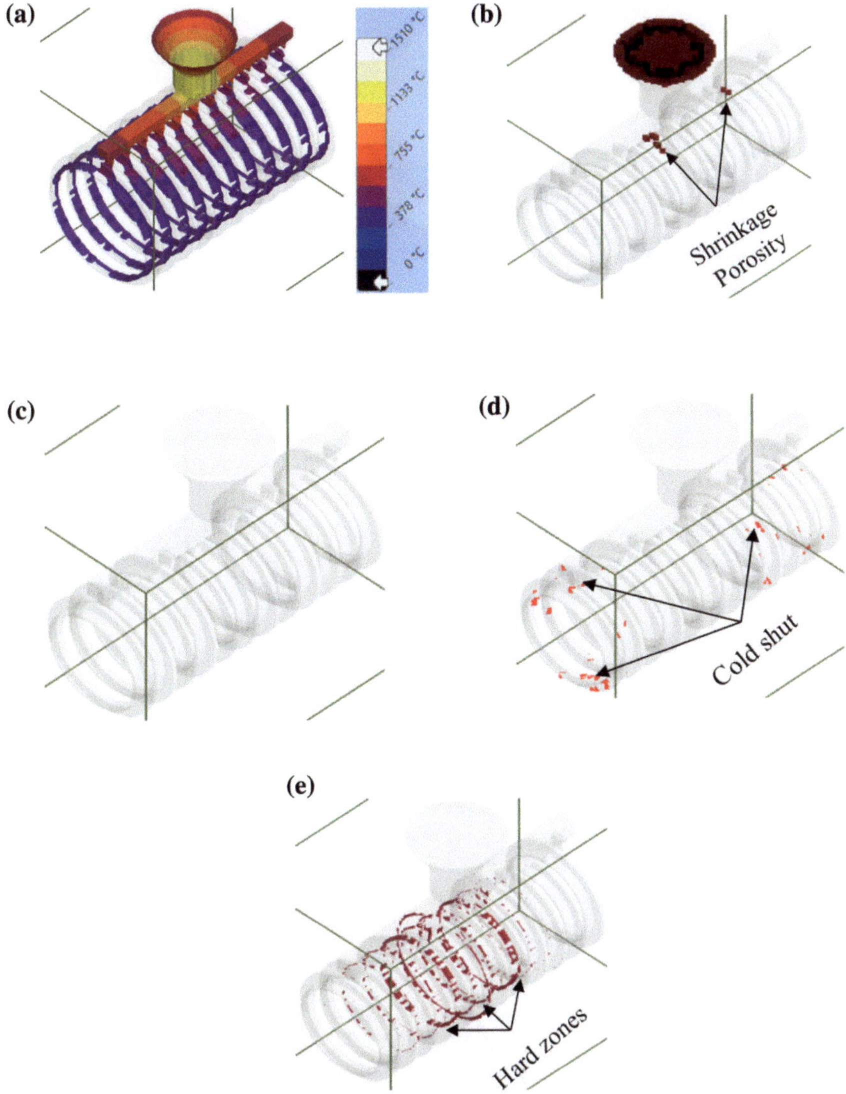

Fig. 12.5 Simulated casting defects for taper in-gate of size (20 × 20 × 300)

but no blowhole is there. Figure 12.5a shows that shrinkages are nearer to the riser which implies that the riser design is not proper to meet the requirement of feeding. Figure 12.6 shows the effect of square ring in-gate of size (15 × 15 × 300) on several casting defects. The probability of shrinkage porosity and blowholes are there, but result of simulation shows that probability of cold shut is minimum and probability of hard zones is maximum. Figure 12.6(a) shows the solidification

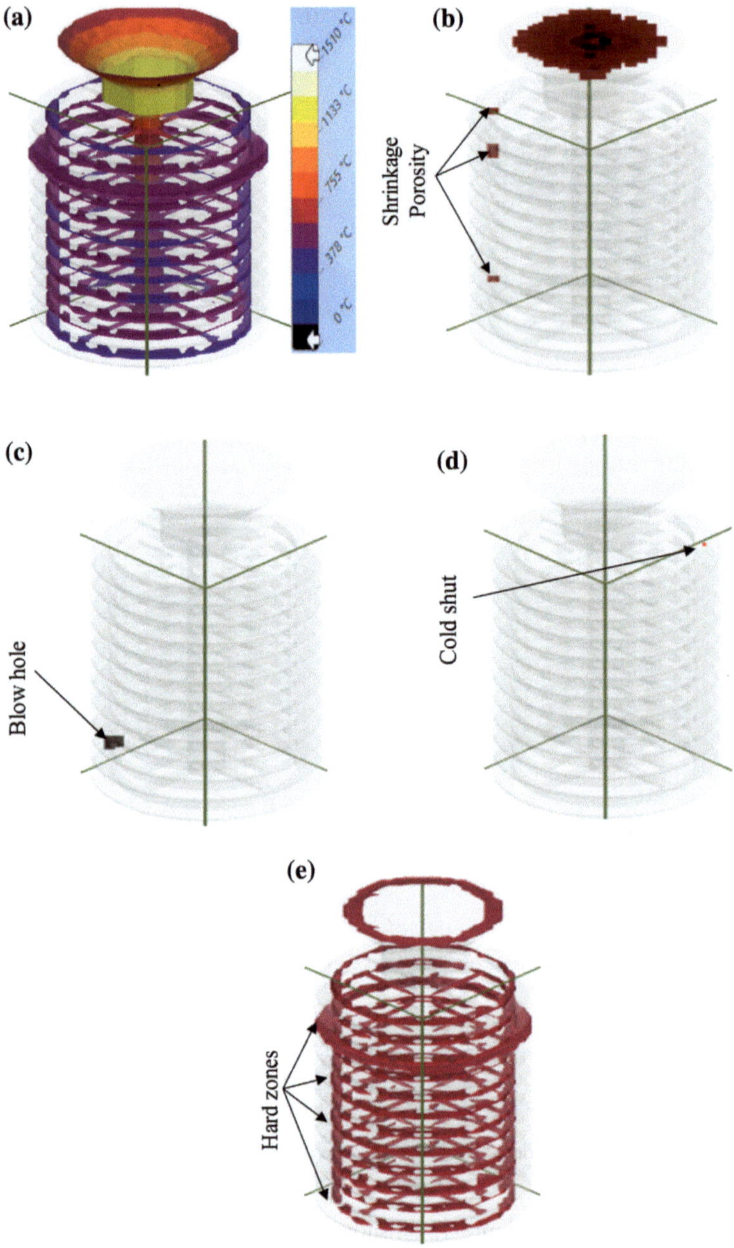

Fig. 12.6 Simulated casting defects for ring in-gate of size (15 × 15 × 300)

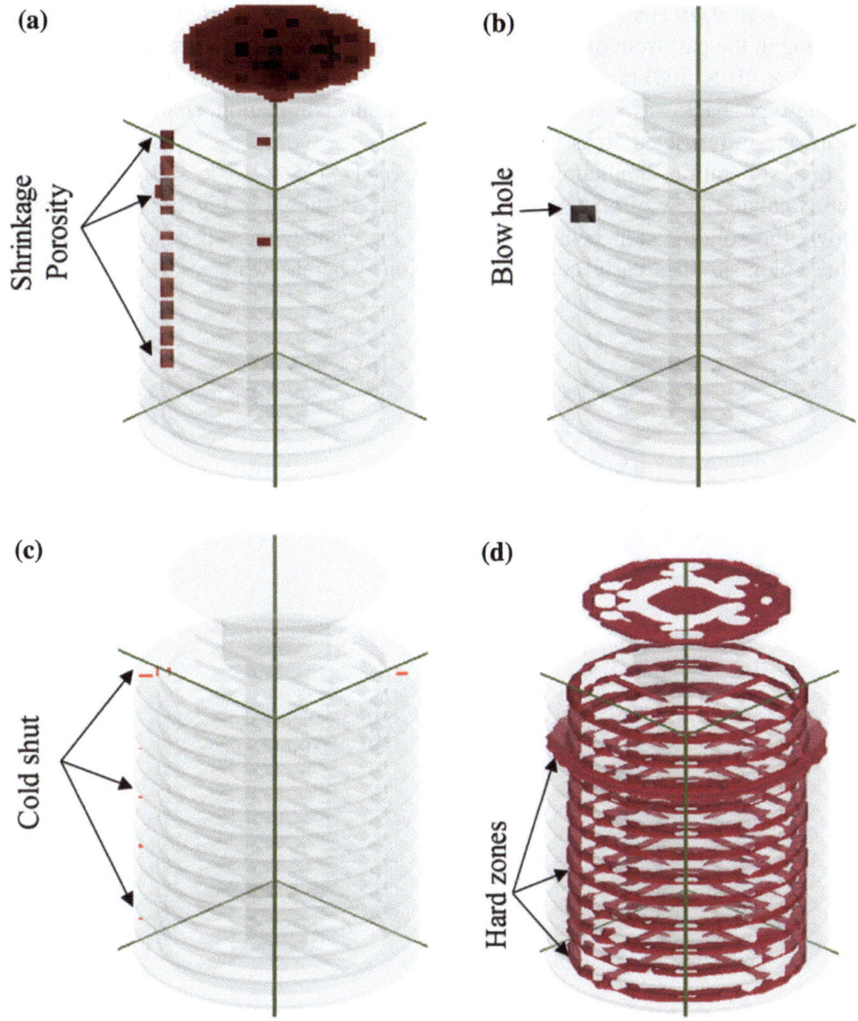

Fig. 12.7 Simulated casting defects for ring in-gate of size (20 × 20 × 300)

nature of the casting. It shows that molten metal will enter from different directions; thus, there is probability of gas entrapment. As no proper venting mechanism is there, the entrapment will happen at the middle of entrances. Thus, shrinkages are there in the simulation result. But the directional solidification will lead the entrapped air toward in-gate which will reduce the probability of blowholes. Further simulation of square ring in-gate of size (20 × 20 × 300) is done to check its effect on several casting defects. The simulation result is shown in Fig. 12.7. It is seen that the percentage of probable shrinkage porosity and blowhole has increased along with cold shut and hard zones compared to ring square in-gate

(15 × 15 × 300). Here, metal is falling freely thus creating turbulence, and melt is entering in the part from different directions; thus, air entrapment issue will be more for (20 × 20 × 300)-type in-gate. When the jetted metal entered into the flanges, a rolling-type back-wave was formed. Hence, the simulation result shows more shrinkage in restricted areas.

In next iteration, simulation of flat in-gate of size (15 × 15 × 300) is performed. The simulated view of post-solidification figure is shown in Fig. 12.8a. The figure shows the solidification behavior of the component with respect to temperature. Blue color signifies early coagulation region, whereas white or yellowish white

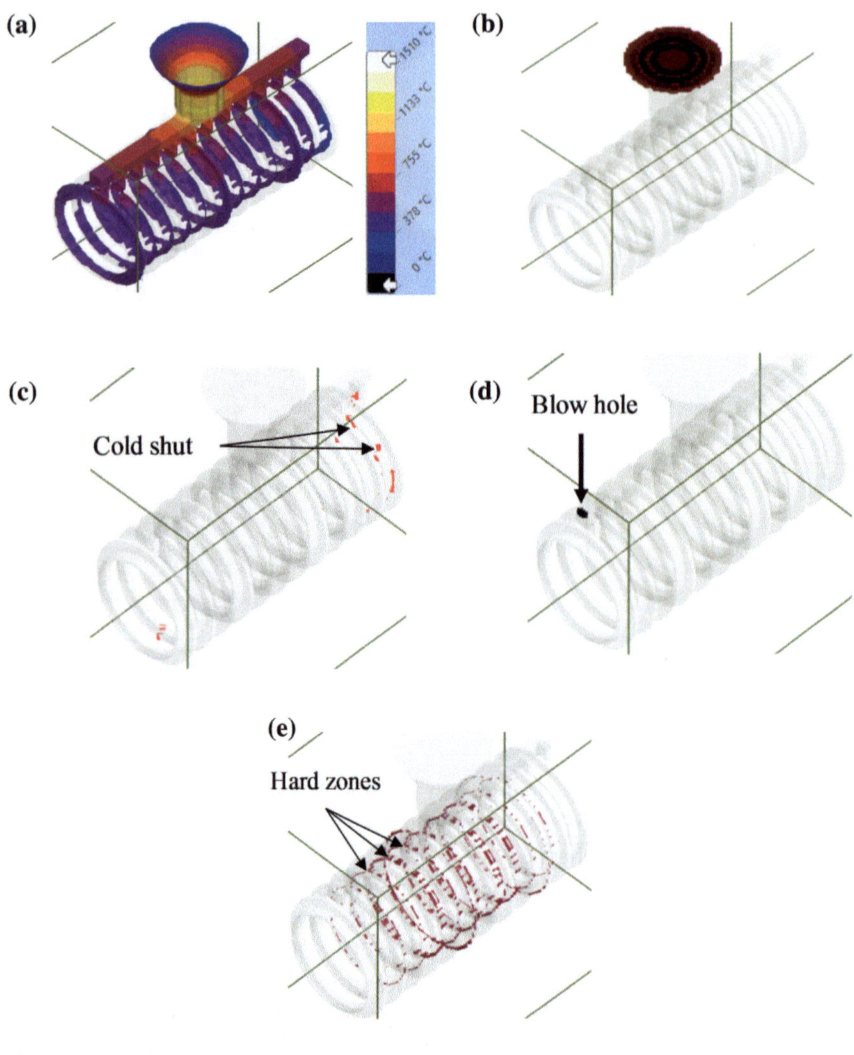

Fig. 12.8 Simulated casting defects for flat in-gate of size (15 × 15 × 300)

signifies the slowest freezing region. Here, the in-gate region is hotter compared to the flange body. Thus, the in-gate part will remunerate the volumetric shrinkages which help to avert porosity or under-fill-related defects. Results of simulation are shown in Fig. 12.8. Figure 12.8 shows that no probable shrinkage porosity is present in the casting and blowhole is there in the in-gate, whereas few cold shuts are present along with some hard zones. Cold shuts can be eliminated by increasing pouring rate but that will cause air entrapment. Thus, further iteration is done to minimize cold shut and hard zone if possible. In this iteration, simulation is performed by taking in-gate (20 × 20 × 300). Results of this iteration are shown in Fig. 12.9 which shows that percentage of probable shrinkage porosity, blowholes, cold shut, and hard zones has increased compared to the previous iteration. Thus, the OEM is advised to use flat in-gate of size (15 × 15 × 300) for shop-floor trial. They have got satisfactory result by having shrinkage and blowhole-free casting with minimum rejection. Thus, further iterations are not needed.

To study the solidification behavior of the component having in-gate size (15 × 15 × 300), cooling curve is plotted and examined by taking different points at different thickness and location of the in-gate tree. Different points and cooling curve are shown in Fig. 12.10a, b. In Fig. 12.10b, all curves are showing distinct variation law. After completion of filling, temperature declines rapidly from 1520 to 1410 °C. After that, the melt is entering into the liquidus region; hence, the slope of curve is less steep than the previous region. This phenomenon continues up to 1350 °C. Then, nature of curves is nearly horizontal but of different degrees.

12.2.2 Case Study II: 32 NB Swivel Flange

In this case study, the foundry is producing 32 NB swivel flange of A304 stainless steel by investment casting method. Effect of pouring temperature and method design on shrinkage porosity is discussed in this study. Composition of A304 stainless steel is shown in Table 12.2. Here, simulation is done by using Z-Cast and AutoCAST-XI casting simulation software. During simulation, some parameters are kept constant; they are tabulated in Table 12.3. The de-waxed tree of the considered component is shown in Fig. 12.11. For the first iteration, pouring temperature is considered as 1615 °C and pouring time is 20 s.

Figure 12.12 shows filling condition of this flange at different fill rate (10, 50 and 100%). Figure 12.12a shows that melt is entering in the tree through in-gate. Mold filling is an important issue because melt temperature will start to decrease when it comes in contact with cooler surface. Figure 12.12a shows that molten metal will start to fill from bottom parts. Figure 12.12b shows that at 50% filled condition the components placed at bottom are already getting solidified, which implies that melt temperature has dropped drastically. So the thinner portion of the component will have lesser shrinkage compared to the thicker part as thicker part may not get proper backup.

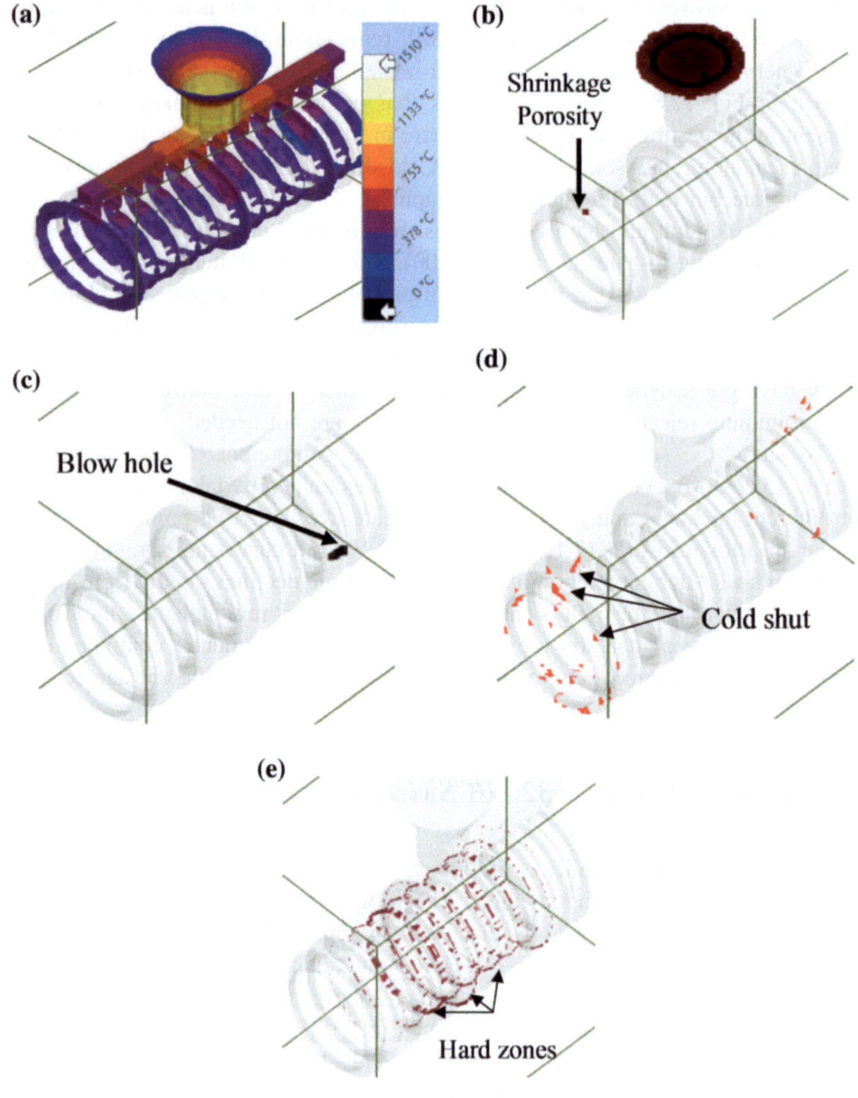

Fig. 12.9 Simulated casting defects for flat in-gate of size (20 × 20 × 300)

The simulation result is shown in Fig. 12.13. The simulation result shows that shrinkages are present on each component and volume of shrinkages at bottom part of a column is more compared to the volume of shrinkage at top part of that column. The sizes of shrinkages of bottom part are shown in Fig. 12.14. For reducing the shrinkage size, second iteration is needed. It is well known that liquid melt can retain more resolved gas compared to solid state. As a result, solidification shrinkage like porosities is formed. So increased temperature will lead to increased

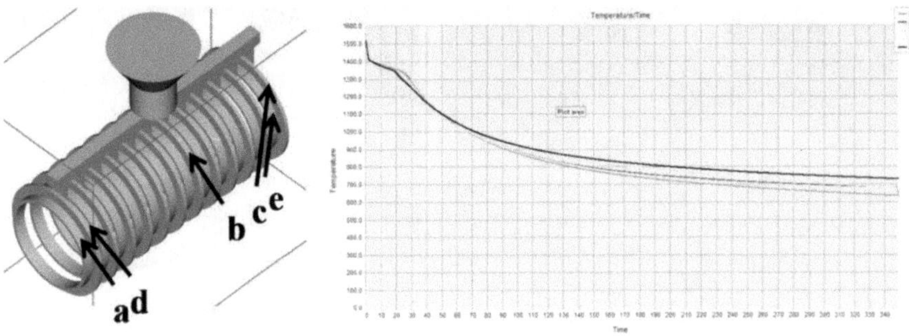

(a) Different points (b) Cooling curves

Fig. 12.10 Cooling curve of solidification process

Table 12.2 Composition of A304 stainless steel

Element	Ni	Cr	Si	Mn	C	P	S	Fe
wt%	10.20	18.30	1.00	2.00	0.080	0.045	0.030	Balance

Table 12.3 Constant parameters during simulation

Parameter	Value
Material	A304 stainless steel
Pouring time	20 s
Shrinkage allowance for material	6%
Mold material	Shell sand
Shell thickness	6 mm
Mold temperature	600 °C
Flow velocity	40–100 m/s

Fig. 12.11 De-waxed tree

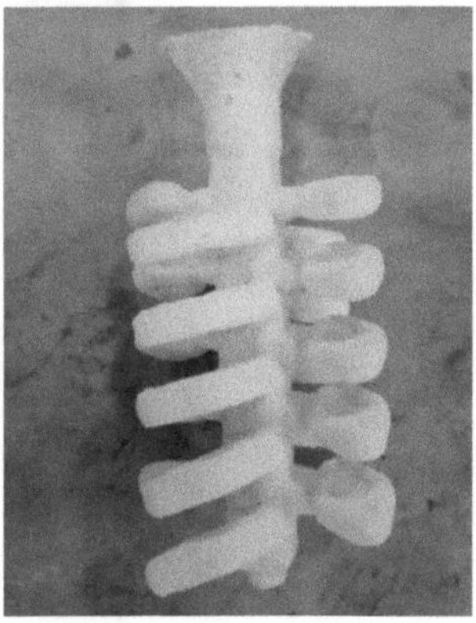

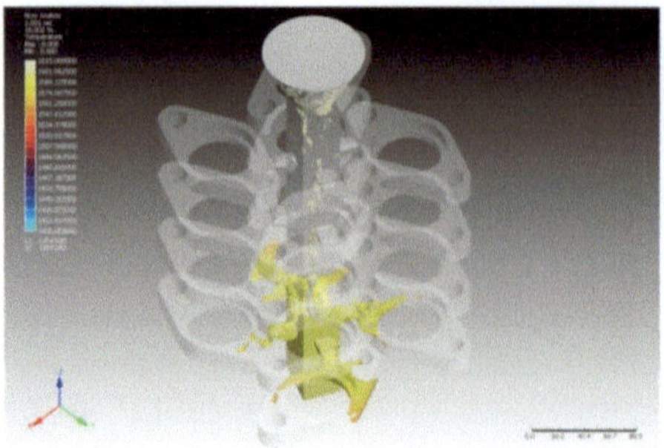

(a) 10% Filled

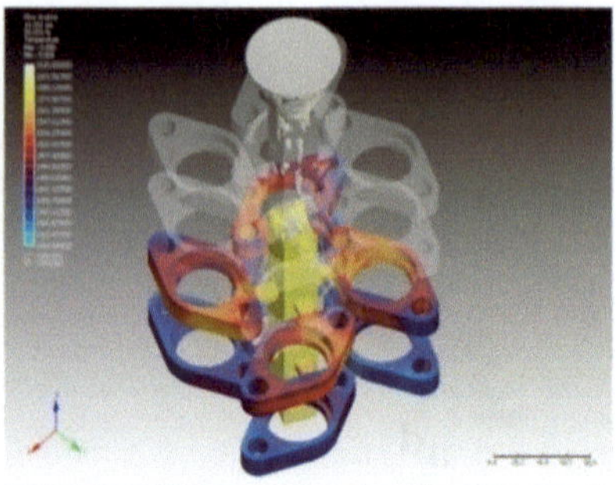

(b) 50% Filled

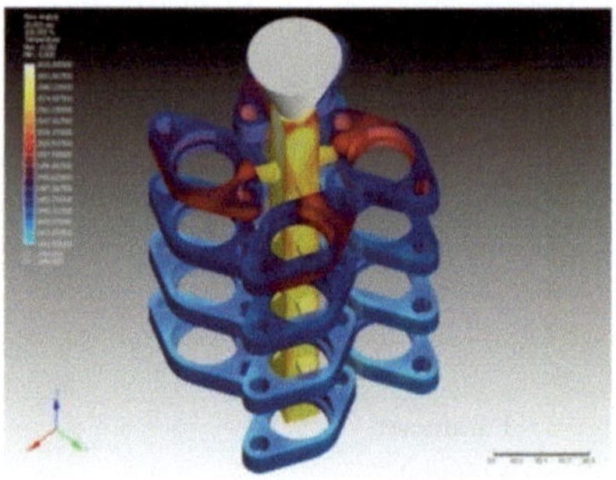

(c) 100% Filled

Fig. 12.12 Filling condition at different fill rate

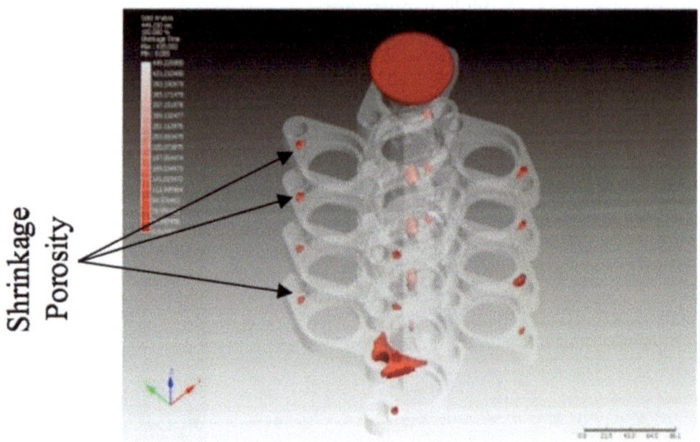

Fig. 12.13 Simulation result of shrinkage

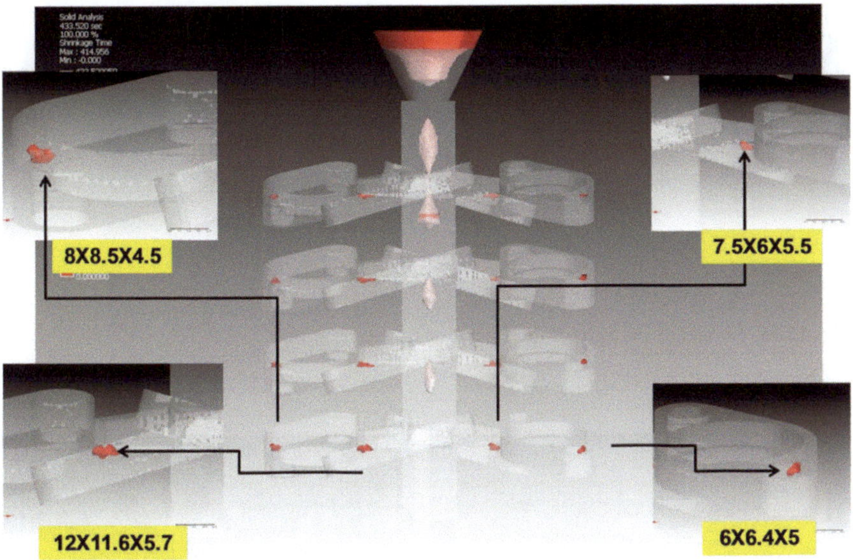

Fig. 12.14 Shrinkage size at bottom part

shrinkage porosity. Hence, pouring temperature of second iteration is reduced to 1550 °C. Simulation results show that shrinkage size has reduced. The top view of shrinkage is shown in Fig. 12.15.

For further improvement, methoding is little bit changed by placing the components at 30° with the horizontal and pouring temperature is kept at 1550 °C. Simulation result of this iteration is shown in Fig. 12.16a, b. Figure 12.16b shows

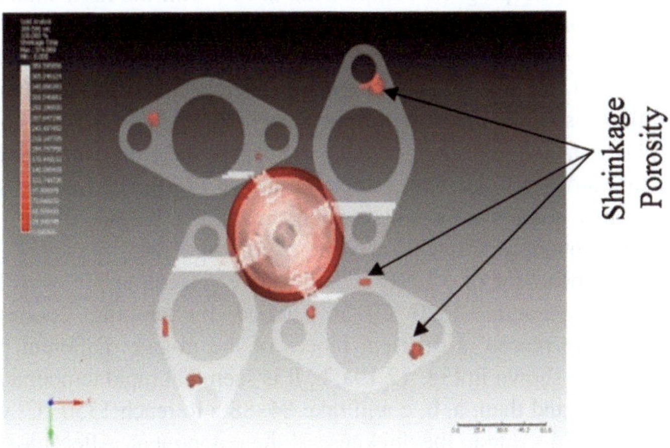

Fig. 12.15 Top view of shrinkage

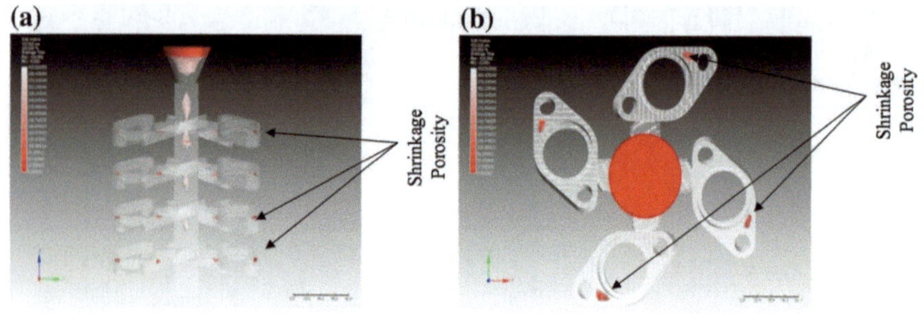

Fig. 12.16 Shrinkage porosity of third iteration

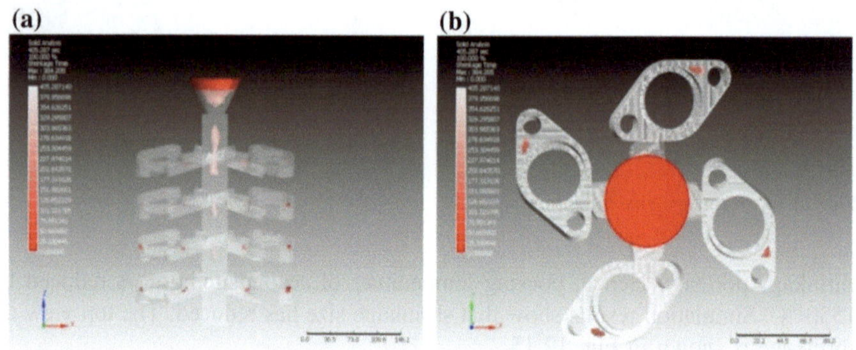

Fig. 12.17 Shrinkage porosity of fourth iteration

that the shrinkage size has reduced again. Finally, pouring temperature has been reduced to 1520 °C keeping the components at 30° with the horizontal to see the combined effect of pouring temperature and proper methoding. Figure 12.17a, b shows the amount of shrinkages present after this iteration. From Fig. 12.17b, it can be concluded that the size of shrinkages has been reduced further. Now, taking same conditions as the fourth iteration (pouring temperature 1520 °C and components are placed at an angle of 30° with the horizontal), the simulation is again performed in AutoCAST-XI. The results are shown in Fig. 12.18. The simulation result shows that some components which are at the bottom of the tree have shrinkage, while other components are free from shrinkage.

Solidification behavior of 32 NB swivel flange (placed at 30° with horizontal and pouring temperature 1520 °C) is studied by plotting cooling curves. Thermocouples are placed at different places of the in-gate. Different points and Cooling curves are shown in Fig. 12.19a, b. It is seen that rapid temperature drop is there up to 1400°, and then, a, b, c will take 84–88 s to reach 1350 °C, whereas d, e, f will take only 24 s due to the thickness of the component. After that, the curves will decline to 900–1000 °C linearly in different degrees.

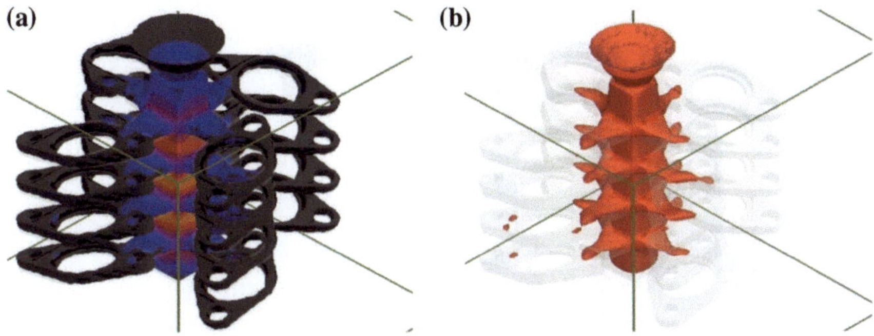

Fig. 12.18 Simulation result of fourth iteration by using AutoCAST-XI

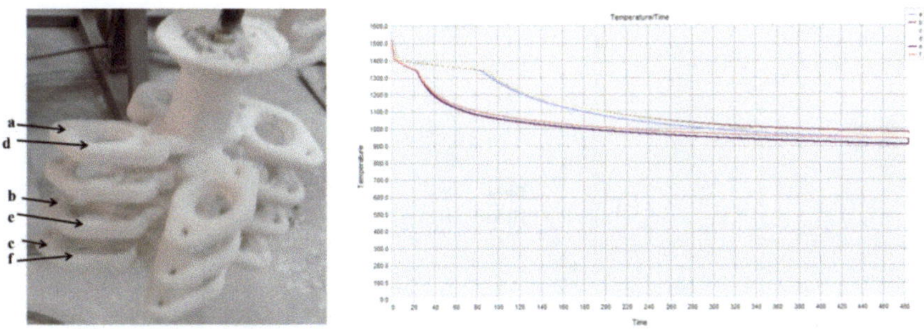

(a) Different points

(b) Cooling curves

Fig. 12.19 Cooling curve of solidification process

12.2.3 Case Study III: Pistol Body

In this study, a foundry is producing pistol body of cast steel-ss-austenitic (composition: C = 0.019%, Si = 0.27%, Mg = 1.4%, Cr = 18.46%, Ni = 10.20%, and Cu = 0.276%) by investment casting. Effect of pouring temperature and method design is studied here to develop defect-free casting. In the first iteration, eight pistol bodies are attached with in-gate to check the simulation result keeping pouring temperature at 1550 °C and pouring time as 22 s. Results of simulation are shown in Fig. 12.20. Figure 12.20 shows that no probable shrinkage is found, but blowholes, cold shuts, and hard zones are present. So, further iteration is needed. In second iteration, pouring temperature is set at 1600 °C. Results of second iteration are shown in Fig. 12.21. Figure 12.21 shows that no shrinkage is present and blowhole is present in the in-gate. It is also seen that cold shut, misrun, and hard zones are present in the casted components. Thus, further iteration is needed. In third iteration, method design is changed by attaching four pistol bodies with the in-gate, while the pouring temperature is 1600 °C. The result of simulation is shown in Fig. 12.22. Figure 12.22 shows that no shrinkage; blowhole; and cold shut are present in pistol body.

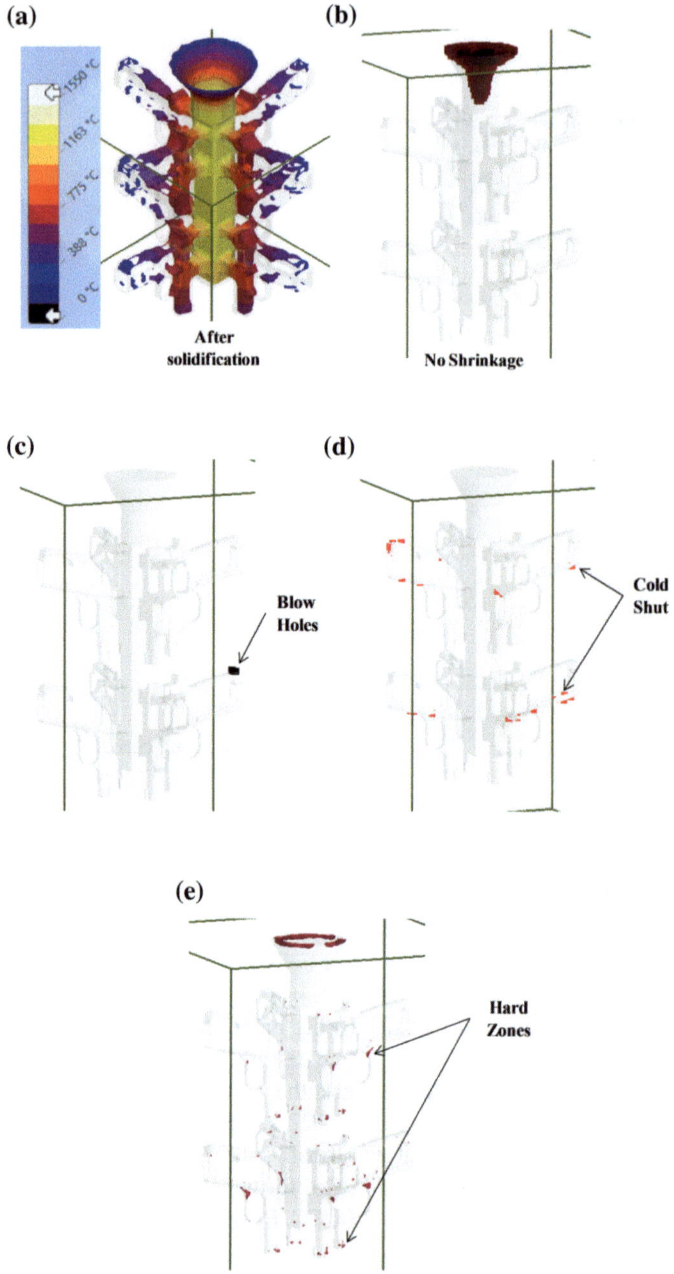

Fig. 12.20 Simulated casting defects for eight pistol bodies, pouring temperature 1550 °C

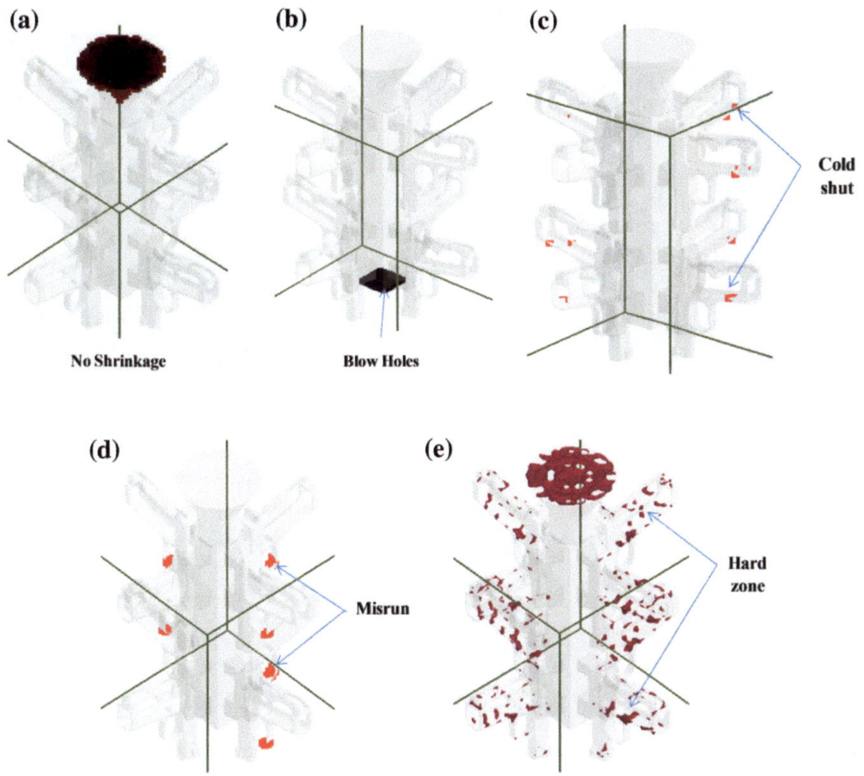

Fig. 12.21 Simulated casting defects for eight pistol bodies, pouring temperature 1600 °C

Selected points on in-gate of four pistol bodies and its cooling curve are shown in Fig. 12.23a, b. Thermocouples are placed at four different positions (a, b, c, d) as shown in Fig. 12.23a. Figure 12.23b shows the cooling curve which shows that after pouring melt temperature will increase to 900 °C in 5 s and melt temperature will decrease to 1400 °C in 4 s for thinner sections and in 10 s for thicker section. After 1400 °C, the slope of curve will change. For thin sections, slope is steeper as these parts will solidify quickly, whereas thick part ‘c’ shows different nature. Thick part ‘c’ will take time to get solidified compared to thin zones; thus, probability of shrinkage was there that is why modification in method design is made by attaching the pistols angularly with the in-gate. Thus, the probable shrinkages are avoided by proper method design. After 180 s, all three curves (a, c and d) are showing linear relation but of different degrees.

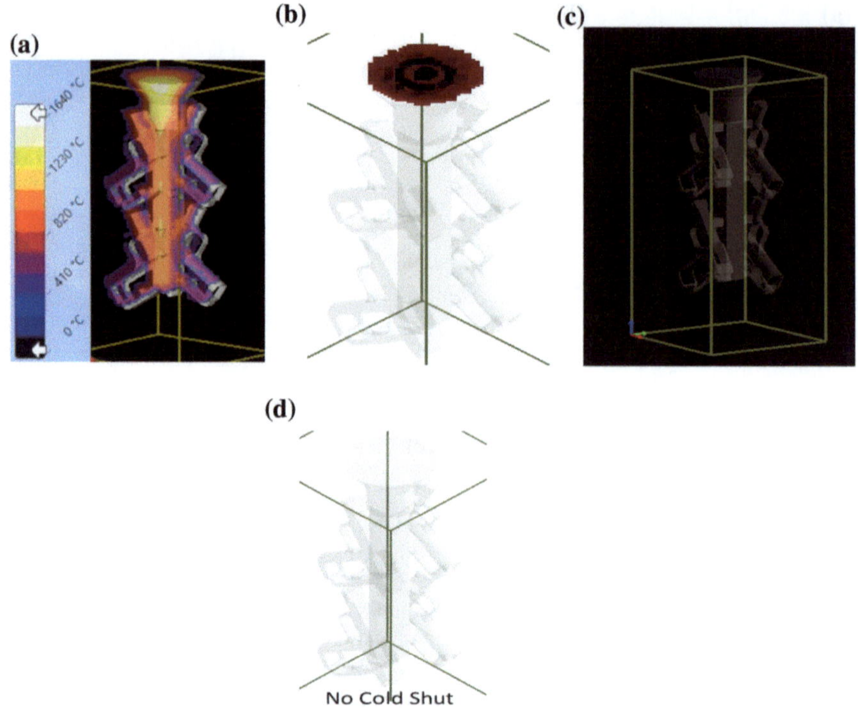

Fig. 12.22 Simulation result of four pistol bodies, pouring temperature 1600 °C

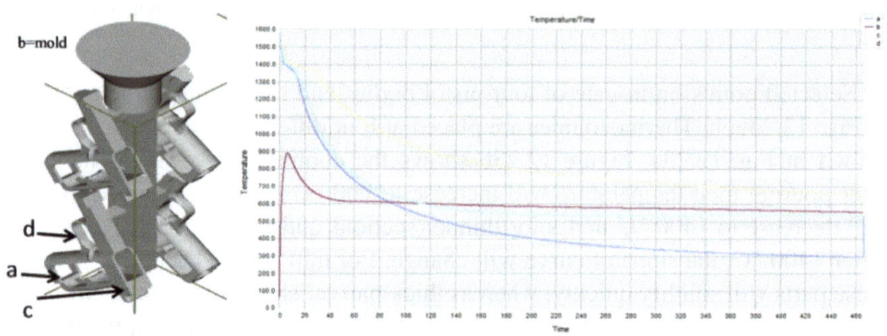

(a) Different points (b) Cooling curves

Fig. 12.23 Cooling curve of solidification process

12.3 Conclusion

Casting simulation has become sufficiently matured to diagnose the shape, appearance, and location of defects of the components. .STL file of the 3D solid model is taken as the input of simulation softwares (Z-Cast and AutoCAST-XI), and simulation is performed to determine casting defects. First cast study shows the effect of several in-gates on 100NB flange, and it is seen that probability of defects is minimized by the help of simulation to make the casting right first time, every time. So casting simulation is essential to produce optimized, high yield, and economical casting without any shop-floor trial for new components under development. This investigation also suggests using flat in-gate of size (15 × 15 × 300) for 100NB flange. In second case, effect of pouring temperature is examined and proper methoding is done depending on the requirement of the system form minimizing shrinkage size. Third case study shows the effect of pouring temperature and method design on position of several casting defects of pistol body casted by investment casting. It is seen that simulated result of the component having four pistol bodies has less probable defects compared to all other conditions. The accuracy of the simulated model is compared with experimental model. It shows that the experimental model completely validates the computer-aided simulated model. So it could be concluded that if the process variables like pouring temperature, shell thickness, and preheating temperature of the shell are properly found in a systematic manner, then better judgment could be taken at the shop-floor level. It is expected that shop-floor foundry men as well as academicians and researchers will get immense benefit from this paper.

Acknowledgements The authors gratefully acknowledge the research support provided by Vikas Hitech Castings, Sankrail Industrial Park, Howrah, 3D Foundry Tech, Mumbai, DHIO Research & Engineering Pvt. Ltd, Bangalore, and financial support of 'SMART FOUNDRY-2020,' DST, Government of India.

References

1. Ravi, B., Datta, G.L.: Metal casting: back to future. In: Proceedings of the 52nd Indian Foundry Congress, Hyderabad, February 2004 (2004)
2. Datta, G.L.: Reinventing investment casting process for sustainable development. Indian Foundry J. **60**, 21–27 (2014)
3. Ravi, B.: Investment casting development: ancient and modern approaches. In: National Conference on Investment Casting Central Mechanical Engineering Research Institute, Durgapur (2003)
4. Chalekar, A.A., Daphal, S.A., Somatkar, A.A., Chinchanikar, S.S.: Minimization of investment casting defects by using computer simulation—a case study. J. Mech. Eng. Autom. **5**, 43–46 (2015)
5. Pal, S.: Ancient indian art casting. Foundry **13**, 13–23 (2001)

6. Eddy, W.P., Barbero, R.J., Dieters, W.I., Esarey, B.J., Frey, L., Gros, J.R., Hall, J.C., Hockin, J., Nelligan, P.J., Powell, R.C., Watts, C.H., Wenner, J.: Investment casting. In: Metals Handbook, American Society for Metals, pp. 237–261 (1974)
7. Pal, D.K., Ravi, B., Bhargava, L.S., Chandrashekhar, U.: Investment casting one-off intricate part using rapid prototyping technology. In: National Conference on Investment Casting, CMERI, Durgapur, pp. 22–23 (2003)
8. Santra, S., Sengupta, G., Bhattacharya, D.: Recent bronze hoard from West Bengal: analytical studies. Indian J. History Sci. **43**, 29–42 (2008)
9. Lal, M., Khanna, O.P.: A Text Book of Foundry Technology. Dhanpat Rai & Sons
10. Mukherjee, M.: Metal Craftsmen of India, Anthropological Survey of India, Govt. of India (1978)
11. Simpson, B.L.: History of the Metal Casting Industry. American Foundrymen's Society, Illinois, USA (1969)
12. Ravi, B.: Computer-aided casting design-past, present and future. Indian Foundry J. **45**, 65–74 (1999)
13. Kochan, D., Chee Kai, C., Zhaohui, D.: Rapid prototyping issues in the 21st century. Comput. Ind. **39**, 3–10 (1999)
14. Deepan, D., Piyush, T.: Advanced in investment casting—India. In: 13th World Conference on Investment Casting, Japan, 2012 (2012)
15. Hazra Choudhury, S.K., Hazra Choudhury, A.K., Roy, N.: Elements of Workshop Technology-I Manufacturing Process. Media Promoters and Publishers Private Limited (1964)
16. 51st Census of World Casting Production: Modest growth in worldwide casting market. Modern Cast. 1–55 (2016)
17. Vinayak, M.: Foundry Industries Development Prospects in the Sixth Plan Period. Foundry Directory, IIF (1982)
18. Oudhia, S.P.: An overview of Indian foundry industry. Metalworld 22–24 (2015)
19. Metalworld Research Team: Indian Foundry industry on growth path. Metalworld 20–23 (2013)
20. SataA, SutariaM: Scope of investment castings supported by survey of foundries in Rajkot cluster. Indian Foundry J. **60**, 42–46 (2014)
21. Das, S., Upadhya, G.K., Paul, A.J., Chandra, U.: Modelling the investment casting process: a novel approach for view factor calculations and defect prediction. Appl. Math. Modelling **19**, 354–362 (1995)
22. Uday, A.D., Rahul, C.B.: Casting defect analysis using design of experiments (DoE) and computer aided casting simulation technique. Procedia CIRP **7**, 616–621 (2013)
23. Amin, L.D., Patel, S., Mishra, P., Joshi, D.: Rapid development of industrial castings using computer simulation. Indian Foundry J. **60**, 40–42 (2014)
24. JadhaoVS, SalunkeJJ: Review paper on simulation based casting. Int. J. Innov. Res. Sci. Eng. Technol. **3**, 17370–17374 (2014)
25. Dong, Y.W., Li, X.L., Zhao, Q., Yang, J., Dao, M.: Modelling of shrinkage during investment casting of thin-walled hollow turbine blades. J. Mater. Process. Technol. **244**, 190–203 (2017)
26. Jabur, A.S., Kushnaw, F.M.: Casting simulation and prediction of shrinkage cavities. J. Appl. Comput. Math. **6**, 1–7 (2017)
27. Shamasundar, S., Manjunatha, T.M.: Computer simulation and analysis of investment casting process. Foundry 29–44 (2004)
28. Yong, M.S., Fu, M.W.: Simulation-enabled casting product defect prediction in die casting process. Int. J. Prod. Res. **47**, 5203–5216 (2009)
29. Tiberto, D., Klotz, U.E.: Computer simulation applied to jewellery casting: challenges, results and future possibilities. IOP Conf. Ser. Mater. Sci. Eng. **33**, 1–10 (2012)
30. Beffel, M.J., Yu, K.O., Robinson, M., Schneider, K.R.: Computer simulation of investment casting process. Design 27–30 (1989)
31. Sabau, A.S.: Numerical simulation of the investment casting process. AFS Trans. **160**, 1–11 (2005)

32. Norouzi, S., Shams, A., Farhangi, H., Darvish, A.: The temperature range in the simulation of residual stress and hot tearing during investment casting. World Acad. Sci. Eng. Technol. **3**, 558–564 (2009)
33. Hang, Z., Qingyan, X., Baicheng, L.: Numerical simulation and optimization of directional solidification process of single crystal superalloy casting. Materials **7**, 16–25 (2014)
34. Chattopadhyay, H.: Estimation of solidification time in investment casting process. Int. J. Adv. Manuf. Technol. **55**, 35–38 (2011)
35. Zhang, D., Jiang, R., Li, J., Wang, W., Bu, K.: Cavity optimization for investment casting die of turbine blade based on reverse engineering. Int. J. Adv. Manuf. Technol. **48**, 839–846 (2010)
36. Guan, J., Dieckhues, G.W., Sahm, P.R.: Analysis of residual stresses and cracking of γ-TiAl castings. Intermetallics **2**, 89–94 (1994)
37. Sung, P.K., Poirier, D.R., Felicelli, S.D., Poirier, E.J., Ahmed, A.: Simulations of micro-porosity in IN718 equiaxed investment castings. J. Cryst. Growth **226**, 363–377 (2001)
38. Blair, M., Monroe, R., Beckermann, C., Hardin, R., Carlson, K., Monroe, C.: Predicting the occurrence and effects of defects in castings. JOM 29–34 (2005)
39. Ravi, B.: Casting simulation—best practice. In: Transaction of 58th IFC, pp. 19–29 (2010)
40. Shamasunder, S.: To believe or not to believe results of casting simulation software. ALUCAST 62–67 (2012)
41. Pattnaik, S., Karunakar, D.B., Jha, P.K.: Developments in investment casting process—a review. J. Mater. Process. Technol. **212**, 2332–2348 (2012)
42. Ravi, B.: Casting simulation and optimization: benefits, bottlenecks, and best practices. Technical Paper for Indian Foundry J., Special Issue 1–12 (2008)
43. Chakaraverthi, G., Prabhakararao, P.: Application of casting simulation. Int. J. Thermal Technol. **1**, 107–113 (2011)
44. Ding, M., Song, J., Honghui, L.: Effect of pouring temperature on typical structure of thin-walled ZL105A alloy casting. Mater. Manufact. Process **29**, 853–863 (2014)

Chapter 13
Experimental Investigation of Fiber Laser Micro-Marking on Aluminum 6061 Alloy

Debal Pramanik, Srinath Das, Soumya Sarkar, Subir Kumar Debnath, Arunanshu Shekhar Kuar and Souren Mitra

13.1 Introduction

Precision of operation, high-speed processing, and low cost are the significant advantages of laser cutting of sheet metals over the conventional cutting methods. Various manufacturing processes are using laser beams for the last four decades. In the field of aeronautics, electronics, semiconductor, automobile, and biomedical industries requiring turbine blades of aircraft engine, integrated circuits, combustion chambers, automotive fuel filters, surgical needles or stent cutting and micro-fluidic devices laser micro-machining including micro-cutting, micro-marking, micro-drilling, micro-channeling, micro-turning has emerged in a large scale during the present decade. Fiber laser technology combining of multi-diode pump solid-state laser and fiber technology is the prime factor for the successful application of laser micro-machining in the present era. Due to its simplicity, compact design, cost-effectiveness, low maintenance, high efficiency, high stability, and smaller spot size, it has become the most promising substitute to the high power, bulk solid-state laser, and gas lasers. Short pulse ranges from millisecond to picosecond, and even femtosecond for precise micro-machining of different materials is the main characteristic of fiber lasers. An extensive range of flexible processes for removing molten material in the micron range is included in the laser micro-machining process [1]. Broad classification of the term of micro-machining is made through the ablation, micro-cutting, turning, drilling, and marking in the ranges of micron. The various fields of production, metrological, medical, and military applications earlier controlled by conventional solid-state lasers and gas lasers have come into the domain of fiber lasers which appears to be most flexible and rapid increasing laser systems during the most recent decade. Fine and precise micro-machining involving an arrangement of both continuous and pulse mode with the help of

D. Pramanik (✉) · S. Das · S. Sarkar · S. K. Debnath · A. S. Kuar · S. Mitra
Department of Production Engineering, Jadavpur University, Kolkata, India
e-mail: debpramanik18@yahoo.com

P. Sahoo and J. P. Davim (eds.), *Advances in Materials, Mechanical and Industrial Engineering*, Lecture Notes on Multidisciplinary Industrial Engineering,
https://doi.org/10.1007/978-3-319-96968-8_13

fusion cutting (inert gas) to sublimation cutting (oxygen cutting) is a good marketable field for fiber lasers (average power of 5–50 W) [2, 3].

13.2 Benefits of Fiber Lasers Over Other Lasers (Gas and Solid State)

a. Light joined with a flexible fiber: It is a fact that light already in a fiber permits its easy transfer to a variable focusing element. Laser cutting, bending welding and folding of metals and polymers bear the importance of this.
b. High output power: Active regions of fiber lasers can be several meters long, for which it can give very high optical gain. Fibers' high surface area proportionate to volume ratio, allowing cooling efficiently is the cause of supporting of kilowatt levels of continuous output powers by lasers.
c. High optical quality: Waveguiding properties of fiber decrease or remove thermal deformation of the optical path, which creates a high-quality optical beam with limited diffraction.
d. Compact size: In comparison with rod or gas lasers, fiber lasers remain to be compact and to save space. No cooling or chilling unit is required for such type of fiber laser machine.
e. Reliability: Vibrational stability at high level with extension of lifetime and maintenance-free turnkey procedure is seen in fiber laser.

13.3 Applications of Fiber Laser in the Area of Micro-Machining

Fiber laser micro-cutting, micro-marking, micro-drilling are more widely used than the lamp current pulsed Nd:YAG lasers in the field of medical sciences. As the medical devices are small in size, expensive in material and able to be fitted into small areas production of them is a very difficult task. To overcome such problems, operation of fiber laser is a fruitful solution for costly and smaller implements in areas tolerating the power of only a few microns. Use of many conventional methods like ink marking, electrochemical marking, stamp marking, and engraving is made to mark the manufactured product with logs and related information for identification and theft prevention. Having a lot of advantageous over the conventional marking methods laser marking has been used increasingly due to the following benefits like no wear of tool, high degree of automation, free programming, and choice of features. Fast marking with precision in automated manufacturing environment requires the process for its most suitable nature. ID marks like QR codes, data codes, barcodes part numbers, 2D matrix codes and graphics, business logos and laser marking of the aesthetic nature are included in the various application fields, but it is not only confined to precision marking related to callipers. Pulse frequency, laser power, duty cycle, scanning speed are the process parameters of laser marking to be optimized for ensuring the acceptable mark quality for almost all the elements of material.

The identification number marking or QR code at the backside of the wafer in the die level is the most significant purpose of laser making, being used increasingly in micro-machining. Identification numbers on silicon wafers by laser marking help to stress out the manufacturing process related to fault analysis. Besides, the qualities like machine readability, minimization, and unfavorable outcome on the onward processing step the laser mark must possess the quality of distinct classification at the end of the production process. Laser marking requirements are turning to be more severe strict as a result of increasingly trend toward, very thin wafers (150 μm) in electronics consumer goods. One of the key areas emerging as the marking of electronic goods of daily household articles remains to be fiber laser marking process. Several manufacturers have made use of hard coated plastic cases to enable scratch resistance. It is very hard task to make laser marking on such surfaces without causing any damage to it because maintenance of a precise stability between the peak power and average power along with the pulse energy has to be made. Resulting fiber lasers are more and more used in place of green and other ultrashort laser for its cost-effective quality. Also, increasingly use of lasers is observed to fabricate medical devices like stents, catheters, and micro-fluidic devices. On the basis of the properties of the materials, selection of laser parameters should be made so that the laser can be used for production of variety of medical devices.

13.4 Fundamental Method of Laser Marking

A considerable quantity of energy in a small predetermined area is produced by laser. At the time of a beam of photons entering into a material surface, transfer of energy appears for intensification of the vibration of the molecules, causing heating to be macroscopic phenomena. Through heating, melting, or vaporizing in selected areas the focused energy achieves in a greater way than by any other process. A schematic representation of surface interaction with laser is shown in Fig. 13.1.

There are significant contrasting achievements of lasers as it can perform melting, annealing, or vaporization of the material, all of which are permanently unique. Modification of the optical appearance of a surface, which a laser beam heats, is the principle of laser marking. A variety of mechanisms are there for the occurrence of this following are the two broad categories for such mechanism:

(a) Marking by way of material removal from the upper surface of material and
(b) Marking by material surface modification.

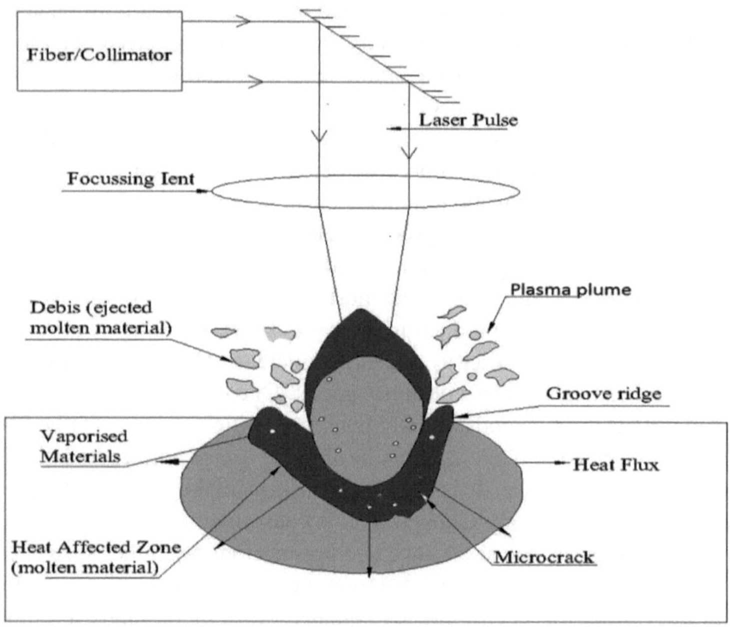

Fig. 13.1 Schematic representation of interaction of material

13.4.1 Marking by Way of Material Removal from the Upper Surface of the Material

Focusing on laser beam to a small spot greatly increases the density of the energy. As the energy becomes sufficiently high, it causes the surface temperature to rise. A visible mark is achieved by laser through either by melting, vaporization, or photochemical decomposition or ablation and engraving or scribing through the mechanism of removal of material.

13.4.1.1 Etching

The process laser etching applies heat to the surface of a material increasingly up to a level causing surface melting. The surface of a material is caused to be heated by the laser beam when cooling is done a different finish is found on the material. For creating a mark, removal of small quantity of material from the metal portion is done by the etching process. When a solid in connection with gas and liquid, a gas in connection a with solid, or by both a solid and gas at a time absorb laser light the processes of etching starts, causing reaction among the solid, liquid, or gas. The light wavelength remains to be the guiding factor for absorption, resulting in electronic excitation which causes melting, heating, or immediate direct heating [4].

13.4.1.2 Surface Melting

Laser beam helps to melt the material. Removal of coated layer like black oxide, anodization, paint for revel of the base material is achieved by the application of the laser beam. For the high contrast in-between the coating and the base material, very sharp and detailed characters arise out of this method. Analytical model studies for the laser melting of inhomogeneous workpieces having parabolic melt pool geometry are made for finding out the melt pool depth during laser marking for finding out the resulting surface layer thickness after solidification, related to the quality of the mark.

13.4.1.3 Ablation (Layer Removal)

High energy pulse irradiation is used by laser ablation for achieving strong electronic bonds breaking and resultantly decomposition of materials. Expansion of the resultant molecular fragments in plasma plume appears, rapidly builds up pressure and quickly ejects it. Resultant effect of the high-power laser pulse related to the material is to eject the material, and the process is known as ablation. Often it is conducted in both vacuum and air conditions. Absorption of a portion of incident beam energy is caused to the material, and subsequently, it is altered into heat energy. Rise of bulk material temperature depends on material properties and thermal conductivity. Melting and vaporization start as temperature rises to a definite threshold value. Vaporization and vapor cause ejection of melt to remove sections of the heated material, reaching at elevated temperatures. A recent layer on the wall of crater is formed by the solidified melt (the sides and under the surface of a layer mark), and deposition of condensed vapor is done as debris on the upper layer around the interaction area. Adverse influence of debris and recast layer may affect the laser mark visibility for which proper choice of processing parameters should be made to minimize them [1, 2].

13.4.1.4 Engraving

The upper layer of the material is penetrated by the laser beam to remove material in the laser path by way of displacement of molten material and/or evaporation. More heat is generated in laser engraving than laser etching, removing substrate material. Localized rising temperatures higher than the melting point of the material result in melting. After re-solidification of the material, an altered surface formation appears in the shape of a mark. Laser engraving cannot attain high contrast as is done by laser coloring or etching as the vaporization and ejection of material are caused during the marking process [5].

13.4.2 Marking by Material Surface Modification

This process causes radiation of laser to affect the composition of material for creation of a mark with high contrast devoid of any material removal. It helps the material to melt locally through oxidization or chemical alteration for forming a noticeable mark.

13.4.2.1 Foaming

Sufficiently high local temperature around the absorption place causes generation of gases through burning or evaporation in foaming process. Molten material surrounded the hot gases which form bubbles through expansion. Controlled laser energy helps to result in bubbles out of foaming, scattering light and resultantly marking white or light on dark marking contrast. High pulse frequencies are the prerequisite setting for the laser for obtaining these type of result, as a result high heat transmission to the compound occurs. Foaming giving very high contrast helps to achieve color markings on dark surfaces [6, 7].

13.4.2.2 Carbonization

A laser heating helps to achieve a dark or black marking through localized polymer matrix carbonization. In the method, local temperature of the material around the absorption place is raised by absorbed energy for causing polymer thermal degradation. Oxidized blackening of the surface material by laser gives a smooth surface. Application of this marking is done to lubricious parts. Different type of lasers may help laser marking of plastic by the use of laser sensitive pigments coming out of certain particle substrates having certain metal oxide coating. For optimization of laser sensitivity, selection of the components of the coating layer is done.

13.4.2.3 Annealing

When laser causes metal to be heated nearly to melting points, a change of color appears at the top layer of material. Laser beam heat causes annealing to achieve a dark, burr free, permanent mark without significant penetration of surface. Heat and oxygen induce an indelible mark coming out of laser annealing without observable material ablation. For all metals showing a color change due to exposure to heat and oxygen, especially stainless steel laser annealing may be done. A dark iridescent appearance, having a light rainbow of greens, blues, and pinks sometimes seen in the content or graphic, is generally given by annealing. Heating temperature is the strong determining factor for the color. No cut or shallow marking can be seen in annealing which typically achieves the darkest as no material removal is done from the metal. Applications of medical devices within the human body require such technique for their precision.

13.4.2.4 Coloring

Exposer to laser radiation of a specific wavelength causes chemical changes in materials when change of color occurs at certain energy densities (below melting point). Discoloration of some metallic substrate materials is done by laser coloring without melting, flaming, or vaporization of the material. Passing of a considerably low power laser beam on the surface material at low speed discolors the mark area. This method of laser marking achieves high-quality, high contrast marking, without any disruption of the surface. All materials are not good fields of action of the process as it possesses unfavorable effect on materials with previous heat treatment and by reduction of the corrosion-resistant properties of some stainless steel alloys. Specific colorant stabilities are the dependent factors for these mixed colorant systems when possibility of all color changes is remote. Toxicity or adverse affection of the product appearance physical properties or functional properties are absent in laser formulations.

13.5 Major Parameters in Laser Beam Machining

A quality machining operation of any material is an intricate method in which a number of parameters are to be followed at the same time. Laser machining process requires the most important parameters like peak power or threshold intensity, pulse width, cutting speed, pulse repetition frequency, cutting speed, and focal length which is elaborated below:

(a) Peak Power or Threshold Intensity

As the metal target is heated up by a laser without gas jet, the absorbed energy is conducted into surrounding colder metal. Threshold intensity is the minimum amount of power impact necessary for initiating evaporation as it is exposed to laser radiation [8]. Melting of the surface region of the metal will be done at higher abortion rates. The dominant mechanism of material removal from the target remains to be vaporization even higher absorption rate. At too high-power density, plasma comes out of the instant transformation of the gas near the workpiece interactions.

(b) Pulse Width

The definition of the time required for vaporization of the material is known to be pulse duration or pulse width. The penetration time of the laser beam should be shorter than the pulse width. The pulse energy and the penetration time have an inverse relation, i.e., when one increases the after decreases [9].

(c) Pulse Frequency

The mechanism of pulsed laser cutting differs from continuous wavelength laser cutting due to the periodic nature of heating. Overlapping of a series of marking operations is similar to the overall effect of laser cutting in pulsed mode. An accumulation of the action of a series of single pulses remains to be the process of cutting action.

(d) Cutting Speed

Feed rates varying from 80 to 90% of the maximum possible cutting speed make sound and safe cutting results practicable. Reduction in speed is done to have certain quality demands. Dross formation during fusion cutting and burnouts during oxidation cutting may occur when the speed is too low. Pulsing of the laser may help to avoid the defect.

(e) Focal Length

It is the distance between the workpiece and the focusing lens. It determines the diameter of focused spot and therefore the light concentration on the work surface. It is important parameter to obtain kerf widths.

13.6 Literature Review on Mark Width and Mark Depth

Researchers have made their study on micro-machining of laser beam and micro-machining of laser-assisted studies with various alternatives of process parameters as to optimize their machining output responses depending upon their applications. Qi et al. [10] have made their research work on the laser beam pulse frequency to find out their influence on the mark width, depth, and mark contrast. Use of Q-switched Nd:YAG laser has been made in the stainless steel laser marking process. The ratio of the apartment brightness between the with mark and without marked areas showing the allowance degree of the mark remains to be the mark contrast. A significant effect of the pulse frequency of a Q-switched Nd:YAG laser on the mark quality has been investigated by them. When mark width appears to be almost constant at different pulse frequency maximum depth of mark is found when the pulse frequency remains to be about 3 kHz. There is a decrease in evaporation of the material when the pulse frequency increases and simultaneously oxidization appears to be more important, leading to improvement in mark contrast. Leone et al. [11] have made use of a Q-switched multi-diode-pumped Nd:YAG laser to carry out laser marking test on AISI 304 steel. Determination of the co-relation appearing between working parameters, like pulse frequency, scanning speed, pulse intensity, and resulting mark visibility is the main objective of the study. Evaluation of the mark visibility has been made on the basis of contrast index calculated from digital images of the marks. For characterization of mark features, its width and roughness have been estimated. Optical and scanning electron microscopy along with energy dispersive X-ray technique is employed for analysis. The experimental results show that both surface roughness and oxidation rise up as a function of frequency, which leads to an enhancement in contrast up to a characteristic value. Ahmed et al. [12] have presented their study results of micro-channel fabrication relating to Nickel-based superalloy (INCONEL 718) through Nd:YAG laser beam machining. Evaluation of the influences of laser parameters upon the characteristic of the machining performance relating to dimensional sizes was made. Laser intensity, scanning speed, and pulse frequency were the three laser parameters considered to be predictors. Process responses were the channel's bottom width, top width, depth, and taperness. For studying and measuring micro-size dimension of the fabricated channels, micrograph by SEM was recorded. To achieve desired machining results, it was critical on the part of the selection of channel size. Study of microstructures of the machined channel, revealing the recast layer with lumber grain structure and phase transformation close to the edges of micro-channels was also made. Peter et al. [4] have studied on the basis of artificial neural network (ANN) intelligence tool and the response surface methodology (RSM), mathematical modeling and an optimization of marking characteristics like mark depth, mark width, and mark intensity by Nd:YAG laser micro-marking on zirconia ceramic. The RSM optimal data output is justified through experiments and compared with predictive model of ANN. Javale et al. [13] have looked into the result of the various process parameters on mark characteristics, i.e., width, depth, contrast using fiber laser on stainless steel, anodized aluminum, phenol formaldehyde. It is observed that with increase in pulse frequency, evaporation of material comes down, while simultaneously oxidization is more important, which bring about mark contrast improvement.

To obtain clear and unique mark, the most essential factors require adoption of appropriate laser marking techniques and proper process parameters are for a specific material, to improve the marking quality and productivity. In this experimental study, micro-marking on Al-6061 alloy has been done using multi-diode pump fiber laser. Aluminum has its remarkable mechanical properties and thermal properties, and it is considerable challenge for machining using laser beam machine due to its high thermal conductivity and weldability. Few research works had been reported so far in the field of modeling and optimization of micro-marking on Al-6061 alloy using diode pumped fiber laser. Nevertheless, no research work has been reported on influence of a relatively uncommon process variable, i.e., defocus height during laser marking operation. The aim of the present research study to develop a mathematical model that can foretell the response characteristics like mark width and mark depth for various input parameter settings in order to maximize productivity.

13.7 Fiber Laser Marking Quality Characteristic

Estimation of mark width and mark depth helps to evaluate laser marking quality characteristics. Legibility characteristics help to assess the quality of marking. Mark width refers to the line segment width formed by laser marking. The laser beam spot size controls the marking width. The line width is affected by other parameters like scanning speed, material absorbity, laser power density, and duty cycle. Normally, an optical microscope helps to measure the mark width. Occasionally to find out the width measurement, use of surface texture measurement equipment has been made. The density of laser energy, the various types of materials, and the material and laser beam interaction time generally affect the depth of marking. A surface profile instrument or optical microscope helps to measure marking depth. Variation from a few microns to several tens of millimeters can be achieved with the help of depth of penetration in the material. The material evaporation from marking region is enhanced by the use of the assist compress gas, resulting in more marking depth. The supply of gases like oxygen, argon, helium, nitrogen, or compressed air, assisting material removal helps to realize further increase in the reaction on the material.

13.8 Experimental Plan on Laser Micro-machining of Al 6061 Alloy

Details of multi-diode fiber laser machining system setup and experimental plan along with result have been discussed thoroughly.

13.8.1 Fiber Laser Machining System Components

Optical fiber, diode pump sources, rare earth material dopant ions, different types of mirrors, collimator, and fiber coupler remain to be the key elements of fiber laser. Several parameters including laser source, (mainly wavelength and emission regime) and characteristic of the material itself are the depending factors for laser interaction with the optical fiber material. As a low loss material silica glass constitute cylindrical waveguide which possesses to be optical fiber as a main component. Doping with ytterbium yb^{3+} and neodium Nd^{3+} as rare earth elements help to improve the optical properties. Comparison with other rare earth elements Yb^{3+} possesses several advantages like its long upper state lifetime, a minor quantum defect (decreasing thermal load per unit length of pump power) and lack of the excited state absorption. Different types of mirrors are utilized in fiber lasers to form cavity. Semiconductor, diode lasers predominantly perform the pumping operation in most of the cases. The diode laser output is directly pulsed into the laser fiber.

Multiple diode lasers are utilized for increasing the pump power reaching to the fiber. To prevent going back of emitted light toward the source, use of fiber couplers is made either at one side of input fibers or at one side of output fibers [8].

The fiber laser key elements relating to marching system are laser source, collimator, beam delivery system, laser beam focusing control system, CNC controller, and axis movement.

a. Laser Source: Pumping of fiber laser is done with the help of special high-power multi-mode diodes through cladding of surrounding as single mode core. In comparison with conventional diode pump, solid-state laser (DPSSL) life of this individual multi-mode is quite long, when a single diode bar pumps DPSSL multiple identical diodes, all of which feed the same gain medium, pump fiber laser.
b. Collimator: Once the laser is generated within the optical fiber, the laser is transferred into a collimator. In the collimator, two prisms are fixed at the two ends of it.
c. Beam Delivery System: The most favorable advantage of fiber laser is that the gain medium remains in the fiber and the delivery is also made through the fiber. As a result, chances of coupling tip failure between gain medium and delivery extending to the workpiece remain to be less.
d. Laser beam focusing control system: To alter the power density and the depth of focus made by the laser of a given beam diameter, the focal length of the focusing lens may be changed. A lower focal length lens achieves higher power density. The importance of the alignment of the focusing lens greatly regulates the coincidence of the beam center with center of the lens and the beam after the lens if it not straight it will drastically lower down the efficiency of the cutting. For getting proper focus on the workpiece, this fiber laser system is provided with CNC interface, CCD camera, and CCTV [14].
e. CNC controller for axis movement: For having appropriate focusing for laser beam through focusing lens, a CNC controller unit guides *X*-*Y* table and the movement of *X*-*Y*-axis. The focus lens of the job is controlled by *Z*-axis.

13.8.2 Procedure and Planning of Laser Marking

Use of pulsed Multi-mode diode pumped fiber laser system and Industrial Mark (I-mark) software manufactured by M/s Sahajanand Laser Technology, India, has been made for the experimental purpose of the study (Fig. 13.2). Table 13.1 shows the specification of the machine setup in details. Experiments have been planned on response surface methodology (RSM), central composite rotatable second-order design (CCD) based. Total 52 runs have been conducted and replicate with three times. On experimental completion, microscopic images of mark width have been taken at 10x magnification and mark depth by 50x magnification by using an optical measuring microscope (Olympus STM 6), which is later measured by image

Table 13.1 Detailed specification of the machine setup

Specification	Description
Laser type	Pulsed fiber laser
Wavelength	1064 nm
Mode of operation	Pulsed mode
Beam diameter $1/e^2$	9 mm (after collimator)
Beam spot diameter	21 μm
Max average power	50 W
Pulse duration	120 ns
Max peak power	7.5 kW
Pulse repetition rate	50–120 kHz

Table 13.2 Controllable parameters and their limits

Parameters	Symbol	Level				
		−2	−1	0	1	2
Power setting (%)	A	30	35	40	45	50
Duty factor (%)	B	45	55	65	75	85
Frequency (kHz)	C	50	60	70	80	90
Speed (mm/s)	D	0.7	1.4	2.1	2.8	3.5
Defocus height (mm)	E	−0.7	−0.35	0	0.35	0.7

analysis software. The controllable process parameter in the present research has been chosen as power setting, duty cycle, pulse frequency, scanning speed, and defocus height at 3 bar constant air pressure. The range of process parameters are selected by basic experiments. The range of each controllable process parameters as per RSM and CCD techniques is listed in Table 13.2.

13.9 Experimental Results and Discussion

From the literature review, it is concluded that certain parameters have a great influence on manufacturing. CCD is a very capable method to fit a second-order model. Each factor ranges have been decided on the basis of some pilot experiments. Each controllable process parameter as per RSM CCD technique is listed in Table 13.2.

Experimental plan with coded values for setting of process parameter and measurement of responses is shown in Table 13.3.

13.9.1 Developing of Mathematical Model

Development of a second-order polynomial equation has been made to carry out further analysis. Minitab 17 statistical software is used for analysis the responses and determining the mathematical models with best fits are given below in Eqs. 13.1 and 13.2.

$$\begin{aligned}\text{Mark Width} = {} & 56.676 + 4.263\text{A} + 1.598\text{B} - 3.517\text{C} - 1.118\text{D} - 6.049\text{E} - 1.146\text{A} \times \text{A} \\ & - 0.902\text{B} \times \text{B} + 0.228\text{C} \times \text{C} + 0.307\text{D} \times \text{D} + 2.950\text{E} \times \text{E} + 0.380\text{A} \times \text{B} \\ & + 0.864\text{A} \times \text{C} + 0.144\text{A} \times \text{D} - 1.020\text{A} \times \text{E} - 0.196\text{B} \times \text{C} + 0.074\text{B} \times \text{D} \\ & + 0.600\text{B} \times \text{E} + 0.419\text{C} \times \text{D} - 1.653\text{C} \times \text{E} - 1.042\text{D} \times \text{E}\end{aligned} \tag{13.1}$$

$$\begin{aligned}\text{Mark Depth} = {} & 13.126 + 1.939\text{A} + 0.329\text{B} - 1.900\text{C} - 1.041\text{D} + 0.263\text{E} - 0.226\text{A} \times \text{A} \\ & + 1.149\text{B} \times \text{B} + 0.599\text{C} \times \text{C} + 0.599\text{D} \times \text{D} - 0.839\text{E} \times \text{E} - 0.224\text{A} \times \text{B} \\ & + 0.663\text{A} \times \text{C} - 1.324\text{A} \times \text{D} - 0.360\text{A} \times \text{E} + 0.213\text{B} \times \text{C} + 0.351\text{B} \times \text{D} \\ & - 0.010\text{B} \times \text{E} + 1.013\text{C} \times \text{D} + 1.002\text{C} \times \text{E} + 0.652\text{D} \times \text{E}\end{aligned} \tag{13.2}$$

13.9.2 Influence of Parameters

For determining the relative significance and importance of the machining process parameters during the experimentation, the analysis of variance has been performed. The analysis of variance (ANOVA) and the F-ratio test has been conducted to evaluate the adequacy of the models as well as the importance of the individual model coefficient. Lack-of-fit value of the model shows non-significance, which is desirable. Quadratic models with other adequacy measure R^2 and adjusted R^2 are shown in Tables 13.4 and 13.5. From Table 13.4, it is noticed that the related p value of less than 0.05 for the model signifies that the model terms are statistically significant. From Table 13.4, it is observed that for mark width, defocus height (51.58%) is the most important factor followed by power setting (25.61%), pulse frequency (17.43%), duty factor (3.6%), and scanning speed (1.76%).

It is also observed that from Table 13.5 for mark depth, power setting (43.5%) is the most important factor followed by pulse frequency (41.8%), scanning Speed (12.5%), duty factor (1.25%), and defocus height (0.802%).

The combined effects of defocus height and power setting on laser mark width have been analyzed through response surface plot obtained from developed mathematical models, where duty factor (65%) frequency (70 kHz), speed (2.1 mm/s) are kept constant. It is observed from Fig. 13.3 that mark width is more in defocus up rather than in focus point or defocus bottom position. The mark width increases with the increase in defocus height up, and with respect to power, it

Table 13.3 Experimental plan showing with coded values of parameter settings and measured responses

Experimental information						Results	
S. No.	Machining parameters (coded value)					Mark width (μm)	Depth (μm)
	A	B	C	D	E		
1.	−1	−1	−1	−1	−1	58.760	16.45
2.	1	−1	−1	−1	−1	68.990	23.80
3.	−1	1	−1	−1	−1	61.686	18.70
4.	1	1	−1	−1	−1	69.916	22.60
5.	−1	−1	1	−1	−1	55.626	9.400
6.	1	−1	1	−1	−1	64.494	15.20
7.	−1	1	1	−1	−1	58.500	7.400
8.	1	1	1	−1	−1	70.138	16.80
9.	−1	−1	−1	1	−1	62.450	14.00
10.	1	−1	−1	1	−1	67.158	16.10
11.	−1	1	−1	1	−1	63.774	15.20
12.	1	1	−1	1	−1	72.536	13.10
13.	−1	−1	1	1	−1	52.450	7.500
14.	1	−1	1	1	−1	70.242	12.60
15.	−1	1	1	1	−1	52.760	8.900
16.	1	1	1	1	−1	68.606	14.50
17.	−1	−1	−1	−1	1	56.932	12.67
18.	1	−1	−1	−1	1	62.394	21.50
19.	−1	1	−1	−1	1	58.650	14.00
20.	1	1	−1	−1	1	65.234	19.70
21.	−1	−1	1	−1	1	41.194	10.90
22.	1	−1	1	−1	1	48.288	17.40
23.	−1	1	1	−1	1	44.688	9.700
24.	1	1	1	−1	1	54.028	14.80
25.	−1	−1	−1	1	1	47.850	15.90
26.	1	−1	−1	1	1	52.026	12.60
27.	−1	1	−1	1	1	53.736	15.00
28.	1	1	−1	1	1	61.520	13.30
29.	−1	−1	1	1	1	41.584	11.20
30.	1	−1	1	1	1	46.924	13.80
31.	−1	1	1	1	1	44.320	14.70
32.	1	1	1	1	1	51.972	16.60
33.	−2	0	0	0	0	45.620	7.900
34.	2	0	0	0	0	61.126	15.30
35.	0	−2	0	0	0	52.040	14.80
36.	0	2	0	0	0	56.656	19.40
37.	0	0	−2	0	0	64.584	18.10
38.	0	0	2	0	0	53.150	11.70

(continued)

Table 13.3 (continued)

Experimental information						Results	
S. No.	Machining parameters (coded value)					Mark width (μm)	Depth (μm)
	A	B	C	D	E		
39.	0	0	0	−2	0	62.962	16.30
40.	0	0	0	2	0	55.408	13.50
41.	0	0	0	0	−2	83.560	6.900
42.	0	0	0	0	2	55.952	11.40
43.	0	0	0	0	0	55.210	14.90
44.	0	0	0	0	0	55.000	13.10
45.	0	0	0	0	0	57.198	11.40
46.	0	0	0	0	0	55.206	13.90
47.	0	0	0	0	0	56.206	13.20
48.	0	0	0	0	0	55.028	12.60
49.	0	0	0	0	0	54.002	12.40
50.	0	0	0	0	0	57.546	12.80
51.	0	0	0	0	0	59.480	13.30
52.	0	0	0	0	0	59.330	14.90

Table 13.4 ANOVA for mark width

Source	DF	Adj SS	Adj MS	*F* value	*P* value
Model	20	3394.34	169.72	28.31	0.000
Linear	5	2837.40	567.48	94.66	0.000
A	1	726.91	726.91	121.26	0.000
B	1	102.19	102.19	17.05	0.000
C	1	494.67	494.67	82.52	0.000
D	1	49.99	49.99	8.34	0.007
E	1	1463.64	1463.64	244.15	0.000
Square	5	353.76	70.75	11.80	0.000
2-way interaction	10	203.17	20.32	3.39	0.004
Error	31	185.84	5.99	–	–
Lack-of-fit	22	153.35	6.97	1.93	0.154
Pure error	9	32.49	3.61	–	–
Total	51	3580.18			
S = 2.44844 R-sq = 94.81% R-sq (adj) = 91.46%					

increases more steeply at higher power. It is observed that relatively uncommon process variable, i.e., defocus height plays a very significant role determining the mark width, from the experimental result it is evident that with increase in defocus height (from negative to positive) the mark width decreases.

The combined effects of power setting and frequency on laser mark depth have been evaluated through response surface plot obtained from developed

Table 13.5 ANOVA for mark depth

Source	DF	Adj SS	Adj MS	*F* value	*P* value
Model	20	596.69	29.83	12.39	0.000
Linear	5	345.35	69.07	28.67	0.000
A	1	150.46	150.46	62.46	0.000
B	1	4.34	4.34	1.80	0.189
C	1	144.47	144.47	59.98	0.000
D	1	43.30	43.30	17.98	0.000
E	1	2.76	2.76	1.15	0.292
Square	5	91.35	18.27	7.59	0.000
2-way interaction	10	159.97	15.99	6.64	0.000
Error	31	74.67	2.40	–	–
Lack-of-fit	22	64.01	2.91	2.46	0.082
Pure error	9	10.66	1.18	–	–
Total	51	671.37			
S = 1.55207 R-sq = 88.88% R-sq (adj) = 81.70%					

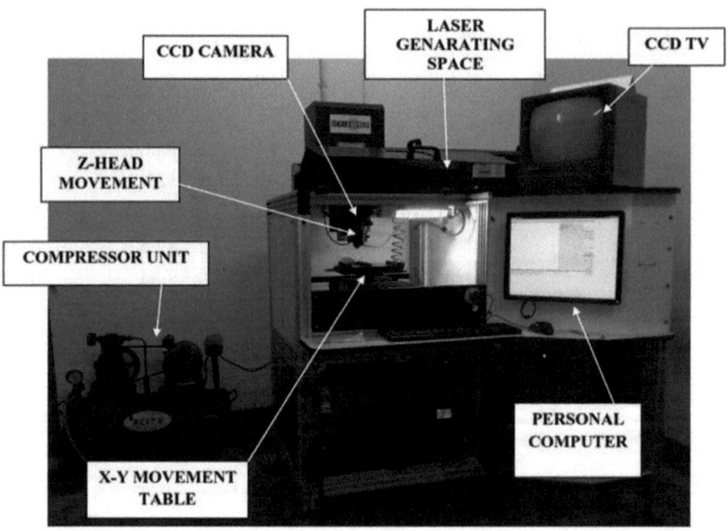

Fig. 13.2 Photographic view of fiber laser machining system

mathematical models, where duty factor (65%), speed (2.1 mm/s), and defocus height (0 mm) are kept constant. It is observed from Fig. 13.4 that mark depth is more in low frequency as at low-frequency peak power is high, which leads in high penetration. And with increase in frequency, the mark depth decreases. It is also observed that with the increase in power setting mark depth also increases.

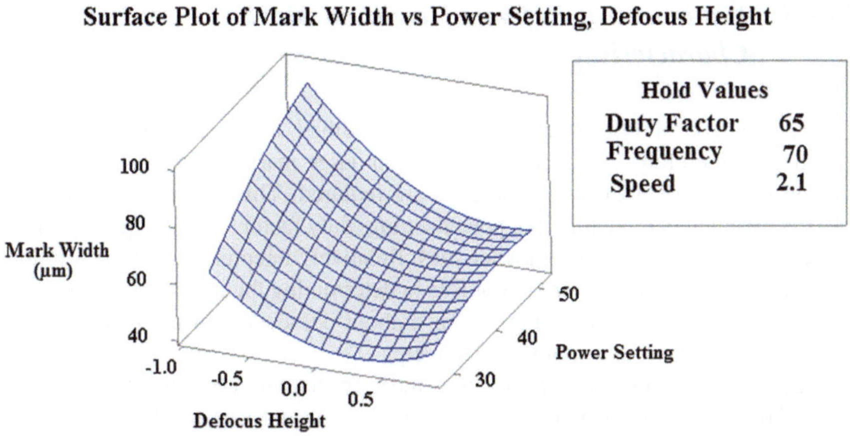

Fig. 13.3 Surface plot of mark width versus power setting, defocus height

Table 13.6 Confirmation testing

Responses	Actual value	Predicted value	% error
Width	38.94	40	2.65
Depth	15.58	15	3.86

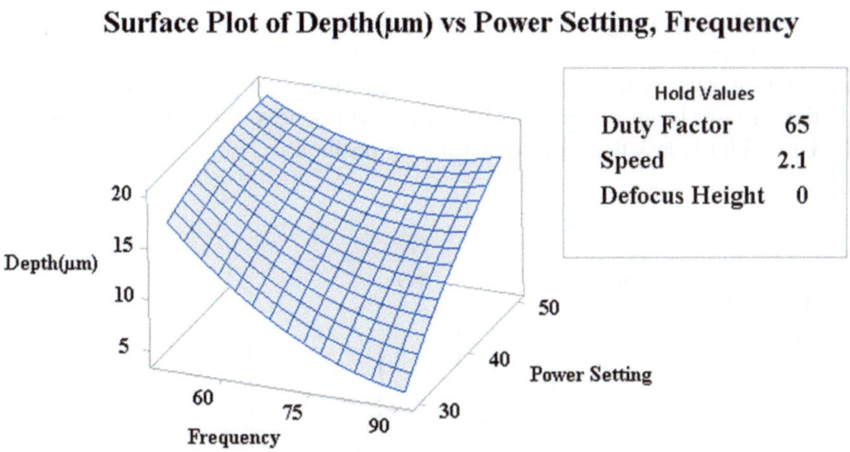

Fig. 13.4 Surface plot of mark depth versus power setting, frequency

13.9.3 Influences of Defocus Height on Mark Characteristics

Characterization of laser marking may be modified for change in the marking parameters; therefore, choosing of the proper laser marking parameter and controlling of the laser marking parameters are required. This paper deals with laser process parameters modeling like power setting, duty factor, pulse frequency, speed defocus height on the mark characteristics. Relatively uncommon process variable, i.e., defocus height, acts a very major role in determining the mark characteristics. It is clearly explained below in Fig. 13.5a–c about defocusing in up, focusing position, and defocusing in bottom, respectively.

Figure 13.5 shows that kerf width is more in defocus up rather than in focus point. The kerf width increases with the increase in defocus height up, and with respect to power, it increases more steeply at higher power. It is also observed that, for the same parameter setting, in defocus up (−), the kerf width is more than defocus in bottom (+). At defocus bottom, where kerf width increases slightly or almost same for higher power also.

Laser beam machining is all about thermal energy transformation, which causes melting and direct vaporization of material. So, more disorder takes place during material removal from the work surface.

When laser beam is in defocus up, laser beam in focusing somewhere over the work surface, after that it diverges toward the work surface and does machining with more beam diameter. So the kerf width of the marking is more but the depth or penetration is less because the energy intensity is less for defocusing position as the beam is falling on more area than focusing position.

While laser beam is in focusing position on the job surface, the energy intensity is so high that its penetration more that in defocusing up position. So when penetration of laser beam is more, the kerf width will be less. So in focusing position, the kerf width is less than the defocusing up position.

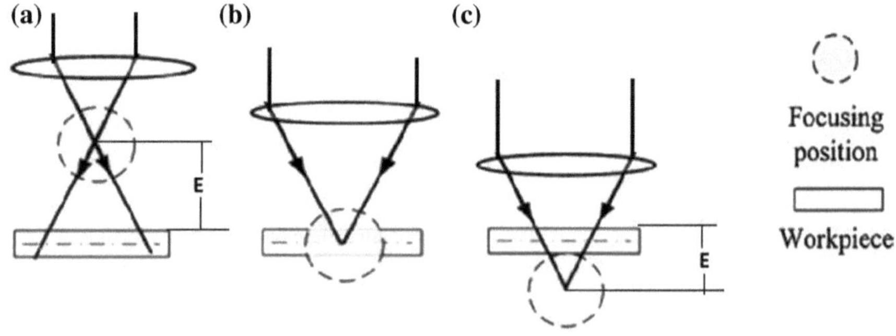

Fig. 13.5 **a** Defocusing in up, **b** focusing position, and **c** defocusing in bottom

Now when the laser beam is in defocus bottom position, a laser beam in focusing somewhere below the work surface. So laser beam is convergent towards the job surface, so it's energy intensity is increase towards the work surface, do machining according to the energy intensity and it is observed that the kerf width in defocus bottom is quite same or slightly less than the kerf width in focusing position.

When the energy intensity is more the depth will be more and kerf width will be less, as the machining is about to material removal from the work surface, but for micro-marking application depth of penetration is less desirable rather than kerf width, so by varying defocus height we can vary the energy intensity. So kerf width will also vary accordingly.

Now it is also observed that kerf width increase with increase in power and after certain value its tendency to slightly decrease. This is because power generates high thermal energy, which outcome in the melting and immediate vaporization of material from the job surface. It is already mention that to get certain average power, and the most important parameter is power to generate thermal energy for melting and vaporization of materials from the work surface. So kerf width increases with increase in power. And after certain value, the tendency of graph is to slightly decrease because after certain power energy intensity is so high that it penetrates more.

13.9.4 Verification Experiments

To further validate the planned mathematical model, a new set of experiments have been done according to the parameter settings, power setting 47.59%, duty factor 45%, frequency 74.1 kHz, speed 3.5 mm/s, defocus height 0.7 mm obtained from the optimization result, shown in Fig. 13.6.

It is noticed from the validation experiments (Table 13.6) that there is a little % of error between the predicted and the experimental values. So it is marked that using the developed model, it is quite possible to achieve the desired mark width and mark depth with good accuracy.

13.10 Conclusion

From the experimental inspection and analysis, it is clear that multi-diode fiber laser marking seems to be a useful potential marking process on Al-6061. Based on the present study and experimental analysis, the following conclusion may be drawn:

(i) From the research study, it is observed that relatively uncommon process variable, i.e., defocus height plays a very significant role determining the mark characteristics. From the experimental result, it is evident that with increase in defocus height (from negative to positive) the kerf width

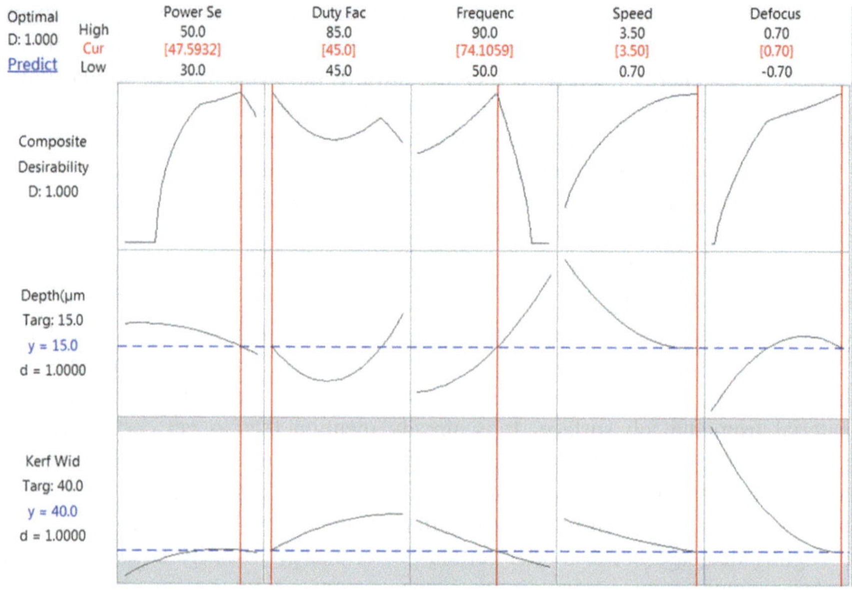

Fig. 13.6 Optimization of desired mark response

decreases. Hence, it is obvious that defocus height can be tactfully utilized for efficient control of laser marking operation.

(ii) It is noticed that the kerf width increases with increase in power setting. This is due to increase in average power with increase in power setting. It is further observed that the impact of defocus height is less pronounced when the power setting is on lower side during laser marking operation. This aspect can play a very important role while marking on slightly curved or uneven surface. To obtain a relatively consistent uniform kerf, low power setting is recommended during such machining operation.

(iii) From the response plots and ANOVA, it is obvious that the defocus height, power, speed, pulse frequency are significant process parameters in determining the mark characteristics.

Thus, it is clear that using the developed model it is quite possible to achieve the desired mark characteristics with good accuracy while maintaining the speed at its highest possible limit (Fig. 13.7). The present research findings is expected to provide a systematic guideline to shop floor engineers for optimum marking of Al-6061 using fiber diode laser setup.

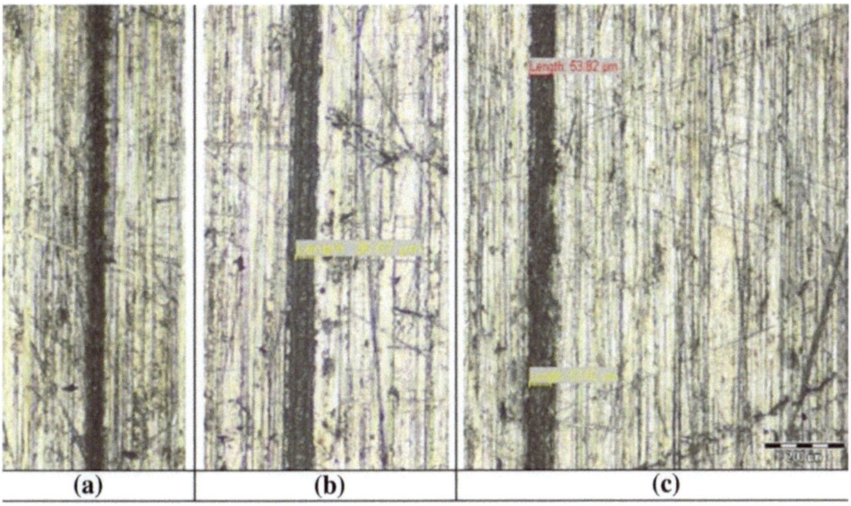

Fig. 13.7 Experimental result of kerf width **a** 30.74 μm, **b** 35.12 μm, **c** 53.82 μm

References

1. Dahotre, N.B., Harimkar, S.P.: Laser fabrication and machining of materials. ISBN 978-0-387-72343-3 (2007)
2. Roy, N., Kuar, A.S., Mitra, S., Acherjee, B.: Nd: YAG laser micro drilling of Sic-30 BN Nanocomposite: experimental study and process optimization, laser based manufacturing. In: Topics in mining, metallurgy and materials engineering, pp. 317–341. Springer. ISBN 978-81-322-2351-13 (2015)
3. Sen, A., Doloi, B., Bhattacharaya, B.: Fiber laser micromachining Ti-6Al-4 V laser based manufacturing. In: Topics in mining, metallurgy and materials engineering, pp. 255–281, 317–341. Springer. ISBN 978-81-322-2351-13 (2015)
4. Peter, J., Doloi, B., Bhattacharya, B.: Nd:YAG laser marking on zirconia ceramic, laser based manufacturing. In: Topics in mining, metallurgy and materials engineering, pp. 282–316, 317–341. Springer. ISBN 978-81-322-2351-13 (2015)
5. Meijer, J.: Laser beam machining (LBM), state of the art and new opportunities. J. Mater. Process. Technol. **149**, 2–17 (2004)
6. Dubey, A., Yadava, V.: Experimental study of Nd:YAG laser beam machining—an overview. Int. J. Machine Tools Manufact. **48**, 609–628 (2008)
7. Velotti, C., Astarita, A., Leone, C., Genna, S., Minutolo, S.F.M.: Study of the laser marking process of cold sprayed titanium coatings on aluminium substrates. Optics Laser Technol. **83**, 168–176 (2016)
8. Jiang, C.Y., Lau, W.S., Yue, T.M., Chiang, L.: On maximum depth and profile of cut in pulsed Nd:YAG laser machining. Ann. CIRP **42**(1), 223–226 (1993)
9. Yilbas, B.S.: Parametric study to improve laser hole drilling process. J. Mater. Process. Technol. **70**, 264–273 (1997)
10. Qi, J., Wang, K.L., Zhu, Y.M.: J. Mater. Process. Technol. **139**, 273–276 (2003)
11. Leone, C., Genna, S., Caprino, G., Lorio, I.D.: AISI 304 stainless steel marking by a Q-switched diode pumped Nd:YAG laser. J. Mater. Process. Technol. **210**, 1297–1303 (2010)

12. Ahmed, N., Darwish, S., Alahmari, A.M., Shar, M.A.: Micro channels by Nd: YAG laser beam machining: fabrication, microstructures, and micro hardness profiles. Int. J. Adv. Manuf. Technol. **85**, 1955–1968 (2016)
13. Javale, Y.V., Nair, N.V.: A review on laser marking by Nd:YAG laser and fiber laser. IJSRD **1**(9), 75 (2013)
14. Biswas, R., Kuar, A.S., Sarkar, S., Mitra, S.: A parametric study of pulsed Nd:YAG laser micro drilling of gamma-titanium aluminide. Opt. Laser Technol. **42**, 23–31 (2010)

Chapter 14
An Approach to Numerical Modeling of Temperature Field in Direct Metal Laser Sintering

Mihir Samantaray, Dhirendra Nath Thatoi and Seshadev Sahoo

14.1 Introduction

Aluminum alloys and composites have been broadly used in the aerospace, automotive and marine industries due to their lightweight, high strength, corrosion resistant, and good tribological properties [1, 2]. Among different types of aluminum alloys and composites, some of them are the complex structure which is difficult to fabricate using conventional manufacturing processes. In conventional processes, coarse-grain structures are formed because of its slow cooling rate [3]. Therefore, novel manufacturing processes are extremely necessitated for the manufacturing of complex components with fine and uniform microstructures. Nowadays, additive manufacturing grabs the attention of the manufacturing industry to produce complex shapes. Direct metal laser sintering is one of the newly developed additive manufacturing technologies that can manufacture three-dimensional metallic components with complex structures directly from the metal powders. In the DMLS process, a high-intensity laser beam scans the powder bed and simultaneously it melts and fuses the powders in layer by layer manner, thus producing high-density parts [4–6]. The process is so rapid that a high-temperature gradient is observed in the powder bed, which results in a rapid cooling rate. The microstructures and the mechanical properties of the build component are directly influenced by the cooling rate. This process involves multiple modes of heat and mass transfer phenomena due to the interaction of the laser beam and powder bed, which make it complicated. The key benefits of the DMLS process are high degree of design freedom, minimal wastage of powder material, and cost-effective than conventional manufacturing processes [7, 8]. The graphical representation of DMLS process is shown in Fig. 14.1.

M. Samantaray · D. N. Thatoi · S. Sahoo (✉)
Department of Mechanical Engineering, Institute of Technical Education and Research, Siksha O Anusandhan (Deemed to Be University), Bhubaneswar 751030, Odisha, India
e-mail: seshadevsahoo@soa.ac.in

P. Sahoo and J. P. Davim (eds.), *Advances in Materials, Mechanical and Industrial Engineering*, Lecture Notes on Multidisciplinary Industrial Engineering,
https://doi.org/10.1007/978-3-319-96968-8_14

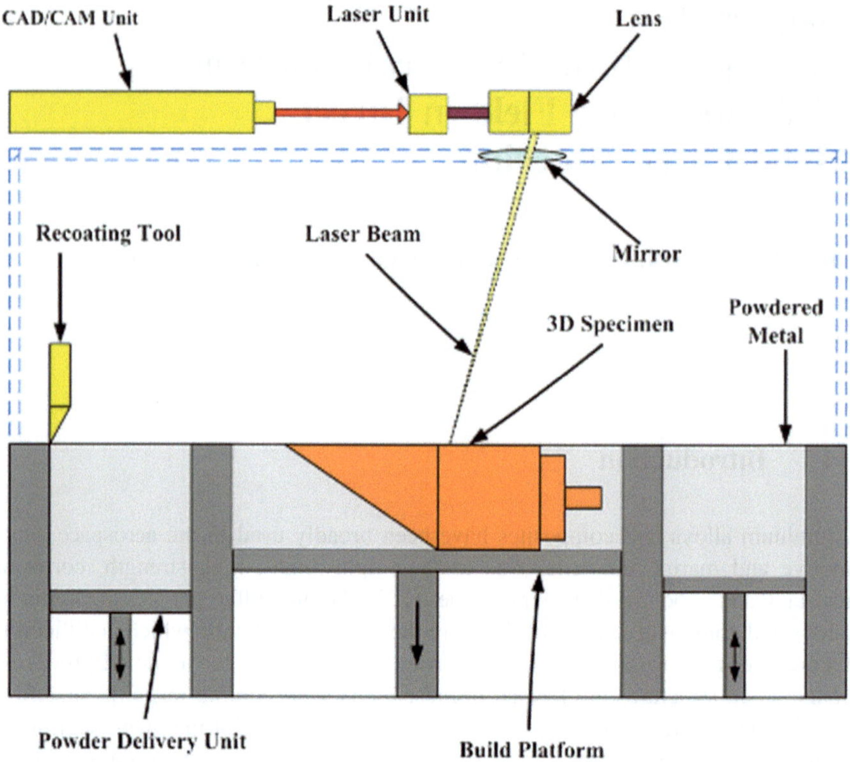

Fig. 14.1 Schematic diagram of DMLS process

As reported by previous researchers [9–12], the sintering phenomena of metallic powders in additive manufacturing processes are solid phase sintering or liquid phase sintering, which depends on the energy source, type of powders, particle size, and processing parameters. In this process, a high-intensity laser beam scans the powder layer resulting in rapid heating of a small region of powders. This will lead to a significant change in temperature distribution, and large temperature gradient in the powder layer generates residual stress which affects the part quality. Over past decades, some of the studies focused on modeling of laser sintering and laser melting processes with different alloy system. Only a few attempts have been conducted to simulate the DMLS process to study the thermal behavior during sintering. A comprehensive thermal model for the selective laser sintering process was developed by Zeng et al. [13] to examine the temperature distribution on the build part. Tang et al. [14] have studied the thermal behavior during processing of copper-based alloy and the quality and accuracy of the structure with different processing condition. A time-dependent numerical model based on finite element approach has been proposed by Dong et al. [15] to understand the phase transformation behavior during the laser sintering process by considering the thermal and the sintering phenomena. Yuan and Gu [16] developed a numerical model for selective laser melting of TiC/AlSi10Mg nano-composites and investigated the temperature variation with respect to process parameters. Simchi [17] had analyzed the significance of processing parameters on the densification of metal powders in the direct metal laser sintering process. He established a correlation between the density of build parts and laser energy during direct metal laser sintering, which was found to be very useful for the metals or alloys. Jian et al. [18] had presented a three-dimensional transient model for simulating the temperature in selective laser sintering process for Al_2O_3 coated with ceramic powder by using a moving CO_2 laser beam. A relationship between process parameter and thermal behavior in selective laser melting process was established by Li and Gu [19] based on a finite element model. The authors used ANSYS 12.0 platform to simulate the developed model and numerically analyzed the thermal behavior and molten pool formation at different laser power and scan speed. Shi et al. [20] numerically analyzed the melting/solidification of Ti-based composite powders in selective laser melting process under different laser power and scan speed. In their study, the authors optimized the processing parameters based on the formation of defects. Also, experiments were conducted to validate the model. Hu et al. [21] developed a finite element model for selective laser melting of multilayer AlSi10Mg powder and investigated the effect of input laser power on temperature distribution, depth of the molten pool, cooling rate, and microstructure of the build part. The finding of this study delivers a valuable theoretical guidance for selective laser melting of different alloy system. Sahoo et al. [22] had developed a thermal and microstructure model for electron beam additive manufacturing of Ti alloy powder by considering moving heat source. The temperature variation in the powder bed with respect to scan speed, laser power, and hatch spacing in DMLS process was investigated by Lee et al. [23] using a particle-based discrete element model. From their investigation, the authors found that laser power has greatly influenced the temperature variation in the powder bed, whereas scan speed and hatch spacing have negligible effect. Zhao et al. [24] numerically studied the mechanism of heat transfer and residual stress in the direct metal laser sintering of Ti alloy using COMSOL multiphysics environment. Their simulation results give information on molten pool size, temperature distribution, and change of residual stresses of the powder layer with the variation of process parameters. The simulation results provide a better understanding of the complex thermo-mechanical mechanisms of laser sintering processes. A finite element model was developed by Dong et al. [25] to simulate the laser sintering of Ti powder bed fusion using the software package ABAQUS. Based on their simulation results, the process parameters were optimized for the manufacturing of different components. Kundakcioglu et al. [26] simulated 3D transient temperature field for additive manufacturing of complex structures in powder bed systems and laser heat source. Ojha et al. [27] developed a mathematical model for direct metal laser sintering of AlSi10Mg alloy by considering fluid flow, heat transfer, and solidification characteristics in the liquid pool. Their simulation results give the fundamental understanding of the densification of powder particles in the DMLS process. Panda and Sahoo [28] developed a thermo-mechanical model to investigate the residual stress formation in the

AlSi10Mg powder bed with respect to scan speed and laser power. Based on the residual stress, they optimized the process parameters to minimize the residual stress in the built part.

Regardless of several advantages of this rapid manufacturing process over conventional processes, still few drawbacks hinder its application for fabrication of various alloys and composites. As this process is applied to an extensive range of powder materials and involves rapid melting and sintering phenomena, the scientific and technical aspects of this processes are still not well understood. To overcome these difficulties, significant research efforts are required to scalable of this process. In the present study, a numerical model has established to investigate temperature distribution, thermal history, molten pool profile, sintering depth, and the sintering mechanism of AlSi10Mg alloy powder bed.

14.2 Establishment of the Model

When the laser beam interacts with the powder layer, the heat energy generated due to the laser beam absorbed by the powder particle results localized melting of powders in the powder bed. The laser beam is considered as a moving Gaussian heat source. When it leaves the melt region, rapid cooling of the melted region takes place leading to the integration of the molten metal powder particles. During this process, the heat transfer mechanism includes multiple modes of heat transfer in the powder bed and the surrounding environment [29], which makes the process complicated. The heat transfer mechanism is shown in Fig. 14.2.

Considering the practical process of heat flow and sintering phenomena in the DMLS process, following physical assumptions are considered for the establishment of the numerical model. The physical assumptions are:

(i) The powder particles are assumed to be spherical in shape.
(ii) The laser beam is considered as a moving Gaussian heat source.
(iii) The laser beam striking perpendicular to the powder bed.

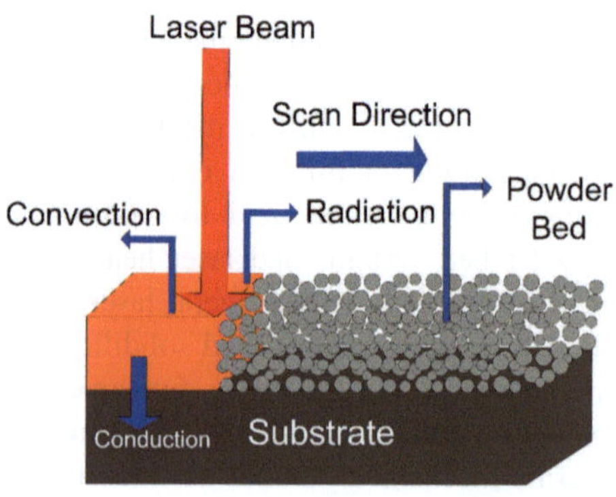

Fig. 14.2 Schematic of the heat transfer mechanism in DMLS process

(iv) The powder particles have consistent physical properties.
(v) A constant absorption rate is assumed in the modeling scheme.
(vi) Thermo-physical properties of the powders are temperature-dependent.
(vii) The heat transfer phenomena are dominated by conduction, convection, and radiation.
(viii) A constant heat transfer coefficient is considered between powder bed and the surrounding environment.

Governing Equation

Numerical simulations of thermal behavior in DMLS processes is carried out by solving three-dimensional Fourier heat conduction equation, which describes temperature as a function of space (*x*, *y*, *z*) and time (*t*) coordinate. The general form of the governing differential equation to simulate the temperature distribution in the DMLS process is given as follows:

$$\frac{\partial}{\partial x}\left(k_x\frac{\partial T}{\partial x}\right)+\frac{\partial}{\partial y}\left(k_y\frac{\partial T}{\partial y}\right)+\frac{\partial}{\partial z}\left(k_z\frac{\partial T}{\partial z}\right)+\dot{Q}(x,y,z,t)=\rho c_p\frac{\partial T}{\partial t} \tag{14.1}$$

where

k_x, k_y, and k_z	are the thermal conductivities in the *x*, *y*, and *z* directions (W/m K).
T	is the temperature (K).
$\dot{Q}(x,y,z,t)$	is the rate of the internal heat generation per unit volume (W/m^3).
ρ	is the density of the metal powder (kg/m^3).
c_p	is the specific heat capacity (J/kg K).
t	is the interaction time between the laser beam and powder material

When the laser beam scans the powder bed, heat generation takes place and it can be modeled in the form of heat flux. In this model, the input heat source follows a Gaussian distribution [30]. The heat flux q with respect to position is defined as

$$q(r)=\frac{2AP}{\Pi r_0^2}\mathrm{e}^{-2r^2/r_0^2} \tag{14.2}$$

where

A laser absorptance of the powder system;
P laser power;
r_o radius of the laser beam;
r radial distance between the laser beam and the center of the spot generated on the top surface of the powder bed.

Initial and Boundary Conditions

In this process, the laser beam interacts with the powder bed and the heat transfer takes place within the powder bed as well as the substrate. Figure 14.3 shows the computational domain.

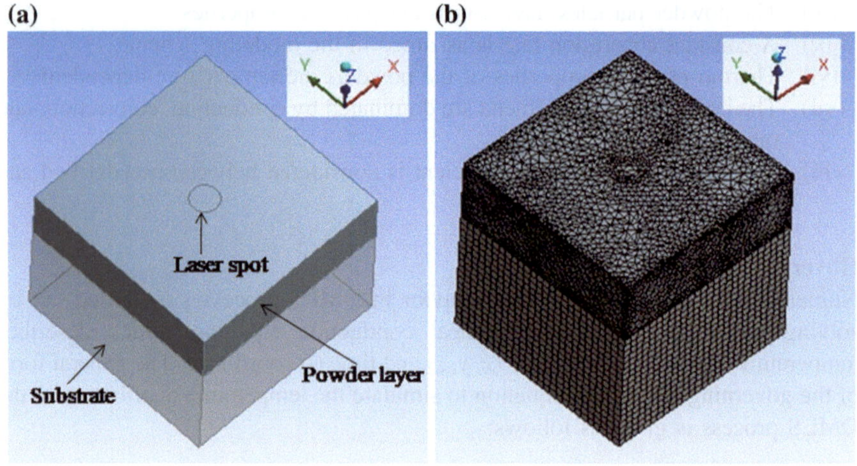

Fig. 14.3 **a** Computational domain, **b** finite element model with meshing used for simulation

Initially, the powder bed is assumed to be at room temperature, i.e., at $t = 0$,

$$[T(x, y, z, t)]_{t=0} = T_0 \tag{14.3}$$

where T_0 is considered as surrounding temperature.

Also, the input laser energy from the moving laser beam at the initial stage is assumed to be zero.

When the sintering process starts, the rate of absorption of laser energy by the powder material also increases. Thus, the heat flux is generated, and heat transfer takes place due to the temperature difference. From each surface of the powder bed, heat loss takes place by convection and radiation mode. Within the powder bed and substrate, heat transfer takes place by conduction mode. But the bottom of the surface is considered as adiabatic. There is no loss of heat from the bottom surface.

The boundary conditions are as follows:

(i) Heat transfer by convection: The amount of energy lost to the surrounding from the powder bed is by convection mode, which can be calculated from Eq. (14.4) written below:

$$q = h(T - T_a) \tag{14.4}$$

where

q heat lost by convection;
h convective heat transfer coefficient in W/m^2K;
T temperature of the top surface of the powder bed;
T_o ambient temperature at $t = 0$.

(ii) Heat transfer by radiation: The amount of heat radiated from the top surface of the powder bed to the surrounding environment can be given by Eq. (14.5) as follows:

$$q = \varepsilon\sigma\left(T^4 - T_0^4\right) \tag{14.5}$$

where

ε emissivity;
σ Stefan–Boltzmann constant for radiation.

The net amount of heat transfer from each surface except the bottom surface is given by Eq. (14.6):

$$-\left(\frac{\partial T}{\partial Z}\right) = \varepsilon\sigma\left(T^4 - T_0^4\right) + h(T - T_0) \tag{14.6}$$

(iii) Heat lost from the bottom surface: The net amount of heat loss from the bottom surface is as follows:

$$\left[k\left(\frac{\partial T}{\partial Z}\right)\right] at(Z = 0) = 0 \tag{14.7}$$

where k = thermal conductivity in W/mK.

Numerical Procedure and Thermo-physical Properties
The transient heat transfer model is developed on the ANSYS 15.0 platform by using finite element approach for direct metal laser sintering of AlSi10Mg alloy powder. The geometry taken for simulation consists of a substrate having the dimension of 3 mm × 3 mm × 2 mm and powder layer having the dimension of 3 mm × 3 mm × 1 mm. The laser beam focuses on the top surface of the powder bed with the diameter of 0.2 mm. The model takes into account the combined heat transfer phenomena, i.e., conduction, convection, and radiation in the powder bed as well as the substrate. Simulations are carried out for sintering of AlSi10Mg in a powder bed as well as the substrate. The grid for computation domain is finalized after a grid-independent test. Based on the grid-independent test, the computational model has meshed into 23,149 nodes and 76,121 elements. The temperature-dependent thermo-physical properties of AlSi10Mg, thermo-physical properties of steel, and processing parameters used for simulations are given in Tables 14.1, 14.2, and 14.3.

Table 14.1 Thermo-physical properties of AlSi10Mg alloy [31]

Temperature	20 °C	100 °C	200 °C	300 °C	400 °C
Thermal conductivity(k) (W/m K)	147	155	159	159	155
Specific heat capacity (c_p) (J/kg K)	739	755	797	838	922
Heat transfer coefficient (h) (W/m^2 K)	80				
Density (ρ) (g/m^3)	2.67				
Emissivity (ε)	0.3				

Table 14.2 Process parameters of DMLS process

Parameters	Values
Laser power (W)	50, 70, 100, 130
Scanning speed (mm/s)	100, 200, 300, 400
Laser spot size (mm)	0.2, 0.3, 0.4, 0.5
Thickness of layer (mm)	1
Laser absorptivity	0.9

Table 14.3 Thermo-physical properties of structural steel [32]

Properties of steel	Values
Thermal conductivity (k) (W/m K)	60.5
Density (ρ) (kg/m^3)	7850
Specific heat (c_p) (J/kg K)	434

14.3 Results and Discussions

14.3.1 Thermal Behavior

For obtaining quality components in the DMLS process, it is necessary to understand the temperature variation in the powder bed as well as in the substrate. The thermal phenomena in the powder bed and the substrate during the densification process provide useful information for determining the thermal and residual stress in the build part. The simulations are carried out for different scan speed, i.e., 100, 200, 300, and 400 mm/s by keeping fixed laser power, i.e., 100 W. The temperature distribution of the powder bed at different scan speed on the top surface of the powder bed is shown in Fig. 14.4.

The temperature distribution on the surface of the powder bed is shown by different colors. The temperature contours on the top surface of the powder bed look like a series of elliptical shape, and these elliptical shapes are more intensive in the scanning direction. From the temperature profile, it is observed that the temperature at specific locations increases rapidly as the laser beam scans that location and decreases sharply as it moves away from that location. This is because of change in thermal conductivity from powder to the solid layer which transfers heat quickly to

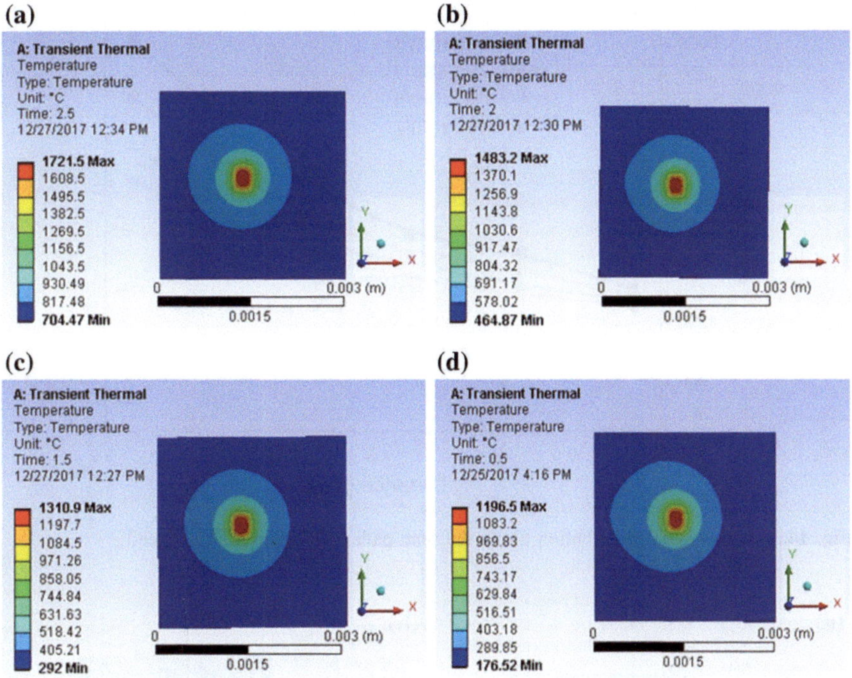

Fig. 14.4 Thermal profile of powder bed at different scan speed **a** 100 mm/s, **b** 200 mm/s, **c** 300 mm/s, and **d** 400 mm/s

the other regions of the powder bed. Also, the substrate plays a significant role in heat dissipation. It is found that, with an increase in scan speed, the laser energy density gradually decreases which result in the decrease in temperature.

When the scan speed is 100 mm/s, the temperature underneath the laser spot on the powder bed is 1721 °C, and at 200 mm/s scan speed, the temperature is 1483 °C. Subsequently, the temperature on the powder bed decreases with an increase in scan speed which is shown in the thermal profile. With the variation in scan speed and keeping the laser power and spot size constant, the interaction time between the laser energy and powder particle changes. For a slower scan speed, the interaction time is longer which results in higher sintering temperature because of more heat accumulation. On the contrary, the interaction time is less at higher scan speed and the temperature is relatively lower around the powder particles. Figure 14.5 shows the change in temperature at different scan speed in the powder bed.

Figure 14.6 illustrates the thermal profile of the powder bed along *XY* plane with varying laser power, i.e., 50, 70, 100, and 130 W while keeping the laser scan speed 100 mm/s constant in all cases. From the thermal profile, it is observed that there is an enhancement of temperature with increasing the laser power. At 50 W laser power, the average temperature in the powder bed is approximately 400 °C, which is below the melting point of AlSi10Mg. When the laser power is increased from

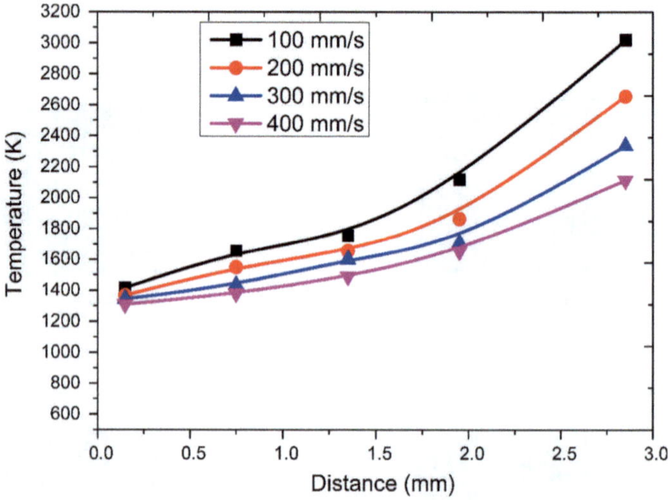

Fig. 14.5 Temperature distribution along the scan path with different scan speed

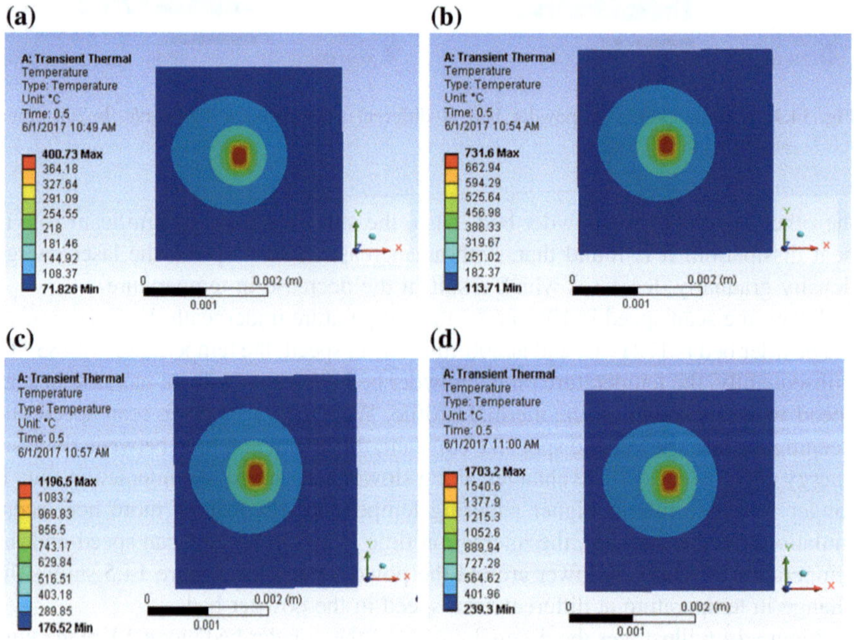

Fig. 14.6 Temperature contour of the powder bed at different laser power **a** 50 W, **b** 70 W, **c** 100 W, and **d** 130 W

50 to 130 W, the average temperature in the powder bed is changed from 400 to 1703 °C and the temperature gradient is significant increased along the scan path, which is clearly observed in the thermal profile. As the laser power increases, the beam intensity increases, which increases the heat flux. As per Eq. (14.2), the heat flux is directly proportional to laser power. So, the higher rate of heat transfer will occur from the laser beam to the powder surface that increases the temperature of the powder layer. The bottom surface of the powder bed is in an adiabatic condition, and there will be no loss of heat from the bottom surface to the outside as the thermal gradient vanished gradually. Hence, no temperature rise can be seen at the bottom wall. This can be clearly evident from the simulation results. Figure 14.7 shows the variation of temperature along the scan path with different laser power. It represents that there is a variation in temperature gradients with the laser power.

Figure 14.8 represents the temperature distribution on the powder bed with the variation of laser spot size by keeping scan speed 100 mm/s and laser power 100 W constant in all cases. From the temperature profile, it is found that with the increase in laser spot size, the maximum temperature at the top surface of the powder bed decreases. At 0.2 mm laser spot size, the temperature is 3181 °C which is well above the melting point of the alloy. When the laser spot size is 0.3 mm, the temperature is 1403 °C, and subsequently, it decreases with increase in spot size. When the laser spot size is so small, the heat is concentrated on the powder bed in a small area of cross section and less amount of heat is dissipated by convection and radiation. With an increase in spot size, the same amount of heat is concentration with the large area of cross section and more amount of heat is transferred by convection and radiation mode to the surrounding. That is why, the temperature decreases. This is also confirmed by Fig. 14.9 which gives the variation of temperature with the different spot sizes on the powder bed.

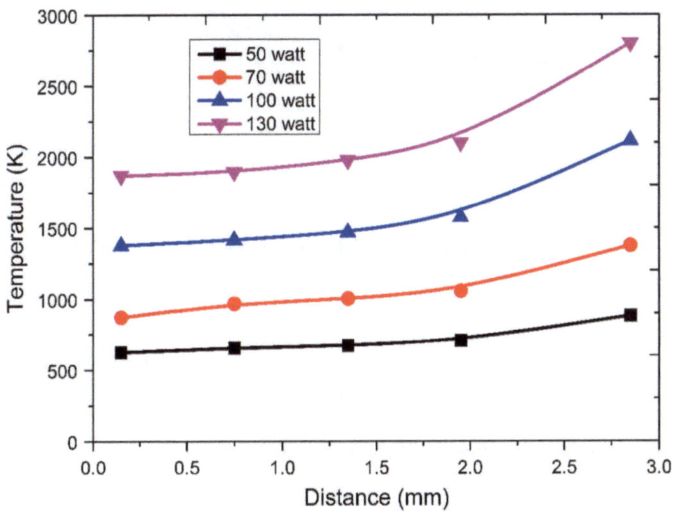

Fig. 14.7 Variation of temperature along the scan path with different laser power

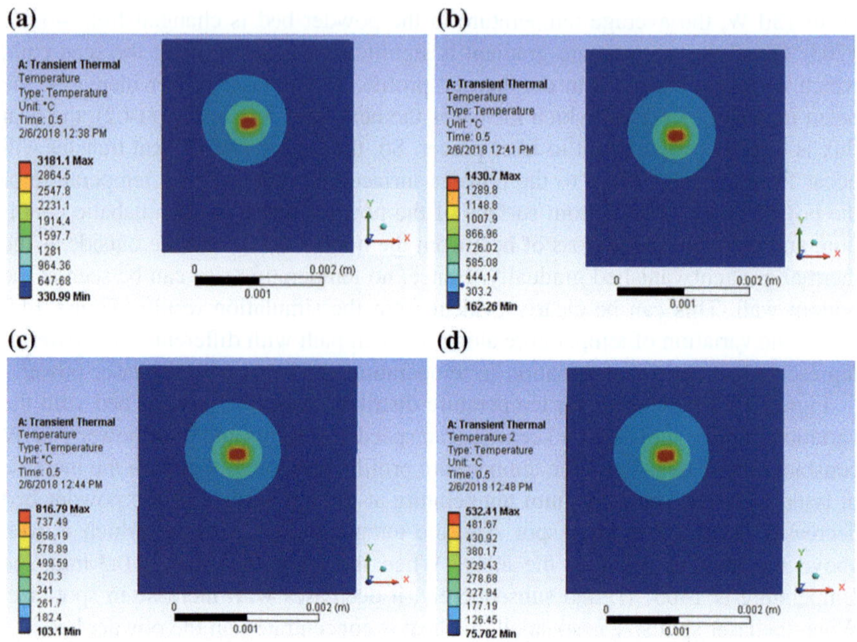

Fig. 14.8 Thermal profile at different laser spot size **a** 0.2 mm, **b** 0.3 mm, **c** 0.4 mm, and **d** 0.5 mm

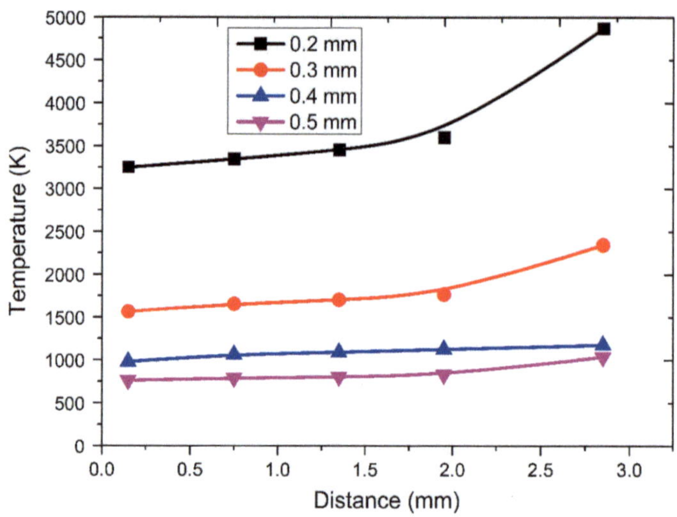

Fig. 14.9 Temperature variation with different laser spot size

14.3.2 Molten Pool Profile

In the DMLS process, when the laser beam strikes at a particular location, the temperature of the powders at that location increases significantly. If the temperature at that location is more than the melting point of the alloy, then a molten pool will be formed at that location. When the laser beam moves from one precise location to nearby location, the temperature and configuration of the molten pool vary significantly. This is significantly attributed to the fact that the heat energy stored in the previous precise location has an influence on the next location which leads to heat accumulation. The repetitive scan of the powder layers results in rapid melting and solidification which accomplish a sound metallurgical bonding between the powder particles and adjacent powder layers. So, the liquid-state sintering of the AlSi10Mg powder particles will take place, and as a result, completely dense build part is produced in the DMLS process. Figure 14.10 represents the change in molten pool size, i.e., length and width with respect to scan speed. From the graph, it is found that the molten pool length and width decrease as the scan speed increases. This is because of the short interaction time between the heat source and the powder bed. At 100 mm/s scan speed, the molten pool length and width are 3 and 3 mm. When the scan speed increases from 100 to 400 mm/s, the length and width of the molten pool are changed to 0.506 and 0.572 mm, respectively.

The dimensions of molten pool variation at different laser power are shown in Fig. 14.11. It is observed that, as the laser power increases, there is a significant change in molten pool length and width, respectively. When the laser power is 50 W, the temperature underneath the laser beam on the powder bed is 400 °C. This temperature is well below the melting point of the AlSi10Mg alloy. So molten

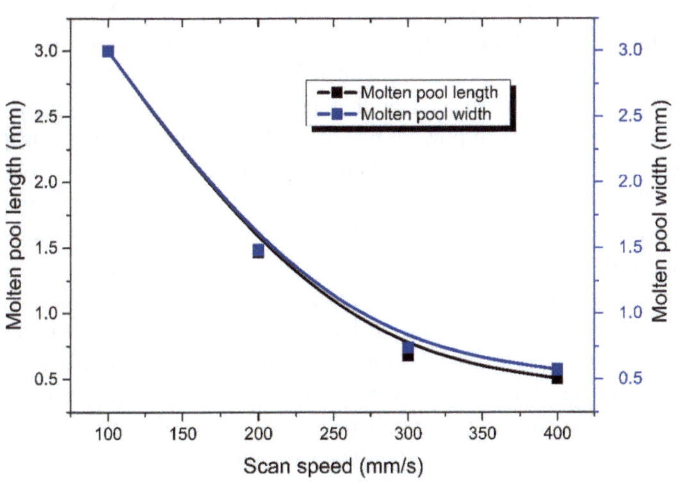

Fig. 14.10 Variation of molten pool dimension with scan speed

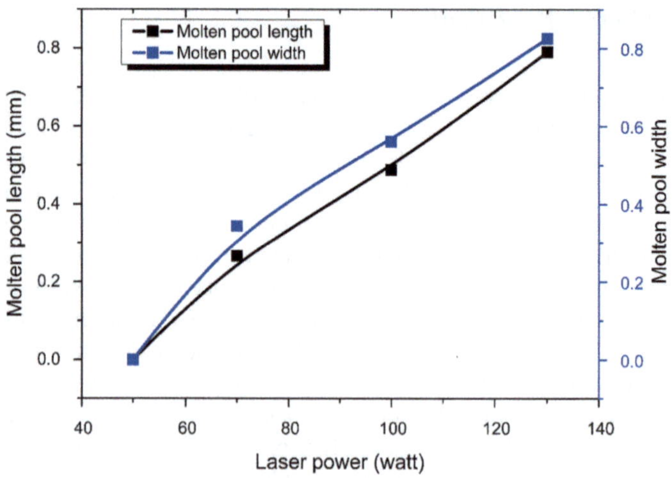

Fig. 14.11 Variation of molten pool dimensions with laser power

pool is not formed at this laser power. When the laser power increases from 50 to 70 W, the temperature of the powder bed increased from 400 to 731 °C which is above the melting point (T_m = 600.2 °C) of the alloy. So the molten pool is formed. The length and width of the molten pool at 70 W are 0.266 and 0.344 mm. By increasing the laser power subsequently up to 130 W, the molten pool dimensions are changed significantly; i.e., length and width are 0.791 and 0.825 mm, respectively.

It is found that the dimensions of the molten pool increase with an increase in laser power and decrease with an increase in scan speed. However, the dimensions like length and width of the molten pool differ from each other. This is because of the interaction time of the laser beam with the powder particle. When the scan speed increases, the interaction time decreases which results in decreases in the melt pool dimension. A high temperature is obtained in the molten pool region as the laser directly touches the powder bed at that point. More amount of heat gradually transfers to other surfaces of powder bed as the laser beam penetrates deeper into it. It is observed that the size of the molten pool in lengthwise is more than the laser spot size. This is because of the scanning direction as the laser beam moves along the x-axis so that the length of the molten pool increases along the x-direction and more amount of heat accumulation helps to increase the molten pool size.

Similarly, Fig. 14.12 shows the molten pool dimension with different laser spot sizes. It has been found that, with an increase in laser spot size, the molten pool length and width decrease. These simulations are done with varying laser spot size by keeping scan speed and laser power fixed. When the spot size changes, the same amount of heat flux strikes the powder bed, but the area of exposure will increases. So the temperature decreases significantly, and the molten pool dimension decreases. The length and width of the molten pool are 1.618 and 1.604 mm at 0.2 mm laser spot size. When the spot size increases from 0.2 to 0.4 mm, the melt

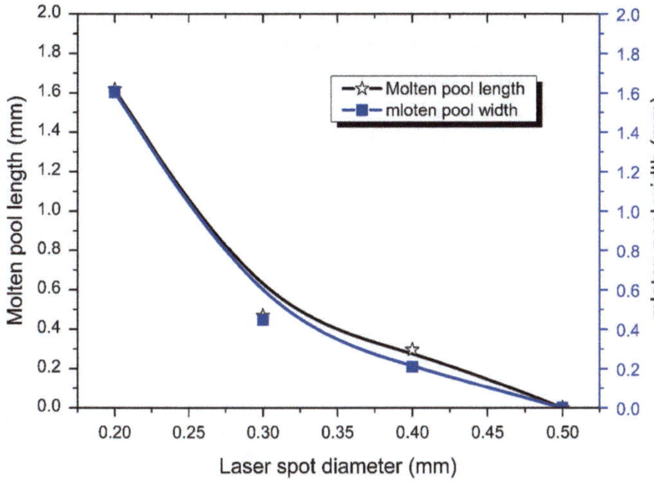

Fig. 14.12 Variation of molten pool dimensions with laser spot size

pool length and width are 0.296 and 0.21 mm. At 0.5 mm laser spot size, there is no existence of the molten pool. Because at 0.5 mm spot size, the temperature at the powder bed is 532 °C which is below the melting point of the alloy. So no molten pool is formed. This is also confirmed from the thermal profile.

14.3.3 Sintering Depth

In the DMLS process, a high-energy heat source is utilized to sinter the powder particles, which melts and fuses the powders with a fraction of second. So liquid-state sintering takes place during this process. Liquid-state sintering occurs when the temperatures are high enough for the powdered metal to melt and fuse in. In solid-state sintering, surface and bulk transport mechanisms play an important role for sintering, whereas in liquid-state sintering shrinkage and densification result from bulk transport mechanisms.

Figure 14.13 shows the plot of sintering depth in the powder layer with laser power. From the graph, it is observed that, with an increase in laser power, laser beam intensity increases and thus sintering depth increases. The sintering depth is measured by using ImageJ software. When the laser power is 50 W, the maximum temperature obtained is 400 °C. So liquid-state sintering does not take place, and sintering depth is 0 mm. When the laser power is 70 W, the temperature is 731 °C and depth is 0.06 mm. With an increase in laser power from 70 to 130 W, the temperature increases from 731 to 1703 °C and sintering depth increases from 0.06 to 0.381 mm, respectively. So, it is found that the laser power strongly affects the temperature and sintering depth values in a positive way. Based on the sintering depth, the thickness of the powder layer is optimized as per the requirement for a particular processing condition. So, the wastage of the powders in the DMLS process can be minimized.

Similarly, Figs. 14.14 and 14.15 represent the sintering depth with the scan speed and the laser spot size. From the graph, it is observed that, when the laser power is constant, i.e. 100 W, by increasing the scan speed, the sintering depth

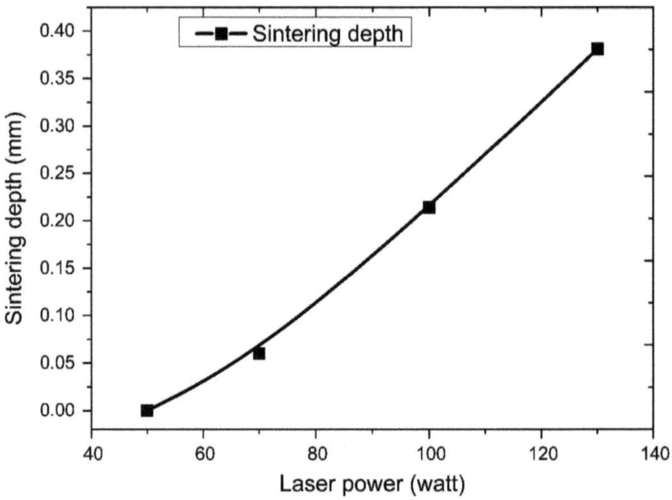

Fig. 14.13 Variation of sintering depth with laser power

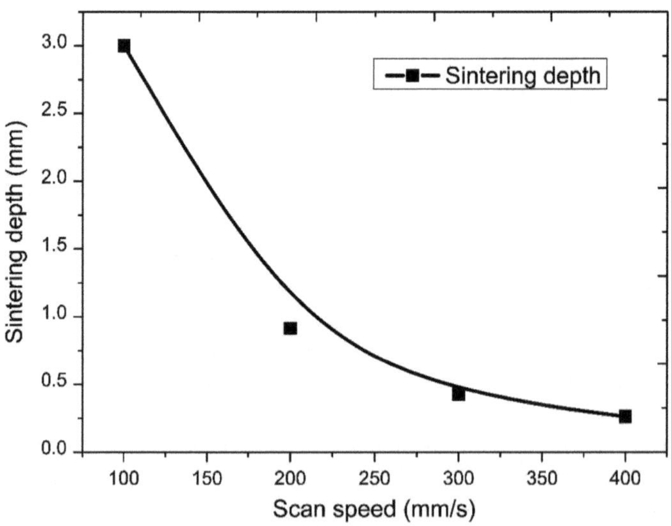

Fig. 14.14 Variation of sintering depth with scan speed

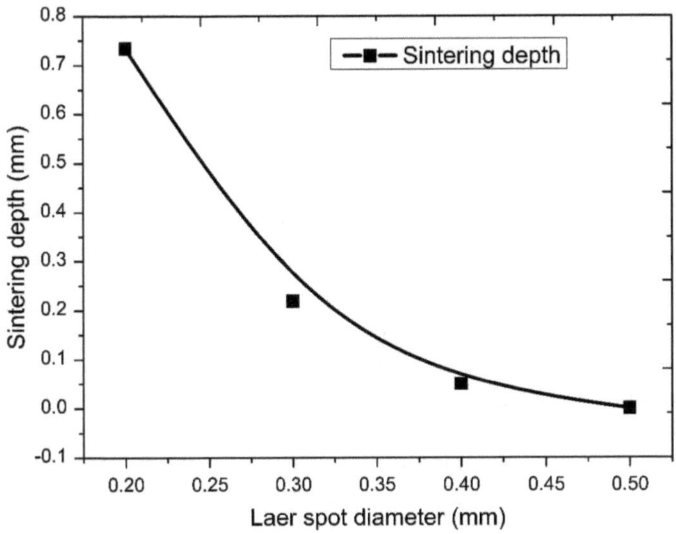

Fig. 14.15 Variation of sintering depth with laser spot size

increases. Further, the quantitative results evidence that the decreasing of sintering depth becomes less prominent as the scan speed is increased above 100 mm/s. As the scanning speed is dependent on laser energy density, with an increase in scan speed, the laser energy density decreases and hence the sintering depth is also decreased. This can be clearly seen in Fig. 14.14. So, from the simulation results, it is found that sintering depth decreases from 0.973 to 0.261 mm with the decrease in temperature value; i.e. depth and temperature are directly proportional to each other at different scanning speed. This is happened because of the interaction time between the powder particle and laser beam. At a constant laser power, by increasing scan speed, the laser energy density decreases. At a high scan speed, the interaction time is less and the powder particles do not get sufficient time to melt and fuse in. But the temperature increases rapidly at low scan speed, which leads to more heat accumulation and results in a larger temperature gradient. From the results, it is concluded that the sintering depth must be greater than the powder layer thickness to ensure the complete sintering.

Similarly, with an increase in laser spot size, the sintering depth also decreases which is shown in Fig. 14.15. This is due to the decrease in temperature which is clearly observed in thermal profile.

14.4 Conclusion

A three-dimensional transient thermal simulation is carried out for the DMLS using finite element approach. The model considers the temperature-dependent thermo-physical properties of the powder materials along with multiple modes of heat transfer phenomena. From the simulation results, it is quite evident that the process parameters and material properties have a great influence on the quality and property of the sintered part and the following conclusions are obtained.

(i) A comprehensive three-dimensional transient model is developed to know the thermal behavior in the DMLS process. This model can be utilized for other alloy systems in the DMLS process.
(ii) From the simulation results, it is quite evident that the process parameters and material properties have a great influence on the quality and property of the sintered part. With increasing in laser power, the molten pool temperature increases from 400 to 1703 °C and the length and width of the molten pool also increase significantly.
(iii) As the scan speed increases from 100 to 400 mm/s, the temperature of the molten pool decreases from 1483 to 1196 °C because of little interaction time of the laser beam with the powder bed.
(iv) The sintering depth of the powder bed increases from 0.061 to 0.872 mm with an increase in laser power, and a reverse phenomenon is observed with increase in scan speed. This will help to optimize the thickness of the powder layer for specific processing parameters and minimize the wastage of the excess powders.
(v) With the increase in laser spot size, the molten pool temperature decreases from 3181 to 532 °C and molten pool dimension and sintering depth decrease significantly.
(vi) The developed model will help to optimize the process parameters for fabricating quality components in DMLS process.

References

1. Olakanmi, E.O., Cochrane, R.F., Dalgarno, K.W.: A review on selective laser sintering/melting (SLS/SLM) of aluminum alloy powders: processing, microstructure, and properties. Prog. Mater. Sci. **74**, 401–477 (2015)
2. Nandy, J., Sarangi, H., Sahoo, S.: Microstructure evolution of Al-Si-10Mg in direct metal laser sintering using phase field modeling. Adv. Manufact. **6**, 107–117 (2018)
3. Atzeni, E., Salmi, A.: Study on unsupported overhangs of AlSi10Mg parts processed by direct metal laser sintering (DMLS). J. Manufact. Process. **20**, 500–506 (2015)
4. Wong, K.V., Hernandez, A.: A review of additive manufacturing. ISRN Mech. Eng. **2012**, 1–10 (2012)

5. Tolochko, N.K., Mozzharov, S.E., Yadroitsev, I.A., Laoui, T., Froyen, L., Titov, V.I., Ignatiev, M.B.: Balling processes during selective laser treatment of powders. Rapid Prototyping J. **10**, 78–87 (2004)
6. Kruth, J.P., Wang, X., Laoui, T., Froyen, L.: Binding mechanisms in selective laser sintering and selective laser melting. Assembly Autom. **23**, 357–371 (2003)
7. Romano, J., Ladani, L., Sadowski, M.: Thermal modeling of laser based additive manufacturing processes within common materials. Procedia Manufact. **1**, 238–250 (2015)
8. Verma, A., Tyagi, S., Yang, K.: Modeling and optimization of direct metal laser sintering process. Int. J. Adv. Manuf. Technol. **77**, 847–860 (2015)
9. Kruth, J.P., Mercelis, P., Vaerenbergh, J.V., Froyen, L., Rombouts, M.: Binding mechanism in selective laser sintering and selective laser melting. Rapid prototyping J. **11**, 26–36 (2005)
10. Chen, T., Zhang, Y.: A particle shrinkage model for selective laser sintering of two-component metal powder layer. Int. J. Heat Mass Transf. **49**, 1489–1492 (2006)
11. Gu, D.D., Meiners, W., Wissenbach, K., Poprawe, R.: Laser additive manufacturing of metallic components: materials, processes, and mechanisms. Int. Mater. Rev. **57**(3), 133–164 (2012)
12. Frazier, W.E.: Metal additive manufacturing: a review. J. Mater. Eng. Perform. **23**(6), 1917–1928 (2014)
13. Zeng, K., Pal, D., Stucker, B.: A review of thermal analysis methods in laser sintering and selective laser melting. In: Proceedings of Solid Freeform Fabrication Symposium Austin, TX, vol. 60, pp. 796–814 (2012)
14. Tang, Y., Loh, H.T., Wong, Y.S., Fuh, J.Y.H., Lu, L., Wang, X.: Direct laser sintering of a copper-based alloy for creating three-dimensional metal parts. J. Mater. Process. Technol. **140**, 368–372 (2003)
15. Dong, L., Makradi, A., Ahzi, S., Remond, Y.: Three-dimensional transient finite element analysis of the selective laser sintering process. J. Mater. Process. Technol. **209**, 700–706 (2009)
16. Yuan, P., Gu, D.: Molten pool behavior and its physical mechanism during selective laser melting of TiC/AlSi10Mg nanocomposites: simulation and experiments. J. Phys. D Appl. Phys. **48**, 1–16 (2015)
17. Simchi, A.: Direct laser sintering of metal powders: mechanism, kinetics and microstructural features. Mater. Sci. Eng. **428**, 148–155 (2006)
18. Jian, X., Weimin, S., Rana, S.R.: 3D modeling and testing of transient temperature in selective laser sintering (SLS) process. Optik **124**, 301–304 (2013)
19. Li, Y., Gu, D.: Parametric analysis of thermal behavior during selective laser melting additive manufacturing of aluminum alloy powder. Mater. Des. **63**, 856–867 (2014)
20. Shi, Q., Gu, D., Xia, M., Cao, S., Rong, T.: Effects of laser processing parameters on thermal behavior and melting/solidification mechanism during selective laser melting of TiC/Inconel 718 composites. Opt. Laser Technol. **84**, 9–22 (2016)
21. Hu, H., Ding, X., Wang, L.: Numerical analysis of heat transfer during multi-layer selective laser melting of AlSi10Mg. Optik **127**, 8883–8891 (2016)
22. Sahoo, S., Chou, K.: Phase-field simulation of microstructure evolution of Ti-6Al-4 V in electron beam additive manufacturing process. Addit. Manufact. **9**, 14–24 (2016)
23. Lee, W.H., Zhang, Y., Zhang, J.: Discrete element modeling of powder flow and laser heating in direct metal laser sintering process. Powder Technol. **315**, 300–308 (2017)
24. Zhao, X., Iyer, A., Promoppatum, P., Yao, S.C.: Numerical modeling of the thermal behavior and residual stress in the direct metal laser sintering process of titanium alloy products. Addit. Manufact. **14**, 126–136 (2017)
25. Dong, L., Correia, J.P.M., Barth, N., Ahzi, S.: Finite element simulations of temperature distribution and of densification of a titanium powder during metal laser sintering. Addit. Manufact. **13**, 37–48 (2017)
26. Kundakcioglu, E., Lazoglu, I., Rawal, S.: Transient thermal modeling of laser-based additive manufacturing for 3D freeform structures. Int. J. Adv. Manuf. Technol. **85**, 493–501 (2016)

27. Ojha, A., Samantaray, M., Thatoi, D.N., Sahoo, S.: Continuum simulation of heat transfer and solidification behavior of AlSi10Mg in direct metal laser sintering process. IOP Conf. Ser. Mater. Sci. Eng. **338**, 1–6 (2018)
28. Panda, B.K., Sahoo, S.: Numerical simulation of residual stress in laser based additive manufacturing process. IOP Conf. Ser. Mater. Sci. Eng. **338**, 1–6 (2018)
29. Labudovic, M., Hu, D., Kovacevic, R.: A three dimensional model for direct metal laser powder deposition and rapid prototyping. J. Mater. Sci. **38**, 35–49 (2003)
30. Roberts, I.A., Wang, C.J.: A three-dimensional finite element analysis of the temperature field during laser melting of metal powders in additive layer manufacturing. Int. J. Mach. Tool Manuf. **49**, 916–923 (2009)
31. Nandy, J., Sarangi, H., Sahoo, S.: Modeling of microstructure evolution in direct metal laser sintering: a phase field approach. IOP Conf. Ser. Mater. Sci. Eng. **178**, 1–8 (2017)
32. Bogaard, R.H., Desai, P.D., Li, H.H., Ho, C.Y.: Thermophysical properties of stainless steels. Thermochim. Acta **218**, 373–393 (1993)

Chapter 15
Parametric Optimization of TIG Welding Process on Mechanical Properties of 316L Stainless Steel Using RSM

Subhas Chandra Moi, Asish Bandyopadhyay and Pradip Kumar Pal

Nomenclature

A	Welding current (A)
B	Welding speed (cm/min)
C	Gas flow rate (l/min)
AISI	American Iron and Steel Institute
ANOVA	Analysis of variance
DF	Degree of freedom
GFR	Gas flow rate
PE	Percentage elongation
RSM	Response surface methodology
TIG	Tungsten inert gas
UTS	Ultimate tensile strength

15.1 Introduction

Austenitic stainless steel is an important material and very frequently used in industries like automobile, thermal power plant, nuclear power plant, chemical and pharmaceutical industries due to its superior corrosion resistance, high strength, high ductility, formability and weldability. Out of 300 series grade of austenitic stainless steels, AISI 316L is an extra low carbon version of austenitic chromium-nickel stainless steel [1]. It contains molybdenum materials that improve corrosion resistance, strength at elevated temperatures and improve resistance to pitting from chloride ion solutions. The extra low carbon content in 316L austenitic stainless steels minimizes harmful carbide precipitation due to welding. Some equipment of above said industries is manufactured by fabrication (welding).

S. C. Moi (✉) · A. Bandyopadhyay · P. K. Pal
Mechanical Engineering Department, Jadavpur University, Kolkata, India
e-mail: sc_moi1@rediffmail.com

P. Sahoo and J. P. Davim (eds.), *Advances in Materials, Mechanical and Industrial Engineering*, Lecture Notes on Multidisciplinary Industrial Engineering,
https://doi.org/10.1007/978-3-319-96968-8_15

Tungsten inert gas (TIG) welding is frequently used for fabrication of austenitic stainless steel. TIG welding is basically an electric arc welding process which uses non-consumable tungsten electrode, and arc is established between the tip of an electrode and the workpiece. Inert gas (argon, helium, etc.) is used to avoid contamination of the weldment with air. Sometimes a mixture of any two of the gases mentioned above is also used. TIG welding is used in modern industries for high quality.

The mechanical property of TIG weld is significantly influenced by the parameters of weld pool geometry, namely weld bead width, weld bead height and depth of penetration. And the weld pool geometry is characterized by the levels of different parameters which are used during welding. Therefore, the proper choice of the welding input parameters and its levels are very important for obtaining desired optimal weld pool geometry [2]. Welding current, arc voltage, electrode size, arc travel speed, electrode stick out, etc., are the important welding process parameters. Further, the quality of weld in TIG welding may be influenced due to welding position, electrode composition, edge preparation, the type of gas used and its flow rate. There are different optimization techniques that can be used to determine the required output responses through the development of mathematical equations to postulate the relationships between the input parameters and output variables. One of the most widely used statistical methods of optimization is the response surface methodology (RSM), in which the researchers try to approximate the unknown mechanism with an appropriate empirical model [3–5]. Murugan and Parmer [6] established quadratic relationship between the parameters of MIG welding and parameters of weld pool geometry on 316L austenitic stainless steel. Benyounis et al. [7] used both numerical and graphical optimization techniques to optimize the laser welding parameters on medium carbon steel using RSM. Zhao et al. [8] developed a mathematical model that relates the relationship between the process parameters and weld pool dimensions of laser beam welding of thin gauge galvanized steel. Olabi et al. [9] applied response surface methodology to optimize the laser welding process parameters for joining of dissimilar materials. Ahn et al. [10] optimized laser welding process and reported the effect of welding process parameters on microstructure and mechanical property. Ragavendran et al. [11] studied the hybrid laser and TIG welding process on 316LN stainless steel and developed mathematical models to correlate between input variables and output responses. Padmanaban et al. [12] investigated laser beam welding process through experiment in butt join configuration made by AZ31B magnesium alloy and optimized the process to obtain highest tensile strength. Elatharasan et al. [13] used central composite design technique of response surface methodology to develop a mathematical model to estimate the performance characteristics: ultimate tensile strength, yield strength and elongation in friction stir welding of AA 6061-T6 aluminium alloy material. Srivastava and Garg [14] had carried out experimental analysis to study the effects of the welding process parameters on welding of mild steel plates using gas metal arc welding process by using Box–Behnken design of response surface methodology. Lakshminarayanan et al. [15] had conducted comparative analysis among shielded metal arc welding, gas metal arc welding and gas tungsten arc welding processes to study the tensile and impact properties of the ferritic stainless steel of thickness 4 mm. Researchers had been found from their study was that gas tungsten arc welding technique was found to be advantageous compared to other welding operations for joining ferritic stainless steels, due to the presence of finer grains in fusion zone and heat-affected zone. Kumar et al. [16] experimentally analysed that parameters of weld bead geometry are largely dependent on incident angle for laser welding. They also optimized the welding process for obtaining maximum value of tensile strength by RSM. Ghosh et al. [17] applied grey-based Taguchi method on the experiment of MIG welding on 316L stainless steel. They concluded that the welding current is the most significant factor for obtaining highest joint strength.

From the literature survey, it is found that various aspects of TIG welding like parametric studies, mathematical modelling and process optimization have been explored by the investigators. But work done by them appears not sufficient, as reliable databank has not yet been established and that can be obtained through many more extensive researches involving parametric studies, analysis of the response in terms of quality like joint strength, microstructures particularly in the field of TIG welding of 316L stainless steel. This sound knowledge base and precise understanding will help in controlling the TIG welding process more reliably and predicatively. In practice, one will be able to select proper parametric combination depending upon the level of quality required. This will also help automatic/ intelligent control of the process. Hence, the present work is one attempt towards the improvement of weld joint quality by selection of proper process parameters and to enrich the existing knowledge base in the area of TIG welding.

15.2 Experimental Procedures

In order to achieve desired aim, the research work has been considered in the following sequence:

(i) Selecting the base metal and filler metal.
(ii) Identifying the most significant TIG welding parameters and their levels.
(iii) Selecting the design of experiment and constructing experimental design matrix.
(iv) Performing the experiments according to design matrix.
(v) Testing of welded samples through non-destructing and destructive testing methods and recording the responses like UTS, PE.
(vi) Developing a regression model using RSM and identifying the significant factors.
(vii) Verifying the predicted results for optimization through confirmatory test.

The base material has been selected as 316L austenitic stainless steel of dimension 50 mm × 80 mm × 3 mm. ER316L austenitic stainless steel of diameter 1.6 mm has been taken as a filler material. The chemical composition of the base material

Table 15.1 Composition of the base material and filler material

Weight %	C	Mn	Si	S	P	Cr	Ni	Mo	Cu	Fe
AISI 316L	0.026	0.97	0.26	0.012	0.043	16.12	10.08	2.03	0.15	70.31
ER 316L	0.040	1.50	0.45	0.03	0.030	18.20	12.00	2.30	0.75	64.70

Table 15.2 Mechanical properties of base material

Ultimate tensile strength (MPa)	Yield strength (MPa)	% elongation	Hardness (VHN)
608.88	306.92	12.02	212

and filler material and its mechanical properties are given in Tables 15.1 and 15.2, respectively.

The most influencing parameters which affect weld pool geometry as well as mechanical properties of TIG welded joint are identified through extensive literature survey and previous work done. The important identified process parameters are welding current, gas flow rate and welding speed. A large number of trials have been performed on 3-mm-thick AISI 316L stainless steel plates to find out the efficient and practicable working limits of TIG welding parameters, and this has been done by altering one of the selected parameters while others remain constant. The visual inspection and non-destructive tests have been used to identify the working limits of the welding process parameters. Table 15.3 shows the identified welding parameters along with their three levels.

The experimental design matrix has been developed according to the Box–Behnken design of response surface methodology (RSM). RSM is a collection of mathematical and statistical techniques used for modelling and optimizing engineering problems where response is influenced by several input variables. In Box–Behnken design of RSM, there should be at least three factor and three levels for each factor. The upper limit of each parameter is coded as '+1', and the lower bound is coded as '−1'. The design of experiment software, Minitab v17, has been applied to build up the experimental design matrix for RSM. Table 15.4 represents the experimental plan.

The experiments have been planned and conducted according to a Box–Behnken design of response surface methodology. Figure 15.1 shows the photographic view of a semi-automatic TIG welding machine (Model—SUPERGEN320, M/C SR.

Table 15.3 Process parameters and their levels

Parameters	Units	Notation	Levels		
			−1	0	+1
Welding current	*A*	*A*	100	125	150
Travel speed	cm/min	*B*	12	15	18
Gas flow rate	l/min	*C*	6	9	12

Table 15.4 Experimental design matrix as per RSM

	Welding current		Travel speed		Gas flow rate	
Exp. no.	Coded	Uncoded (*A*)	Coded	Uncoded (cm/min)	Coded	Uncoded (l/min)
1	−1	100	−1	12	0	9
2	1	150	−1	12	0	9
3	−1	100	1	18	0	9
4	1	150	1	18	0	9
5	−1	100	0	15	−1	6
6	1	150	0	15	−1	6
7	−1	100	0	15	1	12
8	1	150	0	15	1	12
9	0	125	−1	12	−1	6
10	0	125	1	18	−1	6
11	0	125	−1	12	1	12
12	0	125	1	18	1	12
13	0	125	0	15	0	9
14	0	125	0	15	0	9
15	0	125	0	15	0	9

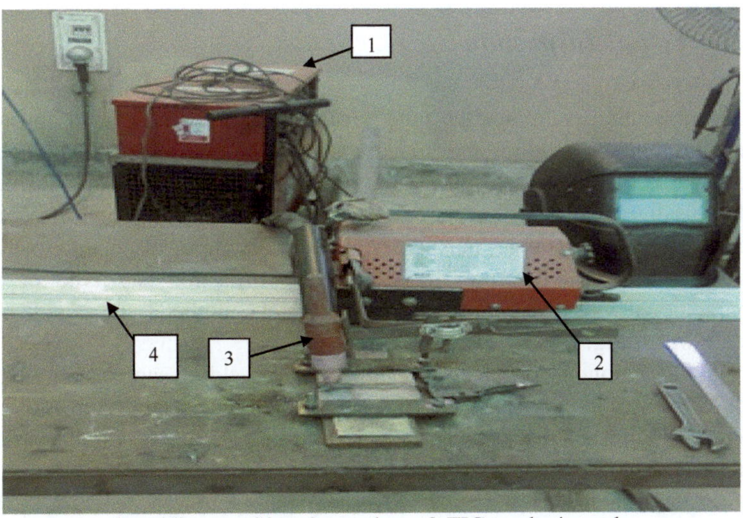

1-welding machine, 2-travel car, 3-TIG torch, 4- track

Fig. 15.1 Photographic view of experimental set-up

No-12111097670 and Make—Ador Welding Ltd.) Here, the TIG torch has been fixed to the travel car to confirm the torch is set at a predetermined angle. The travel car travels on the defined path and can move only straight line direction. Control of current, speed and gas flow rate is done manually for each run. All plates of size 50 mm × 80 mm × 3 mm and filler rods are cleaned by stainless steel wire brush with acetone for removing the dust, oil, grease and thin oxide coating before welding. Square butt joint configurations have been prepared by tungsten inert gas welding using argon with purity 99.99%, as shielding gas. The welding direction has been taken as normal to the direction of rolling. During welding, the workpieces have been placed on copper plate and mechanical clamps are also used for avoiding distortion.

15.3 Results and Discussions

Square butt welded joints are being made under varied welding input factors, and then quality of weld is being judged, first by visual inspection and then by non-destructive and destructive testing methods. In so far a non-destructive test is concerned, X-ray radiography has been taken up. Tensile test specimens have been prepared, and tensile test has been performed for all the samples.

15.3.1 Visual Inspection

A visual inspection examines the quality of weldment such as bead appearance, width and thickness, welding defects like undercut, cracks, pits and slag inclusions in the surfaces of the welded joints, and whether the throat is as thick as desired, and the misalignment is within the allowable allowance. The visual inspection of welded joints is easy, simple, inexpensive and is capable of inspecting many weld zones at one time. Therefore, it is generally used for any type of welds. Photographic view of two welded specimens, sample number 1 and sample number 14, is shown in Fig. 15.2.

Visual inspection has been carried out for all the welding samples after welding. The observed results are given in Table 15.5. The defects observed in the visual inspection in some of the samples include: little undercut at the end, large reinforcement height, lack of penetration. Under certain parametric conditions, almost no defects are found in a few samples like sample numbers 2, 5, 8, 9, 12, 13, 14 and 15. The possible reasons for the defects found in several samples are due to lack of environmental suitability, fluctuation of voltage and due to semi-automatic machine set-up, etc. Further, the skill of the operator or technician is very crucial for weld quality. Defects may come from any type of irregularities/inhomogeneity in the base metal as well as filler wire. Undercut defect may be caused by inappropriate joint geometry in some of the samples. It may have been resulted because of

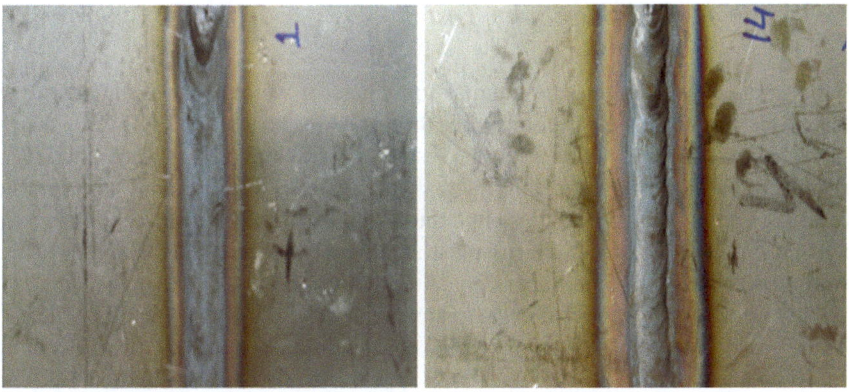

Fig. 15.2 Photographic view of welded sample number 1 and 14

Table 15.5 Results of visual inspection

Exp. no.	Current (*A*)	Speed (cm/min)	GFR (l/min)	Observations
1	100	12	9	Large reinforcement height at the starting, good penetration
2	150	12	9	Uniform penetration and uniform weld bead
3	100	18	9	Lack of penetration at the end
4	150	18	9	Little undercut at the end
5	100	15	6	Good penetration throughout
6	150	15	6	Large reinforcement height
7	100	15	12	Lack of penetration at the end
8	150	15	12	Uniform HAZ width, no defect
9	125	12	6	Good penetration throughout
10	125	18	6	Minimum porosity but good penetration throughout
11	125	12	12	Lack of penetration at the end
12	125	18	12	Uniform penetration and uniform weld bead
13	125	15	9	Good penetration throughout
14	125	15	9	Uniform penetration and uniform weld bead
15	125	15	9	Good penetration throughout

excessive welding current and voltage. The combined effects of various welding input parameters can also form undercut defect on weldment. Lack of penetration is one of the main defects which will affect the joint strength of the welded sample. The reasons for lack of penetration are faster travel speed, incorrect selection of welding currents levels, improper cleaning of welding zone, and presence of oxides, scale and other impurities which do not allow the deposited metal to fuse accurately with the base metal.

15.3.2 X-Ray Radiography Test

X-ray is used to determine the internal soundness of welds. This examination has been done for all 15 welded joints after the visual inspection by XXQ-2005 X-ray flaw detector. Table 15.6 illustrates the results of X-ray radiography examination. It is noticed that almost all the samples pass through X-ray radiography test with no significant defect remarks. Only sample number 11 is found with defect like lack of penetration. Lack of penetration is most significantly affecting the strength of the weld. Lower heat input, improper choice of current or less skill of the welder, etc., [18] as pointed out earlier, are the major causes for lack of penetration defect. Hence, this sample has been replaced by a new one done by same welding condition.

Qualities of the joints are further evaluated by other tests and studies, as well. If the results of visual inspection and the results of X-ray radiography test are compared, then some consistency can be observed. It is found from the two tables (Tables 15.5 and 15.6) that low current is not the primary cause of lack of penetration, because for sample numbers 1, 3, 5 and 7 also have the current level 100A, but for these samples, no defect is found. The skill of the welding operator is high depending on weld quality. Lack of penetration defect may also come from any irregularities in the base metal or filler wire as mentioned earlier in case of visual inspection. Large electrode angle may also be a possible reason for this type of defect. Any improper welding environment can also affect the weld quality. This may have occurred due to entrapped gas in the weld pool, which could have been

Table 15.6 Results of X-ray radiography test

Exp. no.	Current (*A*)	Speed (cm/min)	GFR (l/min)	Observations	Remarks
1	100	12	9	No significant defect	Acceptable
2	150	12	9	No significant defect	Acceptable
3	100	18	9	No significant defect	Acceptable
4	150	18	9	No significant defect	Acceptable
5	100	15	6	No significant defect	Acceptable
6	150	15	6	No significant defect	Acceptable
7	100	15	12	Minimum porosity	Acceptable
8	150	15	12	No significant defect	Acceptable
9	125	12	6	No significant defect	Acceptable
10	125	18	6	Minimum porosity	Acceptable
11	125	12	12	Lack of penetration	Not satisfactory
12	125	18	12	No significant defect	Acceptable
13	125	15	9	No significant defect	Acceptable
14	125	15	9	No significant defect	Acceptable
15	125	15	9	No significant defect	Acceptable

avoided by proper selection of gas flow rate and arc gap. A faster travel speed and too small root gap may also be the reasons for such defect. Therefore, defects may possibly appear from inappropriate choice of different welding parameters, namely current, voltage, arc gap, including their ranges also. It is noticed from the tables of visual as well as radiography examination that almost no welding defects are observed, when welding with middle value of current, speed and gas flow rate. It is also found from visual inspection and X-ray radiographic examination that no significant defect presences in sample numbers 2, 5, 9 and 12. Thus, this is the reason for obtaining higher value tensile properties of the samples under tensile testing.

15.3.3 Tensile Test

After X-ray radiography test, tensile test samples have been prepared in accordance with ASTM E8 M-04. Tensile specimens of the weld joint are obtained in taking specimen from transverse direction of weld joint. Top view of weldment in butt joint configuration is shown in Fig. 15.3. Figure 15.4 represents typical diagram showing the dimensions of the standard sub-size tensile specimen. Tensile test of all welded samples has been conducted in universal testing machine (Model—8801, Make—INSTRON and capacity—100 kN) at strain rate of 0.001/s. The tensile properties mainly UTS, yield strength, percentage elongation, of base metal as well as TIG welded joints have been evaluated. The measured responses (UTS and PE) are listed in Table 15.7.

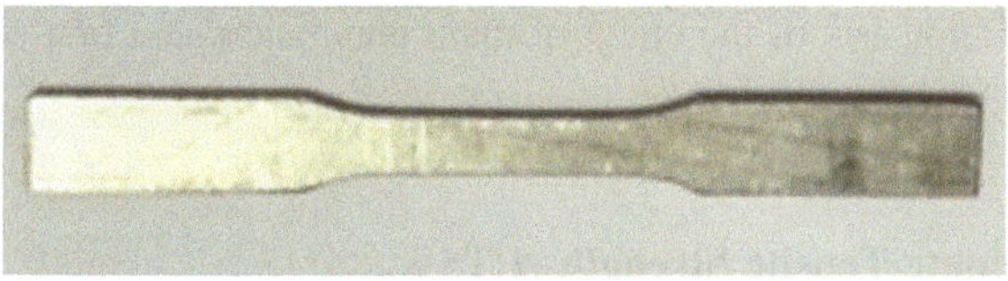

Fig. 15.3 Top view of welded sample in butt joint configuration

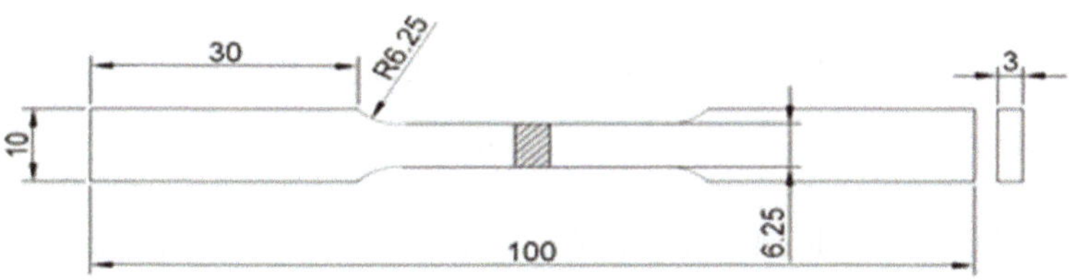

Fig. 15.4 Schematic diagram of TIG welded sample for tensile test

Table 15.7 Design matrix and experimental results

Run	Current (*A*)	Speed (cm/min)	GFR (l/min)	UTS (MPa)	% elongation
1	100	12	9	612.36	08.53
2	150	12	9	616.68	11.94
3	100	18	9	609.94	10.24
4	150	18	9	630.55	11.05
5	100	15	6	621.46	10.45
6	150	15	6	612.89	13.47
7	100	15	12	580.61	11.45
8	150	15	12	607.75	13.18
9	125	12	6	624.09	10.16
10	125	18	6	616.98	09.60
11	125	12	12	592.44	10.12
12	125	18	12	615.92	11.56
13	125	15	9	639.84	14.68
14	125	15	9	645.31	14.44
15	125	15	9	642.46	14.58

15.3.4 Developing the Mathematical Models: Regression Analysis

In the present study, Minitab v17 software has been applied to the experimental data to obtain the quadratic mathematical equations for ultimate tensile strength (UTS) and percentage elongation (PE). This regression equations/ mathematical model is used for predicting responses (UTS and PE) in terms of current, speed and gas flow rate. It consists of the effect of main parameters and first-order interaction of all parameters.

15.3.4.1 Ultimate Tensile Strength (UTS)

In the present context, the response i.e. ultimate tensile strength (σ) is function of TIG welding current (A), travel speed (B) and gas flow rate (C) and it can be expressed as

$$\sigma = f(\text{welding current, travel speed, gas flow rate})$$

$$\sigma = f(A, B, C) \tag{15.1}$$

Typically, a second-order polynomial equation is used in RSM that is provided below:

$$Y = b_0 + \sum_{i=1}^{k} (b_i x_i) + \sum_{i=1}^{k} \left(b_{ii} x_i^2\right) + \sum \sum \left(b_{ij} x_i x_j\right) + € \tag{15.2}$$

where

Y response variables, i.e. dependent variables
x_i predicted variables, i.e. independent variables
b_0 model constant
€ random error.

Parameters b_i, b_{ii}, b_{ij} are known as regression coefficient, where i = 1, 2, 3, … k and j = 1, 2, 3, …k.

In the study, three parameters have been considered. Therefore, for three factors, the selected polynomial equations can be provided as:

$$\sigma = b_0 + b_1(A) + b_2(B) + b_3(C) + b_{11}A^2 + b_{22}B^2 + b_{33}C^2 + b_{12}(AB) + b_{13}(AC) + b_{23}(BC) \tag{15.3}$$

where σ is the predicted response, b_0 model constant, b_1, b_2 and b_3 linear coefficients, b_{11}, b_{22} and b_{33} quadratic coefficients and b_{12}, b_{13} and b_{23} cross-interaction coefficients.

The second-order regression equation for the response UTS in terms of coded factors is given below:

$$\begin{aligned} \text{UTS} = {} & 160.6 + 4.699\,A + 17.51\,B + 10.98\,C - 0.02547\,A*A - 1.026\,B*B - 2.327\,C*C \\ & + 0.0543\,A*B + 0.1190\,A*C + 0.850\,B*C \end{aligned} \tag{15.4}$$

The adequacy of the developed model for UTS has been tested using the statistical analysis of variance (ANOVA) technique. Table 15.8 shows the adequate ANOVA test results for UTS. From table, it is found that the regression is significant with linear and quadratic terms for UTS at 95% confidence level as its p value is less than 0.05. The lack-of-fit value of the model is not significant which is desirable. The model indicates that the welding process parameters: current (*A*), speed (*B*) and gas flow rate (*C*) and its squared terms [current*current (A^2), speed*speed (B^2) and gas flow rate*gas flow rate (C^2)] and interaction effects [current and speed ($A \times B$), current and gas flow rate ($A \times C$) and speed and gas flow rate ($B \times C$)] have significant effect on UTS. The 'R^2' value is 0.9895, and 'Pred R^2' of 0.8805 is in reasonable agreement with the 'Adj R^2' of 0.9707. The 'R^2' value is higher and nearer to 1, that is desirable.

The ANOVA result indicates that the gas flow rate is most significantly effects ultimate tensile strength followed by welding current and speed.

Table 15.8 ANOVA results for testing adequacy of the UTS

Source	DF	Adj SS	Adj MS	*F*-value	*P* value	
Model	9	4268.93	474.33	52.59	0.000	Significant
Linear	3	1107.49	369.16	40.93	0.001	Significant
A—current	1	236.53	236.53	26.23	0.004	Significant
B—travel speed	1	96.74	96.74	10.73	0.022	Significant
C—gas flow rate	1	774.21	774.21	85.84	0.000	Significant
Square	3	2542.36	847.45	93.96	0.000	Significant
*A***A*	1	935.46	935.46	103.72	0.000	Significant
*B***B*	1	315.04	315.04	34.93	0.002	Significant
*C***C*	1	1619.34	1619.34	179.54	0.000	Significant
Two-way interaction	3	619.08	206.36	22.88	0.002	Significant
*A***B*	1	66.34	66.34	7.36	0.042	Significant
*A***C*	1	318.80	318.80	35.35	0.002	Significant
*B***C*	1	233.94	233.94	25.94	0.004	Significant
Error	5	45.10	9.02			
Lack of fit	3	30.13	10.04	1.34	0.454	Not significant
Pure error	2	14.97	7.48			
Total	14	4314.02				

Model summary
S R^2 R^2 (adj) R^2 (pred)
3.00320 98.95% 97.07% 88.05%

15.3.4.2 Percentage Elongation (PE)

The second-order regression equation for the response, percentage elongation is developed by using RSM, in terms of coded factors is given below:

$$\begin{aligned}\mathrm{PE} = & -121.13 + 0.6834\,A + 10.494\,B + 2.323\,C - 0.001879\,A*A - 0.3280\,B*B \\ & - 0.1394\,C*C - 0.00867\,A*B - 0.00430\,A*C + 0.0556\,B*C\end{aligned} \tag{15.5}$$

Table 15.9 shows the adequate ANOVA test results for percentage elongation (PE), and it is found that model is significant ($P < 0.05$) at 95% confidence level or 5% significant level. The lack-of-fit value of the model is not significant which is desirable. The model F-value is 119.66 which indicates that the model is significant. Here A, B, C, A^2, B^2, C^2, $A \times B$, $A \times C$, and $B \times C$ are the significant model terms. The 'R^2' value is 0.9954, and 'Pred R^2' of 0.9335 is in reasonable agreement with the 'Adj R^2' of 0.9871. The 'R^2' value is higher and nearer to 1, that is desirable.

According to the mathematical models, it is understood that gas flow rate is found to be the most important factor affecting UTS and current is the most significant parameter on percentage elongation. But welding speed is insignificant factor for ultimate tensile strength and percentage elongation.

Table 15.9 ANOVA results for testing adequacy of percentage elongation (PE)

Source	DF	Adj SS	Adj MS	*F*-value	*P* value	
Model	9	53.3411	5.9268	119.66	0.000	Significant
Linear	3	11.2835	3.7612	75.94	0.000	Significant
A—current	1	10.0576	10.0576	203.07	0.000	Significant
B—travel speed	1	0.3613	0.3613	7.29	0.043	Significant
C—gas flow rate	1	0.8646	0.8646	17.46	0.009	Significant
Square	3	38.9516	12.9839	262.15	0.000	Significant
*A***A*	1	5.0941	5.0941	102.85	0.000	Significant
*B***B*	1	32.1777	32.1777	649.68	0.000	Significant
*C***C*	1	5.8116	5.8116	117.34	0.000	Significant
Two-way interaction	3	3.1060	1.0353	20.90	0.003	Significant
*A***B*	1	1.6900	1.6900	34.12	0.002	Significant
*A***C*	1	0.4160	0.4160	8.40	0.034	Significant
*B***C*	1	1.0000	1.0000	20.19	0.006	Significant
Error	5	0.2476	0.0495			
Lack of fit	3	0.2186	0.0729	5.01	0.171	Not significant
Pure error	2	0.0291	0.0145			
Total	14	53.5887				

Model summary
S R^2 R^2(adj) R^2(pred)
0.222550 99.54% 98.71% 93.35%

The soundness of the developed model has also been checked by the normal probability plot of the residuals for UTS and PE as shown in Figs. 15.5 and 15.6, respectively. From figure, it is seen that the residuals are falling on the straight line, which means the errors are distributed normally and the mathematical relationship is correctly developed.

15.3.5 Effect of TIG Welding Process Parameters on UTS

The surface and contour plots have been drawn from Eq. 15.4 to study the interaction effects of process parameters on UTS and shown in Figs. 15.7, 15.8 and 15.9. It is observed that UTS first increases gradually then decreases with the increase in the levels of the welding process parameters. From Fig. 15.7a, b, it is noticed that UTS increases with gas flow rate (GFR) and then stars decreasing to its minimum value. UTS is maximum at medium values of GFR and welding current. Figure 15.8a, b shows the combined effect of welding speed and welding current on UTS. It is found that UTS is maximum at the medium values of welding speed and current. The interaction between GFR and welding speed on UTS is shown in

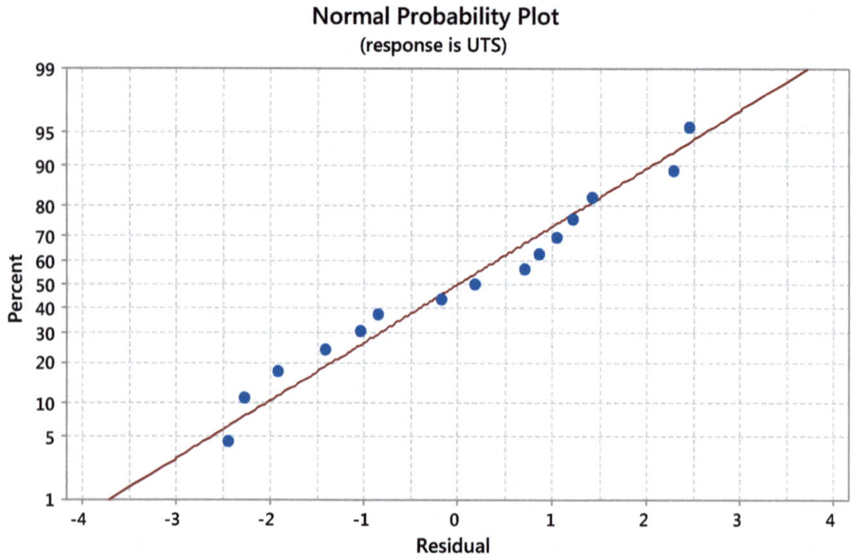

Fig. 15.5 Residual plot of tensile strength

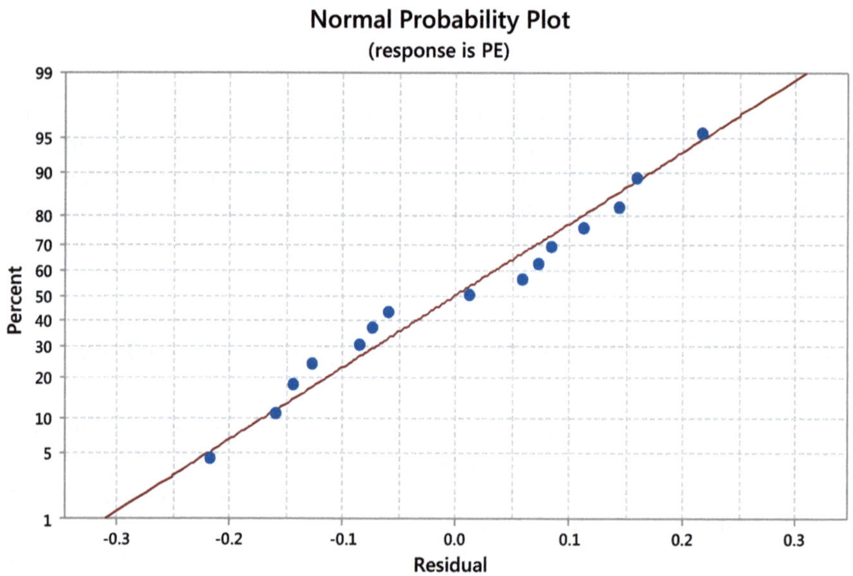

Fig. 15.6 Residual plot of percentage elongation

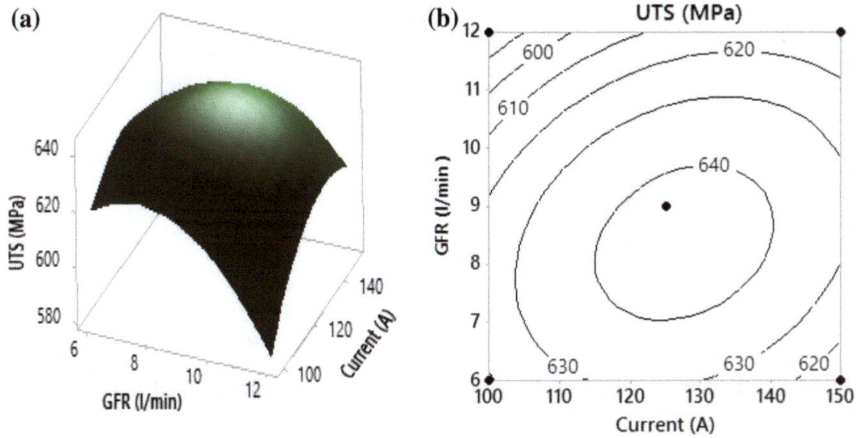

Fig. 15.7 **a** Response surface plot and **b** contour plot showing the effect of GFR and current on the UTS at welding speed of 15 cm/min

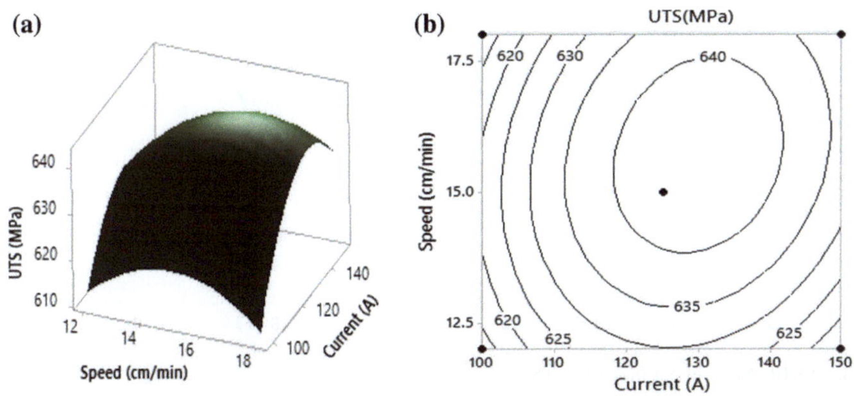

Fig. 15.8 **a** Response surface plot and **b** contour plot showing the effect of welding speed and current on the UTS at GFR of 9 l/min

Fig. 15.9a, b. It is observed that the UTS is optimum at the middle value of GFR and welding speed.

The main effect plots for UTS with process parameters like welding current, travel speed and GFR are shown in Fig. 15.10. The graph indicates that UTS increases from low value to maximum value, when current increases from 100 to 130 A as depth of penetration attains its maximum value. Similar types of patterns are also seen in case of welding speed and GFR. As welding speed changes from 12 to 15 cm/min, UTS increases from 630 to 643 MPa. GFR increases from 6 to 9 l/min UTS increases from 628 to 643 MPa due to decrease in porosity level of weld metals.

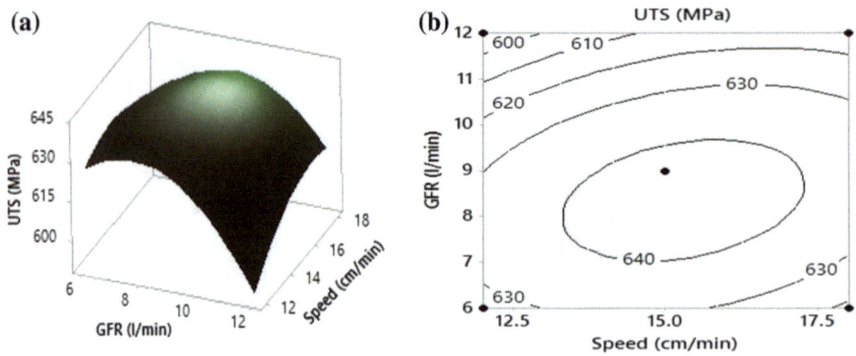

Fig. 15.9 **a** Response surface plot and **b** contour plot showing the effect of GFR and speed on the UTS at welding current of 125 A

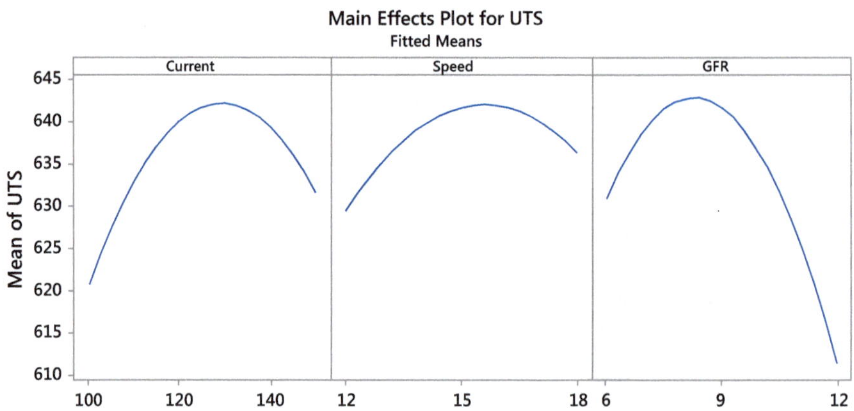

Fig. 15.10 Main effect plot for UTS

15.3.6 Effect of TIG Welding Process Parameters on PE

Figures 15.11, 15.12 and 15.13 illustrate the response surface plot and contour plot for the response of percentage elongation. Similar type of patterns is also observed in case of percentage elongation. The interactive effects of welding speed and GFR on percentage elongation are shown in Fig. 15.11a, b. The value of percentage elongation is maximum at the medium levels of both welding speed and GFR. Cooling rate decreases with increase in heat input at low welding speed. This in turn will take longer time for solidification. As a result, it produces coarse grain, hence low tensile strength and low percentage elongation. The combined effect of GFR and welding current on percentage elongation is shown in Fig. 15.12a, b. Welding current has a significant effect on PE. At lower welding current, the percentage

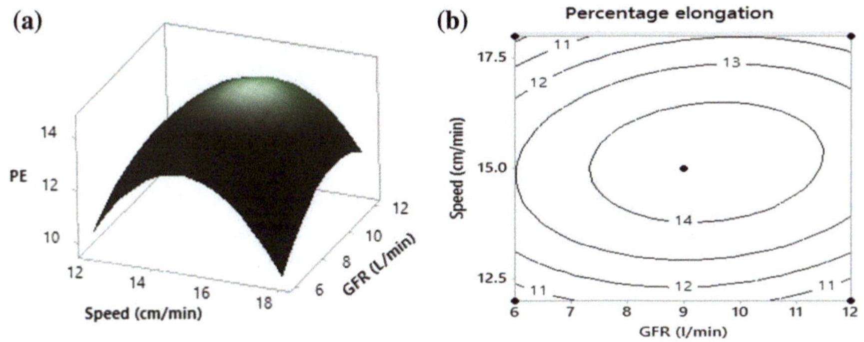

Fig. 15.11 **a** Response surface plot and **b** contour plot showing the effect of welding speed and GFR on the PE at current of 125 A

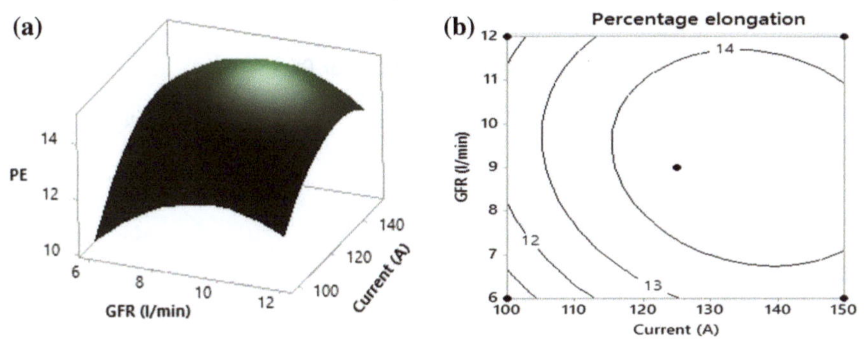

Fig. 15.12 **a** Response surface plot and **b** contour plot showing the effect of GFR and current on the PE at welding speed of 15 cm/min

elongation of the welded joint is lower. When current is increased, the PE is also increased and reached its maximum value as it forms uniform penetration. Figure 15.13a, b demonstrates the interaction effect of welding speed and welding current on percentage elongation at GFR of 9 l/min.

Figure 15.14 represents the main effect plot of percentage elongation with welding current, welding speed and GFR. The value of percentage elongation increases as welding current increases from 100 to 130 A due to proper penetration. As travel speed changes from 12 to 15 cm/min, PE achieves its greatest value. For further increase in the level of welding speed, the PE starts to decrease. PE reaches its optimum value when GFR changes from 6 to 9 l/min.

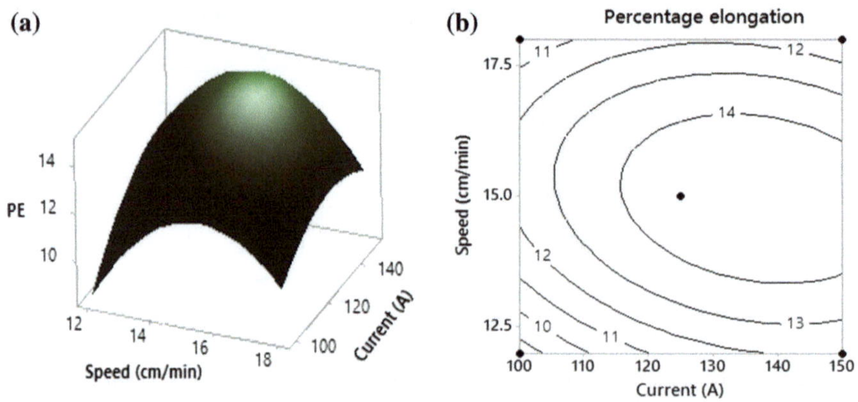

Fig. 15.13 **a** Response surface plot and **b** contour plot showing the effect of welding speed and current on the PE at GFR of 9 l/min

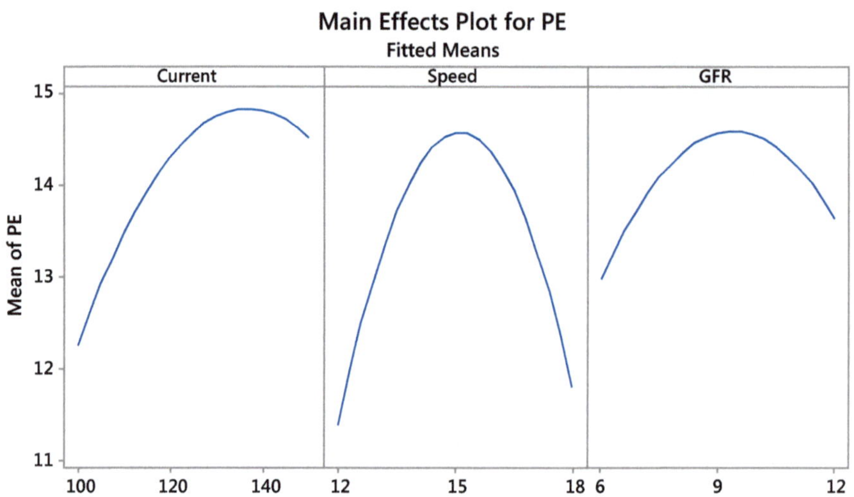

Fig. 15.14 Main effect plot for percentage elongation

15.3.7 Optimization of Ultimate Tensile Strength

One of the main targets of experimental work related to welding is to attain higher tensile strength and as well as percentage elongation. In the study, the aim is to achieve maximum ultimate tensile strength and maximum percentage elongation of welded specimens simultaneously which is desirable for good quality of welded joint. The results optimization done by RSM for UTS and PE is shown in Fig. 15.15. In this figure, each column represents one of the TIG welding process parameters and each row of the graph represents a response like UTS, PE and how

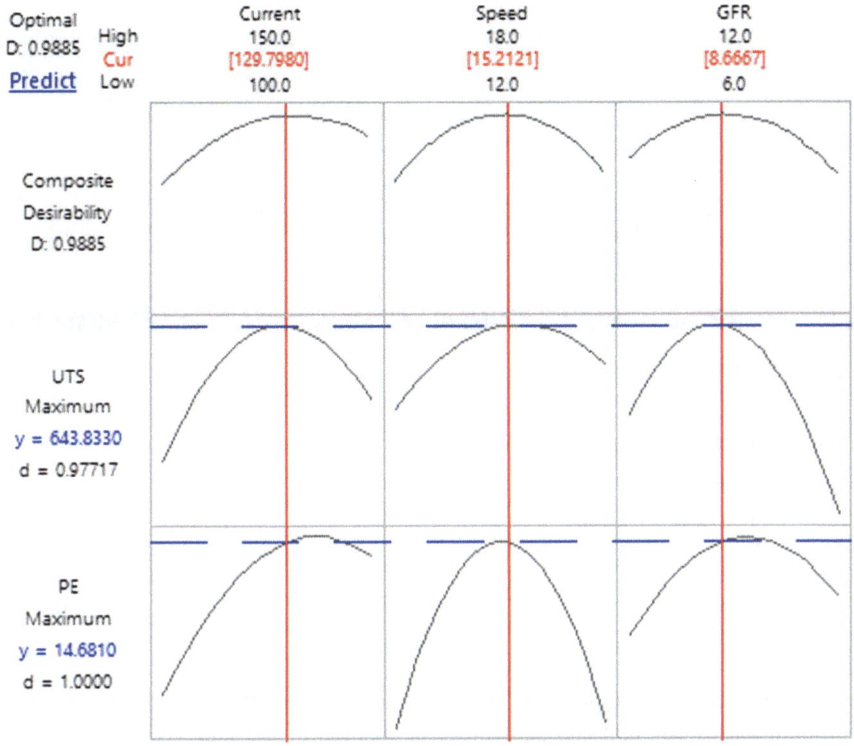

Fig. 15.15 Optimization results of ultimate tensile strength and percentage elongation

this response variable alters as a function of the welding process parameters, while maintaining other process parameters unchanged. The optimum predicted response values are shown by horizontal dotted lines, and corresponding optimum process parameters are designated by vertical line inside the graph. At the top of the column shows the values of optimum process parameters, and left-hand side shows the predicted responses values. Individual desirability values are also given in this figure. The design of experiment software, Minitab v17, has been applied for optimization of TIG welding process. The most favourable ultimate tensile strength, 643.83 MPa, and optimum percentage elongation, 14.68, have been obtained at welding current of 129.8 A, welding speed of 15.2 cm/min and gas flow rate of 8.7 l/min. The value of composite desirability factor, $D = 0.9885$.

Microstructural examination has been carried out by scanning electron microscopy (SEM) of base material and weld material of the welded joint that has been prepared with optimum parametric setting as mentioned earlier. Figure 15.16a, b represents typical SEM micrographs of base and weld metal, respectively. Base material mostly consists of austenitic structure along with little annealing twins. From Fig. 15.16b, it is found that very fine skeletal δ-ferrite is produced in weld

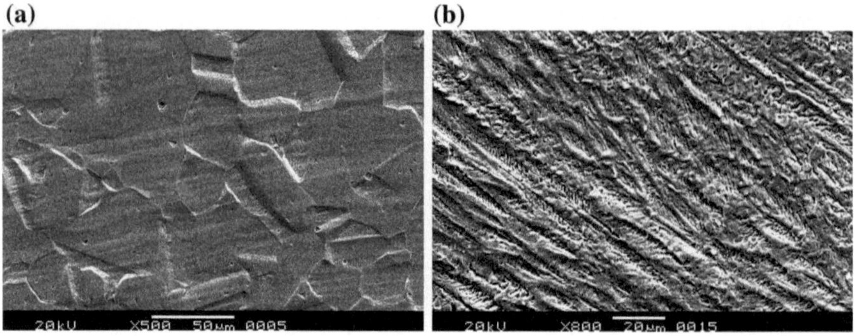

Fig. 15.16 SEM micrographs of **a** base metal and **b** weld zone

Table 15.10 Multi-objective optimization results

Optimum condition					
A (A)	*B* (cm/min)	*C* (l/min)		UTS (MPa)	PE
129.8	15.2	8.7	Average actual predicted |Error %|	650.45 643.83 1.02	14.75 14.68 0.47

area and reasonable precipitation. These are the most important causes for obtaining optimum tensile strength and percentage elongation of the said TIG welded joint.

15.4 Confirmation Test

The results of optimization obtained have been validated by performing confirmatory experiments. Table 15.10 represents the results of confirmatory tests that are conducted at optimum conditions. It is noticed from table that the error in terms of percentage between the predicted and experimental results is very small. This indicates that the optimized TIG welding process parameters can be considered to obtained higher ultimate tensile strength and percentage elongation of 316L stainless steel.

15.5 Conclusions

The effect of welding process parameters: current, speed and gas flow rate on ultimate tensile strength and percentage elongation in TIG welding of AISI 316L stainless steel materials have been studied. The following points can be concluded from the experiments and analyses:

1. Response surface methodology (RSM) is found to be very helpful in the process of optimization carried out in the present study. Here the predicted value obtained from the models is very near to the experimental value.
2. Gas flow rate is the most important welding parameter which affecting the ultimate tensile strength, and it is followed by welding current and welding speed.
3. The most significant factor affecting percentage elongation is welding current and the next important factor in this context is gas flow rate followed by welding speed.
4. Welding speed is not found to be most significant at all in affecting either UTS or percentage elongation.
5. It is observed that both UTS and percentage elongation increase with increasing in the levels of welding process parameters up to its centre value; thereafter, the responses start decreasing with increasing in the levels of process parameters.
6. An optimum tensile strength of 643.83 MPa and percentage elongation of 14.68 are obtained under the welding conditions of current at 129.8 A, speed at 15.2 cm/min and gas flow rate at 8.7 l/min.
7. The error between predicted and confirmatory test is obtained approximately 1%, which validates the used optimization technique.

Acknowledgements The authors would like to acknowledge Dr. D. Bose, ME Department, NITTTR, Kolkata, West Bengal, for providing welding set-up for experiment for the present research work.

References

1. Ghosh, N., Pal, P.K., Nandi, G., Rudrapati, R.: Parametric optimization of gas metal arc welding process by PCA based Taguchi method on austenitic stainless steel AISI 316L. Mater. Today Proc. **5**, 1620–1625 (2018)
2. Tarng, Y.S., Yang, W.H.: Optimisation of the weld bead geometry in gas tungsten arc welding by the Taguchi method. Int. J. Adv. Manufact. Technol. **14**, 549–554 (1998)
3. Benyounis, K.Y., Olabi, A.G.: Optimization of different welding processes using statistical and numerical approaches—a reference guide. Adv. Eng. Softw. **39**, 483–496 (2008)
4. Gunaraj, V., Murugan, N.: Application of response surface methodology for predicting weld bead quality in submerged arc welding of pipes. J. Mater. Process. Technol. **88**(1–3), 266–275 (1999)
5. Kumar, A., Sundarrajan, S.: Optimization of pulsed TIG welding process parameters on mechanical properties of AA 5456 aluminum alloy weldments. Mater. Des. **30**, 1288–1297 (2009)
6. Murugan, N., Parmer, R.S.: Effects of MIG process parameters on the geometry of the bead in the automatic surfacing of stainless steel. J. Mater. Process. Technol. **41**(4), 381–398 (1994)
7. Benyounis, K.Y., Olabi, A.G., Hashmi, M.S.J.: Optimizing the laser-welded butt joints of medium carbon steel using RSM. J. Mater. Process. Technol. **164–165**, 986–989 (2005)
8. Zhao, Y., Zhang, Y., Hu, W., Lai, X.: optimization of laser welding thin-gage galvanized steel via response surface methodology. Opt. Lasers Eng. **50**, 1267–1273 (2012)

9. Olabi, A.G., Alsinani, F.O., Alabdulkarim, A.A., Ruggiero, A., Tricarico, L., Benyounis, K. Y.: Optimizing the CO_2 laser welding process for dissimilar materials. Opt. Lasers Eng. **51**, 832–839 (2013)
10. Ahn, J., Chen, L., Davies, C.M., Dear, J.P.: Parametric optimisation and microstructural analysis on high power Yb-fibre laser welding of Ti-6Al-4V. Opt. Lasers Eng. **86**, 156–171 (2016)
11. Ragavendran, M., Chandrasekhar, N., Ravikumar, R., Saxena, R., Vasudevan, M., Bhaduri, A.K.: Optimization of hybrid laser-TIG welding of 316LN steel using response surface methodology (RSM). Opt. Lasers Eng. **94**, 27–36 (2017)
12. Padmanaban, G., Balasubramanian, V.: optimization of laser beam welding process parameters to attain Maximum tensile strength in AZ31B magnesium alloy. Opt. Laser Technol. **42**, 1253–1260 (2010)
13. Elatharasan, G., Senthil Kumar, V.S.: An experimental analysis and optimization of process parameter on friction stir welding of AA 6061-T6 aluminum alloy using RSM. Procedia Engineering **64**, 1227–1234 (2013)
14. Srivastava, S., Garg, R.K.: Process parameter optimization of gas metal arc welding on IS:2062 mild steel using response surface methodology. J. Manufact. Process. **25**, 296–305 (2017)
15. Lakshminarayanan, A.K., Shanmugam, K., Balasubramanian, V.: Effect of autogenous arc welding processes on tensile and impact properties of ferritic stainless steel joints. J. Iron Steel Res. Int. **16**(1), 62–68 (2009)
16. Kumar, N., Mukherjee, M., Bandyopadhyay, A.: Study on laser welding of austenitic stainless steel by varying incident angle of pulsed laser beam. Opt. Laser Technol. **94**, 296–309 (2017)
17. Ghosh, N., Pal, P.K., Nandi, G.: Parametric optimization of MIG welding on 316L austenitic stainless steel by Grey-Based Taguchi method. Procedia Technol. **25**, 1038–1048 (2016)
18. Ghosh, N., Rudrapati, R., Pal, P.K., Goutam, N.: Parametric optimization of gas metal arc welding process by using Taguchi method on ferritic stainless steel AISI 409. Mater. Today Proc. **4**, 2213–2221 (2017)

Chapter 16
A Study on the Structural Behaviour of AFG Non-uniform Plates on Elastic Foundation: Static and Free Vibration Analysis

Hareram Lohar, Anirban Mitra and Sarmila Sahoo

Nomenclature

A_0	Cross-sectional area of the plate at root side
a	Length of the plate
b	Breadth of the plate
d	Unknown coefficients
E_0	Elastic modulus of the plate material at root side
$\{f\}$	Load vector
I_0	Moment of inertia of the plate at root side
D	Flexural rigidity of plate
$[K]$	Stiffness matrix
$[M]$	Mass matrix
nf, nw, nv	Number of constituent functions for w, u and v, respectively
α_i	Set of orthogonal functions for u
β_i	Set of orthogonal functions for v
ϕ_i	Set of orthogonal functions for w
K_f	Foundation stiffness
ω_1	Fundamental linear frequency
nfg	Number of Gauss points
q	External uniformly distributed load
t_0	Thickness of the plate at root side
T	Kinetic energy of the system
u	Displacement field in x-axis
U	Strain energy stored in the system
v	Displacement field in y-axis
V	Potential energy of the external forces
w	Displacement field in z-axis

H. Lohar (✉) · A. Mitra
Department of Mechanical Engineering, Jadavpur University, Kolkata, India
e-mail: hr.lohar343@gmail.com

S. Sahoo
Department of Civil Engineering, Heritage Institute of Technology, Kolkata, India

P. Sahoo and J. P. Davim (eds.), *Advances in Materials, Mechanical and Industrial Engineering*, Lecture Notes on Multidisciplinary Industrial Engineering,
https://doi.org/10.1007/978-3-319-96968-8_16

δ	Variational operator
μ	Poisson's ratio
ρ_0	Density of the plate material at ($\xi = 0$)
τ	Time coordinate
ξ, η	Normalized axial coordinates
α	Taper parameter
ω_{nf}	Nonlinear frequency

16.1 Introduction

Functionally graded materials (FGMs) are new class of advance materials. The interface problems in layered composite material are avoided in FGM through functional and continuous gradation of material properties. Due to excellent thermo-mechanical properties FGM finds huge applications in certain engineering fields and also catches the attention of the researchers to explore its behaviours quiet often. In FGM, functional gradation of material properties may be considered in two orthogonal directions, either in the transverse or in axial direction. A huge number of papers are available in the literature in the domain of inhomogeneity in the thickness/transverse direction. A critical review was presented by Jha et al. [1] to identify the most recent and relevant aspects of FGM plate. Birman and Byrd [2] presented a review on most recent development of FG materials since 2000. Various theoretical aspects and applications of FGM were reflected upon in this paper. Kennedy et al. [3] equivalently modelled a FG plate as a plate of sequentially stacked multiple isotropic layers. Classical Plate theory (CPT), First-Order Shear Deformation Theory (FSDT) and Higher Order Shear Deformation Theory (HSDT) were used to obtain the governing equations for separate models to study buckling and dynamic behaviour. Singha et al. [4] utilized high precision FE method to study the nonlinear behaviours of FGM plates. The formulation was based on FSDT and solution was obtained through Newton–Raphson iteration technique. Panyatong et al. [5] used second-order shear deformation theory (SDT) to determine the natural frequencies of Functionally Graded nanoplate which was embedded in elastic medium. The formulation was on the basis of nonlocal elasticity and the derivation of governing equations was performed through Hamilton's principle to investigate the influences of the medium stiffness and temperature on natural frequencies. Benferhat et al. [6] performed free vibration study on graded plate supported on Winkler–Pasternak type of elastic foundation on the basis of the neutral surface concept. Sharma and Parashar [7] employed generalized differential quadrature method to study the free vibration of FG piezoelectric annular plate on the basis of modified Mindlin plate theory. Abrate [8] proposed a new approach to analyse FG plate by using no special tools as he postulated that FG plates behaved same as homogeneous plates. In the model, the variation in material properties of the plate was performed by introducing a coupling between the in-plane and transverse deformations. Keleshteri et al. [9] utilized generalized differential quadrature method to study nonlinear vibration of FG carbon nanotube reinforced composite annular sector plates with piezoelectric layers on the basis of FSDT. Von Karman type of geometrical nonlinearity was considered. The governing set of equations was derived through Hamilton principle and solved using direct iterative method and GDQM to highlight the variation of frequency versus amplitude plot. Chi and Chung [10] investigated the mechanical behaviour of a simply supported FGM plate of moderate thickness loaded under transverse loading. The solution was based on the CPT and Fourier series expansion. Akbarzadeh et al. [11] represented an analytical solution to investigate the behaviour of FGM rectangular plates based on FSDT and 3rd order SDT. The mathematical formulation was displacement based and derived governing sets of equations were solved by the Fourier series expansion to obtain the natural frequencies.

On the other hand, research work involving material inhomogeneity in the in-plane/axial direction are hardly available in the literature, although they have potential for applications in civil, mechanical, aerospace and marine engineering. Uymaz et al. [12] carried out vibration analysis of FG plates with in-plane material gradation with different boundary conditions. Formulation was on the basis of shear deformable plate theory and assumed displacement field. Xiang et al. [13] used scaled boundary FE approach to study free vibration and buckling of FG plates with in-plane material imperfection. Two-dimensional higher order spectral element was considered to model the in-plane dimensions of the plate. The stiffness matrix was derived directly from spectral element. Liu et al. [14] presented free vibration of a FG rectangular plate with in-plane material imperfection. The edge condition was considered as simply supported and a Levy-type solution was formulated. The differential equations were solved considering a particular integration method. Hussein and Mulani [15] dealt with the optimization of in-plane FG nano-reinforced panels for buckling load. For that purpose, different types of panels with or without stiffener and cut-outs were considered. The main objective was to minimize the nano-inforcement to satisfy the desired buckling constraints. Recently, Kumar et al. made an effort to explore the static and dynamic behaviour of in-plane/axial inhomogeneous FG plate by using energy principle based on displacement field. The authors [16, 17] performed nonlinear forced vibration on non-uniform AFG plates. For the formulation purpose, Hamilton's principle was used to obtain the set of governing equations and the solution of equations was performed through Broyden method and direct iterative method. The influences of taper parameter and excitation amplitude on forced vibration frequency response of the plates were observed. The authors also studied [18] free vibration of axial inhomogeneous plate through backbone curves to show the geometric nonlinearity effects.

Beside the classical end conditions, plate structures are often connected to other members, supported by elastic restraints or supported by elastic foundation. Application of such type of supported structure can be found in railroad, biomechanics, road, marine, geo-technics and engineering. So, studies should not be always limited to classical edges which are ideal clamped, simply supported and free edges. In reality, such type of structures is modelled by a number of distributed linear spring at the boundaries. On the other hand, plates supported on foundations can be idealized through an array of springs attached to the face of the plate. Due to their critical behaviour, investigations involving plates on elastic foundation are of crucial significance and have received sufficient attention recently. Taczała et al. [19] performed a post-buckling study of FG plates on elastic foundation. FSDT was used for the mathematical formulation, and the results were obtained numerically by the FE method. Ebrahimi et al. [20] proposed a four-variable refined shear deformation theory to perform free vibration of porous magneto-electro-elastic FG plates. The set of differential equations were obtained with the help of Hamilton's principle. Kutlu et al. [21] presented a mixed FE method and utilized a boundary element approach to study the dynamics of thick plates on elastic foundation. The plate was considered to be interacting with a quiescent fluid on the other side. To represent the plate-foundation system, a two-field mixed FE formulation, on the basis of Hellinger–Reissner variational principle was used. A boundary element solution was incorporated for the fluid–structure interaction. Gupta et al. [22] performed the dynamic analysis of FG simply supported plates on elastic foundation. The foundation was considered as two-parameter Pasternak type. The formulation was displacement based and HSDT was used for derivation purpose. Mohammadzadeh and Noh [23] presented an analytical approach for obtaining nonlinear frequency response of sandwich plates. The equations of motion were obtained by using the HSDT along with Hamilton's principle. Navier's solution and Runge-Kutta numerical scheme was used to solve the problem. Zamani et al. [24] investigated free vibration of laminated viscoelastic composite plate with simply supported edge resting on Pasternak viscoelastic foundation. Based on 3rd order SDT, the partial differential equations were attained by Hamilton principle and Laplace transformation. The equations were solved by iterative methods which were weighted residual method and Fourier transform. Najafi et al. [25] presented the impact analysis of FGM plates which were resting on elastic support with simply supported edge condition. The elastic foundation was considered as a nonlinear three-parameter one. The governing equations were derived based on a Reddy's HSDT and were solved by using an analytical procedure. Von Karman type of kinematic nonlinearity was considered. Wattanasakulpong and Chaikittiratana [26] investigated static and dynamic behaviour of CNR composite plates resting on elastic foundation. The elastic foundation was considered to be Pasternak type which also included shear layer and springs. The formulation was on the basis of generalized shear deformation theory. Barati et al. [27] examined the buckling behaviour of FG piezoelectric porous plates on basis of a refined four-variable plate theory. Derivation of the governing set of equations was performed through Hamilton's principle and implementing an analytical approach the equations were solved. Shahsavari et al. [28] analysed the free vibration behaviour of FG porous plates which were supported on elastic foundations. Formulation was carried out on the basis of quasi-3D hyperbolic plate model and solution was performed using Galerkin method.

From the literature review, it can be concluded that an enormous number of literature are available in the domain of FGM where inhomogeneity is considered in the thickness direction. At the same time, in-plane inhomogeneity is newer domain and a few literatures are available. Its also comprehensible that research works involving in-plane material inhomogeneity (AFG) are quite rare. So, in the present study the structural behaviour of AFG non-uniform plates on elastic foundation is considered.

16.2 Mathematical Formulation

In the present formulation, a semi-analytical technique (displacement field based) is adopted for carrying out the analysis. At first, the static problem is taken up and the governing set of equations is derived through application of minimum potential energy principle following variational method [29]. After that the subsequent dynamic problem is solved using Hamilton principle from the known displacement field which is derived previously from the static analysis. Geometric nonlinearity is incorporated in the system following nonlinear strain-displacement relations.

A tapered axially functionally graded (AFG) square plate is considered to be rested on elastic foundation as shown in Fig. 16.1a. The length and width of the plate are a and b, respectively. Formulation is performed in normalized domain which are characterized by $\xi = x/a$ and $\eta = y/b$ and their values vary from 0 to 1. The thickness of the plate is varying from the root side along axial direction following the relation $t(\xi) = t_0(1 - \alpha\xi^2)$. Here, t_0 is the root thickness and α is taper parameter. Axial variation of material properties is considered in a similar way to the thickness variation described above. These variations [30] for elastic modulus and density are expressed as, $E(\xi) = E_0(1 + \xi)$ and $\rho(\xi) = \rho_0(1 + \xi + \xi^2)$. Here E_0 and ρ_0 are values at the root side. All ends of the plate are considered to be

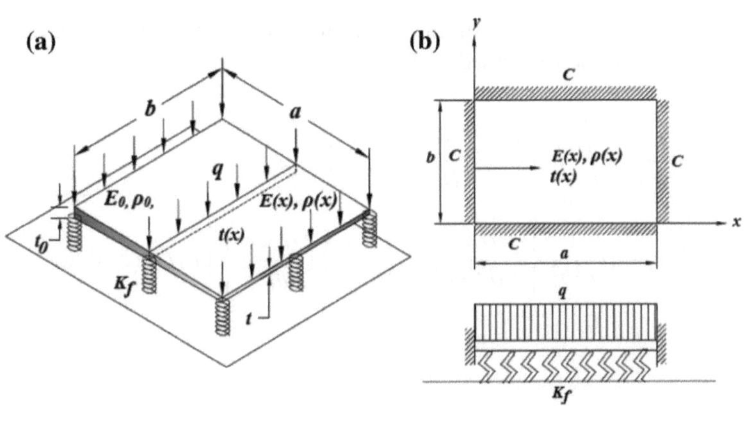

Fig. 16.1 **a** AFG plate on elastic foundation. **b** End condition and loading pattern

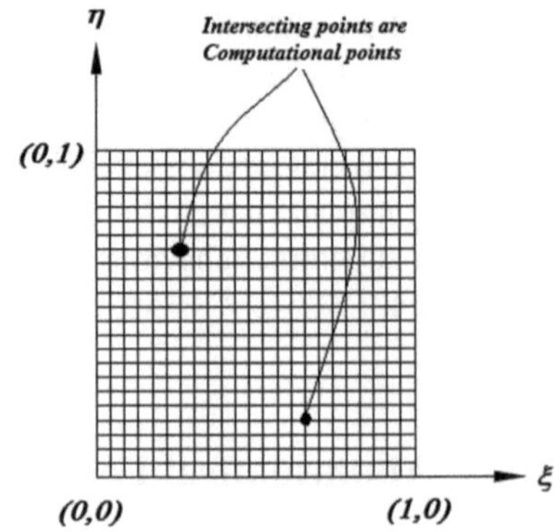

Fig. 16.2 Computational points in the domain

clamped (C) or simply supported (S), and they are represented as CCCC and SSSS end conditions. The stiffness of the elastic foundation is considered as K_f. It is important to note that linear behaviour of the foundation is assumed by attaching a series of parallel linear spring at the bottom face of the plate. The plate is loaded under evenly distributed pressure type load of magnitude q as shown in Fig. 16.1b. To create a computational domain in the present study, Gauss points are created with the help of a numerical module along normalized ξ and η directions and hence constant ξ and constant η lines are formed as shown in Fig. 16.2. At these intersection points of the lines, all relevant numerical computations of the problem are performed.

Due to large deflection and elastic foundation, the calculation of total strain energy of the present system has three distinct parts. The first part corresponds to strain energy due to pure bending (U_b), the second one is due to stretching of mid-plane of the plate (U_s) and the third part arises because of inclusion of elastic foundation (U_f). So, total strain energy (U) of the system is represented as,

$$U = U_b + U_s + U_f \tag{16.1}$$

By substituting appropriate strain displacement relations and loading condition, U_b, U_s, and U_f can be expressed as [16],

$$\begin{aligned} U_b = \frac{a}{2(1-\mu^2)} \int_0^1 \int_0^1 \Bigg[& \left\{ \frac{1}{a^2}\left(\frac{\partial^2 w}{\partial \xi^2}\right) + \frac{1}{b^2}\left(\frac{\partial^2 w}{\partial \eta^2}\right) \right\}^2 \\ & + 2(1-\mu)\frac{1}{a^2 b^2}\left\{ \left(\frac{\partial^2 w}{\partial \xi \partial \eta}\right)^2 - \left(\frac{\partial^2 w}{\partial \xi^2}\right)\left(\frac{\partial^2 w}{\partial \eta^2}\right) \right\} \Bigg] E(\xi) I(\xi)\, \mathrm{d}\xi \mathrm{d}\eta \end{aligned} \tag{16.2a}$$

$$U_s = \frac{a}{2(1-\mu^2)}\int_0^1\int_0^1 \left[\frac{1}{a^2}\left(\frac{\partial u}{\partial \xi}\right)^2 + \frac{1}{a^3}\left(\frac{\partial u}{\partial \xi}\right)\left(\frac{\partial w}{\partial \xi}\right)^2 + \frac{1}{b^2}\left(\frac{\partial v}{\partial \eta}\right)^2\right.$$
$$+\frac{1}{b^3}\left(\frac{\partial v}{\partial \eta}\right)\left(\frac{\partial w}{\partial \eta}\right)^2 + \frac{1}{4}\left\{\frac{1}{a^2}\left(\frac{\partial w}{\partial \xi}\right)^2 + \frac{1}{b^2}\left(\frac{\partial w}{\partial \eta}\right)^2\right\}^2$$
$$+2\mu\left\{\frac{1}{ab}\left(\frac{\partial u}{\partial \xi}\right)\left(\frac{\partial v}{\partial \eta}\right) + \frac{1}{2a^2 b}\left(\frac{\partial v}{\partial \eta}\right)\left(\frac{\partial w}{\partial \xi}\right)^2 + \frac{1}{2b^2 a}\left(\frac{\partial u}{\partial \xi}\right)\left(\frac{\partial w}{\partial \eta}\right)^2\right\}$$
$$+\frac{1-\mu}{2}\left\{\frac{1}{b^2}\left(\frac{\partial u}{\partial \eta}\right)^2 + \frac{2}{ab}\left(\frac{\partial u}{\partial \eta}\right)\left(\frac{\partial v}{\partial \xi}\right) + \frac{1}{a^2}\left(\frac{\partial v}{\partial \xi}\right)^2\right.$$
$$\left.\left.+\frac{2}{ab^2}\left(\frac{\partial u}{\partial \eta}\right)\left(\frac{\partial w}{\partial \xi}\right)\left(\frac{\partial w}{\partial \eta}\right) + \frac{2}{a^2 b}\left(\frac{\partial v}{\partial \xi}\right)\left(\frac{\partial w}{\partial \xi}\right)\left(\frac{\partial w}{\partial \eta}\right)\right\}\right] E(\xi)A(\xi)\, \mathrm{d}\xi \mathrm{d}\eta \tag{16.2b}$$

$$U_f = (ab)K_f \int_0^1\int_0^1 (w\partial w)\, \mathrm{d}\xi \mathrm{d}\eta \tag{16.2c}$$

The potential energy due to external load (V) of the system can be expressed as,

$$V = (ab)\int_0^1\int_0^1 (qw)\, \mathrm{d}\xi \mathrm{d}\eta \tag{16.3}$$

Also, kinetic energy (T) of the system can be expressed as,

$$T = \frac{a}{2}\int_0^1\int_0^1 \left\{\left(\frac{\partial w}{\partial \tau}\right)^2 + \left(\frac{\partial u}{\partial \tau}\right)^2 + \left(\frac{\partial v}{\partial \tau}\right)^2\right\}\rho(\xi)A(\xi)\, \mathrm{d}\xi \mathrm{d}\eta \tag{16.4}$$

where, w, u and v are approximated displacement field which are assumed in the transverse (w) and in plane (u and v) directions. τ is the time coordinate.

16.2.1 Static Analysis

In static analysis, to derive the governing set of equations, minimum potential energy principle is utilized, which is expressed as,

$$\delta(U + V) = 0 \tag{16.5}$$

The assumed static displacement field (w, u and v) are generated as linear combinations of orthogonal admissible functions (ϕ, α, β) and unknown coefficients (d) as shown,

$$w(\xi,\eta) = \sum_{i=1}^{nw} d_i\phi_i(\xi,\eta) \tag{16.6a}$$

$$u(\xi,\eta) = \sum_{i=nw+1}^{nw+nu} d_i\alpha_{i-nw}(\xi,\eta) \tag{16.6b}$$

$$v(\xi,\eta) = \sum_{i=nw+nu+1}^{nw+nu+nv} d_i\beta_{i-nw-nu}(\xi,\eta) \tag{16.6c}$$

where, nw, nu and nv are number of orthogonal functions for each of the displacement field w, u and v respectively. Utilizing the boundary conditions, appropriate start function ($\phi_1, \alpha_1, \beta_1$) are selected which are shown in Table 16.1. Then generation of higher order functions (ϕ, α, β) are performed by using Gram–Schmidt orthogonalization procedure [16, 18].

Substituting the expressions from Eqs. (16.2a) and (16.3) along with the Eqs. (16.6a) into Eq. (16.5), the governing equations are found as,

$$[K]\{d\} = \{f\} \tag{16.7}$$

Equation (16.7) here is nonlinear in nature. This is due to the reason that the stiffness matrix, $[K]$, is function of unknown coefficient (d). Hence, exact solution is hardly possible. The approximate solution can be done with the help of a well-established iterative method [31]. Here, direct substitution with relaxation method is utilized to solve the unknown coefficients on the basis of some initial guess.

16.2.2 Dynamic Analysis

Hamilton's principle is employed to derive the governing set of equations in the dynamic analysis, which is expressed as,

Table 16.1 Start functions for different boundary conditions

End conditions	Start functions
CCCC	$\{\xi(1-\xi)\}^2 * \{\xi(1-\xi)\}^2$
SSSS	$\sin(\pi\xi) * \sin(\pi\xi)$
Inmovable	$\xi(1-\xi) * \eta(1-\eta)$

$$\delta\left(\int_{\tau 1}^{\tau 2} (T - U - V)\mathrm{d}\tau\right) = 0 \tag{16.8}$$

It needs to be mentioned here that the present free vibration study is performed on a pre-stressed plate, whose static solution has already been obtained in the previous step. So, the potential energy of the external load (V) can be set to zero. The assumed dynamic displacement field (w, u and v) are expressed as,

$$w(\xi, \eta, \tau) = \sum_{i=1}^{nw} d_i \phi_i(\xi, \eta) e^{j\omega\tau} \tag{16.9a}$$

$$u(\xi, \eta, \tau) = \sum_{i=nw+1}^{nw+nu} d_i \alpha_{i-nw}(\xi, \eta) e^{j\omega\tau} \tag{16.9b}$$

$$v(\xi, \eta, \tau) = \sum_{i=nw+nu+1}^{nw+nu+nv} d_i \beta_{i-nw-nu}(\xi, \eta) e^{j\omega\tau} \tag{16.9c}$$

Substituting the expressions from Eqs. (16.2), (16.3) and (16.4) along Eqs. (16.9) into Eq. (16.8), the set of the governing equations are found as,

$$-\omega^2[M]\{d\} + [K]\{d\} = 0 \tag{16.10}$$

where, $[K]$ and $[M]$ indicate stiffness matrix and mass matrix, respectively, and which are now completely known. $[K]$ is already solved by static analysis and $[M]$ is known from the problem definition. So, the unknown frequency parameter (ω) is solved through an Eigen value analysis, whereas solution of the unknown coefficients is associated with Eigen vectors.

16.3 Results and Discussions

In the present chapter, nonlinear structural behaviour of non-uniform AFG plate on elastic foundation is carried out considering clamped (CCCC) and simply supported (SSSS) end condition and pressure type of loading. The geometric dimensions of the plate are taken as, $a = b = 0.4$ m, $t_0 = 0.0025$ m and $\alpha = 0.2$. The material properties value are considered as, $E_0 = 210$ Gpa, $\rho_0 = 7850$ Kg/m^3 and $\mu = 0.3$, which resemble the material properties of mild steel. The non-dimensional elastic foundation stiffness $\left[K_f = a^4 k_f / D\right]$ values for each result are taken as, $K_f = 0, 10^2, 10^3$ and 10^4, respectively. Here, k_f is dimensional value of stiffness and $D = E_0 t_0^3 / 12(1 - \mu^2)$. The number of Gauss points (ngp) and number of orthogonal functions $(nw = nu = nv)$ are selected as 24 and 5, respectively. These values are selected by performing appropriate convergence studies. As a result, the number of computational points in the domain becomes 24 × 24 and total number of function for displacement become 5 × 5. The start functions due to transverse and in-plane displacement are shown in Table 16.1. The present study results are generated when dimensionless maximum amplitude (w_{max}/t_0) is less than 2.0 to minimize the computation time.

The present methodology and solution procedure is compared with the results of previously published articles of Timoshenko and Woinowsky-Krieger [32], Leissa [33] and Saha et al. [34] for fully homogeneous and uniform plate without considering foundation support. The comparison of linear dimensionless natural frequencies for first six modes are tabulated in Table 16.2 for different end conditions. In Fig. 16.3a, the comparison of the load versus deflection is shown for CCCC plate and in Fig. 16.3b, the comparison of the first backbone curve is shown for the same plate. From the above comparison, it can be conclusive that the current results satisfy the established results.

In Fig. 16.4 the load versus deflection curves are shown in dimensionless load and deflection plane for CCCC and SSSS end conditions, respectively. The effects of the foundation stiffness on the deflection amplitude of the system can be found through this plot. Four curves are considered corresponding to different spring stiffness values (K_f = 0, 10^2, 10^3 and 10^4). It can be observed that with the increase of the foundation stiffness value the slope of the curve is decreasing in nature. Such type of results is found due to the reason that increase of stiffness makes the system more stiff to deflection under the application of same load. As a usual consequence of the increase in stiffness the load-bearing capacity of the system also increases. It is also observed that for the same load application the deflection of the SSSS plate is more than the CCCC plate. This type of occurrence happens due to the reason that the CCCC plate is stiffer to deflect than SSSS plate.

Figure 16.5 represents the deflected shape of the plates centreline along axial (ξ) direction for four foundation stiffness value. These plots are obtained taking particular external load value of 100 KN/m^2 and 15 KN/m^2 for CCCC and SSSS plate, respectively, to ensure best fit. From the plot, it is seen that deflection is maximum near the middle of the plate. It is also observed that at zero value of stiffness the

Table 16.2 Comparison of first six linear non-dimensional natural frequency square plate

Boundary condition	Literature	Linear dimensionless frequencies					
		ω_1	ω_2	ω_3	ω_4	ω_5	ω_6
CCCC	Leissa [33]	35.992	73.407	73.407	108.261	131.599	132.195
	Saha et al. [34]	35.992	73.407	73.407	108.235	131.755	132.402
	Present study	35.865	73.192	73.192	107.942	131.371	132.002
SSSS	Leissa [33]	19.743	49.344	49.344	78.944	98.687	98.687
	Saha et al. [34]	19.743	49.369	49.369	78.996	99.282	99.282
	Present study	19.712	49.317	49.317	78.873	99.103	99.103

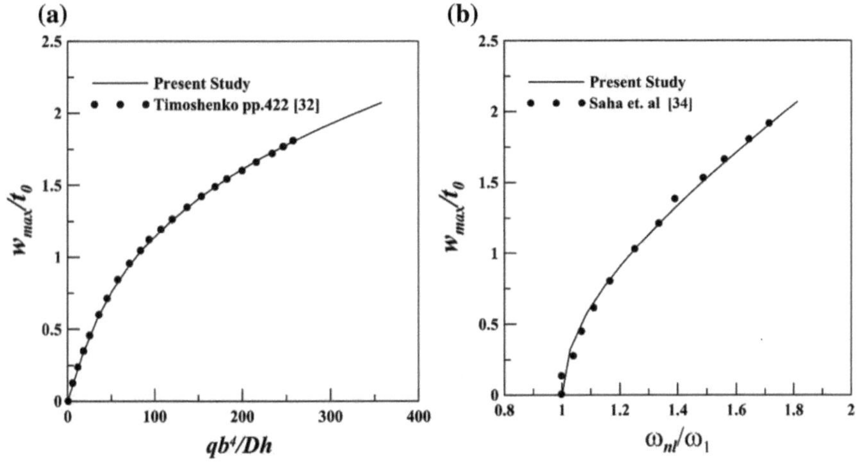

Fig. 16.3 **a** Comparison plot of load versus deflection curve. **b** Comparison plot of backbone curve

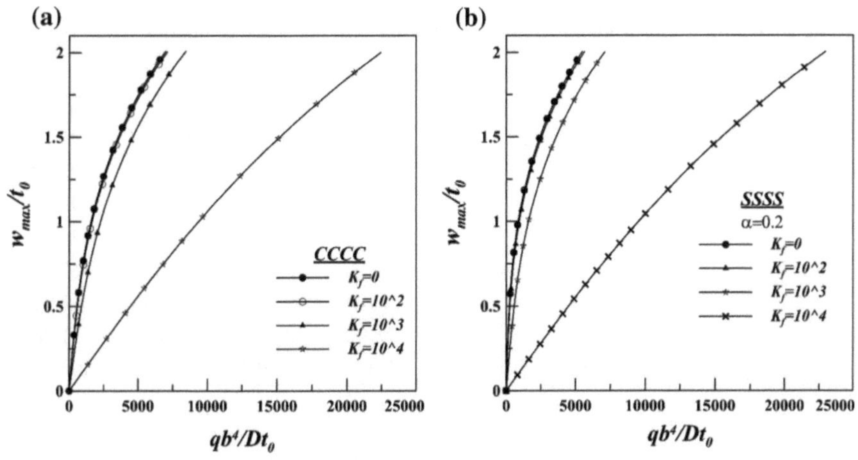

Fig. 16.4 Load versus deflection plot for CCCC and SSSS boundary conditions, respectively

deflection is maximum, and it is minimum when stiffness is maximum. Again it is also clear from the plots that SSSS plate deflect more than the CCCC plate under same load application.

Linear dimensionless frequencies are tabulated in Table 16.3 to show the effects of the end conditions and the foundation stiffness values on natural frequencies. From the table, it is clear that in each case higher the stiffness value of the foundation increasing the natural frequency and the reason behind this occurrence is due to the fact that the system becomes more rigid for stiffer foundation. It is also observed that due to rigidity of the plate edge, boundary condition also affects the

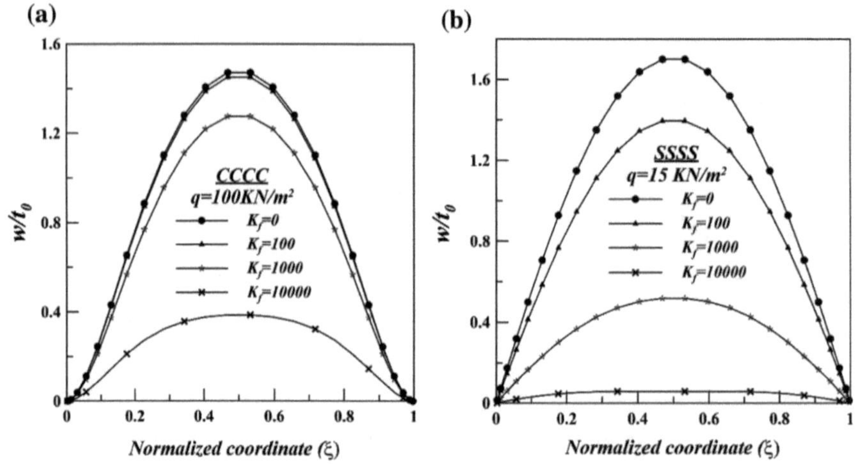

Fig. 16.5 Deflected shape of the CCCC and SSSS plate in centreline along axial direction for different foundation stiffness

Table 16.3 Linear dimensionless frequencies for different boundary conditions

Boundary condition	Stiffness of foundation	Linear dimensionless frequencies			
		ω_1	ω_2	ω_3	ω_4
CCCC	0	30.365	61.904	62.427	92.153
	100	31.326	62.398	62.889	92.489
	1000	38.895	66.692	66.89	95.4594
	10,000	81.535	97.779	100.337	121.376
SSSS	0	16.732	41.854	42.136	67.449
	100	18.419	42.541	42.87	67.912
	1000	29.452	48.252	49.016	71.961
	10,000	76.002	85.392	90.661	104.499

natural frequency and CCCC plate will obtain highest frequency value for a constant value of foundation stiffness.

The dynamic response of the system is shown through backbone curves in dimensionless frequency amplitude plane. In Fig. 16.6, backbone curves are plotted to show the effect of the foundation stiffness value in separate two diagrams for CCCC and SSSS plate, respectively. It is cleared from the figure that slope of the curves are incremental with foundation stiffness value and hence, proving the system is getting stiffer to vibrate/deflect.

The effect of the end conditions of the plates on backbone curve is shown in Fig. 16.7. Two separate plots are there to accommodate first four backbone curves at each foundation stiffness value to analyse the effect of the end conditions of the plates. Hence, for four foundation stiffness value, i.e. 0, 10^2, 10^3 and $10^{4,}$

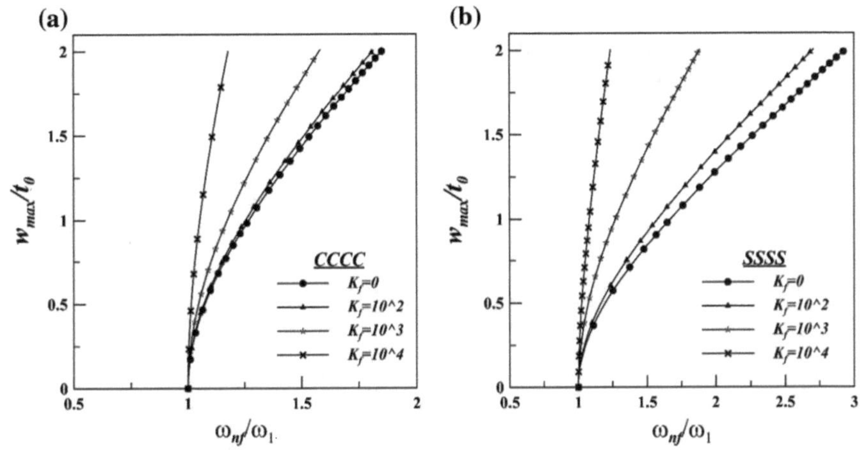

Fig. 16.6 Effect of foundation stiffness on backbone curve 1 for CCCC and SSSS boundary conditions, respectively

respectively, eight plot will be obtained. Here also from the figures, it is also observed that for a particular value of foundation stiffness, the backbone curves obtained in case of SSSS end condition have lesser slope. This type of phenomenon happens due to the reason that the rigidity of the plate edges conditions significantly affect the free vibration response.

In Fig. 16.8, through the representation of backbone curve, the effect of the taper parameter and the material gradation are shown. These plots are obtained in case of CCCC plate taking foundation stiffness, $K_f = 10^3$. It is clear from the figure that with the introduction of taper parameter ($\alpha = 0.2$) and material gradation the slope of the backbone curves will be less than from the ideal case (uniform and homogeneous material). This type of phenomenon will happen due to the reason that in the present system the thickness is gradually decreasing $\left[t(\xi) = t_0\left(1 - \alpha\xi^2\right)\right]$ and material properties are gradually increasing $[E(\xi) = E_0(1+\xi)$ and $\rho(\xi) = \rho_0\left(1+\xi+\xi^2\right)]$ in nature. So at the root side of the plate thickness is maximum, but material property values are minimum, and these values are considered in the ideal case. So the softening effect will be predominate and that's why decreasing slope of the backbone curve will be observed in the present scenario.

Dynamic behaviour of FGM plates on elastic foundation are presented more accurately through the representation of mode shape plots. Both linear $[(w_{max}/t_0) = 2.0]$ and nonlinear $[(w_{max}/t_0) = 2.0]$ mode shapes for the first six modes for CCCC and SSSS plates are highlighted in Figs. 16.9 and 16.10, respectively, at $K_f = 1000$. There are observable changes between the two sets (linear and nonlinear) of mode shapes, which points towards the effect of nonlinearity. For higher order mode shape, these effects appear to be more prominent.

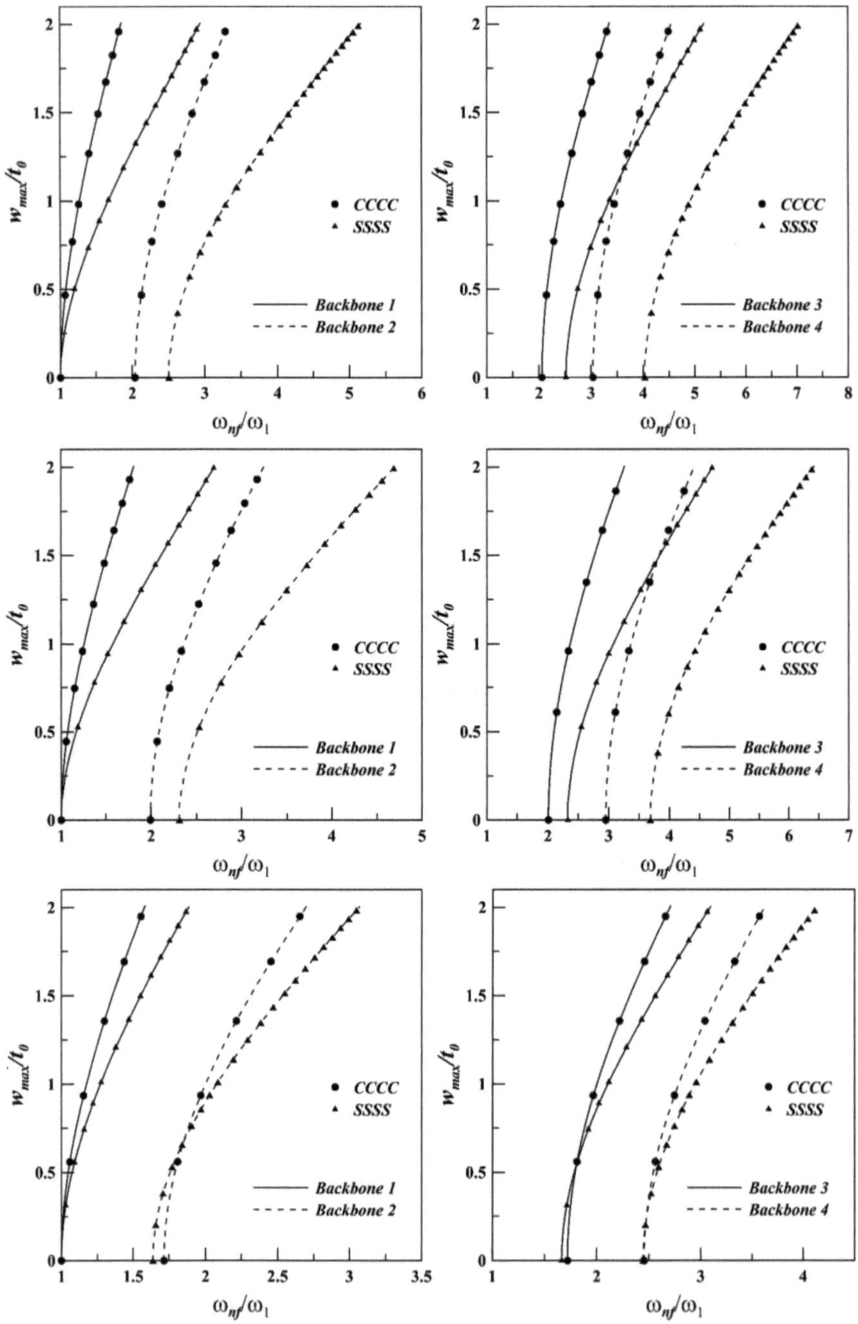

Fig. 16.7 Backbone curves for different boundary conditions at $K_f = 0$, $K_f = 10^2$, $K_f = 10^3$ and $K_f = 10^4$ respectively

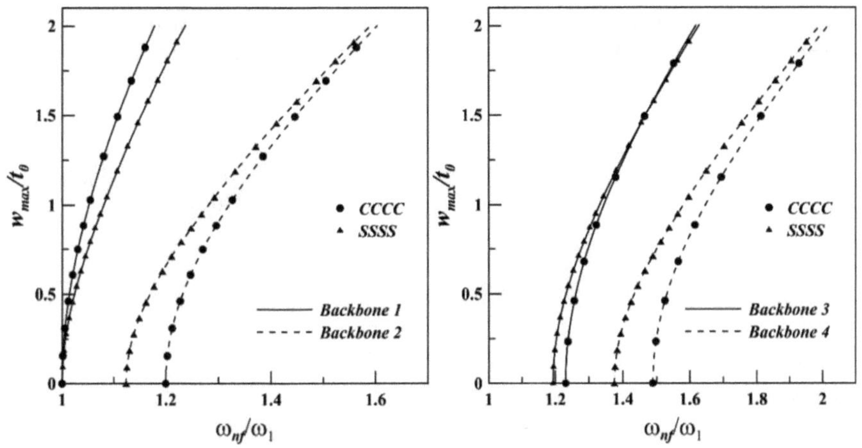

Fig. 16.7 (continued)

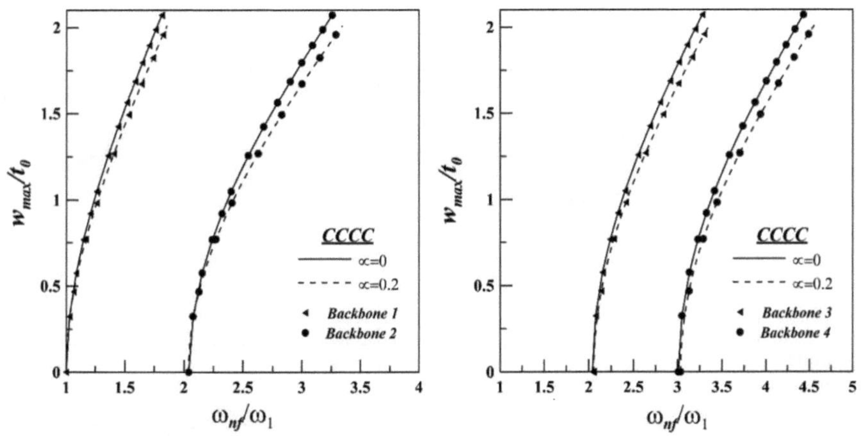

Fig. 16.8 Effect of the tapering and material gradation at $K_f = 10^3$

Mode shape	**Linear Mode shape** $(w_{max}/t_0) = 0$	**Non Linear Mode Shape** $(w_{max}/t_0) = 2.0$
Mode shape 1	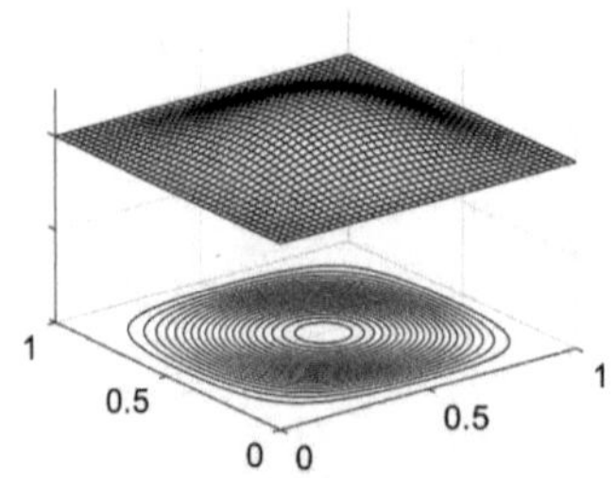	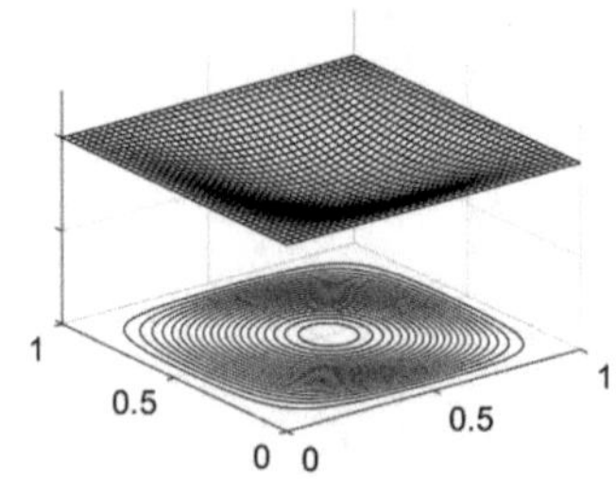
Mode shape 2	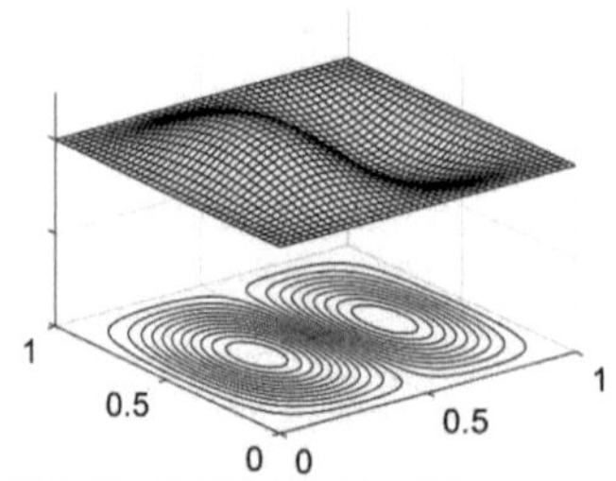	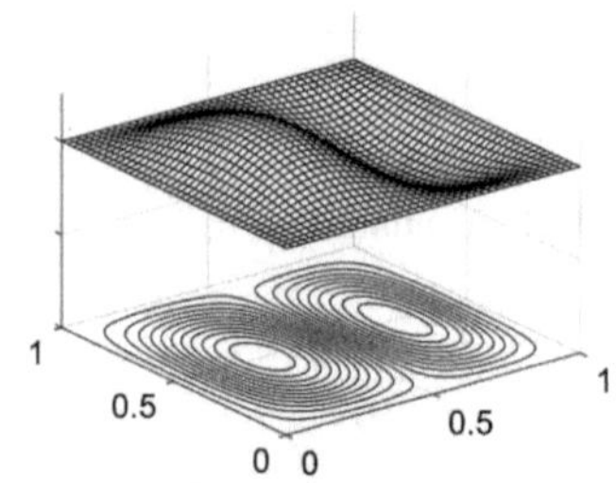
Mode shape 3	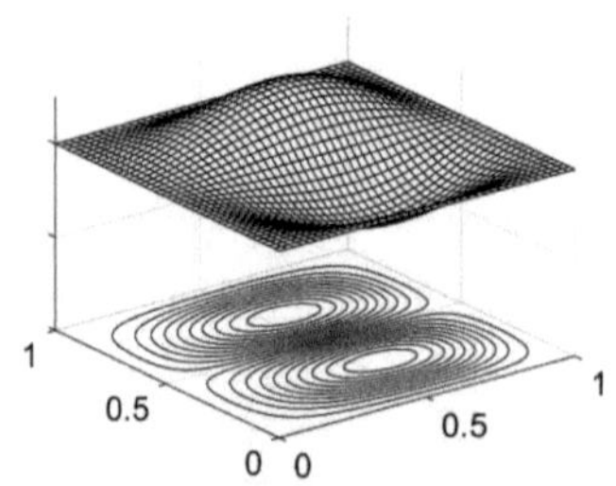	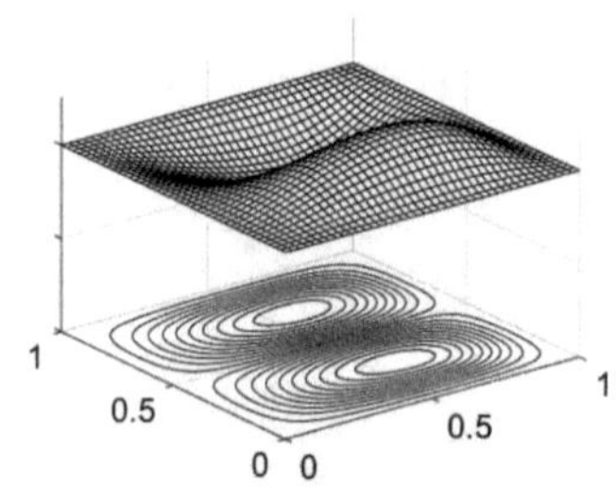
Mode shape 4	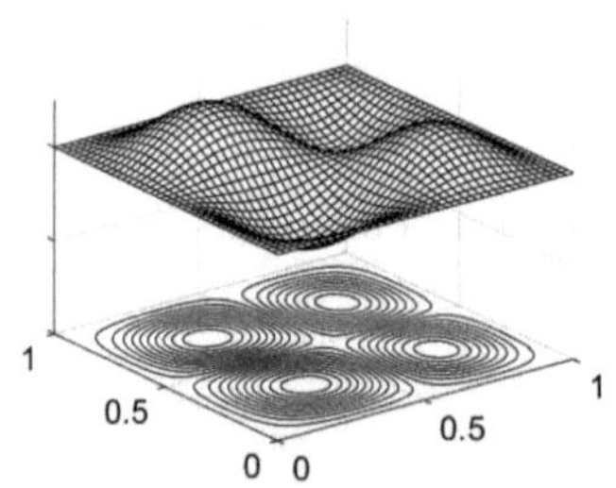	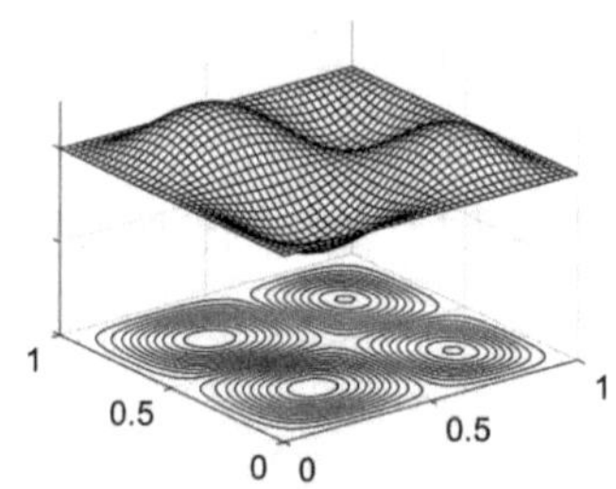

Fig. 16.9 Mode shape for CCCC boundary condition at $K_f = 10^3$

Mode shape 5	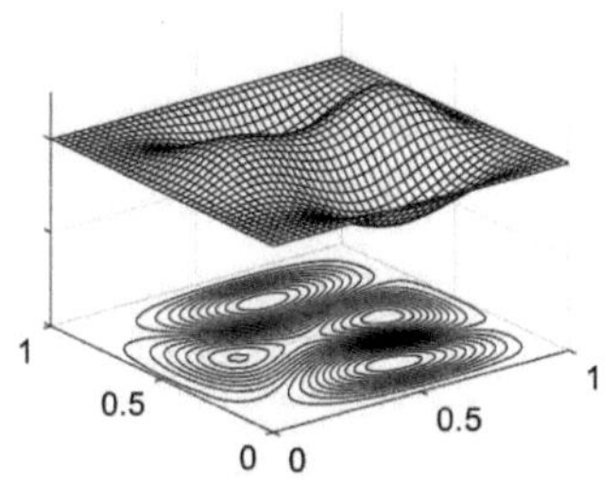	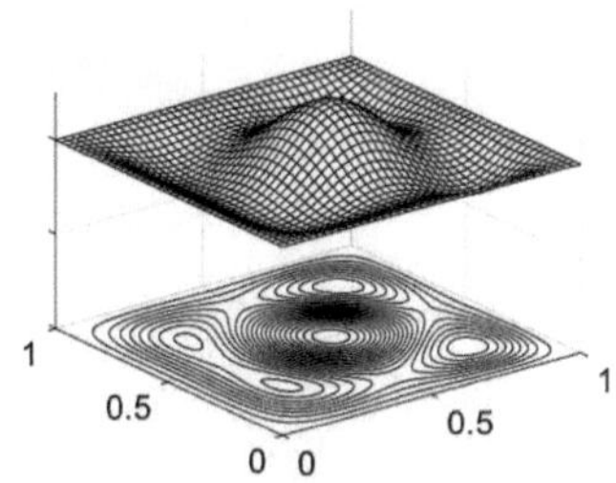
Mode shape 6	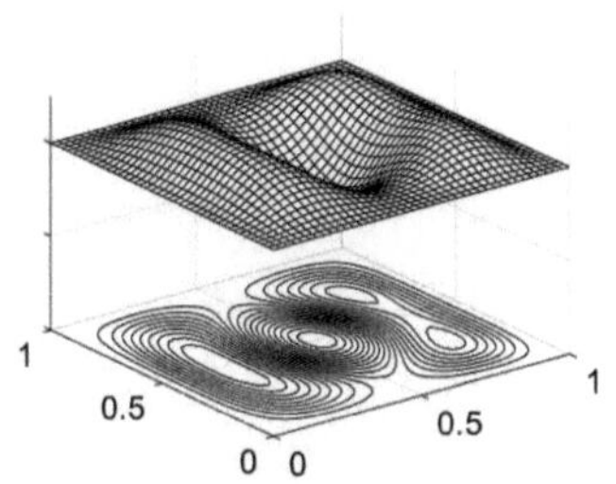	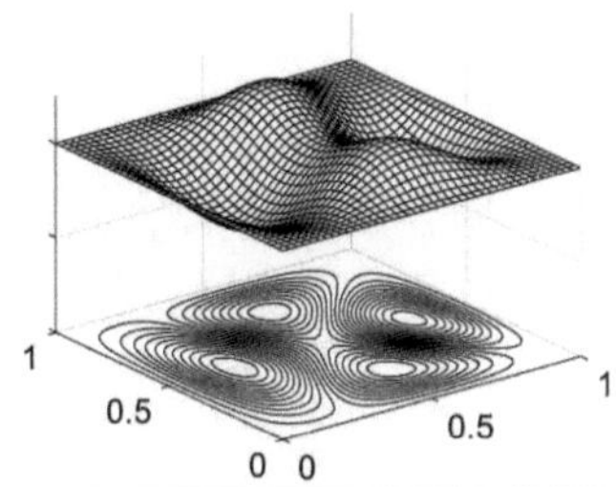

Fig. 16.9 (continued)

Mode shape	**Linear Mode shape** $(w_{max}/t_0) = 0$	**Non Linear Mode Shape** $(w_{max}/t_0) = 2.0$
Mode shape 1	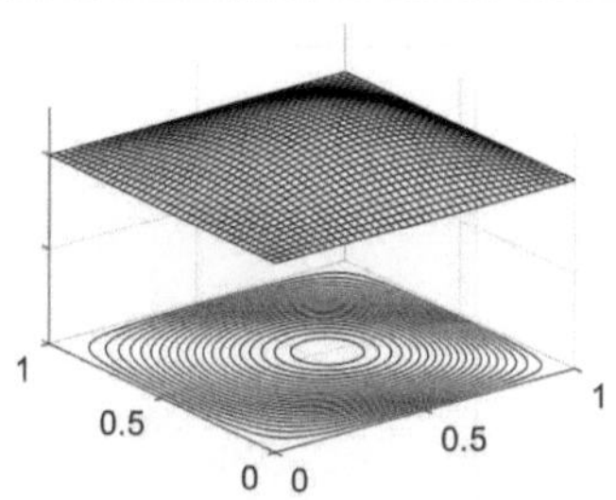	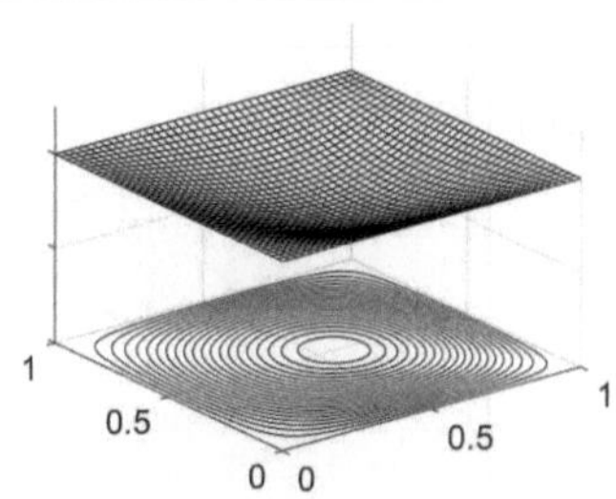
Mode shape 2	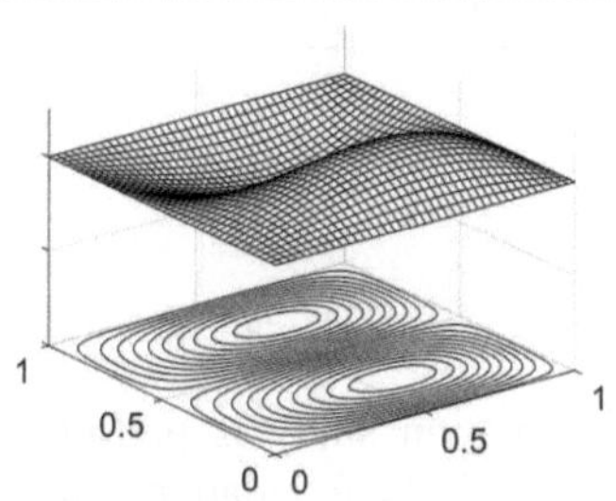	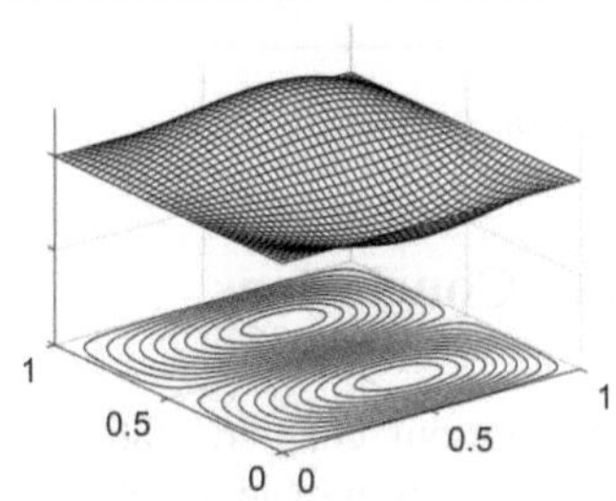

Fig. 16.10 Mode shape for SSSS boundary condition at $K_f = 10^3$

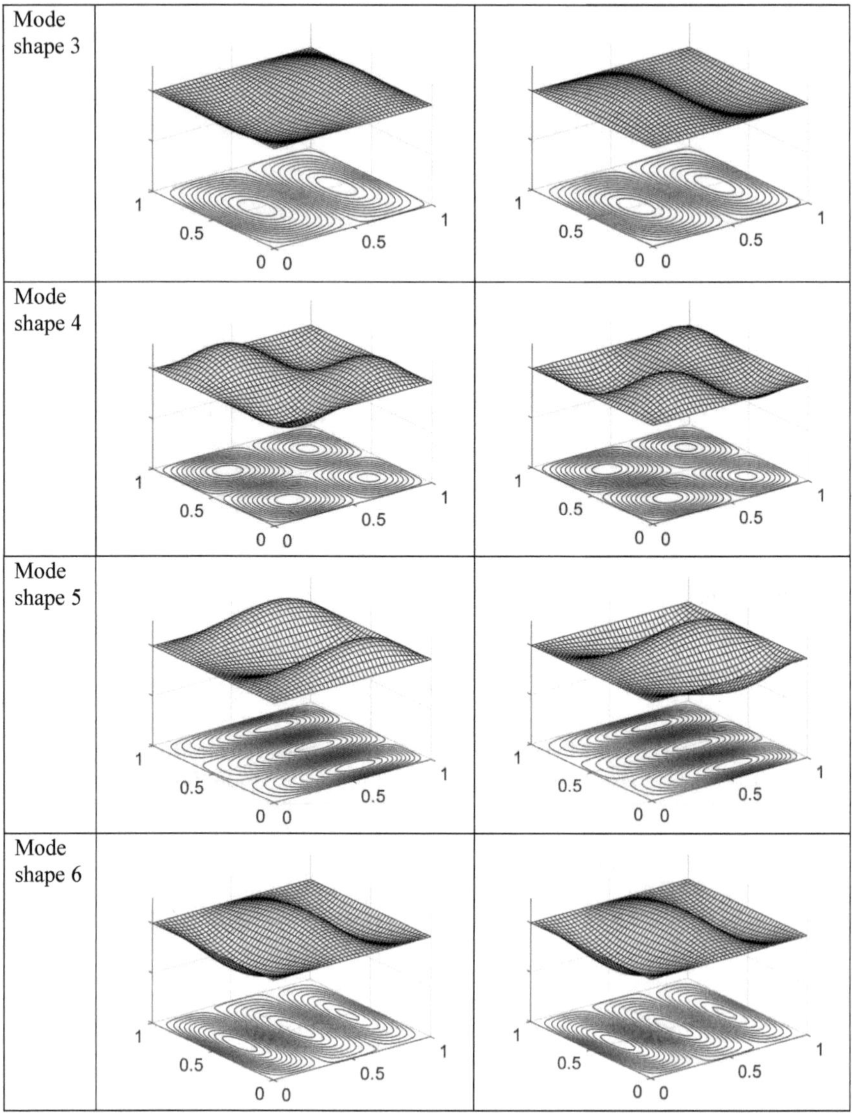

Fig. 16.10 (continued)

16.4 Conclusions

In the present chapter, static and free vibration study is performed on a AFG non-uniform thin plate which is supported on elastic foundation of different stiffness with CCCC and SSSS end conditions, respectively. Minimum potential energy principle is utilized for static case, whereas in case of free vibration Hamilton's principle is used to obtain the governing sets of equations. An iterative method (direct substitution with relaxation) is numerically implemented to solve of the nonlinear equations. The methodology is flexible so that other type of loading pattern, gradation parameter, taper pattern and foundation behaviour can be incorporated. The generated results are successfully validated with the previously published results. The effect of the foundation is found out through load versus amplitude plot and deflected shape plot in static analysis, whereas, in free vibration analysis backbone curves for five foundation stiffness values are furnished.

References

1. Jha, D.K., Kant, T., Singh, R.K.: A critical review of recent research on functionally graded plates. Compos. Struct. **96**, 833–849 (2013)
2. Birman, V., Byrd, L.W.: Modeling and analysis of functionally graded materials and structures. Appl. Mech. Rev. **60**, 195–216 (2007)
3. Kennedy, D., Cheng, R.K.H., Wei, S., AlcazarArevalo, F.J.: Equivalent layered models for functionally graded plates. Comput. Struct. **174**, 113–121 (2016)
4. Singha, M.K., Prakash, T., Ganapathi, M.: Finite element analysis of functionally graded plates under transverse load. Finite Elem. Anal. Des. **47**, 453–460 (2011)
5. Panyatong, M., Chinnaboon, B., Chucheepsakul, S.: Free vibration analysis of FG nanoplates embedded in elastic medium based on second-order shear deformation plate theory and nonlocal elasticity. Compos. Struct. **153**, 428–441 (2016)
6. Benferhat, R., Daouadji, T.H., Mansour, M.S.: Free vibration analysis of FG plates resting on an elastic foundation and based on the neutral surface concept using higher-order shear deformation theory. C. R. Mecanique **344**, 631–641 (2016)
7. Sharma, P., Parashar, S.K.: Free vibration analysis of shear-induced flexural vibration of FGPM annular plate using generalized differential quadrature method. Compos. Struct. **155**, 213–222 (2016)
8. Abrate, S.: Functionally graded plates behave like homogeneous plates. Compos. Part B **39**, 151–158 (2008)
9. Mohammadzadeh-Keleshteri, M., Asadi, H., Aghdam, M.M.: Geometrical nonlinear free vibration responses of FG-CNT reinforced composite annular sector plates integrated with piezoelectric layers. Compos. Struct. **171**, 100–112 (2017)
10. Chi, S.H., Chung, Y.L.: Mechanical behavior of functionally graded material plates under transverse load—part I: analysis. Int. J. Solids Struct. **43**, 3657–3674 (2006)
11. Akbarzadeh, A.H., Hosseinizad, S.K., Eslami, M.R., Sadighi, M.: Mechanical behaviour of functionally graded plates under static and dynamic loading. Proc. Inst. Mech. Eng. Part C. J. Mech. Eng. Sci. **203–210**, 1989–1996 (2011)
12. Uymaz, B., Aydogdu, M., Filiz, S.: Vibration analyses of FGM plates with in-plane material inhomogeneity by Ritz method. Compos. Struct. **94**, 1398–1405 (2012)
13. Xiang, T., Natarajan, S., Man, H., Song, C., Gao, W.: Free vibration and mechanical buckling of plates with in-plane material inhomogeneity—A three dimensional consistent approach. Compos. Struct. **118**, 634–642 (2014)
14. Liu, D.Y., Wang, C.Y., Chen, W.Q.: Free vibration of FGM plates with in-plane material inhomogeneity. Compos. Struct. **92**, 1047–1051 (2010)
15. Hussein, O.S., Mulani, S.B.: Optimization of in-plane functionally graded panels for buckling strength: Unstiffened, stiffened panels, and panels with cutouts. Thin-Walled Struct. **122**, 173–181 (2018)

16. Kumar, S., Mitra, A., Roy, H.: Forced vibration response of axially functionally graded non-uniform plates considering geometric nonlinearity. Int. J. Mech. Sci. **128–129**, 194–205 (2017)
17. Kumar, S., Mitra, A., Roy, H.: Forced vibration analysis of functionally graded plates with geometric nonlinearity. In: Proceedings of ASME 2015 Gas Turbine India Conference - GTINDIA2015, Hyderabad, India, pp. 1–8 (2015)
18. Kumar, S., Mitra, A., Roy, H.: Large amplitude free vibration analysis of axially functionally graded plates. In: Proceedings of ASME 2014 Gas Turbine India Conference - GTINDIA2014, New Delhi, India, pp. 1–8 (2014)
19. Taczała, M., Buczkowski, R., Kleiber, M.: Postbuckling analysis of functionally graded plates on an elastic foundation. Compos. Struct. **132**, 842–847 (2015)
20. Ebrahimi, F., Jafari, A., Barati, M.R.: Vibration analysis of magneto-electro-elastic heterogeneous porous material plates resting on elastic foundations. Thin-Walled Struct. **119**, 33–46 (2017)
21. Kutlu, A., Ugurlu, B., Omurtag, M.H.: A combined boundary-finite element procedure for dynamic analysis of plates with fluid and foundation interaction considering free surface effect. Ocean Eng. **145**, 34–43 (2017)
22. Gupta, A., Talha, M., Chaudhari, V.K.: Natural frequency of functionally graded plates resting on elastic foundation using finite element method. Procedia Technol. **23**, 163–170 (2016)
23. Mohammadzadeh, B., Noh, H.C.: Analytical method to investigate nonlinear dynamic responses of sandwich plates with FGM faces resting on elastic foundation considering blast loads. Compos. Struct. **174**, 142–157 (2017)
24. Zamani, H.A., Aghdam, M.M., Sadighi, M.: Free vibration analysis of thick viscoelastic composite plates on visco-Pasternak foundation using higher-order theory. Compos. Struct. **182**, 25–35 (2017)
25. Najafi, F., Shojaeefard, M.H., Googarchin, H.S.: Nonlinear low-velocity impact response of functionally graded plate with nonlinear three-parameter elastic foundation in thermal field. Compos. B **107**, 123–140 (2016)
26. Wattanasakulpong, N., Chaikittiratana, A.: Exact solutions for static and dynamic analyses of carbon nanotube-reinforced composite plates with Pasternak elastic foundation. Appl. Math. Model. **39**, 5459–5472 (2015)
27. Barati, M.R., Sadr, M.H., Zenkour, A.M.: Buckling analysis of higher order graded smart piezoelectric plates with porosities resting on elastic foundation. Int. J. Mech. Sci. **117**, 309–320 (2016)
28. Shahsavari, D., Shahsavari, M., Li, L., Karami, B.: A novel quasi-3D hyperbolic theory for free vibration of FG plates with porosities resting on Winkler/Pasternak/Kerr foundation. Aerosp. Sci. Technol. **72**, 134–149 (2018)
29. Reddy, J.N.: Energy Principles & Variational Methods in Applied Mechanics, 2nd edn. Wiley, New Jersey, USA (2002)
30. Lohar, H., Mitra, A., Sahoo, S.: Geometric nonlinear free vibration of axially functionally graded non-uniform beams supported on elastic foundation. Curved Layer. Struct. **3**, 223–239 (2016)
31. Mitra, A., Sahoo, P., Saha, K.N.: Nonlinear vibration analysis of simply supported stiffened plate by a variational method. Mech. Adv. Mater. Struct. **20**, 373–396 (2013)
32. Timoshenko, S., Woinowsky-Krieger, S.: Theory of Plates and Shells, 2nd edn. McGraw-Hill Classic Textbook Reissue Series, New York, USA (1964)
33. Leissa, A.W.: The free vibration of rectangular plates. J. Sound Vib. **3**, 257–293 (1973)
34. Saha, K.N., Misra, D., Pohit, G., Ghosal, S.: Large amplitude free vibration study of square plates under different boundary conditions through a static analysis. J. Sound Vib. **10**, 1009–1028 (2004)

Chapter 17
Free Vibration of Rotating Twisted Composite Stiffened Plate

Mrutyunjay Rout and Amit Karmakar

Nomenclature

L	Length of the plate (m)
b	Width of the plate (m)
h	Thickness of the plate (m)
b_{st}	Width of stiffener (m)
d_{st}	Depth of stiffener (m)
ϕ	Pretwist angle of the plate (deg.)
ω_n	Fundamental natural frequency of the stiffened plate without rotation (rad/s)
Ω'	Actual rotational speed (rad/s)
Ω	Non-dimensional rotational speed ($\Omega = \Omega'/\omega_n$) (dimensionless)

17.1 Introduction

Earlier, tourbomachinery blades are assumed as twisted beam wherein chordwise bending is found missing particularly in moderate-to-low aspect ratio models and the analysis becomes inaccurate. Thereafter, these blades are modelled as twisted cantilever composite plates. In general, these blades fail due to flutter, which in turn induces high value of repeated stresses. In order to have a safety of operation, these thin plates are very often attached with rib-like structures called stiffeners at suitable orientation to increase the overall stiffness, thereby increasing the fundamental frequency. The high-speed rotation of the turbomachinery blades is very much related to their frequencies and often prone to failure due to the centrifugal force. The deformation of the geometry due to centrifugal force is represented by geometric stiffness called centrifugal stiffening, which is the source of initial stresses.

M. Rout (✉) · A. Karmakar
Department of Mechanical Engineering, Jadavpur University, Kolkata, India
e-mail: kulu2670371@gmail.com

P. Sahoo and J. P. Davim (eds.), *Advances in Materials, Mechanical and Industrial Engineering*, Lecture Notes on Multidisciplinary Industrial Engineering,
https://doi.org/10.1007/978-3-319-96968-8_17

Hence, an intensive study of the stiffened blades considering the effects of different parameters will be extremely useful for design engineers.

Leissa and Ewing [1] presented the free vibration results of turbomachinery blades considering both beam and shallow shell theories and finally reported that beam theory was found inadequate for moderate-to-low aspect ratio blades because chordwise bending was found missing. Kielb et al. [2] carried out the theoretical and numerical analysis of twisted cantilever plates wherein finite elements' results were obtained considering plate, shell and solid elements. In theoretical method, they considered both shell and beam theory to carry out the investigation. Qatu and Leissa [3] were the first investigators to report the effects of twist angles of the composite plates on the natural frequencies and mode shapes. Liew et al. [4] used Ritz method to study the effects of twist angle on the vibration response and mode shapes of composite conical shells. Kuang and Hsu [5] studied the free vibration results of tapered pretwisted orthotropic composite plate employing differential quadrature method (DQM). Lee et al. [6] investigated the vibration response of composite twisted plates, cylindrical and conical shells with cantilevered boundary conditions employing finite element method (FEM). Sreenivasamurthy and Ramamurti [7] studied the effects of Coriolis component on the natural frequencies of the plates, rotating at different speeds, and reported that its effect is found marginal at low and moderate speed of rotation. Ramamurti and Kielb [8] reported the eigenfrequencies of twisted rotating isotropic plate wherein the effects of rotation were included by using a stress smoothing technique. Karmakar and Sinha [9] worked to investigate the free vibration response of laminated twisted plate at moderate speed neglecting Coriolis effect. They used three-dimensional finite element method to perform the parametric studies. Rao and Gupta [10] presented the fundamental frequencies of twisted and tapered Timoshenko beams rotating at different speeds. Hu et al. [11] computed the free vibration response of cantilevered twisted conical shell subjected to axial and centrifugal force by Rayleigh–Ritz method, while Kee and Kim [12] worked on the free vibration response of rotating twisted thick cylindrical shell structures employing FEM wherein Coriolis acceleration was considered with centrifugal forces. Joseph and Mohanty [13] reported a finite element numerical method to compute the natural frequency of the rotating sandwich plates.

The remarkable researchers [14–17] investigated the free vibration characteristics of stationary untwisted laminated eccentrically and concentric stiffened plates/ shells using FEM based on the kinematics of FSDT. Sadek and Tawafik [18] used higher-order shear deformation theory (HSDT), which eliminates the use of shear correction factor to study the flexural response of the stiffened plates. Qing et al. [19] worked on the free vibration of stiffened plates based on the theory of state-vector equation, wherein the compatibility of stresses and displacement at the interface of stiffener and plate were satisfied. Yuan and Dawe [20] studied the stability and vibration characteristics of sandwich plates stiffened eccentrically employing spline finite strip method, while Guo et al. [21] presented a finite element model based on zigzag theory of the laminated stiffened composite plates, wherein the inter-layer continuity was maintained by employing bilinear in-plane displacement constraints. Li et al. [22] united two theories such as layerwise laminated theory and traditional finite element theory to investigate the static and free vibration response of laminated stiffened cylindrical shells, while Bhar et al. [23] presented a comparative study of laminated stiffened plates employing both the kinematics of FSDT and HSDT. Zaho and Kapania [24] presented an efficient finite element method to investigate the vibration response of laminated composite stiffened panel, wherein curved composite stiffeners were considered. Castro and Donadon [25] developed a semi-analytical model to investigate the buckling and vibration response of laminated composite stiffened panel with inclusion of the debonding problem. Rout et al. [26, 27] presented the free vibration response of rotating stiffened cylindrical shells with preexisting delamination, while Damnjanović et al. [28] used the dynamic stiffness method to investigate the free vibration responses of composite stiffened and cracked plate assemblies based on the HSDT and FSDT theories.

Though plenty of literature is available in the theme of free vibration of stiffened plates, the investigators have not studied the effects of rotational speed on the free vibration characteristics of the initially twisted laminated composite eccentrically stiffened plate. Hence, the present work aims at studying the vibration response of rotating twisted composite stiffened plate employing finite element method, wherein the composite plate is modelled with an eight-noded isoparametric element and the stiffener is with a three-noded isoparametric beam element. The compatibility between the plate element and the stiffener element is established by considering the eccentricity of the stiffener element. The degrees of freedom of the stiffener element at each node are transferred to the corresponding nodes of the plate element. The initial stiffening due to rotation is manifesting itself through geometric stiffness, considering Green–Lagrangian strain components for initial stresses. The effect of fibre orientation angle, twist angles, stiffener depth-to-plate thickness ratio and rotational speeds on the free vibration response of the stiffened plate is furnished. Finally, the influence of twist angles and moderate speed of rotation on the mode shapes of the eccentrically stiffened composite plate is presented.

17.2 Theoretical Formulation

The laminated composite crossed-stiffened plate of uniform thickness h, curvature of twist R_{xy}, length L is shown in Fig. 17.1 with the global coordinate system. The twist angle of the stiffened plate is expressed as,

$$\tan\phi = -\frac{L}{R_{xy}} \tag{17.1}$$

The isoparametric eight-noded plate element consisting of five degrees of freedom including three translations (u, v, w) and two rotations (α, β) per node is

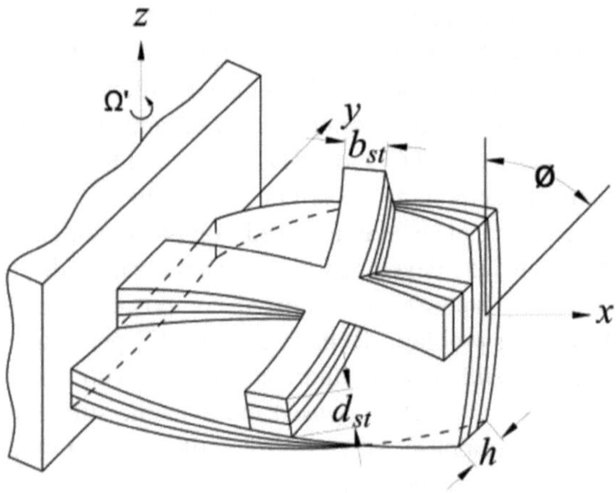

Fig. 17.1 Typical twisted laminated composite stiffened plate

considered for discretization of the entire plate. The generalized strain vector of the plate based on the kinematics of FSDT is expressed as,

$$\begin{aligned}\left[\varepsilon_x \quad \varepsilon_y \quad \gamma_{xy} \quad \gamma_{xz} \quad \gamma_{yz}\right]^T = \left[\varepsilon_x^0 \quad \varepsilon_y^0 \quad \gamma_{xy}^0 \quad \gamma_{xz}^0 \quad \gamma_{yz}^0\right]^T \\ + z\left[\kappa_x \quad \kappa_y \quad \kappa_{xy} \quad \kappa_{xz} \quad \kappa_{yz}\right]^T\end{aligned} \tag{17.2}$$

where $\varepsilon_x^0, \varepsilon_y^0, \gamma_{xy}^0, \gamma_{xz}^0, \gamma_{yz}^0$ are the mid-surface strains and $\kappa_x, \kappa_y, \kappa_{xy}, \kappa_{xz}, \kappa_{yz}$ corresponds to the curvatures of the plate.

$$\begin{Bmatrix} \varepsilon_x^0 \\ \varepsilon_y^0 \\ \gamma_{xy}^0 \\ \gamma_{xz}^0 \\ \gamma_{yz}^0 \end{Bmatrix} = \begin{Bmatrix} \partial u/\partial x \\ \partial v/\partial y \\ \partial u/\partial x + \partial v/\partial x + 2w/R_{xy} \\ \alpha + \partial w/\partial x \\ \beta + \partial w/\partial y \end{Bmatrix} \tag{17.3}$$

and

$$\begin{Bmatrix} \kappa_x \\ \kappa_y \\ \kappa_{xy} \\ \kappa_{xz} \\ \kappa_{yz} \end{Bmatrix} = \begin{Bmatrix} \partial \alpha/\partial x \\ \partial \beta/\partial y \\ \partial \alpha/\partial y + \partial \beta/\partial x \\ 0 \\ 0 \end{Bmatrix} \tag{17.4}$$

The mid-plane strain field and curvatures can be written as,

$$\{\bar{\varepsilon}\} = [B]\{d_e\} \tag{17.5}$$

where

$$\{d_e\} = [\,u_1 \quad v_1 \quad w_1 \quad \alpha_1 \quad \beta_1 \quad \cdots \quad u_8 \quad v_8 \quad w_8 \quad \alpha_8 \quad \beta_8\,]^T \tag{17.6}$$

$$[B] = \begin{bmatrix} N_{i,x} & 0 & 0 & 0 & 0 & \cdots \\ 0 & N_{i,y} & 0 & 0 & 0 & \cdots \\ N_{i,y} & N_{i,x} & 2N_i/R_{xy} & 0 & 0 & \cdots \\ 0 & 0 & 0 & N_{i,x} & 0 & \cdots \\ 0 & 0 & 0 & 0 & N_{i,y} & \cdots \\ 0 & 0 & 0 & N_{i,y} & N_{i,x} & \cdots \\ 0 & 0 & N_{i,x} & N_i & 0 & \cdots \\ 0 & 0 & N_{i,y} & 0 & N_i & \cdots \end{bmatrix}_{i=1,8} \tag{17.7}$$

The constitutive relation of the plate can be expressed as,

$$\begin{Bmatrix} N \\ M \\ Q \end{Bmatrix} = \begin{bmatrix} A_{ij} & B_{ij} & 0 \\ B_{ij} & D_{ij} & 0 \\ 0 & 0 & S_{ij} \end{bmatrix} \begin{Bmatrix} \varepsilon_p \\ \kappa_b \\ \gamma_s \end{Bmatrix} = [D]\{\bar{\varepsilon}\} \tag{17.8}$$

where A_{ij}, B_{ij}, D_{ij} and S_{ij} are the stiffness coefficients, while ε_p, κ_b and γ_s represent the in-plane strains, the curvatures and shear strains in transverse directions, respectively.

Based on the finite element method, the standard formulations used to compute the element stiffness and mass matrices are expressed as,

$$[K_p] = \int_{-1}^{1}\int_{-1}^{1} [B]^T[D][B]|J|\,\mathrm{d}\xi\mathrm{d}\eta \tag{17.9}$$

$$[M_p] = \int_{-1}^{1}\int_{-1}^{1} [N]^T[m][N]|J|\,\mathrm{d}\xi\mathrm{d}\eta \tag{17.10}$$

where $[B]$ is the strain–displacement matrix, $[D]$ is the elasticity matrix, $[N]$ is the shape function matrix and $[m]$ is the inertia matrix per unit area of the plate element, respectively.

The stiffener element is considered as a one-dimensional three-noded isoparametric beam element consisting of four degrees of freedom per node, which includes two translations and two rotations. The shape functions considered for the stiffener element are given by,

$$
\begin{aligned}
N_i^{sx} &= 0.5\xi\xi_i(1+\xi\xi_i) \quad \text{for } i = 1,3 \\
N_i^{sx} &= (1-\xi^2) \quad \text{for } i = 2
\end{aligned} \tag{17.11}
$$

The strain–displacement relations of a stiffener, whose axis is parallel to x-axis, is given by,

$$
\begin{Bmatrix} \varepsilon_x^{sx} \\ \gamma_{xy}^{sx} \\ \gamma_{xz}^{sx} \end{Bmatrix} = \begin{Bmatrix} \frac{\partial u^{sx}}{\partial x} + z\frac{\partial \alpha^{sx}}{\partial x} crz \frac{\partial \beta^{sx}}{\partial x} \\ \alpha^{sx} + \frac{\partial w^{sx}}{\partial x} - y\frac{\partial \beta^{sx}}{\partial x} \end{Bmatrix} \tag{17.12}
$$

The stress resultants developed in the cross section of the x-directional stiffener are computed and arranged as,

$$
\begin{Bmatrix} N_x^{sx} \\ M_x^{sx} \\ T_x^{sx} \\ Q_{xz}^{sx} \end{Bmatrix} = \begin{bmatrix} A_{11}^{sx} b_{st} & B_{11}^{sx} b_{st} & B_{16}^{sx} b_{st} & 0 \\ B_{11}^{sx} b_{st} & D_{11}^{sx} b_{st} & D_{16}^{sx} b_{st} & 0 \\ B_{16}^{sx} b_{st} & D_{16}^{sx} b_{st} & \frac{1}{6}\left(\bar{Q}_{66}^{sx} + \bar{Q}_{44}^{sx}\right) d_{st} b_{st}^3 & 0 \\ 0 & 0 & 0 & k_s^{sx} A_{44}^{sx} b_{st} \end{bmatrix} \begin{Bmatrix} \frac{\partial u^{sx}}{\partial x} \\ \frac{\partial \alpha^{sx}}{\partial x} \\ \frac{\partial \beta^{sx}}{\partial x} \\ \alpha^{sx} + \frac{\partial w^{sx}}{\partial x} \end{Bmatrix}
$$

$$
= [E_{sx}]\{\varepsilon_{sx}\} \tag{17.13}
$$

The nodes of the plate and the stiffener elements are assumed to be collinear in thickness direction of the global coordinate system. The relation between the nodal displacement vector of the stiffener element and that of the mid-surface of the plate element is expressed as,

$$
\begin{Bmatrix} u^{sx} \\ w^{sx} \\ \alpha^{sx} \\ \beta^{sx} \end{Bmatrix} = \begin{Bmatrix} 1 & 0 & 0 & e & 0 \\ 0 & 0 & 1 & 0 & 0 \\ 0 & 0 & 0 & 1 & 0 \\ 0 & 0 & 0 & 0 & 1 \end{Bmatrix} \begin{Bmatrix} u \\ v \\ w \\ \alpha \\ \beta \end{Bmatrix} = \left[T_{ce}^{sx}\right]\{\delta^p\} \tag{17.14}
$$

In consequence, the elemental nodal degrees of freedom of the x-directional stiffener can be presented in terms of nodal degrees of freedom of the plate element as,

$$
\sum_{i=1}^{3} \begin{Bmatrix} u_i^{sx} \\ w_i^{sx} \\ \alpha_i^{sx} \\ \beta_i^{sx} \end{Bmatrix} = \sum_{i=1}^{3} \left[T_{ce}^{sx}\right] \sum_{j=1}^{8} \begin{bmatrix} N_{ij} & 0 & 0 & 0 & 0 \\ 0 & N_{ij} & 0 & 0 & 0 \\ 0 & 0 & N_{ij} & 0 & 0 \\ 0 & 0 & 0 & N_{ij} & 0 \\ 0 & 0 & 0 & 0 & N_{ij} \end{bmatrix} \begin{Bmatrix} u_j \\ v_j \\ w_j \\ \alpha_j \\ \beta_j \end{Bmatrix} \tag{17.15}
$$

In Eq. (17.15), N_{ij} is defined as the jth quadratic shape function of plate element computed at the ith node of the stiffener element.

Equation (17.15) can be expressed as,

$$\{d_e^{sx}\} = [T_{ce}^{sx}][T_{sx}^{sh}]\{d_e\} = [T^{sx}]\{d_e\} \tag{17.16}$$

where $[T^{sx}]$ is called as transformation matrix, which is used to transfer the degrees of freedom of the nodes of the stiffener element to the corresponding nodes of the plate element taking eccentricity ($e = (h + d_{st})/2$) of the stiffener into account.

The stiffness and mass matrices of the stiffener element are computed as,

$$[K_{sxe}] = \int_{-1}^{1} [T^{sx}]^T[B_{sx}]^T[E_{sx}][B_{sx}][T^{sx}]|J^{sx}|\,\mathrm{d}\xi \tag{17.17}$$

$$[M_{sxe}] = \int_{-1}^{1} [T^{sx}]^T[N^{sx}]^T[m^{sx}][N^{sx}][T^{sx}]|J^{sx}|\,\mathrm{d}\xi \tag{17.18}$$

where $[B_{sx}]$ is the strain–displacement matrix and $[m^{sx}]$ is the inertia matrices of the one-dimensional stiffener element. The strain–displacement matrix is expressed as,

$$[B_{sx}] = \begin{bmatrix} N_{i,x}^{sx} & 0 & 0 & 0 & \cdots \\ 0 & 0 & N_{i,x}^{sx} & 0 & \cdots \\ 0 & 0 & 0 & N_{i,x}^{sx} & \cdots \\ 0 & N_{i,x}^{sx} & N_i^{sx} & 0 & \cdots \end{bmatrix}_{i=1,3} \tag{17.19}$$

and the inertia matrix of the stiffener is written as,

$$[m^{sx}] = \begin{bmatrix} p_{sx} & 0 & 0 & 0 \\ 0 & p_{sx} & 0 & 0 \\ 0 & 0 & I_{sx} & 0 \\ 0 & 0 & 0 & J_{sx} \end{bmatrix} \tag{17.20}$$

where $p_{sx} = \sum_{k=1}^{nl} b_{st}(z_k - z_{k-1})\rho_k$, $J_{sx} = \frac{1}{3}\sum_{k=1}^{nl} b_{st}^3(z_k - z_{k-1})\rho_k$ and $I_{sx} = \frac{1}{3}\sum_{k-1}^{nl} b_{st}(z_k^3 - z_{k-1}^3)\rho_k$.

The same procedure may be adopted to compute the elasticity stiffness matrix and mass matrix of the stiffener placed along y-axis of the plate.

The stiffness matrix and mass matrix of the stiffened plate element can be computed as,

$$[K_e] = [K_p] + [K_{sxe}] + [K_{sye}] \tag{17.21}$$

$$[M_e] = [M_p] + [M_{sxe}] + [M_{sye}] \tag{17.22}$$

The generalized dynamic equilibrium equation is derived from Lagrange's equation of motion. While deriving the dynamic equilibrium equation, it is assumed that the stiffened plate is rotating at moderate speed. For moderate speed of rotation, the Coriolis effect is neglected and the dynamic equilibrium equation in global form is written as [9],

$$[M]\left\{\ddot{\delta}\right\} + ([K] + [K_\sigma])\{\delta\} = \left\{F(\Omega^2)\right\} \tag{17.23}$$

In Eq. (17.23), $[M]$ is the global mass matrix, $[K]$ is the global elastic stiffness matrix, $[K_\sigma]$ is the global geometric stiffness matrix, $\{\delta\}$ is the global displacement vector and $\{F(\Omega^2)\}$ is the global centrifugal force vector [9]. The computation of the geometric stiffness matrix $[K_\sigma]$ is based on iterative solution [7, 9], because it depends on the values of initial stresses.

$$([K] + [K_\sigma])\{\delta\} = \left\{F(\Omega^2)\right\} \tag{17.24}$$

In the first phase of solution, the initial stresses are equal to zero and the equation becomes,

$$[K]\{\delta\} = \left\{F(\Omega^2)\right\} \tag{17.25}$$

The solution of the above equation gives a stress distribution σ^0. Taking this stress distribution, the geometric stiffness matrix is derived. The equation becomes,

$$([K] + [K_{\sigma^0}])\{\delta\} = \left\{F(\Omega^2)\right\} \tag{17.26}$$

The solution of Eq. (17.26) gives another new stress distribution σ^1. Similarly, the procedure can be repeated to get the converged value of the stresses.

The rotational velocity component matrix contributing for angular acceleration is expressed as [7, 9],

$$[A_x] = \begin{bmatrix} \Omega_y'^2 + \Omega_z'^2 & -\Omega_x'\Omega_y' & -\Omega_x'\Omega_z' \\ -\Omega_x'\Omega_y' & \Omega_x'^2 + \Omega_z'^2 & -\Omega_y'\Omega_y' \\ -\Omega_x'\Omega_z' & -\Omega_y'\Omega_z' & \Omega_x'^2 + \Omega_y'^2 \end{bmatrix} \tag{17.27}$$

The centrifugal force vector of an element can be written as [7, 9],

$$\{F_{\Omega e}\} = \rho \int_{\text{vol}} [N]^T [A_x] \begin{Bmatrix} x \\ y \\ z \end{Bmatrix} \mathrm{d(vol)} \tag{17.28}$$

In Eq. (17.28), ρ and $[N]$ are the density of the composite material and the matrix of shape functions. Considering the Green–Lagrangian nonlinear strain components due to rotation, the geometric stiffness matrix of the stiffened plate element can be computed as [9, 29],

$$\{K_{\sigma e}\} = \int_{\text{vol}} [G]^T [M_\sigma][G] \, \mathrm{d(vol)} \tag{17.29}$$

In the above equation, the matrix $[G]$ contains the derivatives of shape functions and $[M_\sigma]$ represents the matrix of initial in-plane stress resultants developed due to rotation. The expressions of [G] and $[M_\sigma]$ of the stiffened plate are given by,

$$[G] = \begin{bmatrix} N_{i,x} & 0 & 0 & 0 & 0 & \cdots \\ N_{i,y} & 0 & 0 & 0 & 0 & \cdots \\ 0 & N_{i,x} & 0 & 0 & 0 & \cdots \\ 0 & N_{i,y} & 0 & 0 & 0 & \cdots \\ 0 & 0 & N_{i,x} & 0 & 0 & \cdots \\ 0 & 0 & N_{i,y} & 0 & 0 & \cdots \\ 0 & 0 & 0 & N_{i,x} & 0 & \cdots \\ 0 & 0 & 0 & N_{i,y} & 0 & \cdots \\ 0 & 0 & 0 & 0 & N_{i,x} & \cdots \\ 0 & 0 & 0 & 0 & N_{i,y} & \cdots \\ 0 & 0 & 0 & N_i & 0 & \cdots \\ 0 & 0 & 0 & 0 & N_i & \cdots \end{bmatrix}_{i=1,8} \tag{17.30}$$

$$[M_\sigma] = \begin{bmatrix} N_x & N_{xy} & 0 & 0 & 0 & 0 & 0 & 0 & 0 & 0 & 0 & 0 \\ N_{xy} & N_y & 0 & 0 & 0 & 0 & 0 & 0 & 0 & 0 & 0 & 0 \\ 0 & 0 & N_x & N_{xy} & 0 & 0 & 0 & 0 & 0 & 0 & 0 & 0 \\ 0 & 0 & N_{xy} & N_y & 0 & 0 & 0 & 0 & 0 & 0 & 0 & 0 \\ 0 & 0 & 0 & 0 & N_x & N_{xy} & 0 & 0 & 0 & 0 & 0 & 0 \\ 0 & 0 & 0 & 0 & N_{xy} & N_y & 0 & 0 & 0 & 0 & 0 & 0 \\ 0 & 0 & 0 & 0 & 0 & 0 & \frac{N_x h^2}{12} & \frac{N_{xy} h^2}{12} & 0 & 0 & 0 & 0 \\ 0 & 0 & 0 & 0 & 0 & 0 & \frac{N_{xy} h^2}{12} & \frac{N_y h^2}{12} & 0 & 0 & 0 & 0 \\ 0 & 0 & 0 & 0 & 0 & 0 & 0 & 0 & \frac{N_x h^2}{12} & \frac{N_{xy} h^2}{12} & 0 & 0 \\ 0 & 0 & 0 & 0 & 0 & 0 & 0 & 0 & \frac{N_{xy} h^2}{12} & \frac{N_y h^2}{12} & 0 & 0 \\ 0 & 0 & 0 & 0 & 0 & 0 & 0 & 0 & 0 & 0 & 0 & 0 \\ 0 & 0 & 0 & 0 & 0 & 0 & 0 & 0 & 0 & 0 & 0 & 0 \end{bmatrix} \tag{17.31}$$

The QR iteration algorithm [30] is used to compute the natural frequencies of stiffened panel. The solution of governing equation of motion is given as,

$$[A]\{\delta\} = \lambda\{\delta\} \tag{17.32}$$

where $[A] = ([K] + [K_\sigma])^{-1}[M]$ and $\lambda = 1/\omega_n^2$.

17.3 Result and Discussion

For the computational purpose, an in-house computer programme based on the above formulation is developed in MATLAB environment. A mesh convergence study is conducted, and it is found that the mesh size of 8×8 gives the converged results. So for the entire analysis of the present study, this mesh size is used to compute the results. The validation of the formulation with respect to stiffened panel is shown in Table 17.1, which shows the agreement of the computed results with that of Nayak and Bandyopadhyay [16] and Das and Chakravorty [31]. The natural frequencies of antisymmetric cross-ply crossed-stiffened laminated composite plates are furnished in Table 17.1. The non-dimensional fundamental frequencies of an isotropic cantilever plate are furnished in Table 17.2 corresponding to different rotational speeds. The computed results corresponding to various rotational speeds show very good agreement with that of Sreenivasamurthy and Ramamurti [7]. The non-dimensional fundamental frequencies of a twisted composite plate are computed and compared with that of Qatu and Leissa [3] corresponding to different fibre orientation angles. These comparisons of results are shown in Table 17.3. The capability of the present MATLAB code in respect of stiffener formulation, rotation of the panel and pretwist angle is well established and acceptable. Therefore, it is obvious that the MATLAB code can effectively compute the natural frequencies of the pretwisted stiffened panel subjected to various rotational speeds.

Table 17.1 Natural frequencies (Hz) of simply supported cross-ply (0°/90°) crossed-stiffened plate, $L = b = 254$ mm, $h = 12.7$ mm, $b_{st} = 6.35$ mm, $d_{st} = 25.4$ mm, $E_1 = 144.8$ GPa, $E_2 = 9.67$ GPa, $G_{12} = G_{13} = 4.14$ GPa, $G_{23} = 3.45$ GPa, $\nu = 0.3$, $\rho = 1389.23$ kg/m^3 and $n_x = n_y = 1$

Mode number	Nayak and Bandyopadhyay [16]	Das and Chakravorty [31]	Present FEM
1	1141.00	1123.17	1142.03
2	2394.17	2367.77	2398.12
3	2415.82	2407.57	2417.10
4	2646.18	2656.00	2646.31

Table 17.2 Non-dimensional fundamental frequencies $\left(\overline{\omega} = \omega_n L^2 \sqrt{\rho h / D}\right)$ of an isotropic rotating cantilever plate

Non-dimensional speed $\Omega = \Omega' / \omega_n$	Sreenivasamurthy and Ramamurti [7]	Present FEM
0.0	3.43685	3.41748
0.4	3.75280	3.72640
0.8	4.56786	4.51250
1.0	5.09167	5.01066

$L/b = 1$, $h/L = 0.12$, $D = Eh^3/12(1 - \nu^2)$, $\nu = 0.3$

Table 17.3 Non-dimensional fundamental frequencies $\left(\overline{\omega} = \omega_n L^2 \sqrt{\rho / E_{11} h^2}\right)$ of three layer [θ/−θ/θ] graphite–epoxy twisted plates

θ (Deg.)	Qatu and Leissa [3]	Present FEM
0	0.9553	0.9431
15	0.8759	0.8629
30	0.6923	0.6812

$L/b = 1$, $b/h = 20$, twist angle $(\phi) = 30°$, $E_1 = 138.70$ GPa, $E_2 = 8.96$ GPa, $G_{12} = 7.1$ Gpa, $\nu_{12} = 0.3$

The fundamental frequencies of the composite stiffened plate are presented corresponding to different twist angles and rotational speeds. The geometric and material properties of the composite stiffened plate made of graphite–epoxy are as follows:

$$L/b = 2,\ L/h = 100,\ b_{st} = h,\ d_{st} = 4h,\ \nu_{12} = 0.30,\ E_1 = 138.0\ \text{GPa},$$
$$G_{12} = G_{13} = 7.1\ \text{GPa}, E_2 = 8.96\ \text{GPa}, G_{23} = 2.84\ \text{GPa}$$

The entire analysis is based on the laminated composite plate with composite stiffeners placed symmetrically along the nodal lines. The stacking sequence of the plate and the stiffener is always same. The boundary condition considered for the stiffened plate is given below:

$$u = v = w = \alpha = \beta = 0 \text{ at } x = 0. \tag{17.33}$$

The effect of fibre orientation angle on the fundamental natural frequency of an eight-layered laminated composite $[\theta/-\theta/\theta/-\theta]_s$ stiffened plates is presented in Fig. 17.2 for two panels, wherein one panel is attached with a single x-directional stiffener while the second panel is appended with a single y-directional stiffener. At the same time, the stiffened plates of different angles of pretwist are also considered. Form the graph, it may be observed that raise in the value of fibre orientation angle

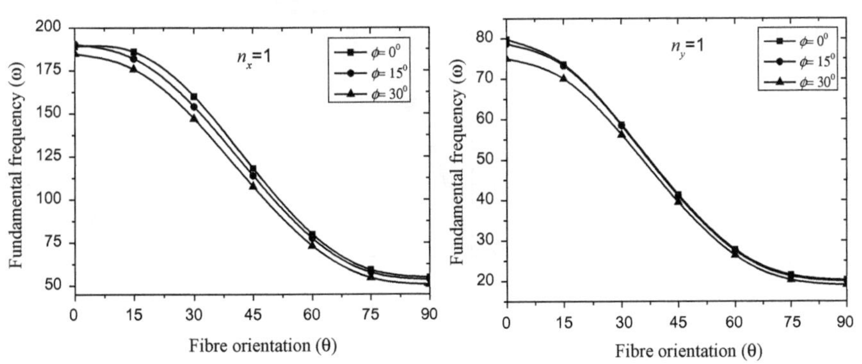

Fig. 17.2 Variation of fundamental frequency with fibre angle of different twisted stiffened plates

decreases the fundamental frequency irrespective of twist angle. When the fibres of stiffener and plate are orthogonal to the clamped edge ($\theta = 0°$), fundamental frequency is obtained maximum and minimum at $\theta = 90°$ irrespective of twist angles. Increase in the value of pretwist angle is found to reduce the value of fundamental frequency because of decrease in structural stiffness, and the results corroborate with the results of Qatu and Leissa [3]. The maximum percentage of reduction in fundamental frequency with increase in twist angle is observed at fibre orientation of 30° while minimum at $\theta = 90°$ for x-directional stiffened plate. In case of y-directional stiffened plate, insignificant variation of fundamental frequency is observed corresponding to 15° twist angle. At 30° pretwist angle of y-directional stiffened plate, maximum reduction in fundamental frequency is depicted at $\theta = 0°$. Comparing x- and y-directional stiffened plates, it may be observed that maximum value of fundamental frequency is obtained with x-directional stiffener. Hence, x-directional stiffener is found to be more efficient in rendering maximum stiffness to the plates, thereby increasing the fundamental frequency. This observation is limited to cantilever boundary condition.

The variation of fundamental frequency with increase in number of x-directional stiffener of stationary composite $[30°/-30°/30°/-30°]_s$ twisted stiffened plate corresponding to twist angle 0°, 15°, 30° is furnished in Fig. 17.3, because from the previous observation it is clear that x-stiffener is more efficient in terms of increasing fundamental frequency. It reveals that increase in number of stiffeners increases the value of fundamental frequency, as normally expected. However, the rate of increase in fundamental frequency, especially at early stage, gradually decreases with increase in number of stiffeners for all the cases. Excellent performance in terms of improving fundamental frequency is achieved by appending maximum three numbers of stiffeners; thereafter, the increase in fundamental frequency is found marginal as the curves gradually become horizontal.

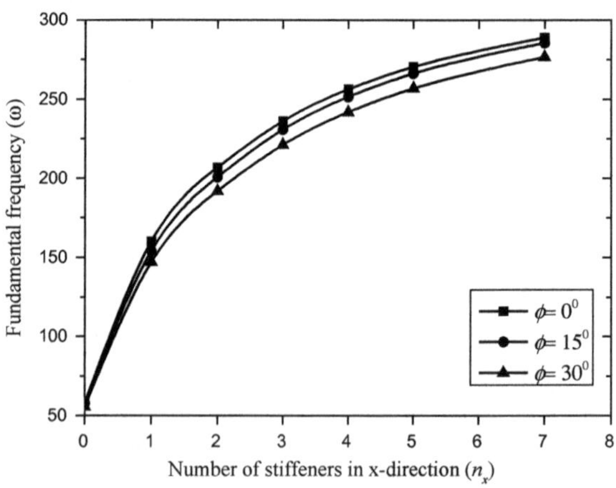

Fig. 17.3 Variation of fundamental frequency with respect to number of stiffeners of various twisted stiffened plates. Stiffeners are symmetrically across the width

The variation of first fundamental frequency with stiffener depth-to-plate thickness ratio (d_{st}/h) of both untwisted and twisted laminated composite [30°/−30°/30°/−30°]$_s$ stiffened plate is illustrated in Fig. 17.4. A single x-directional composite stiffener placed symmetrically is considered for this particular case. It is evident that increase in (d_{st}/h) increases the fundamental frequency, because it increases the eccentricity of the stiffener, which in turn increases the second moment of area. The rate of increase of fundamental frequency is found slow at early stage while its growth is found rapid after $d_{st}/h = 2$ in both the cases. This observation will be helpful for the investigators to select the depth of the stiffener.

Figure 17.5 shows the variation of fundamental frequency of the composite [30°/−30°/30°/−30°]$_s$ stiffened plate with non-dimensional rotational speeds ($\Omega = 0.00, 0.50$ and 1.00). Three cases are considered corresponding to twist angles 0°, 15° and 30°, respectively. The present investigation is performed for two different stiffened panels: one is embedded with a single x-stiffener and other with a single y-stiffener. The fundamental frequency is found minimum at twist angle 30° while found maximum in untwisted stiffened plate as normally expected for both the panels. It is evident that increasing the rotational speed raises the value of fundamental frequency because of centrifugal stiffening. The percentage increase in fundamental frequency due to rotational speed is found maximum in the twisted plate ($\phi = 30°$) than untwisted plate. This observation is found in x-directional stiffened plate, while in y-directional stiffened plate, no such observation is noticed. Hence, the rotating effect is more pronounced for the twisted plate than the untwisted plate, when attached with x-directional stiffener.

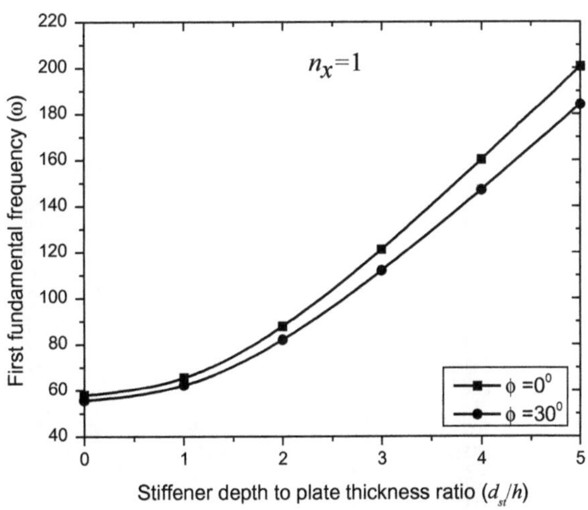

Fig. 17.4 Variation of first fundamental frequency of the stiffened plate with respect to stiffener depth-to-plate thickness ratio

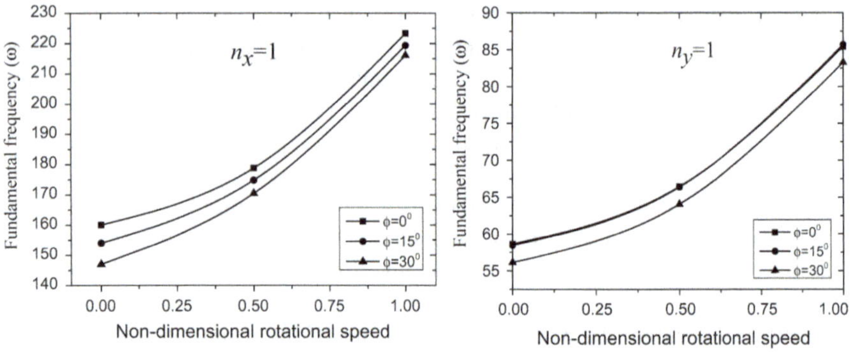

Fig. 17.5 Variation of fundamental frequency with rotational speed of composite $[30/-30/30/-30]_s$ stiffened plates with different angles of pretwist

The mode shapes of the composite stiffened plate for various twist angle and rotational speeds are furnished. First the effect of different twist angles of stiffened plate on its mode shapes is presented in Table 17.4. The continuous and dashed lines denote the upward and downward displacements, respectively. It reveals that the symmetry of mode shapes disappears with increase in twist angles.

The effects of different rotational speeds on the mode shapes of a twisted stiffened plate are shown in Table 17.5. The first five mode shapes of the composite stiffened plate are shown corresponding to non-dimensional rotational speeds 0.0, 0.5 and 1.0, respectively. The first five modes of the non-rotating stiffened plate are, in order, first spanwise bending (1B), first torsional mode (1T), second torsional

Table 17.4 Mode shapes of the cantilevered composite $[30°/-30°/30°/-30°]_s$ stiffened plate corresponding to different twist angles

Twist angle	Mode-1	Mode-2	Mode-3	Mode-4	Mode-5
0°		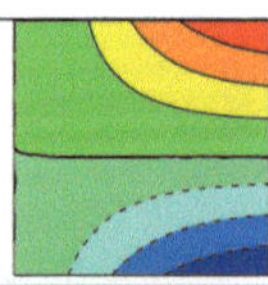	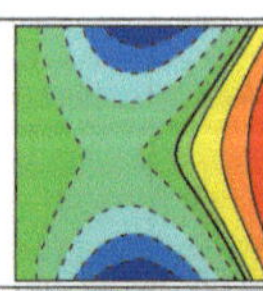	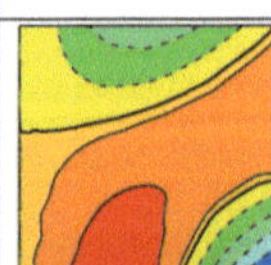	
15°			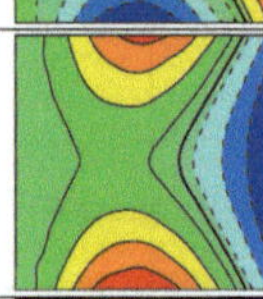		
30°	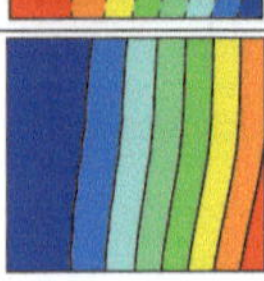			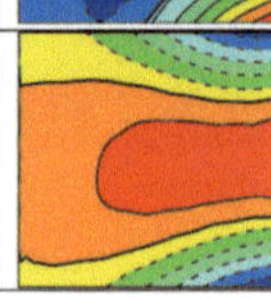	

Table 17.5 Mode shapes of the cantilevered composite [30°/−30°/30°/−30°]$_s$ twisted (30°) stiffened plate corresponding to different rotational speeds

Rot. speed	Mode-1	Mode-2	Mode-3	Mode-4	Mode-5
0.0				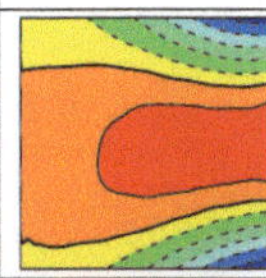	
0.5			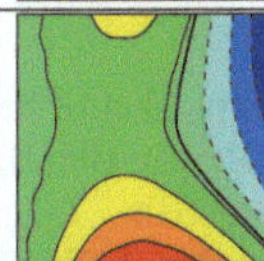	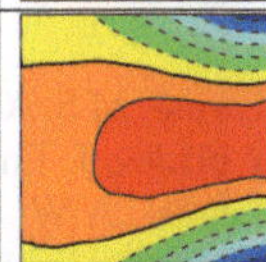	
1.0			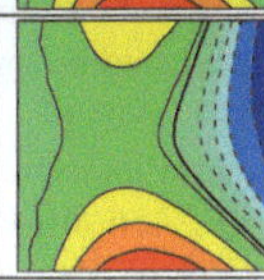	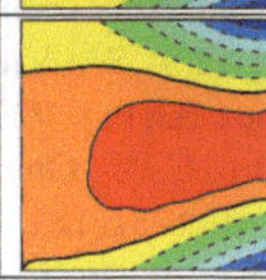	

mode (2T), first chordwise bending mode (1C) and third torsional mode (3T), respectively. The symmetry of mode shapes is not observed due to twist angle and rotational speed. It is observed that the mode-1 is not at all influenced by the rotational speed, while the influence of rotational speed can be seen after mode-2.

17.4 Conclusions

In this investigation, the finite element formulation of the rotating twisted stiffened plate is presented and the accuracy and effectiveness of the formulation are well verified with the results available in the open literature. The major conclusions drawn from the parametric studies are listed below:

1. The fibre orientation of the laminated stiffened plate has large control over the fundamental frequency, wherein maximum and minimum values of fundamental frequency are obtained corresponding to 0° and 90°, respectively.
2. The x-directional stiffener is found to be more efficient in increasing the fundamental frequency of the twisted stiffened plate.
3. The fundamental frequency is noticed to reduce with increase in twist angle of the composite stiffened plate.
4. Increase in stiffener depth-to-plate thickness ratio has also a striking effect on the fundamental frequency.

5. The effect of speed of rotation on the fundamental frequencies is observed to be more prominent in twisted stiffened plates. Increase in the rotational speed of the stiffened plate increases the fundamental frequency due to centrifugal stiffening irrespective of twist angle.

References

1. Leissa, A.W., Ewing, M.S.: Comparison of beam and shell theories for the vibrations of thin turbomachinery blades. J. Eng. Power **105**(2), 383–392 (1983)
2. Kielb, R.E., Leissa, A.W., Macbain, J.C.: Vibrations of twisted cantilever plates—a comparison of theoretical results. Int. J. Numer. Meth. Eng. **21**(8), 1365–1380 (1985)
3. Qatu, M.S., Leissa, A.W.: Vibration studies for laminated composite twisted cantilever plates. Int. J. Mech. Sci. **33**(11), 927–940 (1991)
4. Liew, K.M., Lim, C.W., Ong, L.S.: Vibration of pretwisted cantilever shallow conical shells. Int. J. Solids Struct. **31**(18), 2463–2476 (1994)
5. Kuang, J.H., Hsu, M.H.: The effect of fiber angle on the natural frequencies of orthotropic composite pre-twisted blades. Compos. Struct. **58**(4), 457–468 (2002)
6. Lee, J.J., Yeom, C.H., Lee, I.: Vibration analysis of twisted cantilevered conical composite shells. J. Sound Vib. **255**(5), 965–982 (2002)
7. Sreenivasamurthy, S., Ramamurti, V.: Coriolis effect on the vibration of flat rotating low aspect ratio cantilever plates. J. Strain Anal. Eng. Des. **16**(2), 97–106 (1981)
8. Ramamurti, V., Kielb, R.: Natural frequencies of twisted rotating plates. J. Sound Vib. **97**(3), 429–449 (1984)
9. Karmakar, A., Sinha, P.K.: Finite element free vibration analysis of rotating laminated composite pretwisted cantilever plates. J. Reinf. Plast. Compos. **16**(16), 1461–1491 (1997)
10. Rao, S.S., Gupta, R.S.: Finite element vibration analysis of rotating Timoshenko beams. J. Sound Vib. **242**(1), 103–124 (2001)
11. Hu, X.X., Sakiyama, T., Matsuda, H., Morita, C.: Fundamental vibration of rotating cantilever blades with pre-twist. J. Sound Vib. **271**(1–2), 47–66 (2004)
12. Kee, Y.J., Kim, J.H.: Vibration characteristics of initially twisted rotating shell type composite blades. Compos. Struct. **64**(2), 151–159 (2004)
13. Joseph, S.V., Mohanty, S.C.: Free vibration of a rotating sandwich plate with viscoelastic core and functionally graded material constraining layer. Int. J. Struct. Stab. Dyn. **17**(10), 1750114 (2017)
14. Chandrashekhara, K., Kolli, M.: Free vibration of eccentrically stiffened laminated plates. J. Reinf. Plast. Compos. **16**(10), 884–902 (1997)
15. Prusty, B.G., Satsangi, S.K.: Finite element transient dynamic analysis of laminated stiffened shells. J. Sound Vib. **248**(2), 215–233 (2001)
16. Nayak, A.N., Bandyopadhyay, J.N.: Free vibration analysis of laminated stiffened shells. J. Eng. Mech. **131**(1), 100–105 (2005)
17. Sahoo, S., Chakravorty, D.: Stiffened composite hypar shell roofs under free vibration: Behaviour and optimization aids. J. Sound Vib. **295**(1), 362–377 (2006)
18. Sadek, E.A., Tawfik, S.A.: A finite element model for the analysis of stiffened laminated plates. Comput. Struct. **75**(4), 369–383 (2000)
19. Qing, G., Qiu, J., Liu, Y.: Free vibration analysis of stiffened laminated plates. Int. J. Solids Struct. **43**(6), 1357–1371 (2006)
20. Yuan, W.X., Dawe, D.J.: Free vibration and stability analysis of stiffened sandwich plates. Compos. Struct. **63**(1), 123–137 (2004)

21. Guo, M., Harik, I.E., Ren, W.X.: Free vibration analysis of stiffened laminated plates using layered finite element method. Struct. Eng. Mech. **14**(3), 245–262 (2002)
22. Li, D., Qing, G., Liu, Y.: A layerwise/solid-element method for the composite stiffened laminated cylindrical shell structures. Compos. Struct. **98**, 215–227 (2013)
23. Bhar, A., Phoenix, S.S., Satsangi, S.K.: Finite element analysis of laminated composite stiffened plates using FSDT and HSDT: A comparative perspective. Compos. Struct. **92**(2), 312–321 (2010)
24. Zhao, W., Kapania, R.K.: Vibrational analysis of unitized curvilinearly stiffened composite panels subjected to in-plane loads. In: 57th AIAA/ASCE/AHS/ASC Structures, Structural Dynamics, and Materials Conference, p. 1500 (2016)
25. Castro, S.G., Donadon, M.V.: Assembly of semi-analytical models to address linear buckling and vibration of stiffened composite panels with debonding defect. Compos. Struct. **160**, 232–247 (2017)
26. Rout, M., Bandyopadhyay, T., Karmakar, A.: Free vibration analysis of pretwisted delaminated composite stiffened shallow shells: a finite element approach. J. Reinf. Plast. Compos. **36**(8), 619–636 (2017)
27. Rout, M., Hota, S.S., Karmakar, A.: Free vibration characteristics of delaminated composite pretwisted stiffened cylindrical shell. In: Proceedings of the Institution of Mechanical Engineers, Part C: Journal of Mechanical Engineering Science, doi: 0954406216686389 (2017)
28. Damnjanović, E., Marjanović, M., Nefovska-Danilović, M.: Free vibration analysis of stiffened and cracked laminated composite plate assemblies using shear-deformable dynamic stiffness elements. Compos. Struct. **180**, 723–740 (2017)
29. Cook, R.D.: Concepts and Applications of Finite Element Analysis. Wiley, Hoboken (2007)
30. Bathe, K.J.: Finite Element Procedures in Engineering Analysis. PHI, New Delhi (1990)
31. Das, H.S., Chakravorty, D.: Bending analysis of stiffened composite conoidal shell roofs through finite element application. J. Compos. Mater. **45**, 525–542 (2010)

Chapter 18
Growth of Yield Front of Functionally Graded Non-uniform Bars Under Thermal Load

Priyambada Nayak and Kashi Nath Saha

18.1 Introduction

In the mid-1980s, the Japanese researchers [1] first introduced the idea of functionally graded material, i.e. FGM, as high-temperature-resistant materials for various applications of engineering such as chemical plants and fusion reactors, space vehicles, components of aircraft and different branches of electrical, civil and mechanical engineering. The physical and mechanical properties continuously vary in the desired direction with suitable metallic and ceramic volume fraction mixture in FGM. Despite excellent characteristics of ceramics to withstand severe environmental effect, such as wear, corrosion and high temperature in combination with the toughness of metals, the structures using FGM can withstand severe mechanical and thermal loading conditions.

Many theoretical research works on FGMs for engineering structures have been significantly expanded over the most recent couple of decades. The post-elastic investigation of functionally graded non-uniform bars under thermo-mechanical loading is important for optimum design of mechanical, civil and aerospace structures. The theoretical treatment of elastic behaviour of bar with uniform geometry under uniform temperature field is discussed in the textbook of Timoshenko [2]. The elasto-plastic conduct of different geometries of solid slender bars under various loading types is a fascinating field of research in structural mechanics field to provide the requirement of designers [3]. Niknam et al. [4] studied the nonlinear bending of tapered FG beams by the implementation of analytical and numerical approaches subjected to thermo-mechanical loading. Wattanasakulpong et al. [5] investigated the thermal buckling load of FGM beam under uniform thermal load by applying an improved third-order shear deformation theory. Paul and Das [6] presented nonlinear analysis of FGM Timoshenko beam

P. Nayak (✉) · K. N. Saha
Department of Mechanical Engineering, Jadavpur University, Kolkata 700032, India
e-mail: priyambada53@gmail.com

P. Sahoo and J. P. Davim (eds.), *Advances in Materials, Mechanical and Industrial Engineering*, Lecture Notes on Multidisciplinary Industrial Engineering,
https://doi.org/10.1007/978-3-319-96968-8_18

for post-buckling load along the beam thickness at steady-state condition under non-uniform temperature distribution.

In a recent article, the thermo-mechanical analysis of non-uniform bars for the evaluation of axial strain and stress is carried out by Nayak and Saha [7] by utilizing a variational principle. In a subsequent article, Nayak and Saha [8] investigated the yield front propagation by using an iterative variational method for non-uniform bars subjected to thermo-mechanical load in elasto-plastic regime. The thermo-elastic fracture of functionally graded materials is simulated by Garg and Pant [9] with the implementation of element-free Galerkin method (EFGM). They solved the heat transfer problem to obtain the temperature field distribution. Cho and Oden [10] used finite element method for the investigation of thermal stress for FGMs. Tayyar [11] presented post-elastic analysis of cantilever beams under uniform moment distribution based on kinematic displacement theory (KDT) for the solution of finite strain. In KDT, deflection is generated without assuming geometry and using differential equations of the deflection curve.

Horgan and Chan [12, 13] examined the impacts of material inhomogeneity on the torsion response of linearly elastic isotropic and anisotropic bars. Katsikadelis and Tsiatas [14] formulated the torsion issue in terms of the warping function and also developed a boundary solution to solve the resulting partial differential equation with variable coefficients under Neumann-type boundary condition for non-homogeneous anisotropic bars. Ecsedi [15] presented some analytical solutions to Saint-Venant torsion of solid and hollow cross-sectional non-homogeneous anisotropic bars. Kolodziej and Gorzelanczyk [16] studied the elasto-plastic material behaviour of the torsion of prismatic bars on the basis of the presumption of deformation theory of plasticity and Saint-Venant displacement. Dwivedi et al. [17] employed finite difference method with nonlinear work hardening for the spring-back investigation of thin rectangular bars. Tsiatas and Babouskos [18] solved the post-elastic torsional problem for arbitrary cross-sectional FG bars by employing a new integral solution technique and deformation theory of plasticity. Sapountzakis and Tsipiras [19] presented the boundary element method (BEM) for the elasto-plastic issue of arbitrary cross-sectional composite cylindrical bars under uniform torsional loading, considering the impact of geometric nonlinearity. Huang et al. [20] presented a semi-analytic solution technique to examine the post-elastic buckling behaviour of FG cylindrical shells subjected to torsional loading by assuming multilinear hardening model for materials.

The literature survey reveals that most of the studies are on elastic and thermo-elastic analysis of functionally graded bars and analysis has been done by different analytical and numerical techniques. However, very little work has been carried out on thermo-elasto-plastic investigation of functionally graded bars. Hence in the present analysis, a numerical technique based on variational principle employing Hencky's deformation theory of plasticity and von Mises yield criterion is used to address the thermo-elasto-plastic behaviour of functionally graded non-uniform bars. The material of the functionally graded bar is modelled by continuous distribution of metal–ceramic composite using power law variation for ceramic and metallic constituents along the length. For the prediction of the governing equation's unknown displacement field, an iterative technique has been employed to obtain the solution algorithm which ultimately yields the resulting stress field. The comparison study is carried out for the proposed mathematical model with that of ABAQUS FEM software, revealing coherent correlation. The numerical results for the elasto-plastic field subjected to thermal loading are presented graphically showing the effect of material parameters on clamped–clamped functionally graded non-uniform bars having various geometries subjected to uniform and non-uniform thermal loads.

18.2 Mathematical Formulation

Thermo-elasto-plastic investigation is carried out for solid circular cross-sectional functionally graded taper bars as represented in Fig. 18.1 where the dotted and solid lines demonstrate the parabolic and linear diameter variations. For parabolic geometry bar at location x, the diameter variation is expressed by the relation $d = d_1 + \left((d_0 - d_1)\sqrt{1-\xi}\right)$, where d_0 = larger diameter, d_1 = smaller diameter, and the normalized axial coordinate is $\xi(= x/L)$ where L is the bar length. The diameter for linear taper geometry is expressed by $d = d_0 - \xi(d_0 - d_1)$. Furthermore, as detailed in Fig. 18.1, the aspect ratio, A_R, and slenderness ratio, S_R, are also used to define the bar geometry.

18.2.1 Material Parameters for FGM

A functionally graded bar comprised of a metallic phase and ceramic phase is considered in this investigation. A continuous distribution of ceramic and metallic

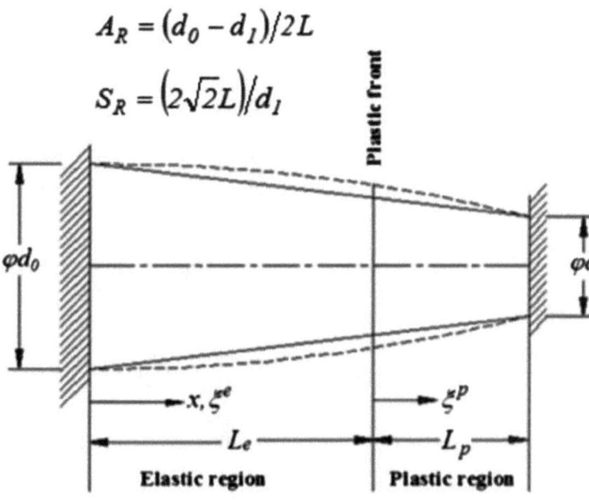

Fig. 18.1 Schematic diagram of geometry of taper bar

volume fractions along the axial direction is assumed for modelling of FGM. The ceramic volume fraction (V_c) and metallic volume fraction (V_m) constituent along the axial direction follow the power law distribution [21],

$$V_c(x) = V_{c0}\left(\frac{x}{L}\right)^n \tag{18.1a}$$

and

$$V_c + V_m = 1 \tag{18.1b}$$

where the constituent volume fraction is denoted by V with subscripts m and c, which correspond to the metallic and ceramic constituents. The volume fraction of the ceramic in the bar's right fixed end is V_{c0}, and n is the volume fraction exponent. The FG bar becomes a pure metallic bar when $V_{c0} = 0$, and the ceramic phase of the bar with uniform distribution is denoted by $n = 0$. For $V_{c0} = 1$, the left and right ends of the bar comprise of purely metallic and purely ceramic parts, respectively.

For the post-elastic investigation of FGM bars, it is assumed that the metal matrix has bilinear elasto-plastic behaviour model with elastic modulus E_m, tangent modulus E_{tm} and yield stress σ_{ym}, whereas the ceramic is assumed to be linear elastic with elastic modulus, E_c, as shown in Fig. 18.2. In the present investigation, for estimating the FGM effective properties for the modelling of elasto-plastic FGMs, TTO homogenization scheme is adopted as defined by Tamura et al. [22]. The model was extended by Bocciarelli [23] for ceramic–metal compound to depict the elasto-plastic behaviour of FGMs.

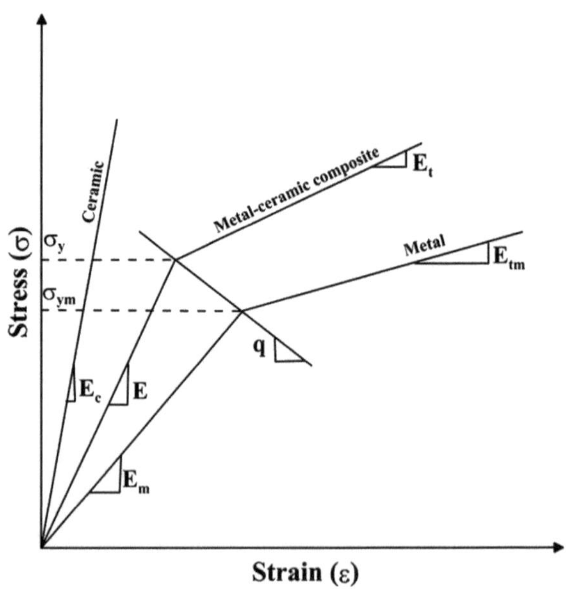

Fig. 18.2 Schematic bilinear stress–strain curve for the FGM

The assumption of the TTO model is that with the yielding of metallic constituents the yielding of composite takes place [24]. Thus, the elasto-plastic behaviour of FGM is obtained by the introduction of the ratio of stress to strain transfer $q(=\tilde{q}E_c)$ where the stress transfer parameter is $\tilde{q}$ and $\tilde{q} \geq 0$. It should be noted that once the elastic limit of metallic constituents is achieved, the plastic flow of FGMs occurs which is represented by $\tilde{q} = 0$. The value of q relies on various factors such as composition, material microstructure, loading condition. However, a constant q value is assumed because of the absence of experimental data beyond the elastic range. The elasto-plastic material properties of FGM bar along the axial location can be defined by Nakamura et al. [25] as

$$E(x) = \left[V_m E_m \left(\frac{q+E_c}{q+E_m}\right) + V_c E_c\right] \Big/ \left[V_m \left(\frac{q+E_c}{q+E_m}\right) + V_c\right] \tag{18.2a}$$

$$\sigma_y(x) = \sigma_{ym}\left[V_m + \left(\frac{q+E_m}{q+E_c}\right)\frac{E_c}{E_m}V_c\right] \tag{18.2b}$$

$$E_t(x) = \left[V_m E_{tm} \left(\frac{q+E_c}{q+E_{tm}}\right) + V_c E_c\right] \Big/ \left[V_m \left(\frac{q+E_c}{q+E_{tm}}\right) + V_c\right] \tag{18.2c}$$

where $E(x)$ is the elastic modulus, $\sigma_y(x)$ is the overall yield stress, and $E_t(x)$ is the tangent modulus of the FGM bar as shown in Fig. 18.2.

The coefficient of linear thermal expansion $\alpha(x)$ of FGM bar can be found out by using the modified rule of mixtures [26] as provided below.

$$\alpha(x) = \alpha_m V_m + \alpha_c V_c \tag{18.2d}$$

However, for both the phases (metal and ceramic) a constant value is assumed for Poisson's ratio in this analysis as there is a little change between ceramic and metallic values; hence, an average value is considered throughout the FGM. However, in elastic and elasto-plastic regions, Poisson's ratios are denoted by v and v_p.

18.2.2 Solution to the Problem

In the present paper, the investigation of clamped–clamped FGM taper bars under thermal load is performed. For the present investigation, two types of material models are considered: bilinear and multilinear. The FGM bar is considered to be locally isotropic, and the governing equation is obtained by using Hencky's total deformation theory of plasticity and von Mises criterion by an energy-based variational technique. As the yield limit and stress–strain relations are temperature dependent, the plastic behaviour investigation turns out to be more complicated due to the consideration of temperature field. The creep effect is neglected in the present analysis because the assumption of temperature rise is not permanent. In view of the presumptions of the present investigation in the total regime of deformation, temperature field T induces the relative volumetric change, which is elastic and is expressed by $\varepsilon = 3k\sigma + 3\alpha T$ where $k(= (1-2v)/E)$ is the volumetric compression coefficient, σ is the mean stress, and α is the thermal expansion coefficient.

As indicated by Hencky's deformation theory, the components of deviatoric strain are made up of elastic and plastic strain components, $e_{ij} = e_{ij}^{e} + e_{ij}^{p}$. The components of the elastic deviatoric strain are related to stress components by Hooke's law $e_{ij}^{e} = s_{ij}/2G$, where G is the modulus of rigidity, while the plastic strain components of strain are related to the deviatoric stress in the form [27] $e_{ij}^{p} = \psi s_{ij}$, where ψ is a scalar function of deviatoric stress. With the aid of these equations, the expression for total strain can be written by:

$$\varepsilon_{ij} = \frac{\varepsilon}{3}\delta_{ij} + e_{ij} = k\sigma\delta_{ij} + \alpha T\delta_{ij} + \psi s_{ij} \tag{18.3}$$

In Hencky's total deformation theory of plasticity, the minimum potential energy principle $\delta(\Pi) = 0$ is applied and the displacement field under thermal load T is obtained. U and V are strain energy and potential energy of the energy principle, $\Pi = U + V$, respectively. The total strain energy of the bar under elastic conditions is obtained with the aid of the potential of work of deformation [28] and expressed as

$$\Pi = \int_{0}^{L} \left[\left(\frac{K(1-2v)^2}{2} + \frac{2G(1+v)^2}{3} \right) (u')^2 - 3K(1-2v)\alpha T u' \right] A(x)\,\mathrm{d}x \tag{18.4}$$

Here, A is the cross-sectional area and $(\ )'$ indicate first derivative w.r.t coordinate variable. Π is the total potential energy which gets composed into two parts: first part is same as that of the strain energy U stored in the bar, and the second part is potential energy due to thermal load. So, the strain energy U is expressed by

$$U = \int_{0}^{L} \left\{ \left(\frac{K(1-2v)^2}{2} + \frac{2G(1+v)^2}{3} \right) (u')^2 \right\} A(x)\,\mathrm{d}x \tag{18.5}$$

where the axial strain $\varepsilon_x = u'$. Due to thermal loading, the potential energy V expression is given by

$$V = -\int_{0}^{L} \{3K(1-2v)\alpha T u'\} A(x)\,\mathrm{d}x \tag{18.6}$$

In the present investigation, the consideration for shear deformation is excluded due to the axisymmetric nature of the bar problem.

For the taper bar's smaller fixed end $(x = L)$, the occurrence of yield initiation takes place at a certain temperature and this specific temperature is named as elastic limit temperature T_y. The yield front slowly moves towards the larger fixed end $(x = 0)$ with further increment in temperature, and the coalescence of elasto-plastic region takes place there at a particular temperature, named as plastic collapse temperature T_c. The separation of the bar domain into two regions takes place beyond the state of yield limit, a higher diameter elastic region (0 to L_e) and a plastic region (L_e to L) with relatively smaller diameter. Consequently, the total strain energy comprises of an elastic (U_e) and a plastic $\left(U_p\right)$ part in the post-elastic region.

The strain energy in the elastic region is expressed by

$$U_e = \int_0^{L_e} \left(\frac{K(1-2v)^2}{2} + \frac{2G(1+v)^2}{3} \right) A(x)(u')^2 \mathrm{d}x \tag{18.7}$$

and for the post-elastic region, the strain energy is expressed as

$$U_p = \int_{L_e}^{L} \left(\frac{K(1-2v)^2}{2} + \frac{2g(\Gamma)\left(1+v_p\right)^2}{3} \right) A(x)(u')^2 \mathrm{d}x \tag{18.8}$$

where the modulus of plasticity is $g(\Gamma)$.

The governing equation is obtained from the energy principle $\delta(U + V) = 0$, on substitution of Eqs. (18.5) to (18.8),

$$\delta \left[\begin{array}{l} \int_0^{L_e} \left(\frac{K(1-2v)^2}{2} + \frac{2G(1+v)^2}{3} \right) A(x)(u')^2 \mathrm{d}x \\ + \int_{L_e}^{L} \left(\frac{K(1-2v)^2}{2} + \frac{2g(\Gamma)\left(1+v_p\right)^2}{3} \right) A(x)(u')^2 \mathrm{d}x \\ - \int_0^{L} \{3K(1-2v)\alpha T u'\} A(x) \mathrm{d}x \end{array} \right] = 0 \tag{18.9}$$

Equation (18.9) is normalized by using length parameters $\xi^e = x/L_e$ in elastic and $\xi^p = (x - L_e)/L_p$, in post-elastic regions, where the elastic and post-elastic domain lengths are L_e and L_p. A linear combination of sets of orthogonal coordinate functions $u(\xi) = \sum c_i \phi_i$, $i = 1, 2, \ldots, n_f$, is used in Eq. (18.9) for the approximation of the displacement function $u(\xi)$, where Gram–Schmidt scheme is used for the development of the set of orthogonal functions ϕ_i, c_i is the unknown coefficients, and n_f is number of functions. The start function ϕ_0 is necessary to satisfy the geometric boundary conditions of the bar, $u = 0$ at $\xi = 0$ and $u = 0$ at $\xi = 1$ which is given by $\phi_0 = \xi(1 - \xi)$. For elastic and post-elastic domains, the expressions used for the displacement functions are $u(\xi^e) \cong \sum c_i \phi_i^e$ and $u(\xi^p) \cong \sum c_i \phi_i^p$, respectively. On substitution of these assumed displacement functions in the governing differential equation as per Galerkin error minimization principle and replacing δ by $\partial/\partial c_j$, $j = 1, 2, \ldots, n,$, we get

$$\sum_{i=1}^{n}\sum_{j=1}^{n} c_i \Bigg[\left(\frac{K(1-2v)^2}{2} + \frac{2G(1+v)^2}{3} \right) \left(\frac{2}{L_e} \right) \int_0^1 \left(A(\xi^e) \phi_i^{e'} \phi_j^{e'} \right) \mathrm{d}\xi^e + \left(\frac{K(1-2v)^2}{2} + \frac{2g(\Gamma)(1+v_p)^2}{3} \right) \left(\frac{2}{L_p} \right) \int_0^1 \left(A(\xi^p) \phi_i^{p'} \phi_j^{p'} \right) \mathrm{d}\xi^p \Bigg] = 3K(1-2v)\alpha T \sum_{j=1}^{n} \int_0^1 \left(A(\xi) \phi_j' \right) \mathrm{d}\xi \tag{18.10}$$

Equation (18.10) is in matrix form, and a single-step matrix inversion process is used to obtain the solution of vector $\{c_i\}$ which needs to be examined for convergence of yield front location. The strain and stress fields for any specified temperature T value can be determined from the axial displacement field (u). For an expansion of elastic solution, the issue is solved in which at the zeroth approximation, $g(\Gamma)$ is set equivalent to G. The iterative scheme is followed as reported in [27, 28] for updating the value of $g(\Gamma)$ until the attainment of final convergence on yield front in subsequent approximations.

18.3 Results and Discussions

The present analysis is carried out for functionally graded non-uniform bars subjected to thermal loading. The material properties of the bar are represented in Table 18.1 that is used in the analyses taken from Farimani and Toussi [21]. The length of the bar L is taken as 1.2 m. In elastic region, Poisson's ratio v value is taken as 0.3 and in post-elastic state v_p is taken as 0.5 for the FGM bar.

The results are presented by considering three different volume fraction values for the ceramic content in the right fixed end, V_{c0}, as 0.5, 0.8 and 1. The volume fraction exponent, n, is taken as 0.5 and 2. For distinct n and V_{c0} values, a plot for

Table 18.1 Material properties

	E (GPa)	σ_y (MPa)	α $(10^{-6}/°\mathrm{C})$	E_t (GPa)	q (GPa)
Metal	208	260	11.5	80	17.2
Ceramic	324	–	7.7	–	

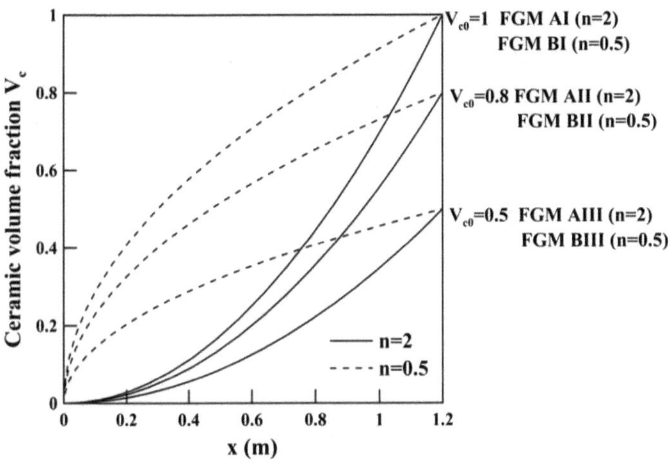

Fig. 18.3 Plots for ceramic volume fraction V_c for six materials, FGM AI, FGM BI, FGM AII, FGM BII, FGM AIII and FGM BIII

Eq. (18.1a) is displayed in Fig. 18.3 for six materials, FGM AI, FGM BI, FGM AII, FGM BII, FGM AIII and FGM BIII.

It is clear from Eqs. 18.2a–18.2c that the material properties of the bar are functions of axial location as the values of V_c and V_m are functions of axial location. The values of modulus of elasticity, tangent modulus and yield stress of the bar are given in detail in Table 18.2 at different normalized co-ordinates. The graphical representation of the variation of the material properties with axial co-ordinate is shown in Fig. 18.4 for six FGM materials.

18.3.1 Temperature Effect on FGM Bar

Three distinct types of temperature distributions are considered for the present investigation such as uniform $T(\xi) = T_1$, linearly decreasing $T(\xi) = T_1 - (\{T_1 - T_0\}\xi)$ and parabolically decreasing $T(\xi) = T_1 - (\{T_1 - T_0\}\xi^2)$ where T_0 and T_1 are ambient and maximum temperatures. In Fig. 18.5, the nature of temperature field distributions is shown in a sample plot by taking $T_0 = 20$ °C and $T_1 = 100$ °C.

The present analysis is carried out on the consideration of temperature effect on yield stress and elasticity modulus of the metallic part of the bar as obtained from Martinez [29], whereas the elasticity modulus value at ambient temperature is considered for the ceramic part although. The plot for normalized yield stress and elasticity modulus with temperature is presented graphically by best fit curve from the experimental data points taken from Martinez [29] in Fig. 18.6, where $\sigma_{yo} =$ 260 MPa and $E_o = 208$ GPa are the ambient temperature values.

Table 18.2 Material properties of the bar at different normalized co-ordinates for FGM AI

Normalized length (ξ)	Elasticity modulus, E (GPa)	Tangent modulus, E_t (GPa)	Yield stress, σ_y (MPa)
0.00	208.00	80.00	260.00
0.01	208.01	80.01	260.00
0.03	208.07	80.06	260.00
0.06	208.24	80.20	260.01
0.09	208.61	80.51	260.02
0.13	209.27	81.07	260.03
0.18	210.35	81.99	260.06
0.23	211.95	83.37	260.10
0.28	214.19	85.35	260.16
0.34	217.18	88.06	260.23
0.40	221.03	91.69	260.32
0.47	225.81	96.42	260.43
0.53	231.60	102.50	260.56
0.60	238.43	110.22	260.70
0.66	246.29	119.97	260.86
0.72	255.12	132.17	261.02
0.77	264.77	147.34	261.19
0.82	275.03	165.99	261.35
0.87	285.53	188.55	261.50
0.91	295.82	215.03	261.64
0.94	305.33	244.61	261.77
0.97	313.45	274.96	261.87
0.99	319.56	301.79	261.94
1.00	324.00	320.00	262.00

18.3.2 Validation Study

In Fig. 18.7, the comparison of the stress fields acquired from the present technique is carried out with the results of ABAQUS FEM software for clamped–clamped linear taper bar of $A_R = 0.1$, $S_R = 20$ and $L = 1.2$ m revealing coherent correlation within the limits of the software. With 2400 elements in ABAQUS, an eight-node thermal coupled linear brick element is considered for the modelling of bar under uniform thermal loading.

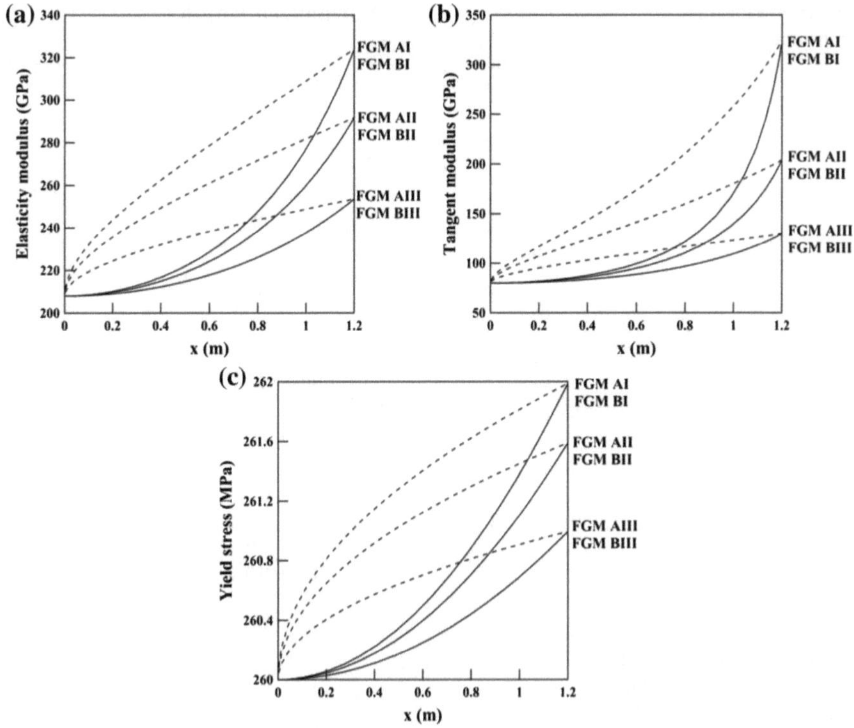

Fig. 18.4 Variation in material properties of the disc: **a** elasticity modulus, **b** tangent modulus and **c** yield stress

18.3.3 Present Results

For clamped FGM taper bar, the study on T_y and T_c is carried out by considering the effect of temperature field distributions, slenderness ratio, aspect ratio, the volume fraction of ceramic in the right end (V_{c0}) and the volume fraction exponent (n) and presented in subsequent sections. Results are exhibited for both the material behaviours: bilinear and multilinear. However, both the bar geometries, parabolic and linear tapers, are taken into consideration for bilinear material behaviour, whereas the results for linear taper geometry are exhibited for multilinear material behaviour.

18.3.3.1 Elasto-plastic Behaviour of FGM Taper Bar

For clamped–clamped FGM bar, the axial distribution of stress fields is reported in Fig. 18.8a–c through waterfall plots at different temperatures for the analysis of the propagation of yield front under thermal loading. The ceramic volume fraction at the right fixed end, $V_{c0} = 1$, and volume fraction exponent, $n = 2$, are considered for this analysis.

Fig. 18.5 Plot for temperature variations

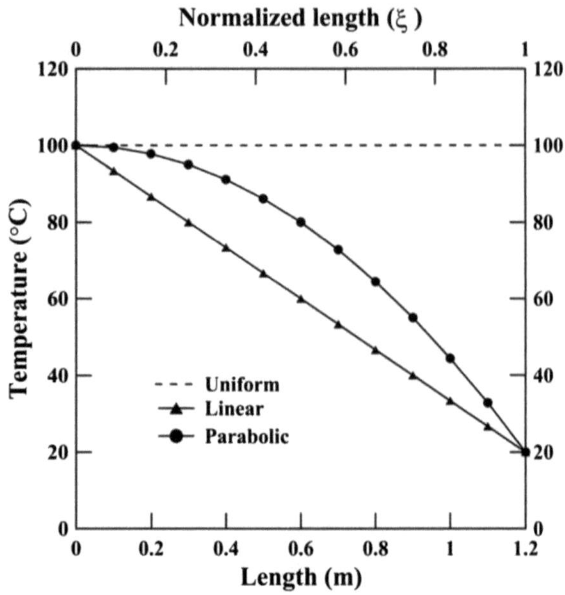

Fig. 18.6 Normalized yield stress and elasticity modulus variation with temperature of mild steel, after Martinez [29]

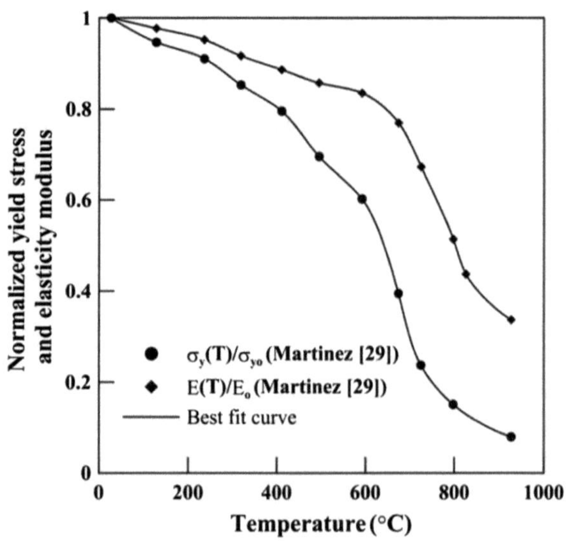

The plots are presented for a specific geometry of bar, $L = 1.2$, $S_R = 20$ and $A_R = 0.1$. These plots are presented for initial elastic limit to final collapse limit temperature range for three types of temperature variations. It is remarked that the increase in compressive stress takes place for the temperature increment for three instances of temperature variations, yielding occurs at the bar's smallest fixed end $(x = L)$, and when the temperature reaches T_c, entire bar becomes elasto-plastic.

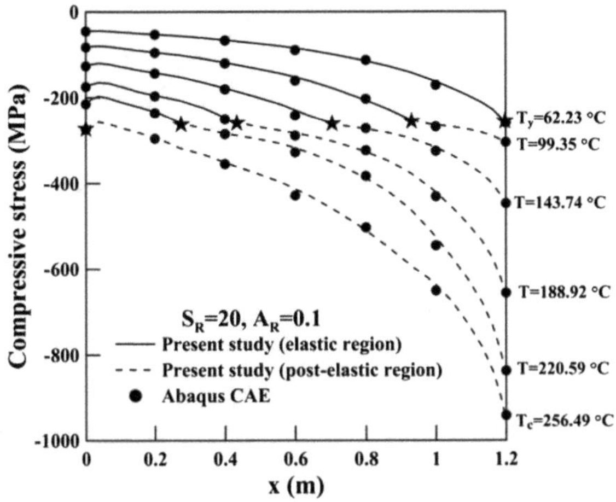

Fig. 18.7 Comparison plot for stress field for clamped–clamped taper bar

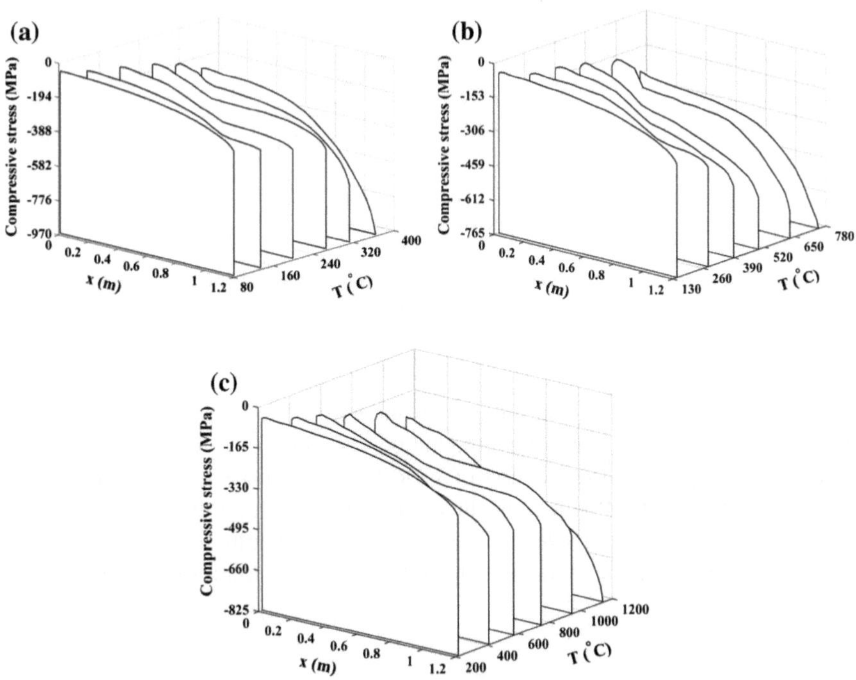

Fig. 18.8 Waterfall plot of stress fields at different temperatures for FGM taper bar of $A_R = 0.1$ and $S_R = 20$, under **a** uniform, **b** linearly decreasing and **c** parabolically decreasing temperatures

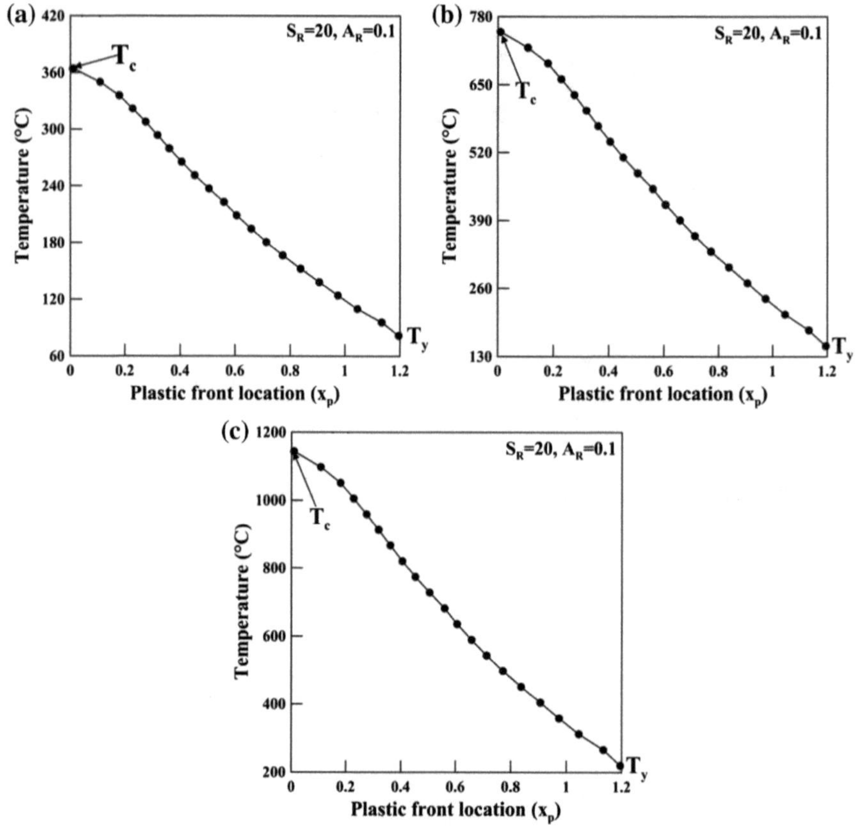

Fig. 18.9 Plot of propagation of yield front location with increase in temperature for FGM taper bar of $A_R = 0.1$ and $S_R = 20$, under **a** uniform, **b** linearly decreasing and **c** parabolically decreasing temperatures

For uniform temperature field, there is an increase in compressive stress in comparison with linear and parabolic temperature variations.

For FGM taper bar, the location of yield front at various temperature levels is captured from Fig. 18.8 with stress field waterfall plots, and the temperature with location is given in Fig. 18.9a–c. These figures demonstrate the advance of locations of yield front on increment in temperatures for three kinds of temperature field variations.

18.3.3.2 Effect of Material Parameters V_{c0} and n

To comprehend the significance of the material parameters on FGM taper bar, a parametric study by changing the values of V_{c0} for a specific volume fraction exponent, n, is performed. Similarly, for a specific ceramic content at the right end, V_{c0}, the results are obtained by changing the values of n.

Effect of the Ceramic Content at the Right Fixed End, V_{c0}

The plot for location of yield front with temperatures is presented in Fig. 18.10a–c for FGM taper bar under different types of temperature distributions. The effect of V_{c0} values is observed for the plots containing three different values for V_{c0} (equals to 0.5, 0.8 and 1) and $n = 2$. It is obvious that T_y and T_c increase with increment in V_{c0} for various temperature field variations. It is apparent from the present investigation that the limit temperatures get improved with the increase in content of ceramic in a ceramic–metal FGM bar.

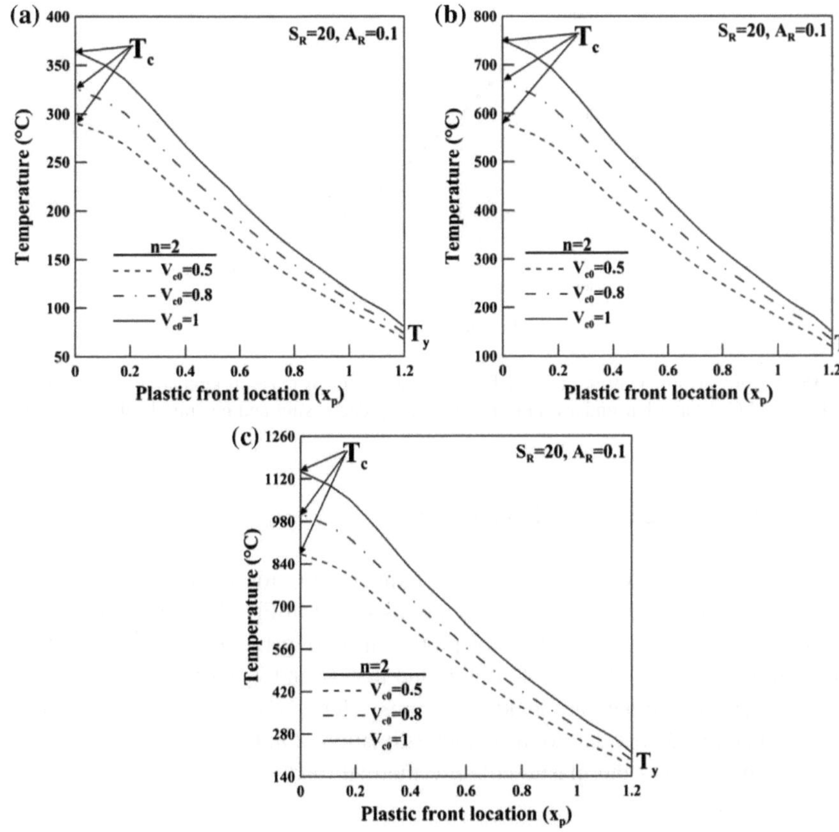

Fig. 18.10 Plot of propagation of yield front location with increase in temperature for different V_{c0} values for FGM taper bar under **a** uniform, **b** linearly decreasing and **c** parabolically decreasing temperatures

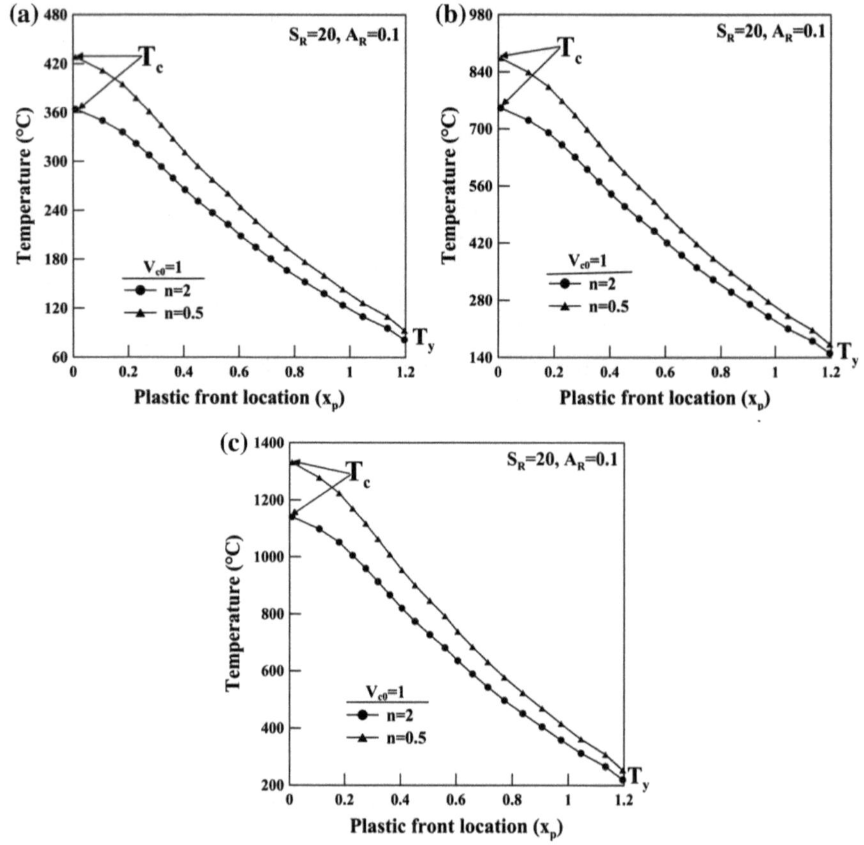

Fig. 18.11 Plot of propagation of yield front location with increase in temperature for different *n* values for FGM taper bar under **a** uniform, **b** linearly decreasing and **c** parabolically decreasing temperatures

Effect of the Volume Fraction Exponent, *n*

The plots for yield front locations with temperatures for FGM taper bar under various types of temperature variations are displayed in Fig. 18.11a–c. The impact of change in n values is observed for plots considering two different values for n (equals to 0.5 and 2) and $V_{c0} = 1$. It is found that T_y and T_c increase with decrease in n for various temperature field variations. Here, it may be presumed for a ceramic–metal FGM bar that for a smaller estimation of n, the ceramic percentage is more which improves the plastic collapse thermal load.

18.3.3.3 Elasto-plastic Behaviour of FGM Parabolic Bar

For various temperature variations, the location of yield front captured at distinct temperatures and the locations has been provided for FGM parabolic bar in Fig. 18.12. These figures also furnish the results of Fig. 18.9 for FGM taper bar in solid lines, and it is observed that T_y and T_c are lower for parabolic bar for all the temperature distributions.

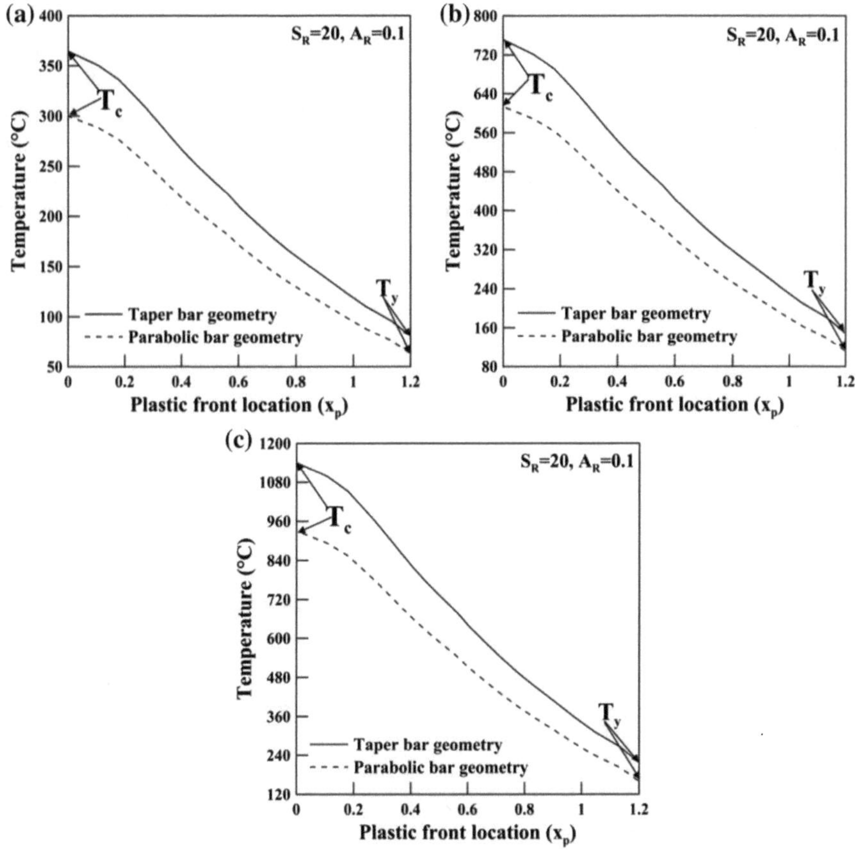

Fig. 18.12 Plot of propagation of yield front location with increase in temperature for FGM taper and parabolic bar under **a** uniform, **b** linearly decreasing and **c** parabolically decreasing temperatures

18.3.3.4 FGM Taper Bar with Multilinear (Four Segment) Material Behaviour

For FGM taper bar with multilinear material model, the results are exhibited under various types of temperature variations. In this investigation, the values of V_{co} and n are taken as 1 and 2, respectively. The waterfall plot of stress fields at different temperatures is presented in Fig. 18.13a–c for multilinear material behaviour. The comparison of these plots is made with the corresponding Fig. 18.8 waterfall plots of bilinear material behaviour. The stress field's nature is found to be identical, but there is an increase in compressive stress considering multilinear model for all the temperature field distributions.

For multilinear model, the location of yield front with temperatures is presented in Fig. 18.14a–c which depicted their difference with the comparing figure for bilinear material behaviour. These figures showed that T_y remains the same for both the material behaviours, bilinear and multilinear. However, for multilinear behaviour, the increase in T_c is more for all instances of temperature field distributions. It is also observed from both the material models that at low temperatures for

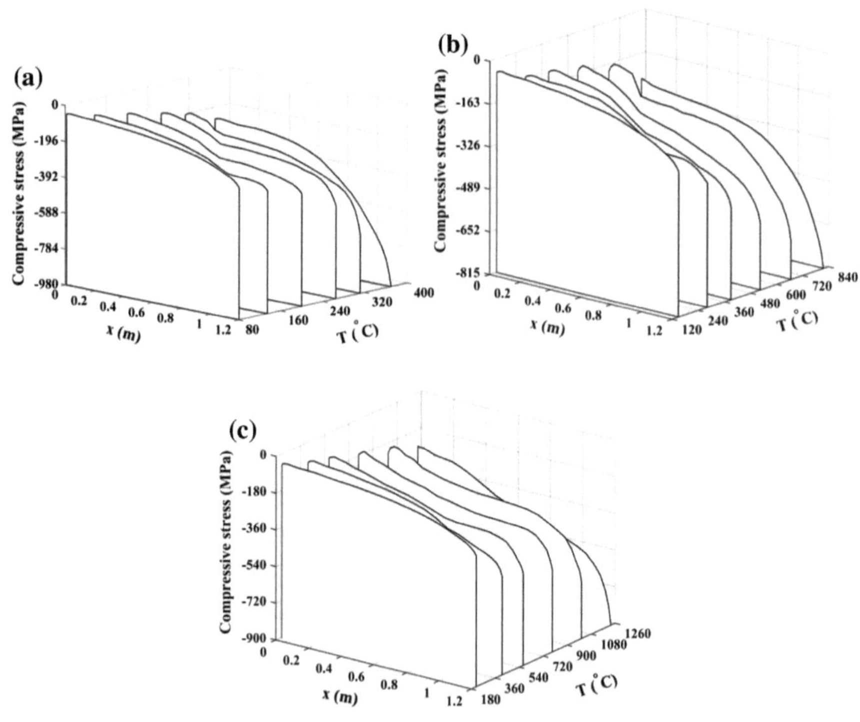

Fig. 18.13 Waterfall plot of stress fields at different temperatures for FGM taper bar of $A_R = 0.1$ and $S_R = 20$, for multilinear material model under **a** uniform, **b** linearly decreasing and **c** parabolically decreasing temperatures

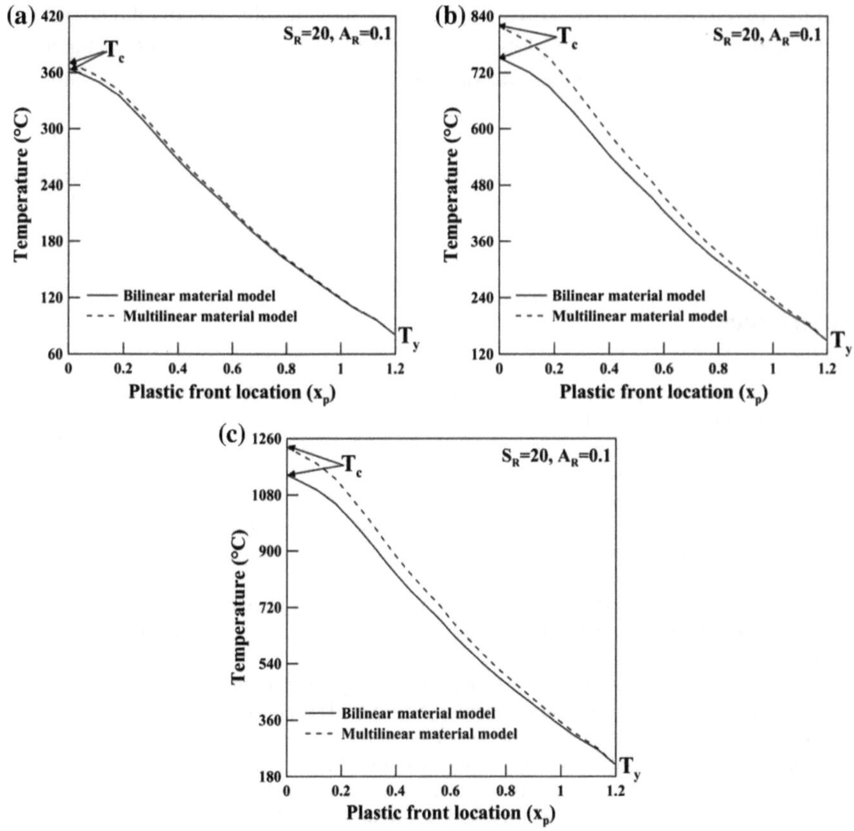

Fig. 18.14 Plot of propagation of yield front location with increase in temperature for FGM taper bar of $A_R = 0.1$ and $S_R = 20$, for both the material models under **a** uniform, **b** linearly decreasing and **c** parabolically decreasing temperatures

uniform temperature distribution, the difference in the location of yield fronts is not much prominent. However, for linear and parabolic temperature field variations, there is a significant difference in the location of yield fronts.

18.4 Conclusion

The thermo-elasto-plastic stress fields and propagation of yield front for clamped FGM non-uniform bar are formulated through a variational method by employing Hencky's deformation theory of plasticity and von Mises yield criterion in an iterative manner and considering bilinear and multilinear material behaviours. The outcomes acquired from the present method are compared with those from ABAQUS FEM software for a clamped taper bar subjected to uniform thermal loading, revealing coherent correlation. Some new results are displayed for the stress fields through waterfall plots and growth of plastic fronts for FGM taper and parabolic bar geometries under various kinds of temperature variations, and it is observed for all the cases that for parabolic temperature variation, the yield and collapse temperature are higher. The effect of material parameters V_{c0} and n is considered for FGM taper bars, and it is observed that the limit temperatures get improved with the increase in content of ceramic in a ceramic–metal FGM bar.

References

1. Yamanoushi, M., Koizumi, M., Hiraii, T., Shiota, I.: Proceedings of the First International Symposium on Functionally Gradient Materials, Japan (1990)
2. Timoshenko, S.: Strength of Materials: Part I Elementary Theory and Problems. D. Van Nostrand Company Inc, New York (1930)
3. Hill, R.: The Mathematical Theory of Plasticity. Oxford University Press, Oxford (1950)
4. Niknam, H., Fallah, A., Aghdam, M.M.: Nonlinear bending of functionally graded tapered beams subjected to thermal and mechanical loading. Int. J. Non-Linear Mech. **65**, 141–147 (2014)
5. Wattanasakulpong, N., Prusty, B.G., Kelly, D.W.: Thermal buckling and elastic vibration of third-order shear deformable functionally graded beams. Int. J. Mech. Sci. **53**, 734–743 (2011)
6. Paul, A., Das, D.: Non-linear thermal post-buckling analysis of FGM Timoshenko beam under non-uniform temperature rise across. Eng. Sci. Technol. Int. J. **19**, 1608–1625 (2016)
7. Nayak, P., Saha, K.N.: Optimized design of thermo-mechanically loaded non-uniform bars by using a variational method. J. Phys. Conf. Ser. **738**, 012002 (2016)
8. Nayak, P., Saha, K.N.: Growth of yield front in thermo-mechanically loaded non-uniform bars using an iterative variational method. J. Strain Anal. **53**, 66–79 (2018)
9. Garg, S., Pant, M.: Numerical simulation of thermal fracture in functionally graded materials using element-free Galerkin method. Sadhana-Acad. Proc. Eng. Sci. **42**, 417–431 (2017)
10. Cho, J.R., Oden, J.T.: Functionally graded material: a parametric study on thermal-stress characteristics using the Crank-Nicolson-Galerkin scheme. Comput. Meth. Appl. Mech. Eng. **188**, 17–38 (2000)
11. Tayyar, G.T.: A new approach for elasto-plastic finite strain analysis of cantilever beams subjected to uniform bending moment. Sadhana-Acad. Proc. Eng. Sci. **41**, 451–458 (2016)
12. Horgan, C.O., Chan, A.M.: Torsion of functionally graded isotropic linearly elastic bars. J. Elast. **52**, 181–199 (1999)
13. Horgan, C.O.: On the torsion of functionally graded anisotropic linearly elastic bars. IMA J. Appl. Math. **72**, 556–562 (2007)
14. Katsikadelis, J.T., Tsiatas, G.C.: Saint-Venant torsion of non-homogeneous anisotropic bars. J. Appl. Comput. Mech. **2**, 42–53 (2016)
15. Ecsedi, I.: Some analytical solutions for Saint-Venant torsion of non-homogeneous anisotropic cylindrical bars. Mech. Res. Commun. **52**, 95–100 (2013)
16. Kolodziej, J.A., Gorzelanczyk, P.: Application of method of fundamental solutions for elasto-plastic torsion of prismatic rods. Eng. Anal. Bound. Elem. **36**, 81–86 (2012)
17. Dwivedi, J.P., Shah, S.K., Upadhyay, P.C., Talukder, N.K.D.: Springback analysis of thin rectangular bars of non-linear work-hardening materials under torsional loading. Int. J. Mech. Sci. **44**, 1505–1519 (2002)
18. Tsiatas, G.C., Babouskos, N.G.: Elastic-plastic analysis of functionally graded bars under torsional loading. Compos. Struct. **176**, 254–267 (2017)

19. Sapountzakis, E.J., Tsipiras, V.J.: Nonlinear inelastic uniform torsion of composite bars by BEM. Comput. Struct. **87**, 151–166 (2009)
20. Huang, H., Chen, B., Qiang, H.: Investigation on buckling behaviours of elastoplastic functionally graded cylindrical shells subjected to torsional loads. Compos. Struct. **118**, 234–240 (2014)
21. Farimani, M.R., Toussi, H.E.: Plastic limit speed of FGM disc due to the variation of temperature and material composition. J. Reinf. Plast. Compos. **32**, 1257–1267 (2013)
22. Tamura, I., Tomota, Y., Ozawa, H.: Strength and ductility of Fe–Ni–C alloys composed of austenite and martensite with various strength. Proc. Third Int. Conf. Strength Met. Alloy., Cambridge Institute of Metals **1**, 611–615 (1973)
23. Bocciarelli, M., Bolzon, G., Maie, G.: A constitutive model of metal–ceramic functionally graded material behaviour: formulation and parameter identification. Comput. Mater. Sci. **43**, 16–26 (2008)
24. Gao, D.Y., Ogden, R.W.: Advances in Mechanics and Mathematics, vol. II. Kluwer Academic publishers, Boston (2003)
25. Nakamura, T., Wang, T., Sampath, S.: Determination of properties of graded materials by inverse analysis and instrumented indentation. Acta Mater. **48**, 4293–4306 (2000)
26. Suresh, S., Mortensen, A.: Fundamentals of functionally graded materials. IOM Communications Ltd, London (1998)
27. Kachanov, L.M.: Foundations of the theory of plasticity. North-Holland Publication Co., Amsterdam (1971)
28. Nayak, P., Saha, K.N.: Analysis of statically indeterminate non-uniform bar problem in post elastic domain by an iterative variational method. Appl. Math. Model. **51**, 86–108 (2017)
29. Martinez, I.: Thermal effects on materials. http://webserver.dmt.upm.es/~isidoro/ot1/Thermal%20effects%20on%20materials.pdf (online article 2016)

Chapter 19
Nonlinear Static Analysis and Superharmonic Influence on Nonlinear Forced Vibration of Timoshenko Beams

Brajesh Panigrahi and Goutam Pohit

Nomenclature

A_j	Temporal coordinates
B_j	Temporal coordinates
b	Width of beam
C_j	Temporal coordinates
E	Young's modulus
F	Amplitude of loading
h	Depth of beam
K_1, K_2, K_3, K_4	Stiffness parameters
k_3, k_4	Stiffness parameter (dimensionless)
L	Length of beam
M_1, M_2, M_3	Inertial parameters
m_3	Inertial parameter (dimensionless)
N	Number of polynomial terms
T^*	Kinetic energy
T^{**}	Kinetic energy (dimensionless)
U	Longitudinal displacement
U^*	Potential energy
U^{**}	Potential energy (dimensionless)
u	Axial displacement (dimensionless)
W	Transverse displacement
w	Transverse displacement (dimensionless)
x_f	Load application point

B. Panigrahi (✉) · G. Pohit
Department of Mechanical Engineering, Jadavpur University, Kolkata 700032, India
e-mail: brajeshpanigrahi248@gmail.com

P. Sahoo and J. P. Davim (eds.), *Advances in Materials, Mechanical and Industrial Engineering*, Lecture Notes on Multidisciplinary Industrial Engineering,
https://doi.org/10.1007/978-3-319-96968-8_19

Greek Symbols

ν Poisson's ratio (dimensionless)
ξ Normalized axial coordinate
ξ_f X_f/L
ρ Mass density
Φ Polynomial functions
Ψ Rotational displacement
ψ Normalized rotational displacement
Ω Frequency of excitation
ω Normalized frequency

19.1 Introduction

Structures in practices, most of the time shows nonlinear characteristics. If these are not considered for the design purpose, it may lead to faulty design and hampers the operation in subsequent process. Therefore, engineers must consider these nonlinearities while setting design parameters of such structural systems. Nature of the static as well as the dynamic response changes considerably due to presence of such nonlinearity. In the analysis of structural components such as beams, two types of nonlinearities are commonly encountered, namely geometric nonlinearity and material nonlinearity. Material nonlinearity leads to a nonlinear stress–strain relation. Large amplitude motion causes a nonlinear strain–displacement relation, which in turn leads to a nonlinear force–deflection relation in static response, whereas in dynamics it leads to amplitude-dependent frequencies (generally referred as nonlinear frequencies). This type of nonlinearity is called as geometric nonlinearity. Therefore, large amplitude vibration of beam-like structure is studied by various authors. Modeling as well as analysis of such system is studied by various authors adopting various methods. Nonlinear static and dynamic analysis of beam-like structures is a well-established area of research. Numbers of theoretical, analytical, and experimental methods, regarding this field, have been proposed as of now. Some of these methods are published in the textbooks [1, 2].

Apart from the textbooks, there are various literatures available on the nonlinear analysis of beams. Nonlinear force–displacement analysis is investigated in various literatures [3–6]. Newmark [7] proposed a computational method based on direct time integration to analyze the dynamic response for structures of any degree of complications. Eisley [8] analyzed the dynamics of different structures in post-buckled region. Herrmann and Chu [9] studied the effect of large amplitudes on vibration of plate with hinged ends. Yamaki [10] proposed a modal analysis by reducing independent set of differential equation of motion with quadratic nonlinearity. Hsu [11] published a detailed study on the systems having Duffing's-type equation of motion, known to respond in the form of elliptic functions during its free vibration. The period of which can be expressed as elliptic integrals. Author analyzed the response of such systems subjected to periodic elliptical forcing functions. Evensen [12] used the perturbation method to analyze amplitude-dependent frequencies of beams.

Tseng and Dugundji [13] analyzed dynamic behavior of beams considering harmonic balance method. Authors also reported the experimental results obtained from a shake test. In their later studies, Tseng and Dugundji [14] extended the previous work for buckled beams. Bennett and Eisley [15] studied the steady-state nonlinear transverse response for free and forced vibrating conditions with elastic restrained ends. For the spatial form of solution, some linear mode shapes are assumed and nonlinear ordinary differential equations are obtained, which are solved in order to obtain time variables. Mei [16] analyzed nonlinear dynamic response of initially excited beams following a finite-element matrix displacement technique. Stupnicka [17–19] used the generalized Ritz approximation and by balancing the different harmonics, primary as well as higher mode of vibration in nonlinear multi-DOF systems and the parametric dynamic systems are studied. Rao and Raju [20] proposed a method, which avoids inclusion of geometric stiffness matrices in the formulation for nonlinear dynamics of initially excited beams. Lau and Cheung [21] studied the steady-state response of different types of nonlinearity by balancing the harmonics followed by the Newton–Raphson method. Later, Lau et al. [22] used perturbation method in conjunction with above method for nonperiodic motion.

Bennouna and white [23] studied the effect of large amplitude vibration on fundamental mode shapes. Experimental mode shapes are used to obtain higher-order differentiation with respect to space coordinates using polynomial fittings of deflection curves. Furthermore, in subsequent studies, Bennouna and White [24] investigated the influence of excitation and the induced harmonic distortion in fatigue life of the structure vibrating nonlinearly. Benamar et al. [25] derived a theoretical modal based on spectral analysis to study dynamics of beam undergoing large amplitude motion. Nonlinear sets of algebraic equations are deduced and subsequently solved using suitable technique. In their later study, Benamar et al. [26] studied the nonlinear behavior of clamped homogenous plates under electrodynamic point excitation at the center. In their studies, it was evident that clearly due to large amplitude motion the responses are quite different from the linear approximations. Following the pervious works, Azrar et al. [27] used Lagrange principle followed by balancing of the harmonics terms to derive nonlinear algebraic equation to analyze the nonlinear dynamic response of externally excited beams with C–C and S–S boundary conditions. Authors have presented a single-mode analysis. Methodology reveals the fact that considering a single mode, the equations may be reduced to multi-dimensional Duffing's equations. In a later study, El Kadiri et al. [28] extended this methodology to analyze second nonlinear vibrational mode for plates. However, in subsequent study, Azrar et al. [29] proposed a multi-modal approach to analyze steady-state dynamic response for externally periodically excited beams. The methodology proposed by Benamar et al. [25] is extended for the analysis of coupled transverse–circumferential mode shapes of shells and composite plates by Harras et al. [30]. El Kadiri and Benamar [31] modified the previous method based on the linearization of the nonlinear algebraic equations, written on the modal basis, in the neighborhood of each resonance for beams. Qaisi [32] proposed an analytical method based on harmonic balance method to analyze the periodic response of nonlinear free vibration of beams. Ribeiro and Petyt [33] used harmonic balance method and hierarchical finite-element method to investigate the geometrically nonlinear free and forced dynamic response of thin beams. In the later studies, Ribeiro [34] clubbed the shooting method with above methodology to analyze nonlinear dynamics of beam and plate. Chen et al. [35] studied the nonlinear forced response of an axially moving beam with internal resonance between the first two transverse modes. Multi-dimensional Lindstedt–Poincare and Galerkin methods are used for analysis purpose. Cheung and Chen [36] used the incremental harmonic balance methods to analyze nonlinear dynamic responses of different structure systems. Ramezani et al. [37] studied nonlinear vibration of C–C microbeams using thick beam theory. They demonstrated that for slender beams, thin beam theory can be adopted instead of thick beam theory.

Luongo et al. [38] studied the frequency–amplitude response and the stability characteristics of thick beams using Timoshenko beam theory. Authors have derived a unique nonlinear equation of motion consisting of only one differential equation and solved them using perturbation technique. Lenci et al. [39] have focused on clearly showing the transition of hardening to softening behavior and vice versa for nonlinear dynamic response of initially excited Euler–Bernoulli and Timoshenko beams. In another study, Clementi et al. [40] extended this discussion on transition to softening or hardening behavior using an analytical and a numerical finite-element technique as well. A relatively easier method for approximate solution of such nonlinear system is assumption of simple harmonic motion, which reduces the equations of motion as nonlinear algebraic equations [41]. Kitipornchai et al. [42] proposed an energy method-based technique followed by an iterative scheme to analyze nonlinear dynamic behavior of initially excited functionally graded thick beams with edge crack.

In this study, a methodology is proposed by balancing the harmonic terms followed by an iterative technique. Present method is computationally easier than incremental harmonic balance method. Unlike the incremental harmonic balance method, in the present method governing differential equations are converted into set of nonlinear algebraic equations directly using harmonic balance method. Nonlinear stiffness matrices are simplified in order to eliminate the temporal terms present within the matrices. These algebraic equations are solved using an iterative technique based on the modified direct substitution (incremental amplitude technique). The methodology is quite simple and computationally efficient. Impact of the higher-order harmonic terms in the solution of such nonlinear problems is shown by considering different number of harmonic terms. Once the accuracy of the method is established, static and dynamic analysis for various structural parameters is obtained.

19.2 Problem Formulation

Figure 19.1 represents a C–C beam of length 'L', depth 'h', and width 'b' under harmonic excitation $F\cos(\Omega t)$. Ω and F be frequency and amplitude of excitation, respectively. Following Timoshenko beam theory and considering Von Kerman's nonlinear strain–displacement relation, expression for kinetic energy (T^*) and potential energy (U^*) can be obtained as mentioned in Eqs. (19.1) and (19.2), respectively.

$$U^* = \frac{1}{2}\int_0^L \int_{-h/2}^{h/2} \frac{E}{1-\upsilon^2}\left\{\left[\frac{\delta U}{\delta x} + z\frac{\delta \Psi}{\delta x} + \frac{1}{2}\left(\frac{\delta W}{\delta x}\right)^2\right]^2 + \frac{E}{2(1+\upsilon)}\left(\frac{\delta W}{\delta x} + \Psi\right)^2\right\} \mathrm{d}z\mathrm{d}x \tag{19.1}$$

$$T^* = \frac{1}{2}\int_0^L \int_{-h/2}^{h/2} \rho\left\{\left(\frac{\delta U}{\delta t} + z\frac{\delta \Psi}{\delta t}\right)^2 + \left(\frac{\delta W}{\delta t}\right)^2\right\} \mathrm{d}z\mathrm{d}x \tag{19.2}$$

In Eqs. (19.1) and (19.2), U, W, and Ψ are longitudinal, transverse, and rotational displacements, respectively. E, ρ, and υ are Young's modulus, mass density, and Poisson's ratio. Taking the stiffness and inertial parameters as given in Eq. (19.3) and considering the dimensionless parameter as mentioned in Eq. (19.4) and dividing energy expressions by K_1h^3/L^2, one may reduce the nondimensional kinetic (T^{**}) and potential (U^{**}) energy expressions in the form of Eq. (19.5)

$$\{K_1, K_2, K_3\} = \int_{-h/2}^{h/2} \frac{E}{1-\upsilon^2}\{1, z, z^2\}\mathrm{d}z,$$

$$K_4 = \int_{-h/2}^{h/2} \kappa\frac{E}{2(1+\upsilon)}\mathrm{d}z,$$

$$\{M_1, M_2, M_3\} = \int_{-h/2}^{h/2} \rho\{1, z, z^2\}\mathrm{d}z \tag{19.3}$$

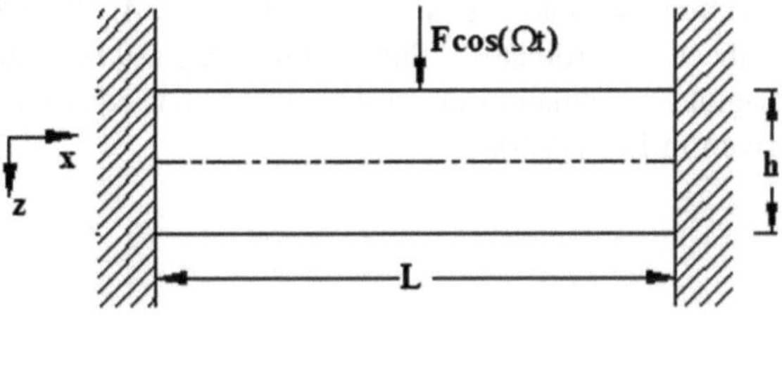

Fig. 19.1 C–C beam under harmonic excitation

$$\xi = \frac{x}{L}, u = \frac{U}{h}, w = \frac{W}{h}, \psi = \Psi, \alpha = \frac{L}{h}$$
$$\{m_3, k_3, k_4\} = \left\{\frac{M_3}{M_1 h^2}, \frac{K_3}{K_1 h^2}, \frac{K_4}{K_1}\right\}, \tau = \Omega t, \omega^2 = \frac{M_1 \Omega^2 L^3}{K_1 h} \tag{19.4}$$

$$U^{**}_{\text{linear}} = \frac{\alpha}{2}\int_0^1 \left\{\left(\frac{\delta u}{\delta \xi}\right)^2 + k_3\left(\frac{\delta \psi}{\delta \xi}\right)^2 + k_4\left(\frac{\delta w}{\delta \xi}\right)^2 + k_4\alpha^2\psi^2 + k_4\alpha\psi\frac{\delta w}{\delta \xi}\right\} \mathrm{d}\xi$$
$$U^{**}_{\text{nonlinear}} = \frac{\alpha}{2}\int_0^1 \left\{\frac{1}{\alpha}\left(\frac{\delta u}{\delta \xi}\right)\left(\frac{\delta w}{\delta \xi}\right)^2 + \frac{1}{4\alpha^2}\left(\frac{\delta w}{\delta \xi}\right)^4\right\} \mathrm{d}\xi$$
$$T^{**} = \frac{\omega^2}{2}\int_0^1 \left\{\left(\frac{\delta u}{\delta \tau}\right)^2 + m_3\left(\frac{\delta \psi}{\delta \tau}\right)^2 + \left(\frac{\delta w}{\delta \tau}\right)^2\right\} \mathrm{d}\xi \tag{19.5}$$

Similarly, dividing load potential by K_1h^3/L^2 potential due to external load (V^*) is derived and mentioned in Eq. (19.6).

$$V^* = \frac{F \cos(\tau) w_{(\xi=\xi_f)} L^2}{K_1 b h^2} = P\cos(\tau) \tag{19.6}$$

One may deduce the functional of the problem as

$$J = T^{**} - \left(U^{**}_{\text{linear}} + U^{**}_{\text{nonlinear}} + V^*\right) \tag{19.7}$$

Displacement field is written in separable form of temporal and spatial coordinates as

$$u = \sum_{j=1}^{N} A_j(\tau)\Phi_{uj},\; w = \sum_{j=1}^{N} B_j(\tau)\Phi_{wj},\; \psi = \sum_{j=1}^{N} C_j(\tau)\Phi_{\psi j} \tag{19.8}$$

In Eq. (19.8), Φ is set of coordinate functions which must satisfy at least the geometric boundary condition [11]. For clamped–clamped case, the trial function Φ must be chosen in such a way that it should satisfy the essential boundary condition, i.e., displacements at both the clamped boundaries must be zero. Therefore, trail functions are taken as $\Phi_{u,w,\psi} = \xi^j(1-\xi)$. A_j, B_j, and C_j are time-dependent generalized coordinates. Substituting Eq. (19.8) into Eq. (19.7) and applying Lagrange's equation governing differential equation of motion can be derived as given in Eq. (19.9).

$$\omega^2 M\ddot{q} + Kq + \{K_{nl1}(w) + K_{nl2}(w^2)\}q = P\cos\ \tau \tag{19.9}$$

In Eq. (19.9), dot denotes differentiation with respect to τ. M and K are mass and linear stiffness matrices. K_{nl1} and K_{nl2} are nonlinear stiffness matrices depending on the displacement field, and q contains temporal coefficients as $q = [A_j, B_j, C_j]^T$. In order to obtain steady-state solution, these time-dependent terms can be eliminated from nonlinear stiffness matrices and Eq. (19.9) can be reduced to a Duffing's equation with quadratic and cubic nonlinearity. P is the load vector containing amplitude of excitation. Governing differential equation of motion can be simplified into a nonlinear set of algebraic equations using harmonic balance method [1]. Considering different numbers of harmonic terms $q = \sum_{i=0}^{i=nc} q_i \cos\,(i\tau)$ and expanding the higher powers of cos (τ) in the form of higher-order harmonics, i.e., cos (τ), cos (2τ), cos (3τ), then equating the coefficients of each order of harmonics, nonlinear set of algebraic equations are derived. In the present work, order of harmonic terms is considered up to three terms. In the first case, a simple harmonic assumption is considered which reduces Eq. (19.9) into

$$-\omega^2 Mq + Kq + \{K_{nl1}(q) + K_{nl1}(q^2)\}q = P \tag{19.10}$$

Considering different numbers of harmonic terms, one may derive the governing algebraic equations as mentioned in [1]. Final set of algebraic equations, which can be derived after the balancing of harmonic terms, are mentioned in Table 19.1. It can be noticed that as the number of harmonic terms is increasing, the number of unknown vectors (q_n, q_m, q_p) is also increasing. However, equating the coefficient of harmonic terms' relations between these vectors can be derived. According to classical harmonic balance method used for a single degree of freedom system Duffing's equation, it is known that the coefficients of harmonics other than the first harmonics are of the order of square of coefficient of first harmonics (q_n, $q_p = O\,(q_m^2)$).

These nonlinear sets of equations are solved using an iterative technique based on modified direct substitution (amplitude incremental method). Solution process begins with the static solution of the nonlinear system. Maximum displacement obtained from the static solution is treated as the response amplitude for zero excitation. All the static analysis in the present paper is done by considering the zero excitation frequency. Increment in the amplitude is given, and the

Table 19.1 Final set of algebraic equations derived for different number of harmonics considered

Number of harmonics considered	Final equations to be solved
$q = q_m \cos\,(\tau)$	$-\omega^2 Mq_m + Kq_m + \frac{3}{4}\{K_{nl2}(q_m^2)\}q_m = P$
$q = q_n + q_m \cos\,(\tau)$	$\{-\omega_e^2 M + K + 2[(K_{nl1})q_n] + \frac{3}{4}[(K_{nl2})q_m^2]\}q_m = P$
$q = q_n + q_m \cos\,(\tau) + q_p \cos\,(2\tau)$	$\left\{\begin{matrix} -\omega_e^2 M + K + 2[(K_{nl1})q_n] \\ + [(K_{nl1})q_p] + \frac{3}{4}[(K_{nl2})q_m^2] \end{matrix}\right\} q_m = P$

displacement field is scaled up according to the incremented amplitude. These equations are solved by taking an assumed amplitude parameter. Displacement fields obtained from the static analysis are scaled up with the amplitude parameter, and convergence is searched for the nonlinear frequency. Once the convergence in nonlinear frequency is obtained, load step is solved and amplitude parameter is incremented for new load step. Therefore, in the method instead of providing a frequency sweep, amplitude is incremented and frequency is searched in subsequent iteration.

19.3 Results and Discussion

Analysis is performed following the methodology presented in above segment. Nonlinear dynamic response is shown as response plots. Ordinates are assigned as maximum normalized amplitude of vibration (W_{max}/h). Abscissas represent nonlinear frequency ratios ($\omega_{excitation}/\omega_{linear\ natural}$). It is to be noticed here that ordinates are plotted for absolute values of maximum amplitudes in response curves and the ordinates in the deformed shape plots are having the exact values with its original sign. In the backward sweep, the excitation frequency is greater than fundamental natural frequency. Therefore, it is obvious that response will be out of phase with that of in the forward sweep. However, all the results presented here are plotted with absolute values of maximum amplitude. Since the methodology captures the steady-state response, it automatically assumes that the nonlinear response frequency is equal to the frequency of excitation. For the static analysis, results are furnished with load–displacement plane.

For insuring preciseness in the results obtained by present methodology, it is mandatory to perform a validation study which sets the minimum required number of terms in the trail functions without changing efficacy of obtained results. In the present methodology, dynamic analysis starts with the assumption of initial static analysis (zero excitation frequency). Therefore, validation study is only focused for the dynamic problem. It can be noted that an accurate dynamic model will automatically lead to an accurate static model. Subsequent to the validation study, static and dynamic responses are captured for various conditions. In order to study the effect of slenderness, load–displacement plots are obtained for different slenderness (L/h ratios) and for different load application points (x_f/L). Lastly, frequency–amplitude responses are obtained in order to show the influence of slenderness and load application points on the nonlinear dynamic behavior of beams.

19.3.1 Validation Study

Before generating results, it becomes necessary to ensure what should be least numbers of terms in the trail functions for small as well as large amplitude of free

Table 19.2 Convergence study of number for polynomial terms for free vibration of beam with (L/h = 6, h = 0.1 m, E_1 = 70 GPa, ρ_1 = 2780 kg/m^3)

Number of polynomial terms	Nonlinear frequency in dimensionless form	
	W_{max}/h = 0.2	W_{max}/h = 1
2	0.494447117	0.631092806
4	0.456829499	0.613334074
6	0.456808633	0.613083836
8	0.456800326	0.613073507
10	0.456800249	0.613072947
12	0.456800247	0.613072831

vibration (F = 0). In this respect, Table (19.2) shows the nonlinear frequencies for different amplitudes of motion (lower as well as higher). Properties used for this particular study are as below:

$$h = 0.1 \text{ m}, L = 0.6 \text{ m}, \rho_1 = 2780; E_1 = 70 \text{ GPa}.$$

Data shown in Table (19.2) clarifies that eight numbers of terms in trail function are sufficient enough to capture the accurate results for the couple of cases. Therefore, eight numbers of terms in trail functions are considered for the analysis purpose throughout this work. To ascertain validity of methodology, free vibration results (F = 0) are compared with that of results obtained in [41]. For validation purpose, a constant along with two harmonic terms are taken as $q = q_n + q_m \cos\,\tau + q_p \cos\,2\tau$. Table 19.3 shows very good agreement with referred values, insuring the method to be valid. Figure 19.2 represents the frequency–amplitude response of beam having L = 0.406 m, b = 0.02 m, h = 0.002 m, E = 71.72 GPa, and ρ = 2800 kg/m^3, amplitude of the load F = 0.134 N with different number of harmonic terms considered. Results obtained by Ribeiro [34] are also plotted with the results to show accuracy of results with increasing numbers of harmonic terms.

Table 19.3 Comparison of nonlinear frequency ratio with [41]

w/radius of gyration	$\omega_{nonlinear}/\omega_{linear}$	
	Present	Reference [41]
0.2	1.0008975	1.0008946
0.4	1.0035984	1.0035740
0.6	1.0081252	1.0080236
0.8	1.0142312	1.0142203
1.0	1.0223512	1.0221322
1.5	1.0514213	1.0491411
2	1.0874562	1.0858255
2.5	1.1332410	1.1312446
3	1.1854123	1.1843938

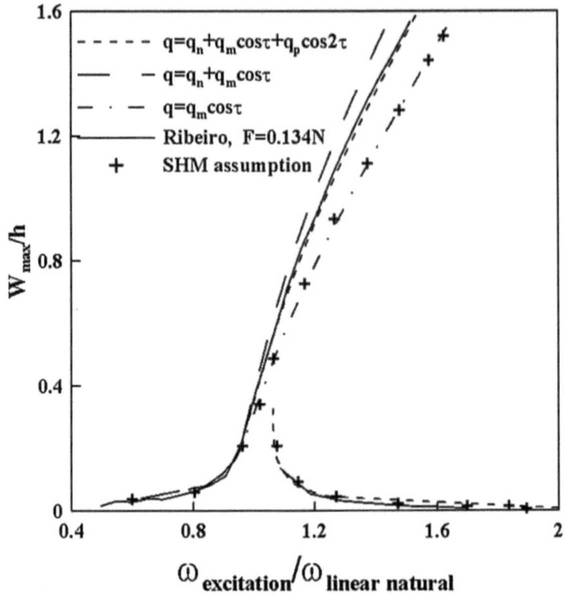

Fig. 19.2 Frequency–amplitude response of beam considering different number of harmonic terms

It is evident from Fig. 19.2 that only one harmonic term yields almost identical results as obtained by assuming simple harmonic motion. With the increase in number of these terms, accuracy of solution also improves and for one constant and two harmonic terms the results are very close to the accurate results obtained by Ribeiro [34]. In Fig. 19.3, similar results are shown with a relatively thicker beam with other properties remain similar as mentioned before, except the thickness is now $L/20$. SHM assumption and one constant with two harmonic terms are plotted in Fig. 19.3. Significant variation is observed for both the cases.

19.3.2 Static Analysis

In Fig. 19.4, nonlinear static analysis is represented as load–deflection curve. Influence of load application position on load–deflection curve is shown in Fig. 19.4 for different slenderness (L/h = 200, 175, 150, 125, 100, 75, 50, and 25). It can be observed that load application points are varied from 0.5 to 0.9 considering the symmetric nature of boundary condition (clamped-clamped) taken in present work. For large deflections, nonlinearity in load–deflection response can be clearly seen. However as the load application point shifts toward any of the fixed ends, large amount of load is required for a large deflection as the concentrated external loadings are most effective when applied at middle of the beam. Similarly as the slenderness (L/h) reduces, response shows lower values of deflection. Therefore, higher loads are required for less slender beams for larger displacements.

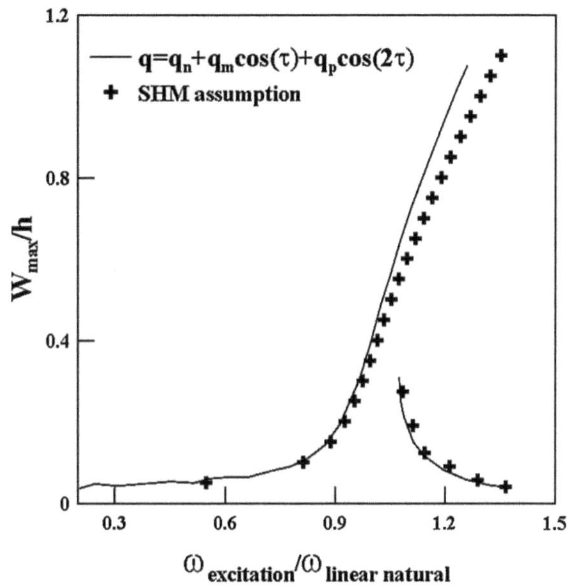

Fig. 19.3 Frequency–amplitude response of thick beam considering SHM assumption and a constant with two harmonic terms

Once the influence of load application points and slenderness is studied, it is required to study the level of loads required to produce some prescribed values of deflection. Figure 19.5 shows the variation in the applied load required to produce a same static deflection with its application points. These variations are shown for different slenderness (*L*/*h*) as well. It can be observed that a very large amount of load is required to produce same amount of deflection when load application point is near to any of the fixed ends. Same is true for less slender beams as well.

19.3.3 Dynamic Analysis

Figure 19.6 represents nonlinear frequency–amplitude response subjected to different loading amplitudes. Backbone curve is generated considering the free vibration characteristics. It can be observed that although static deflections are quite different of various amplitudes of loading but as the excitation frequency increases each of the response curves tends to bend toward right and follows the corresponding backbone curves. Each response curve is comprised of two different branches. Higher branch of response curves are generated during forward sweep, whereas lower branch of response curves are generated during backward sweep. In all the cases, a jump phenomenon can be clearly observed. In addition to that, a hardening type of nonlinearity is also common in all cases. Figure 19.7 shows the frequency–amplitude responses for different slenderness (*L*/*h* = 200, 175, 150, 125, 100, 75, and 50) for an amplitude of excitation *F* = 5 N. It can be stated clearly that frequency–amplitude response become steeper after particular value of excitation

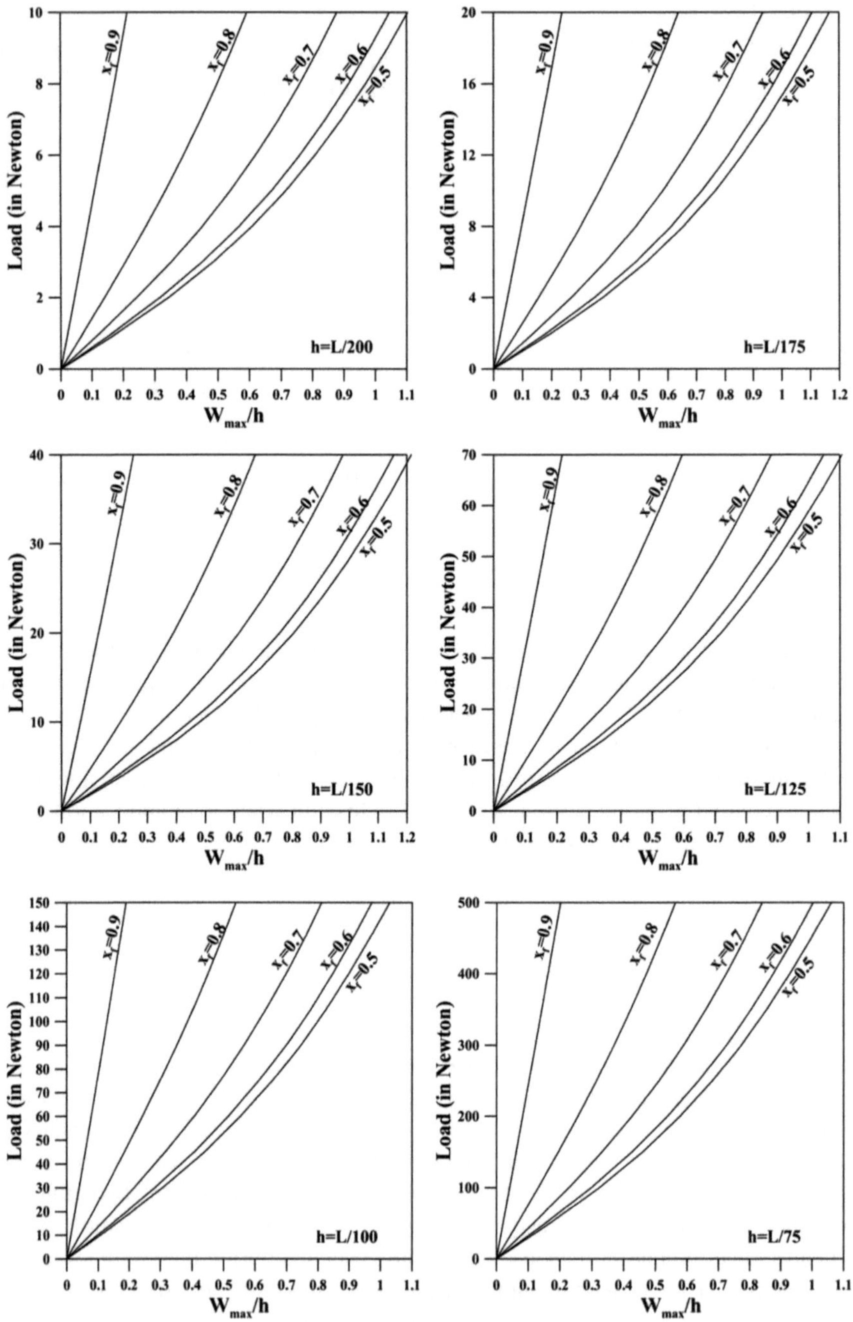

Fig. 19.4 Load–displacement response of beams with $L = 0.402$ and for different slenderness ($L/h = 200, 175, 150, 125, 100, 75, 50$, and 25)

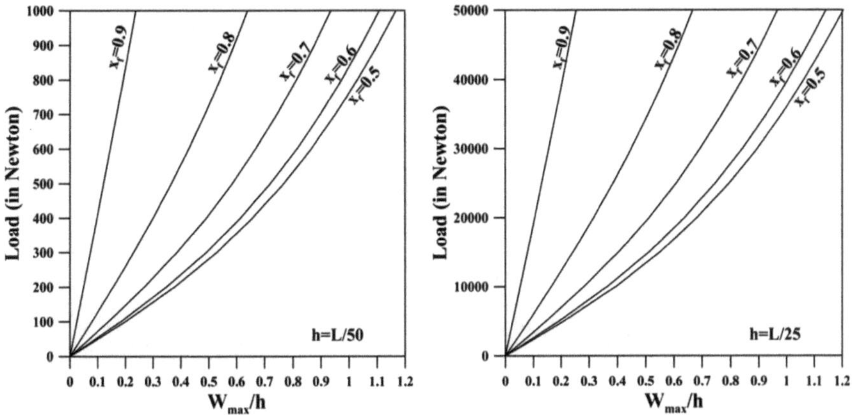

Fig. 19.4 (continued)

frequency for lesser values of *L*/*h*. Response curves increase gradually with increase in slenderness ratios (*L*/*h*). It is clear by observing Fig. 19.7 that higher branch of response curves for *L*/*h* = 75 and 50 have a steep response curves, whereas other response curves are comparatively gradual in nature. Less slender beams have a lower level of static deflection for particular amplitude of loadings compared to higher slenderness (*L*/*h*) beams. This is the reason for steep increment in response curves of less slender (thick) beams. In Fig. 19.8, responses are plotted for various load application points and for different values of slenderness (*L*/*h* = 15, 20, 25, and 30) with amplitude of excitation *F* = 16,000 N. Figure 19.9 shows the operational deformed shapes of a clamped–clamped beam with *h* = *L*/20 and *F* = 16,000 N. It can be noticed that operational deformed shapes are plotted with the original sign of the amplitude value. Negative values indicate that the response is out of phase for lower branch of frequency–amplitude response.

19.4 Conclusions

A relatively easier method is proposed using harmonic balance method and an iterative algorithm for the solution of Timoshenko beam under large amplitude of motion under forced vibration condition. In the present work, harmonic balance method is extended for a multi-DOF system. Solution technique as mentioned in Sect. 19.2 follows an amplitude incremental scheme, and even the multiple solution zones in the frequency–amplitude response can be easily obtained using the direct substitution iterative technique. Incremental harmonic balance method uses Newton–Raphson method in order to solve nonlinear algebraic equations. Furthermore, other methods for obtaining steady-state solution for such nonlinear systems usually adopt a solution scheme such as Newton–Raphson, Broyden's

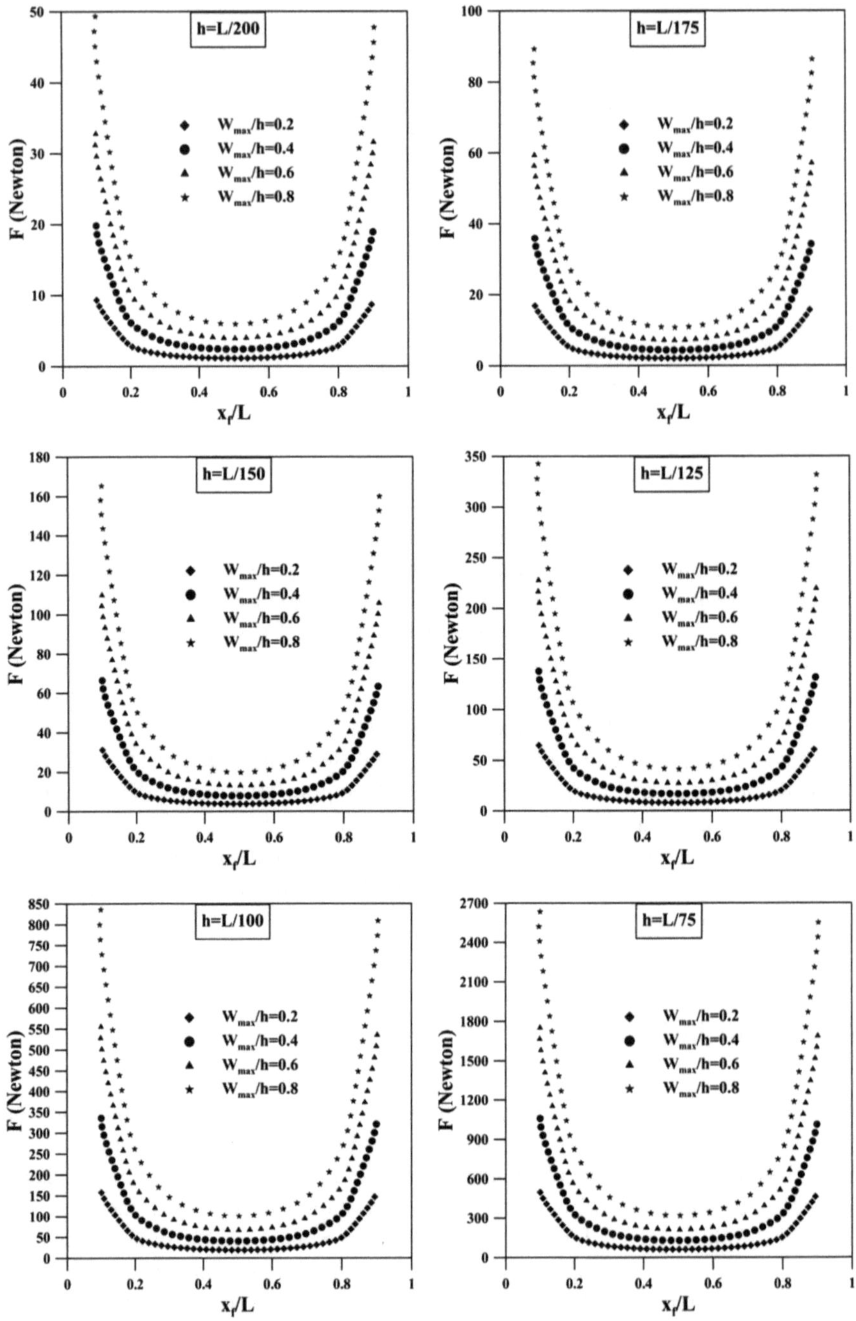

Fig. 19.5 Variation of amplitude of loading with load application points for same maximum static deflections for different slenderness (L/h = 200, 175, 150, 125, 100, 75, 50, and 25)

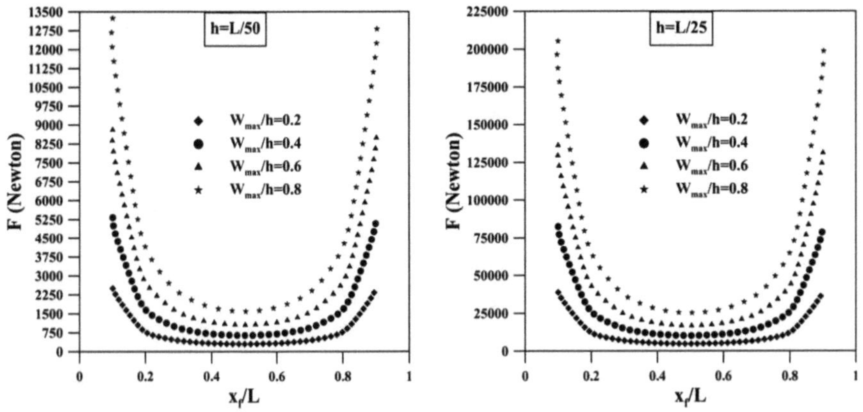

Fig. 19.5 (continued)

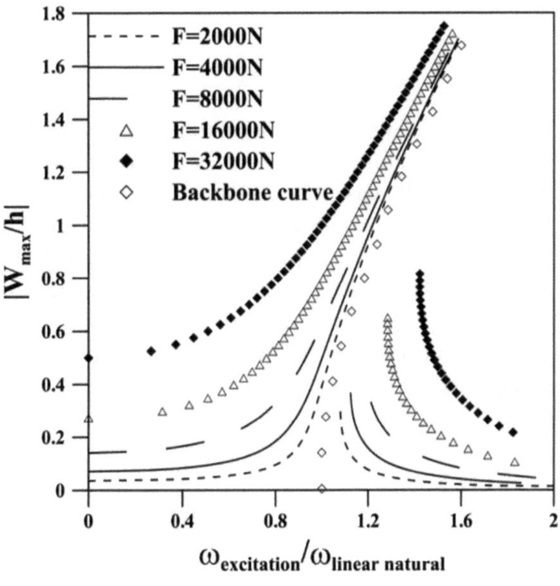

Fig. 19.6 Frequency–amplitude response for different amplitude of excitations with $h = L/20$

method, with the forward or a backward frequency sweep. The procedure becomes time consuming as there are multiple response zones. On the other hand with the present methodology solution becomes simpler as although in the multiple solution zones, the aim is to search a unique response frequency. Therefore, present methodology is simpler than the incremental harmonic balance method and other conventional methods used to obtain steady-state response. Using the proposed

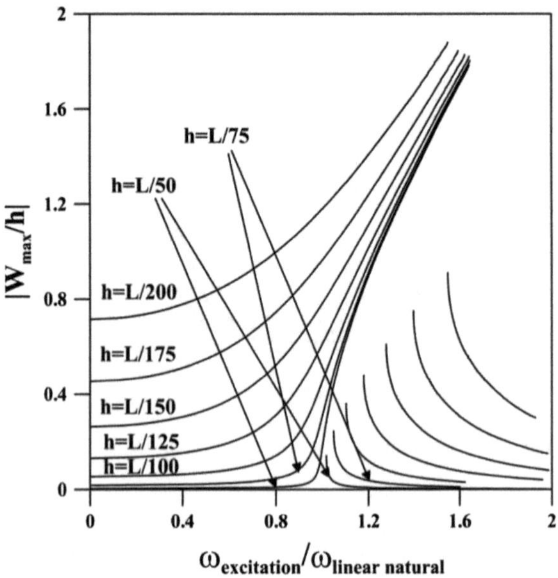

Fig. 19.7 Frequency–amplitude response for different slenderness (*L*/*h*) of with *F* = 5 N

method effect of higher-order harmonics on the frequency–amplitude response is studied successfully. It is found that although SHM assumption makes the problem computationally easier but at higher amplitude of vibration, it may yield erroneous results. Accurate results can be obtained by increasing the numbers of higher-order harmonics terms. A constant and two harmonic terms yield fairly precised solutions. Therefore in the present work, further increment in number of harmonic terms is not considered. Methodology is successfully implemented for nonlinear vibration analysis and nonlinear static response analysis for various geometric and loading parameters.

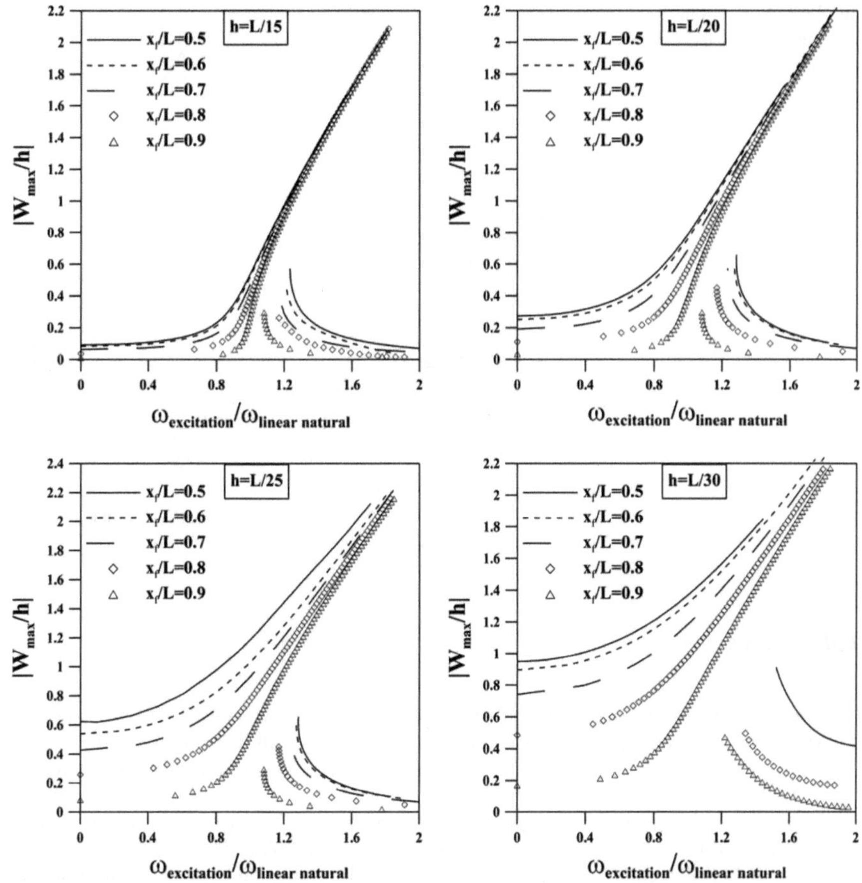

Fig. 19.8 Frequency–amplitude response for different slenderness (L/h = 15, 20, 25, and 30) of with F = 16,000 N

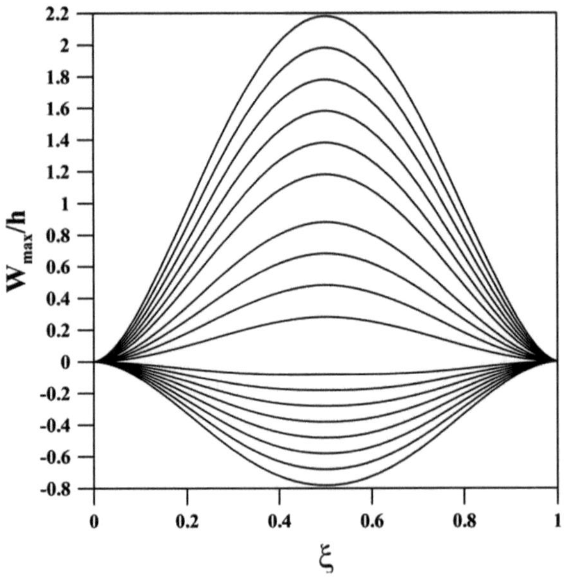

Fig. 19.9 Operational deformed shapes for L/h = 20 with F = 16,000 N

References

1. Nayfeh, A.H., Mook, D.T.: Nonlinear Oscillations. Wiley, New York (1979)
2. Meirovitch, L.: Methods of Analytical Dynamics. McGraw-Hill, New York (1970)
3. Noor, A.K., Peters, J.M.: Reduced basis technique for nonlinear analysis of structures. AIAA J. **18**, 455–462 (1980)
4. Desai, Y.M., Popplewell, N., Shah, A.H., Buragohain, D.N.: Geometric nonlinear static analysis of cable supported structures. Comput. Struct. **29**, 1001–1009 (1988)
5. Bathe, K.J., Bolourchi, S.: Large displacement analysis of three-dimensional beam structures. Int. J. Numer. Meth. Eng. **14**, 961–986 (1979)
6. Mata, P., Oller, S., Barbat, A.H.: Static analysis of beam structures under nonlinear geometric and constitutive behavior. Comput. Meth. Appl. Mech. Eng. **196**, 4458–4478 (2007)
7. Newmark, N.M.: A method of computation for structural dynamics. ASCE Eng. Mech. Div. **85**, 67–94 (1959)
8. Eisley, J.G.: Nonlinear vibration of beams and rectangular plates. Zeitschrift für angewandte Mathematik und Physik ZAMP **15**, 167–175 (1964)
9. Chu, H., Herrmann, G.: Influence of large amplitudes on free flexural vibrations of rectangular elastic plates. ASME J. Appl. Mech. **23**, 532–540 (1956)
10. Yamaki, N.: Influence of large amplitudes on flexural vibrations of elastic plates. J. Appl. Math. Mech. **41**, 501–510 (1961)
11. Hsu, C.S.: On the application of elliptic functions in nonlinear forced oscillations. Q. Appl. Math. **17**, 393–407 (1960)
12. Evensen, D.A.: Nonlinear vibrations of beams with various boundary conditions. AIAA J. **6**, 370–372 (1968)
13. Tseng, W.Y., Dugundji, J.: Nonlinear vibrations of a beam under harmonic excitation. J. Appl. Mech. **37**, 292–297 (1970)
14. Tseng, W.Y., Dugundji, J.: Nonlinear vibrations of a buckled beam under harmonic excitation. J. Appl. Mech. **38**, 467–476 (1971)

15. Bennett, J.A., Eisley, J.G.: A multiple degree-of-freedom approach to nonlinear beam vibrations. AIAA J. **8**, 734–739 (1970)
16. Mei, C.: Nonlinear vibration of beams by matrix displacement method. AIAA J. **10**, 355–357 (1972)
17. Stupnicka, W.: A study of main and secondary resonances in nonlinear multi-degree-of-freedom vibrating systems. Int. J. Nonlinear Mech. **10**, 289–304 (1975)
18. Stupnicka, W.: The generalised harmonic balance method for determining the combination resonance in the parametric dynamic systems. J. Sound Vib. **58**, 347–361 (1978)
19. Stupnicka, W.: Nonlinear normal modes and the generalised Ritz method in the problems of vibrations of nonlinear elastic continuous systems. Int. J. Nonlinear Mech. **18**, 149–165 (1983)
20. Rao, G., Raju, K.: Finite element formulation for the large amplitude free vibrations of beams and orthotropic plates. J. Comput. Struct. **6**, 169–172 (1976)
21. Lau, S.L., Cheung, Y.K.: Amplitude incremental variational principle for nonlinear vibration of elastic systems. J. Appl. Mech. **48**, 59–964 (1981)
22. Lau, S.L., Cheung, Y.K., Wu, S.Y.: Incremental harmonic balance method with multiple time scales for aperiodic vibration of nonlinear systems. J. Appl. Mech. **50**, 871–876 (1983)
23. Bennouna, M.M.K., White, R.G.: The effects of large vibration amplitudes on the fundamental mode shape of a clamped-clamped uniform beam. J. Sound Vib. **96**, 309–331 (1984)
24. Bennouna, M.M.K., White, R.G.: The effects of large vibration amplitudes on the dynamic strain response of a clamped-clamped beam with consideration of fatigue life. J. Sound Vib. **96**, 281–308 (1984)
25. Benamar, R., Bennouna, M.M.K., White, R.G.: The effects of large vibration amplitudes on the mode shapes and natural frequencies of thin elastic structures. Part I: simply supported and clamped-clamped beams. J. Sound Vib. **149**, 179–195 (1991)
26. Benamar, R., Bennouna, M.M.K., White, R.G.: The effects of large vibration amplitudes on the mode shapes and natural frequencies of thin elastic structures, part III: fully clamped rectangular isotropic plates-measurements of the mode shape amplitude dependence and the spatial distribution of harmonic distortion. J. Sound Vib. **175**, 377–424 (1994)
27. Azrar, L., Benamar, R., White, R.G.: Semi-analytical approach to the nonlinear dynamic response problem of S–S and C–C beams at large vibration amplitudes, part I: general theory and application to the single mode approach to free and forced vibration analysis. J. Sound Vib. **224**(2), 183–207 (1999)
28. El Kadiri, M., Benamar, R., White, R.G.: The non-linear free vibration of fully clamped rectangular plates: second nonlinear mode for various plate aspect ratios. J. Sound Vib. **228**, 333–358 (1999)
29. Azrar, L., Benamar, R., White, R.G.: A semi-analytical approach to the nonlinear dynamic response problem of beams at large vibration amplitudes, part II: multimode approach to the steady state forced periodic response. Journal Sound Vib. **255**(1), 1–41 (2002)
30. Harras, B., Benamar, R., White, R.G.: Geometrically nonlinear free vibration of fully clamped symmetrically laminated rectangular composite plates. J. Sound Vib. **251**, 579–619 (2002)
31. El Kadiri, M., Benamar, R., White, R.G.: Improvement of the semi-analytical method for determining the geometrically nonlinear response of thin straight structures Part I: application to clamped-clamped and simply supported-clamped beams. J. Sound Vib. **249**, 263–305 (2002)
32. Qaisi, M.I.: Application of the harmonic balance principle to the nonlinear free vibration of beams. Appl. Acoust. **40**, 141–151 (1993)
33. Ribeiro, P., Petyt, M.: Non-linear vibration of beams with internal resonance by the hierarchical finite element method. J. Sound Vib. **224**, 591–624 (1999)
34. Ribeiro, P.: Non-linear forced vibrations of thin/thick beams and plates by the finite element and shooting methods. Comput. Struct. **82**, 1413–1423 (2004)
35. Chen, S.H., Haung, J.L., Sze, K.Y.: Multidimensional Lindstedt–Poincare′ method for nonlinear vibration of axially moving beams. J. Sound Vib. **306**, 1–11 (2007)

36. Cheung, Y.K., Chen, S.H.: Application of the incremental harmonic balance method to cubic non-linearity systems. J. Sound Vib. **140**, 273–286 (1990)
37. Ramezani, A., Alasty, A., Akbari, J.: Effects of rotary inertia and shear deformation on nonlinear free vibration of micro beams. J. Vib. Acoust. **128**, 611–615 (2006)
38. Luongo, A., Rega, G., Vestroni, F.: On nonlinear dynamics of planar shear indeformable beams. J. Appl. Mech. **53**, 619–624 (1986)
39. Lenci, S., Clementi, F., Rega, G.: A comprehensive analysis of hardening/softening behaviour of shearable planar beams with whatever axial boundary constraint. Meccanica **51**, 2589–2606 (2016)
40. Clementi, F., Lenci, S., Rega, G.: Cross-checking asymptotics and numerics in the hardening/softening behaviour of Timoshenko beams with axial end spring and variable slenderness. Arch. Appl. Mech. **87**, 865–880 (2017)
41. Gupta, R.K., Babu, G.J., Janardhan, G.R., Venkateswara Rao, G.: Relatively simple finite element formulation for the large amplitude free vibrations of uniform beams. Finite Elem. Anal. Des. **45**, 624–631 (2009)
42. Kitipornchai, S., Ke, L.L., Yang, J., Xiang, Y.: Non-linear vibration of edge cracked functionally graded Timoshenko beams. J. Sound Vib. **324**, 962–982 (2009)

Chapter 20
Identification of Crack of Cantilever Beam Using Experimental Results and a Hybrid Neuro-GA Optimization Technique

Amit Banerjee and Goutam Pohit

20.1 Introduction

Structural element beams are largely applied in machinery production and steel manufacture. The presence of crack in structures may cause hampered its accomplishment considerably. Therefore, the identification of crack depth and its location in beam-like structures is a crucial manner of structural health checking and securing their period security. Dynamic response-based damage identification method attracts most attention due to its integrity for application and capability of collecting the global as well as local report of design. Accordingly, diversity of analytical, numerical, and experimental study presently subsists.

Chandros and Dimarogonas [1] presented an analysis of flaws in structures in the form of cracks which produce local change in stiffness altering their dynamic response. Sekhar and Prabhu [2] presented a technique to measure the vibration character with the help of FEM. Saavedra and Cuitino [3] introduced a cracked finite element stiffness matrix for FEM analysis of crack systems. Sinha and Friswell [4] studied dynamic nature of a breathing cracked free–free beam by experimental study under sinusoidal input force with the help of cracked beam finite element model. This method used finite element (FE) modelling with standard two node Euler–Bernoulli beam elements, except for those elements containing a crack. Zheng and Kessissoglou [5] presented a finite element analysis of cracked beam through measured natural frequencies with its corresponding mode shapes. Nahvi and Jabbari [6] had presented an analytical with experimental study to crack detection of cantilever beams by dynamic analysis. Patil and Maiti [7] presented experimental approach to verify the certainty and capability of their numerical method. The beam response of cantilever beam with multiple cracks was analyzed by the FFT analyzer.

A. Banerjee (✉) · G. Pohit
Department of Mechanical Engineering, Jadavpur University, Kolkata 700032, India
e-mail: mr.amit_banerjee@yahoo.com

P. Sahoo and J. P. Davim (eds.), *Advances in Materials, Mechanical and Industrial Engineering*, Lecture Notes on Multidisciplinary Industrial Engineering,
https://doi.org/10.1007/978-3-319-96968-8_20

Choubey et al. [8] proposed crack detection method of pressure vessel with the help of artificial neural networks. Different crack depths at a distinct location were considered as input data sets of an ANN. Orhan [9] presented crack detection of cantilever beam by both free and forced vibration analysis. Circumferentially, torsional vibration of a cracked cylindrical shaft by an analytical solution called 'exact' and a numerical FEM analysis was presented by Chondros and Labeas [10]. Yoon et al. [11] studied an experimental analysis on double crack simply supported beam. Saridakis et al. [12] presented bending vibrations of a cracked shaft causing coupling effect. For detection of crack, three optimizations techniques, ANN, fuzzy logic, and GA, were presented. Perk et al. [13] presented sequential way for damage identification in beams applying time-modal features with neural networks controller. Singh and Tiwari [14] developed a two-step multi-crack detection algorithm established on forced responses of a non-rotating shaft. The multi-objective genetic algorithm is applied to investigate the cracks size and their location. Parhi and Dash [15] presented the dynamics analysis of multiple transverse cracked beams by ANN optimization. Buezas et al. [16] determined crack depth and crack location identification using genetic algorithms considering a crack contact model. Al-Ansari et al. [17] performed a comparative study between experimental result and ANSYS program results of cracked simple supported beam. Universal vibration approaches were employed in that study. Mazanoglu and Sabuncu [18] presented an algorithm for identifying cracks on the beams and minimizing the measurement errors a statistical method. Jena et al. [19] presented defect identification of a beam with single crack with the help of theoretical and experimental investigation applying vibration signatures. Thatoi et al. [20] presented of three type of ANN for crack identification of cantilever beam. A hybrid optimization with response surface methodology with genetic algorithm to identifying crack on cantilever beam structure is developed by Garg et al. [21]. Jena et al. [22] studied dynamic analysis composite cracked beam through angle change of bidirectional fibers. Recently, two optimization techniques for detection of crack in Timoshenko beam are presented by Banerjee et al. [23].

In the present paper, evaluate first four free vibrational modes of a beam having transverse crack having clamped-free boundary condition over experiment. The vibration of modal data obtained from experimental analysis is compared with finite element analysis. Identification of crack on FGM beams is attempted by two techniques cascade artificial neural network (CANN) and genetic algorithm (GA). Identify unknown crack parameter is evaluated with the help of known vibrational parameters obtained from experimental investigation. Lastly, a new neuro-genetic base hybrid technique is proposed, and the results obtained are compared.

20.2 Methodology

In this section, vibration parameter of cracked cantilever beams by is determined through experiment. Then, FEM analysis is also presented on those cracked beams. Based on the result generated by experimentation, detection of crack size and its

Table 20.1 Geometric and physical properties of beam

Beam	Density (kg/m^3)	Poisson ratio	Elasticity modulus (GPa)	Length (*L*) (mm)	Thickness (*W*) (mm)	Depth (*h*) (mm)
Cantilever	2645	0.33	70	850	50	10

location is determined by CANN, GA, and a new hybrid neuro-GA technique. Analysis is done on MATLAB platform. Table 20.1 shows material properties of the beam.

20.2.1 Experimental Details and Procedure

Specimens formed an aluminum metal having cracks consisting different sizes and locations along the length of the beam. EDM wire cut machine is used to make crack on beam having 0.6 mm width for all cases. Some specimens are shown in Fig. 20.1.

Figure 20.2 shows entire experimental setup. Figure 20.2 also indicates the major components, hydraulic fixture, pressure gauge, accelerometer, coupler for accelerometer, oscilloscope. The blow of a soft rubber hammer is applied to excite the system. As a result, deflected shape produced free vibration response of the intact and series of cracked beams. The schematic layout of experimental setup is shown in Fig. 20.3.

An accelerometer [manufacturer: Kistler Instrument Corporation, type: 8728A500, acceleration range: ±500 g ($g = 9.80665\ m/s^2$), range of frequency: 1 Hz–10 kHz (±5%)] with a mass of 1.6 g, is attached at a suitable position on the top surface of the beam using Petro-Wax adhesive material. Mass of accelerometer

Fig. 20.1 Different crack positions of sample test

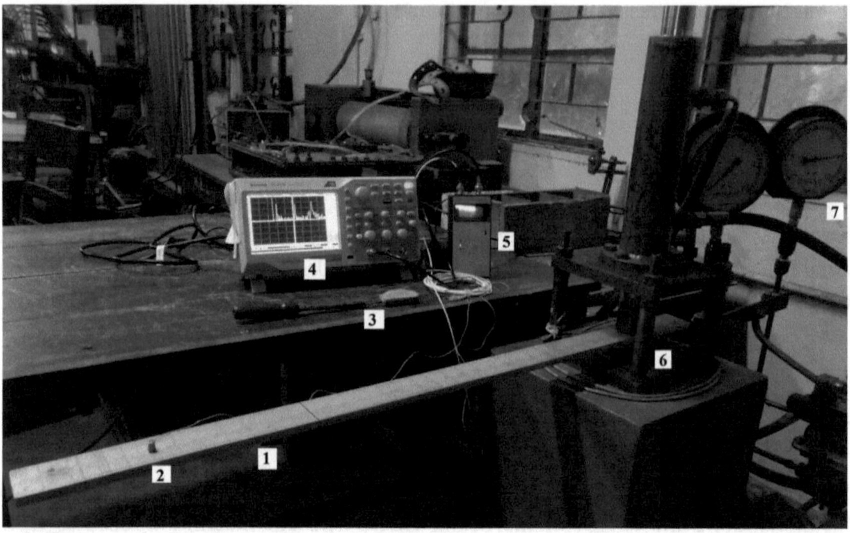

1. Beam 2. Accelerometer 3. Hammer 4. Oscilloscope 5. Coupler 6. Hydroulic Fixture 7. Pressure Gauge

Fig. 20.2 Photograph of experimental setup

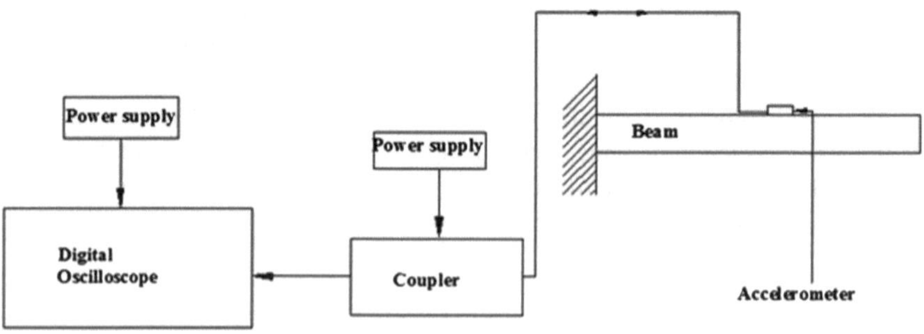

Fig. 20.3 Schematic diagram of experimental setup with oscilloscope

for affecting dynamic response of beam may be ignored because mass of the accelerometer is very less compare to beam. Accelerometer is placed on the top of the beam near anti-nodal point to avoid minimum amplitude displacement point. Accelerometer has two parts: impedance converter and sensing assembly. Impedance converter is a scale-down universal electronic circuit which converts low-impedance output voltage signal from high-impedance charge signal by the piezoelectric element. An external power source name coupler is required. Second component sensing assembly can run without any external power source. It has two parts: a piezoelectric crystal and a seismic mass.

The accelerometer is connected to a coupler [manufacturer: Kistler Instrument Corporation, frequency response: 0.07 Hz–60 kHz (±5%), type: 5114], which decouples output signal into DC bias voltage and arranges the constant current power support to the accelerometer impedance converter. An AC–DC power adapter source is connected to the coupler. DC bias voltage decouples by coupler from the output signal. It operates as the electrical interface between low-impedance, recording device or display and voltage mode piezoelectric sensor. In the present scenario, display device is a digital oscilloscope. The accelerometer and coupler is connected by a two-wire cable. Signal and power both are carried by this cable.

The coupler output is connected to two channel digital real time oscilloscope display device (Tektronix TBS 1072B) with the specification list rate range of sample: 50 samples/s–1 Gigasamples/s, peak detect bandwidth: 70 MHz, record length: 2500 samples, and frequency limit (lower): 10 Hz. Oscilloscope is used to performed FFT operation. It transformed signal frequency domain from time domain through a fast Fourier transform (FFT) module. The analyzer shows a digital readout for the peaks.

20.2.2 Finite Element Simulation

Free vibration finite element analysis is carried out first on a cracked free beam and then on a series of cracked beams. An open transverse cracked beam having crack depth '*a*' at location L_1 from fixed end is shown in Fig. 20.4. *L* and *h* denote beam length and depth, respectively. Finite element model of a cantilever beam with single open transverse crack is developed in ANSYS environment. For mashing purpose, two types of elements are picked: One is quad element 8-node Plane 183, and other is 20-node SOLID 186. A special mashing arrangement is implemented at crack trip to outcome singularity effect. A mesh model of cracked beam is shown in Fig. 20.5. Natural frequencies of the beam are obtained by block Lanczos mode extraction method. First, four natural frequencies in transverse direction are evaluated for intact and set of cracked cantilever beams.

Fig. 20.4 Cantilever beam with a crack

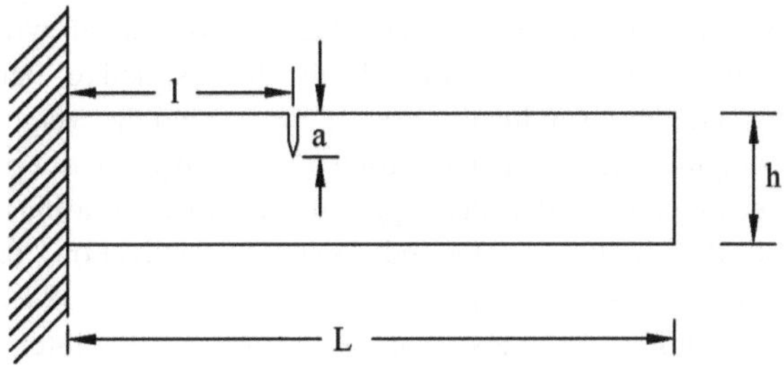

Fig. 20.5 Mesh model with singular element

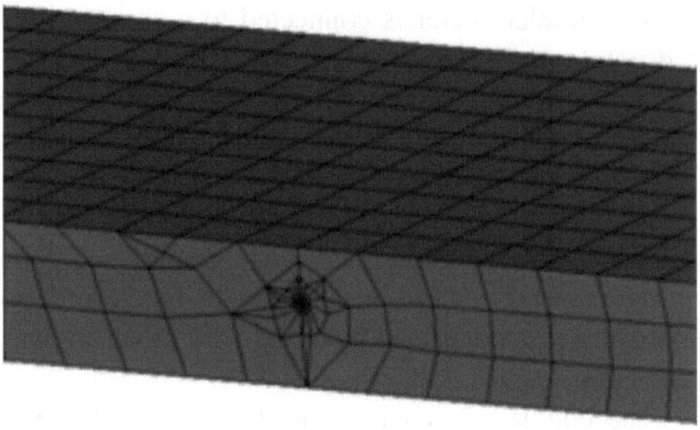

20.2.3 Cascade Artificial Neural Network

Cascade artificial neural networks (CANN) has risen as an auspicious device for structural health controlling and damage diagnosis on machining equipment. This optimization is healthily equipped for clarifying vibrational inverse problems in the damage identification and situation of monitoring because of their model acceptance and interpolation competence. ANN is also a favorable way to categorize the problems correlated with nonlinearities, supported they are healthily depicted by input patterns. It can stay away from the intricacy received by traditional computational approach. It consists of a given set of inputs for which desired outputs are determined by establishing proper and desired relationship between the inputs and there outputs. The calibrating among the input and the output is hindered but must be learned, and after the calibrating is learned or trained, the coveted outputs can be collected. It guides to raise the design method capability.

In this present paper, an artificial neuron is presented in the vibration surroundings. A scalar function n is the summation of the bias b and weighted input w_p. This summation is the altercation of the transfer function. The transfer function denoted by f is generally a sigmoid function or a step function that catching the altercation n and out comes the output a. Alterable scalar parameters of the neuron are denoted as w and b. The central idea of neural networks is that such parameters can be adjusted so that the network exhibits some desired or interesting behavior. Thus, you can train the network to do a particular job by adjusting the weight or bias parameters, or perhaps the network itself will adjust these parameters to achieve some desired end. Here, the distinctive element inputs p_1, p_2 ... p_R are extended by weights $w_{1,\,1}$, $w_{1,\,2}$...$w_{1,\,R}$, and the weighted values are delivered to the computing junction. Their total is scalar Wp, the dot product of the matrix $W$and the vector p. Weighted net inputs n are added by a bias b. The total internal sum, n, of the inputs is moved over an activation or transfer function, f. So, $n = w_{1,1}p_1 + w_{1,2}p_2 + \cdots + w_{1,R}p_R + b$ or, $n = W * p + b$.

A nonlinear transformation expresses the intention of using a neural filter application. A nonlinear transfer function which is used in network has output limit capacity of each approach of a layered network towards a described field and also limits the amplitude range as well as limits the output strength. All the neural network constructions operate the transfer function. It is prescribed as activity level of output neuron at its input (bounds from 0 to 1 or −1 to 1). In this present approach, activation functions are the sigmoid functions (Eq. 20.1). This is because they are differentiable and probabilistic interpretation which simplifies the classification task.

$$f(x) = \frac{1}{1 + \mathrm{e}^{-\alpha x}}, \quad \alpha > 0 \tag{20.1}$$

A learning rule defines a method for revising the network weights and biases. There are three main learning criterions, each equivalent to a particular theoretical learning assignment. Names of these learning assignments are unsupervised, supervised, and reinforcement. Unsupervised learning neural network simply required an input data set, and it is the network controller's authority to find part of arrangement within the inputs without given any external support. A set of input–output pairs is required for instructed supervised learning. The objective is to train the network to understand between known input and wanted output. There is a similarity of outcome feedback of learning process between supervised and reinforcement. One main difference is reinforcement supplying a target output score established on how healthy network is achieved. Supervised learning is applied for crack detection due to this process more efficient than other learning process.

The backpropagation algorithm of CANN models is executed better than backpropagation algorithm of feed-forward ANN models, particularly in lowering the disperse of forecasting. In this thesis, cascade-forward backpropagation is applied for crack detection. A weight connection is associated from every layer to the following layers.

Backpropagation algorithm is most generally applied in the multilayer cascade-forward network. A gradient descent algorithm is applied in common backpropagation. In the gradient descent algorithm, weights of network are carried on the adverse of performance activity gradient. Gradient is calculated for multilayer networks with nonlinear environment, so it is called backpropagation algorithm. For crack detection method back propagation technique is employed to reduce the error with optimize the inputs relative natural frequencies to achieve the target that is normalized crack location and crack depth ration. Neurons are originated by the number of training data and the convergence of the error during training to a minimum threshold error to control the crack location and crack depth. Network performance is measured by mean squared difference between $\psi_{\text{desired},\,n}$ and $\psi_{\text{actual},\,n}$ for supplied training arrangement data list

$$\text{Error} = \frac{1}{2} \sum_{\substack{\text{all training} \\ \text{patterns}}} \left(\psi_{\text{desired},m} - \psi_{\text{actual},m}\right)^2 \tag{20.2}$$

where m = output parameters

Error of training of ANN network is accomplished by error backpropagation method. In order to reduce error, appropriate weight corrections are required. Appropriate weight is produced by calculating neighboring error gradients. Gradient of error of output layer δ^i (*i*th layer) is

$$\delta^i = f'\left(N^i\right)\left(\psi_{\text{desired},n} - \psi_{\text{desired},n}\right) \tag{20.3}$$

where $N_j^i = \sum_{n=1}^{z} W_{jn}^i y_n^{i-1} + W_j^i u_j$ and $W_j^i = f\left(N_j^i\right)$

i = layer number, j = hidden layer jth neuron label, W_{jn}^i = connection weight of neuron n in $i - 1$ th layer to neuron jth layer i and u = input neuron.

Hidden layer local gradient of neuron is given by:

$$\delta_j^i = f'\left(N_j^i\right)\left(\sum_k \delta_k^{i+1} W_{kj}^{i+1}\right) \tag{20.4}$$

Neurons weight is updated as:

$$W_{jn}(t+1) = W_{jn}(t) + \Delta W_{jn}(t+1) \tag{20.5}$$

where $\Delta W_{jn}(t+1) = \alpha \Delta W_{jt}(t) + \beta \delta_j^i y_n^{i-1}$

where, α = momentum coefficient is chosen empirically as 0.2 in this computational analysis; β = learning rate is chosen empirically as 0.35 in this computational work, t = iteration number, the final neural network output is,

$$\psi_{\text{actual},m} = f\left(N_m^{i_{\text{output}}}\right) \tag{20.6}$$

where $f_m^{i_{\text{output}}} = \sum_{n=1} W_{mn}^{i_{\text{output}}} y_n^{i_{\text{output}-1}}$

Each activation consists of the adjustment of learning sample with neuron weight interchange.

20.2.4 Genetic Algorithm

In this study, GA method is used to forecast crack depth and its location of a structure utilizing data set of natural frequency. The first four natural frequencies in transverse direction of the beams are applied to create the objective function. The genetic algorithm is used to forecast the global optimal crack depth and its location with the help of minimizing the objective function which is found on experimental result frequencies (f_n) and calculated frequencies by regression equation (f_r).

The process used for prediction of crack in GA as follows:

- First select all variable. The GA starts by representing two variables, crack depth ratio and normalized location of crack whose values should be optimized.

- The following equation is represented the cost function. This function should be minimized:

$$\text{Cost}\,(l,d) = \sum_{i=1}^{4} (f_{\text{ei}} - f_{\text{ri}})^2 \tag{20.7}$$

where

d_c	crack depth ratio
l_c	crack location (normalized)
f_{ni}	normalized first three natural frequencies from experimental result
f_{ri}	normalized first three natural frequencies from regression field.

- Twenty numbers of initial population are developed.
- Initial population is collected from numerical or finite element analysis.
- Initial population parent set is applied.
- The chosen parents are transformed decimal values to binary string
- Cost functions are picked.
- During crossover binary string is randomly incision at any point.
- The children of the parents are determined.
- The fitness value is figured out for each variable.
- Stochastic uniform is considered for selection. After eliminating worsted chromosomes only the best chromosomes are picked.
- After achieved a new population, the selection process and crossover are redone with mutation which is applied to all of the data expect the best children. For best fitness value, mutation is applied with uniform rate 0.01.
- The method is re-evaluated and performed repeatedly the algorithm just before the fitness value extents confident level. Maximum generation number and function tolerance for every run are decided at 500 and 1e−12, respectively.

20.3 Neuro-Genetic Technique

In order to get more accurate results, a new hybrid technique is proposed. First, genetic algorithm model is prepared from experimental result and general regression equation as described in previous section. Then, optimized output results (crack location and depth) from GA are used as input neurons of neuron network. So this hybrid neural network technique has six input parameters and four natural frequencies obtained from experiment with two optimized data output of GA crack location and its depth (Fig. 20.6). These training and learning processes are same as described in Sect. 20.2.3.

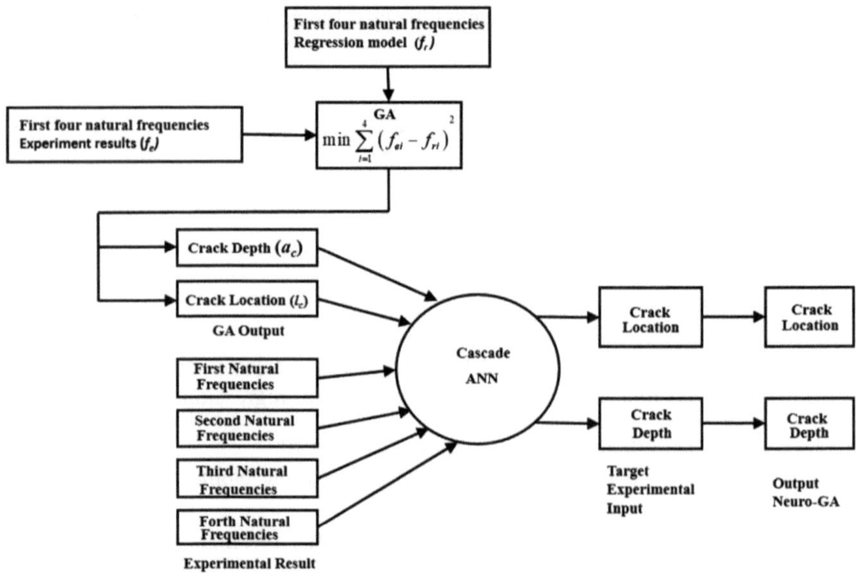

Fig. 20.6 Block diagram of neuro-GA technique

20.4 Result and Discussion

Free vibration of a large set of cantilever beams with different sizes of crack at different locations is executed by experiment. In order to determinate natural frequencies, FEM analysis also done same kind of cracked cantilever beam. Three different crack depth fractions (a/h) 0.2, 0.4, and 0.6 of cracked beam are considered for FEM analysis and experiment. Normalizing crack location (l/L) values is diverse from 0.0588 to 0.8824. The relative frequencies of different cracked beams are received from the following equation:

$$\text{Relative Natural Frequency} = \frac{\text{Natural frequency of beam having crack}}{\text{Natural frequency of intract beam}} \quad (20.8)$$

20.4.1 Comparison Results of Experimentation with FEM Results

Table 20.2 shows a comparison result. Results show that the experimental value and FEM analysis values are equated quite well.

Table 20.2 **a** Comparison of first natural frequency of cracked cantilever beam obtained through experimental investigation and ANSYS simulation. **b** Comparison of second natural frequency of cracked cantilever beam obtained through experimental investigation and ANSYS simulation. **c** Comparison of third natural frequency of cracked cantilever beam obtained through experimental investigation and ANSYS simulation. **d** Comparison of forth natural frequency of cracked cantilever beam obtained through experimental investigation and ANSYS simulation

Specimen crack location and depth (mm)	Free vibration frequency (Hz)	
	Experimental	ANSYS
a		
l = 50 and a = 2	10.5	11.27
l = 150 and a = 2	10.6	11.3
l = 250 and a = 2	10.6	11.32
l = 350 and a = 2	10.7	11.33
l = 450 and a = 2	10.7	11.33
l = 550 and a = 2	10.8	11.348
l = 650 and a = 2	10.8	11.351
l = 750 and a = 2	10.8	11.353
l = 50 and a = 4	10.2	11.01
l = 150 and a = 4	10.3	11.11
l = 250 and a = 4	10.5	11.2
l = 350 and a = 4	10.6	11.27
l = 450 and a = 4	10.6	11.31
l = 550 and a = 4	10.8	11.34
l = 650 and a = 4	10.8	11.35
l = 750 and a = 4	10.8	11.356
l = 50 and a = 6	9.1	10.3
l = 150 and a = 6	9.7	10.6
l = 250 and a = 6	10	10.87
l = 350 and a = 6	10.4	11.07
l = 450 and a = 6	10.6	11.22
l = 550 and a = 6	10.7	11.31
l = 650 and a = 6	10.7	11.35
l = 750 and a = 6	10.8	11.359
b		
l = 50 and a = 2	67.8	70.77
l = 150 and a = 2	69	71.06
l = 250 and a = 2	68	71.04
l = 350 and a = 2	67.9	70.85
l = 450 and a = 2	67.7	70.29
l = 550 and a = 2	68	70.857
l = 650 and a = 2	69	70.99
l = 750 and a = 2	69.1	71.08
l = 50 and a = 4	66.9	69.79

(continued)

Table 20.2 (continued)

Specimen crack location and depth (mm)	Free vibration frequency (Hz)	
	Experimental	ANSYS
l = 150 and a = 4	68.8	71
l = 250 and a = 4	67.5	70.87
l = 350 and a = 4	66.9	70.14
l = 450 and a = 4	66.4	69.77
l = 550 and a = 4	67.7	70.08
l = 650 and a = 4	68.8	70.68
l = 750 and a = 4	68.8	71.05
l = 50 and a = 6	65.8	67.45
l = 150 and a = 6	68.6	70.83
l = 250 and a = 6	68	70.4
l = 350 and a = 6	66.2	68.12
l = 450 and a = 6	65.2	66.95
l = 550 and a = 6	66.3	67.85
l = 650 and a = 6	68.3	69.75
l = 750 and a = 6	69.1	70.96
c		
l = 50 and a = 2	193	198.37
l = 150 and a = 2	192	198.77
l = 250 and a = 2	193	198.22
l = 350 and a = 2	192	198.58
l = 450 and a = 2	191	198.73
l = 550 and a = 2	193	198.15
l = 650 and a = 2	191	198.1
l = 750 and a = 2	192	198.72
l = 50 and a = 4	192	196.84
l = 150 and a = 4	190	198.41
l = 250 and a = 4	191	195.92
l = 350 and a = 4	191	197.58
l = 450 and a = 4	189	198.64
l = 550 and a = 4	193	195.69
l = 650 and a = 4	188	195.36
l = 750 and a = 4	189	198.21
l = 50 and a = 6	192	193.43
l = 150 and a = 6	185	197.37
l = 250 and a = 6	191	189.87
l = 350 and a = 6	190	194.92
l = 450 and a = 6	186	198.17
l = 550 and a = 6	193	189.32
l = 650 and a = 6	183	187.64

(continued)

Table 20.2 (continued)

Specimen crack location and depth (mm)	Free vibration frequency (Hz)	
	Experimental	ANSYS
l = 750 and a = 6	187	196.61
d		
l = 50 and a = 2	378	388.88
l = 150 and a = 2	377	388.36
l = 250 and a = 2	378	388.75
l = 350 and a = 2	378	388.85
l = 450 and a = 2	375	385.65
l = 550 and a = 2	378	389.32
l = 650 and a = 2	375	387.73
l = 750 and a = 2	377	388.58
l = 50 and a = 4	376	387.52
l = 150 and a = 4	373	385
l = 250 and a = 4	375	386.78
l = 350 and a = 4	376	387.15
l = 450 and a = 4	371	383.32
l = 550 and a = 4	379	389.3
l = 650 and a = 4	370	382.33
l = 750 and a = 4	372	385.95
l = 50 and a = 6	377	384.57
l = 150 and a = 6	365	375.87
l = 250 and a = 6	375	381.91
l = 350 and a = 6	374	382.65
l = 450 and a = 6	366	371.35
l = 550 and a = 6	379	389.25
l = 650 and a = 6	361	369.15
l = 750 and a = 6	368	377.71

20.4.2 Identification Crack Depth and Location

Total 24 beams with different types of crack are examined having different crack schemes. Now, those natural frequencies are used for input optimization technique for the evaluation of the crack. Comparison studies are performed between experimental results and optimization results of CANN, GA, and cascade neuro-GA.

20.4.2.1 Identification Crack Parameter by CANN

Cascade algorithm starts with a single-layer network to which hidden units are then added one by one. Total 24 data sets acquired from experimental study with four inputs (four natural frequencies) with two targets (location and depth of crack) are spat between training, validation, and testing scheme. Data set ratio for training, validation, and testing is 4:1:1. Gradient descent with bias and momentum weight and Levenberg–Marquardt optimization are chosen for training function and learning function, respectively. In order to achieve optimized value, tansigmoid hyperbolic activation function is applied. Neuron numbers in a specific hidden layer with total hidden layer numbers are changed to produce best target efficiency. Best MSE validation performance value 0.00987 at 47 epochs is achieved at the time of forecasting of crack depth and location in CANN. Regression value for this training is 0.9143. Neural network model for this analysis consists of 3 inputs, 2 output and 2 hidden layer having three neurons each.

Forecasting results are correlated with experimental result, and percentage errors are calculated with the help of the following equation,

$$E_h^* = \left|\frac{a_\mathrm{c} - a_\mathrm{c}^*}{h}\right| \times 100; E_L^\# = \left|\frac{l_\mathrm{c} - l_\mathrm{c}^*}{L}\right| \times 100 \qquad (20.9)$$

Outcome results of CANN optimization process are presented on Table 20.3. For all 24 crack scenarios, maximum errors of 15.91 and 19.21% are noticed for crack location and depth, respectively, although average errors are 4.21 and 2.99%. Results shows that four cases have large amount of error (greater than 10%) in case of prediction of location where as only one case has large amount of error in case of prediction of depth. Predicted results of crack depth are found to be more accurate than location. Over all crack prediction is reasonably well with few amount of experimental dada using CANN.

20.4.2.2 Identification Crack Parameter by GA

GA is applied to predict the global optimal depth and location of crack by minimizing the cost function. It is established from experimental results *(f_e)* with general regression *(f_r)*. The following equation is applied to minimize the fitness function.

$$\text{fitness}\ (l_\mathrm{c}, a_\mathrm{c}) = \sum_{i=1}^{4} (f_{\mathrm{e}i} - f_{\mathrm{r}i})^2 \qquad (20.10)$$

where f_e is the first four natural frequencies which are used for crack identification structure considered as inputs obtained from experimental results. f_r is the first four

Table 20.3 CANN optimization results of cantilever beam with error

Case	Normalized natural frequencies				Experiment		CANN		Error (%)	
	First	Second	Third	Forth	l_c/L	a_c/h	l_c^*/L	a_c^*/h	E_L^*	E_h^*
1	0.821	0.667	0.495	0.950	0.059	0.2	0.123	0.211	6.387	1.127
2	0.866	0.971	0.600	0.902	0.176	0.2	0.180	0.210	0.348	1.030
3	0.911	0.715	0.495	0.950	0.294	0.2	0.297	0.208	0.285	0.823
4	0.643	0.429	0.493	0.852	0.059	0.4	0.067	0.414	0.777	1.449
5	0.732	0.928	0.710	0.654	0.176	0.4	0.135	0.392	4.109	0.841
6	0.821	0.595	0.391	0.802	0.294	0.4	0.304	0.303	0.959	9.727
7	0.000	0.142	0.210	0.902	0.059	0.6	0.059	0.600	0.000	0.014
8	0.375	0.880	0.594	0.213	0.176	0.6	0.217	0.598	4.063	0.227
9	0.554	0.715	0.000	0.754	0.294	0.6	0.306	0.593	1.141	0.729
10	0.955	0.694	0.570	0.944	0.412	0.2	0.458	0.212	4.588	1.160
11	0.955	0.643	0.600	0.796	0.529	0.2	0.689	0.273	15.918	7.329
12	1.000	0.715	0.495	0.964	0.647	0.2	0.597	0.206	4.990	0.649
13	0.866	0.424	0.706	0.816	0.412	0.4	0.355	0.373	5.705	2.745
14	0.911	0.310	0.810	0.556	0.529	0.4	0.514	0.430	1.514	3.009
15	1.000	0.643	0.594	1.000	0.647	0.4	0.542	0.208	10.469	19.207
16	0.777	0.258	0.511	0.722	0.412	0.6	0.347	0.597	6.511	0.252
17	0.911	0.000	0.810	0.261	0.529	0.6	0.524	0.594	0.500	0.582
18	0.955	0.286	0.087	1.000	0.647	0.6	0.647	0.594	0.037	0.571
19	1.000	0.976	0.702	0.758	0.765	0.2	0.754	0.264	1.111	6.351
20	1.000	1.000	0.890	0.910	0.882	0.2	0.860	0.225	4.236	2.456
21	1.000	0.930	0.501	0.516	0.765	0.4	0.866	0.450	10.082	4.987
22	1.000	0.930	0.691	0.633	0.882	0.4	0.833	0.387	4.963	1.269
23	0.957	0.790	0.000	0.000	0.765	0.6	0.882	0.600	11.763	0.001
24	1.000	1.000	0.421	0.363	0.882	0.6	0.878	0.548	0.460	5.233

natural frequencies, which are functions of crack location (l_c) and its depth (a_c) and are calculated from the general regression cracked beam model.

Stochastic uniform and adaptive feasible mutations are used for selection and mutation, respectively. Maximum generation number and function tolerance for every run are decided at 500 and 1e−12, respectively.

All 24 crack scenarios are optimized for cost function of GA, and the value of optimum fitness forward along comparable crack depth and its location is preferred being best result. Results of GA are shown in Table 20.4. The following equation represents the percentage of errors.

$$E_h^{**} = \left|\frac{a_c - a_c^{**}}{h}\right| \times 100; E_L^{**} = \left|\frac{l_c - l_c^{**}}{L}\right| \times 100 \tag{20.11}$$

Table 20.4 GA optimization of cantilever beam with error

Case	Normalized natural frequencies				Experiment		GA		Error (%)	
	First	Second	Third	Forth	l_c/L	a_c/h	l_c^{**}/L	a_c^{**}/h	E_L^{**}	E_h^{**}
1	0.821	0.667	0.495	0.950	0.059	0.2	0.068	0.233	0.941	3.3
2	0.866	0.971	0.600	0.902	0.176	0.2	0.120	0.139	5.647	6.1
3	0.911	0.715	0.495	0.950	0.294	0.2	0.288	0.225	0.588	2.5
4	0.643	0.429	0.493	0.852	0.059	0.4	0.061	0.377	0.235	2.3
5	0.732	0.928	0.710	0.654	0.176	0.4	0.153	0.398	2.353	0.2
6	0.821	0.595	0.391	0.802	0.294	0.4	0.300	0.426	0.588	2.6
7	0.000	0.142	0.210	0.902	0.059	0.6	0.059	0.600	0.000	0
8	0.375	0.880	0.594	0.213	0.176	0.6	0.182	0.586	0.588	1.4
9	0.554	0.715	0.000	0.754	0.294	0.6	0.306	0.600	1.176	0
10	0.955	0.694	0.570	0.944	0.412	0.2	0.412	0.267	0.000	6.7
11	0.955	0.643	0.600	0.796	0.529	0.2	0.444	0.252	8.588	5.2
12	1.000	0.715	0.495	0.964	0.647	0.2	0.614	0.240	3.294	4
13	0.866	0.424	0.706	0.816	0.412	0.4	0.428	0.364	1.647	3.6
14	0.911	0.310	0.810	0.556	0.529	0.4	0.521	0.415	0.824	1.5
15	1.000	0.643	0.594	1.000	0.647	0.4	0.613	0.294	3.412	10.6
16	0.777	0.258	0.511	0.722	0.412	0.6	0.614	0.543	20.235	5.7
17	0.911	0.000	0.810	0.261	0.529	0.6	0.562	0.600	3.294	0
18	0.955	0.286	0.087	1.000	0.647	0.6	0.680	0.669	3.294	6.9
19	1.000	0.976	0.702	0.758	0.765	0.2	0.771	0.227	0.588	2.7
20	1.000	1.000	0.890	0.910	0.882	0.2	0.891	0.178	0.824	2.2
21	1.000	0.930	0.501	0.516	0.765	0.4	0.722	0.329	4.235	7.1
22	1.000	0.930	0.691	0.633	0.882	0.4	0.809	0.426	7.294	2.6
23	0.957	0.790	0.000	0.000	0.765	0.6	0.751	0.592	1.412	0.8
24	1.000	1.000	0.421	0.363	0.882	0.6	0.894	0.607	1.176	0.7

where l_c and a_c are represented as experimental location and depth of the crack, respectively, and $a_c^{\#\#}$ and $l_c^{\#\#}$ represent the forecasting depth and location of the crack, respectively evaluated by GA.

Table 20.4 presents the predicted results of GA optimization technique of cantilever beam. For all 24 cracks scenarios, maximum errors of 20.6 and 10.6% are noticed for crack location and depth, respectively, although average errors are 3.01 and 3.27%. Accuracy of prediction of location and depth is nearly equal in GA method. Overall average value of prediction error is very similar to CANN method.

20.4.2.3 Identification Crack Parameter by Neuro-GA

In order to get more accurate results, neuro-GA hybrid technique is proposed. First genetic algorithm model is prepared from experimental result and general regression equation as per described in the previous section.

All 24 optimized output results of experimental different crack scenarios are used as input parameters of neuro-GA optimization. Optimized output results (crack location and depth) from GA are also used as input neurons of this neuro-GA neuron network. Best MSE validation performance value 0.00907 at 57 epochs is achieved at the time of forecasting of crack depth and location in CANN. Regression value for this training is 0.9433. Neural network model present optimization has of six inputs, two outputs, and three hidden layers with three neurons each.

Table 20.5 Neuro-GA optimization of cantilever beam with error

Case	Normalized natural frequencies				Experiment		CANN		Error (%)	
	First	Second	Third	Forth	l_c/L	a_c/h	l_c^{***}/L	a_c^{***}/h	E_L^{***}	E_h^{***}
1	0.821	0.805	0.751	0.974	0.059	0.2	0.079	0.200	2.062	0.040
2	0.866	0.983	0.801	0.949	0.176	0.2	0.174	0.200	0.245	0.007
3	0.911	0.833	0.751	0.974	0.294	0.2	0.307	0.201	1.267	0.078
4	0.643	0.666	0.750	0.923	0.059	0.4	0.063	0.401	0.427	0.100
5	0.732	0.958	0.850	0.820	0.176	0.4	0.258	0.344	8.126	5.572
6	0.821	0.763	0.700	0.897	0.294	0.4	0.302	0.405	0.791	0.454
7	0.000	0.498	0.607	0.949	0.059	0.6	0.059	0.600	0.006	0.000
8	0.375	0.930	0.800	0.590	0.176	0.6	0.179	0.600	0.267	0.007
9	0.554	0.833	0.500	0.872	0.294	0.6	0.242	0.600	5.252	0.001
10	0.955	0.821	0.788	0.971	0.412	0.2	0.565	0.206	15.335	0.629
11	0.955	0.791	0.801	0.894	0.529	0.2	0.512	0.202	1.748	0.229
12	1.000	0.833	0.751	0.981	0.647	0.2	0.631	0.204	1.572	0.376
13	0.866	0.663	0.855	0.904	0.412	0.4	0.403	0.400	0.913	0.041
14	0.911	0.596	0.900	0.769	0.529	0.4	0.531	0.402	0.174	0.153
15	1.000	0.791	0.800	1.000	0.647	0.4	0.710	0.321	6.291	7.857
16	0.777	0.566	0.759	0.855	0.412	0.6	0.435	0.600	2.331	0.002
17	0.911	0.415	0.900	0.615	0.529	0.6	0.561	0.600	3.136	0.027
18	0.955	0.582	0.550	1.000	0.647	0.6	0.655	0.600	0.797	0.012
19	1.000	0.986	0.853	0.874	0.765	0.2	0.770	0.200	0.539	0.041
20	1.000	1.000	0.946	0.953	0.882	0.2	0.873	0.201	0.906	0.069
21	1.000	0.959	0.754	0.748	0.765	0.4	0.832	0.313	6.737	8.698
22	1.000	0.959	0.848	0.809	0.882	0.4	0.867	0.402	1.536	0.204
23	0.957	0.877	0.504	0.480	0.765	0.6	0.753	0.599	1.141	0.114
24	1.000	1.000	0.715	0.668	0.882	0.6	0.882	0.599	0.025	0.112

Forecasting results are correlated with experimental result, and percentage errors are calculated with the help of Eq. 20.12.

$$E_h^{***} = \left|\frac{a_c - a_c^{***}}{h}\right| \times 100; E_L^{***} = \left|\frac{l_c - l_c^{***}}{L}\right| \times 100 \quad (20.12)$$

Results of neuro-GA are shown in Table 20.5. For all 24 crack scenarios, maximum errors of 15.35 and 8.7% are noticed for crack location and depth, respectively, although average errors are 2.57 and 1.45%. Result shows that accuracy prediction trough neuro-GA is better than homogeneous optimization of CANN or GA techniques.

20.5 Conclusion

In the present paper, natural frequencies of intact and series of beam having cracked are evaluated by the experiment followed by FEM analysis, and comparison study is performed with measured frequencies. The second part of this paper represents crack parameter prediction NDT technique using three optimization techniques. A neuro-GA base hybrid optimization is also performed. It is noticed that forecasting results by optimization techniques are quite satisfactory with limited number of the experimental values. However, better predictions results are shown in cases of hybrid technique. It can be concluded that these optimization methods are fairly suitable for structural health monitoring with practical significance.

References

1. Chondros, T.G., Dimarogonas, A.D.: Identification of cracks in welded joints of complex structures. J. Sound Vib. **69**, 531–538 (1980)
2. Sekhar, A.S., Prabhu, B.S.: Crack detection and vibration characteristics of cracked shaft. J. Sound Vib. **157**, 375–381 (1992)
3. Saavedra, P.N., Quitino, L.A.: Crack detection and vibration behavior of cracked beams. Comput. Struct. **79**, 1451–1459 (2001)
4. Sinha, J.K., Friswell, M.I.: Simulation of the dynamic response of a cracked beam. Comput. Struct. **80**, 1473–1476 (2002)
5. Zheng, D.Y., Kessissoglou, N.J.: Free vibration analysis of a cracked beam by finite element method. J. Sound Vib. **273**, 457–475 (2004)
6. Nahvi, H., Jabbari, M.: Crack detection in beams using experimental modal data and finite element model. Int. J. Mech. Sci. **47**, 1477–1497 (2005)
7. Patil, D.P., Maiti, S.K.: Experimental verification of a method of detection of multiple cracks in beams based on frequency measurements. J. Sound Vib. **281**, 439–451 (2005)
8. Choubey, A., Sehga, D.K., Tandon, N.: Finite element analysis of vessels to study changes in natural frequencies due to cracks. Int. J. Press. Vessels Pip. **83**, 181–187 (2006)
9. Orhan, S.: Crack detection in beams using experimental modal data and finite element model. NDT&E Int. **40**, 443–450 (2007)

10. Chondros, T.G., Labeas, G.N.: Torsional vibration of a cracked rod by variational formulation and numerical analysis. J. Sound Vib. **301**, 994–1006 (2007)
11. Yoon, H.I., Son, I.S., Ahn, S.J.: Free vibration analysis of Euler Bernoulli beam with double cracks. J. Mech. Sci. Technol. **21**, 476–485 (2007)
12. Saridakis, K.M., Chasalevris, A.C., Papadopoulos, C.A., Dentsoras, A.J.: Applying neural networks, genetic algorithms and fuzzy logic for the identification of cracks in shafts by using coupled response measurements. Comput. Struct. **86**, 1318–1338 (2008)
13. Park, J.H., Kim, J.T., Hong, D.S., Ho, D.D., Yi, J.H.: Sequential damage detection approaches for beams using time-modal features and artificial neural networks. J. Sound Vib. **323**, 451–474 (2009)
14. Singh, S.K., Tiwari, R.: Identification of multiple crack in a shaft system using transverse frequency response functions. Mech. Mach. Theor. **45**, 1813–1827 (2010)
15. Parhi, D.R., Dash, A.K.: Proc. Inst. Mech. Eng. Part C: J. Mech. Eng. Sci. **225**, 1329–1339 (2011)
16. Buezas, F.S., Rosales, M.B., Filipich, C.P.: Damage detection with genetic algorithms taking into account a crack contact model. Eng. Fract. Mech. **78**, 695–712 (2011)
17. Al-Ansari, L.S.: Calculating of natural frequency of stepping cantilever beam. Int. J. Mech. Mech. Eng. **12**(5), 59–68 (2012)
18. Mazanoglu, K., Sabuncu, M.: A frequency based algorithm for identification of single and double cracked beams via a statistical approach used in experiment. Mech. Syst. Signal Process. **30**, 168–185 (2012)
19. Jena, P.C., Parhi, D.R., Pohit, G.: Faults detection of a single cracked beam by theoretical and experimental analysis using vibration signatures. J. Mech. Civil Eng. **4**(3), 1–18 (2012)
20. Thatoi, D., Guru, P., Jane, P.K., Choudhury, S., Das, H.C.: Modelling and Simulation in Materials Science and Engineering, 1–13 (2014)
21. Garg, S., Chawla, R., Singh, B.: Crack detection in cantilever beams using a new hybrid approach. Int. J. Mech. Prod. Eng. **2**(4), 22–27 (2014)
22. Jena, P.C., Parhi, D.R., Pohit, G.: Dynamic study of composite cracked beam by hanging the angle of bi-directional fibres. Iran. J. Sci. Technol. **40**(1), 27–37 (2016)
23. Banerjee, A., Panigrahi, P., Pohit, G.: Nondestr. Test. Eval. **31**, 142–164 (2016)

[illegible]

Chapter 21
Experimental Investigation on Static and Free Vibration Behavior of Concentrically Stiffened Plates

Sayantan Mandal, Anirban Mitra and Prasanta Sahoo

21.1 Introduction

Stiffened plates are structural elements that are fabricated with reinforcement of a single or a set of beams located at discrete spacing on one or both sides of the plate to increase the load-bearing capability. These plates can be broadly classified into two categories depending on the presence of stiffeners on one or both sides of the plate. These two categories are, namely, eccentric stiffening, where stiffeners are fixed to only a single side of the plate and concentric stiffening, where the stiffeners are attached to both sides of the plate. The material used for the stiffener and plate specimens can be different or identical. The benefit of stiffened plates over unstiffened ones is that they offer substantially greater strength-to-weight ratio. These stiffeners enhance the bending stiffness of the structure with minimum addition of material. It also provides better vibration and stability characteristics along with economic and easy fabrication. In the present scenario, considering these numerous advantages, there are wide-ranging applications of stiffened plates in marine structures, aerospace industry, highway bridges, elevated roadways, etc.

A significant amount of research work has been documented in this domain through different analytical and numerical techniques by several researchers. Recent advancement in simulation techniques and analysis methods coupled with an exponential increase in computational capabilities has opened new avenues of study in the field. It has also caused a shift in focus on simulation studies in determining mechanical behavior of stiffened plates. In the following section, a brief literature survey, with emphasis on experimental works in recent years, has been presented. In static analysis, both the maximum deflection and the defected profile of the

S. Mandal · A. Mitra (✉) · P. Sahoo
Department of Mechanical Engineering, Jadavpur University,
Kolkata 700032, West Bengal, India
e-mail: anirban.mitra@jadavpuruniversity.in

P. Sahoo and J. P. Davim (eds.), *Advances in Materials, Mechanical and Industrial Engineering*, Lecture Notes on Multidisciplinary Industrial Engineering,
https://doi.org/10.1007/978-3-319-96968-8_21

deformed system under loading are of importance and these two parameters depend on several factors, for example, boundary conditions, loading pattern.

An extensive review work on various aspects of studies performed on stiffened plates was presented by Bedair [1]. His emphasis was mainly on stability of stiffened plates subjected to uniform compression. Experiments were performed on clamped square plates under uniformly distributed periodic load with different edge displacements by Yamaki et al. [2] in order to determine their nonlinear response. Bau-Madsen et al. [3] performed experimental investigations into clamped sandwich plates under external lateral loading. The authors measured surface strains and deflections (large) to establish relations between the measured parameters and the external transverse load. Stiffened plates with square cut out were investigated on by Kumar et al. [4] under out of plane and in-plane loading in order to determine the ultimate strength. Xu et al. [5] carried out comparisons for ultimate strength of wide stiffened panels obtained from simulation as well as experimentation. Pedram and Khedmati [6] studied the influence of welding on the stiffened plates (material was considered to be aluminum), which were under combined loading of lateral pressure and uniaxial compression. Experimental work on stiffened plates subjected to similar kind of loading condition as mentioned above, i.e., a combination of transverse pressure and in-plane compression, was also performed by Shanmugam et al. [7]. Experimental study and numerical simulation (considering the effect of strain rate) of laterally impacted stiffened plates were carried out by Liu et al. [8]. Zheng et al. [9] investigated clamped stiffened plates subjected to confined blast load for large deflection behavior by conducting experiments, analytical calculations, and numerical simulations. Morin et al. [10] tested and simulated aluminum stiffened panels under the conditions of quasi-static and low-velocity impact loads.

Majority of the research papers, mentioned in the above paragraph, deal with static analysis of stiffened plates. However, there exist in the literature various contributions on free vibration studies of these structural elements. Fahy and Wee [11] measured average acoustically induced strains developed in stiffened plates under acoustic excitation. These stiffened plates were constructed by attaching, through various methods, identical beams to mild steel plates. Yurkovich et al. [12] performed dynamic experiments on rectangular stiffened panels with simply supported boundaries. Using real-time laser holography, Olson and Hazell [13] determined 24 free vibration frequencies and mode shapes of four stiffened square plates with fixed boundary conditions. Wilson et al. [14] designed experiments to analyze free vibration of thin square plates with edge support. A spline finite strip method was applied to linear and nonlinear transient vibration analysis of stiffened plates by Sheikh and Mukhopadhyay [15]. Amabili [16] experimentally tested plates with geometrical imperfections for large amplitude forced vibration behavior under harmonic excitation. The author considered simply supported clamped plates with and without in-plane edge displacements. Qing et al. [17] proposed a novel method for analyzing stiffened laminated plates under free vibration condition. Hu et al. [18] put forward fatigue behavior study of fully clamped aluminum stiffened plates subjected to random vibration loading. Zhou et al. [19] investigated thin plate with periodic stiffening and viscoelastic damping material utilizing simplified superelement method to find out the flexural free vibration characteristics. Mindlin plate and Timoshenko beam theories were implemented by Cho et al. [20] for formulating the free vibration problem of panels with stiffness attachments and lumped mass.

Current research paper presents an effort to experimentally study static and free vibration behavior of concentrically stiffened plates subjected to lateral uniformly distributed loading. Laboratory experiments are conducted on different types of concentric stiffened plate specimens along with simulation study to determine the static deflection and loaded natural frequencies.

21.2 Experimentation Description

Test specimens, laboratory experimental setup with detailed description of various components of the setup, and sequential test procedure are described in the following subsections.

21.2.1 Test Specimen Description

Experiments are performed on test specimens constructed out of uniform thickness square plates and rectangular cross-sectional flat bars. These bars are attached as stiffeners to both sides of the plate at different locations and orientations through spot welding process. Mild steel is the material chosen for both the plates and the stiffeners. There are five different types of concentric stiffened plates that are fabricated as described above. It should be mentioned here that these five categories are separate in terms of number of stiffeners and their corresponding orientation of the face of the plate. Dimensions of the stiffeners, such as length, width, and thickness, are same throughout all of the specimens, while the variation is being made to the attaching position of the stiffener in the plate. These different types of specimen are, namely, uniaxially single stiffened (UASS), uniaxially double stiffened (UADS), biaxially cross stiffened (BACS), diagonally cross stiffened (DCS), biaxially double stiffened (BADS) plates. The abbreviations for different types of specimen are created by taking the first letter of the words in their names. Schematic diagrams with symbols and photographs of the above-mentioned types of stiffened specimens are depicted in Figs. 21.1 and 21.2. The figures also include an unstiffened specimen, which is included in order to provide a reference to comparison.

The basic geometric parameters of the specimen are as follows: length of plate (L) = 480 mm, clamped length of plate (L_c) = 400 mm, thickness of plate (t_p) = 2 mm, length of stiffener (L_s) = 400 mm, thickness of stiffener (t_s) = 25 mm, breadth of stiffener (b_s) = 5 mm. The dimensions relating to the position of the stiffeners corresponding to different types of plates are provided in Table 21.1. From Figs. 21.1 and 21.2, it is clearly visible that there is a series of circular holes

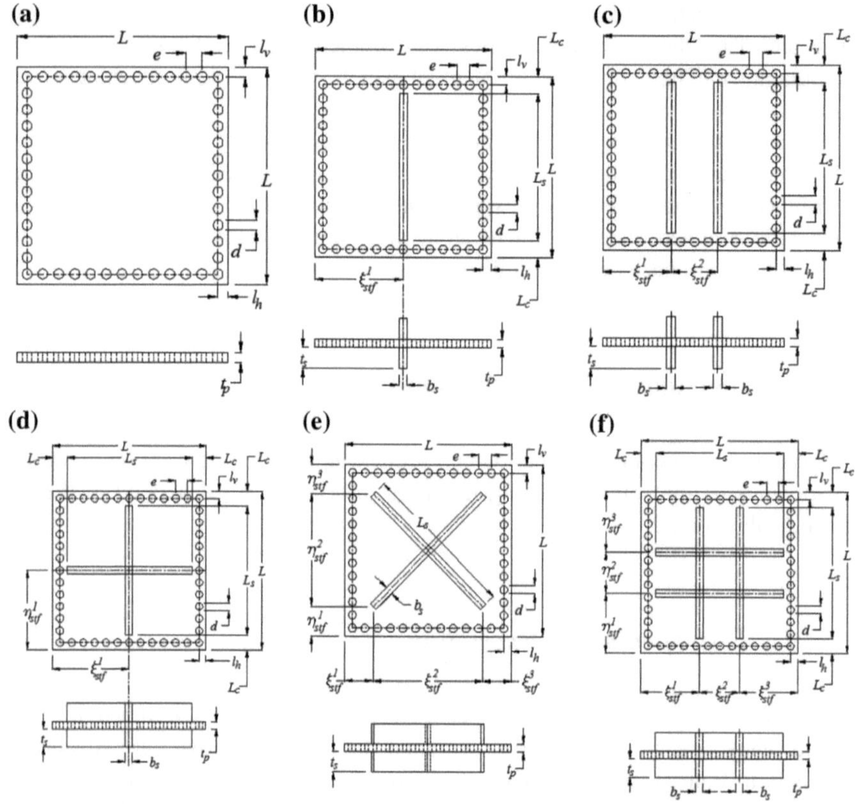

Fig. 21.1 Schematic diagrams of **a** unstiffened, **b** uniaxially single stiffened, **c** uniaxially double stiffened, **d** biaxially cross stiffened, **e** diagonally cross stiffened, **f** biaxially double stiffened plates

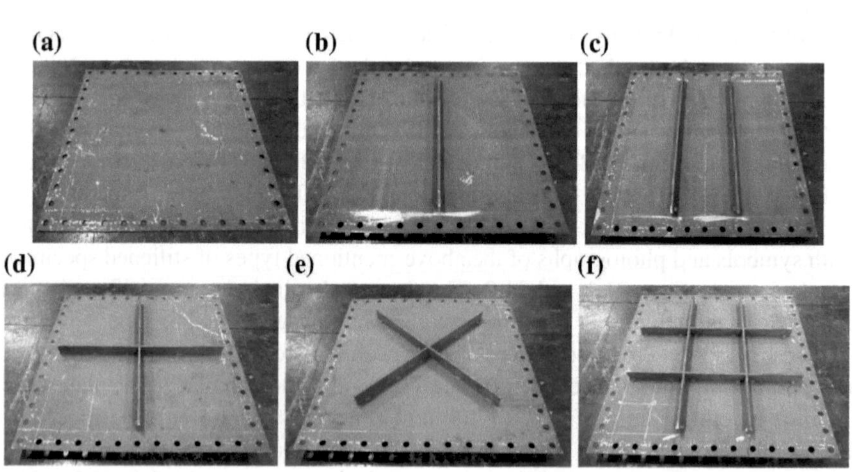

Fig. 21.2 Images of **a** unstiffened, **b** uniaxially single stiffened, **c** uniaxially double stiffened, **d** biaxially cross stiffened, **e** diagonally cross stiffened, **f** biaxially double stiffened plates

Table 21.1 Stiffener position for various types of specimens (mm)

Sl. no.	Stiffened plate type	ξ^{1}_{stf}	ξ^{2}_{stf}	ξ^{3}_{stf}	η^{1}_{stf}	η^{2}_{stf}	η^{3}_{stf}
1	UASS	240	–	–	–	–	–
2	UADS	170	140	170	–	–	–
3	BACS	240	–	–	240	–	–
4	DCS	98.6	282.8	98.6	98.6	282.8	98.6
5	BADS	170	140	170	170	140	170

(constant diameter) present along the outer circumference of the plates. These drilled holes correspond a set of holes present on the edge of the holding arrangement in the experimental setup and are used to clamp the specimen in position. The distance between the centerline of the holes and the plate edge in both orthogonal directions ($l_v = l_h$) is 14.5 mm. Center distance between two adjacent holes (e) is 37.5 mm, while the diameter of these clamping holes (d) is 11 mm. For bolting the plate securely, a certain amount of plate dimension around the periphery of the plate is considered on all the sides. Therefore, the effective dimension of the clamped plate is (L–2L_c).

21.2.2 Experimental Setup

Static deflection and free vibration experiments on concentrically stiffened plates under uniform transverse loading are carried out on an experimental setup as shown in Fig. 21.3. The main components of the experimental setup are also marked in the figure, and these are compressors for supplying compressed air, channel frame having an inlet with bypass mechanism for application of pneumatic pressure, gauges to measure pressure and deflection, accelerometer along with coupler and digital oscilloscope. Uniform transverse pneumatic pressure is applied to induce static deflection of the specimen. Subsequently, the free vibration experiment is carried out by exciting the system about its deflected position. The instantaneous excitation is provided from the blow of a soft rubber hammer. The schematic diagram of the experimental setup is shown in Fig. 21.4.

It has already been mentioned that the experiments are performed under pneumatic pressure. Hence, the first and foremost requirement of the setup is an enclosed/sealed chamber that can retain high-pressure air. An open box with side walls made of thick metal plates is used for the purpose (shown as item 8 in Fig. 21.3). There is provision for attaching plates at the top and bottom through a sequence of drilled holes on the flanges of the box. At the bottom, a thick backing plate is secured, while the test specimen is attached at the top, thus creating an enclosed chamber. This holding arrangement simulates clamped boundary conditions at all the edges of the specimen. On one of the side panels, there is an inlet that introduces high-pressure air from the compressor into the sealed space. In this inlet

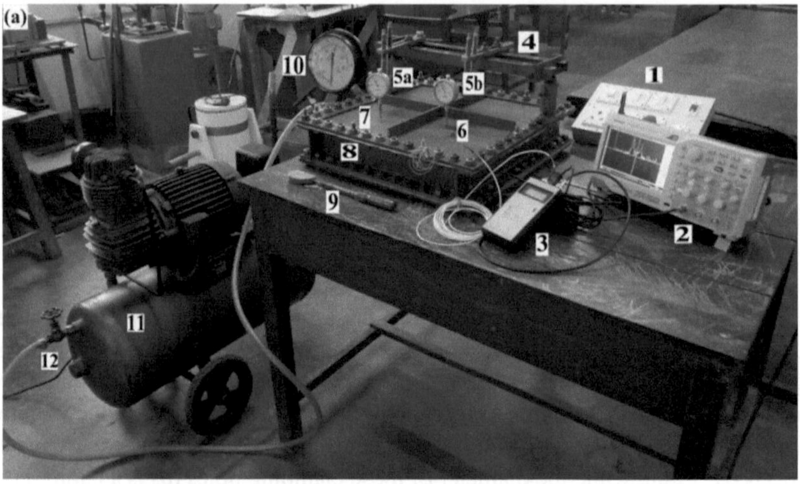

1. External power supply
2. Digital osciiloscope
3. Power supply coupler
4. Fixed arm for positioning dial gauges
5. (a), (b) Dial gauges
6. Accelerometer
7. Stiffened plate specimen
8. Channel frame
9. Rubber hammer
10. Pressure gauge
11. Air compressor
12. Air compressor outlet

Fig. 21.3 Photograph of experimental setup with notations for different components

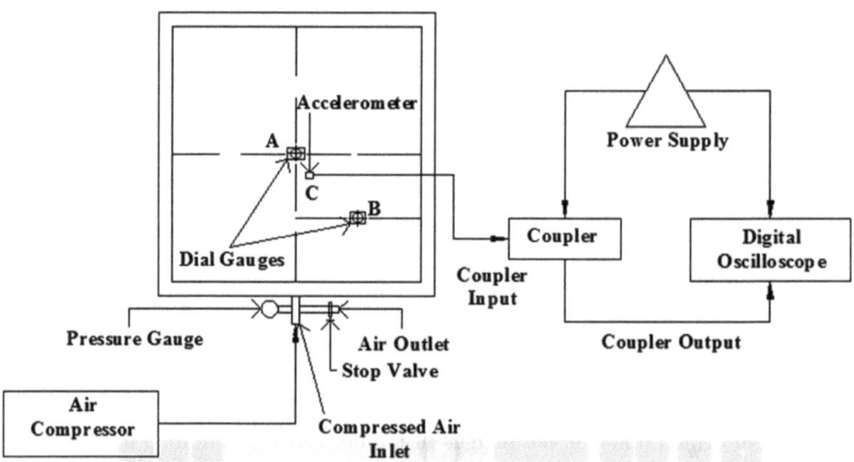

Fig. 21.4 Schematic diagram of experimental setup

assembly, there is provision for attaching a pressure gauge and a bypass valve that can be used to control the pressure level in the chamber. It is important to take preventive measures against leakage from various joints and the chamber itself. To ensure that there is no escape of pressurized air, rubber gaskets and sealing tapes are used in between the frame and the top and bottom plates.

Defections under transverse pressure loading are recorded with the help of two dial gauges which are positioned at locations A and B as depicted in Fig. 21.4. These points indicate that deflection data are collected at two predefined locations and the gauges are positioned with the help of an overhanging arm supported on vertical columns fixed to opposite sides of the frame. A shear mode accelerometer is fixed to the vibrating specimen at an appropriate location with petro-wax adhesive. Extreme care is necessary in locating the accelerometer so as to avoid nodal points and place it approximately near the point of maximum amplitude. However, the location of the point of maximum displacement is dependent on the placement of stiffeners. Hence, for a different type of stiffened plate specimen, the position of the accelerometer is varied. The accelerometer used for the present experimentation is a miniature one and weighs 1.6 g due to significantly low dynamic mass, thus eliminating any chance of mass loading effect on the response. It consists of an impedance converter, which is a miniature electronic circuit, and a sensing assembly made up of a piezoelectric crystal and a seismic mass. Motion of the vibrating system normal to the base is sensed by the piezoelectric element, which generates a high impedance charge. It is then converted into a low impedance voltage output signal by the electronic circuitry. It should be mentioned here that the sensing assembly does not require any external power source to operate, while a coupler (power source) is necessary to power the circuit. An AC–DC power adapter is required to externally power the coupler itself. This instrument decouples the output signal and DC bias voltage and provides an interface between the sensor and the recording/display device (digital oscilloscope). A two-wire cable which carries both the signal and the power connects the coupler and the accelerometer. The transmitted signal of the vibrating system is received by the digital storage oscilloscope, which has the capability to display, post-process, and store the signal. The instrument is equipped with a Fast Fourier Transform (FFT) extension module that can convert the time domain signal into a frequency domain one. Specifications of different instruments used in the laboratory experiments are detailed in Table 21.2.

21.2.3 Test Procedure

The present laboratory experiment has the objective of determining the static deflection at two predefined points and the natural frequency at statically deflected configuration under transverse pneumatic pressure loading for stiffened as well as unstiffened plate specimens. Preparing the setup to conduct experiments involves certain preprocessing tasks. First of all, the specimen under consideration is tightly clamped in position with gaskets (rubber) and sealing tapes in place. Initially,

Table 21.2 Specifications of different instruments used in the laboratory experiments

Instruments	Specifications
Air compressor	Manufacturer: Elgi Equipments Limited, India Model no: LG 100 P; tank capacity: 45 L Maximum working pressure: 9 kg/cm^2
Pressure gauge	Manufacturer: Karnataka Instruments, India Working range: 0–10 kg/cm^2; resolution: 0.2 kg/cm^2
Dial gauge	Manufacturer: Baker, India; type: JO 8; least count: 0.01 mm
Accelerometer	Manufacturer: Kistler Instrument Corporation, India Type: 8728A500; measuring range: 500 g (g = 9.80665 m/s^2); sensitivity: 10.03 mV/g; weight: 1.6 g
Power supply/coupler for accelerometer (externally powered AC–DC power adapter)	Manufacturer: Kistler Instruments Corporation, India Type: 5114; frequency response (±5%): 0.07 Hz–60 kHz
Digital oscilloscope	Manufacturer: Tektronix Inc., USA; model no: TBS 1072B Peal detect bandwidth: 70 MHz; sample rate range: 1 gigasamples/s; voltage range: 100–240 V, 115 V; frequency: 50–60 Hz, 400 Hz; power maximum: 30 W

only the dial gauges are positioned at their respective locations with the help of an arm/bridge supported on two posts. Outlet of the compressor is attached to the inlet of the enclosed chamber through a flexible pipe, and the pressure gauge is also attached to the inlet section. Instruments (compressor, oscilloscope, and coupler) requiring electric power are connected to the power source/supply. Connection between the coupler and the accelerometer is through a two-wire cable, which is shared by power for the impedance circuit and response signal from the accelerometer. The accelerometer is located on the specimen with adhesives at appropriately selected position. The coupler output is taken to the digital oscilloscope through one of the channels of the device. FFT mode is selected on the oscilloscope by pushing the appropriate buttons to make it ready for receiving and capturing the signal.

At first, the static experiment is performed by increasing the air pressure inside the chamber by supplying compressed air gradually. To avoid sudden impact loading, rate of increase of pressure is kept low. After a brief settling period, the readings from the dial gauges are tabulated against the pressure gauge reading. Once these readings are noted, the dial gauge probes are lifted up and kept at a distance with the help of plastic stops. This is necessary in order to let the specimen vibrate without interference from the gauge probes. A rubber hammer is used to strike the deflected plate to initiate free vibration pulses. The signal from the oscillating specimen is captured by the digital oscilloscope and subsequently plotted in frequency domain. These frequency amplitude plot images, in jpg format, are stored and saved by pressing the save button in the oscilloscope and finally taken from it by a pen drive. This completes a single round of data collection corresponding to a specific pressure level. At this stage, the dial gauge probes are set down onto the plate surface by removing the plastic stops. Now, the pressure level is adjusted to a new value and the corresponding data for static deflection and loaded natural frequencies of the system are collected following the same procedure as described above. The entire experimental process is repeated for various concentric stiffened plates along with the unstiffened one.

21.3 Finite Element Modeling and Analysis Using ANSYS

Objective of performing finite element simulation is to provide independent validation of the experimental results. This analysis is undertaken to generate the results for the same parameters as found out from the experiments (static deflection and loaded natural frequencies) and perform a comparison. The entire finite element studies are carried out with the help of finite element software ANSYS Mechanical APDL 15.0. The analysis performed in ANSYS can be conveniently divided into three stages, namely preprocessing, solution, and post-processing.

21.3.1 Selection of Element Type

In the finite element procedure, the main task is to replicate all the physical conditions that are employed in the laboratory experiment. As the basic structural element under consideration is a plate with stiffeners, so in the preprocessing stage of ANSYS the element is chosen to have similar features with the test specimens. In this regard, SHELL 181 element is used to model the plate and SOLID 186 is utilized for the stiffeners. SHELL 181 is an element with four nodes, and each of these nodes has six degrees of freedom. This element is equipped to handle linear, large rotation, and/or large strain nonlinear applications and can be used for analyzing thin to moderately thick shell structures. On the other hand, SOLID 186 is a 3-D solid element with 20 nodes, each with three degrees of freedom. This element exhibits quadratic displacement behavior. All the material properties are provided for mild steel for both the plate and the stiffeners. These material properties are elastic modulus (E) = 211 GPa, Poisson's ratio (ν) = 0.3, density (ρ) = 7830 kg/m^3, while shell thickness is supplied as t_p = 0.002 m. Property values for both the elements are kept same as the plate and stiffeners are made out of the same material.

21.3.2 Modeling and Meshing

The next task is to create a realistic model of the specimens and meshing the model on the basis of selected element types. Modeling is done by creating areas as per the previously mentioned plate dimensions. The stiffeners are created by drawing the appropriate area according to dimensions and extruding it in the normal direction. For different types of stiffened plates, different stiffener orientations are fashioned following appropriate measure. After modeling is completed, meshing is performed for unstiffened and each type of concentrically stiffened plates. Meshing is the process of dividing the physical domain into smaller elements upon which the analysis is carried out. First of all, the element size is fixed and this selection is on the basis of number of divisions that are intended on the plate and stiffeners. As by increasing the number of divisions (i.e., by creating a finer mesh) the accuracy of the results increases, it is necessary to create sufficient divisions to obtain better results. Figures 21.5 and 21.6 show (a) created model and (b) meshing of model in ANSYS for uniaxially single stiffened (UASS) and biaxially cross stiffened (BACS) plates, respectively.

(a) (b)

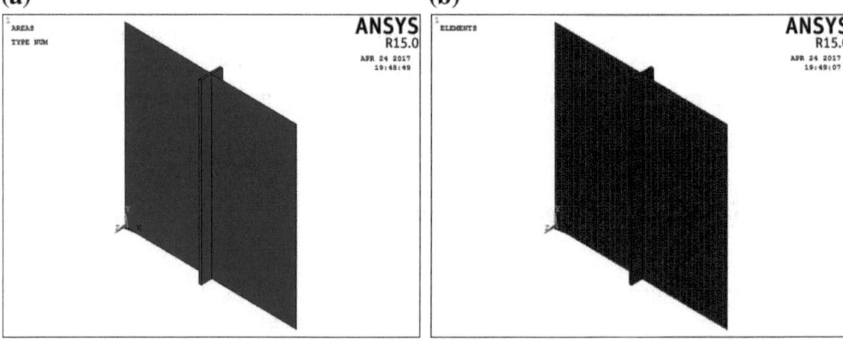

Fig. 21.5 **a** Created model and **b** meshing of model in ANSYS for uniaxially single stiffened (UASS) plate

(a) (b)

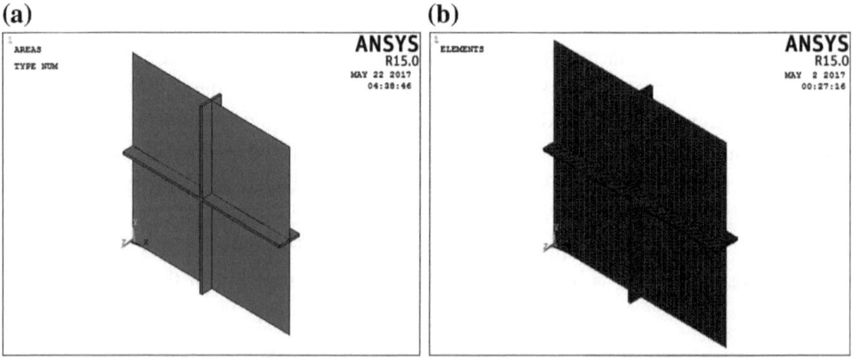

Fig. 21.6 **a** Created model and **b** meshing of model in ANSYS for biaxially cross stiffened (BACS) plate

21.3.3 Contact of Stiffener with Plate

Another important aspect of modeling the specimen is to define the type of contact between the plate and the stiffeners. Physically, the stiffener is attached to the plate by spot welding process. It should be mentioned here that as the plate thickness is small it is difficult to perform a continuous welding of the stiffener to the plate. Due to excessive heat generation, there is a chance of distortion and bending of the plate. So, the stiffener in reality is rigidly connected to the plate at the locations of spot welding. In order to model the contact nature at the interface of the plate and stiffener, a contact pair consisting of a contact surface and target surface is defined. The stiffener element is chosen for target surface, and plate element is chosen as contact element using contact manager option in ANSYS. Here, certain optional settings are to be set before the contact pair is created. First, multi-point constraint (MPC) algorithm is selected as the formulation for contact region. The formulation for the contact region is the mathematical method that the code uses to enforce the contact compatibility conditions such as no penetration allowed, no separation, no sliding. The multi-point constraint (MPC) connection uses rigid constraint equations between the solid elements on the contact and target faces for a truly bonded connection. The contact detection is assumed to be on nodes normal to target, and behavior of contact surfaces is always bonded. There is no initial penetration, while the type of constraints remains in auto assembly detection and automatic contact adjustment is set as close gap.

21.3.4 Finite Element Solution

Once the preprocessing is finished with setting up the problem accurately, solutions can be obtained by running the specialized solvers. Before the solution is performed in static analysis, some solution parameters need to be predefined. In solution control, the option is set as large displacement static analysis, where the loads are imparted in certain substeps. Next, the boundary conditions are imposed by providing constraints (zero displacement in all three directions) at the edges of the plate so that the entire boundary remains in clamped condition. Then, a uniformly distributed transverse load of a certain magnitude is applied to the specimen. Finally, the solution is obtained by clicking the run option. Once converged results are obtained, one can move onto the post-processing segment for documentation of the obtained data. The same procedure is repeated from loading onward with a modified (increased) value of the transverse load. After a few iterations of the same process, results for the desired load range are obtained.

In order to obtain the loaded natural frequencies of the system, a prestressed modal analysis needs to be performed. Here as a prerequisite to the modal analysis, large amplitude static analysis is to be carried out with prestress effects taken into account. Apart from that, the other steps are same as the normal static analysis.

So, application of boundary conditions, transverse loading, and running the solution remains the same as before, but once the solution is obtained it is necessary to check the nodal deflection value. After that, one needs to restart the process as a modal analysis, which is tackled through perturbation technique. For modal solution, some analysis options are to be set before moving on to solving. Block Lanczos mode extraction method is used for the analysis. Block Lanczos is an extraction method to solve the modal problem of a large model of many modes. The method works well when the model consists of shells or a combination of solids and shells. Once the converged solution is obtained, it is necessary to retrieve the identified modal analysis data from the perturbation result file. The method is repeated for increasing values of transverse uniformly distributed loads, so that the variations in loaded natural frequencies can be plotted.

21.3.5 Post-processing

Once the solution is obtained for static analysis, it has been checked for the z component nodal solution value for all the load conditions. The data are utilized to plot the normalized pressure–deflection curves for the system. In the modal analysis, natural frequencies are extracted to plot the backbone curves.

21.4 Results and Discussions

The present experimental analysis of concentrically stiffened plates has two main objectives: determination of static deflection and finding out the loaded natural frequencies of the system under uniform transverse loading. It is also intended to observe the effect of number, positioning, and orientation of the stiffeners on the static and free vibration characteristics of the system. All the results, static as well as dynamic, are normalized with respect to appropriate parameters and plotted in non-dimensional planes using Grapher 8.0 software. Load–deflection curves are obtained from the static experiments, and these curves are exhibited in a non-dimensional plane, where the abscissa is normalized intensity of uniformly distributed load ($Pa^4/E_p t_p^4$), while the ordinate represents normalized deflection (w/t_p). The experimental data of the free vibration tests are plotted as backbone curves. In this plot, the abscissa represents normalized frequency (ω/ω_1) and normalized deflection (w/t_p) is shown along the ordinate.

21.4.1 Validation Study

Validation of the present experimental and finite element analysis is carried out with respect to previously published and established results. The comparisons are done with respect to the results furnished by Mitra et al. [21], who conducted experimental investigations of static and dynamic behavior of eccentrically stiffened plates. In this regard, it should be mentioned here that eccentric stiffening is associated with stiffeners fixed to only one face of the plate. Geometric details of the specimens in that study were as follows: $L = 480$ mm, $t_p = 0.002$ mm, $L_s = 400$ mm, $b_s = 0.005$ mm, $t_s = 0.025$ mm. Here, the same experiments are performed in the laboratory along with simulations in ANSYS and the results are plotted together in Fig. 21.7 (static deflection results) and Fig. 21.8 (backbone curves for free vibration analysis). Each of the two figures contains two separate cases corresponding to unstiffened plate and uniaxially single stiffened (UASS) eccentric plates. It is seen from the comparisons that the general trend of the curves is quite similar. Experimental results and finite element simulation results generated in ANSYS are verified against numerical results of a semi-analytical displacement-based analysis put forward by Mitra et al. [21]. From the validation study, it can be said that the present experimental procedure and FE analysis generate authentic results within the limitations of experimental errors and differences in different types of modeling.

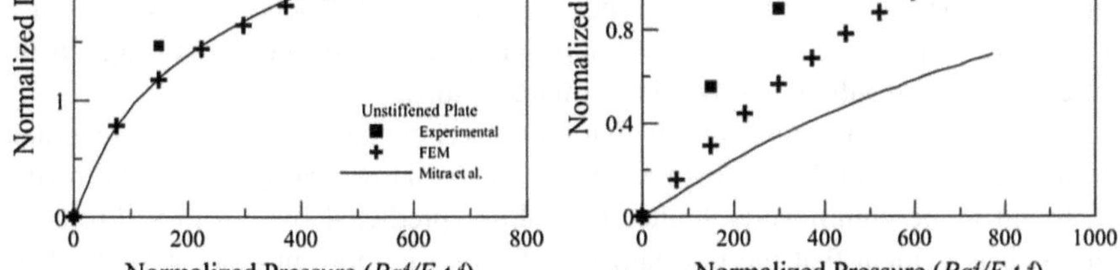

Fig. 21.7 Comparison of static results for **a** unstiffened and **b** uniaxially single stiffened eccentric plates

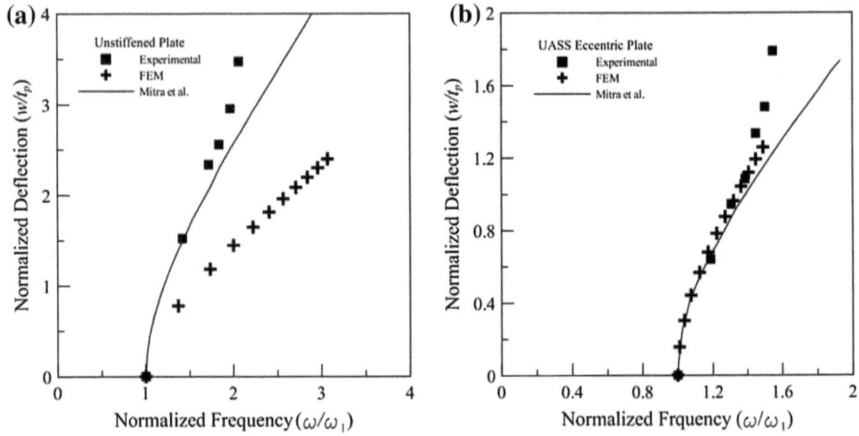

Fig. 21.8 Comparison of free vibration results for **a** unstiffened and **b** uniaxially single stiffened eccentric plates

21.4.2 *Static Analysis*

The experimental results acquired from static deflection tests are matched against those generated by finite element (FE) simulation of ANSYS 15.0. The plate and stiffener dimensions of the specimens used in the experiment are supplied to the ANSYS simulation, and the boundary condition is assumed as all edges clamped. The deflection values obtained for both types of analysis are normalized by thickness of plate, and the normalized load–deflection plots for UASS plates are furnished in Fig. 21.9. Normalized load–deflection plots (obtained experimentally and through ANSYS) for uniaxially double stiffened (UADS), biaxially cross stiffened (BACS), diagonally cross stiffened (DCS), and biaxially double stiffened (BADS) plates are presented in Fig. 21.10. It should be mentioned here that in these figures the normalized deflection on the vertical axis corresponds to either point A or point B, depending on whichever is higher. As mentioned previously, in the static deflection results the x-axis is normalized load.

It is observed from these figures that the experimental curve and that generated by finite element simulation display similar trends. However, experimental results are found to be significantly higher than those predicted by ANSYS. From the figures, it is evident that the normalized deflection is more in case of DCS plates while the values of deflection for BADS plates are quite low. This is an expected outcome as with increase in stiffener number the overall stiffness of the system increases giving rise to lower deflection at the same load level.

The difference between the simulated (ANSYS) and experimental load–deflection curves can be attributed to variation in the boundary conditions in the two situations. In case of experimentation, the plate is bolted securely to the frame, thus nullifying any vertical displacement. However, the axial displacement is restricted

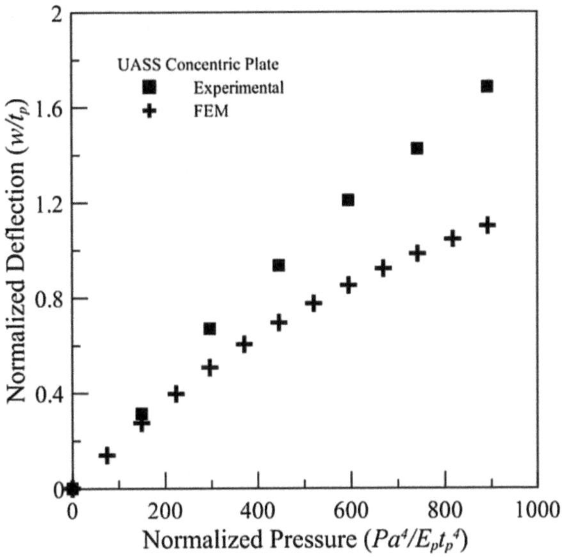

Fig. 21.9 Normalized load–deflection plot for uniaxially single stiffened concentric plate

only by the frictional force generated by the clamping pressure between the plate and the frame. At low pressure, intensity friction may be sufficient to stop any kind of in-plane displacement, but at higher pressure the edges of the plate may be rendered movable. On the other hand, in ANSYS the boundary condition provided at the edges of the plates completely restricts movement in transverse as well as axial direction. It is noticeable that the variation between experimentation and simulation results increases with increase in pressure. It is indicative of the fact that at high transverse pressure, stretching forces come into picture. It is also important to point out that the interaction between the stiffeners while crossing each other may contribute toward determining the overall behavior of the system. In finite element modeling, the interaction between the stiffeners at this interface has been neglected. This may also be one of the reasons behind the divergences in the deflection values obtained from experimental and ANSYS results.

21.4.3 Free Vibration Analysis and Backbone Curves

In the present analysis, variation of natural frequencies of free vibration with respect to increase in transverse loading on the system has been studied for different concentrically stiffened plates. Large amplitude dynamic behavior of a system is represented by backbone curves. Finite element simulation results are compared with the experimentally obtained ones in terms of the backbone curve (first mode of vibration) as shown in Fig. 21.11 for UASS plates. This backbone curve is depicted in a non-dimensional plane (normalized frequency versus normalized

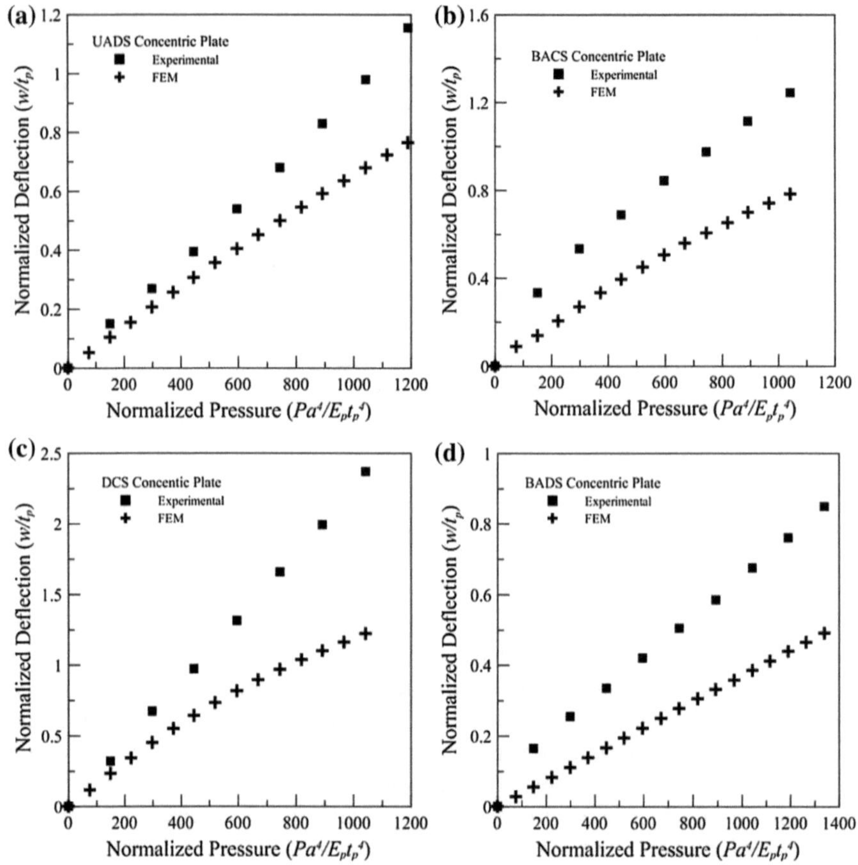

Fig. 21.10 Normalized load–deflection curve for **a** uniaxially double stiffened, **b** biaxially cross stiffened, **c** diagonally cross stiffened, **d** biaxially double stiffened concentric plates subjected to transverse uniformly distributed load

displacement). The normalization of the frequency axis is done with respect to the linear fundamental frequency (ω_1) (at zero load condition) of the specimen. Hence, it is important to individually provide the normalization values for each of the plate specimens. In Table 21.3, the first natural frequency (ω_1) at no-load condition is supplied for various concentrically stiffened plates corresponding to experimental and finite element analysis. Utilizing the values provided in the table, the non-dimensional backbone curves can be readily transformed into dimensional curves and data corresponding to any load value can be extracted for practical use.

Experimentally obtained backbone curves for the fundamental mode along with those simulated in ANSYS for UADS, BACS, DCS, and BADS plates are furnished in Fig. 21.12. It is visible from the figure that in the non-dimensional plane the experimentally generated backbone curve for UADS and BACS plate is in close proximity to those generated from ANSYS. However, it does not indicate that the

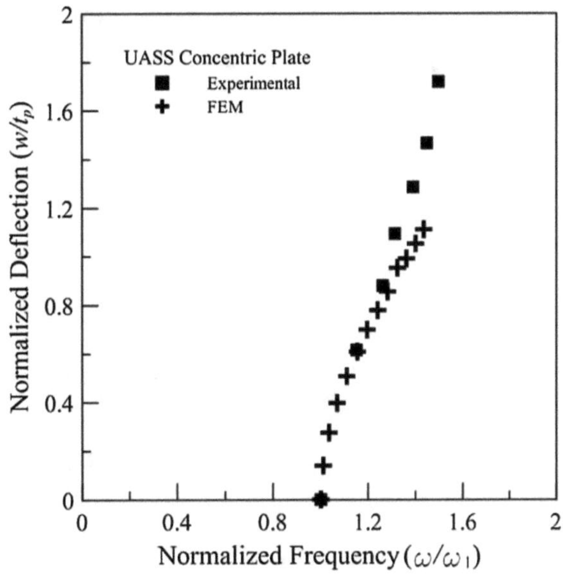

Fig. 21.11 Backbone curve for uniaxially single stiffened concentric plate

Table 21.3 Natural frequency for first mode of vibration at no-load condition for different types of concentrically stiffened plates

Sl. no.	Plate type	Pressure or load (kg/cm^2)	Natural frequency (Hz)	
			Experimental	Finite elements
1	UASS	0	230	256.91
2	UADS	0	368	426.45
3	BACS	0	354	417.07
4	DCS	0	215.2	202.96
5	BADS	0	430	552.60

actual values of the frequencies are matching closely; but the trends of both the curves are almost identical. The results corresponding to BADS plate exhibit slight variation, whereas results for DCS plate show a significant difference from the experimental curves. However, hardening-type (increase in load/deflection causes loaded natural frequency to increase) nonlinearity is exhibited in all the cases.

The free vibration results evidently demonstrate that for each of the types of stiffened plates at lower end of pressure, difference between finite element simulation and experimental results is minor. But the two curves tend to move apart from each other with increase in pressure. In order to get to the root cause of the deviations, further detailed studies are necessary. But at this point, it can be quite safely inferred that better matching between the two sets of data would be obtained through realistic modeling of boundary conditions of the plates in ANSYS.

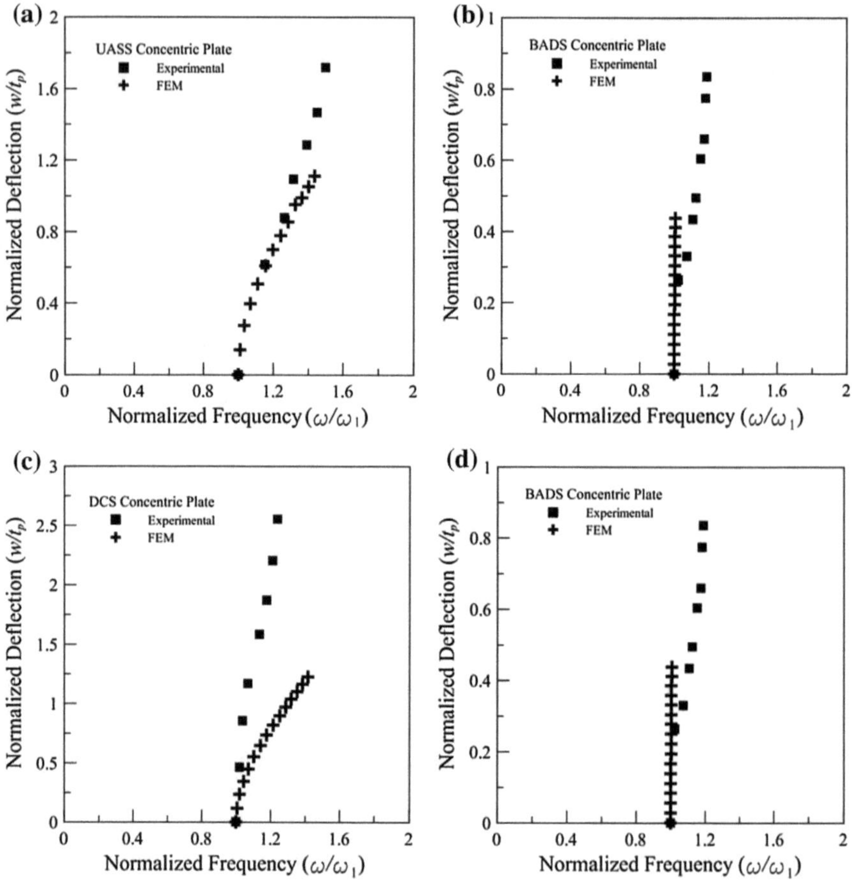

Fig. 21.12 Backbone curves corresponding to fundamental mode of frequency for **a** uniaxially double stiffened, **b** biaxially cross stiffened, **c** diagonally cross stiffened, **d** biaxially double stiffened concentric plates

Instead of providing no displacement at the edges of the plates, a finite yet small amount of in-plane displacement reduces the overall stiffness of the system, hence bringing down the natural frequency values. Another source of deviation could be in the nature of interaction at the plate–stiffener interface. In case of the physical specimens, the stiffeners are attached to the plate by welding at some predefined positions, but in the simulation 3-D model the stiffener is attached to the plate throughout by bonded condition.

21.5 Conclusions

The present paper conducts an experimental study into static and free vibration behavior of different types of concentrically stiffened plates subjected to uniform transverse pressure. The laboratory tests are conducted by utilizing an experimental setup with provision for bolting at all edges of the plate specimen to simulate clamped boundaries. The setup also has provision for introducing high-pressure air from a compressor into an enclosed chamber such that uniformly distributed transverse loading can be obtained. Experiments are carried out for an unstiffened and five separate types of concentric stiffened plates. These are uniaxially single stiffened (UASS), uniaxially double stiffened (UADS), biaxially cross stiffened (BACS), diagonally cross stiffened (DCS), biaxially double cross (BADS) plates. The objective of the present work is the determination of maximum static deflection of the plate under external uniform transverse loading, while the loaded natural frequencies are obtained from the free vibration tests. A validation study has been conducted by comparison with previously published results to authenticate the data obtained from experiments as well as finite element analysis. Normalized load–deflection plots are presented as the results of static experiment, whereas frequency data are provided as backbone curves in a non-dimensional plane. The results show hardening-type nonlinearity for the system, which indicates that the loaded natural frequency increases with increasing pressure. The experimental results for both static and free vibration studies are compared with those generated by the finite element simulation in ANSYS 15.0 software. Although in terms of magnitude the experimental results are found to be higher than the finite element simulations, the trend of the two sets of curves is quite similar.

References

1. Bedair, O.K.: A contribution to the stability of stiffened plates under uniform compression. Comput. Struct. **66**(5), 535–570 (1998)
2. Yamki, N., Otomo, K., Chiba, M.: Nonlinear vibrations of a clamped rectangular plate with initial deflection and initial edge displacement – part II: Experiment. Thin-Walled Struct. **1**(2), 101–119 (1983)
3. Bau-Madsen, N.K., Svendsen, K.H., Kildegaard, A.: Large deflections of sandwich plates-an experimental investigation. Compos. Struct. **23**(1), 47–52 (1993)
4. Kumar, M.S., Alagusundaramoorthy, P., Sundaravadivelu, R.: Ultimate strength of stiffened plates with a square opening under axial and out-of-plane loads. Eng. Struct. **31**(11), 2568–2579 (2009)
5. Xu, M.C., Soares, C.G.: Comparisons of calculations with experiments on the ultimate strength of wide stiffened panels. Mar. Struct. **31**, 82–101 (2013)
6. Pedram, M., Khedmati, M.R.: The effect of welding on the strength of aluminium stiffened plates subject to combined uniaxial compression and lateral pressure. Int. J. Naval Arch. Ocean Eng. **6**(1), 39–59 (2014)
7. Shanmugam, N.E., Dongqi, Z., Choo, Y.S., Arockiaswamy, M.: Experimental studies on stiffened plates under in-plane load and lateral pressure. Thin-Walled Struct. **80**, 22–31 (2014)

8. Liu, K., Wang, Z., Tang, W., Zhang, Y., Wang, G.: Experimental and numerical analysis of laterally impacted stiffened plates considering the effect of strain rate. Ocean Eng. **99**, 44–54 (2015)
9. Zheng, C., Kong, X.S., Wu, W.G., Liu, F.: The elastic-plastic dynamic response of stiffened plates under confined blast load. Int. J. Impact Eng. **95**, 141–153 (2016)
10. Morin, D., Kaarstad, B.L., Skajaa, B., Hopperstad, O.S., Langseth, M.: Testing and modelling of stiffened aluminium panels subjected to quasi-static and low-velocity impact loading. Int. J. Impact Eng. **110**, 97–111 (2017)
11. Fahy, F.J., Wee, R.B.S.: Some experiments with stiffened plates under acoustic excitation. J. Sound Vib. **7**(3), 431–436 (1968)
12. Yurkovich, R.N., Schmidt, J.H., Zak, A.R.: Dynamic analysis of stiffened panel structures. J. Aircr. **8**(3), 149–155 (1971)
13. Olson, M.D., Hazell, C.R.: Vibration studies on some integral rib-stiffened plates. J. Sound Vib. **50**(1), 43–61 (1977)
14. Wilson, J.F., Henry, J.K., Clark, R.L.: Measured free vibrations of partially clamped square plates. J. Sound Vib. **231**(5), 1311–1320 (2000)
15. Sheikh, A.H., Mukhopadhyay, M.: Linear and nonlinear transient vibration analysis of stiffened plate structures. Finite Elem. Anal. Des. **38**(6), 477–502 (2002)
16. Amabili, M.: Theory and experiments for large-amplitude vibrations of rectangular plates with geometric imperfections. J. Sound Vib. **291**(3), 539–565 (2006)
17. Qing, G., Qiu, J., Liu, Y.: Free vibration analysis of stiffened laminated plates. Int. J. Solids Struct. **43**(6), 1357–1371 (2006)
18. Hu, H.T., Li, Y.L., Tao, S.U.O., Feng, Z.H.A.O., Miao, Y.G., Pu, X.U.E., Qiong, D.E.N.G.: Fatigue behavior of aluminum stiffened plate subjected to random vibration loading. Trans. Nonferrous Met. Soc. China **24**(5), 1331–1336 (2014)
19. Zhou, X.Q., Yu, D.Y., Shao, X., Wang, S., Zhang, S.Q.: Simplified-super-element-method for analyzing free flexural vibration characteristics of periodically stiffened-thin-plate filled with viscoelastic damping material. Thin-Walled Struct. **94**, 234–252 (2015)
20. Cho, D.S., Kim, B.H., Kim, J.H., Choi, T.M., Vladimir, N.: Free vibration analysis of stiffened panels with lumped mass and stiffness attachments. Ocean Eng. **124**, 84–93 (2016)
21. Mitra, A., Sahoo, P., Saha, K.N.: In: Proceedings of the IASTED International Conference Engineering and Applied Science - EAS'2012, Colombo, Sri Lanka, pp. 6–13 (2012)

Chapter 22
A Shaft Finite Element for Analysis of Viscoelastic Tapered and Hollow Tapered Rotors

Amit Bhowmick, Arghya Nandi, Sumanta Neogy and Smitadhi Ganguly

22.1 Introduction

A load deformation characteristic of solid material is explained by Hooke's law for linear elastic material, where energy is stored when load is applied and released on unloading without any energy dissipation. In presence of relative velocity in-between fluid layers energy dissipates, where dissipated energy is directly proportional to the rate of shear strain and shear stress is quantified by Newton's law of viscosity. In reality, solid materials not only store energy but also dissipate energy on application of external load. In free vibration analysis of a damped flexible rotor, several models explaining both storing and dissipation of energy are used to incorporate this energy dissipation, like two-, three- or four-element spring damper model, Maxwell model, Maxwell–Wiechert model and Kelvin–Voigt model.

In the literature of vibration analysis of rotor-bearing system, several works have been done using finite element approach. Finite element model for rotor-bearing system considering Euler–Bernoulli beam theory for taper and cylindrical shaft is analysed by Nelson [1, 2]. Similarly, a comparative study on dynamic analysis of rotor is done [4] in between Timoshenko and Euler–Bernoulli theory for different slenderness ratio of rotor. These analyses are performed considering the shaft material as elastic. In free vibration analysis, gyroscopic force makes natural frequency to be dependent of spin speed and real part remains zero, but if damping is present in material, then the real part becomes nonzero, and after certain spin speed, this value becomes positive which makes system unstable. Hence, the study of rotor

A. Bhowmick (✉) · A. Nandi · S. Neogy
Department of Mechanical Engineering, Jadavpur University, Kolkata 700032, India
e-mail: amitbhowmick555@gmail.com

S. Ganguly
Department of Mechanical Engineering, Hooghly Engineering and Technology College, Hooghly 712103, India

P. Sahoo and J. P. Davim (eds.), *Advances in Materials, Mechanical and Industrial Engineering*, Lecture Notes on Multidisciplinary Industrial Engineering,
https://doi.org/10.1007/978-3-319-96968-8_22

considering damping is important to find out the spin speed beyond which real part of eigenvalue becomes positive.

Detailed FEM formulation of rotor-bearing system on spinning reference frame is done by Ganguly et al. [3] considering Euler–Bernoulli beam theory for cylindrical rotor element. In this current work, stability analysis of a tapered viscoelastic rotor is investigated using Maxwell–Wiechert model and a comparison is carried out in between stepped cylindrical rotor element and tapered rotor element in terms number of element required to converge using Euler–Bernoulli beam theory.

22.2 Maxwell–Wiechert Model

Maxwell–Wiechert model used to describe the behaviour of linear viscoelastic materials. In this model a spring is connected to one or more then one number of Maxwell Branch (Spring and damper connected in series). Figure 22.1 represents Maxwell–Wiechert model with three numbers of Maxwell branch. In this current work, we have considered this model.

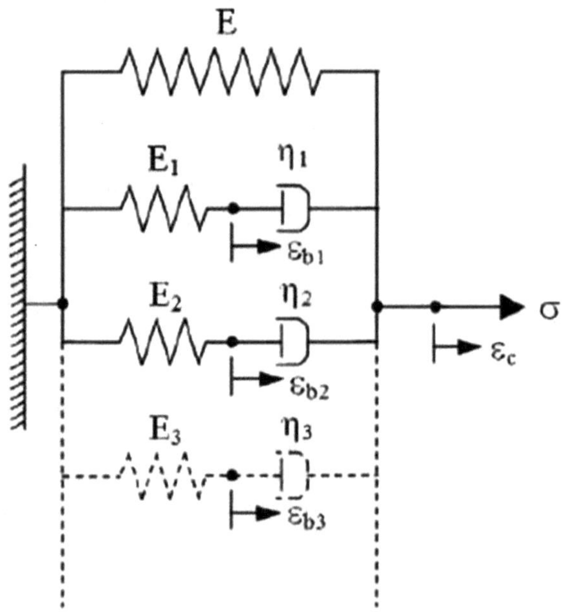

Fig. 22.1 Maxwell–Wiechert model with '*n*' number of Maxwell branch

22.3 Formulations

22.3.1 Equation of Motion

Finite element model is formulated using virtual work principle. Expression for virtual strain energy and virtual work is given by Eqs. (22.1) and (22.2), respectively.

$$\int_{1} \{\delta \varepsilon_c\}^{\mathrm{T}} \{\sigma\} \mathrm{d}x$$

$$= \{\delta U_c\}^{\mathrm{T}} \int_{1} E [B]^{\mathrm{T}} [B] \mathrm{d}x \{U_c\} + \sum_{i=1}^{n} \{\delta U_c\}^{\mathrm{T}} \int_{1} \eta_i [B]^{\mathrm{T}} [B] \mathrm{d}x \{\dot{U}_c\}$$
$$- \sum_{i=1}^{n} \{\delta U_c\}^{\mathrm{T}} \int_{1} \eta_i [B]^{\mathrm{T}} [B] \mathrm{d}x \{\dot{U}_{\mathrm{bi}}\}$$

Or,

$$\int_{V} \{\delta \varepsilon_c\}^{\mathrm{T}} \{\sigma\} \mathrm{d}V = \{\delta U_c\}^{\mathrm{T}} E [K_e^e] \{U_c\} + \sum_{i=1}^{n} \{\{\delta U_c\}^{\mathrm{T}} \eta_i [K_e^e] \{\dot{U}_c\} - \{\delta U_c\}^{\mathrm{T}} \eta_i [K_e^e] \{\dot{U}_{\mathrm{bi}}\}\} \tag{22.1}$$

$$\partial w = -\{\delta U_c\}^{\mathrm{T}} \left(\left([M_{\mathrm{T}}^e] + [M_{\mathrm{R}}^e] \right) \{\ddot{U}_c\} + 2\omega [\hat{M}_{\mathrm{T}}^e] \{\dot{U}_c\} - \omega^2 \left([M_{\mathrm{T}}^e] - [M_{\mathrm{R}}^e] \right) \{\dot{U}_c\} \right) \tag{22.2}$$

Strain energy will be equal to virtual work, and hence, the equation of motion in spinning frame is presented by Eq. (22.3)

$$\left([M_{\mathrm{T}}^e] + [M_{\mathrm{R}}^e] \right) \{\ddot{U}_c\} + 2\omega [\hat{M}_{\mathrm{T}}^e] \{\dot{U}_c\} - \omega^2 \left([M_{\mathrm{T}}^e] - [M_{\mathrm{R}}^e] \right) \{U_c\} + E [K_e^e] \{U_c\} + \sum_{i=1}^{n} \eta_i [K_e^e] \{\dot{U}_c\}$$
$$- \sum_{i=1}^{n} \eta_i [K_e^e] \{\dot{U}_{\mathrm{bi}}\} = \{f^e\} \tag{22.3}$$

And corresponding equation in state space is given below (Eq. 22.4).

$$[A_n] \{\dot{U}_n\} + [B_n] \{U_n\} = \{f_n\} \tag{22.4}$$

Details of expression for all element matrices are shown in Appendix 1.

22.3.2 Formulation of Element Matrices

For numerical analysis of viscoelastic tapered rotor, two types of rotor elements are formulated

(i) Solid tapered rotor element.
(ii) Hollow tapered rotor element.

Details of expression of formulated element matrices are given in Appendix 1.

22.4 Validation

For validation of formulated code, the results achieved by current analysis are compared with ANSYS and Ref. [2] as described below. Five different cases are considered

a. Solid elastic tapered rotor without disc (Fig. 22.2a).
b. Solid viscoelastic uniform rotor with eccentric disc (Fig. 22.2b).
c. Hollow elastic tapered rotor without disc (Fig. 22.2c).
d. Solid viscoelastic tapered rotor with eccentric disc (Fig. 22.2d).

22.4.1 Validation of Tapered Solid Rotor Element

First of all, frequency response is computed for a mild steel, solid tapered rotor without disc (Case 'a'). The results are in good agreement with those found using ANSYS (Fig. 22.3). Now a uniform (non-tapered) PVC rotor (Case 'b') is considered for which the stability limit is obtained. The results are identical with those presented in Ref. [3] and are shown in Fig. 22.4.

22.4.2 Validation of Tapered Hollow Rotor Element

Frequency response is computed for a mild steel, hollow tapered rotor without disc (Case 'c'). The results match those found using ANSYS (Fig. 22.5). A viscoelastic, solid tapered rotor (Case 'd') is considered next. The stability limit for this rotor is obtained using hollow tapered element with a zero value for the inner radius. The results are compared with the stability limits obtained using solid tapered rotor element (Fig. 22.6) and found to be in good agreement.

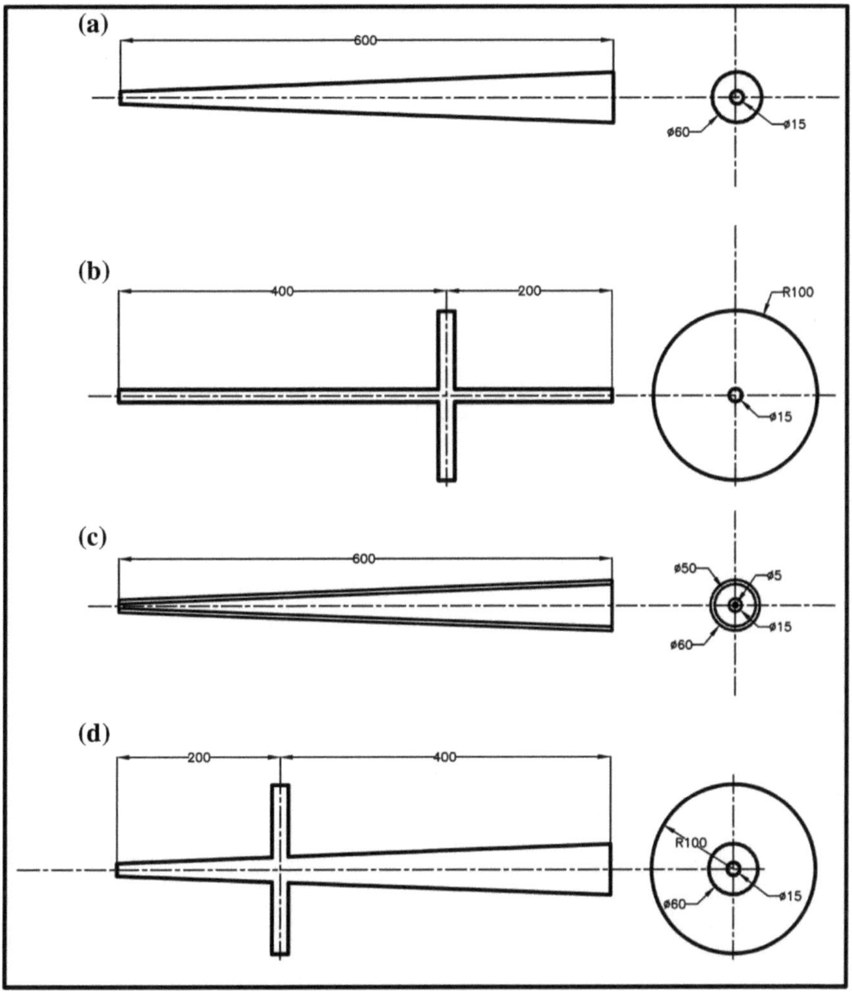

Fig. 22.2 Rotor model for numerical analysis—**a** solid tapered beam, **b** solid uniform rotor with eccentric disc, **c** hollow tapered beam and **d** solid tapered rotor with eccentric disc

22.5 Results and Analysis

22.5.1 Stability Analysis

A simply supported solid tapered rotor and a hollow tapered rotor have been taken (Figs. 22.2d and 22.7, respectively). A rigid disc of mass 1 kg and radius of gyration 100 mm is placed 100 mm from the centre towards smaller diameter for both the rotors. Both PVC and mild steel have been used as rotor materials.

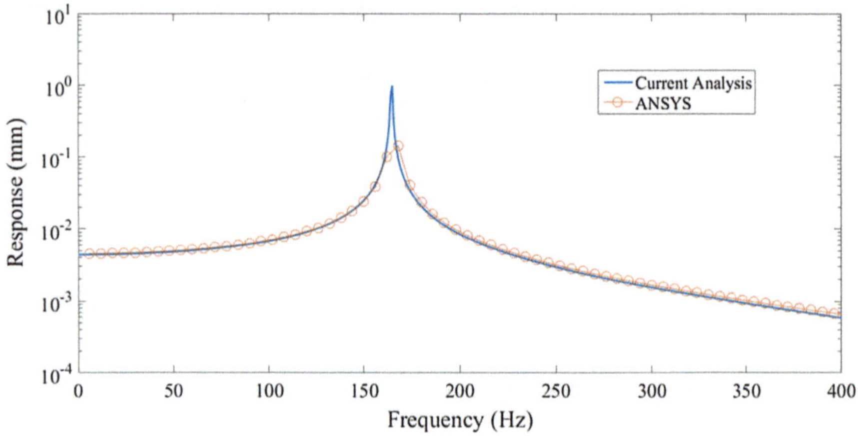

Fig. 22.3 Comparison of frequency response for solid tapered rotor obtained by current analysis and ANSYS

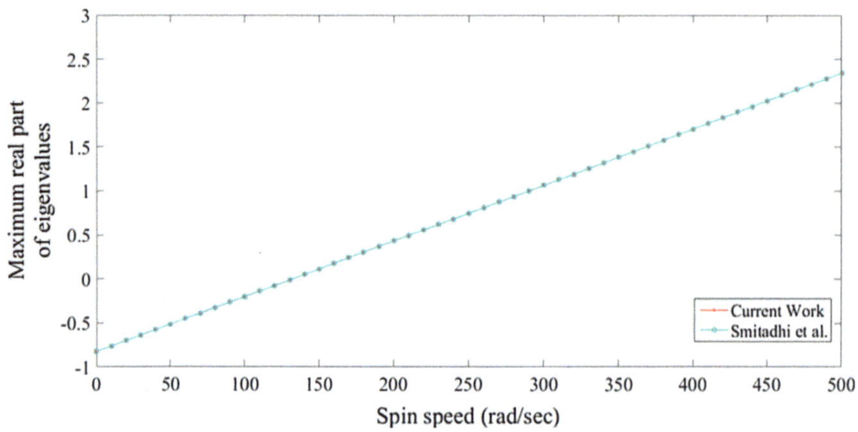

Fig. 22.4 Comparative plot of maximum real part versus spin speed for solid straight rotor

22.5.1.1 Example 1: Solid Tapered Rotor

The rotor configuration is shown in Fig. 22.2d. For both PVC and mild steel, a Maxwell–Wiechert material model with three branches is considered. The branch properties are presented in Table 22.1. The stability limit is computed using the proposed tapered solid element and stepped cylindrical element (Ref. [3]). The results are compared in Table 22.2. For solid tapered shaft made up of PVC, taper element model takes 9 elements (Fig. 22.8), whereas stepped cylindrical element model requires more than 24 elements to converge (Fig. 22.9), and a comparative plot is shown in Fig. 22.10. For mild steel, solid tapered shaft, taper element takes 9

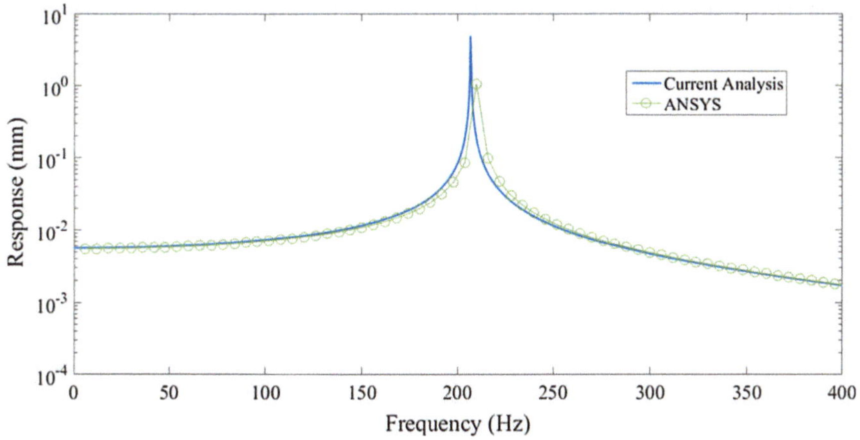

Fig. 22.5 Comparison of frequency response for hollow tapered rotor obtained by current analysis and ANSYS

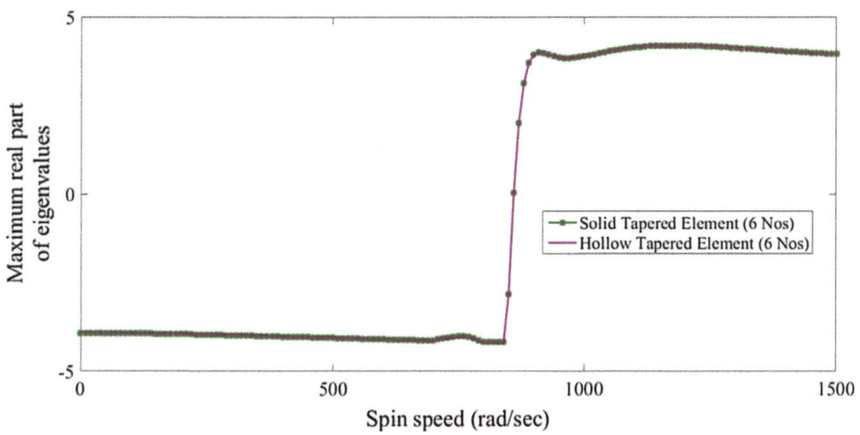

Fig. 22.6 Comparative plot of maximum real part versus spin speed for solid straight rotor

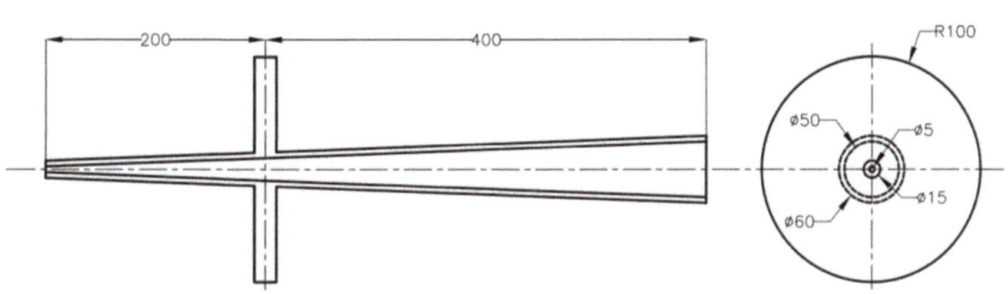

Fig. 22.7 Tapered hollow rotor with eccentric disc for numerical analysis

Table 22.1 Material properties of PVC and mild steel, taken from Refs. [3, 5], respectively

Material	PVC	Mild steel
E	2.46×10^{10} N/m^2	$2.0 \times 10^{11} \frac{\text{N}}{\text{m}^2}$
E_1	0.858×10^{10} N/m^2	$3.407 \times 10^{9} \frac{\text{N}}{\text{m}^2}$
E_2	0.494×10^{10} N/m^2	$2.651 \times 10^{9} \frac{\text{N}}{\text{m}^2}$
E_3	0.285×10^{10} N/m^2	$3.407 \times 10^{9} \frac{\text{N}}{\text{m}^2}$
η_1	0.4×10^{6} Ns/m^2	$1.136 \times 10^{8} \frac{\text{Ns}}{\text{m}^2}$
η_2	0.752×10^{6} Ns/m^2	$8.836 \times 10^{6} \frac{\text{Ns}}{\text{m}^2}$
η_3	$1.32 \times 10^{6} \frac{\text{Ns}}{\text{m}^2}$	$1.136 \times 10^{6} \frac{\text{Ns}}{\text{m}^2}$
ρ	200 kg/m^3	7800 kg/m^3

Table 22.2 Comparison of shaft spin speed of stable operation for solid tapered shaft for mild steel and PVC material

Stability frequency limit					
	Material	Mild steel		PVC	
	Element type	Taper	Cylindrical	Taper	Cylindrical
No. of element	3(2 + 1)	848.86	746.53	523.06	452.35
	6(4 + 2)	860.04	828.89	517.52	496.89
	9(6 + 3)	860.02	846.00	516.74	507.85
	12(8 + 4)	859.94	852.16	516.57	511.73
	15(10 + 5)	859.89	855.00	516.52	513.50
	18(12 + 6)	859.88	856.51	516.51	514.45
	21(14 + 7)	859.87	857.42	516.51	515.01
	24(16 + 8)	859.87	858.01	516.50	515.37

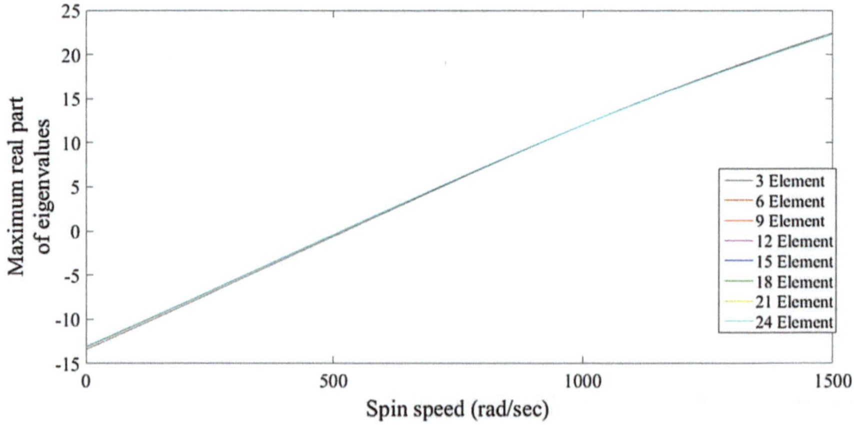

Fig. 22.8 Element-wise stability plot for solid tapered PVC rotor obtained with solid tapered element

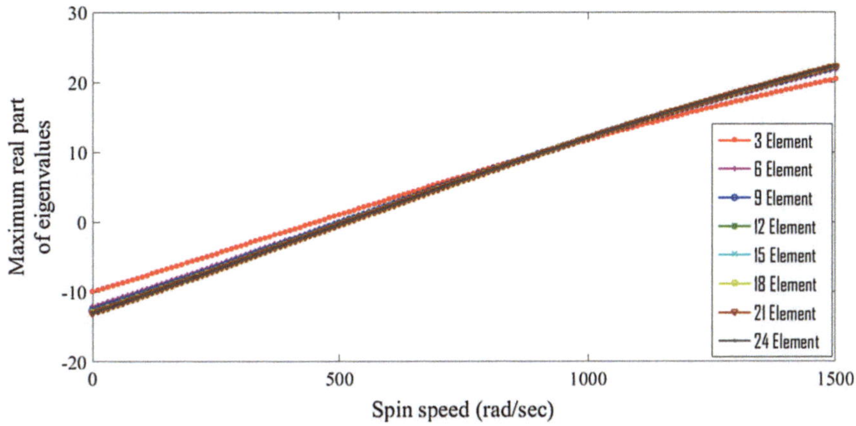

Fig. 22.9 Element-wise stability plot for solid tapered PVC rotor obtained with stepped cylindrical element

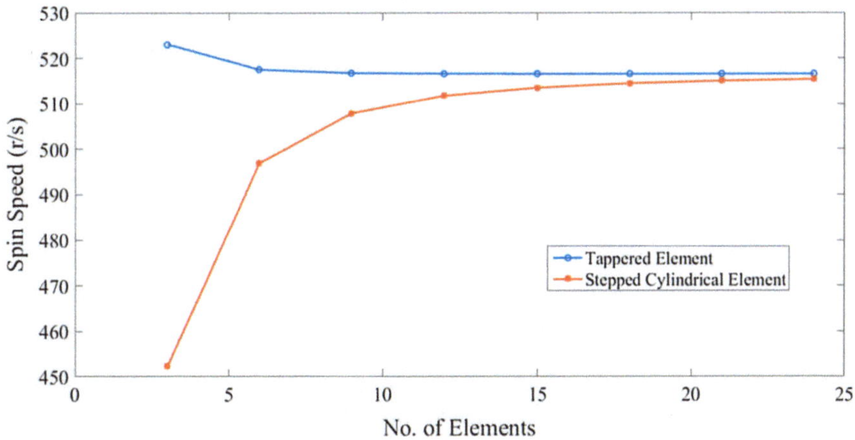

Fig. 22.10 Element-wise comparative plot of frequency limit of stable operation for solid tapered PVC rotor

elements (Fig. 22.11), whereas stepped cylindrical element needs more than 24 elements to converge (Fig. 22.12). A comparative plot is shown in Fig. 22.13.

22.5.1.2 Example 2: Hollow Tapered Rotor

The rotor configuration is shown in Fig. 22.7. The material properties are same as those considered in the previous example. The computed stability limits using the proposed element and stepped cylindrical element are compared in Table 22.3.

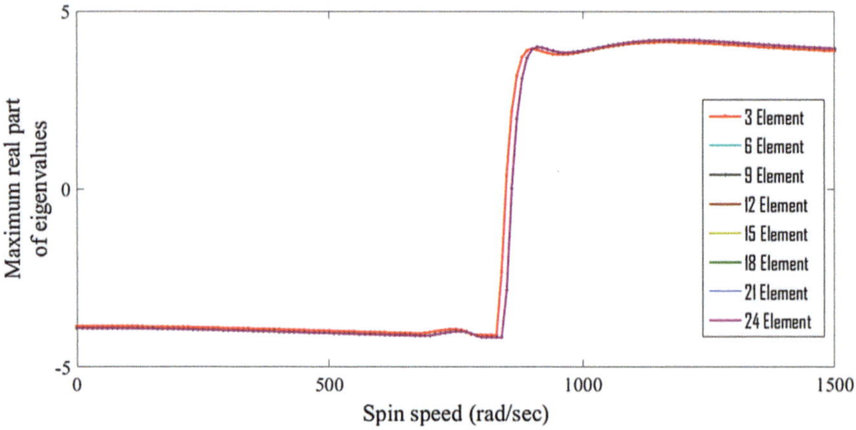

Fig. 22.11 Element-wise stability plot for solid tapered mild steel rotor obtained with solid tapered element

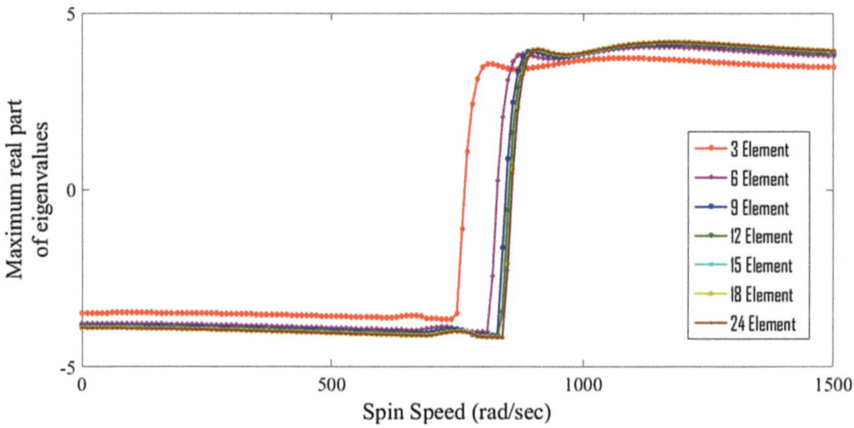

Fig. 22.12 Element-wise stability plot for solid tapered mild steel rotor obtained with stepped cylindrical element

Relevant plots for PVC hollow tapered rotor are shown in Figs. 22.14, 22.15 and 22.16. For mild steel, hollow tapered rotor, the comparative plots are presented in Figs. 22.17, 22.18 and 22.19.

22.5.2 *Frequency Response Analysis*

Comparative study of frequency responses of the above-mentioned rotors is performed next.

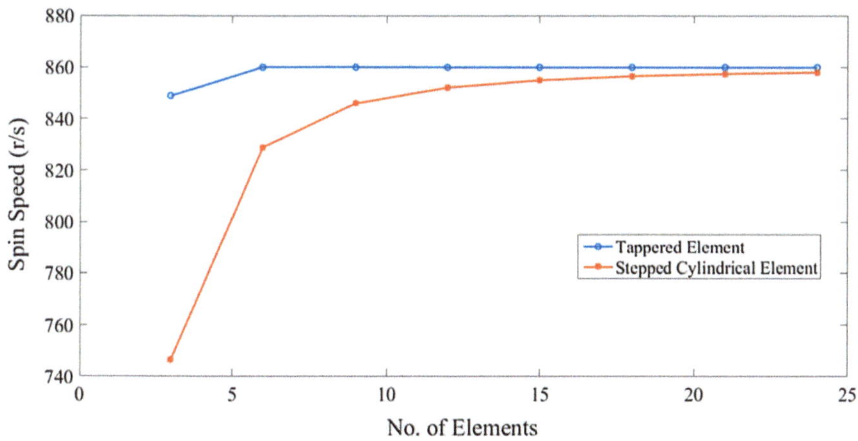

Fig. 22.13 Element-wise comparative plot of frequency limit of stable operation for solid tapered mind steel rotor

Table 22.3 Comparison of shaft spin speed of stable operation for hollow tapered shaft for mild steel and PVC material

Stability frequency limit					
	Material	Mild steel		PVC	
	Element type	Taper	Cylindrical	Taper	Cylindrical
No. of element	3(2 + 1)	918.92	876.91	453.62	425.79
	6(4 + 2)	937.63	937.04	461.69	460.29
	9(6 + 3)	945.90	952.01	465.53	468.27
	12(8 + 4)	950.26	957.43	467.67	471.05
	15(10 + 5)	952.97	959.88	469.01	472.33
	18(12 + 6)	954.81	961.19	469.92	473.01
	21(14 + 7)	956.14	961.97	470.57	473.41
	24(16 + 8)	957.12	962.48	471.06	473.67

22.5.2.1 Example 3: Solid Tapered Rotor

The rotor of Example 1 is now considered as a beam (non-rotating). A harmonic excitation is applied at the location of disc, and the response is computed at the same location. First resonant frequency with increasing number of elements is shown in Figs. 22.20 and 22.21. Figures 22.22 and 22.23 show the frequency responses of solid tapered rotor of mild steel and PVC material, respectively.

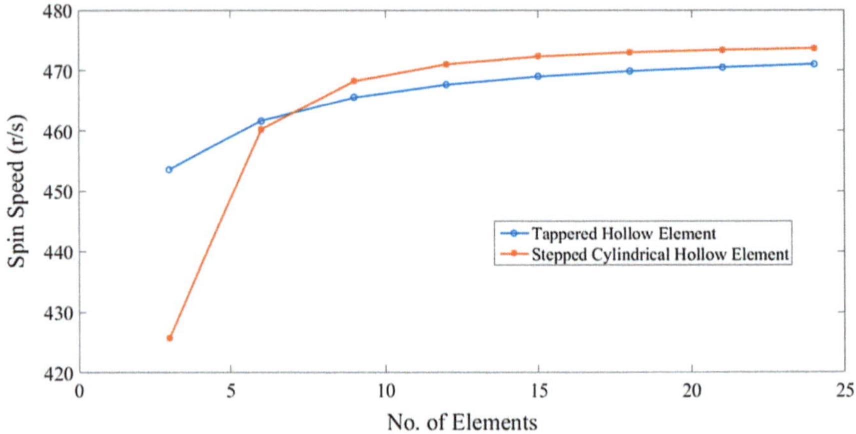

Fig. 22.14 Element-wise comparative plot of frequency limit of stable operation for hollow tapered PVC rotor

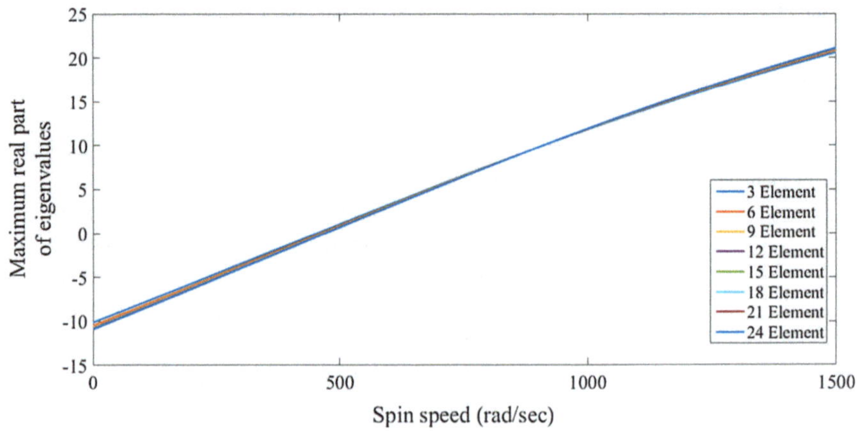

Fig. 22.15 Element-wise stability plot for hollow tapered PVC rotor obtained with hollow tapered element

22.5.2.2 Example 4: Hollow Tapered Rotor

Just like Example 3, the rotor of Example 2 is considered as a beam with harmonic excitation at the location of disc. Figures 22.24 and 22.25 show the variation of first resonant frequency with increasing number of elements. The frequency responses of solid tapered rotor of mild steel and PVC material are plotted in Figs. 22.26 and 22.27, respectively.

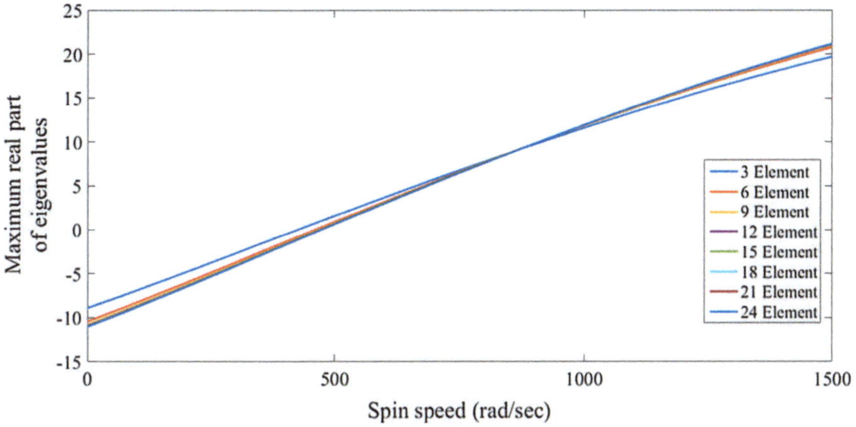

Fig. 22.16 Element-wise stability plot for hollow tapered PVC rotor obtained with stepped cylindrical element

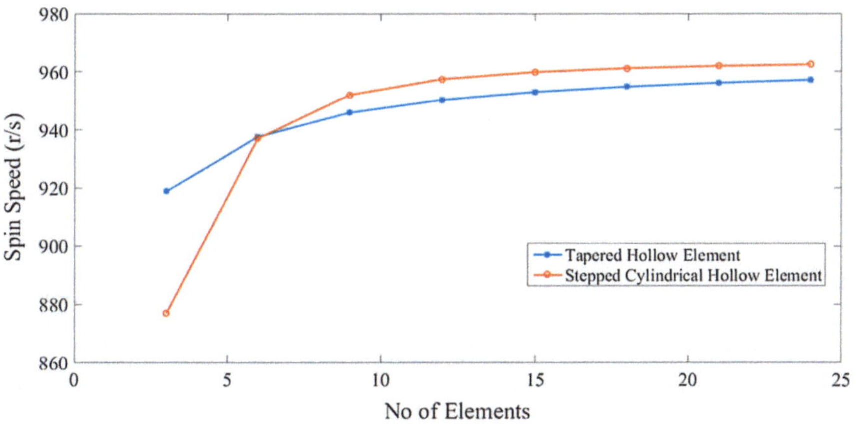

Fig. 22.17 Element-wise comparative plot of frequency limit of stable operation for hollow tapered mild steel rotor

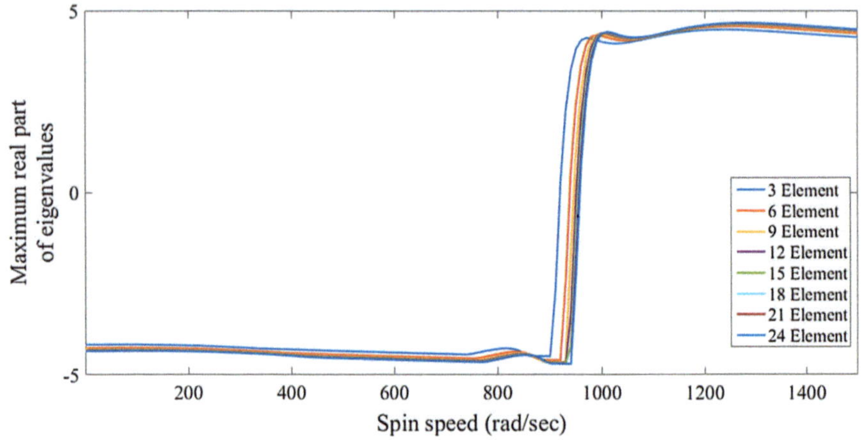

Fig. 22.18 Element-wise stability plot for hollow tapered mild steel rotor obtained with hollow tapered element

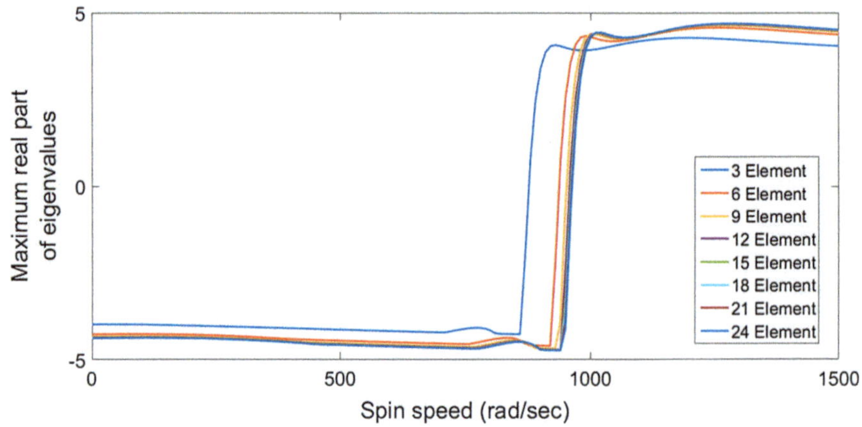

Fig. 22.19 Element-wise stability plot for hollow tapered mild steel rotor obtained with stepped cylindrical element

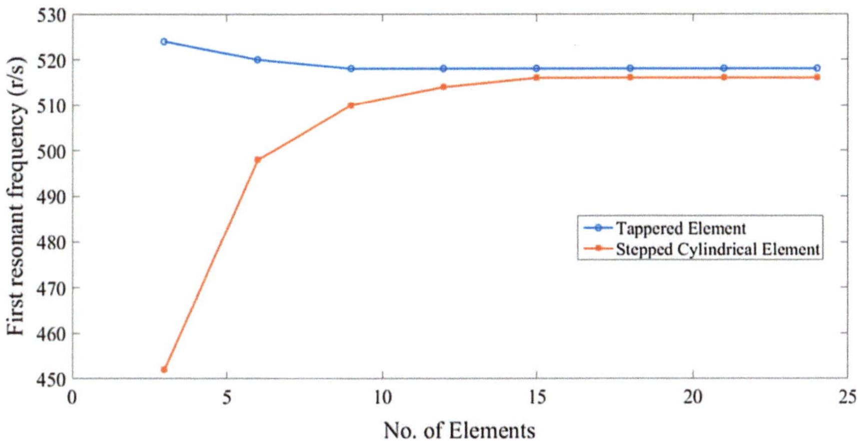

Fig. 22.20 Element-wise comparative plot of first resonant frequency for solid tapered PVC rotor

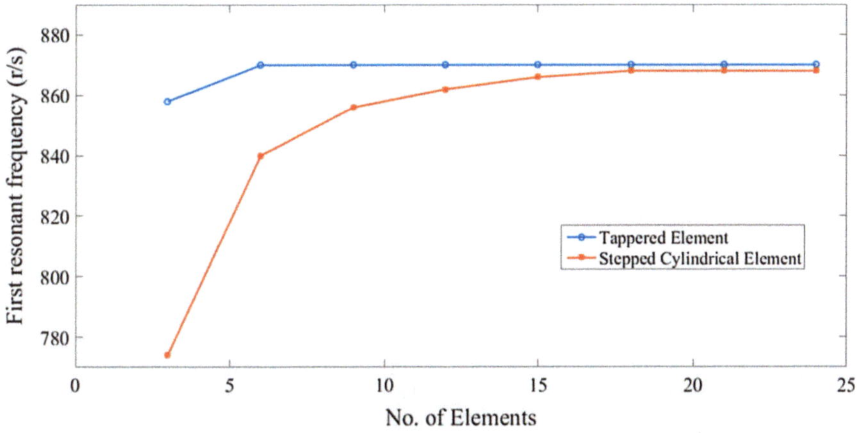

Fig. 22.21 Element-wise comparative plot of first resonant frequency for solid tapered mild steel rotor

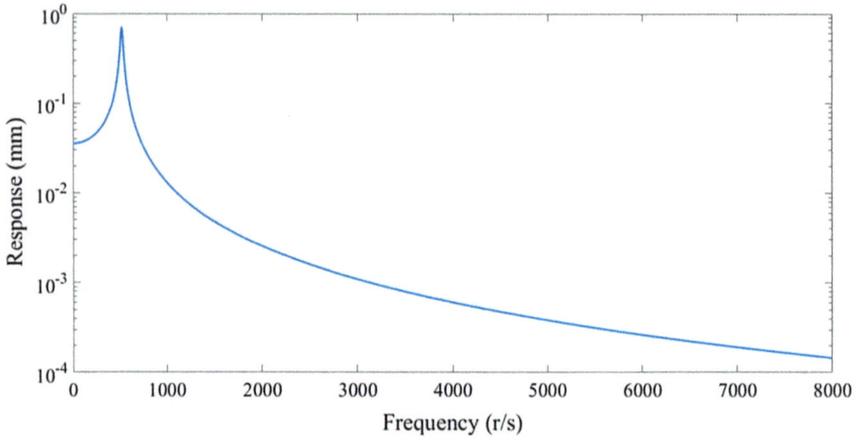

Fig. 22.22 Frequency response of solid tapered PVC rotor

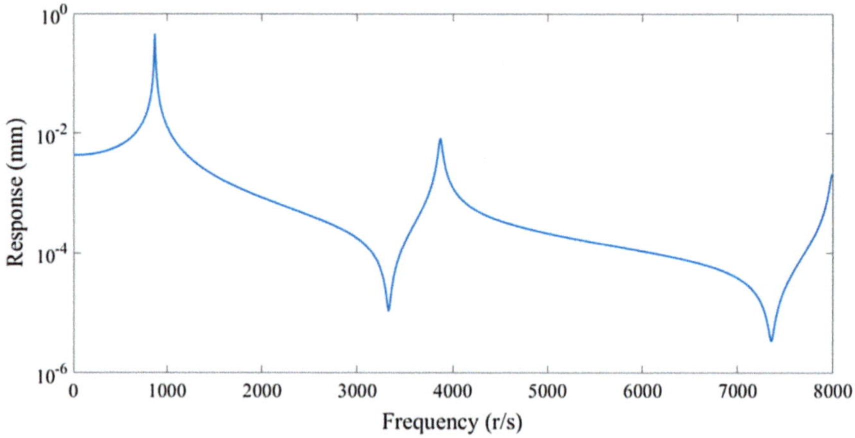

Fig. 22.23 Frequency response of solid tapered mild steel rotor

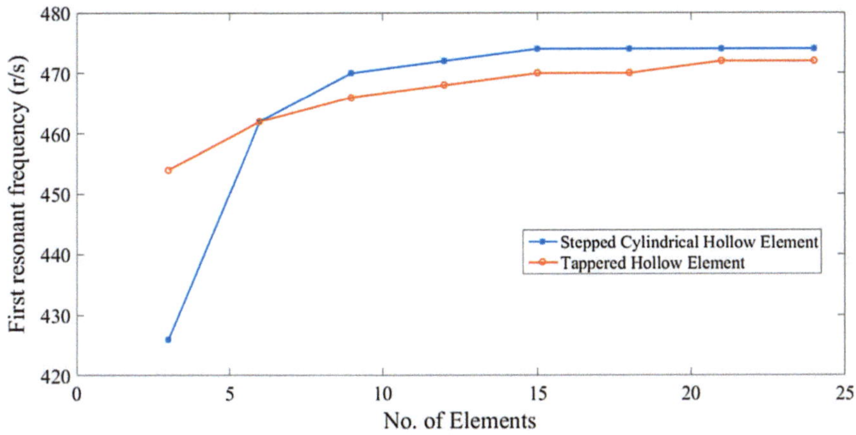

Fig. 22.24 Element-wise comparative plot of first resonance frequency for hollow tapered PVC rotor

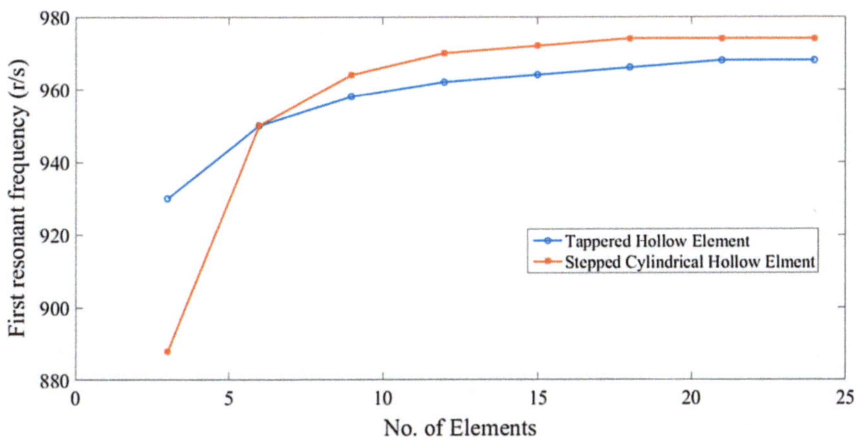

Fig. 22.25 Element-wise comparative plot of first resonance frequency for hollow tapered mild steel rotor

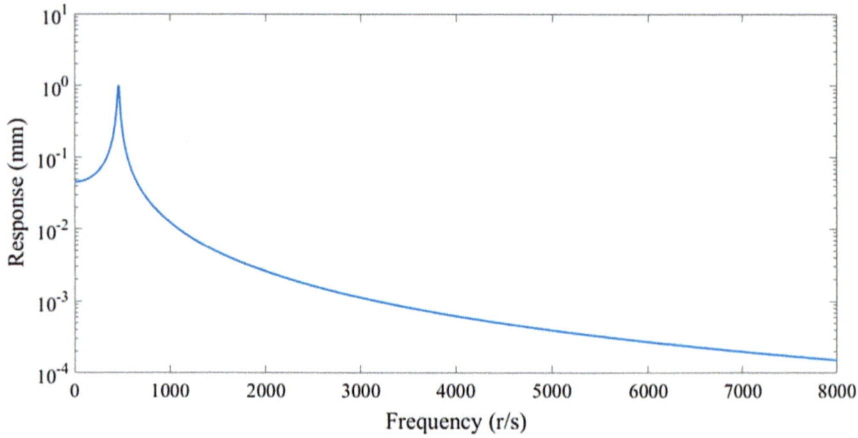

Fig. 22.26 Frequency response of hollow tapered PVC rotor

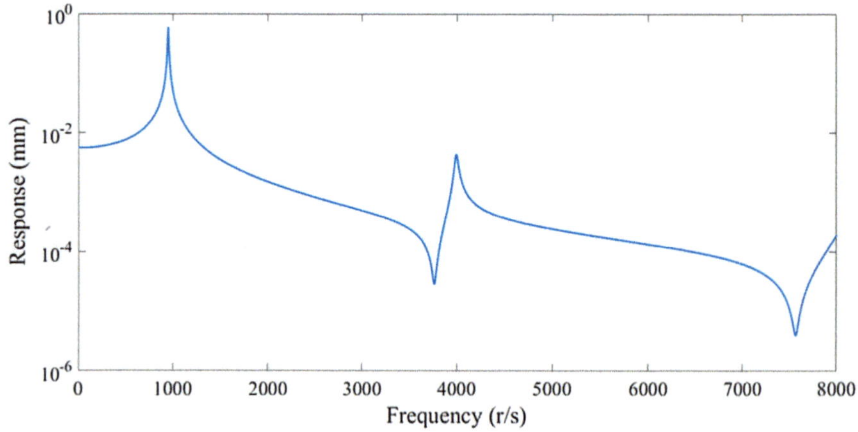

Fig. 22.27 Frequency response of hollow tapered mild steel rotor

22.6 Conclusions

From the analysis, it has been found that tapered rotor element produces equally accurate results with smaller number of elements than stepped cylindrical rotor element. The results using tapered rotor element are expected to match with closed-form solution for linearly tapered rotors. However, for rotors tapered other way the tapered rotor element is expected to produce converged results with much lesser elements than stepped cylindrical rotor elements.

Appendix 1

(i) Translation mass matrix

$$[M_{\mathrm{T}}^{e}] = \int_{0}^{L} \mu(x)[\psi]^{\mathrm{T}}[\psi]\mathrm{d}x$$

$$[M_{\mathrm{T}}^{e}] = \frac{\mu_0 L}{420}\begin{bmatrix} M11 & 0 & 0 & M12 & M13 & 0 & 0 & M14 \\ & M11 & -M12 & 0 & 0 & M13 & -M14 & 0 \\ & & M22 & 0 & 0 & -M23 & M24 & 0 \\ & & & M22 & M23 & 0 & 0 & M24 \\ & & & & M33 & 0 & 0 & M34 \\ & & & & & M33 & -M34 & 0 \\ \text{Symmetric} & & & & & & M44 & 0 \\ & & & & & & & M44 \end{bmatrix}$$

where

$$M11 = (156 + 72NL + 12.67N^2L^2)$$
$$M12 = (22L + 14NL^2 + 2.83N^2L^3)$$
$$M13 = (54 + 54NL + 15.33N^2L^2)$$
$$M14 = (-13L - 12NL^2 - 3.17N^2L^3)$$
$$M22 = (4L^2 + 3NL^3 + 0.67N^2L^4)$$
$$M23 = (13L + 14NL^2 + 4.17N^2L^3)$$
$$M24 = (-3L^2 - 3NL^3 - 0.83N^2L^4)$$
$$M33 = (156 + 240NL + 96.67N^2L^2)$$
$$M34 = (-22L - 30NL^2 - 10.83N^2L^3)$$
$$M44 = (4L^2 + 5NL^3 + 1.67N^2L^4)$$

(ii) Rotational mass matrix

$$\left[\mathbf{M}_{\mathrm{R}}^{e}\right] = \int_{0}^{L} i_D(x)[\varphi]^{\mathrm{T}}[\varphi]\mathrm{d}x$$

$$\left[M_{\mathrm{R}}^{e}\right] = \frac{\mu_0 r_0^2}{120L}\begin{bmatrix} m11 & 0 & 0 & m12 & m13 & 0 & 0 & m14 \\ & m11 & -m12 & 0 & 0 & m13 & -m14 & 0 \\ & & m22 & 0 & 0 & -m23 & m24 & 0 \\ & & & m22 & m23 & 0 & 0 & m24 \\ & & & & m33 & 0 & 0 & m34 \\ & & & & & m33 & -m34 & 0 \\ \text{Symmetric} & & & & & & m44 & 0 \\ & & & & & & & m44 \end{bmatrix}$$

where

$$\begin{aligned}
m11 &= \left(36 + 72NL + 67.7N^2L^2 + 25.7N^3L^3 + 4.29N^4L^4\right)\\
m12 &= \left(3L + 12NL^2 + 12.86N^2L^3 + 6N^3L^4 + 1.1N^4L^5\right)\\
m13 &= \left(-36 - 72NL - 61.7N^2L^2 - 25.7N^3L^3 - 4.29N^4L^4\right)\\
m14 &= \left(3L - 5.14N^2L^3 - 4.29N^3L^4 - 1.1N^4L^5\right)\\
m22 &= \left(4L^2 + 4NL^3 + 3.43N^2L^4 + 1.57N^3L^5 + 0.29N^4L^6\right)\\
m23 &= \left(-3L - 142NL^2 - 12.86N^2L^3 - 6N^3L^4 - 1.1N^4L^5\right)\\
m24 &= \left(-L^2 - 2NL^3 - 2.57N^2L^4 - 1.57N^3L^5 - 0.36N^4L^6\right)\\
m33 &= \left(36 + 72NL + 61.71N^2L^2 + 25.7N^3L^3 + 4.29N^4L^4\right)\\
m34 &= \left(-3L + 5.14N^2L^3 + 4.29N^3L^4 + 1.1N^4L^5\right)\\
m44 &= \left(4L^2 + 12NL^3 + 15.4N^2L^4 + 9.29N^3L^5 + 2.14N^4L^6\right)
\end{aligned}$$

(iii) Bending stiffness matrix

$$[K^e] = \int_0^L I(x)[\psi'']^{\mathrm{T}}[\psi'']\mathrm{d}x$$

$$[K^e] = \frac{I_0}{L^3}\begin{bmatrix} K11 & 0 & 0 & K12 & K13 & 0 & 0 & K14 \\ & K11 & -K12 & 0 & 0 & K13 & K14 & 0 \\ & & K22 & 0 & 0 & -K23 & -K24 & 0 \\ & & & K22 & K23 & 0 & 0 & K24 \\ & & & & K33 & 0 & 0 & K34 \\ & & & & & K33 & -K34 & 0 \\ \text{Symmetric} & & & & & & K44 & 0 \\ & & & & & & & K44 \end{bmatrix}$$

where

$$K11 = \left(12 + 24NL + 28.8N^2L^2 + 16.8N^3L^3 + 3.8N^4L^4\right)$$
$$K12 = \left(6L + 8NL^2 + 8.4N^2L^3 + 4.8N^3L^4 + 1.1N^4L^5\right)$$
$$K13 = -\left(12 + 24NL + 28.8N^2L^2 + 16.8N^3L^3 + 3.8N^4L^4\right)$$
$$K14 = \left(6L + 16NL^2 + 20.4N^2L^3 + 12N^3L^4 + 2.7N^4L^5\right)$$
$$K22 = \left(4L^2 + 4NL^3 + 3.2N^2L^4 + 1.6N^3L^5 + 0.34N^4L^6\right)$$
$$K23 = -\left(6L + 8NL^2 + 8.4N^2L^3 + 4.8N^3L^4 + 1.1N^4L^5\right)$$
$$K24 = \left(2L^2 + 4NL^3 + 5.2N^2L^4 + 3.2N^3L^5 + 0.74N^4L^6\right)$$
$$K33 = \left(12 + 24NL + 28.8N^2L^2 + 16.8N^3L^3 + 3.8N^4L^4\right)$$
$$K34 = -\left(6L + 16NL^2 + 20.4N^2L^3 + 12N^3L^4 + 2.7N^4\right)$$
$$K44 = \left(4L^2 + 12NL^3 + 15.2N^2L^4 + 8.8N^3L^5 + 1.94N^4L^6\right)$$

(iv) Translation mass matrix for tapered and hollow element

$$\left[M_{\mathrm{T}}^{e}\right]=\int_{0}^{L}\mu(x)[\psi]^{\mathrm{T}}[\psi]\mathrm{d}x$$

$$\left[M_{\mathrm{T}}^{e}\right]=\frac{\mu_0 L}{420}\begin{bmatrix} M11 & 0 & 0 & M12 & M13 & 0 & 0 & M14 \\ & M11 & -M12 & 0 & 0 & M13 & -M14 & 0 \\ & & M22 & 0 & 0 & -M23 & M24 & 0 \\ & & & M22 & M23 & 0 & 0 & M24 \\ & & & & M33 & 0 & 0 & M34 \\ & & & & & M33 & -M34 & 0 \\ \text{Symmetric} & & & & & & M44 & 0 \\ & & & & & & & M44 \end{bmatrix}$$

where

$$M11=\left(\frac{38}{3}\right)(N1^2L^2-N2^2L^2)+72(N1L-N2fL)+156\left(1-f^2\right)$$

$$M12=\left(\frac{17}{6}\right)(N1^2L^3-N2^2L^3)+14\left(N1L^2-N2fL^2\right)+22\left(L-Lf^2\right)$$

$$M13=\left(\frac{46}{3}\right)(N1^2L^2-N2^2L^2)+54(N1L-N2fL)+54\left(1-f^2\right)$$

$$M14=-\left(\frac{19}{6}\right)(N1^2L^3-N2^2L^3)-12\left(N1L^2-N2fL^2\right)-13\left(L-Lf^2\right)$$

$$M22=\left(\frac{2}{3}\right)(N1^2L^4-N2^2L^4)+3\left(N1L^3-N2fL^3\right)+4\left(L^2-L^2f^2\right)$$

$$M23=-\left(\frac{25}{6}\right)(N1^2L^3-N2^2L^3)-14\left(N1L^2-N2fL^2\right)-13\left(L-Lf^2\right)$$

$$M24=-\left(\frac{5}{6}\right)(N1^2L^4-N2^2L^4)-3\left(N1L^3-N2fL^3\right)-3\left(L^2-L^2f^2\right)$$

$$M33=\left(\frac{290}{3}\right)(N1^2L^2-N2^2L^2)+240(N1L-N2fL)+156\left(1-f^2\right)$$

$$M34=-\left(\frac{65}{6}\right)(N1^2L^3-N2^2L^3)-30\left(N1L^2-N2fL^2\right)-22\left(L-Lf^2\right)$$

$$M44=\left(\frac{5}{3}\right)(N1^2L^4-N2^2L^4)+5\left(N1L^3-N2fL^3\right)+4\left(L^2-L^2f^2\right)$$

(v) Rotational mass matrix for tapered and hollow element

$$\left[\mathrm{M}_{\mathrm{R}}^{e}\right] = \int_{0}^{L} i_D(x)[\varphi]^{\mathrm{T}}[\varphi]\mathrm{d}x$$

$$\left[M_R^e\right] = \frac{\mu_0 r_0^2}{120L}\begin{bmatrix} m11 & 0 & 0 & m12 & m13 & 0 & 0 & m14 \\ & m11 & -m12 & 0 & 0 & m13 & -m14 & 0 \\ & & m22 & 0 & 0 & -m23 & m24 & 0 \\ & & & m22 & m23 & 0 & 0 & m24 \\ & & & & m33 & 0 & 0 & m34 \\ & & & & & m33 & -m34 & 0 \\ \text{Symmetric} & & & & & & m44 & 0 \\ & & & & & & & m44 \end{bmatrix}$$

where

$$m11 = 36(1-f^4) + 72(N1L - N2f^3L) + \left(\frac{432}{7}\right)(N1^2L^2 - N2^2f^2L^2) + \left(\frac{180}{7}\right)(N1^3L^3 - N2^3fL^3) + \left(\frac{30}{7}\right)(N1^4L^4 - N2^4L^4)$$

$$m12 = 3(L - Lf^4) + 12(N1L^2 - N2f^3L^2) + \left(\frac{90}{7}\right)(N1^2L^3 - N2^2f^2L^3) + 6(N1^3L^4 - N2^3fL^4) + \left(\frac{15}{14}\right)(N1^4L^5 - N2^4L^5)$$

$$m13 = -36(1-f^4) - 72(N1L - N2f^3L) - \left(\frac{432}{7}\right)(N1^2L^2 - N2^2f^2L^2) + \left(\frac{180}{7}\right)(N1^3L^3 - N2^3fL^3) - \left(\frac{30}{7}\right)(N1^4L^4 - N2^4L^4)$$

$$m14 = 3(L - Lf^4) - \left(\frac{36}{7}\right)(N1^2L^3 - N2^2f^2L^3) - \left(\frac{30}{7}\right)(N1^3L^4 - N2^3fL^4) - \left(\frac{15}{14}\right)(N1^4L^5 - N2^4L^5)$$

$$m22 = 4(L^2 - L^2f^4) + 4(N1L^3 - N2f^3L^3) + \left(\frac{24}{7}\right)(N1^2L^4 - N2^2f^2L^4) + \left(\frac{11}{7}\right)(N1^3L^5 - N2^3fL^5) + \left(\frac{2}{7}\right)(N1^4L^6 - N2^4L^6)$$

$$m23 = -3(L - Lf^4) - 12(N1L^2 - N2f^3L^2) - \left(\frac{90}{7}\right)(N1^2L^3 - N2^2f^2L^3) - 6(N1^3L^4 - N2^3fL^4) - \left(\frac{15}{14}\right)(N1^4L^5 - N2^4L^5)$$

$$m24 = -(L^2 - L^2f^4) - 2(N1L^3 - N2f^3L^3) - \left(\frac{18}{7}\right)(N1^2L^4 - N2^2f^2L^4) - \left(\frac{11}{7}\right)(N1^3L^5 - N2^3fL^5) - \left(\frac{5}{14}\right)(N1^4L^6 - N2^4L^6)$$

$$m33 = 36(1-f^4) + 72(N1L - N2f^3L) + \left(\frac{432}{7}\right)(N1^2L^2 - N2^2f^2L^2) + \left(\frac{180}{7}\right)(N1^3L^3 - N2^3fL^3) + \left(\frac{30}{7}\right)(N1^4L^4 - N2^4L^4)$$

$$m34 = -3(L - Lf^4) + \left(\frac{36}{7}\right)(N1^2L^3 - N2^2f^2L^3) + \left(\frac{30}{7}\right)(N1^3L^4 - N2^3fL^4) + \left(\frac{15}{14}\right)(N1^4L^5 - N2^4L^5)$$

$$m44 = 4(L^2 - L^2f^4) + 12(N1L^3 - N2f^3L^3) + \left(\frac{108}{7}\right)(N1^2L^4 - N2^2f^2L^4) + \left(\frac{65}{7}\right)(N1^3L^5 - N2^3fL^5) + 15(N1^4L^6 - N2^4L^6)$$

(vi) Bending stiffness matrix for tapered and hollow element

$$\left[K_e^e\right] = \int_0^L I(x)[\psi'']^{\mathrm{T}}[\psi'']\mathrm{d}x$$

$$\left[K_e^e\right] = \frac{I_0}{L^3}\begin{bmatrix} K11 & 0 & 0 & K12 & K13 & 0 & 0 & K14 \\ & K11 & -K12 & 0 & 0 & K13 & -K14 & 0 \\ & & K22 & 0 & 0 & -K23 & K24 & 0 \\ & & & K22 & K23 & 0 & 0 & K24 \\ & & & & K33 & 0 & 0 & K34 \\ & & & & & K33 & -K34 & 0 \\ \text{Symmetric} & & & & & & K44 & 0 \\ & & & & & & & K44 \end{bmatrix}$$

where

$$K11 = 12\left(1 - f^4\right) + 24\left(N1L - N2f^4L\right) + \left(\frac{144}{5}\right)\left(N1^2L^2 - N2^2f^4L^2\right) + \left(\frac{84}{5}\right)\left(N1^3L^3 - N2^3f^4L^3\right) + \left(\frac{132}{35}\right)\left(N1^4L^4 - N2^4f^4L^4\right)$$

$$K12 = 6\left(L - Lf^4\right) + 8\left(N1L^2 - N2f^4L^2\right) + \left(\frac{42}{5}\right)\left(N1^2L^3 - N2^2f^4L^3\right) + \left(\frac{24}{5}\right)\left(N1^3L^4 - N2^3f^4L^4\right) + \left(\frac{38}{35}\right)\left(N1^4L^5 - N2^4f^4L^5\right)$$

$$K13 = -12\left(1 - f^4\right) - 24\left(N1L - N2f^4L\right) - \left(\frac{144}{5}\right)\left(N1^2L^2 - N2^2f^4L^2\right) - \left(\frac{84}{5}\right)\left(N1^3L^3 - N2^3f^4L^3\right) - \left(\frac{132}{35}\right)\left(N1^4L^4 - N2^4f^4L^4\right)$$

$$K14 = 6\left(L - Lf^4\right) + 16\left(N1L^2 - N2f^4L^2\right) + \left(\frac{102}{5}\right)\left(N1^2L^3 - N2^2f^4L^3\right) + 12\left(N1^3L^4 - N2^3f^4L^4\right) + \left(\frac{94}{35}\right)\left(N1^4L^5 - N2^4f^4L^5\right)$$

$$K22 = 4\left(L^2 - L^2f^4\right) + 4\left(N1L^3 - N2f^4L^3\right) + \left(\frac{16}{5}\right)\left(N1^2L^4 - N2^2f^4L^4\right) + \left(\frac{8}{5}\right)\left(N1^3L^5 - N2^3f^4L^5\right) + \left(\frac{12}{35}\right)\left(N1^4L^6 - N2^4f^4L^6\right)$$

$$K23 = 6\left(L - Lf^4\right) + 8\left(N1L^2 - N2f^4L^2\right) + \left(\frac{42}{5}\right)\left(N1^2L^3 - N2^2f^4L^3\right) + \left(\frac{24}{5}\right)\left(N1^3L^4 - N2^3f^4L^4\right) + \left(\frac{38}{35}\right)\left(N1^4L^5 - N2^4f^4L^5\right)$$

$$K24 = 2\left(L^2 - L^2f^4\right) + 4\left(N1L^3 - N2f^4L^3\right) + \left(\frac{26}{5}\right)\left(N1^2L^4 - N2^2f^4L^4\right) + \left(\frac{16}{5}\right)\left(N1^3L^5 - N2^3f^4L^5\right) + \left(\frac{26}{35}\right)\left(N1^4L^6 - N2^4f^4L^6\right)$$

$$K33 = 12\left(1 - f^4\right) + 24\left(N1L - N2f^4L\right) + \left(\frac{144}{5}\right)\left(N1^2L^2 - N2^2f^4L^2\right) + \left(\frac{84}{5}\right)\left(N1^3L^3 - N2^3f^4L^3\right) + \left(\frac{132}{35}\right)\left(N1^4L^4 - N2^4f^4L^4\right)$$

$$K34 = -6\left(L - Lf^4\right) + 16\left(N1L^2 - N2f^4L^2\right) - \left(\frac{102}{5}\right)\left(N1^2L^3 - N2^2f^4L^3\right) - 12\left(N1^3L^4 - N2^3f^4L^4\right) - \left(\frac{94}{35}\right)\left(N1^4L^5 - N2^4f^4L^5\right)$$

$$K44 = 4\left(L^2 - L^2f^4\right) + 12\left(N1L^3 - N2f^4L^3\right) + \left(\frac{76}{5}\right)\left(N1^2L^4 - N2^2f^4L^4\right) + \left(\frac{44}{5}\right)\left(N1^3L^5 - N2^3f^4L^5\right) + \left(\frac{68}{35}\right)\left(N1^4L^6 - N2^4f^4L^6\right)$$

$$[A_n]\begin{bmatrix} [M^e] & [0] & -\eta_1[\overline{K^e}] & -\eta_2[\overline{K^e}] & -\eta_3[\overline{K^e}] \\ [0] & [I] & [0] & [0] & [0] \\ [0] & [0] & \eta_1[\overline{K^e}] & [0] & [0] \\ [0] & [0] & [0] & \eta_2[\overline{K^e}] & [0] \\ [0] & [0] & [0] & [0] & \eta_3[\overline{K^e}] \end{bmatrix} \quad \{U\} = \begin{Bmatrix} \{\dot{U}_c\} \\ \{U_c\} \\ \{U_{b1}\} \\ \{U_{b2}\} \\ \{U_{b3}\} \end{Bmatrix}$$

$$[B_n] = \begin{bmatrix} (\eta_1+\eta_2+\eta_3)[\overline{K^e}] + [G^e] & E[\overline{K^e}] + [K_1^e] & [0] & [0] & [0] \\ -[I] & [0] & [0] & [0] & [0] \\ -\eta_1[\overline{K^e}] & [0] & E_1[\overline{K^e}] & [0] & [0] \\ -\eta_2[\overline{K^e}] & [0] & [0] & E_2[\overline{K^e}] & [0] \\ -\eta_3[\overline{K^e}] & [0] & [0] & [0] & E_3[\overline{K^e}] \end{bmatrix} \{f\} = \begin{Bmatrix} \{f^e\} \\ \{0\} \\ \{0\} \\ \{0\} \end{Bmatrix}$$

Appendix 2

Nomenclature

$[B], [\psi'']$	Strain displacement relation matrix
E	Young's modulus (N/mm^2)
$\{U_c\}$	Nodal displacement
η_1, η_2, η_3	Maxwell branch parameter (Ns/mm^2)
E_1, E_2, E_3	Maxwell branch parameter (N/mm^2)
$\{U_b\}$	Additional nodal variable
$[M_T^e]$	Translation mass matrix
$[M_R^e]$	Rotational mass matrix
ω	Spin speed (Rad/s)
$[\hat{M}_T^e]$	Transformed translation mass matrix in spinning frame
$[M^e]$	Mass matrix, $([M_T^e] + [M_R^e])$
$[K^e]$	$\int_0^1 I(x)[\psi'']^T[\psi'']dx$
$[G^e]$	$2\omega[\hat{M}_T^e]$
$[K_1^e]$	$-\omega^2([M_T^e] - [M_R^e])$
f_e	Unbalance force vector
α_s	Half taper angle
d_s	Smaller diameter of rotor element (mm)
N	$(2\tan\alpha_s/d_s)$
α_{hi}	Half taper angle of inner material line
d_{hi}	Inner diameter of rotor element (mm) at smaller end
$N2$	$(2\tan\alpha_{hi}/d_{hi})$
α_{ho}	Half taper angle of outer material line
d_{ho}	Outer diameter of rotor element (mm) at smaller end

(continued)

(continued)

$N1$	$(2\tan\alpha_{ho}/d_{ho})$
f	d_{hi}/d_{ho}
L	Element length (mm)
$\mu(x)$	Mass of rotor per unit length
$i_d(x)$	Mass moment of inertia per unit length
$[\psi]$	Translation displacement function matrix
$[\varphi]$	Rotational displacement function matrix
$I(x)$	Area moment of inertia at x
I_0	Smaller area moment of inertia of rotor element (mm^4)
μ_0	Mass per unit length at smaller end of rotor element
r_0	Radius of rotor element at smaller end (mm)

References

Journals

1. Nelson, H.D., MeVaugh, J.M.: The dynamics of rotor-bearing systems using finite elements. J. Eng. Ind., pp. 593–600 (1976)
2. Greenhill, L.M., Bickford, W.B., Nelson, H.D.: A conical beam finite element for rotor dynamics analysis. J. Vib. Acoust. Stress Reliab **107**(4), 421–430 (Oct 01, 1985)
3. Ganguly, S., Nandi, A., Neogy, S.: A state space viscoelastic shaft finite element for analysis of rotors. Procedia Eng. **144**, 374–381 (2016)
4. Nelson, H.D.: A finite rotating shaft element using timoshenko beam theory. ASME J. Mech. Des. **102**, 793–803 (1980)

Book

5. Genta, G.: Vibration Dynamics and Control. Springer, New York, USA, XXIV (2009)

Chapter 23
Numerical Analysis of Heat Transfer and Entropy Generation for Natural Convection in a Quadrantal Cavity with Non-uniform Heating at the Bottom Wall

Shantanu Dutta, Arup Kumar Biswas and Sukumar Pati

Nomenclature

c_p	Specific thermal capacity (J kg^{-1} K^{-1})
g	Acceleration owing to gravity (m s^{-2})
h	Heat transfer coefficient (W m^2 K^{-1})
k	Thermal conductivity (W m^{-1} K^{-1})
L	Enclosure length (m)
$\overline{Nu}$	Average Nusselt number
Nu	Nusselt number (dimensionless)
P	Dimensionless pressure
p	Pressure (N m^2)
Pr	Prandtl number (dimensionless)
Ra	Rayleigh number (dimensionless)
T	Temperature (K)
U, V	Dimensionless velocity component in the X- and Y-directions
u, v	Velocity component in the x- and y-directions (m s^{-1})
X, Y	Dimensionless coordinates
x, y	Cartesian coordinate system

S. Dutta (✉) · A. K. Biswas
Department of Mechanical Engineering, National Institute of Technology,
Durgapur 713209, India
e-mail: shantanu.dutta@nshm.com

S. Pati
Department of Mechanical Engineering, National Institute
of Technology Silchar, Silchar 788010, India

P. Sahoo and J. P. Davim (eds.), *Advances in Materials, Mechanical and Industrial Engineering*, Lecture Notes on Multidisciplinary Industrial Engineering,
https://doi.org/10.1007/978-3-319-96968-8_23

Greek symbols

α	Thermal diffusivity ($m^2\ s^{-1}$)
β	Coefficient of thermal expansion (K^{-1})
θ	Dimensionless temperature
ν	Kinematic viscosity ($m^2\ s^{-1}$)
ρ	Density (kg m^{-3})
ψ	Stream vorticity (m^2 s)
Ψ	Dimensionless stream
Vorticity	$(=\psi/\alpha)$

Subscripts

Avg	Average
c	Cold wall
h	Hot, bottom wall
max	Maximum
min	Minimum
num	Number

23.1 Introduction

Natural convection in an enclosure continues to be a hot topic of research for the last few decades because of its importance in various engineering applications. A host of articles both numerical and experimental are available in the literature pertaining to attributes of fluid flow, thermal transport, and irreversibility analysis in regular geometries. Davis and De [1] were among the first who presented a benchmark explanation and results for natural convection considering air as the working fluid within a square enclosure considering the upright walls being maintained at unequal temperatures. Hamadi et al. [2] executed a laboratory experiment and implemented numerical investigation of free convection of air in a square inclined cavity. From the results presented, they came to conclusions that the laboratory experiment and numerical investigation were in very good compliance. It is important to mention here that the geometry of the enclosure considered in most of the previous studies was regular in nature and accordingly extensive investigations were carried out on natural convection inside enclosures by considering different fluids and choosing distinctive boundary conditions, such as in square enclosures [3–6] and rectangular enclosures [7, 8]. A comprehensive list of investigation, based on natural convection heating strategy, has also been reported on triangular [9–11], parallelogrammic [12–14], trapezoidal [15, 16] enclosures in the literature, in the past too.

Generation of irreversibility in an enclosure can be attributed due to transfer of heat and fluid flow with distinct boundary conditions causing a difference of temperature and is an irretrievable process by nature. In order to utilize the energy resources efficiently and hence for a thermodynamically efficient thermal systems, the entropy generation associated with the processes should be minimized. Bejan [39] was the first researcher who popularized the academic fraternity by considering the principle of minimization of entropy generation. Then onward, several authors have investigated the effects of several influencing criteria on the entropy generation characteristics. Many researchers in the past undertook numerical experiments in studying thermal transport and irreversibility for buoyancy-induced convection in enclosures. These studies are well documented in the literature and can be referred here [17–31]. Abu-Hijileh et al. [17] analyzed the irreversibility associated with buoyancy-induced convection in a horizontal cylinder. The authors came to the conclusion that increasing the radius of cylinder resulted in a decrease of irreversibility. Interested readers working on trapezoidal and triangular enclosures can refer to articles [18, 19]. Analysis of irreversibility has also been carried out in complicated enclosures like wavy wall and Γ-shaped enclosures too and can be referred to articles [21, 22]. Datta et al. [23] theoretically analyzed buoyancy-induced convective thermal transport and irreversibility in a square porous cavity with different block sizes maintained in insulated condition inside the enclosure. The study gave a clear insight into the readers that the presence of insulated blocks affected the transfer of heat inside the enclosure. The authors also pointed out in their study about detailed heatline analysis and multiple regions of heat transfer, like region of heat transfer enhancement, region of heat transfer enhancement together with a decrease in entropy production region, and a third region where heat transfer together with entropy generation abatement was noticed. The authors also highlighted about a critical insulated block size; over and above, this limit resulted in heat transfer abatement. Chakravarty et al. [24] numerically investigated the buoyancy-induced non-equilibrium thermal transport and irreversibility analysis in a porous cylindrical cavity. The numerical study was achieved for a large range of Ra (10^7–10^{11}) and Darcy number (10^{-6}–10^{-10}) and clearly disclosed that transfer of heat in the porous zone could be classified into conduction-dominated and convection-dominated regimes. The authors also analyzed the irreversibilities in the system by carrying out an entropy generation analysis too and deduced the fact that entropy generation has a close resemblance to the above-mentioned two porous zone demarcations. In another article, Mani Rathnam et al. [25] did a finite element numerical analysis on irreversibility study inside a right-angled porous triangular cavity considering vertex angles of 15° and 45°. The authors concluded that a triangular enclosure of 15° vertex angle is better for thermal design since it involves higher rate of transfer of heat and lesser irreversibility. Das and Basak [26] carried out several entropy generation comparative analysis studies in square and triangle porous enclosures which were heated along the side boundary walls and concluded that triangular enclosure is better for thermal analysis design since it involves higher rate of transfer of heat and lesser irreversibility. Abu-Nada and Oztop [27] did a finite volume numerical analysis of Al_2O_3water nanofluid natural convection in a wavy-walled complicated geometry and found out that heat transfer enhancements could be attributed to larger Rayleigh number chosen for the study and higher volume fraction of nanofluids considered. The waviness amplitude considered for the study also played a major role in heat transfer. In another study, Oztop and Abu-Nada [28] carried out buoyancy-induced convection in partial rectangular enclosures comprised of nanofluids subjected to flush-mounted heater, heat source. They analyzed different nanofluids and also deduced that the location of heater was primarily responsible for affecting temperature and flow fields.

The literature study reveals the presence of a wide spectrum of practical utilization involving natural convection method in fluid-filled enclosures which are enforced to non-isothermal boundary conditions. Accordingly, a host of articles can be found out in the literature survey. Meshram et al. [29] analyzed natural convection thermal transport along with entropy generation in a porous square cavity, subjected to non-uniform heating at various inclination angles, and came to the conclusion that the angle of inclination affects the heat transfer rate in the enclosure. In another article, Bhardwaj et al. [30] carried out an in-depth numerical investigation in a porous triangular enclosure subjected to non-uniform heating. They came to the conclusions that the undulation parameters in the left upright wall not only resulted in an increase of heat transfer in enclosure, but also produced more entropy inside the enclosure. In another article, Dutta et al. [31] performed a numerical analysis pertaining to heat transfer along with irreversibility subjected to free convection inside rhombic enclosures considering sinusoidal heating appearing in both top and bottom walls, with cold inclined side walls. It was also evaluated and found by the authors that a significant contributor of entropy generation is heat transfer and the positional region of maximum entropy rested with the bearing of phase shift angle of the top wall.

The study of natural convection in irregular geometry, such as quadrantal cavity, is gaining interest from the research community because of the applications of such geometry in the domain of electronics, cooling system, heating coil, etc. Aydin and Yesiloz [32] were the first researchers who carried out a laboratory experiment and numerical investigation within quadrantal enclosure subjected to uniform heating from the bottom boundary, along with upright cold wall, considering Ra in the span of 10^3–10^7. In another article, Yesiloz and Aydin [33] analyzed the thermal transport in an inclined quadrantal cavity by changing the angle of inclination in the span of $0° \leq \phi \leq 360$ and varying the Ra from 10^5 to 10^7. Sen et al. [34, 35] theoretically analyzed the buoyancy-induced thermal on quadrantal cavity subjected to a constant temperature at bottom wall, along with upright cold wall with adiabatic arched or curved wall and combined heater on bottom and upright walls with arched or curved wall maintained cold. Natural convection in a quadrantal cavity was also analyzed by Dutta and Biswas [36] where they analyzed heat transfer and entropy generation in a porous quadrantal enclosure. They came to the conclusion that the total entropy production was not very high even considering higher Darcy number, and it could be highly recommended in enclosures pertaining to circular design. Recent numerical investigations were also performed by Dutta et al. [37] where the authors investigated the natural convection phenomenon heat transfer along with entropy generation considering quadrantal enclosure, filled with porous media subjected to non-isothermal heating strategy applied to bottom wall.

The objective of the current work is to carry out a numerical investigation on buoyancy-induced convective heat transfer along with entropy generation aspects inside a quadrantal cavity. The boundary conditions adapted for the present study include a heated bottom wall subjected to non-uniform heating strategy. The left upright wall is maintained cold throughout, and also there is a presence of insulated arched wall from the surroundings. The results are depicted through streamline contour, isothermal contour, local heat transfer rate along the heated as well as cold upright wall, average heat transfer rate. The authors have also discussed entropy generation owing to transfer of heat as well as fluid flow friction.

23.2 Problem Description

The present investigation is carried out in quadrantal enclosure filled with air and can be found in Fig. 23.1. The bottom wall is heated non-uniformly as per article [37] and can be found out in Eq. (23.1). The left upright wall is kept cold along with right curved wall being maintained in adiabatic condition.

$$T(x) = T_{\mathrm{c}} + \frac{T_{\mathrm{h}} - T_{\mathrm{c}}}{2}\left(1 - \cos\left(\frac{2\pi x}{L}\right)\right) \tag{23.1}$$

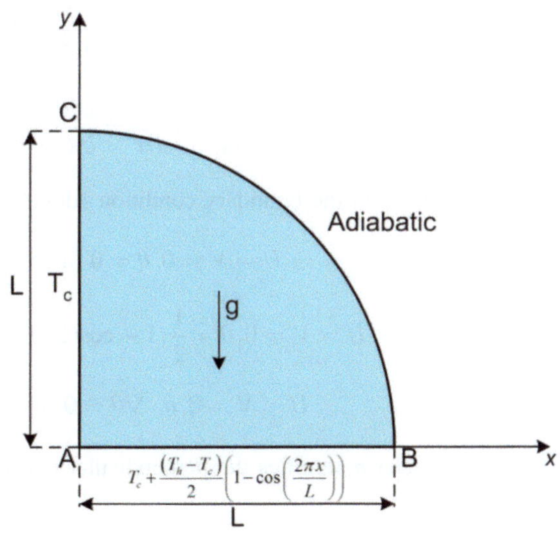

Fig. 23.1 Schematic diagram of the physical model

Here, T_{h} is an extreme (max.) temperature corresponding to bottom wall which is non-uniformly heated. T_{c} is corresponding to cold upright wall temperature. Also, L corresponds to length for the quadrantal model adopted for our study.

23.2.1 Governing Equations and Boundary Conditions

We consider laminar, steady, incompressible flow of Newtonian fluid (air) inside the cavity. All the properties are treated to be constant exempting the density appearing in y-momentum equation in the form of the body force term. The well-known Boussinesq approximation is harnessed to model density variations due to change in temperature. The heat transfer due to radiation is neglected because of very small temperature difference. Viscous dissipation's presence in the fluid is also ignored.

We first present the resulting dimensionless variables: $X = \frac{x}{L}$, $Y = \frac{y}{L}$, $U = \frac{uL}{\alpha}$, $V = \frac{vL}{\alpha}$, $\theta = \frac{T - T_{\mathrm{c}}}{T_{\mathrm{h}} - T_{\mathrm{c}}}$, $P = \frac{pL^2}{\rho\alpha^2}$, $\mathrm{Pr} = \frac{\nu}{\alpha}$, $Ra = \frac{g\beta(T_{\mathrm{h}} - T_{\mathrm{c}})L^3\,\mathrm{Pr}}{\nu^2}$.

The transport equations appropriate to the problem under consideration can be expressed, after due consideration of the dimensionless variables

$$\frac{\partial U}{\partial X} + \frac{\partial V}{\partial Y} = 0 \tag{23.2}$$

$$U\frac{\partial U}{\partial X} + V\frac{\partial U}{\partial Y} = -\frac{\partial P}{\partial X} + Pr\left(\frac{\partial^2 U}{\partial X^2} + \frac{\partial^2 U}{\partial Y^2}\right) \tag{23.3}$$

$$U\frac{\partial V}{\partial X} + V\frac{\partial V}{\partial Y} = -\frac{\partial P}{\partial Y} + Pr\left(\frac{\partial^2 V}{\partial X^2} + \frac{\partial^2 V}{\partial Y^2}\right) + Ra\,Pr\theta \tag{23.4}$$

$$U\frac{\partial \theta}{\partial X} + V\frac{\partial \theta}{\partial Y} = \frac{\partial^2 \theta}{\partial X^2} + \frac{\partial^2 \theta}{\partial Y^2} \tag{23.5}$$

We now refer to the boundary condition adopted in this chapter.

$$U = V = 0, \theta = 0 \text{ (along } AC) \tag{23.6a}$$

$$U = V = 0, \theta = \frac{1}{2}(1 - \cos(2\pi X)) \text{ (along } AB) \tag{23.6b}$$

$$U = V = 0, n \cdot \nabla\theta = 0 \text{ (along } BC) \tag{23.6c}$$

The parameter n signifies the perpendicular normal to the corresponding wall.

23.2.2 Nusselt Number

The heat transfer rate (*Nu*) along any domain of the enclosure is conceivably obtained locally as follows:

$$Nu = -\frac{\partial \theta}{\partial n} \tag{23.7}$$

The average heat transfer rate ($\overline{Nu}$) at the bottom wall is given by:

$$\overline{Nu_b} = \frac{\int_0^1 Nu_b \mathrm{d}X}{X|_0^1} = \int_0^1 Nu_b \mathrm{d}X \tag{23.8}$$

23.2.3 Entropy Generation

While designing an effective thermal enclosed system, care must be taken to minimize the energy losses arising out of irreversibilities inside the enclosure. Any thermal enclosed system having a definite/indefinite temperature gradient will definitely undergo some losses. Some losses will also occur because of inherent fluid flow too. This energy losses primarily attributed due to irreversibilities in the system can be measured in expressions of entropy generation inside the enclosure. Designing an efficient thermal enclosed system means the entropy generation associated with the processes should be kept to a minimum. The mathematical equation relating to entropy distribution and generation for heat and fluid flow for a two-dimensional thermal enclosed system subjected to thermodynamic equilibrium for coupled heat and mass transport in Cartesian coordinates is penned down in definite form as:

$$S_\theta = \left(\frac{\partial \theta}{\partial X}\right)^2 + \left(\frac{\partial \theta}{\partial Y}\right)^2 \tag{23.9}$$

$$S_\varphi = \phi\left\{2\left(\left(\frac{\partial U}{\partial X}\right)^2 + \left(\frac{\partial V}{\partial Y}\right)^2\right) + \left(\frac{\partial U}{\partial Y} + \frac{\partial V}{\partial X}\right)^2\right\} \tag{23.10}$$

Here, S_θ and S_ψ refer to local entropy produced on account of irreversibility on account of heat transfer (thermal gradient) and fluid friction (fluid flow), respectively. Here, φ refers to the irreversibility dissemination ratio and is determined accordingly.

$$\phi = \frac{\mu T_0}{k}\left(\frac{\alpha}{L\Delta T}\right)^2 \quad (23.11)$$

The present numerical simulations of this chapter are calculated considering ϕ to be 10^{-4}.

The sum total of entropy generation in the presence of transfer of heat ($S_{\theta,\text{total}}$) and fluid frictional effects ($S_{\psi,\text{total}}$) can be estimated from the dissemination of local entropy generation as follows:

$$S_{\theta,\text{total}} = \int_{\Omega} S_{\theta} \mathrm{d}\Omega \quad (23.12a)$$

$$S_{\psi,\text{total}} = \int_{\Omega} S_{\psi} \mathrm{d}\Omega \quad (23.12b)$$

The comprehensive entire total entropy production (S_{T}) in the quadrantal enclosure can be found to be

$$S_{\text{T}} = S_{\theta,\text{total}} + S_{\psi,\text{total}} \quad (23.13)$$

Another way of looking into irreversibility distribution parameter is the Bejan number (Be), which is the ratio of rate of entropy generation on account of transfer of heat in the enclosure to the comprehensive entire entropy production rate (refer Bejan [38]).

$$Be = \frac{S_{\theta}}{S_{\theta} + S_{\psi}} \quad (23.14)$$

Average Bejan number (Be_{avg}) is referred to quantify the pertinent global influence of entropy generation in relation to transfer of heat in the enclosure and is expressed by:

$$Be_{\text{avg}} = \frac{S_{\theta,\text{total}}}{S_{\theta,\text{total}} + S_{\psi,\text{total}}} \quad (23.15)$$

When the $Be_{\text{avg}} \gg 0.5$, it signifies the dominance of irreversibility on account of transfer of heat. When $Be_{\text{avg}} \ll 0.5$, the limit is just the opposite and it signifies the dominance of irreversibility on account of fluid friction effects.

23.3 Numerical Solution Strategy with Code Validation

The transport equations after converting into non-dimensional form (Eqs. 23.2–23.5) together with the boundary conditions enumerated in (Eqs. 23.6a–23.6c) are solved by a finite element numerical strategy. We employ the well-known Galerkin weighted strategy to remodel the controlling mathematical equations into an arrangement of mathematical integral equations. The exhaustive technique and approach of this method can be found in the book by Zienkiewicz [39]. Further, the Gauss's quadrature method also referred to the book by Zienkiewicz [39] is utilized to accomplish the numerical integration. The algebraic equations are taken care of by employing the Newton–Raphson iteration method. The solution is assumed to be converged when the relative residual is less than 10^{-6}.

A grid independency test considering the quadrantal enclosure with the following boundary conditions of uniform heating strategy on bottom wall, cold upright wall, and insulated curved wall had been carried out to establish vividly to prove these results do not depend on the grid used for this numerical simulation. The values of average Nusselt number ($\overline{Nu}$) for the bottom wall considering distinctive grid layout are represented in Table 23.1. We use 26,330 elements for numerical simulation, because of the fact that when we increase the number of elements from 26,330 to 35,714, the maximum relative error is 0.5%.

The numerical results obtained in the present study were validated for the published results of Aydin and Yesiloz [32] who had undertaken a numerical study too on quadrantal enclosure considering water to be the working fluid (Pr = 6.62) with the boundary conditions of uniformly heated bottom wall, along with upright wall maintained cold and arched wall being insulated. Figure 23.2 of this chapter demonstrates a correlation of the published results of streamline vortices and isotherms of Aydin and Yesiloz [32] for $Ra = 1.7 \times 10^5$. The results apparently displayed in Fig. 23.2 are recognized to be in close correlation with the previous results already available in the literature [32].

Table 23.1 Comparison of average Nusselt number on the bottom wall with uniform heating for different grid systems with Pr = 0.71

Ra	No. of elements		
	9719	26,330	35,714
10^3	1.3585	1.3591	1.3591
10^4	1.9083	1.9114	1.9117
10^5	3.8250	3.8398	3.8412
10^6	6.9458	7.0229	7.0301

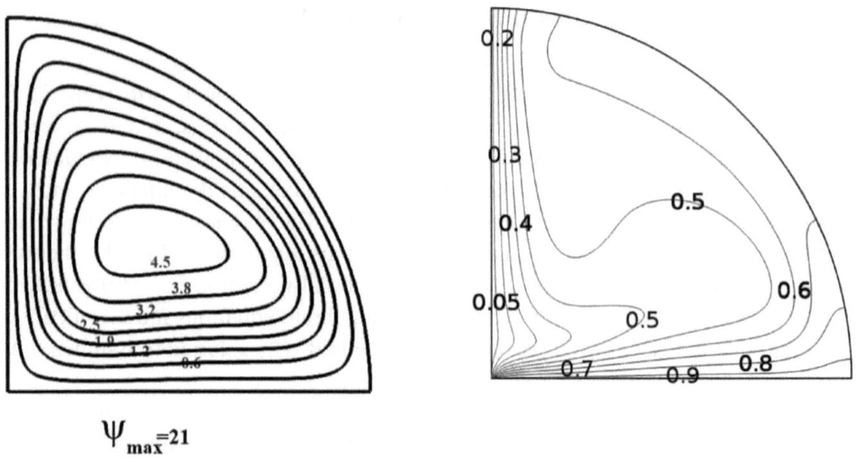

Fig. 23.2 Streamline and isotherm contours for a quadrantal enclosure filled with a fluid (Pr = 6.62) where bottom wall is heated uniformly, upright side wall is cooled at a constant temperature, while arched wall is insulated at $Ra = 1.7 \times 10^5$

23.4 Results and Discussion

The main purpose of the current numerical exploration is to examine the fluid flow and heat transfer along with irreversibility characteristics due to natural convection in a quadrantal cavity. The quadrantal enclosure is heated non-uniformly along the bottom wall, while the perpendicular vertical wall is maintained cold all along. The working fluid for the present investigation is chosen to be air (Pr = 0.71). The resulting flow features are demonstrated through streamline contour, isothermal contour, local and average Nusselt number on hot bottom, cold upright walls, along with illustration of entropy generation. Average Bejan number (Be_{avg}) has also been discussed in this chapter too.

23.4.1 Streamlines and Isotherms

The streamlines and isotherms in the enclosure are represented in the left and right portion in Fig. 23.3 for four distinctive values of Rayleigh number (Ra = 10^3, 10^4, 10^5, and 10^6). It should also be recognized that we allocate positive symbol for ψ, which expresses counterclockwise circulation, while the clockwise circulation is expressed by a negative ψ symbol. It is observed from Fig. 23.3(left) that apart from a primary counterclockwise circulation cell occupying the major portion of the enclosure, there is a formation of secondary cell (clockwise direction) with a lesser vortex strength in the right-hand bottom side of the quadrantal enclosure for the corresponding numerical values of Rayleigh number (Ra) appropriately assigned for this study. It can be observed from Fig. 23.3a that at *Ra* = 10^3, the strength of the vortex of main circulatory cell has a very low value (ψ_{max} = 1.37) because of high resistance to flow. When *Ra* is changed from 10^3 to 10^4, the strength of the vortex has changed to an appreciable value ψ_{max} = 15.5 (counterclockwise circulation) (see Fig. 23.3b). The corresponding clockwise circulation strength changes from a magnitude of 0.49 to 6.65. With further enhancement of *Ra* subsequently from 10^4 to 10^5, we observe a marked increase of circulation strength to ψ_{max} = 70.8 (counterclockwise) signifying dominance of convective effects in the enclosure. Further increase of *Ra* from 10^5 to 10^6 signifies an even greater effect of buoyancy forces with complete dominant of buoyancy effects. The corresponding value of the strength of the vortex is ψ_{max} = 183 (anticlockwise circulation) and ψ_{max} = 55 (clockwise circulation) as can be seen in Fig. 23.3d. It can be further observed from streamline contour in Fig. 23.3d that the nature of main circulation strength vortices is no longer smooth and has undergone twist and curving in the centric portion of the enclosure.

From the isotherm contour plot as depicted in the right portion in Fig. 23.3, it can be inferred that there is a continuous pattern of isothermal contour in the enclosure as there is an absence of finite discontinuity in this study with the chosen boundary conditions depicted for this study. The isotherm patterns revealed in the quadrantal enclosure for the case *Ra* = 10^3 are found to be profoundly compressed in a zone having a close proximity from the bottom heated wall ($0.25 \leq X \leq 0.75$), considering the non-uniform boundary conditions enforced at this wall (Fig. 23.3a). With a change of *Ra* from 10^3 to10^4 as depicted in Fig. 23.3b, the isotherms undergo a twist in the region ($0.25 \leq X \leq 0.75$) signifying that convection effects have just set in inside the enclosure. With further increase of *Ra* = 10^4 to 10^5, resistance force due to buoyancy overcomes fluid friction with an increase of distortion of isotherms observed in the region discussed above (Fig. 23.3c), indicating that the prevalent advancement for heat transfer is buoyant convection in the enclosure at *Ra* = 10^5. On further increase in *Ra* from 10^5 to 10^6, as observed from Fig. 23.3d, the flow moves away from bottom wall and impinges on the left wall as observed in the pattern of distorted isotherms with strong convection effects.

23.4.2 Nusselt Number

The pattern of local Nusselt number (*Nu*) on the heated wall is plotted in Fig. 23.4a for distinctive values of *Ra*. One can see from Fig. 23.4a that the variation of *Nu* alongside the length of the bottom wall is almost sinusoidal. This is for the non-uniform boundary condition enforced on bottom wall. The values of *Nu* are positive from X = 0.45 to 0.95 and have negative values in other regions of the enclosure. The negative value of *Nu* physically connotes that we are transferring the heat from the working fluid, i.e., heated air to the enclosure walls. The maximum value of Nu_{max} is very close to 15 for *Ra* = 10^6.

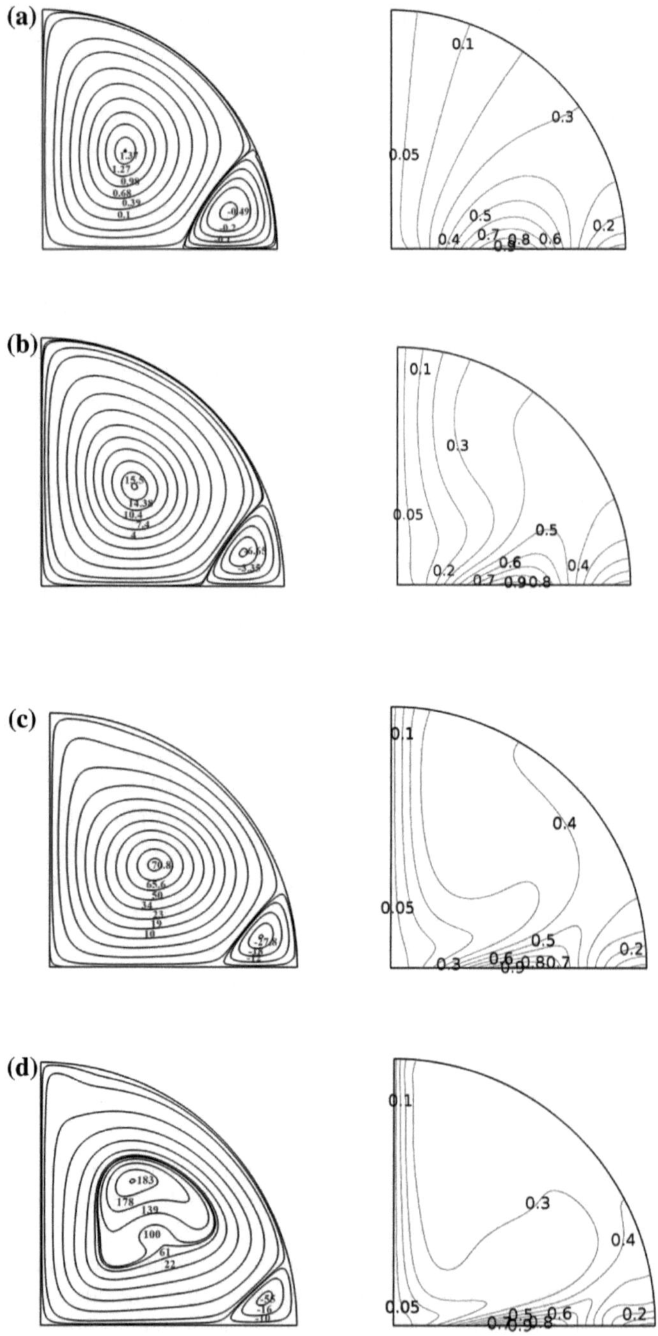

Fig. 23.3 Streamlines and isotherms with non-uniform heating for different Rayleigh numbers: **a** $Ra = 10^3$, **b** $Ra = 10^4$, **c** $Ra = 10^5$, **d** $Ra = 10^6$

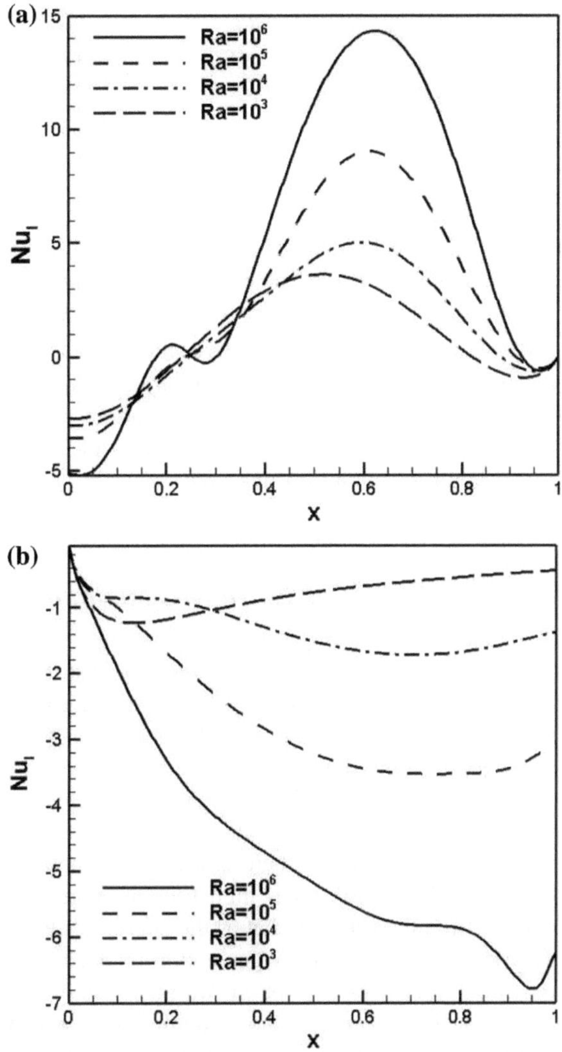

Fig. 23.4 Variation of local Nusselt number on **a** heated bottom wall and **b** cold vertical wall for different values of Rayleigh number

The local Nusselt number (*Nu*) for vertical cold wall is depicted in Fig. 23.4b and is negative for all sets of *Ra* at all locations on the wall. The numerical value of *Nu* escalates with an increase in *Ra* similar to the case of bottom wall. The position of the maximum value of *Nu* is dependent on the *Ra* which can be explained from the isotherm contour. For example, when the *Ra* is small (10^3), *Nu* initially increases to a small length of the wall, attains its pinnacle, and henceforth decreases. On the contrary, for larger values of *Ra*, *Nu* increases almost monotonically except at the corner, where there is a decrease in its value to satisfy the boundary condition.

Table 23.2 Average Nusselt number for bottom and vertical wall for different values of Rayleigh number

Ra	$\overline{Nu_b}$	$\overline{Nu_v}$
10^3	0.796	−0.79258
10^4	1.317	−1.3187
10^5	2.676	−2.6809
10^6	4.645	−4.6661

The numerical values of average Nusselt number ($\overline{Nu}$) on heated bottom wall with Ra are presented in Table 23.2. It is self-evident from Table 23.2 that there is a significant enhancement in $\overline{Nu}$ for the bottom wall with an increase in Ra. Further, $\overline{Nu}$ for cold upright wall is negative and shows a downward trend (negative) with an increase of Ra. We can summarize our discussions, by saying that $\overline{Nu}$ is maximum when heat is transmitted from the hot bottom wall in direction of enclosed fluid or from the heated fluid to the adjoined cold wall, whichever the case may be when Ra is maximum pertaining to our study.

23.4.3 Analysis of Entropy Generation

Noting the points of maximum entropy distribution and generation inside the quadrantal enclosure is of special interest and gives an idea about how to optimize the effects of entropy and irreversibilities. Researchers, working on this area, should also keep in mind that minimum entropy generation with maximum heat transfer holds the key to an efficient design of an enclosure. Figure 23.5 shows the distribution of local entropy generation for three different Ra (=10^4−10^6). For Fig. 23.5, it can be visualized that there is a complete absence of entropy generation in the core of enclosure for either fluid friction or heat transfer. This is quite logical to explain from the situation that as the curved wall is adiabatic the entropy bearing is very less there and entropy generation on account of heat transfer is maximum in the close vicinity of hot bottom wall because of imposed temperature profile. It is also realized that, at the conduction dominant occupancy at lesser Ra ($\leq 10^4$), the entropy generation can be attributed to heat transfer irreversibilities, with the maximum value of entropy generation on account of heat transfer ($S_{\theta\ max}$ = 29.5 and corresponding $S_{\psi\ max}$ = 20). With an increase of Ra to 10^5, the maximum value of entropy generation as a result of fluid friction is increased manifold to a value of $S_{\psi\ max}$ = 908.9. On the contrary, although the maximum numerical value of entropy production on account of heat transfer increases, the enhancement is not to that extent ($S_{\theta max}$ = 85.9). Appreciable amount of entropy generates on account of fluid friction. This entropy distribution is also found in close proximity to curved wall. When the Ra is increased further to 10^6, the value of entropy generation on account of fluid friction throughout the enclosure increases significantly having the maximum value of $S_{\psi\ max}$ = 23874.7, whereas the corresponding value of entropy generation on account of heat transfer is $S_{\theta\ max}$ = 210. We can summarize our

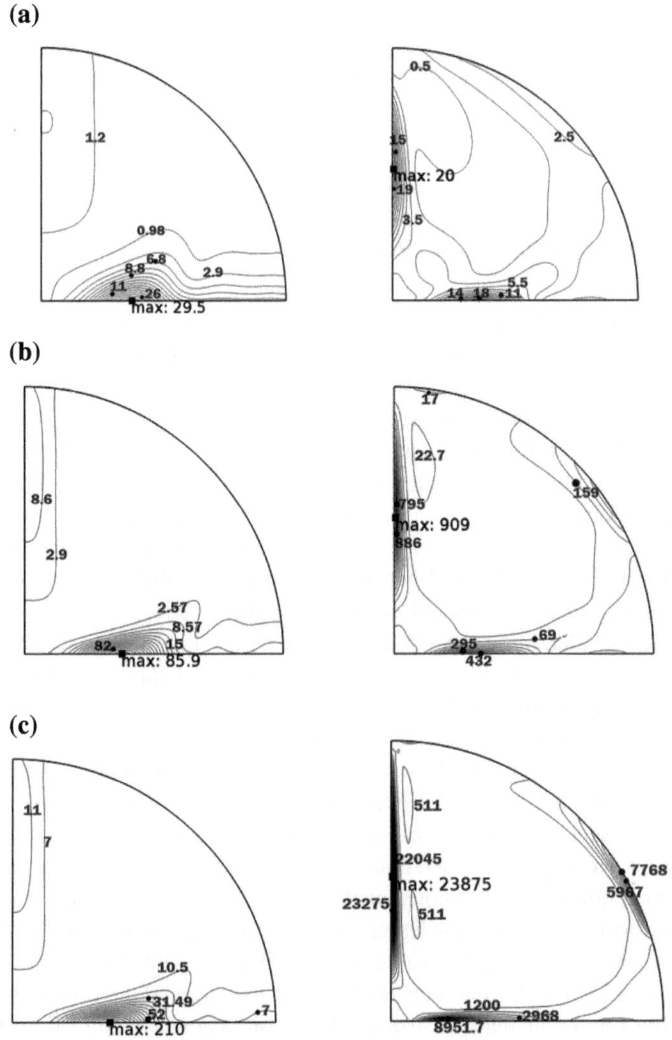

Fig. 23.5 Local entropy generation due to heat transfer (S_θ) and fluid friction (S_ψ) for **a** $Ra = 10^4$ case, **b** $Ra = 10^5$, and **c** $Ra = 10^6$

discussions, by saying that both the heat transfer and the frictional irreversibility escalation occur at high *Ra* for the present nonlinear heating strategy adopted. It is also ascertained that, beyond a critical Rayleigh number, fluid friction irreversibility dictates the total entropy generation inside the quadrantal enclosure.

Table 23.3 Total entropy generation rate and average Bejan number for different Rayleigh numbers

Ra	$S_{\theta,max}$	$S_{\psi,max}$	S_T	Be_{avg}
10^3	14.1	0.1556	14.2556	0.989
10^4	29.5	19.970	49.47	0.596
10^5	85.9	908.92	994.82	0.085
10^6	210	23874.7	24084.7	0.008

23.4.4 Analysis of Bejan Number

As already mentioned, one can predict the global dominance of irreversibility after knowing the average Bejan number (Be_{avg}). Accordingly, Be_{avg} is evaluated and reported in Table 23.3. From Table 23.3, it is clear that when Ra is changed from 10^3 to 10^5, the change in the magnitude of $S_{\theta max}$ is almost negligible. However, the magnitude of $S_{\psi max}$ varies significantly and it records a very less value for $Ra = 10^3$ ($S_\psi = 0.1556$) and has a very high value for $Ra = 10^5$($S_\psi = 908.92$) signifying that entropy generation on account of fluid friction escalation occurs with an increase of Ra. The corresponding average Bejan number (Be_{avg}) noted at $Ra = 10^3$ is 0.989, while at $Ra = 10^5$ it is 0.085. When Ra is increased further to 10^6, the total irreversibility in the presence of heat transfer (S_θ) is increased significantly to $S_\theta = 210$ compared to its value of $S_\theta = 85.9$ at $Ra = 10^5$ signifying that irreversibility due to heat transfer increases to a large value with an increase of Ra from 10^5 to 10^6. Also, Be_{avg} values clearly reflect that entropy generation due to heat transfer is the dominating parameter for $Ra = 10^3$ and thereafter at $Ra = 10^4$ the effect of irreversibility on account of fluid flow and heat transfer is of equal significance. The average Bejan number (Be_{avg}) noted at $Ra = 10^4$ is 0.596 which clearly proves our discussion in Sect. 2.3. With further increase of Ra to 10^5 and 10^6, the irreversibility on account of fluid friction is a commanding factor. The corresponding average Bejan number (Be_{avg}) noted at $Ra = 10^6$ is even lesser which is 0.008. This is attributed to the fact that, for lesser values of Ra, heat transfer is dominated by conduction and accordingly the value of Be_{avg} is large. However, with an increase in Ra, both the irreversibility on account of heat transfer and the fluid flow increase, although the rate of increase of irreversibility on account of fluid flow friction is more a representative factor than due to heat transfer because of intensified effects of buoyancy force strength above fluid viscous force as already reiterated earlier.

23.5 Conclusions

The buoyancy-induced fluid flow, thermal transport, and irreversibility characteristics in a quadrantal cavity have been numerically analyzed. The following boundary conditions are adopted for this chapter. The bottom wall is subjected to non-uniform heating, while the upright left wall is kept cold and at the same time the arched wall is maintained at insulated condition to surroundings. The final conclusions are:

The local Nusselt number at the heated wall follows a sinusoidal variation. At lower values of Ra (=10^3,10^4), the mechanism of thermal transport is conduction, while thermal transport is due to convection at $Ra = 10^6$.

At low values of Ra (10^3), the entropy generation is essentially generated owing to heat transfer irreversibility (I_{HT}), while for larger range of Ra (10^5 and 10^6) fluid friction irreversibility (I_{FF}) is predominant over I_{HT}. For intermediate values of Ra, both I_{HT} and I_{FF} are comparable.

References

1. Davis, V., De, G.: Natural convection of air in a square cavity: a bench mark numerical solution. Int. J. Numer. Methods Fluids **3**(3), 249–264 (1983). https://doi.org/10.1002/fld.1650030305
2. Hamadi, F.J., Lloyd, J.R., Yang, H.Q., Yang, K.T.: Study of local natural convection heat transfer in an inclined enclosure. Int. J. Heat Mass Transfer **32**(9), 1697–1708 (1989). https://doi.org/10.1016/0017-9310(89)90052-5
3. Roy, S., Basak, T.: Finite element analysis of natural convection flows in a square cavity with non-uniformly heated wall(s). Int. J. Eng. Sci. **43**, 668–680 (2005). https://doi.org/10.1016/j.ijengsci.2005.01.002
4. Basak, T., Roy, S., Balakrishnan, A.R.: Effects of thermal boundary conditions on natural convection flows within a square cavity. Int. J. Heat Mass Transf. **49**, 4525–4535 (2006). https://doi.org/10.1016/j.ijheatmasstransfer.2006.05.015
5. Aminossadati, S.M., Ghasemi, B., Kargar, A.: Computational analysis of magneto-hydrodynamic natural convection in a square cavity with a thin fin. Eur. J. Mech. B-Fluids **46**, 154–63 (2014). https://doi.org/10.1016/j.euromechflu.2014.03.002
6. Yousaf, M., Usman, S.: Natural convection heat transfer in a square cavity with sinusoidal roughness elements. Int. J. Heat Mass Transf. **90**, 180–190 (2015). https://doi.org/10.1016/j.ijheatmasstransfer.2015.06.049
7. Ganzarolli, M.M., Milanez, L.F.: Natural convection in rectangular enclosures heated from below and symmetrically cooled from the sides. Int. J. Heat Mass Transf. **38**, 1063–1073 (1995). https://doi.org/10.1016/0017-9310(94)00217-J
8. Aydin, O., Unal, A., Ayhan. T.: Natural convection in rectangular enclosures heated from one side and cooled from the ceiling, Int. J Heat Mass Transf. **42**, 2345–2355 (1999). https://doi.org/10.1016/S0017-9310(98)00319-6
9. Varol, Y., Oztop, H.F., Yilmaz, T.: Two-dimensional natural convection in a porous triangular enclosure with a square body. Int. Commun. Heat Mass Transf. **34**, 238–247 (2007b. https://doi.org/10.1016/j.icheatmasstransfer.2006.10.001
10. Basak, T., Roy, S., Babu, S.K., Balakrishnan, A.R.: Finite element analysis of natural convection flow in a isosceles triangular enclosure due to uniform and non- uniform heating at the side walls. Int. J. Heat Mass Transf. **51**, 4496–4505 (2008). https://doi.org/10.1016/j.ijheatmasstransfer.2007.12.018
11. Basak, T., Roy, S., Babu, S.K., Pop, I.: Finite element simulations of natural convection flow in an isosceles triangular enclosure filled with a porous medium: effects of various thermal boundary conditions. Int. J. Heat Mass Transf. **51**, 2733–2741 (2008b). https://doi.org/10.1016/j.ijheatmasstransfer.2007.10.009
12. Seki, N., Fukusako, S., Yamaguchi, A.: An experimental study of free convective heat transfer in a parallelogramic enclosure. J. Heat Transf. ASME, **105**, 433–439 (1983). https://doi.org/10.1115/1.3245603

13. Moukalled, F., Acharya, S.: Laminar natural convection heat transfer in an eccentric rhombic annulus. Numer. Heat Transf. A **26**, 551–568 (1994). https://doi.org/10.1080/10407789408956009
14. Aldridge, K.D., Yao, H.: Flow features of natural convection in a parallelogramic enclosure. Int. Commun. Heat Mass Transf. **28**, 923–931 (2001). https://doi.org/10.1016/S0735-1933(01)00296-2
15. Salah, H., Roslan, R., Hashim, I.: Natural convection heat transfer in a nanofluid–filled trapezoidal enclosure. Int. J. Heat Mass Transf. **54**, 194–201 (2011). https://doi.org/10.1016/j.ijheatmasstransfer.2010.09.053
16. Esfe, M.H., Arani, A.A.A., Yan, W.M., Ehteram, H., Aghaie, A., Afrand, M.: Natural convection in a trapezoidal enclosure filled with carbon nanotube-EG- water nanofluid. Int. J. Heat Mass Transf. **92**, 76–82 (2016). https://doi.org/10.1016/j.ijheatmasstransfer.2015.08.036
17. Abu-Hijleh, B.A., Abu-Qudais, M., Abu-Nada, E.: Numerical prediction of entropy generation due to natural convection from a horizontal cylinder. Energy **24**, 327–333 (1999). https://doi.org/10.1016/S0360-5442(98)00103-0
18. Varol, Y., Oztop, H.F., Pop, I.: Entropy analysis due to conjugate-buoyant flow in a right-angle trapezoidal enclosure filled witha porous medium bounded by a solid vertical wall. Int. J. Therm. Sci. **48**(6), 1161–1175 (2009). https://doi.org/10.1016/j.ijthermalsci.2008.08.002
19. Varol, Y., Oztop, H.F., Pop, I.: Entropy generation due to natural convection in non-uniformly heated porous isosceles triangular enclosures at different positions. Int. J. Heat Mass Transfer **52**(5–6), 1193–1205 (2009). https://doi.org/10.1016/j.ijheatmasstransfer.2008.08.026
20. Pal G.C., Goswami N., Pati S.: Numerical investigation of unsteady natural convection heat transfer and entropy generation from a pair of cylinders in a porous enclosure, Numer. Heat Transfer Part A: Appl. (2018). https://doi.org/10.1080/10407782.2018.1507887
21. Mahmud, S., Sadrul Islam, A.K.M.: Laminar free convection and entropy generation inside an inclined wavy enclosure. Int. J. Therm. Sci. **42**(11), 1003–1012 (2003). https://doi.org/10.1016/S1290-0729(03)00076-0
22. Dagtekin, I., Oztop, H.F., Bahloul, A.: Entropy generation for natural convection in Γ-shaped enclosures. Int. Commun. Heat Mass Transfer. **34**, 502–510 (2007). https://doi.org/10.1016/j.icheatmasstransfer.2007.01.003
23. Datta, P., Mahapatra, S., Ghosh, K., Manna, N.K., Sen, S.: Heat transfer and entropy generation in a porous square enclosure in presence of an adiabatic block. Transp. Porous Med. **111**, 305–329 (2016). https://doi.org/10.1007/s11242-015-0595-5
24. Chakravarty, A., Datta, P., Ghosh, K., Sen, S., Mukhopadhyay, A.: Thermal non-equilibrium heat transfer and entropy generation due to natural convection in a cylindrical enclosure with a truncated conical, heat-generating porous bed. Transp. Porous Med. **116**, 353–377 (2017). https://doi.org/10.1007/s11242-016-0778-8
25. Mani, Rathnam V., Biswal, P., Basak, T.: Analysis of entropy generation during natural convection within entrapped porous triangular cavities during hot or cold fluid disposal. Numer. Heat Transfer Part A: Appl. **69**(6), 931–956 (2016). https://doi.org/10.1080/10407782.2015.1109362
26. Das, D., Basak, T.: Analysis of entropy generation during natural convection in discretely heated porous square and triangular enclosures. Numer. Heat Transfer Part A: Appl. **71**(10), 979–1003 (2017). https://doi.org/10.1080/10407782.2017.1326785
27. Abu-Nada, E., Oztop, H.F.: Numerical analysis of Al_2O_3/water nano-fluids natural convection in a wavy walled cavity. Numer. Heat Transfer Part A: Appl. **59**(5), 403–419 (2008). https://doi.org/10.1080/10407782.2011.552363
28. Oztop, H.F., Abu-Nada, E.: Numerical study of natural convection in partially heated rectangular enclosures filled with nanofluids. Int. J. Heat Fluid Flow **29**, 1326–1336 (2008). https://doi.org/10.1016/j.ijheatfluidflow.2008.04.009
29. Meshram, P., Bhardwaj, S., Dalal, A., Pati, S.: Effects of inclination angle on natural convection heat transfer and entropy generation in square porous enclosure. Numer. Heat Transfer Part A: Appl. **70**, 1271–1296 (2016). https://doi.org/10.1080/10407782.2016.1230433

30. Bhardwaj, S., Dalal, A., Pati, S.: Influence of wavy wall and non-uniform heating on natural convection heat transfer and entropy generation inside porous complex enclosure. J. Energy **79**, 467–481 (2014). https://doi.org/10.1016/j.energy.2014.11.036
31. Dutta, S., Biswas, A.K., Pati, S.: Analysis of heat transfer and entropy generation during natural convection within rhombic enclosures with non-uniform heating on both top and bottom walls, and cold side walls. In: Proceedings of the 2th International and 24th National ISHMT-ASTFE Heat and Mass transfer conference, pp. 1–9, 27–30, 2017, ISHMTC-2017-07-0487 (2017)
32. Aydin, O., Yesiloz, G.: Natural convection in a quadrantal cavity heated and cooled on adjacent walls. J. HeatTransfer **133**, 052501–052507 (2011). https://doi.org/10.1115/1.4003044
33. Yesiloz, G., Aydin, O.: Natural convection in an inclined quadrantal cavity heated and cooled on adjacent walls. Exp. Therm. Fluid Sci. **35**, 1169–1176 (2011). https://doi.org/10.1016/j.expthermflusci.2011.04.002
34. Sen, D., Bose, P.K., Panua, R., Das, A.K.: Numerical analysis of laminar natural convection in a quadrantal cavity with a hot bottom and cold curved walls. Heat Transfer Res. **46**(7), 631–641 (2015). https://doi.org/10.1615/heattransres.2015004629
35. Sen, D., Bose, P.K., Panua, R., Das, A.K., Sen, P.: Laminar natural convection study in a quadrantal cavity using heater on adjacent walls. Frontiers Heat Mass Transfer (FHMT) **4**(1), 1–7 (2013). https://doi.org/10.5098/hmt.v4.1.3005
36. Dutta, S., Biswas, A.: Entropy generation due to natural convection with non -uniform heating of porous quadrantal enclosure—a numerical study. Frontiers Heat Mass Transfer (FHMT) **10** (8), 1–12 (2018). https://doi.org/10.5098/hmt.10.8
37. Dutta, S., Biswas, A., Pati, S.: Natural convection heat transfer and entropy generation inside porous quadrantal enclosure with non-isothermal heating at the bottom wall. Numer. Heat Transfer, Part A: Appl. **73**(4), 222–240 (2018). https://doi.org/10.1080/10407782.2018.1423773
38. Bejan, A.: Entropy Generation Minimization. The method of thermodynamic optimization of finite size system and finite size processes. CRC, Boca Raton (2013)
39. Zienkiewicz, O.C., Taylor, R.L.: The Finite Element Method, 4th edn. McGraw-Hill, London (1991)

Chapter 24
Impact of Magnetic Field on Thermal Convection in a Linearly Heated Porous Cavity

Aakash Gupta, Sayanta Midya, Nirmalendu Biswas and Nirmal K. Manna

Nomenclature

B	Uniform magnetic field (tesla)
Da	Darcy number
H	Cavity height/length scale, m
Ha	Hartmann number
K	Porous medium permeability, m^2
L	Cavity length, m
Nu	Average Nusselt number
p	Pressure, Pa
Pr	Prandtl number
Ra	Rayleigh number
T	Temperature, K
u, v	Velocity components, m/s
U, V	Dimensionless velocity components
x, y	Cartesian coordinates, m
X, Y	Non-dimensional coordinates

Greek symbols

α	Thermal diffusivity, m^2/s
β	Volumetric expansion coefficient, K^{-1}
γ	Inclination angle of the magnetic field
θ	Non-dimensional temperature
ε	Porosity
υ	Kinematic viscosity, m^2/s
Π	Non-dimensional heatfunction

A. Gupta (✉) · S. Midya · N. Biswas · N. K. Manna
Department of Mechanical Engineering, Jadavpur University, Kolkata 700032, India
e-mail: aakashgupta068@gmail.com

P. Sahoo and J. P. Davim (eds.), *Advances in Materials, Mechanical and Industrial Engineering*, Lecture Notes on Multidisciplinary Industrial Engineering,
https://doi.org/10.1007/978-3-319-96968-8_24

ρ Density, kg/m^3
κ Electrical conductivity (μS cm^{-1})
ψ Non-dimensional stream function

Subscripts

a Ambient
c Cooling
h Heating

24.1 Introduction

Before formulating the present problem, a thorough literature survey is conducted. From the vast pool of literatures, it is noted that magnetohydrodynamics coupled with buoyancy-driven flow is an interesting area of current research because of its wide varieties of applications. Magnetohydrodynamic (MHD) convection finds numerous applications in medical science, engineering and technology and in various thermal flow devices [1–3]. In a fluid with electrical conduction when magnetic force is applied, it develops Lorentz force. The buoyancy results from temperature gradient in the working fluid. A thermal system working along with magnetic fields is governed by Lorentz force as well as buoyancy force. Both are body forces in nature, and through their mutual interaction affects both thermal and fluid flow structures in the cavity. Thus, for the better visualization of thermal characteristics, insight knowledge about the flow structure and heat transport phenomena is required. Many researchers have investigated the MHD natural convection experimentally, analytically and numerically. Sivasankaran and Ho [4] studied a square cavity MHD problem using water ($Pr = 11.6$) with temperature-dependent fluid properties and Hartmann number (Ha). This study indicates the reduction in average Nusselt number (Nu) at increased Ha. The orientation of magnetic field also affects Nu. Similar observations were noted by Yu et al. [5] for a rectangular cavity with external uniform magnetic field for $Pr = 0.025$. In addition, they found that when cavity aspect ratio is less or more than 1.0, the field angle has strong impact on thermo-fluid flow characteristics.

Sathiyamoorthy and Chamkha [6] conducted MHD study numerically in a cavity with uniform bottom heating and linear heating through sidewalls using liquid gallium ($Pr = 0.025$). Here also, Hartmann number nonlinearly reduces heat transfer rate, irrespective of the angle of applied magnetic fields. Nanofluid-based MHD natural convection study in a differentially heated cavity was presented by Ghasemi et al. [7]. With Al_2O_3–water, they observed the marked effect of Hartmann number as well as nanoparticle concentration. Very recently, Mliki et al. [8] reported MHD nanofluid convection with heat absorption or generation using

Cu–water in a square cavity, heated linearly at left wall, right wall cold, bottom and top adiabatic. The observation of reducing heat transfer at increased *Ha* is maintained, although the rise of volume fraction of nanoparticles and Rayleigh number enhances the rate of heat transfer.

Mahmoudi et al. [9] through a numerical study assessed the influence of nanoparticles and Hartman number on heat transfer process in a trapezoidal cavity using uniform magnetic field and Cu–water nanofluid. They found decreased entropy generation with nanoparticles, but more entropy generation with magnetic fields. The effect of thermal boundaries was investigated numerically by Mansour and Bakier [10] through five different thermal configurations of a square cavity, keeping bottom wall condition fixed. The magnetic field effect on free convection in cavity was analysed by Kefayati [11] applying linear heating on left wall and power-law fluids. It shows at $Ha = 60$ when the index of power law increases, it enhances heat transfer. However, the negative impact of Hartmann number on the heat transfer persists for all values of the index. In his another work [12], similar heating configuration was applied but ferrofluid of cobalt–kerosene was considered to address MHD. Linearly heated square cavity (heating through sidewalls) undergoing natural convection was studied by Sathiyamoorthy et al. [13] without the consideration of MHD. Considering Cu–water nanofluid as flowing medium in [14], the numerical investigation on the entropy generation has been conducted using uniform heat sink and source configuration in an inclined porous enclosure during MHD natural convection. This study reveals that Nusselt number reduces with the decrease in Hartmann number or volume fraction of nanofluid. In addition, the optimum configuration of source–sink arrangement for the best thermal performance was obtained. Very recently, Emami et al. [15] numerically studied natural convection in an inclined porous cavity using nanofluid of Cu–water for laminar flow ($10^4 \leq Ra \leq 10^6$). The results revealed that the configuration of wall heater has an adverse effect on heat transfer. Heat transport in a porous cavity was analysed in [16] using sinusoidal and uniform heating on the bottom wall. The analysis was carried out considering same mean temperature of heating, irrespective of the mode of heating. The study indicates marked high heat transfer from the sinusoidal heating. Nemati et al. [17] examined CuO nanoparticle-based MHD in square enclosure. The observations are same as others; Nusselt number becomes higher as the nanoparticles increase, but *Nu* decreases with the magnetic field. Ghasemi and Siavashi [18] investigated a nanofluid saturated porous cavity applying various linear temperature distribution of sidewalls.

Main objective of this work is to address fundamental aspects of thermo-fluid flow by investigating the impacts of linear heating, magnetic field and its inclination on natural convection in a cavity. The heating elements are located on both the sidewalls; however, its temperature varies linearly with minimum temperature at top and maximum temperature at bottom. Within the laminar flow, analysis is conducted considering clear domain as well as porous medium. The flow physics along with the thermal phenomenon is illustrated through heatlines.

24.2 Problem and Solution Method

The problem considers a two-dimensional cavity of length L and height H as described in Fig. 24.1. The electrical conductivity of the working fluid negotiates with the magnetic fields. The magnetic source is applied externally and produces uniform fields in the cavity. The top wall is directly exposed to the ambient condition. Its temperature is taken constant at ambient temperature ($T_a = T_c$). While the bottom wall is insulated and adiabatic, the heating element is located along the sidewalls with linearly increasing temperature in downward direction. The top-end temperature of the sidewalls is chosen same as the temperature of the top wall (to avoid sudden jump in temperature). The working fluid ($Pr = 0.71$) is saturated with a porous medium. The buoyant fluid flow and the resulting heat transfer are analyzed under the influence of magnetic field (of amplitude B).

The present analysis is performed numerically for the condition of steady laminar flow. The buoyancy effect is considered within the Boussinesq approximation. The fluid is assumed as Newtonian and incompressible. The minor variation in thermo-physical properties of the working fluid is neglected along with the induced magnetic effect. Joule heating, Hall effect, and dissipation and pressure works are also ignored. The governing equations for the present problem are obtained from the fundamental conservation principles (mass, momentum and energy), which are given in non-dimensional form as

$$\frac{\partial U}{\partial X} + \frac{\partial V}{\partial Y} = 0 \tag{24.1}$$

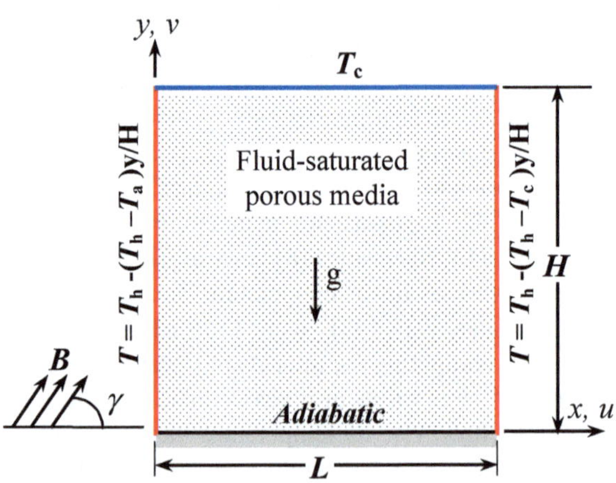

Fig. 24.1 Description of the problem

$$\frac{1}{\varepsilon^2}\left(U\frac{\partial U}{\partial X}+V\frac{\partial U}{\partial Y}\right)=-\frac{\partial P}{\partial X}+\frac{Pr}{\varepsilon}\left(\frac{\partial^2 U}{\partial X^2}+\frac{\partial^2 U}{\partial Y^2}\right)-\frac{Pr}{Da}U-\frac{F_C\sqrt{U^2+V^2}}{\sqrt{Da}}U$$
$$+Ha^2Pr(V\sin\gamma\cos\gamma-U\sin^2\gamma) \quad (24.2)$$

$$\frac{1}{\varepsilon^2}\left(U\frac{\partial V}{\partial X}+V\frac{\partial V}{\partial Y}\right)=-\frac{\partial P}{\partial Y}+\frac{Pr}{\varepsilon}\left(\frac{\partial^2 V}{\partial X^2}+\frac{\partial^2 V}{\partial Y^2}\right)-\frac{Pr}{Da}V-\frac{F_C\sqrt{U^2+V^2}}{\sqrt{Da}}V$$
$$+Ha^2Pr(U\sin\gamma\cos\gamma-V\cos^2\gamma)+RaPr\theta \quad (24.3)$$

$$U\frac{\partial\theta}{\partial X}+V\frac{\partial\theta}{\partial Y}=\left(\frac{\partial^2\theta}{\partial X^2}+\frac{\partial^2\theta}{\partial Y^2}\right) \quad (24.4)$$

In the above, U and V stand for velocity components (in non-dimensional form), respectively, along the dimensionless coordinates X and Y. Dimensionless temperature and pressure are indicated by θ and P, respectively. These quantities are detailed as

$$\begin{aligned}&(X,Y)=(x,y)/H,\ (U,V)=(u,v)/(\alpha/H),\\&P=p/\rho(\alpha/H)^2,\theta=(T-T_c)/(T_h-T_c)\end{aligned} \quad (24.5)$$

The dimensionless numbers namely Prandtl, Rayleigh, Hartmann and Darcy number (*Pr*, *Ra*, *Ha* and *Da*, respectively) are expressed by

$$\begin{aligned}&Ra=g\beta(T_h-T_c)H^3/\alpha\upsilon,\ Ha=BH\sqrt{\kappa/\rho\upsilon}\\&Pr=\upsilon/\alpha, Da=K/H^2, F_C=1.75/\sqrt{150\varepsilon^3}\end{aligned} \quad (24.6)$$

The following boundary conditions are applied: $U=V=0$ on all the walls as they are stationary, $\theta=1-Y$ on both the sidewalls, and $\partial\theta/\partial Y=0$ on the top and bottom walls.

The flow physics and heat transfer phenomenon are characterized by streamlines, isotherms, heatlines and Nusselt number. In the present problem, since identical linear heaters are placed along both the sidewalls, the Nusselt number on the top wall will have contribution from both the left ($X=0$) and right wall ($X=1$). Moreover, the top wall is maintained at constant ambient temperature (T_c), so the average Nusselt number along this isothermal wall is assessed as defined below.

$$Nu=\int_0^1\left(-\left.\frac{\partial\theta}{\partial Y}\right|_{Y=1}\right)\mathrm{d}X \quad (24.7)$$

The thermal energy transport phenomenon is visualized through heatlines which represent constant heatfunction at different values. It is formulated from the steady-state energy equation and takes care of both conductive and convective heat flux. For the present problem of natural convection, the heatfunction (Π) is expressed as

$$-\frac{\partial \Pi}{\partial X} = V\theta - \frac{\partial \theta}{\partial Y} \text{ and } \frac{\partial \Pi}{\partial Y} = U\theta - \frac{\partial \theta}{\partial X} \tag{24.8}$$

The above heatfunction equations are solved by defining $\Pi = 0$ at the mid-point of the bottom wall. For fluid flow visualization through streamlines, the stream function equations are solved which are given by

$$-\frac{\partial \psi}{\partial X} = V \text{ and } \frac{\partial \psi}{\partial Y} = U \tag{24.9}$$

The above equations are solved with $\psi = 0$ at the cavity walls. Negative sign of ψ indicates clockwise rotation and the positive for counterclockwise rotation.

24.3 Numerical Scheme

The governing equations of the present problem on the constraint of imposed boundary conditions (Fig. 24.1) are solved adapting the finite volume method. Equations (24.1)–(24.4) are discretized after linearizing the advection terms of the momentum equations using staggered grids. The discretization results into linear equations which are solved by tri-diagonal matrix algorithm. The solution is performed iteratively using SIMPLE algorithm [19] in an indigenous code. The convergence is marked by setting maximum residuals of all the governing equations below 10^{-7}.

24.4 Grid Independence Study

A grid independent study is conducted considering staggered grids of 70 × 70, 100 × 100, 130 × 130, 160 × 160 and 200 × 200 with non-uniform distribution. The average Nusselt number (*Nu*) and the errors in the percentage of *Nu* with respect to the immediate grid are calculated and given in brackets in Table 24.1. It indicates that 160 × 160 grids are sufficient for the current problem.

Table 24.1 Grid independence study using average Nusselt number (*Nu*) at $Pr = 0.71$, $Ha = 30$

Ra	*Nu* (% error with respect to immediate coarser grid)				
	70 × 70	100 × 100	130 × 130	160 × 160	200 × 200
10^3	0.929	0.929 (0.01%)	0.929 (0.01%)	0.929 (0.03%)	0.929 (0.03%)
10^4	0.924	0.922 (0.02%)	0.922 (0.12%)	0.922 (0.01%)	0.922 (0.01%)
10^5	0.964	0.963 (0.18%)	0.961 (0.14%)	0.961 (0.09%)	0.959 (0.11%)

24.5 Code Validation

For the solution of present MHD convection, the accuracy level of the present code is also checked as indicated in Fig. 24.2. The validating results are generated on the MHD convection in a cavity using the prediction of Ghasemi et al. [7] for $Ra = 10^5$ and fluid Prandtl number (*Pr*) equal to 6.2. Local distributions of thermo-fluid flow fields of both the results are also compared for various *Ha* values that reveal close matching (as could be realized from a sample result in Fig. 24.2 pertaining to $Ha = 30$). The isotherms and streamlines of published and present results are found consistent.

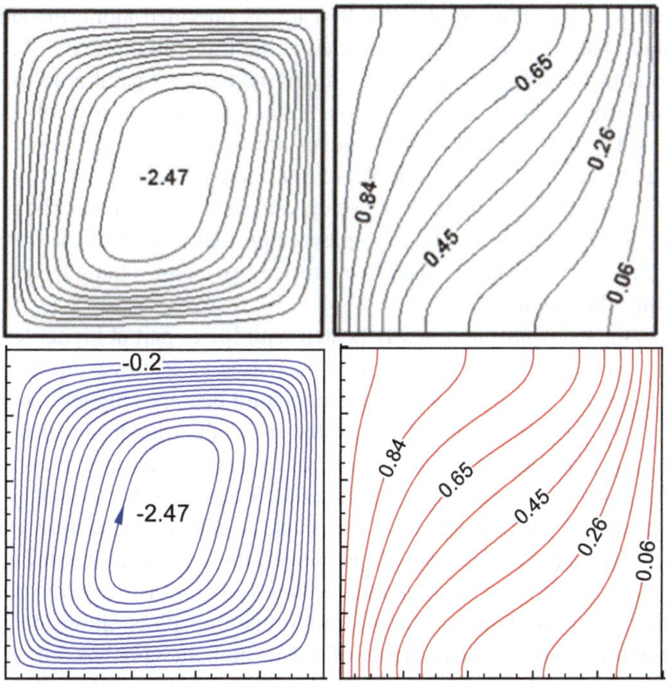

Fig. 24.2 Code validation: the present predictions in the top panel and the published results in Ref. [7] in the bottom panel, consisting of streamlines (first column) and isotherms (second column) for $Pr = 6.2$, $Ra = 10^5$ and $Ha = 30$ as considered by Ghasemi et al. [7]. The present results agree well with the published results

24.6 Results and Discussion

The present work is carried out for three *Ra* values (*Ra* = 10^3, 10^4, 10^5) with a range of *Ha* values (*Ha* = 0, 10, 20, 30, 40, 50, 60, 70, 80, 90, 100) and the magnetic field inclination angle (γ = 0°, 20°, 30°, 40°, 50°, 65°, 90°, 120°, 150°, 165°, 180°) using clear domain and porous domain. The different values of both porosity (ε = 0.1, 0.3, 0.5, 0.6, 0.8, 1.0) and Darcy number (*Da* = 10^{-3}, 10^{-4}, 10^{-5}, 10^{-6}, 10^{-7}) of porous substance are also considered. The results are systematically presented for different parametric investigation in subsequent sections.

24.6.1 Impact of Rayleigh Number

The thermal and fluid flow structures are presented in terms of isotherms (in the top row), streamlines (in the middle row) and heatlines (in the bottom row) in Figs. 24.3a, b, respectively, for both the domains (clear and porous). The values of parameters for these plots are *Ra* = 10^3, 10^4, 10^5 keeping Hartmann number constant at *Ha* = 30 and horizontal magnetic field (inclination angle γ = 0°). The linear heating with $\theta = 1$ at bottom and $\theta = 0$ at top is provided along the sidewalls in same manner. The main cooling is provided at the top wall of cavity. Accordingly, the isotherms in top region of cavity are distributed almost parallel to the top wall, and the isotherms with increasing contour values are located towards the bottom ends of the sidewalls. The heated fluid moves upward and the relatively colder fluid is driven downward along the middle of the cavity. It happens due to density change. The hot fluid being lighter moves up, and on the other hand, the colder fluid being heavier descents down. For the present problem, as both the sidewalls are identically hot, it is assumed that two vortices of same strength will circulate upward along the sidewalls.

Again, as the sidewalls are linearly heated with lower part at higher temperature, the heating rate of fluid inside the cavity decreases as it moves up along the hot sidewalls. However, when the hot fluid temperature becomes higher than that of the sidewalls particularly at the upper part of the sidewalls (due to linear heating with zero temperature at the top ends), instead of heating, some heat may be rejected through the sidewalls, particularly at higher Rayleigh numbers. That is, the sidewalls can provide both heating and cooling due to linear temperature variation, in addition to the cooling at the top wall. However, the results shown in Fig. 24.3 are free from such mixed up of heating and cooling as *Ra* $\leq 10^5$, where conduction is more predominant. Both the isotherm and streamline plots show the symmetric pattern about the mid-vertical plane of clear domain (Fig. 24.3a) and porous domain (Fig. 24.3b). The maximum value of stream function (an indicator of circulation strength) is mentioned on the individual plots.

Heatline plots for clear and porous domain (in Fig. 24.3) depict the corridors for source-to-sink heat flow. It presents how heat energy transfers take place in the

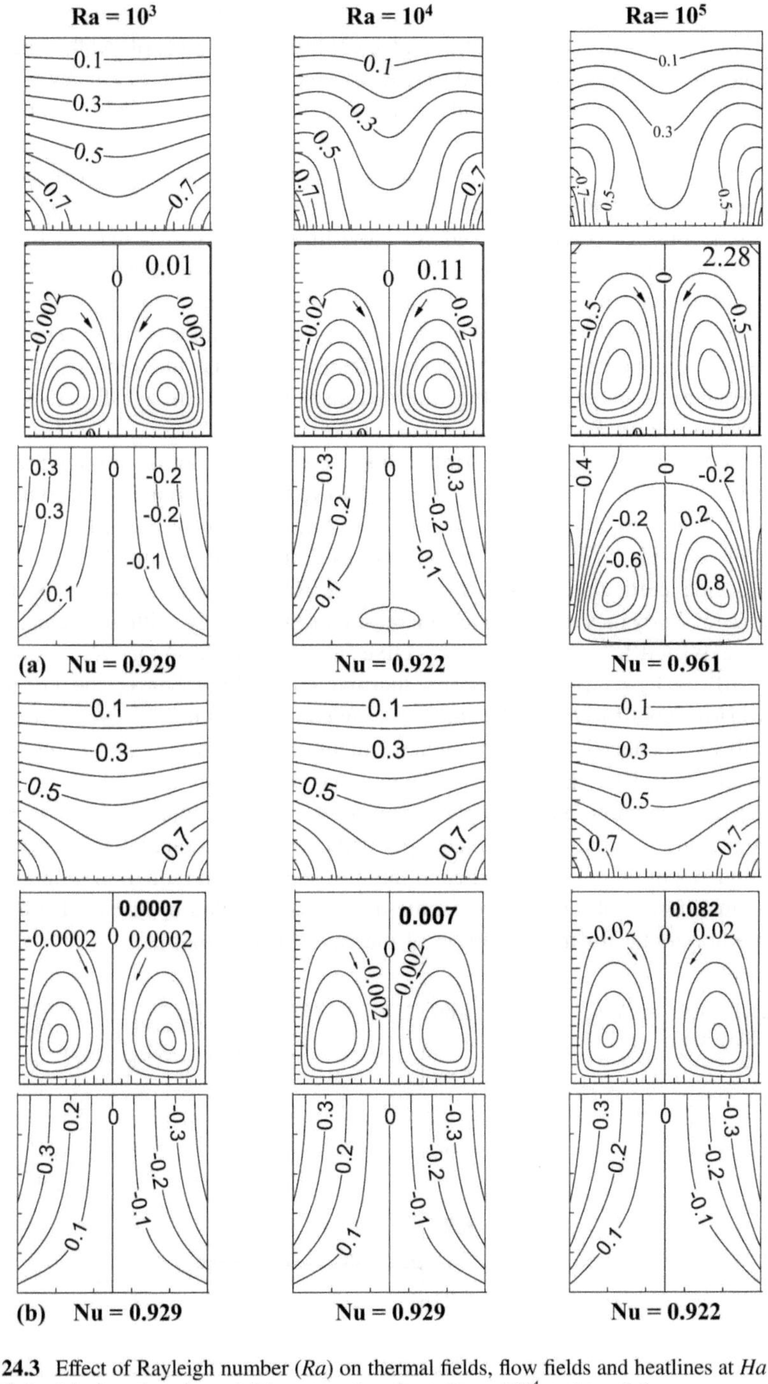

Fig. 24.3 Effect of Rayleigh number (*Ra*) on thermal fields, flow fields and heatlines at Ha = 30, γ = 0° for (**a**) clear domain and (**b**) porous domain at Da = 10^{-4} and ε = 0.6

cavity. The heat/energy flow originates from the hot wall, then travels through the cavity fluid and finally disposes through the cold (or colder parts of) walls. The fluid is subjected to both buoyancy and magnetic forces. The heatlines emanating from the source walls terminates normally at the sink walls almost parallel to neutral line (through the mid-vertical plane). The heatlines are symmetric across the mid-vertical plane due to identical linear heating along the sidewalls. For the clear domain considering $Ra = 10^3$ and 10^4, there is very less or no circulation of heat lines on account of conduction and very low flow velocity as is evident from the maximum stream function values. The heatlines imitate the stream function to a great extent, conforming more heat transfer by convection (*Nu* increases). For the porous domain, heatline plots are more or less identical with less energy circulation due to the hindrance offered by porous medium against the flow. Conduction predominates as observed by the minimal rise in *Nu* (heat transfer) even with significant value of *Ra*.

Due to mixed-up or partial sharing of both heating and cooling through the same sidewalls, *Nu* of the top wall is processed to assess heat transfer rate, which is indicated below by the heatline plots in Fig. 24.3a, b, and in Fig. 24.4. The variations of *Nu* with *Ra* in both clear and porous domains are nominal, and it is also not expected particularly when $Ra = 10^5$. The Rayleigh number could be representative of buoyancy force which increases proportionally with *Ra* value. It, in turn, is expected to cause more convective heat transfer at higher *Ra*. Flow with $Ra \leq 10^4$ is basically governed by thermal conduction; hence, the change is minor. However, for $Ra = 10^5$ thermal convection becomes apparent, which is reflected by enhanced maximum stream function (200 times greater compared to that with $Ra = 10^3$). But, no significant change in heat transfer is observed. Enhanced circulation means higher convective velocity that sag the isotherm contours along the mid-vertical plane, particularly in clear domain (Fig. 24.3a). As the porous medium offers resistance to the flow, substantial reduction in maximum stream function is observed. The isotherms (being dictated by the cold top wall) become horizontal for major portion of the cavity (Fig. 24.3b).

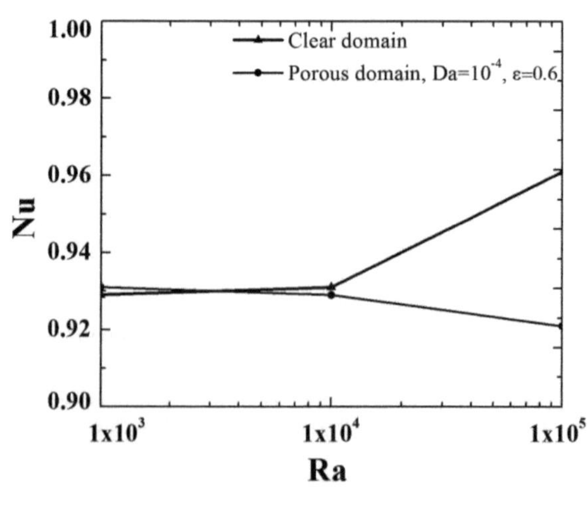

Fig. 24.4 Effect of *Ra* at $Ha = 30$, $\gamma = 0°$

24.6.2 *Effect of Magnetic Fields*

Buoyancy-induced flow inside the cavity is affected by magnetic field. Its effect at $Ra = 10^5$ is demonstrated in Figs. 24.5 and 24.6 for various *Ha* values and inclination angles, respectively. In Fig. 24.5a, the curly isotherms tend to straighten up more as *Ha* increases. The fluid circulation becomes weaker with *Ha* for clear domain, but for porous domain, no such impact is found in Fig. 24.5b. The stream function plots are more or less identical in both clear and porous domains with the maximum stream function value decreasing with increase in Hartmann value. For $Ha = 10$, the high stream function reflects the presence of thermal aspect with minimal magnetic effect, and corresponding flow physics resembling the flow phenomenon as without magnetic field. Hartmann number imposes a damping force on the flow fields, which retards circulation velocity, in turn, reduces convective heat transfer (*Nu*). However, contrary to the usual notion, Nusselt number increases (though very minimal rise) for both clear and porous domain.

With increase in Hartmann number for clear domain, the energy circulation reduces as observed from the heatline plots in Fig. 24.5a. The highest energy circulation occurs at $Ha = 10$. It happens as the Lorentz force adversely damps fluid velocity, thus retarding the flow. It is also reflected from the decrease in the maximum stream function values. For $Ra = 10^3$ and 10^4, the circulation of energy is seen, and thus, heat transfer is more at low *Ha* value. However, for porous domain in Fig. 24.5b, the heatline plots are identical to each other.

The thermo-fluid dynamics in the cavity is influenced by the orientation of imposed magnetic field. The impact of field inclination angle on the structures of isotherms and streamlines is illustrated in Fig. 24.6a, b for both clear and porous domain. Through the inclination angle, the sharing of the magnetic force takes place between two momentum equations that in turn contribute to individual velocity components (*U* and *V*). It results in inclined and distorted isothermal curves for clear domain. The curve sags down along the inclination of magnetic field. Thus, the inclination angle decides in which direction the fluid would spread. However, for porous medium the impact is minimal. Similar to the isothermal plots in clear domain, inclination angle also affects the fluid circulation. The vortices are oriented along the direction of magnetic field. However, for porous domain, no such effect is observed. In the clear domain, heatlines get distorted in the same way of the fluid vortices by the direction of the magnetic field. It is also reflected in the energy circulation as its transport is governed by the flow vector. So as the flow vector is oriented with the magnetic field, the energy lines get adjusted accordingly in that direction. For the porous medium, flow velocity is very less, and hence, the change in heatlines is not marked.

The heat transfer rate for clear and porous domain is indicated in Fig. 24.7a considering various Hartmann numbers. The figure reflects that for clear domain, *Nu* increases up to $Ha = 20$ then decreases till $Ha = 40$, afterwards *Nu* gains marginally as *Ha* increases. Since the wall offers linear heating (with higher temperature at bottom, and at top, zero equal to top wall temperature), the higher

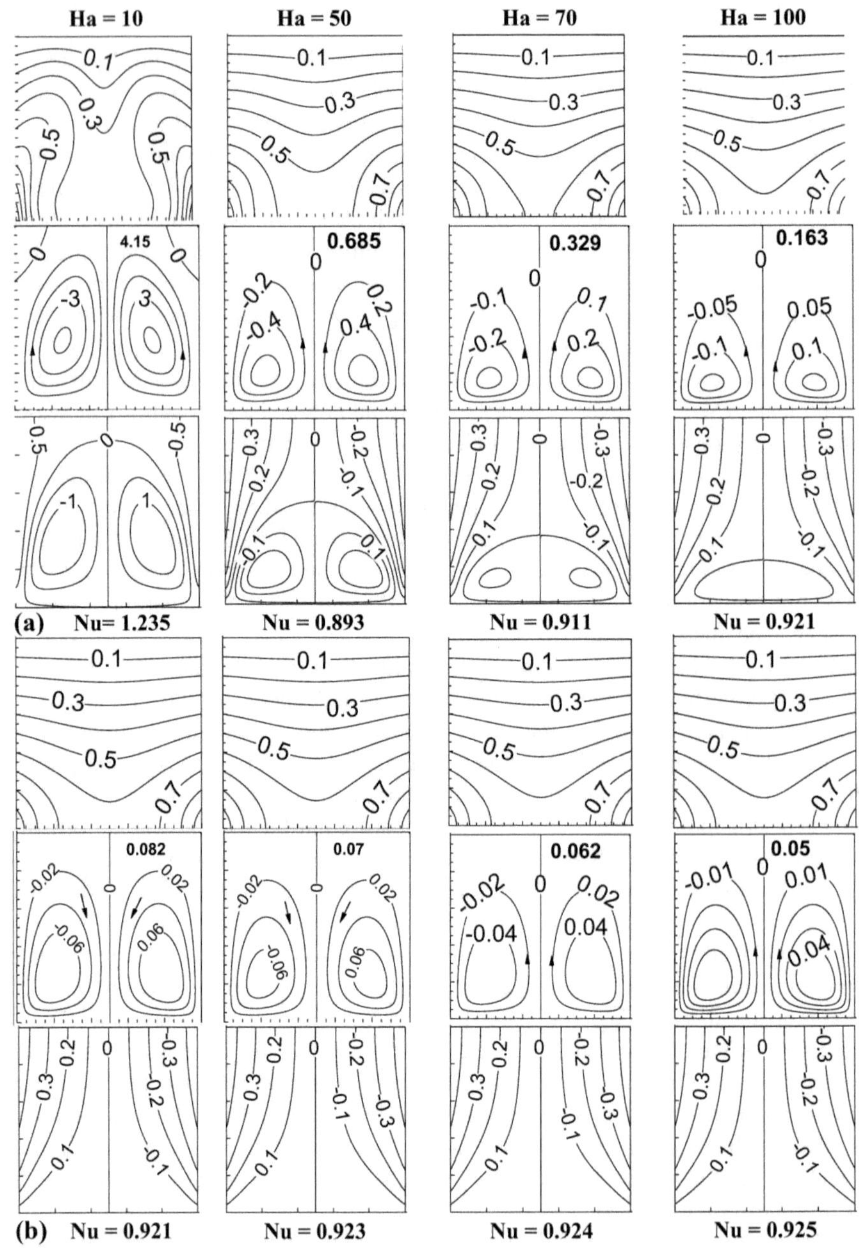

Fig. 24.5 Effect of Hartmann number (*Ha*) on thermal fields and flow fields and heatlines at $Ra = 10^5$, $\gamma = 0°$ for **a** clear domain and **b** porous domain at $Da = 10^{-4}$ and $\varepsilon = 0.6$

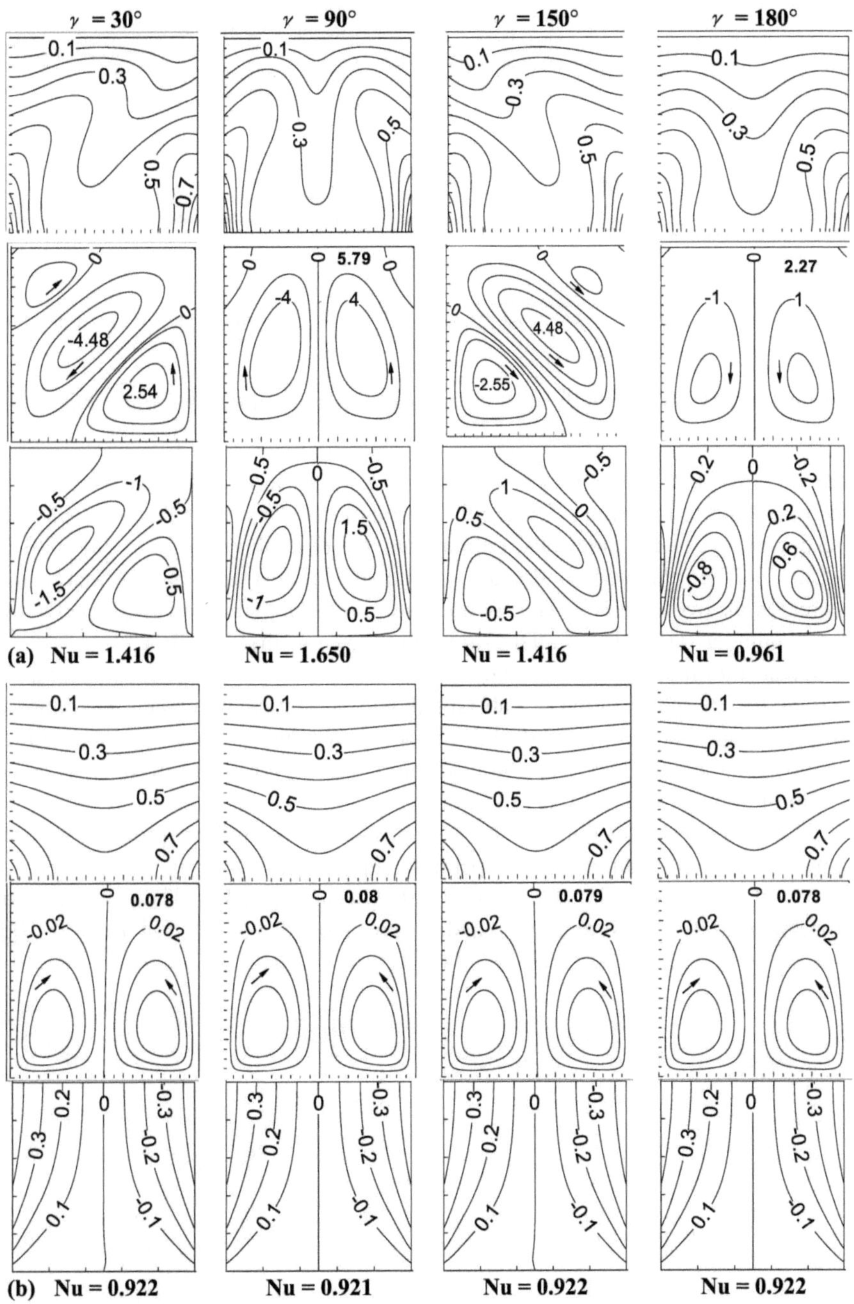

Fig. 24.6 Effect of magnetic field inclination angle (γ) on thermal fields, flow fields and heat line at $Ra = 10^5$, $Ha = 30$ for **a** clear domain and **b** porous domain at $Da = 10^{-4}$, $\varepsilon = 0.6$

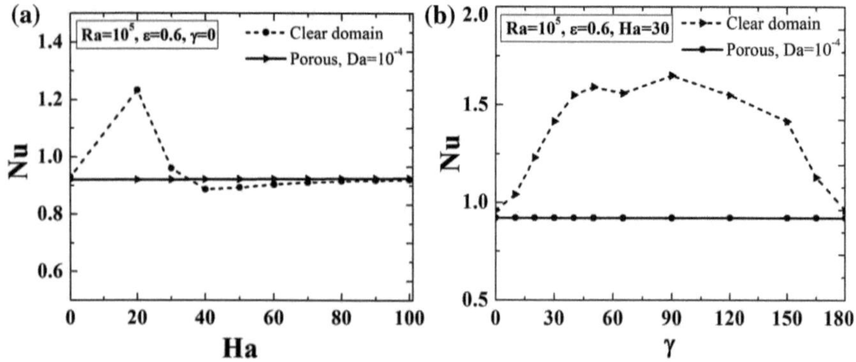

Fig. 24.7 **a** Effect of *Ha* with $Ra = 10^5$, $\gamma = 0$ and **b** effect of γ at $Ra = 10^5$, $Ha = 30$, $\gamma = 0$

temperature fluid moving up along the sidewalls, its temperature continually decreases due to rejecting all heat either through the top wall or that top wall plus the upper part of the sidewalls, culminating in some abnormality at $Ha = 20$ in Fig. 24.7a. For the porous domain, it remains almost constant or marginally decreases as *Ha* increases. This is because porous domain with moderately low Darcy number ($Da = 10^{-4}$) offers strong hindrance to the fluid flow.

When the inclination of magnetic fields has been changed as shown in Fig. 24.7b, for clear domain, substantial rise of *Nu* is observed. A continual increasing and decreasing trend of *Nu* is noted as γ increases. *Nu* is maximum at $\gamma = 90°$. On the other hand, for porous medium, almost no change in *Nu* is observed due to aforementioned flow resistance.

To explore more insight into the flow physics of the present problem in clear domain, mid-horizontal plane temperature and vertical velocity component are processed at $\gamma = 0°$ with $Ra = 10^5$ as shown in Fig. 24.8. The variations of non-dimensional temperature (θ) and *V* are plotted with *X* for $Ha = 0$–100.

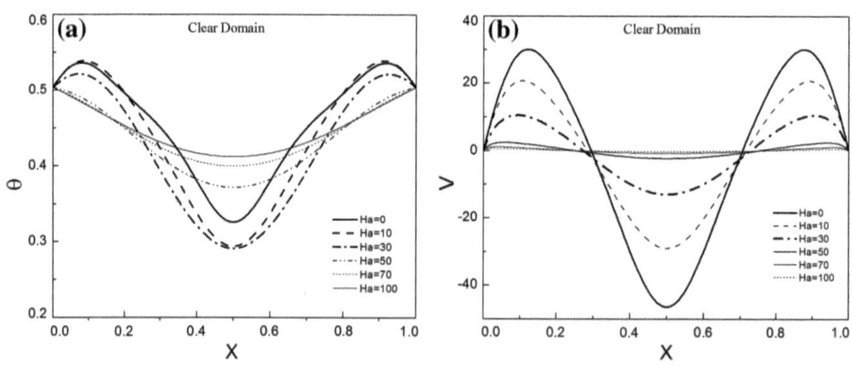

Fig. 24.8 Mid-horizontal plane temperature (**a**) and *V*-velocity profile (**b**) at $Ra = 10^5$ and different *Ha* with $\gamma = 0°$ in clear domain

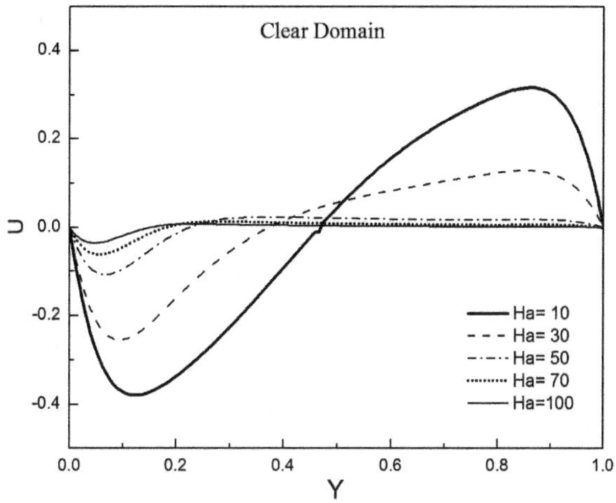

Fig. 24.9 Mid-vertical plane *U*-velocity profile at $Ra = 10^5$ and different *Ha* with $\gamma = 0°$ in clear domain

The formation of both thermal and velocity boundary layer at the sidewalls is found at the lower range of *Ha* values ($Ha \leq 30$). Hence, the curves attain local maxima near the sidewalls. But at the higher *Ha*, the flow velocity becomes very weak resulting in a continuous sagging down of temperature curve from the both sidewalls. The variances in θ and V are in compliance with the isotherm plot and the circulation pattern discussed earlier. At lower *Ha*, in the flow structure two vortices in each halves of the mid-vertical plane exist with clockwise and counterclockwise rotation, respectively. Accordingly, in the respective halves, the direction of the velocity changes. With the formation of velocity boundary layer, the velocity along the mid-vertical plane increases, resulting in a significant downward flow. The magnitude of the velocity reduces with the increase of *Ha* due to stronger damping effect at higher *Ha*.

For the completeness, the horizontal velocity component at the mid-vertical plane is also processed. Corresponding plot for $Ha = 0–100$ is shown in Fig. 24.9. The profile resembles the stream function plot for the respective *Ha* variations. Also, the enhanced damping from higher *Ha* is clearly seen from the decrease in magnitude of horizontal velocity.

24.6.3 *Effect of Porous Medium*

Influence of different resistance levels of flow medium is also explored in Figs. 24.10 and 24.11 by changing the pertinent parameters, Darcy number and porosity, respectively. Usually as Darcy number increases, the value of average Nusselt number becomes higher indicating better and more heat transfer. In contrast, for the present case, as shown in Fig. 24.10, *Nu* reduces with *Da*. When flow structures are scrutinized, the isotherms near the top cold wall is found curving

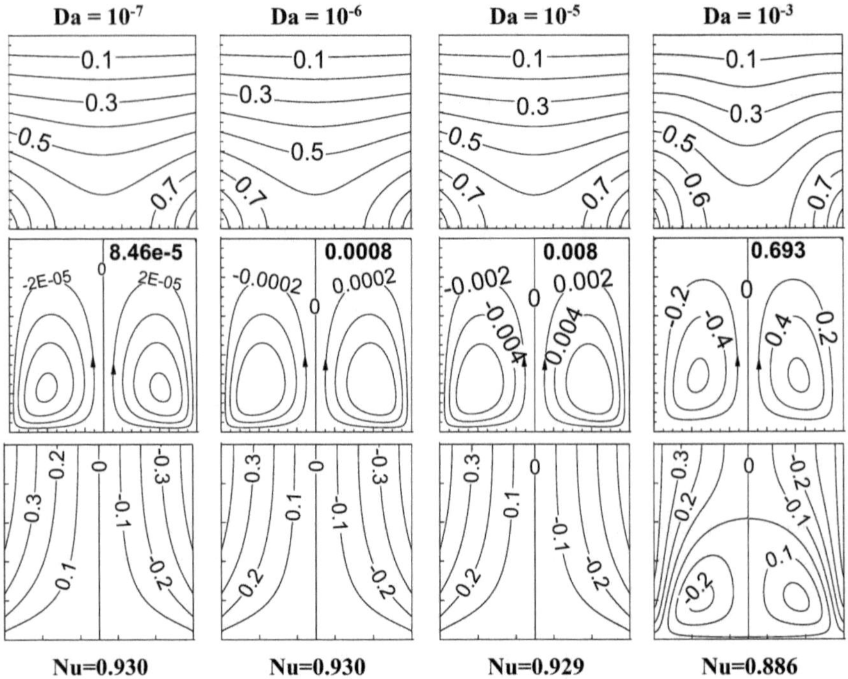

Fig. 24.10 Effect of Darcy number (*Da*) on thermal fields and flow fields at $Ra = 10^5$, $Ha = 30$, $\gamma = 0$ for porous domain at $\varepsilon = 0.6$

away more and more as *Da* increases. It indicates thickening of thermal boundary layer that results in lesser heat transfer, though the flow strength increases with higher Darcy value following the common notion. As heat transfer decreases with *Da*, corresponding *Nu* decreases. It should be mentioned here that for this investigation, porosity ε is kept at 0.6 fixed along with other fixed parameters ($Ra = 10^5$, $Ha = 30$, $\gamma = 0°$). The cooling through the top wall reduces at higher *Da* as the upper part of the linearly heated sidewalls rejects heat with simultaneous heating at their lower portion. This is the root cause behind the deviation from the usual observation. With the increase in *Da* value, flow circulation (maximum stream function) increases but overall heat transfer decreases. So, for $Da < 10^{-3}$ heatlines are predominantly conduction flux. However, the energy circulation increases with rise in *Da* value, with symmetric vortices observed at $Da = 10^{-3}$.

When porosity ε is varied from 0.1 to 1.0 in Fig. 24.11 at $Da = 10^{-4}$ and the same set of other parameters, almost no change in *Nu* is observed as porosity increases. It is more or less consistent with the existing findings. Similar results are obtained for the flow curves although a slight increase in maximum stream function

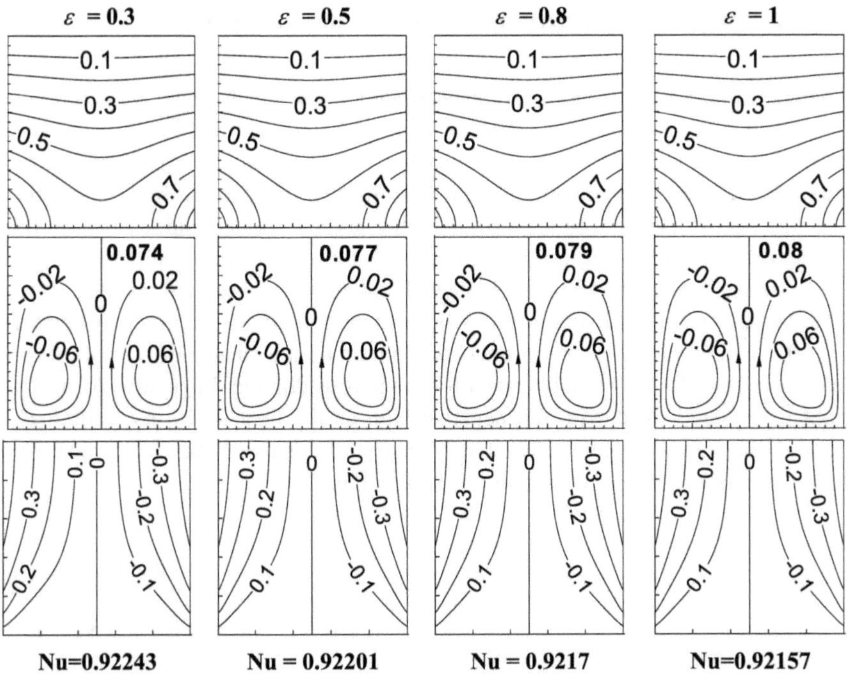

Fig. 24.11 Effect of porosity (ε) on thermal fields and flow fields and heat line plots at $Ra = 10^5$, $Ha = 30$ and $\gamma = 0$, for porous domain at $Da = 10^{-4}$

value is noted. The heatlines follow the same notion with minimal change, with the heat flowing symmetrically from sidewalls and terminating normally at the cold top wall.

The above observations are clearly reflected in the trend curve of *Nu* as shown in Fig. 24.12.

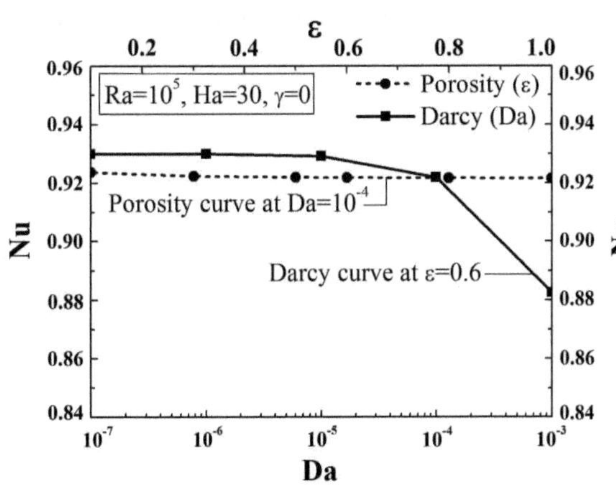

Fig. 24.12 Effect of *Da* and ε on average Nusselt number (*Nu*) with $Ra = 10^5$, $Ha = 30$ and $\gamma = 0$

24.6.4 *Impact of Hartmann Number on Local* Nu

To understand heat flow along the linear heated sidewalls, the local Nusselt number is computed for $Ra = 10^5$ as shown for the clear domain in Fig. 24.13a and the porous domain in Fig. 24.13b. The local *Nu* with various *Ha* at the left and right sidewalls indicates a symmetrical heat flow pattern through the both sidewalls in both the flow domains. The local *Nu* is higher at the bottom of sidewalls ($Y = 0$) as it interacts the circulating fluid first with maximum temperature difference. Local *Nu* becomes zero at the top of the sidewalls as there imposed temperature is zero. The interesting point of simultaneous heating and cooling is observed at lower *Ha* (≤ 30, as mentioned earlier) for the clear domain by the change of sign of local *Nu* in either sidewalls. It is totally absent for the porous domain. In general, local *Nu* of right wall is indicated with the negative sign as the heat flows inside the cavity through the right sidewall (in negative direction of positive *X*-axis). The local *Nu* is found insensitive to any change in *ha* value (Fig. 24.13b).

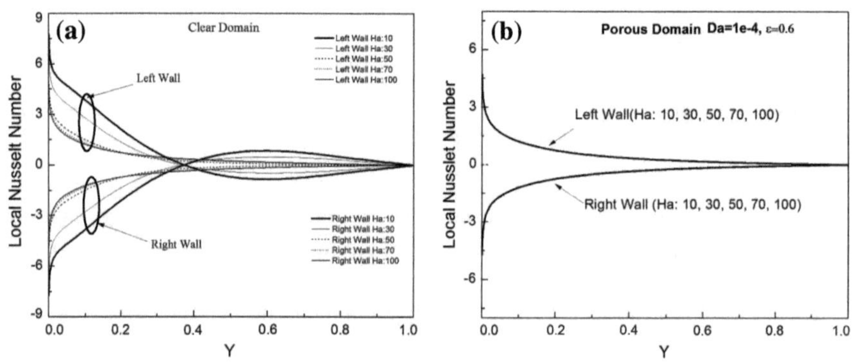

Fig. 24.13 Impact of Hartmann number on the distribution of local Nusselt number along the both sidewalls at $Ra = 10^5$ for clear domain (**a**) and porous domain (**b**)

24.7 Conclusions

The impact of magnetic fields and its orientation with respect to the cavity base on the thermal-fluid flow cavity undergoing natural convection is analysed applying linear heating at both the sidewalls, keeping bottom and top walls, respectively, adiabatic and cold. The major observations are concluded as follows:

- Increased Nusselt number is noted with higher *Ra* for clear domain, which is consistent with the existing findings and theoretical formulation.
- With increase in Hartmann number (*Ha*), *Nu* does not decrease monotonously. It is because the sidewalls act as both heater and absorber of heat, simultaneously.
- The strong effect of *Ha* is revealed with the change of inclination angle of magnetic fields. It is observed at an angle of $\gamma = 50°$ *Nu* minimum, and at $\gamma = 90°$ *Nu* maximum.
- For porous domain, the effect of increase in Rayleigh number strongly depends on permeability or Darcy number of porous substance. *Nu* trends are flat at lower *Da* and of similar pattern as found for clear domain but reduced value at higher *Da*.
- *Nu* reduces with enhanced *Da* value, contrary to the usual notion. The increase in thermal boundary leads to less heat transfer rate despite strong flow strength.

References

1. Rashidi, S., Esfahani, J.A., Maskaniyan, M.: Applications of magneto-hydrodynamics in biological systems-a review on the numerical studies. J. Magn. Magn. Mater. **439**, 358–372 (2017)
2. Sivaraja, C., Sheremet, M.A.: MHD natural convection in an inclined square porous cavity with a heat conducting solid block. J. Magn. Magn. Mater. **426**, 351–360 (2017)
3. Nayak, A.K., Malik, S., Venkateshwarlu, K., Jena, P.K.: Magneto-convection and its effect on partially active thermal zones in a porous square domain. Int. J. Heat Mass Transf. **95**, 913–926 (2016)
4. Sivasankaran, S., Ho, C.J.: Effect of temperature dependent properties on MHD convection of water near its density maximum in a square cavity. Int. J. Thermal Sci. **47**, 1184–1194 (2008)
5. Yu, P.X., Qiu, J.X., Qin, Q., Tian, Z.F.: Numerical investigation of natural convection in a rectangular cavity under different directions of uniform magnetic field. Int. J. Heat Mass Transfer **67**, 1131–1144 (2013)
6. Sathiyamoorthy, M., Chamkha, A.: Effect of magnetic field on natural convection flow in a liquid gallium filled square cavity for linearly heated side wall(s). Int. J. Thermal Sci. **49**(9), 1856–1865 (2010)
7. Ghasemi, B., Aminossadati, S.M., Raisi, A.: Magnetic field effect on natural convection in a nanofluid-filled square enclosure. Int. J. Thermal Sci. **50**(9), 1748–1756 (2011)
8. Mliki, B., Abbassi, M.A., Omri, A., Zeghmati, B.: Augmentation of natural convective heat transfer in linearly heated cavity by utilizing Nanofluid in the presence of magnetic field and uniform heat generation/absorption. Powder Tech. **284**, 312–325 (2015)

9. Mahmoudi, A.H., Pop, I., Shahi, M., Talebi, F.: MHD natural convection and entropy generation in a trapezoidal enclosure using Cu–water nanofluid, Computer. Fluids **72**, 46–62 (2013)
10. Mansour, M.A., Bakier, M.A.Y.: Influence of thermal boundary conditions on MHD natural convection in square enclosure using Cu–water nanofluid. Energy Reports **1**, 134–144 (2015)
11. Kefayati, G.H.R.: FDLBM simulation of magnetic field effect on natural convection of non-Newtonian power-law fluids in a linearly heated cavity. Powder Tech. **256**, 87–99 (2014)
12. Kefayati, G.H.R.: Natural convection of ferrofluid in a linearly heated cavity utilizing LBM. J. Mol. Liq. **191**, 1–9 (2014)
13. Sathiyamoorthy, M., Basak, T., Roy, S., Pop, I.: Steady natural convection flows in a square cavity with linearly heated side wall(s). Int. J. Heat Mass Transf. **50**, 766–775 (2007)
14. Rashad, A.M., Armaghani, T., Chamkha, A.J., Mansoure, M.A.: Entropy generation and MHD natural convection of a nanofluid in an inclined square porous cavity: effects of a heat sink and source size and location, Chinese. J. Phys. **56**, 193–211 (2018)
15. Emami, R.Y., Siavashi, M., Moghaddam, G.S.: The effect of inclination angle and hot wall configuration on Cu-water nanofluid natural convection inside a porous square cavity. Adv. Powder Technol. **29**, 519–536 (2018)
16. Biswas, N., Manna, N.K., Mahapatra, P.S.: Merit of non-uniform over uniform heating in a porous cavity. Int. Commun. Heat Mass Transf. **78**, 135–144 (2016)
17. Nemati, H., Farhadi, M., Sedighi, K., Ashorynejad, H.R., Fattahi, E.: Magnetic field effects on natural convection flow of nanofluid in a rectangular cavity using the Lattice Boltzmann model. Sci. Iranica B **19**, 303–310 (2012)
18. Ghasemi, K., Siavashi, M.: Lattice Boltzmann numerical simulation and entropy generation analysis of natural convection of nanofluid in a porous cavity with different linear temperature distributions on side walls. J. Mol. Liq. **233**, 415–430 (2017)
19. Patankar, S.V.: Numerical Heat Transfer and Fluid Flow. Hemisphere, New York (1980)

Chapter 25
Effect of Turbulence on Stability of Journal Bearing with Micropolar Lubrication: Linear and Non-linear Analysis

Subrata Das and Sisir K. Guha

Nomenclature

A_θ	Constant parameter of turbulent shear coefficient for circumferential flow (dimensionless)
B_θ	Exponential constant parameter of turbulent shear coefficient for circumferential flow (dimensionless)
C	Radial clearance (m)
$C_{\bar{z}}$	Constant parameter of turbulent shear coefficient for axial flow (dimensionless)
D	Journal diameter, (m)
D_{ij}	Damping coefficients of micropolar fluid film, $i = r, \phi$ and $j = r, \phi$ (N s/m)
$\overline{D}_{ij}$	Dimensionless damping coefficients of micropolar fluid film, $\overline{D}_{ij} = \frac{2D_{ij}C^3}{\mu\Omega R^3 L}$, $i = r, \phi$ and $j = r, \phi$ (dimensionless)
$D_{\bar{z}}$	Exponential constant parameter of turbulent shear coefficient for axial flow (dimensionless)
F_i	Fluid film force components along r- and ϕ-directions, $i = r$ and ϕ (N)
$\overline{F}_i$	Non-dimensional force components along r- and ϕ-directions, $\bar{F}_i = F_i C^2 / \left(\mu\Omega^2 R^3 L\right)$, $i = r$ and φ (dimensionless)
h	Local film thickness (m)
$\overline{h}$	Non-dimensional film thickness, $\overline{h} = h/C$ (dimensionless)
$k_\theta, k_{\bar{z}}$	Turbulent shear coefficient along circumferential direction and axial directions, respectively, (dimensionless)
L	Length of bearing (m)
l_{m}	Non-dimensional characteristics length of micropolar fluid (dimensionless)
M	Rotor mass per bearing (kg)

S. Das (✉) · S. K. Guha
Department of Mechanical Engineering, Indian Institute of Engineering Science and Technology, Shibpur, Howrah, India
e-mail: mechsubrata@gmail.com

P. Sahoo and J. P. Davim (eds.), *Advances in Materials, Mechanical and Industrial Engineering*, Lecture Notes on Multidisciplinary Industrial Engineering,
https://doi.org/10.1007/978-3-319-96968-8_25

$\overline{M}$ Non-dimensional mass parameter, $\overline{M} = MC\Omega^2/W_0$ (dimensionless)
$\overline{M}_{cr}$ Critical value of non-dimensional mass parameter (dimensionless)
N Coupling number (dimensionless)
p Local micropolar film pressure in the film region, (N/m^2)
$\overline{p}$ Non-dimensional local micropolar film pressure in the film region, $\overline{p} = pC^2/(\mu\Omega R^2)$ (dimensionless)
P_i Local micropolar film pressure in the film region, i = 0, 1 and 2 for the steady-state and first-order perturbed film pressures along r- and ϕ-directions, (N/m^2)
$\overline{p}_i$ Local micropolar film pressure in the film region, $\overline{p}_i = \frac{p_i C^2}{\mu\Omega R^2}$, i = 0, 1 and 2 for the steady-state and first-order perturbed film pressures along r- and ϕ-directions (dimensionless)
R Radius of the journal (m)
Re Mean or average Reynolds number defined by radial clearance, C, $\mathrm{Re} = \rho\Omega RC/\mu$ (dimensionless)
S_{ij} Stiffness coefficients of micropolar fluid film, $i = r, \phi$ and $j = r, \phi$ (N/m)
$\overline{S}_{ij}$ Stiffness damping coefficients of micropolar fluid film, $\overline{S}_{ij} = \frac{2S_{ij}C^3}{\mu\Omega R^3 L}$, $i = r, \phi$ and $j = r, \phi$ (dimensionless)
t Time (s)
U U velocity of journal, $U = \Omega R$, (m/s)
W Load in bearing, (N)
W_0 Steady-state load in bearing (N)
$\overline{W}_0$ Non-dimensional steady-state load in bearing, $\overline{W}_0 = W_0/\left(\mu\Omega^2 R^3 L\right)$ (dimensionless)
x Cartesian coordinate axis in the circumferential direction, $x = R\theta$, (m)
z Cartesian coordinate axis along the bearing axis (m)
$\overline{z}$ Non-dimensional Cartesian coordinate axis along the bearing axis, $\overline{z} = 2z/L$ (dimensionless)

Greek Symbols

ε Eccentricity ratio (dimensionless)
ε_0 Steady-state eccentricity ratio (dimensionless)
ε_1 Perturbed eccentricity ratio (dimensionless)
λ Whirl ratio, $\lambda = \omega_p/\Omega$ (dimensionless)
ϕ Attitude angle (rad)
ϕ_0 Steady-state attitude angle, (rad)
$\Phi_{x,z}$ Micropolar fluid functions along circumferential and axial directions
$\Phi_{\theta,\overline{z}}$ Non-dimensional micropolar fluid functions along circumferential and axial directions (dimensionless)
Λ Characteristics length of the micropolar fluid, $\Lambda = (\gamma/4\mu)^{1/2}$
γ Viscosity coefficient of micropolar fluid, (N s)
μ Newtonian viscosity coefficient, (Pa s)

ω_p Angular velocity of the orbital motion of the journal centre, (rad/s)
Ω Angular velocity of journal (rad/s)
ω_p Angular velocity of the orbital motion of the journal centre (rad/s)
θ Circumferential coordinate (rad)
θ_c Circumferential coordinate where the film cavitates, (rad)
τ Non-dimensional time, $\tau = \Omega t$ (dimensionless)

25.1 Introduction

The problem of turbulent flow in bearing is very much relevant in modern-day industry because of the high speed of rotating machines and application of low viscosity lubricants. There are many research work carried in past decades addressing the problem of turbulence in journal bearings. The effect of turbulence in high-speed bearing operations was first considered by Wilcock [1]. In his work, he has made an attempt to develop a reasonable engineering theory for turbulent lubrication. Constantinescu [2–4] has worked on the theory of turbulent lubrication in detail. In his work, Prandtl's mixing length concept was applied to derive the modified Reynolds equation applicable to turbulent hydrodynamic lubrication. Ng and Pan [5] and Elrod and Ng [6] used the concept of Reichardt's eddy diffusivity for analysis of bearings operating in turbulent regime. After rigorous study on different turbulent models applied to bearings, Taylor and Dawson [7] suggested the application of the turbulent model proposed by Ng and Pan. Soni et al. [8] carried out non-linear transient analysis of finite journal bearings operating in laminar and turbulent flow regimes. Zhang et al. [9] presented a brief review on different turbulent models that are applied in the analysis of different hydrodynamic lubrication problems.

Most of the research works incorporating turbulence in bearings considered so far has assumed the lubricant as Newtonian fluid. But practically, the lubricants used in the industry are often blended with additives to improve different lubricant properties. After long use in industry the lubricants also get contaminated with worn out particles. Under such conditions, these lubricants do not behave as Newtonian fluid and the predictions of different performance characteristics give erroneous results if conventional continuum theorem is applied in the analysis. Eringen [10] proposed micropolar fluid theory that can be used to analyse performance of bearings lubricated by such non-Newtonian fluid without serious error. According to this theory, the microstructures present in the lubricant have linear as well as micro-rotational motion while the fluid flows through narrow gap, e.g. the lubricant flow through the annular space between the journal and bearing. In such cases, additional viscosity coefficient due the micro-rotation of the microstructure has to be incorporated in the governing equation. Allen and Kline [11] applied the theory of micropolar fluid in bearing lubrication. Prakash and Sinha [12] carried out a theoretical study on the application of the micropolar fluid to a journal bearing considering a two-dimensional incompressible steady laminar flow between two eccentric cylinders in relative rotary motion. Khonsari and Brewe [13] studied the lubricating effectiveness of micropolar fluid for a finite journal bearing.

The analysis of bearings with turbulent flow of micropolar lubricant has only started in the current century when Faralli and Belfiore [14] carried out steady-state analysis of worn spherical bearings. Rana et al. [15] conducted dynamic analysis on short journal bearings in micropolar lubrication with turbulent effect. Ghosh and Guha [16] applied couple stress fluid model in the investigation of linear stability characteristics of journal bearings with turbulent flow of lubricant. Das and Guha [17] carried out linear dynamic analysis to find the effect of turbulence on the dynamic response coefficients of journal bearings of finite width. In continuation to the previous work, Das and Guha [18] carried out the linear stability analysis of journal bearings under turbulent micropolar lubrication. Theoretical analysis of stability characteristics of a micropolar fluid lubricated journal bearing has been carried out by Das et al. [19]. Both linear and non-linear methods have been demonstrated in the analysis. None of the works available in the literature so far that presents non-linear analysis of journal bearings operating with turbulent flow of non-Newtonian lubricant. Hence in the present analysis, an attempt has been made to study the effect of turbulence on the stability of journal bearing under micropolar lubrication following the non-linear technic. A comparative study of the stability characteristics obtained from linear and non-linear study also has been presented.

25.2 Analysis

25.2.1 Governing Equation

A schematic diagram of a hydrodynamic journal bearing with the circumferential coordinate system used in the analysis is shown in Fig. 25.1. The journal rotates with a rotational speed, Ω rad/s about its axis. With the usual assumptions considered for the thin micropolar lubrication film and the assumptions of the absence of the body and inertia forces and body couples, the modified Reynolds equation [14–18] for two-dimensional flow of micropolar lubricant with turbulent effect is written as follows:

$$\frac{\partial}{\partial x}\left[\Phi_x(h,\Lambda,N)\frac{\partial p}{\partial x}\right]+\frac{\partial}{\partial z}\left[\Phi_z(h,\Lambda,N)\frac{\partial p}{\partial z}\right]=\frac{1}{2}\mu U\frac{\partial h}{\partial x}+\mu\frac{\partial h}{\partial t} \tag{25.1}$$

where

$$\Phi_{x,z}(h,\Lambda,N)=\frac{h^3}{k_{x,z}}+\Lambda^2 h-\frac{N\Lambda h^2}{2}\coth\left(\frac{Nh}{2\Lambda}\right)$$

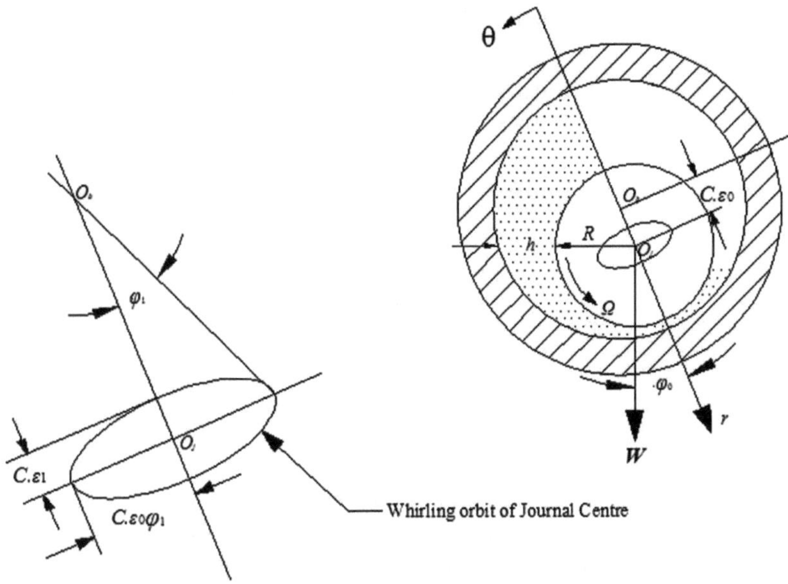

Fig. 25.1 Configuration of the journal bearing showing the whirling orbit of the journal centre

$$N = \left(\frac{\chi}{2\mu + \chi}\right)^{\frac{1}{2}}, \Lambda = \left(\frac{\gamma}{4\mu}\right)^{\frac{1}{2}}, \mu_v = \mu + \frac{1}{2}\chi$$

Here, μ_v is the viscosity of the base fluid, μ is the viscosity of the Newtonian fluid, χ is the spin viscosity, γ is the material coefficient, N is the coupling number, Λ is the characteristics length of the micropolar fluid. h is the film thickness, p is the micropolar film pressure in the bearing and $k_{x,\ z}$ are turbulent constants the value of which is given in Ref. [7]. The coupling number and the micropolar characteristic length are known as micropolar parameters as they signify the presence of microstructure and the interaction of these microstructures with the film gap between the journal and the bearing. The function $\Phi_{x,z}(h, \Lambda, N)$ is known as micropolar fluid function. This function takes into account all the parameters (geometrical, micropolar and flow parameters) that affect the performance of the journal bearing system.

Among the available theories of turbulent lubrication, the theories proposed by Constantinescu [2–4] and Ng, Pan and Elrod [5, 6] are most popular. It is interesting to note that both the theories have resulted in the same form of Reynolds equation for turbulent lubrication.

As per Constantinescu's approach, the derivation of the modified Reynolds equation is based on the concept of Prandtl's mixing length which is a function of the local Reynolds number, Re_h. By adopting the methods of linearization, the following empirical relations for the turbulent coefficients, k_x and k_z are obtained:

Table 25.1 Values of the turbulent shear constants as proposed by Ng and Pan [5]

Reynolds number	A_x	B_x	C_z	D_z
$Re < 5000$	0.0039	1.06	0.0021	1.06
$5000 \leq Re < 10{,}000$	0.0250	0.84	0.0088	0.88
$10{,}000 \leq Re < 50{,}000$	0.0250	0.84	0.0136	0.84
$Re > 50{,}000$	0.0388	0.8	0.0213	0.8

$$k_x = \frac{1}{G_x} = 12 + 0.0260(Re_h)^{0.8265} \tag{25.2a}$$

$$k_z = \frac{1}{G_z} = 12 + 0.0198(Re_h)^{0.741} \tag{25.2b}$$

In a different approach, the derivation of the modified Reynolds equation, Ng, Pan and Elrod [5, 6] used the concept of Reichard's eddy diffusivity to represent the turbulent shearing stress, in terms of the mean velocity gradient. In this approach, both the shearing stresses under the laminar flow condition and that due to the eddy viscosity under the turbulent flow condition contribute to the resultant shearing stress. The expressions for turbulent shear coefficients are given by.

$$k_x = 12 + A_x(Re_h)^{B_x} \tag{25.3a}$$

$$k_z = 12 + C_z(Re_h)^{D_z} \tag{25.3b}$$

Here, A_x, B_x, C_z and D_z are turbulent shear constants and their values, as proposed by Ng and Pan [5] are given in Table 25.1.

Introducing a rotating coordinate system for a journal whose centre is rotating with an angular velocity of ω_p the modified Reynolds equation as in (25.1) can be written as

$$\frac{\partial}{\partial x}\left[\Phi_x(h,\Lambda,N)\frac{\partial p}{\partial x}\right] + \frac{\partial}{\partial z}\left[\Phi_z(h,\Lambda,N)\frac{\partial p}{\partial z}\right] = \frac{1}{2}\mu U\frac{\partial h}{\partial x} + \mu\frac{\partial h}{\partial t} - \mu\omega_p\frac{\partial h}{\partial \theta} \tag{25.4}$$

Using the following substitutions

$$\theta = \frac{x}{R};\quad \bar{z} = \frac{z}{L/2};\quad \bar{h} = \frac{h}{C};\quad \bar{p} = \frac{pC^2}{\mu\Omega R^2};\quad l_m = \frac{C}{\Lambda};\quad \tau = \Omega t.$$

Equation (25.4) can be normalized as follows:

$$\frac{\partial}{\partial\theta}\left[\Phi_\theta\left(\overline{h}, l_{\mathrm{m}}, N\right)\frac{\partial\overline{p}}{\partial\theta}\right] + \left(\frac{D}{L}\right)^2\frac{\partial}{\partial\overline{z}}\left[\Phi_{\overline{z}}\left(\overline{h}, l_{\mathrm{m}}, N\right)\frac{\partial\overline{p}}{\partial\overline{z}}\right] = \frac{1}{2}(1-2\phi')\frac{\partial\overline{h}}{\partial\theta} + \frac{\partial\overline{h}}{\partial\tau} \tag{25.5}$$

where

$$\Phi_{\theta,\overline{z}}\left(\overline{h}, l_{\mathrm{m}}, N\right) = \frac{\overline{h}^3}{k_{\theta,\overline{z}}} + \frac{\overline{h}}{l_{\mathrm{m}}^2} - \frac{1}{2}\frac{\overline{h}^2 N}{l_{\mathrm{m}}}\coth h\left(\frac{\overline{h}\cdot l_{\mathrm{m}}\cdot N}{2}\right);$$
$$k_\theta = 12 + A_\theta\left(Re\overline{h}\right)^{B_\theta} \text{ and } k_{\overline{z}} = 12 + C_{\overline{z}}\left(Re\overline{h}\right)^{D_{\overline{z}}}$$

and

$$\phi' = \frac{\partial\phi}{\partial\tau}$$

25.2.2 *Theoretical Analysis of Linear Stability Characteristics*

25.2.2.1 Perturbed Equations

It is assumed that the journal undergoes a whirling motion in an elliptical orbit about its mean steady-state position (ε_0, ϕ_0) with amplitudes $Re\left(\varepsilon_1 e^{i\lambda\tau}\right)$ and $Re\left(\varepsilon_0\phi_1 e^{i\lambda\tau}\right)$ along the line of centres and perpendicular to the line of centres, respectively. For the first-order perturbation, the non-dimensional pressure and film thickness can be expressed as

$$\left.\begin{aligned} p &= \overline{p}_0 + \overline{p}_1\varepsilon_1 e^{i\lambda\tau} + \overline{p}_2\varepsilon_0\phi_1 e^{i\lambda\tau} \\ h &= \overline{h}_0 + \varepsilon_1 e^{i\lambda\tau}\cos\theta + \varepsilon_0\phi_1 e^{i\lambda\tau}\sin\theta \\ k_\theta &= k_{\theta 0} + k_{\theta 1}\varepsilon_1 e^{i\lambda\tau} + k_{\theta 2}\varepsilon_0\phi_1 e^{i\lambda\tau} \\ k_{\overline{z}} &= k_{\overline{z}0} + k_{\overline{z}1}\varepsilon_1 e^{i\lambda\tau} + k_{\overline{z}2}\varepsilon_0\phi_1 e^{i\lambda\tau} \end{aligned}\right\} \tag{25.6}$$

where

$$\overline{h}_0 = 1 + \varepsilon_0\cos\theta; \quad \varepsilon = \varepsilon_0 + \varepsilon_1 e^{i\lambda\tau}; \quad \phi = \phi_0 + \phi_1 e^{i\lambda\tau} \quad \text{and} \quad \lambda = \omega/\Omega. \tag{25.7a}$$

$$k_{\theta 0} = 12 + a_\theta\left(Re\overline{h}_0\right)^{b_\theta}; \quad k_{\overline{z}0} = 12 + c_{\overline{z}}\left(Re\overline{h}_0\right)^{d_{\overline{z}}} \tag{25.7b}$$

$$k_{\theta 1}=\frac{a_\theta b_\theta\cos\theta\left(Re\overline{h}_0\right)^{b_\theta}}{\overline{h}_0};\quad k_{\theta 2}=\frac{a_\theta b_\theta\sin\theta\left(Re\overline{h}_0\right)^{b_\theta}}{\overline{h}_0} \tag{25.7c}$$

$$k_{\overline{z}1}=\frac{c_{\overline{z}}d_{\overline{z}}\cos\theta\left(Re\overline{h}_0\right)^{d_{\overline{z}}}}{\overline{h}_0};\quad k_{\overline{z}2}=\frac{c_{\overline{z}}d_{\overline{z}}\sin\theta\left(Re\overline{h}_0\right)^{d_{\overline{z}}}}{\overline{h}_0}. \tag{25.7d}$$

Substituting expressions (25.6) into Eq. (25.5) and collecting the zeroth- and first-order terms of ε_1 and $\varepsilon_0\phi_1$ gives the following set of equations in $\overline{p}_0, \overline{p}_1$ and $\overline{p}_2$:

$$\frac{\partial}{\partial\theta}\left[\Phi_{\theta 0}\left(\overline{h}_0,l_{\mathrm{m}},N\right)\frac{\partial\overline{p}_0}{\partial\theta}\right]+\left(\frac{D}{L}\right)^2\frac{\partial}{\partial\overline{z}}\left[\Phi_{\overline{z}0}\left(\overline{h}_0,l_{\mathrm{m}},N\right)\frac{\partial\overline{p}_0}{\partial\overline{z}}\right]=\frac{1}{2}\frac{\partial\overline{h}_0}{\partial\theta} \tag{25.8}$$

$$\begin{aligned}\frac{\partial}{\partial\theta}\left[\Phi_{\theta 0}\left(\overline{h}_0,l_{\mathrm{m}},N\right)\frac{\partial\overline{p}_1}{\partial\theta}\right]+\left(\frac{D}{L}\right)^2\Phi_{\overline{z}0}\left(\overline{h}_0,l_{\mathrm{m}},N\right)\frac{\partial^2\overline{p}_1}{\partial\overline{z}^2}+\frac{\partial}{\partial\theta}\left[\Phi'_{\theta 0}\left(\overline{h}_0,l_{\mathrm{m}},N\right)\cos\theta\frac{\partial\overline{p}_0}{\partial\theta}\right]\\+\left(\frac{D}{L}\right)^2\Phi'_{\overline{z}0}\left(\overline{h}_0,l_{\mathrm{m}},N\right)\cos\theta\frac{\partial^2\overline{p}_0}{\partial\overline{z}^2}=-\frac{1}{2}\sin\theta+i\lambda\cos\theta\end{aligned} \tag{25.9}$$

$$\begin{aligned}\frac{\partial}{\partial\theta}\left[\Phi_{\theta 0}\left(\overline{h}_0,l_{\mathrm{m}},N\right)\frac{\partial\overline{p}_2}{\partial\theta}\right]+\left(\frac{D}{L}\right)^2\Phi_{\overline{z}0}\left(\overline{h}_0,l_{\mathrm{m}},N\right)\frac{\partial^2\overline{p}_2}{\partial\overline{z}^2}+\frac{\partial}{\partial\theta}\left[\Phi'_{\theta 0}\left(\overline{h}_0,l_{\mathrm{m}},N\right)\sin\theta\frac{\partial\overline{p}_0}{\partial\theta}\right]\\+\left(\frac{D}{L}\right)^2\Phi'_{\overline{z}0}\left(\overline{h}_0,l_{\mathrm{m}},N\right)\sin\theta\frac{\partial^2\overline{p}_0}{\partial\overline{z}^2}=\frac{1}{2}\cos\theta+i\lambda\left(\sin\theta-\frac{1}{\varepsilon_0}\frac{\partial\overline{h}_0}{\partial\theta}\right)\end{aligned} \tag{25.10}$$

$$\Phi_{\theta 0}\left(\overline{h}_0,l_{\mathrm{m}},N\right)=\frac{\overline{h}_0^3}{k_{\theta 0}}+\frac{\overline{h}_0}{l_{\mathrm{m}}^2}-\frac{1}{2}\frac{\overline{h}_0^2N}{l_{\mathrm{m}}}\coth\left(\frac{\overline{h}_0\cdot l_{\mathrm{m}}\cdot N}{2}\right) \tag{25.11a}$$

$$\Phi_{\overline{z}0}\left(\overline{h}_0,l_{\mathrm{m}},N\right)=\frac{\overline{h}_0^3}{k_{\overline{z}0}}+\frac{\overline{h}_0}{l_{\mathrm{m}}^2}-\frac{1}{2}\frac{\overline{h}_0^2N}{l_{\mathrm{m}}}\coth\left(\frac{\overline{h}_0\cdot l_{\mathrm{m}}\cdot N}{2}\right) \tag{25.11b}$$

$$\begin{aligned}\Phi'_{\theta 0}\left(\overline{h}_0,l_{\mathrm{m}},N\right)=\frac{\overline{h}_0^2}{k_{\theta 0}}\left\{3-\frac{a_\theta b_\theta\left(Re\overline{h}_0\right)^{b_\theta}}{k_{\theta 0}}\right\}+\frac{1}{l_{\mathrm{m}}^2}-\frac{\overline{h}_0N}{l_{\mathrm{m}}}\cot h\left(\frac{\overline{h}_0\cdot l_{\mathrm{m}}\cdot N}{2}\right)\\+\frac{N^2\overline{h}_0^2}{4}\mathrm{cosech}^2\left(\frac{\overline{h}_0\cdot l_{\mathrm{m}}\cdot N}{2}\right)\end{aligned} \tag{25.11c}$$

$$\Phi'_{\bar{z}0}\left(\bar{h}_0, l_{\mathrm{m}}, N\right) = \frac{\bar{h}_0^{-2}}{k_{\bar{z}0}}\left\{3 - \frac{c_{\bar{z}}d_{\bar{z}}\left(Re\bar{h}_0\right)^{d_{\bar{z}}}}{k_{\bar{z}0}}\right\} + \frac{1}{l_{\mathrm{m}}^2} - \frac{\bar{h}_0 N}{l_{\mathrm{m}}}\coth\left(\frac{\bar{h}_0 \cdot l_{\mathrm{m}} \cdot N}{2}\right) + \frac{N^2\bar{h}_0^{-2}}{4}\mathrm{cosech}^2\left(\frac{\bar{h}_0 \cdot l_{\mathrm{m}} \cdot N}{2}\right) \tag{25.11d}$$

Equations (25.8) to (25.10) have been discretized using finite central difference method and solved using Gauss–Seidel iterative method using the following boundary conditions:

(i) $\bar{p}_i(\theta, \pm 1) = 0$ (ambient pressure at both bearing ends)
(ii) $\frac{\partial \bar{p}_i}{\partial \bar{z}} = 0$ (symmetrical pressure at the midplane)
(iii) $\frac{\partial \bar{p}_i(\theta_c, \bar{z})}{\partial \theta} = 0$, $\bar{p}_i(\theta, \bar{z}) = 0$ for $\theta \geq \theta_c$ (cavitation condition)

where θ_c represents the angular coordinate at which film cavitates and the subscripts i = 0, 1, 2 represents the steady-state and first-order perturbed pressures in Eqs. (25.8), (25.9) and (25.10), respectively.

25.2.2.2 Dynamic Response Coefficients

The dynamic pressures p_1 and p_2 are produced due to the dynamic displacements of the journal centre $Re(C\varepsilon_1 e^{i\tau})$ parallel to and $Re(C\varepsilon_0\phi_1 e^{i\tau})$ perpendicular to the line of centres. The components of the dynamic load due to the dynamic pressure $\bar{p}_1$ along and perpendicular to the line of centres can be written as

$$(W_1)_r = \int_0^1 \int_0^{\theta_c} p_1 R\cos\theta \cdot \mathrm{d}\theta \cdot \mathrm{d}z \tag{25.12a}$$

$$(W_1)_\varphi = \int_0^1 \int_0^{\theta_c} p_1 R\sin\theta \cdot \mathrm{d}\theta \cdot \mathrm{d}z \tag{25.12b}$$

It is found that the fluid film which supports the rotor is equivalent to a spring and dashpot system. Since the journal exerts a small harmonic oscillation about its steady-state position, the dynamic load carrying capacity can be expressed as a spring and a viscous damping force, as given below

$$-(W_1)_r \varepsilon_1 e^{i\tau} = S_{rr} \cdot Y + D_{rr} \cdot \frac{\mathrm{d}Y}{\mathrm{d}t} \tag{25.13a}$$

$$-(W_1)_{\phi}\varepsilon_1 e^{i\tau} = S_{\phi r} \cdot Y + D_{\phi r} \cdot \frac{dY}{dt} \tag{25.13b}$$

where

$S_{rr} =$ Direct spring stiffness in radial direction
$D_{rr} =$ Direct viscous damping in radial direction
$S_{\phi r} =$ Cross-spring stiffness; motion in tangential direction creating spring force in radial direction
$D_{\phi r} =$ Cross-viscous damping; motion in tangential direction creating viscous force in radial direction
$Y =$ Position of the journal centre = $C\varepsilon_1 e^{i\tau}$.

Now,

$$(\overline{W}_1)_r = \frac{(W_1)_r C^2}{\mu\omega R^3 L} = \int_0^1 \int_0^{\theta_c} \overline{p}_1 \cos\theta \cdot d\theta \cdot d\overline{z} \tag{25.14a}$$

$$(\overline{W}_1)_{\phi} = \frac{(W_1)_{\phi} C^2}{\mu\omega R^3 L} = \int_0^1 \int_0^{\theta_c} \overline{p}_1 \sin\theta \cdot d\theta \cdot d\overline{z} \tag{25.14b}$$

From Eqs. (25.13a) and (25.14a), we can write

$$\begin{aligned} &-(W_1)_r \varepsilon_1 e^{i\tau} = S_{rr} \cdot (C\varepsilon_1 e^{i\tau}) + D_{rr} \cdot (iC\varepsilon_1 e^{i\tau}) \cdot \omega_p \\ \text{or,}\quad &-(W_1)_r = S_{rr} \cdot C + i \cdot D_{rr} \cdot C \cdot \omega_p \\ \text{or,}\quad &-\tfrac{\mu\omega R^3 L}{C^2} (\overline{W}_1)_r = S_{rr} \cdot C + i \cdot D_{rr} \cdot C \cdot \omega_p \\ \text{or,}\quad &-(\overline{W}_1)_r = -\int_0^1 \int_0^{\theta_c} \overline{p}_1 \cos\theta \cdot d\theta \cdot d\overline{z} = \overline{S}_{rr} + i \cdot \overline{D}_{rr} \cdot \lambda \end{aligned} \tag{25.15a}$$

Similarly from Eqs. (25.13b) and (25.14b), we obtain

$$-(\overline{W}_1)_{\phi} = -\int_0^1 \int_0^{\theta_c} \overline{p}_1 \sin\theta \cdot d\theta \cdot d\overline{z} = \overline{S}_{\phi r} + i \cdot \overline{D}_{\phi r} \cdot \lambda \tag{25.15b}$$

where

$$\overline{S}_{rr} = \frac{S_{rr} C^3}{\mu\omega R^3 L}; \overline{S}_{\phi r} = \frac{S_{\phi r} C^3}{\mu\omega R^3 L}; \overline{D}_{\mathrm{rr}} = \frac{D_{rr} C^3}{\mu\omega R^3 L} \text{ and } \overline{D}_{rr} = \frac{D_{rr} C^3}{\mu\omega R^3 L}$$

Since $\overline{p}_1$ is complex, the dynamic load $\overline{W}_1$ can be expressed in terms of real and imaginary parts as

$$\overline{W}_1 = \mathrm{Re}\left(\overline{W}_1\right) + i\mathrm{Im}\left(\overline{W}_1\right) \quad (25.16)$$

Hence from Eqs. (25.15a, 25.15b), we can find out the expressions for dimensionless stiffness and damping coefficients. These expressions are written as

$$\overline{S}_{rr} = -\mathrm{Re}\left(\int_0^1\int_0^{\theta_c} \overline{p}_1 \cos\theta \cdot \mathrm{d}\theta \cdot \mathrm{d}\overline{z}\right) \quad (25.17a)$$

$$\overline{S}_{\varphi r} = -\mathrm{Re}\left(\int_0^1\int_0^{\theta_c} \overline{p}_1 \sin\theta \cdot \mathrm{d}\theta \cdot \mathrm{d}\overline{z}\right) \quad (25.17b)$$

$$\overline{D}_{rr} = -\mathrm{Im}\left(\int_0^1\int_0^{\theta_c} \overline{p}_1 \cos\theta \cdot \mathrm{d}\theta \cdot \mathrm{d}\overline{z}\right)\Big/\lambda \quad (25.17c)$$

$$\overline{D}_{\phi r} = -\mathrm{Im}\left(\int_0^1\int_0^{\theta_c} \overline{p}_1 \sin\theta \cdot \mathrm{d}\theta \cdot \mathrm{d}\overline{z}\right)\Big/\lambda \quad (25.17d)$$

Similarly, considering the dynamic displacement of the journal centre along $\phi-$ direction, we obtain the following result

$$\overline{S}_{\phi\phi} = -\mathrm{Re}\left(\int_0^1\int_0^{\theta_c} \overline{p}_2 \sin\theta \cdot \mathrm{d}\theta \cdot \mathrm{d}\overline{z}\right) \quad (25.18a)$$

$$\overline{S}_{r\phi} = -\mathrm{Re}\left(\int_0^1\int_0^{\theta_c} \overline{p}_2 \cos\theta \cdot \mathrm{d}\theta \cdot \mathrm{d}\overline{z}\right) \quad (25.18b)$$

$$\overline{D}_{\phi\phi} = -\mathrm{Im}\left(\int_0^1\int_0^{\theta_c} \overline{p}_2 \sin\theta \cdot \mathrm{d}\theta \cdot \mathrm{d}\overline{z}\right)\Big/\lambda \quad (25.18c)$$

$$\overline{D}_{r\phi} = -\mathrm{Im}\left(\int_0^1\int_0^{\theta_c} \overline{p}_2 \cos\theta \cdot \mathrm{d}\theta \cdot \mathrm{d}\overline{z}\right)\Big/\lambda \quad (25.18d)$$

The stiffness and damping coefficients, also known as dynamic response coefficients, can be used to study the linear stability characteristics of a rigid rotor in terms of critical mass parameter and whirl ratio.

25.2.2.3 Stability Characteristics

The stability of the journal is analysed by combining the equations of motion and the resultant film forces F_r and F_ϕ in the $r-$ and $\phi-$ directions, respectively. Referring to Fig. 25.1, the equation of motion of the rigid journal, assuming the rotor to be rigid, can be written as

$$F_r + W\cos\phi - MC\left[\frac{d^2\varepsilon}{dt^2} - \varepsilon\left(\frac{d\phi}{dt}\right)^2\right] = 0 \tag{25.19}$$

$$F_\phi - W\sin\phi - MC\left[\varepsilon\frac{d^2\phi}{dt^2} + 2.\frac{d\varepsilon}{dt}.\frac{d\phi}{dt}\right] = 0 \tag{25.20}$$

For a journal operating in steady-state condition the equation of motion can be written as

$$F_{r0} + W_0\cos\phi_0 = 0 \tag{25.21a}$$

$$F_{\phi 0} - W_0\sin\phi_0 = 0 \tag{25.21b}$$

Substituting Eqs. (25.13a, 25.13b) and (25.21a, 25.21b) into Eqs. (25.19) and (25.20), neglecting the higher-order terms and normalizing, we obtain the following equations

$$\left(-\overline{M}\overline{W}_0\lambda^2 + \overline{S}_{rr} + i\lambda\overline{D}_{rr}\right)\varepsilon_1 + \left(\overline{W}_0\sin\phi_0 + \overline{S}_{r\phi}\varepsilon_0 + i\lambda\overline{D}_{r\phi}\varepsilon_0\right)\phi_1 = 0 \tag{25.22}$$

$$\left(\overline{S}_{\phi r} + i\lambda\overline{D}_{\phi r}\right)\varepsilon_1 + \left(-\overline{M}\overline{W}_0\lambda^2\varepsilon_0 + \overline{W}_0\cos\phi_0 + \overline{S}_{\phi\phi}\varepsilon_0 + i\lambda\overline{D}_{\phi\phi}\varepsilon_0\right)\phi_1 = 0 \tag{25.23}$$

where

$$\overline{M} = \frac{MC\Omega^2}{W_0};\ \overline{W}_0 = \frac{W_0C^2}{\mu\Omega^2R^3L};\ \overline{F}_r = \frac{F_rC^2}{\mu\Omega^2R^3L}\ \text{and}\ \overline{F}_\phi = \frac{F_\phi C^2}{\mu\Omega^2R^3L}$$

For a non-trivial solution of ε_1 and ϕ_1, the determinant of Eqs. (25.22) and (25.23) must vanish and thus

$$\begin{vmatrix} -\overline{M}\overline{W}_0\lambda^2 + \overline{S}_{rr} + i\lambda\overline{D}_{rr} & \overline{W}_0\sin\phi_0 + \overline{S}_{r\phi}\varepsilon_0 + i\lambda\overline{D}_{r\phi}\varepsilon_0 \\ \overline{S}_{\phi r} + i\lambda\overline{D}_{\phi r} & -\overline{M}\overline{W}_0\lambda^2\varepsilon_0 + \overline{W}_0\cos\phi_0 + \overline{S}_{\phi\phi}\varepsilon_0 + i\lambda\overline{D}_{\phi\phi}\varepsilon_0 \end{vmatrix} = 0 \tag{25.24}$$

Equation (25.24) has been simplified, and the real and imaginary terms have been equated to zero to obtain the following two expressions

$$\bar{M}\bar{W}_0 = \frac{1}{\lambda^2(\bar{D}_{rr} + \bar{D}_{\phi\phi})}\left[\begin{array}{r}(\bar{D}_{rr}\bar{S}_{\phi\phi} + \bar{S}_{rr}\bar{D}_{\phi\phi}) - (\bar{D}_{r\phi}\bar{S}_{\phi r} + \bar{S}_{r\phi}\bar{D}_{\phi r}) \\ + \frac{\bar{W}_0}{\varepsilon_0}(\bar{D}_{rr}\cos\phi_0 - \bar{D}_{\phi r}\sin\phi_0)\end{array}\right] \quad (25.25)$$

$$\begin{aligned}(\overline{MW}_0)^2\lambda^4 - \left[\overline{MW}_0\left(\frac{\overline{W}_0\cos\phi_0}{\varepsilon_0} + \overline{S}_{rr} + \overline{S}_{\phi\phi}\right) + (\overline{D}_{rr}\overline{D}_{\phi\phi} - \overline{D}_{r\phi}\overline{D}_{\phi r})\right]\lambda^2 \\ + (\overline{S}_{rr}\overline{S}_{\phi\phi} - \overline{S}_{r\phi}\overline{S}_{\phi r}) + \frac{\overline{W}_0}{\varepsilon_0}(\overline{S}_{rr}\cos\phi_0 - \overline{S}_{\phi r}\sin\phi_0) = 0\end{aligned} \quad (25.26)$$

Equations (25.25) and (25.26) are solved simultaneously to obtain the values of critical mass parameter and whirl ratio.

25.2.3 Theoretical Analysis of Non-linear Stability Characteristics

25.2.3.1 Fluid Film Forces and Steady-State Load

The components of fluid film forces along the line of centres (F_{r}) and perpendicular to the line of centres (F_{φ}) are written as follows:

$$(F_{\mathrm{r}}) = \int_0^L \int_0^{\theta_{\mathrm{c}}} pR\cos\theta \cdot \mathrm{d}\theta \cdot \mathrm{d}z \quad (25.27)$$

$$(F_{\phi}) = \int_0^L \int_0^{\theta_{\mathrm{c}}} pR\sin\theta \cdot \mathrm{d}\theta \cdot \mathrm{d}z \quad (25.28)$$

Equations (25.27) and (25.28) are normalized using the following substitutions.

$$\overline{p} = \frac{pC^2}{\mu\Omega R^2}, \overline{F}_{\mathrm{r}} = \frac{F_{\mathrm{r}}C^2}{\mu\Omega^2 R^3 L}, \overline{F}_{\phi} = \frac{F_{\phi}C^2}{\mu\Omega^2 R^3 L} \text{ and } \overline{z} = \frac{z}{L/2},$$

The non-dimensional components of fluid film forces obtained along radial and tangential directions are written as follows:

$$\overline{F}_{\mathrm{r}} = \int_0^1 \int_0^{\theta_c} \overline{p} \cos\theta \cdot \mathrm{d}\theta \cdot \mathrm{d}\overline{z} \tag{25.29a}$$

$$\overline{F}_{\phi} = \int_0^1 \int_0^{\theta_c} \overline{p} \sin\theta \cdot \mathrm{d}\theta \cdot \mathrm{d}\overline{z} \tag{25.29b}$$

The steady-state load capacity is obtained as

$$\overline{W}_0 = \sqrt{\left\{\left(\overline{F}_{\mathrm{r}}\right)_0\right\}^2 + \left\{\left(\overline{F}_{\phi}\right)_0\right\}^2} \tag{25.30}$$

where $\left(\overline{F}_{\mathrm{r}}\right)_0$ and $\left(\overline{F}_{\phi}\right)_0$ are the dimensionless steady-state hydrodynamic forces along radial and tangential directions, respectively.

25.2.3.2 Equations of Motion

Considering the equilibrium of forces, the equations of motion of the journal along the line of centres and perpendicular to the line of centres can be written as

$$F_{\mathrm{r}} + W\cos\phi - MC\left[\frac{\mathrm{d}^2\varepsilon}{\mathrm{d}t^2} - \varepsilon\left(\frac{\mathrm{d}\phi}{\mathrm{d}t}\right)^2\right] = 0 \tag{25.31}$$

$$F_{\phi} - W\sin\phi - MC\left[\varepsilon\frac{\mathrm{d}^2\phi}{\mathrm{d}t^2} + 2.\frac{\mathrm{d}\varepsilon}{\mathrm{d}t}.\frac{\mathrm{d}\phi}{\mathrm{d}t}\right] = 0 \tag{25.32}$$

Equations (25.31) and (25.32) have been normalized and written as follows:

$$\overline{MW_0}\left[\frac{\mathrm{d}^2\varepsilon}{\mathrm{d}t^2} - \varepsilon\left(\frac{\mathrm{d}\phi}{\mathrm{d}t}\right)^2\right] = \overline{F}_{\mathrm{r}} + \overline{W}_0\cos\phi \tag{25.33}$$

$$\overline{MW_0}\left[\varepsilon\frac{\mathrm{d}^2\phi}{\mathrm{d}t^2} + 2.\frac{\mathrm{d}\varepsilon}{\mathrm{d}t}.\frac{\mathrm{d}\phi}{\mathrm{d}t}\right] = \overline{F}_{\phi} - \overline{W}_0\sin\phi \tag{25.34}$$

where $\overline{M} = \frac{MC\Omega^2}{W_0}$

25.2.3.3 Methos of Solution

The non-dimensional equations of motions Eqs. (25.33) and (25.34) are second-order differential equations. These two equations have arranged into four first-order differential equation in the following way:

$$\varepsilon' = \frac{\partial \varepsilon}{\partial \tau} \tag{25.35a}$$

$$\phi' = \frac{\partial \phi}{\partial \tau} \tag{25.35b}$$

$$\frac{d\varepsilon'}{d\tau} = \frac{1}{M\overline{W}_0}\left(\overline{F}_r + \overline{W}_0 \cos\phi\right) + \varepsilon.\phi'^2 \tag{25.35c}$$

$$\frac{d\phi'}{d\tau} = \frac{1}{M\overline{W}_0.\varepsilon}\left(\overline{F}_\phi - \overline{W}_0 \sin\phi\right) - \frac{2\varepsilon'\phi'}{\varepsilon} \tag{25.35d}$$

Equation (25.5) is initially solved for steady-state conditions ($\varepsilon' = \phi' = \phi = 0$) using Gauss–Seidel iterative technic applying the boundary conditions to obtain non-dimensional pressure. This non-dimensional pressure has been used to compute the steady-state hydrodynamic forces and steady-state load capacity.

The set of first-order differential Eqs. (25.35a, 25.35b, 25.35c, 25.35d) has been solved simultaneously for ε, ϕ, ε' and ϕ' using fourth-order Runge–Kutta Method. These values are then used in Eqs. (25.33) and (25.34) to obtain new values of pressure distributions which are used to find out the dynamic forces. These are again used to compute the new values of state variables. This process continues until a definite pattern of journal orbit is obtained. This method is similar to that followed by Akers et al. [20].

25.3 Results and Discussions

In the present chapter, stability characteristics of journal bearings under turbulent micropolar lubrication have been analysed using both linear and non-linear techniques. The results have been obtained by using values turbulent shear coefficients as proposed by Ng and Pan. The values of stability parameters in terms of critical mass parameter and whirl ratio have been obtained under certain parametric conditions and plotted in graphical forms to show the effect of turbulence on the stability parameters.

25.3.1 Critical Mass Parameter, $\overline{M}_{cr}$

The influence of turbulence on critical mass parameter has been shown in Figs. 25.2 and 25.3 as a function of micropolar characteristic length for $L/D = 1.0$, $N^2 = 0.3$ and $\varepsilon_0 = 0.4$. Figure 25.2 presents the values of critical mass parameter obtained from linear stability analysis, whereas Fig. 25.3 presents the values of critical mass parameter obtained from non-linear stability analysis. It has been found that the values of the critical mass parameter reduce as the Reynolds number increases. Further, at the lower values of l_m, i.e. when the micropolar effect in the lubricant is high, the reduction in critical mass parameter is more significant. A close observation of Eq. (25.1) shows that value of pressure increases with increase in *Re*.

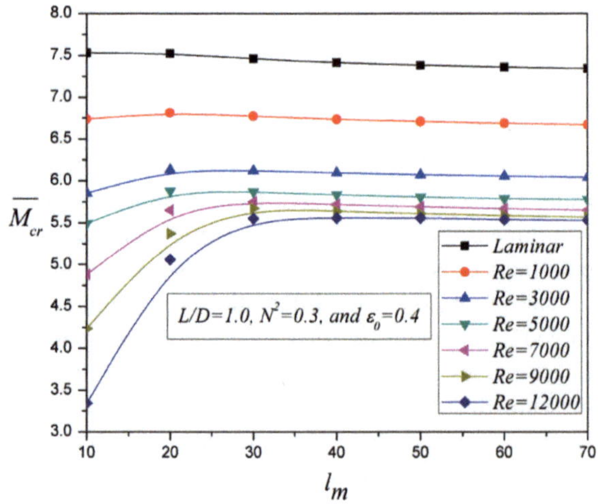

Fig. 25.2 $\overline{M}_{cr}$ versus l_m for different values of *Re* in linear stability analysis

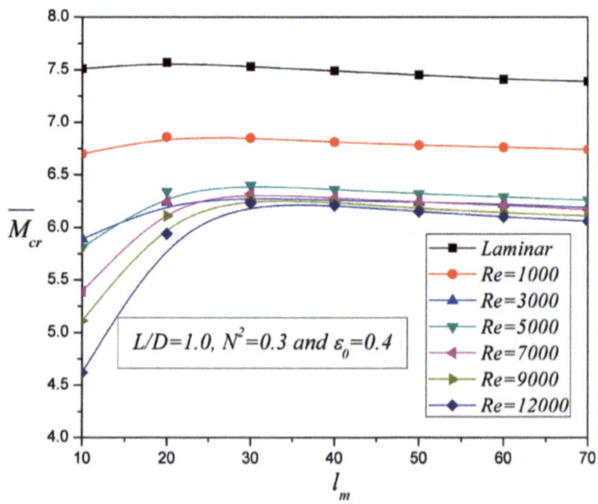

Fig. 25.3 $\overline{M}_{cr}$ versus l_m for different values of *Re* in non-linear stability analysis

Table 25.2 Comparison of linear and non-linear stability analysis at $L/D = 1.0$, lm = 20.0, $N^2 = 0.4$ and $\varepsilon_0 = 0.4$

Flow regime	$\overline{M}_{cr}$ linear	$\overline{M}_{cr}$ non-linear	λ linear	λ non-linear
Laminar	7.61343	7.71	0.53801	0.44917
Turbulent (Re = 10,000)	5.03569	6.49	0.59818	0.48515

This improves the load carrying capacity. Moreover, the non-dimensional mass parameter varies inversely with the load capacity. Hence an increase in steady-state load will decrease the non-dimensional mass parameter. It has been also found that at high values of Reynolds number ($Re \geq 3000$) the critical mass parameter initially increases with increase in l_m up to $l_m \approx 25$ and then remains constant as $l_m \rightarrow \infty$, i.e. the lubricant becomes Newtonian fluid. But at lower values of Reynolds number, the variation of critical mass parameter is almost negligible. The highest value of critical mass parameter is obtained at laminar flow conditions. Similar observations have been reported in [18] for linear stability analysis.

The linear and non-linear stability analyses have yielded similar variation of critical mass parameter. The comparison of results obtained for the stability parameters, viz. critical mass parameter and whirl ratio in linear and non-linear stability analysis, has been shown in Table 25.2 for both laminar and turbulent flow conditions. It can be seen that the values of critical mass parameter obtained in non-linear analysis are more as compared to that obtained in linear stability analysis. In linear stability analysis, only first-order terms of perturbed eccentricity ratio has been considered neglecting the higher-order terms. This makes the linear stability analysis to render approximate values of critical mass parameter. But the values obtained in case of non-linear stability analysis are more accurate as compared to those obtained in linear stability analysis.

A comparison of the values of critical mass parameter obtained using linear stability analysis and non-linear stability analysis is shown in Table 25.2. It can be observed that the value of critical mass parameter obtained by linear stability is less as compared to those obtained using non-linear analysis.

25.3.2 Whirl Ratio, λ

Variation of whirl ratio with respect to l_m is presented in Figs. 25.4 and 25.5 for various values of Reynolds number at $L/D = 1.0$, $N^2 = 0.3$ and $\varepsilon_0 = 0.4$. Figure 25.4 shows the values of whirl ratio obtained from linear stability analysis, and Fig. 25.5 presents the values of whirl ratio obtained in non-linear analysis. It is observed that at laminar flow conditions (i.e. when the values of Re is low), the whirl ratio does not vary significantly with l_m, whereas, for higher values of Re, whirl ratio initially reduces with increase in l_m and almost remains unaltered as l_m is

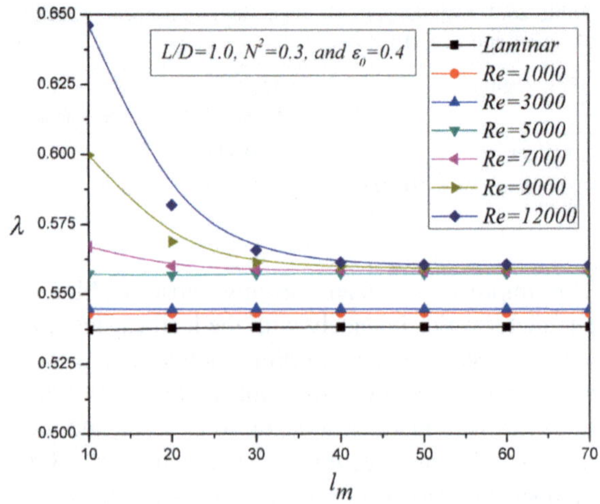

Fig. 25.4 λ versus l_m for different values of *Re* in linear stability analysis

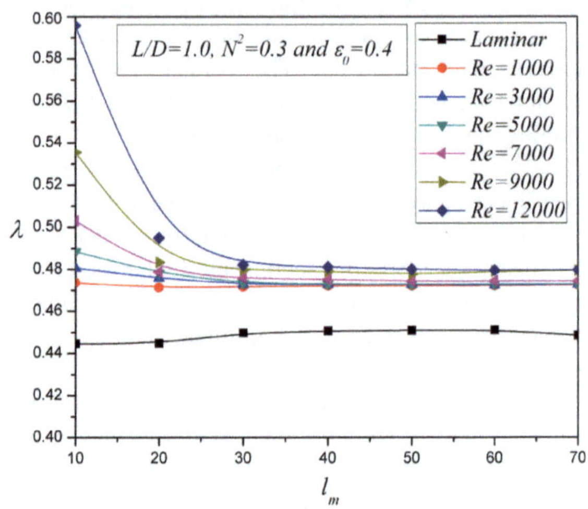

Fig. 25.5 λ versus l_m for different values of *Re* in non-linear stability analysis

further increased. At any value of l_m, the influence of turbulence is to increase the whirl ratio. The effect of Reynolds number on the whirl ratio is more when the micropolar effect is very high, i.e. when the value of l_m is low. This effect is reduced gradually as $l_m \rightarrow \infty$, i.e. the lubricant starts behaving as Newtonian fluid. Hence, the lowest value whirl ratio is observed when the lubricant flow is laminar. Similar results were obtained by Das and Guha [18] for linear stability analysis.

The results obtained from linear and non-linear stability analysis have yielded similar variation of whirl ratio with Reynolds number. The values of whirl ratio obtained from linear and non-linear stability have been presented in Table 25.2.

25.3.3 Journal Orbits

Using the values of the stability parameters, the trajectory of the journal centre has been plotted and is shown in Fig. 25.6a–f at $L/D = 1.0$, $l_m = 20.0$, $\varepsilon_0 = 0.4$, $N^2 = 0.4$ in both laminar and turbulent flow conditions.

The values of the stability parameters have been used to obtain the threshold of stability for which the journal centre executes in a stable cycle called the limit cycle. If the value of the stability mass parameter is less than the threshold value, the journal centre trajectory tends to converge towards the centre of oscillation and the motion of the journal is said to be stable. Whereas, if the mass parameter is higher than the threshold value, the journal centre tends to move away from the centre and the system becomes unstable.

It is observed from Fig. 25.6a, d that the threshold of stability, i.e. critical mass parameter for laminar flow, is higher as compared to that of turbulent flow. As the Reynolds number increases, the formation of eddies starts making the flow turbulent and further increase in *Re* causes more number of eddy formations. This causes the journal to shift away from its stable position causing the journal become unstable. Hence, the critical mass parameter for turbulent flow is less than that of in case of laminar flow under similar operating conditions.

Further, it is also observed from Fig. 25.6b, f that for the similar parametric conditions and same value of mass parameter, a stable orbit is obtained in case of laminar flow but unstable orbit is observed in case of turbulent flow.

25.3.4 Comparison of Different Turbulent Models

The numerical values of stability characteristics were computed using two turbulent models predicted by Constantinescu [2–4] and Ng and Pan [5] for the comparison of two different turbulent models. The results obtained by aforesaid two models of turbulence at various parametric conditions are presented in Tables 25.3, 25.4, 25.5 and 25.6. Tables 25.3 and 25.5 show the values obtained in linear stability analysis, whereas Tables 25.4 and 25.6 show the values of stability parameters computed using non-linear technic.

The tabulated data gives the comparison of the stability characteristics in terms of critical mass parameter and whirl ratio obtained by using Constantinescu's model with those of Ng–Pan when Reynolds number is taken as parameters. The comparison shows a small discrepancy between the two results. This discrepancy may be attributed to the different expressions for turbulent shear coefficients k_x and k_z predicted in two models of turbulence. In fact, Constantinescu proposed the same values of turbulent coefficients at all values of Reynolds number, whereas Ng and Pan have proposed the different values of the turbulent coefficients for various ranges of Reynolds number. On going through the tables, it has been observed that Constantinescu's model predict higher values of stability threshold (critical mass parameter) as compared to those of Ng and Pan models at all parametric conditions.

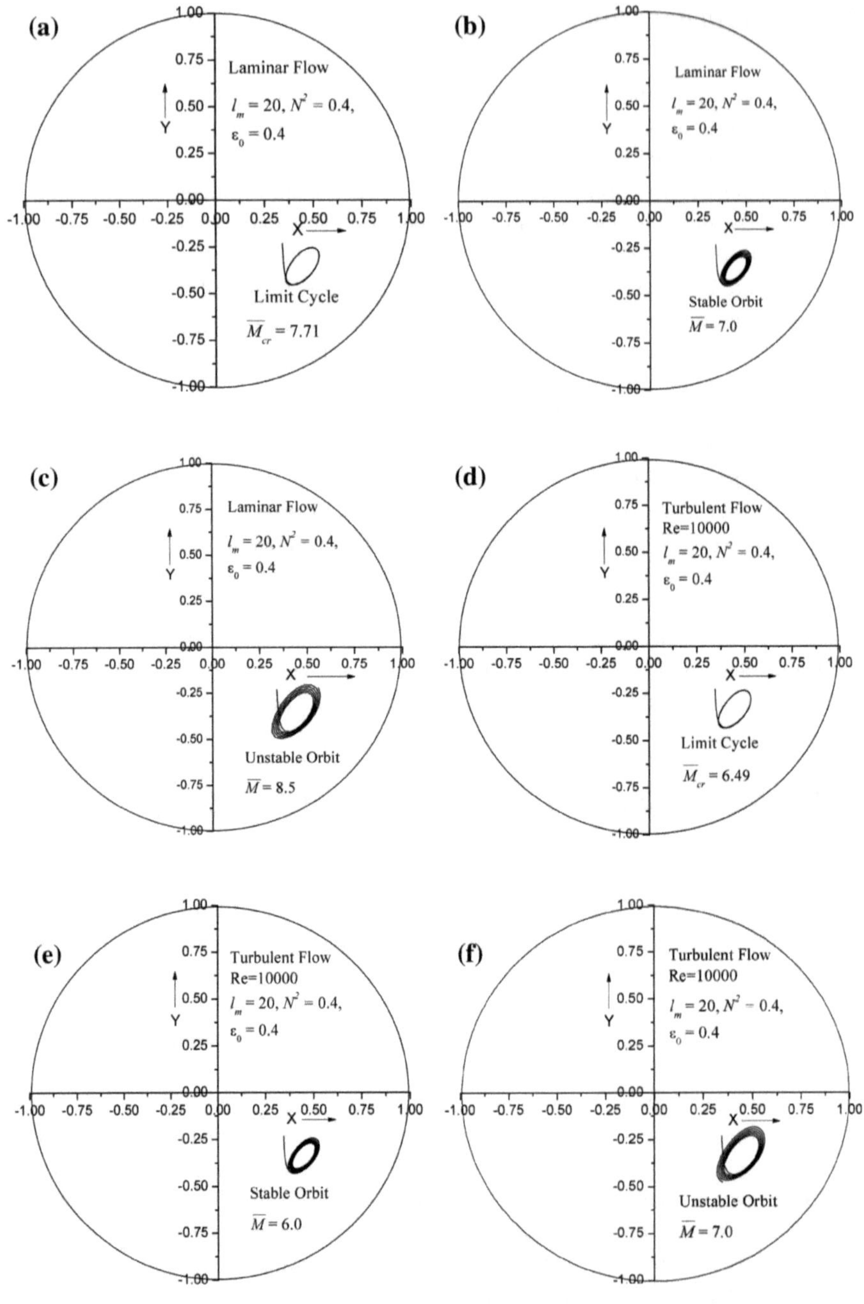

Fig. 25.6 **a** Limit cycle of the journal centre for laminar flow, **b** stable orbit of journal for laminar flow, **c** unstable orbit of journal for laminar flow, **d** limit cycle of the journal centre for turbulent flow, **e** stable orbit of journal for turbulent flow, **f** unstable orbit of journal for turbulent flow

Table 25.3 Comparison of data obtained for critical mass parameter using Ng–Pan model and Constantinescu model at L/D = 1.0, N^2= 0.3 and ε_0 = 0.4 (linear stability)

l_m	Re = 1000.0		Re = 3000		Re = 5000.0		Re = 9000.0	
	$(\overline{M}_{cr})^{ng}$	$(\overline{M}_{cr})^{c}$	$(\overline{M}_{cr})^{ng}$	$(\overline{M}_{cr})^{c}$	$(\overline{M}_{cr})^{ng}$	$(\overline{M}_{cr})^{c}$	$(\overline{M}_{cr})^{ng}$	$(\overline{M}_{cr})^{c}$
10.0	6.7367	6.84118	5.85209	6.10171	5.49081	5.55691	4.236	4.39273
30.0	6.77069	6.85913	6.1213	6.22487	5.86751	6.08421	5.67174	5.63857
50.0	6.70506	6.78765	6.07178	6.4635	5.80715	6.01856	5.6092	5.87342
70.0	6.67024	6.75018	6.04151	6.4226	5.776	5.98028	5.56894	5.8299

$(\overline{M}_{cr})^{ng}$ indicates the value of critical mass parameter obtained using Ng and Pan model
$(\overline{M}_{cr})^{c}$ indicates the value of critical mass parameter obtained using Constantinescu model

Table 25.4 Comparison of data obtained for critical mass parameter using Ng–Pan model and Constantinescu model at L/D = 1.0, N^2= 0.3 and ε_0 = 0.4 (non-linear stability)

l_m	Re = 1000.0		Re = 3000		Re = 5000.0		Re = 9000.0	
	$(\overline{M}_{cr})^{ng}$	$(\overline{M}_{cr})^{c}$	$(\overline{M}_{cr})^{ng}$	$(\overline{M}_{cr})^{c}$	$(\overline{M}_{cr})^{ng}$	$(\overline{M}_{cr})^{c}$	$(\overline{M}_{cr})^{ng}$	$(\overline{M}_{cr})^{c}$
10.0	6.70	6.77	5.89	6.36	5.81	6.17	5.11	5.48
30.0	6.85	6.93	6.28	6.65	6.40	6.58	6.27	6.54
50.0	6.78	6.85	6.24	6.59	6.32	6.49	6.18	6.45
70.0	6.74	6.81	6.19	6.54	6.26	6.44	6.11	6.37

$(\overline{M}_{cr})^{ng}$ indicates the value of critical mass parameter obtained using Ng and Pan model
$(\overline{M}_{cr})^{c}$ indicates the value of critical mass parameter obtained using Constantinescu model

Table 25.5 Comparison of data obtained for Whirl ratio using Ng–Pan model and Constantinescu model at L/D = 1.0, N^2= 0.3 and ε_0 = 0.4 (linear stability)

l_m	Re = 1000.0		Re = 3000		Re = 5000.0		Re = 9000.0	
	$(\lambda)^{ng}$	$(\lambda)^{c}$	$(\lambda)^{ng}$	$(\lambda)^{c}$	$(\lambda)^{ng}$	$(\lambda)^{c}$	$(\lambda)^{ng}$	$(\lambda)^{c}$
10.0	0.54299	0.54361	0.54482	0.55983	0.5571	0.57500	0.5997	0.60481
30.0	0.54333	0.54388	0.54475	0.55979	0.5572	0.55949	0.56123	0.57386
50.0	0.54328	0.54380	0.54466	0.54514	0.55732	0.55932	0.55912	0.55835
70.0	0.54319	0.54370	0.54454	0.54501	0.5573	0.55920	0.5591	0.55828

$(\lambda)^{ng}$ indicates the value of whirl ratio obtained using Ng and Pan model
$(\lambda)^{c}$ indicates the value of whirl ratio obtained using Constantinescu model

Table 25.6 Comparison of data obtained for Whirl ratio using Ng–Pan model and Constantinescu model at L/D = 1.0, N^2= 0.3 and ε_0 = 0.4 (non-linear stability)

l_m	Re = 1000.0		Re = 3000		Re = 5000.0		Re = 9000.0	
	$(\lambda)^{ng}$	$(\lambda)^{c}$	$(\lambda)^{ng}$	$(\lambda)^{c}$	$(\lambda)^{ng}$	$(\lambda)^{c}$	$(\lambda)^{ng}$	$(\lambda)^{c}$
10.0	0.47371	0.47693	0.48067	0.48038	0.50311	0.48201	0.53548	0.49582
30.0	0.47166	0.47256	0.47279	0.47704	0.47571	0.47946	0.47973	0.48035
50.0	0.47205	0.47355	0.47267	0.47685	0.47417	0.47729	0.47758	0.47949
70.0	0.47239	0.47406	0.47258	0.47721	0.47424	0.47734	0.47931	0.48074

$(\lambda)^{ng}$ indicates the value of whirl ratio obtained using Ng and Pan model
$(\lambda)^{c}$ indicates the value of whirl ratio obtained using Constantinescu model

25.4 Conclusions

Based on the ongoing analysis of stability of journal bearing in laminar and turbulent flow regimes, the following conclusions can be drawn:

The influence of turbulence is to reduce the critical mass parameter. This effect is observed for all values of micropolar characteristic length. The highest value of critical mass parameter is found in case of a journal bearing with laminar flow of lubricant. Hence, it can be concluded that the turbulence adversely affects the stability of a journal bearing.

The critical mass parameter initially found to increase with increase in micropolar characteristic length. After achieving an optimum value, the critical mass parameter decreases with further increase in characteristic length when the Reynolds number is taken as a parameter. This optimum value of stability parameter decreases with increase in Reynolds number.

Turbulence in the fluid film causes an enhancement in the whirl ratio. Whirl ratio is found to attain lowest value in case of a journal bearing operating in laminar flow regime.

The stability thresholds obtained using non-linear analysis is more as compared to those obtained using linear analysis for both laminar and turbulent flow. Although the values of stability parameters obtained in case of linear and non-linear analysis are very close to each other, while designing the journal bearing system non-linear analysis may be preferred due to the accuracy of the method.

The parametric study of two different turbulent lubrication models reveals that the difference in values of stability parameters obtained using Ng and Pan models, and Constantinescu model is marginal.

References

1. Wilcock, D.F.: Turbulence in high-speed journal bearings. J. Lubr. Technol. Trans. ASME **72**, 825–834 (1950)
2. Constantinescu, V.N.: On turbulent lubrication. Proc. Inst. Mech. Eng, London **173**, 881–900 (1959)
3. Constantinescu, V.N.: Analysis of bearings operating in turbulent regime. J. Basic Eng. Trans. ASME **84**, 139–151 (1962)
4. Constantinescu, V.N.: Theory of Turbulent Lubrication, p. 159. Int. Symp. Lubr. Wear, Houston (1964)
5. Ng, C.W., Pan, C.H.T.: A linearized turbulent lubrication theory. J. Basic Eng. Trans. ASME **87**, 675–688 (1965)
6. Elrod, H.G., Ng, C.W.: A theory of turbulent fluid films and its application to bearings. J. Lubr. Technol. Trans. ASME. **89**(3), 356–363 (1967)
7. Taylor, C.M., Dawson, D.: Turbulent lubrication theory-application to design. J. Lubr. Technol. Trans. ASME **96**, 36–46 (1974)
8. Soni, S.C., Sinhasan, R., Singh, D.V.: Non-linear analysis of finite hydrodynamic journal bearings in laminar and turbulent flow regimes. Wear **95**, 293–311 (1984)
9. Zhang, Z.M., Wang, X.J., Sun, M.L.: Turbulence models of hydrodynamic lubrication. J. Shanghai Univ. **7**, 305 (2003). https://doi.org/10.1007/s11741-003-0001-3
10. Eringen, A.: Theory of micropolar fluids. J. Math. Mech. **16**, 1–18 (1966)
11. Allen, S., Kline, K.: Lubrication theory of micropolar fluids. J. Appl. Mech. **38**, 646–650 (1971)
12. Prakash, J., Sinha, P.: Lubrication theory of micropolar fluids and its application to a journal bearing. Int. J. Eng. Sci. **13**, 217–232 (1975)
13. Khonsari, M.M., Brewe, D.E.: On the performance of finite journal bearing lubricated with micropolar fluid. STLE. Tribol. Trans. **32**, 155–160 (1989)
14. Faralli, M., Belfiore, N.P.: Steady state analysis of worn spherical bearing operating in turbulent regime with non-newtonian lubricants. In: Proceedings of 2006 International Conference on Tribology—AITC-AIT 2006. Parma, Italy (2006)
15. Rana, N.K., Gautam, S.S., Samanta, S.: Approximate analysis of dynamic characteristics of short journal bearings in turbulent micropolar Lubrication. J. Inst. Eng. (India): Sr. C **95**, 383–388 (2014)
16. Ghosh, A., Guha, S.K.: Linear stability analysis of finite hydrodynamic journal bearing under turbulent lubrication with coupled-stress fluid. Ind. Lubr. Tribol **68**, 386–399 (2016)
17. Das, S., Guha, S.K.: Turbulent effect on the dynamic response coefficients of finite journal bearings lubricated with micropolar fluid. Procedia Technol. **23**, 193–200 (2016)
18. Das, S., Guha, S.K.: Linear stability analysis of hydrodynamic journal bearings operating under turbulent micropolar lubrication. In: IEEE Explore (2017). https://doi.org/10.1109/amiams.2017.8069199
19. Das, S., Guha, S.K., Chattopadhyay, A.K.: Theoretical analysis of stability characteristics of hydrodynamic journal bearings lubricated with micropolar fluids, Inst. Mech. Eng, Part: J. **218**, 45–56 (2004)
20. Akers, A., Michaelson, S., Cameron, A.: Stability contours for a whirling finite journal bearing. J. Lubr. Technol. Trans. ASME **93**, 177–183 (1971)

Chapter 26
MHD Convection with Heat Generation in a Porous Cavity

Soumyodeep Mukherjee, Nirmalendu Biswas and Nirmal K. Manna

Nomenclature

B	Uniform magnetic field (tesla)
Da	Darcy number
G	Ratio of heat generation–absorption
H	Height of the cavity/length scale, m
Ha	Hartmann number
K	Permeability of porous medium, m^2
L	Length of the cavity, m
Nu	Average Nusselt number
p	Pressure, Pa
P	Dimensionless pressure
Pr	Prandtl number
Ra	External Rayleigh number
Ra_{I}	Internal Rayleigh number
Ra_{m}	Modified Darcy–Rayleigh number
T	Temperature, K
g	Velocity components, m/s
U, V	Dimensionless velocity components
x, y	Cartesian coordinates, m
X, Y	Dimensionless coordinates

Greek Symbols

α	Thermal diffusivity, m^2/s
β	Thermal expansion coefficient, K^{-1}
γ	Inclination angle of the magnetic field
θ	Dimensionless temperature
ε	Porosity
υ	Kinematic viscosity, m^2/s

S. Mukherjee (✉) · N. Biswas · N. K. Manna
Department of Mechanical Engineering, Jadavpur University, Kolkata 700032, India
e-mail: soumyodeep1312@gmail.com

P. Sahoo and J. P. Davim (eds.), *Advances in Materials, Mechanical and Industrial Engineering*, Lecture Notes on Multidisciplinary Industrial Engineering,
https://doi.org/10.1007/978-3-319-96968-8_26

ρ Density, kg/m^3
κ Electrical conductivity (μS cm^{-1})
ψ Dimensionless stream function

Subscripts

c, h Cooling, heating

26.1 Introduction

The study on magneto-hydrodynamics (MHD) during natural convection along with heat generation–absorption has relevance in many fields due to its widespread applications, for example, cooling of liquid metal in nuclear reactor system, geophysics, astrophysics, crystal growth, and plasma confinement [1] are few to mention. For establishing relevance and importance of the present study, a literature survey in the context of the present problem is discussed here briefly. Oztop and Bilgen [2] numerically analyzed the effect of heat generation in a square cavity heated differentially. They observed that rate of heat transfer decreases for a specific variation in internal and external Rayleigh numbers ratio, resulting a considerable modification in the flow fields. Considering the effect of magnetic field, a number of studies have been performed for natural convection both numerically and analytically. Conjugate natural convection in a nanofluid-filled porous cavity has been studied by Sheremet and Pop [3], and they observed increasing trend of local Nusselt number at increased Rayleigh number. Amin [4] has investigated internal heat generation during thermal convection of micropolar fluid imposing magnetic field and pointed out that magnetic field parameter changes the overall thermal performance substantially. Role of magnetic field strength during natural convection of nanofluid-saturated cavity has been presented by Ghasemi et al. [5], where it is indicated that depending on magnitude of *Ha* heat transfer changes along with the change of other parameters. Sivaraj and Sheremet [6] have studied an inclined porous cavity subjected to external magnetic field, and they found that effect of magnetic field reduces the rate of heat transfer; furthermore, with the increasing magnetic field inclination angle leads to a nonlinear trend on heat transfer characteristic (average Nusselt number). Selimefendigil and Oztop [7] have investigated MHD in natural convection considering a triangular cavity. The result shows the decreasing trend of heat transfer with the increased magnetic field strength (Hartmann number) and it is more prominent in case of high external Rayleigh numbers, whereas mean heat transfer increases due to the change in magnetic field inclination angle. Bondareva and Sheremet [8] studied the natural convection problem in a square cavity heated locally under the influence of inclined magnetic field. They found that the strength of magnetic field (Hartmann number) and its inclination angle strongly influences the heat transfer characteristic. Mahmoudi et al. [9] examined the natural convection phenomena of nanofluid-filled square cavity using LBM method and found that increase in Rayleigh number leads to the enhanced heat transfer. Further studies reveal the decrease of heat transfer with the increase in strength of applied magnetic field, whereas the angle of magnetic field strongly affects thermal field and fluid flow.

Thermal behavior of a concentric annulus subjected to MHD nanofluid has been analyzed by Sheikholeslami et al. [10]. The results show that with the increase in nanoparticle concentration and Rayleigh number, the rate of heat transfer increases and it is further affected by increment in Hartmann number. Sheikholeslami and Ganji [11] further studied heat transfer phenomena in nanofluid flow through two parallel plates. Their investigation also indicates the similar results as mentioned earlier in [10] along they found a relationship between the heat transfer parameter (Nusselt number) with other governing parameters. A similar study [12] also shows the increment in *Nu* with the increasing nanoparticle concentration. Nanofluid natural convection in a square enclosure subjected to magnetic field and sinusoidal temperature distribution has been investigated numerically by Kefayati [13, 14]. Heat transfer found to decrease with increasing magnetic field strength. Numerical studies performed by Mliki et al. [15] on MHD natural convection using nanofluid-saturated cavity report similar result for the change in magnetic field. Another numerical study presented by Malik and Nayak [16] considering magneto-hydrodynamic (MHD) convection of nanofluid-saturated porous domain found the increment in entropy generation at stronger magnetic field. Their results show that heat transfer rate is more at higher Grashof number.

A number of studies and investigations [17–22] also dealt with the effect of natural convection in different cavities filled with nanoparticles and influenced by MHD. Rashad et al. [23] analyzed an inclined porous-saturated enclosure filled with nanofluid under MHD natural convection phenomena. The study reveals decreasing trend of heat transfer at increasing *Ha*, and peak performance (of heat transfer) takes place between 50° and 310° inclination angles of the enclosure. Gibanov et al. [24] presented that the porous block and magnetic field may contribute as good parametric controlling effect of increase in heat transfer rate as well as the escalation of fluid flow. Yu et al. [25] investigated the impact of uniform magnetic field in an inclined rectangular cavity filled with low Prandtl number ($Pr = 0.025$) fluid under natural convection. The result indicates the significance of cavity aspect ratio as well as its inclination angle, which affects flow structure and heat transfer rate directly. Their results also demonstrated that for higher Hartmann number, the angle of magnetic field inclination contributes as crucial role in the flow structure. Different hydrodynamic effects have been observed with the variation in the cavity aspect ratio. Grosan et al. [26] analytically solved transient free convection in heat-generating porous cavity under magnetic field. Applied horizontal magnetic field suppresses the convective heat flow compared to uniform vertical magnetic field. Similar type of problem has been numerically analyzed and presented by Revnic et al. [27]. Jiang et al. [28] examined free convection phenomenon inside a porous cavity where the problem was defined using Brinkman–Forchheimer–extended Darcy model (BFDM). They considered magnetic quadrupole fields and presented its effect on the cavity flow which demonstrated that flow structure, as well as heat transfer trend, is greatly influenced with magnetic field intensity. Nayak et al. [29] also investigated a porous cavity with two different (constant and periodic temperature) heat sources numerically using BFDM to present the effect of transient MHD natural convection. Fluid having higher Prandtl number at smaller Darcy number ($Da < 1.0$) heat transfer enhances. Several studies addressed the impact of magnetic field for the analysis of heat transfer characteristics [30–35].

Main objective of the present work is to investigate thermal MHD in typical geometry with heat generation and absorption from the fundamental perspective. The impacts of magnetic field strength (in terms of Hartmann number) and its different inclination angle on the heat transfer are explored in porous domain under the variations of relevant parameters. The present study is conducted to supplement the existing knowledge in MHD buoyant flow-area.

26.2 Problem and Solution Method

The studied problem is a two-dimensional geometry having length scale L (= H) as shown in Fig. 26.1. The cavity is packed with electrically-conducting fluid-saturated porous matrix. The right wall is cold (T_c) and the left wall is hot (T_h), whereas the other two walls are considered as adiabatic in nature. For this numerical study, uniformly distributed constant magnetic field of magnitude B with different inclination angle (γ) is considered. The uniform rate of heat generation or absorption (volumetric) is present throughout the flow domain.

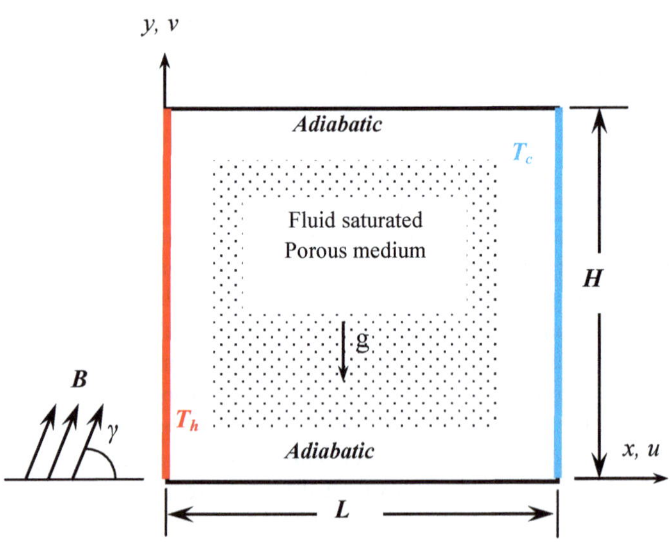

Fig. 26.1 Description of the problem

Considering steady and laminar flow, non-dimensional governing equations for conservation of mass, momentum, and energy are formulated within the validity of Boussinesq approximation [36]. These non-dimensional equations are given below

$$\frac{\partial U}{\partial X}+\frac{\partial V}{\partial Y}=0 \tag{26.1}$$

$$\frac{1}{\varepsilon^2}\left(U\frac{\partial U}{\partial X}+V\frac{\partial U}{\partial Y}\right)=-\frac{\partial P}{\partial X}+\frac{Pr}{\varepsilon}\left(\frac{\partial^2 U}{\partial X^2}+\frac{\partial^2 U}{\partial Y^2}\right)-\frac{Pr}{Da}U-\frac{F_C\sqrt{U^2+V^2}}{\sqrt{Da}}U + Ha^2 Pr(V\sin\gamma\cos\gamma - U\sin^2\gamma) \tag{26.2}$$

$$\frac{1}{\varepsilon^2}\left(U\frac{\partial V}{\partial X}+V\frac{\partial V}{\partial Y}\right)=-\frac{\partial P}{\partial Y}+\frac{Pr}{\varepsilon}\left(\frac{\partial^2 V}{\partial X^2}+\frac{\partial^2 V}{\partial Y^2}\right)-\frac{Pr}{Da}V-\frac{F_C\sqrt{U^2+V^2}}{\sqrt{Da}}V + Ha^2 Pr(U\sin\gamma\cos\gamma - V\cos^2\gamma)+RaPr\theta \tag{26.3}$$

$$\left(U\frac{\partial \theta}{\partial X}+V\frac{\partial \theta}{\partial Y}\right)=\left(\frac{\partial^2 \theta}{\partial X^2}+\frac{\partial^2 \theta}{\partial Y^2}\right)+G \tag{26.4}$$

where G is the ratio of internal to external Rayleigh number, X and Y indicate the dimensionless coordinates, and U and V are the non-dimensional velocity components. Dimensionless temperature and pressure are denoted by θ and P, respectively. The porosity of the porous medium is represented by ε.

The momentum equations are modeled considering buoyancy force evolved due to change in fluid density caused by temperature gradient. Brinkman–Forchheimer–extended Darcy model (BFDM) is used to take care of the effect of flow velocity that causes substantial inertial friction [37] particularly at a higher velocity. This frictional effect is estimated through the Forchheimer coefficient (F_C) expressed by

$$F_C=\frac{1.75}{\sqrt{150}\,\varepsilon^{3/2}}. \tag{26.5}$$

The governing equations are involved with following non-dimensional variables

$$(X,Y)=\frac{(x,y)}{H},\ (U,V)=\frac{(u,v)}{(\alpha/H)},\ P=\frac{(p+\rho g y)-p_a}{\rho(\alpha/H)^2},\theta=\frac{T-T_c}{T_h-T_c} \tag{26.6a}$$

Vital dimensionless parameters (mainly Prandtl number, Darcy number, internal and external Rayleigh number, modified Rayleigh, Hartmann number, and heat generation) are evolved in the process of non-dimensionalization, which are defined by

$$Pr = \frac{\nu}{\alpha}, Da = \frac{K}{H^2}, Ha = BH\sqrt{\kappa/\rho\nu}, G = \frac{Ra_{\rm I}}{Ra} \text{ and}$$
$$Ra = \frac{g\beta(T_{\rm h} - T_{\rm c})H^3}{\nu\alpha}, Ra_{\rm I} = \frac{g\beta\dot{Q}_{\rm V}H^5}{\nu\alpha}, Ra_{\rm m} = \frac{g\beta(T_{\rm h} - T_{\rm c})LK}{\nu\alpha}. \tag{26.6b}$$

In above, $\dot{Q}_{\rm V}$ is internal volumetric heat generation rate (in dimensional form) and is linked to the internal Rayleigh number $Ra_{\rm I}$.

The boundary conditions of the governing equations are set by

$U = 0$, $V = 0$ on all the (stationary) walls,
$\theta = 1$ on the left wall,
$\theta = 0$ on the right wall, and
$\frac{\partial \theta}{\partial Y} = 0$ on the top and bottom adiabatic walls.

The flow fields solved numerically have been used to evaluate the heat transfer trend from the heat source in terms of average Nusselt number and to portray the flow patterns within the enclosure in terms of streamlines. The results have been prepared using the average Nusselt number (Nu) and streamlines. Thus, global heat transfer parameter, average Nusselt number, has been computed by

$$Nu = \int_0^1 \left(- \left. \frac{\partial \theta}{\partial X} \right|_{X=0} \right) \mathrm{d}Y \tag{26.7}$$

26.2.1 Code Validation

The present problem is solved numerically by a well-validated in-house code employing the SIMPLE algorithm and finite volume method [37]. The criterion of mass defect for the maximum value of continuity is set at 10^{-8} for obtaining the convergence of solutions iteratively. The developed code has been validated extensively. A reporting on validation is presented in earlier work of Datta et al. [38] under natural convection flow in a porous cavity. Furthermore, a new validation study is conducted following the problem as reported by Ghasemi et al. [39], by simulating natural convection in a differentially heated square enclosure under the impact of magnetic field as presented in Table 26.1. Variations of average

Table 26.1 Average Nu for various Ha at $Ra = 10^5$ (with maximum stream function in brackets)

Ha	Ghasemi et al. [39]	Present code results
0	4.738 (11.053)	4.776 (11.270)
15	4.143 (8.484)	4.168 (8.612)
30	3.150 (5.710)	3.163 (5.755)
45	2.369 (3.825)	2.377 (3.838)
60	1.851 (2.623)	1.857 (2.626)

Nusselt number and maximum stream function ψ_{max} (in bracket) at different values of *Ha* as computed by using the present code are compared with that of the published results available in Ref. [39] as presented in Table 26.1 for $Ra = 10^5$, solid volume fraction = 0. The comparison as in the table clearly indicates agreement of the present code with the published results [39].

26.2.2 Grid Independence Study

For the influence of grid size, five distinct grid sizes (70 × 70, 100 × 100, 130 × 130, 160 × 160, and 200 × 200) have been considered along with the heat source at left wall and heat sink at the right wall. The study has been executed considering $Ha = 30$ with $\gamma = 0°$, $Da = 10^{-4}$, $\varepsilon = 0.6$, $Pr = 0.71$, and a range of Ra_m values (1–1000). Grids are distributed in a non-uniformed way with the finer grid sizes close to the walls. In case of the higher grid sizes, the size of the finest grid is considered as 0.0009. The details of the grid study which are presented in Table 26.2 is based on the heat transfer rate (average Nusselt number *Nu*) through the hot wall. Thus, scrutinizing the values of average *Nu* for different grid sizes, 160 × 160 grid size is selected for the presented study.

26.3 Results and Discussion

The impact of uniformly distributed magnetic field for a porous-saturated square cavity has been presented by the parametric variations of Ra_m = 1–1000, *Ha* = 10–100 with the magnetic field inclination angle (γ = 0–180°), $Da = 10^{-3}$–10^{-6}, and ε = 0.1–1.0, through extensive computations taking $Pr = 0.71$. Although to identify the influence of magnetic field a constant Ra_m of 100 and $\varepsilon = 0.6$ have been contemplated, different heat-generating boundaries have been also considered to perceive thermal aspects and heat transfer scenarios. Furthermore, some studies on the present cavity are extended considering clear domain inside on the basis same Rayleigh to get more insight. The obtained results are discussed in parametric form (by varying one parameter and keeping others constant) systematically.

Table 26.2 Grid independence study ($Ha = 30$, $\gamma = 0°$, $Da = 10^{-4}$, $\varepsilon = 0.6$)

Ra_m	*Nu* (% error with respect to immediate coarser grid)				
	70 × 70	100 × 100	130 × 1300	160 × 160	200 × 200
1	1.001	1.001 (0%)	1.001 (0%)	1.001 (0%)	1.001 (0%)
10	1.068	1.067 (14%)	1.066 (12%)	1.065 (8%)	1.064 (6%)
100	2.623	2.617 (24%)	2.616 (2%)	2.616 (2%)	2.615 (3%)
1000	8.207	8.163 (54%)	8.165 (−3%)	8.172 (−8%)	8.177 (−7%)

26.3.1 *Porous Domain*

26.3.1.1 Effect of Modified Rayleigh Number

Figure 26.2 (through sub-figures of Fig. 26.2a–d) represents the streamline and isotherm plots for various modified Rayleigh numbers (Ra_m = 1–1000) with fixed Ha = 30 (γ = 0°), $Da = 10^{-4}$, and ε = 0.6. The fluid flows in closed-loop circular form starting from left heat source to the right heat sink. At lower Rayleigh, the magnitude of heat transfer is low due to low buoyancy force. At Ra_m = 1 (Fig. 26.2a), single circular formed in the cavity which is near to the right cold wall for heat generation case at G = +3 and near to the left wall for heat absorption case at G = −3. However, for the case of G = 0, centrally located circulation is formed, and corresponding isotherms are evenly distributed over the entire flow domain vertically. But for nonzero G cases, parallel isotherm lines are clustered to the one side as per the clustering of streamlines. As the value of Ra_m increases, corresponding average Nusselt number also increases due to more heat transfer. However, the change in heat transfer is very negligible at Ra_m = 10 (Fig. 26.2b), but found significant at $Ra_m \geq 100$ (Fig. 26.2c, d). From the isotherms and streamlines, it is noted that contour lines are stretched from the left hot wall lower part to the right cold wall upper part. They are denser adjacent to the heat source and heat sink. As the buoyancy force increases, eddy at the center of the enclosure became elliptical and later on at Ra_m = 1000 (Fig. 26.2d), it becomes almost rectangular. With the increase in Ra_m, it is clearly noticed that both streamlines and isotherms are much more clustered near to the sidewalls, indicating thinner boundary layer thicknesses for velocity and temperature. It is also noticed that at lower Ra_m (=1), the heat transfer through the left is less (zero) due to heat generation. Thus, the temperature adjacent to the left heated wall is same as the imposed wall temperature as could be realized from the isotherm contours. For this, Nu becomes zero. However, this value is nonzero during heat absorption with G = −3. With the heat generation, the rejection of heat through both the sidewalls is possible. However, as Ra_m increases convection becomes stronger that takes place in clockwise direction, and thereby the heat rejection takes place through the right wall only. It leads to increase in Nu as $Ra_m \geq 100$.

For all the cases of heat generation–absorption conditions, the trend of heat transfer at different Ra_m values is depicted in Fig. 26.3. It shows that convective heat transfer from the heated wall is more during heat generation.

26.3.1.2 Effect of Magnetic Field and Inclination Angle

Figure 26.4 (through sub-figures of Fig. 26.4a–d) represents the effect of magnetic field at γ = 0° under the conditions of heat generation and absorption. The top rows of sub-figures indicate streamline plot and the bottom rows of sub-figures indicate isotherm plots. As analyzed, isotherm lines are crowded adjoining to the cold right

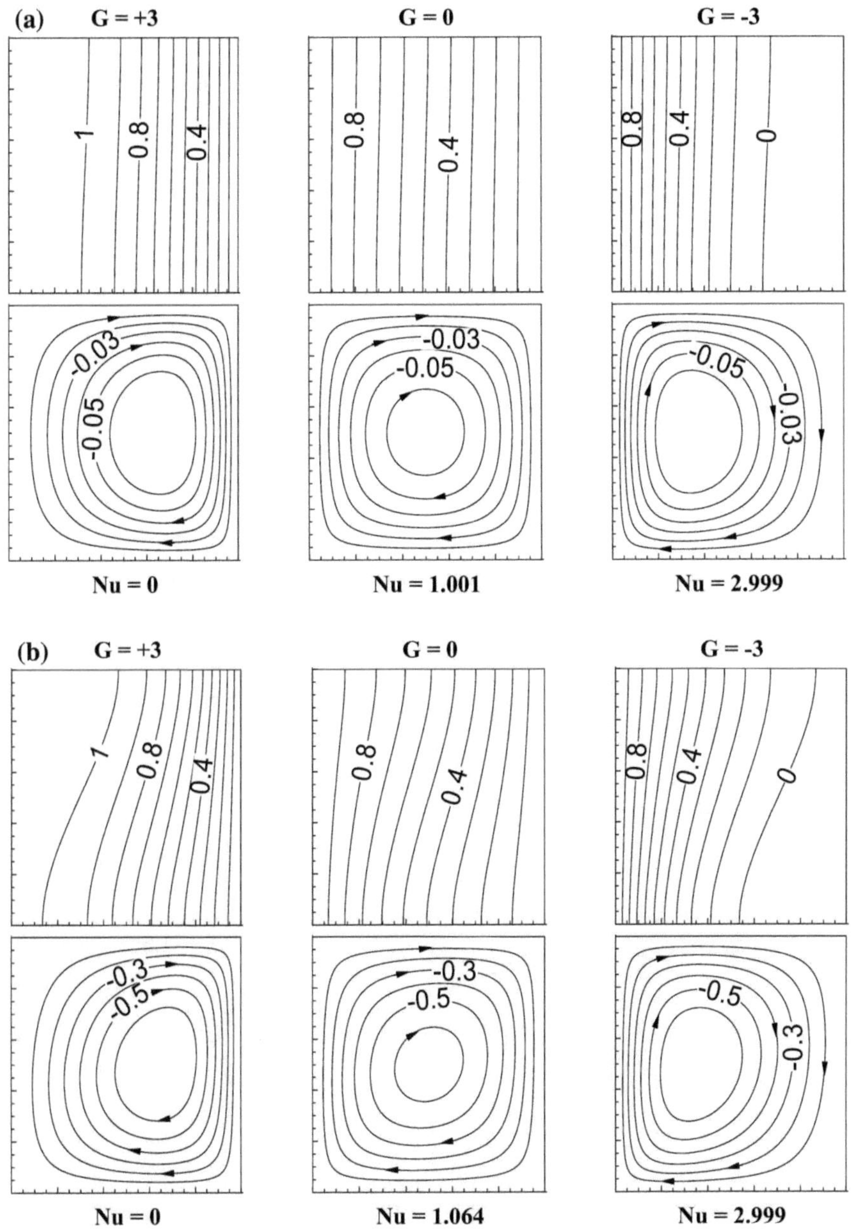

Fig. 26.2 **a** Isotherms (top row) and streamlines (bottom row) in the porous cavity with $Da = 10^{-4}$ and $\varepsilon = 0.6$ at $Ra_{\mathrm{m}} = 1$ in the presence of magnetic field of $Ha = 30$. **b** Isotherms (top row) and streamlines (bottom row) in the porous cavity with $Da = 10^{-4}$ and $\varepsilon = 0.6$ at $Ra_{\mathrm{m}} = 10$ in the presence of magnetic field of $Ha = 30$. **c** Isotherms (top row) and streamlines (bottom row) in the porous cavity with $Da = 10^{-4}$ and $\varepsilon = 0.6$ at $Ra_{\mathrm{m}} = 100$ in the presence of magnetic field of $Ha = 30$. **d** Isotherms (top row) and streamlines (bottom row) in the porous cavity with $Da = 10^{-4}$ and $\varepsilon = 0.6$ at $Ra_{\mathrm{m}} = 1000$ in the presence of magnetic field of $Ha = 30$

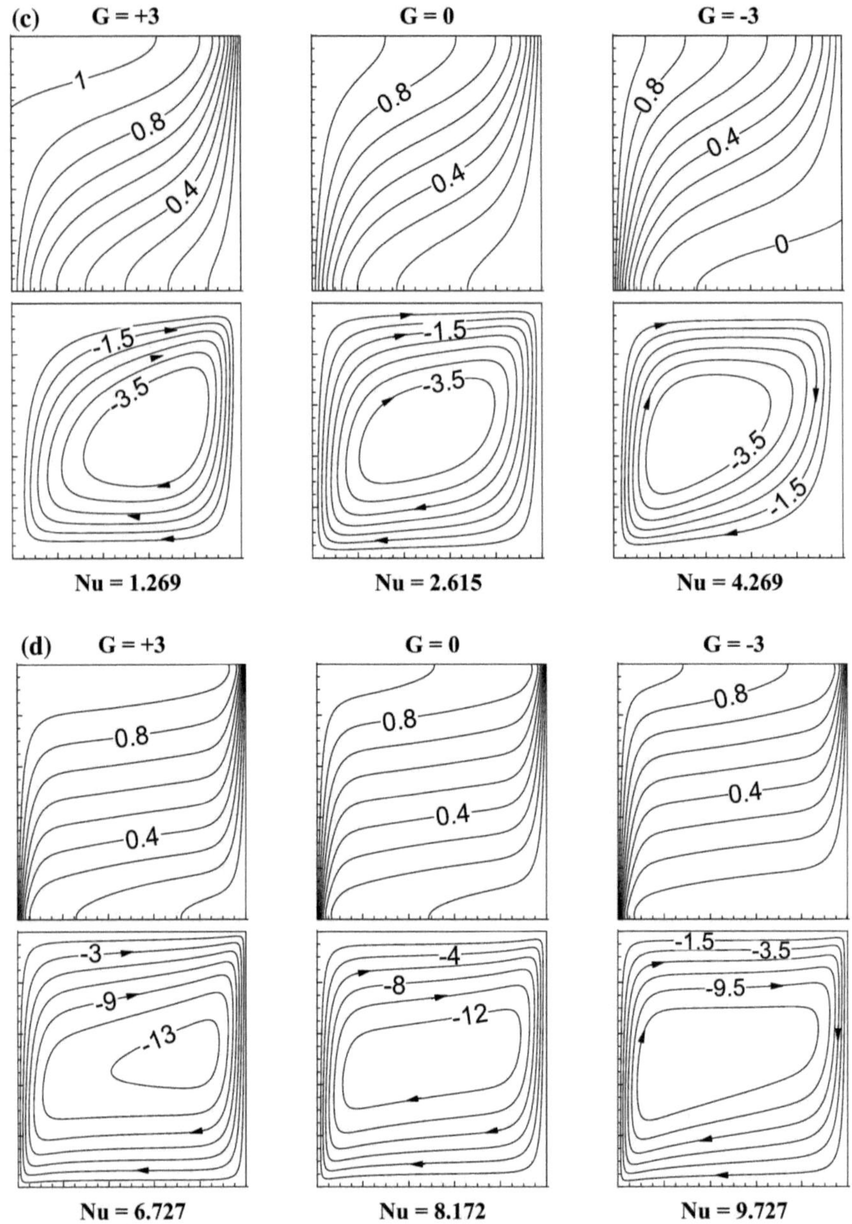

Fig. 26.2 (continued)

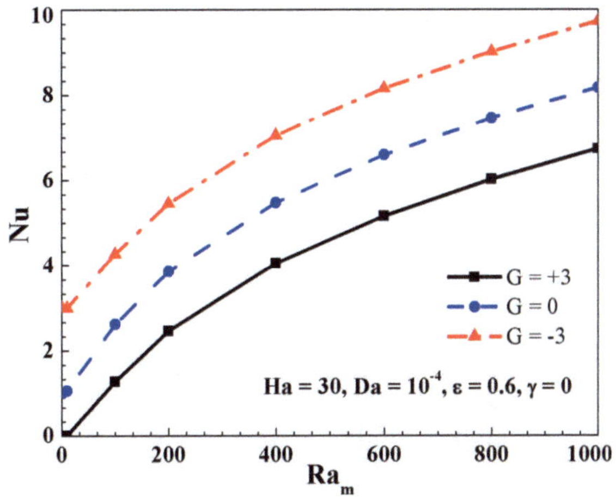

Fig. 26.3 Effect of Ra_m on Nu at $Ha = 30$, $Da = 10^{-4}$, $\varepsilon = 0.6$

top wall, whereas the bottom portion of the left hot wall with heat generation and absorption, respectively. But in case of $G = 0$, they are more distorted in the cavity which does not change much with the change in magnetic field. Overall, with the change in Ha, the change in isotherm contours is very minor and that indicates temperature field is not much affected as the magnetic field increases. Most important changes have been noticed in the flow field. An elliptical cavity has been formed at the center of the enclosure and the elongation of the cavity toward the top right and bottom left corners are observed for heat generation and absorption. The streamlines are also clustered near to the sidewalls with heat generation as well as with heat absorption. The cavity form at the center of the flow field changes it shape and becomes smaller with the increment in Ha. This happens because as magnetic field (in terms of Hartmann number) increases, Lorenz force also increases which restrain the rate of heat transfer. Furthermore, as the strength of magnetic field, the rate of heat transfer reduces by reducing flow velocity. It can also be noted that due to additional heat sinking effect with $G = -3$, apart from the left cooling wall, heat transfer is found lower irrespective of Ha for heat generation.

Figure 26.5 represents the change in Nusselt number with the change in Ha. Decrement in Nusselt number is found irrespective of heat generation–absorption, but the rate of decrement is found higher in the case of heat generation. However, the decrement of rate heat transfer is around 37% for $G = +3$ and 11% for $G = -3$. Fig 26.6 shows the effects of magnetic field inclination angle on the average Nusselt number (Nu) at $Ha = 30$, which indicates almost insignificant effect. At $G = +3$, it is found that up to 90° inclination heat transfer increases by 1%, then start decreasing and become the same as 0° at the end (180°). Same scenario is observed for the other G values ($G = 0$, $G = -3$), but the change in heat transfer is lower in lower G value (that is 0.3% in $G = -3$).

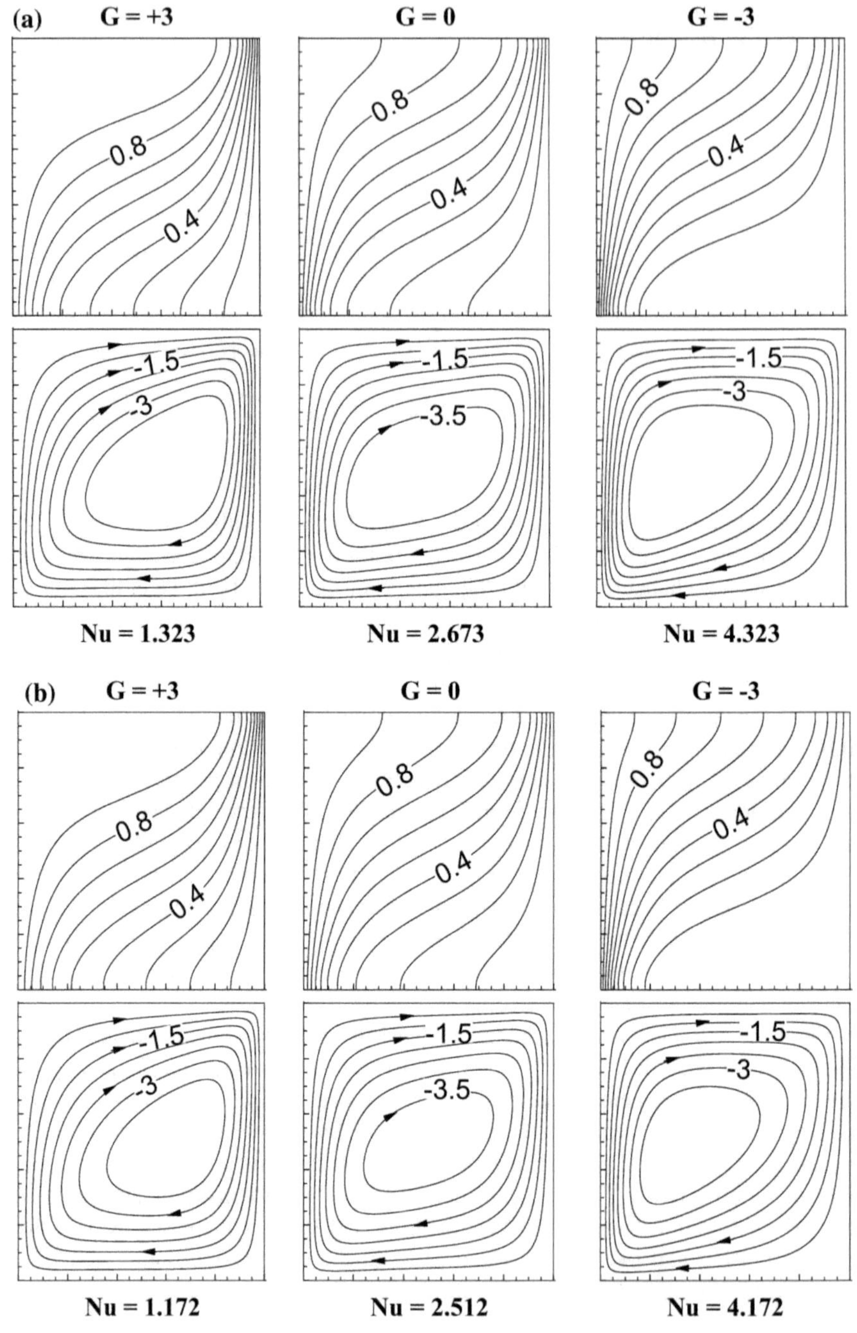

Fig. 26.4 **a** Isotherms (top row) and streamlines (bottom row) in the porous cavity with $Da = 10^{-4}$ and $\varepsilon = 0.6$ at $Ra_m = 100$ in the presence of magnetic field of $Ha = 10$. **b** Isotherms (top row) and streamlines (bottom row) in the porous cavity with $Da = 10^{-4}$ and $\varepsilon = 0.6$ at $Ra_m = 100$ in the presence of magnetic field of $Ha = 50$. **c** Isotherms (top row) and streamlines (bottom row) in the porous cavity with $Da = 10^{-4}$ and $\varepsilon = 0.6$ at $Ra_m = 100$ in the presence of magnetic field of $Ha = 70$. **d** Isotherms (top row) and streamlines (bottom row) in the porous cavity with $Da = 10^{-4}$ and $\varepsilon = 0.6$ at $Ra_m = 100$ in the presence of magnetic field of $Ha = 100$

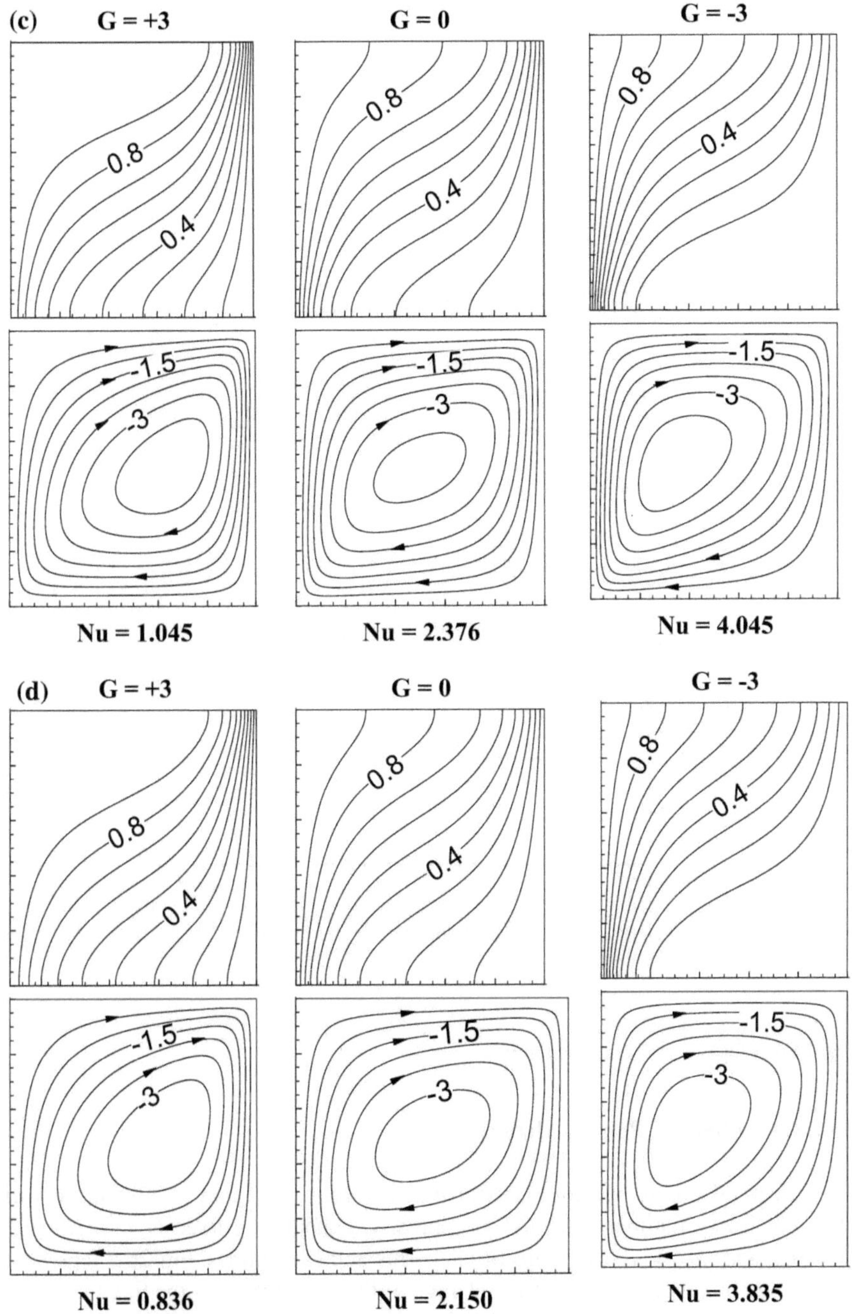

Fig. 26.4 (continued)

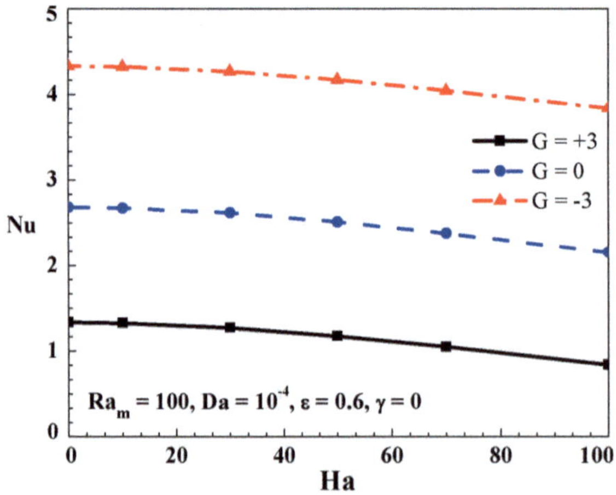

Fig. 26.5 Effect of Ha on Nu at Ra_{m} = 100

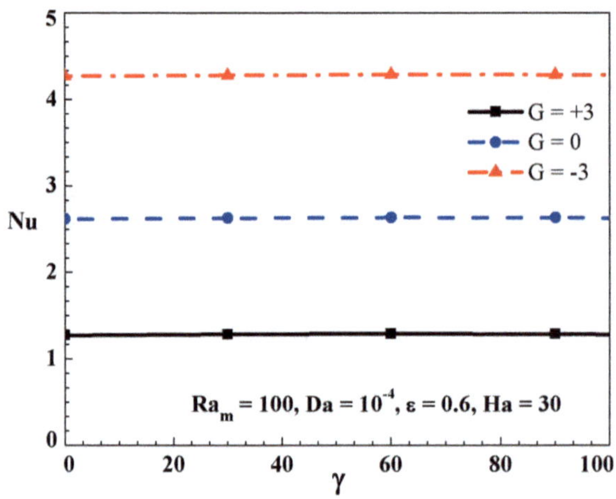

Fig. 26.6 Effect of γ on Nu at Ha = 30

26.3.1.3 Effect of Porous Medium

The effects of porous medium are illustrated in Figs. 26.7 and 26.8 with the variations in Darcy number (Da) and the porosity, considering fixed $G = 0$, $Ha = 30$ ($\gamma = 0°$), and $Ra_{\mathrm{m}} = 100$ in terms of isotherms and streamlines. Fig 26.7 represents the change in heat transfer with the change in Darcy number for the fixed magnetic field. Both thermal and flow field have been found severely affected. At lower Darcy number ($Da = 10^{-6}$ in Fig. 26.7a), isotherm lines are found parallel to each other. Since there is low permeability, fluid velocity is low, and corresponding heat transfer becomes lower. It can also be noted that they are also clustered to a

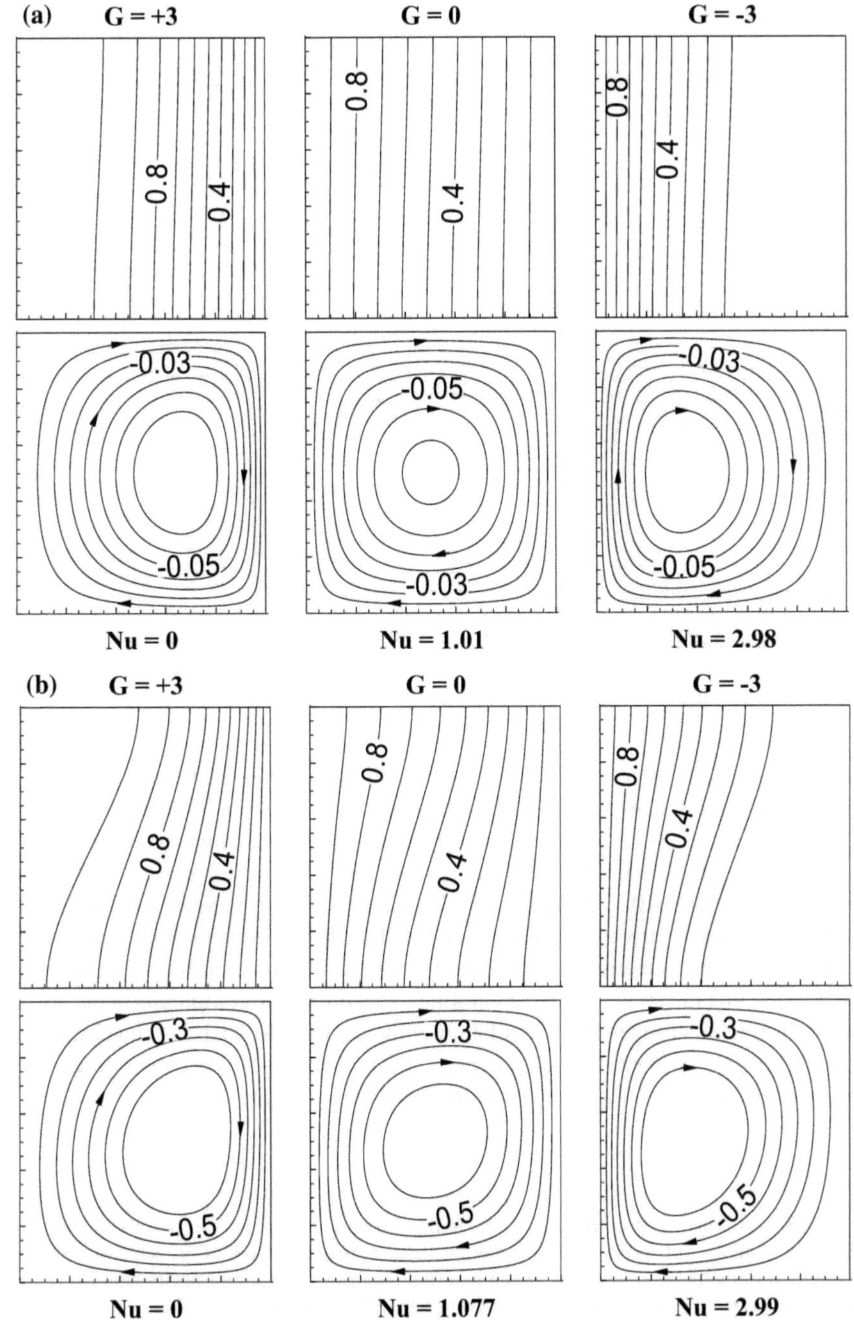

Fig. 26.7 **a** Isotherms (top row) and streamlines (bottom row) in the porous cavity with $Da = 10^{-6}$ and $\varepsilon = 0.6$ at $Ra_{\mathrm{m}} = 100$ in the presence of magnetic field of $Ha = 30$. **b** Isotherms (top row) and streamlines (bottom row) in the porous cavity with $Da = 10^{-5}$ and $\varepsilon = 0.6$ at $Ra_{\mathrm{m}} = 100$ in the presence of magnetic field of $Ha = 30$. **c** Isotherms (top row) and streamlines (bottom row) in the porous cavity with $Da = 10^{-3}$ and $\varepsilon = 0.6$ at $Ra_{\mathrm{m}} = 100$ in presence of magnetic field of $Ha = 30$

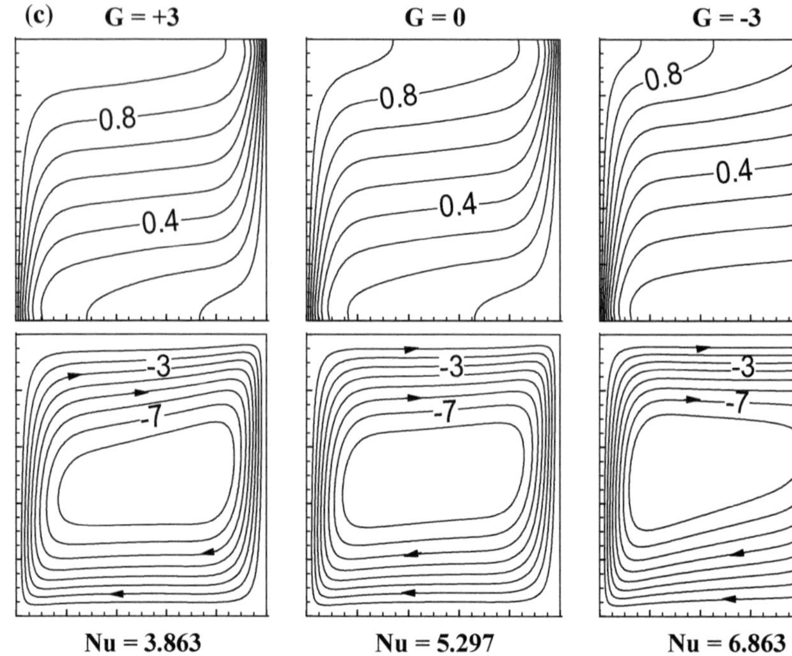

Fig. 26.7 (continued)

particular sidewall at $G = +3$ and $G = -3$. With the increase in *Da* value, isotherm lines become more clustered toward the top right or the bottom left corner. The flow of the fluid becomes strong and develops thin thermal boundary layer near to the sidewalls. In the flow field pattern, a circular cavity is formed in the center of the enclosure at $Da = 10^{-5}$ with $G = 0$ (Fig. 26.7b). With increase in the Darcy number, buoyancy-driven circulation of the fluid flow becomes higher and starts compressing the streamlines toward the sidewalls. It is more prominent at $Da = 10^{-3}$ in Fig. 26.7c where streamlines are too much compressed near to the sidewalls due to stronger fluid circulation in the cavity. The shape of the circulation changes markedly with the increment in Darcy number. It becomes larger and elliptical irrespective of heat generation–absorption.

Fig. 26.8 **a** Isotherms (top row) and streamlines (bottom row) in the porous cavity with $Da = 10^{-4}$ and $\varepsilon = 0.1$ at $Ra_m = 100$ in the presence of magnetic field of $Ha = 30$ as shown in Figure 8. **b** Isotherms (top row) and streamlines (bottom row) in the porous cavity with $Da = 10^{-4}$ and $\varepsilon = 0.3$ at $Ra_m = 100$ in the presence of magnetic field of $Ha = 30$. **c** Isotherms (top row) and streamlines (bottom row) in the porous cavity with $Da = 10^{-4}$ and $\varepsilon = 0.5$ at $Ra_m = 100$ in the presence of magnetic field of $Ha = 30$. **d** Isotherms (top row) and streamlines (bottom row) in the porous cavity with $Da = 10^{-4}$ and $\varepsilon = 1.0$ at $Ra_m = 100$ in the presence of magnetic field of $Ha = 30$ ▶

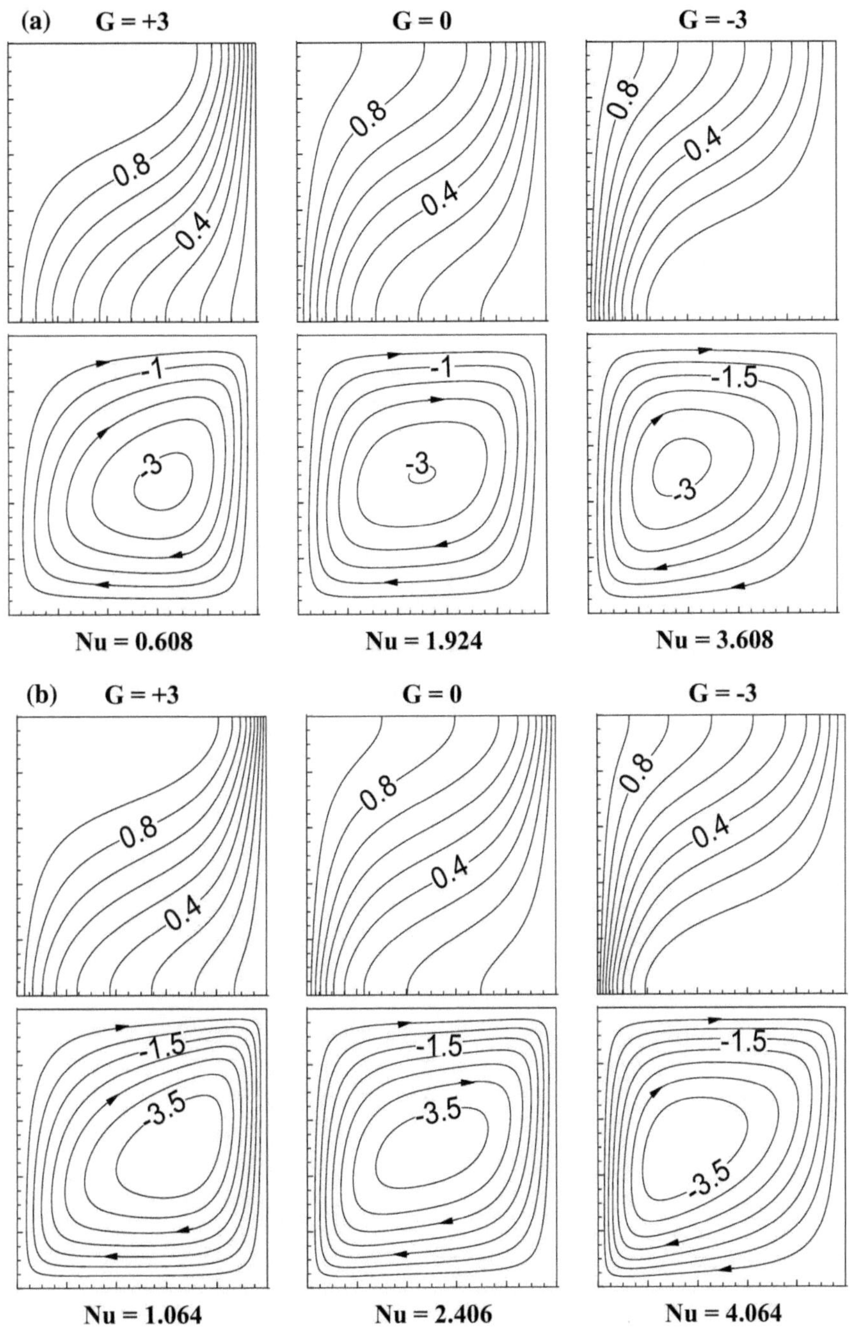
(a)
G = +3
G = 0
G = -3
0.8
0.4
0.8
0.4
0.8
0.4
-1
-3
-1
-3
-1.5
-3
Nu = 0.608
Nu = 1.924
Nu = 3.608
(b)
G = +3
G = 0
G = -3
0.8
0.4
0.8
0.4
0.8
0.4
-1.5
-3.5
-1.5
-3.5
-1.5
-3.5
Nu = 1.064
Nu = 2.406
Nu = 4.064

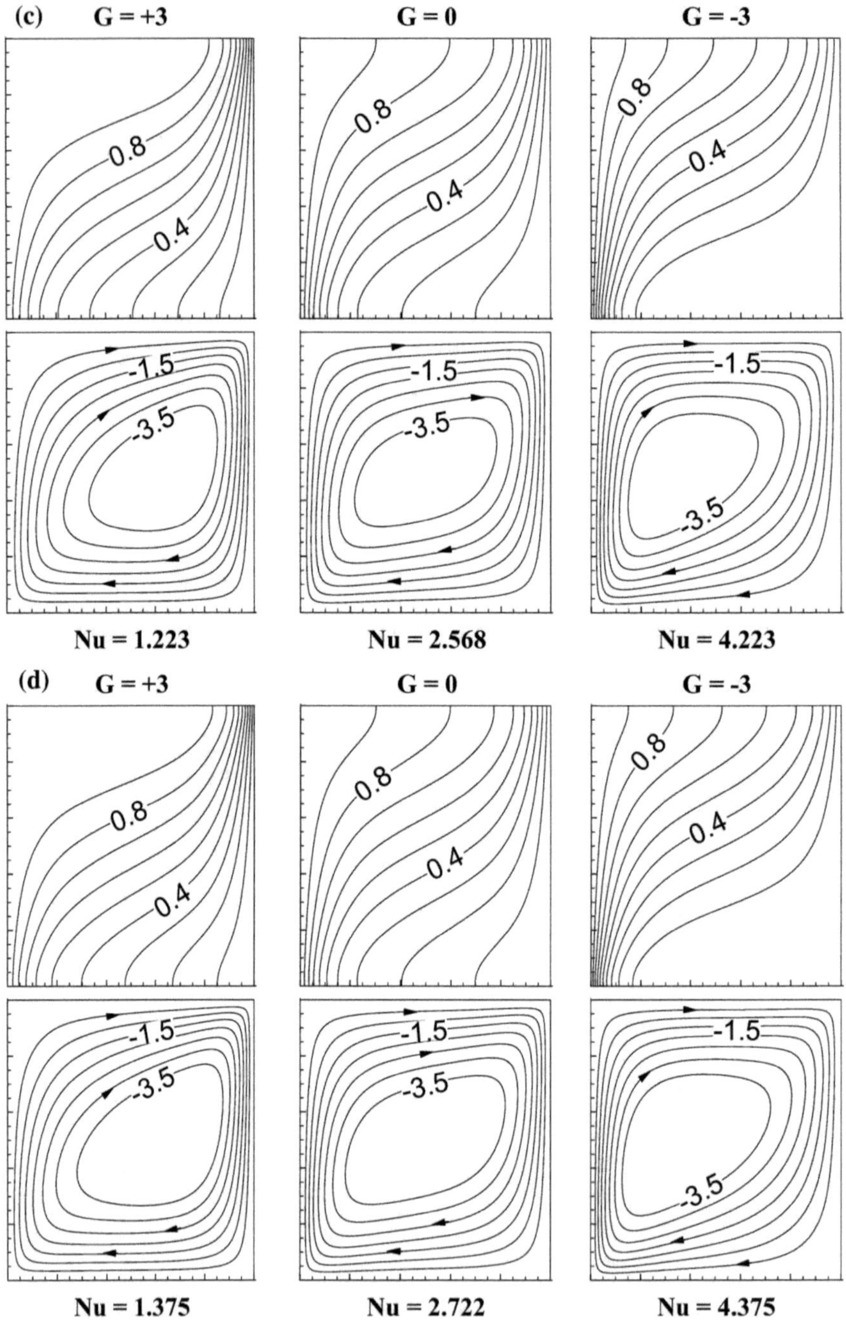

Fig. 26.8 (continued)

The effect of change in porosity (ε) on streamlines and isotherms is displayed in Fig. 26.8. Isotherms are distorted throughout the enclosure for $G = 0$ and for $G = +3$ and $G = -3$, forming thinner thermal boundary wall near to the sidewalls. Changes in isotherm contour plots are prominent as ε changes from 0.1 to 1.0. In the streamline plot, a circular circulation is formed at the center at $G = 0$ in Fig. 26.8a transforms into an ellipse in Fig. 26.8c, as ε increases. Heat generation or absorption leads to shifting of the circulation core to the right or left sidewall.

The parametric effects on the Nusselt number are illustrated in Figs. 26.9 and 26.10. The variation in *Nu* (heat transfer rate) with the change in *Da* is indicated in Fig. 26.9. It is clearly seen that with the increase in *Da*, due to higher permeability and greater fluid velocity, heat transfer increases. With respect to asymptotic value, the maximum heat transfer in terms of *Nu* is found about 81% for $G = 0$. At $G = +3$ and $G = -3$, increment of heat transfer is found 67 and 56%, respectively. Figure 26.10 indicates increased heat transfer at higher porosity. It is observed that rapid change in heat transfer occurs up to $\varepsilon = 0.5$ and *Nu* increases as ε increases and about 29%.

26.3.2 Heat Transfer Assessment in Clear Domain

The effect of fluid-based Rayleigh number and the magnetic field in absence of porous medium is also investigated and is compared with those results presented in Sect. 26.3.1 for $\varepsilon = 0.6$ and $Da = 10^{-4}$, to understand the order of reduction in heat transfer rate. These results have been portrayed in Figs. 26.11 and 26.12 considering fixed angle of inclination of external magnetic field at $\gamma = 0°$ and heat generation $G = 0$. In Fig. 26.11, *Ha* is taken as 30 and it clearly indicates that as Rayleigh number increases heat transfer rate becomes higher. It is because of enhanced thermal convection at higher *Ra*. The reduction of *Nu* with porous

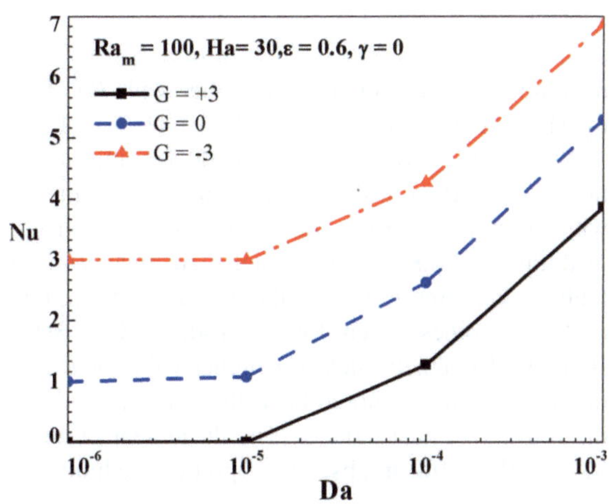

Fig. 26.9 Effect of *Da* on *Nu* at *Ha* = 30

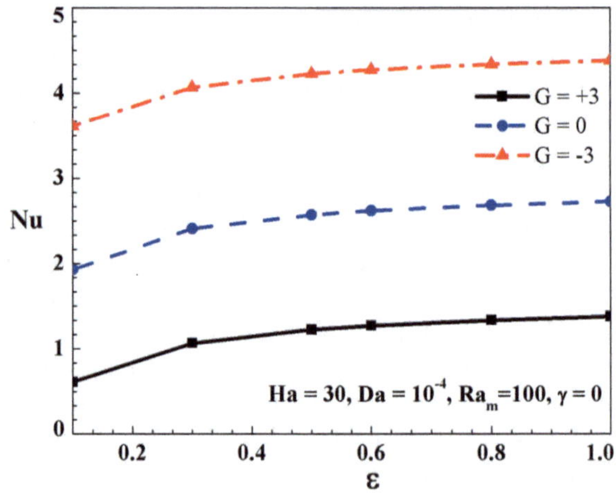

Fig. 26.10 Effect of ε on *Nu* at *Ha* = 30

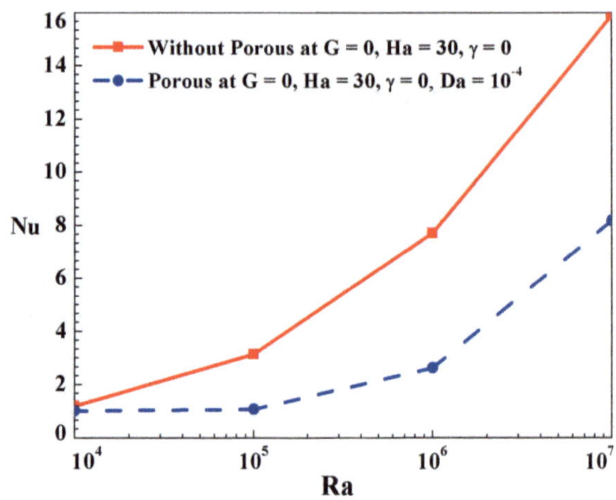

Fig. 26.11 Effect of *Ra* at *Ha* = 30 ($\gamma = 0°$)

medium is found increasingly higher with *Ra*. The maximum increment of heat transfer in clear domain is found 93% at $Ra = 10^7$.

In Fig. 26.12, *Ha* varies up to 100 keeping $Ra = 10^6$ fixed and wider difference in *Nu* is observed in the lower range of *Ha* values. Heat transfer reduces typically with stronger magnetic field but becomes much higher in absence of porous medium (as fluid velocity is high). At lower *Ha*, the heat transfer is around 70% higher comparing to that of the porous medium, but this comparison at higher *Ha* (=100) becomes much lower, around 44%. Besides that, with the increasing strength of magnetic field (or higher value of *Ha*) the heat transfer rate decreases rapidly. Usually, it should be as the magnetic force reduces flow velocity hindering the convective heat transfer. The decrement of heat transfer is found maximum of 57% at *Ha* = 100 in absence of porous medium.

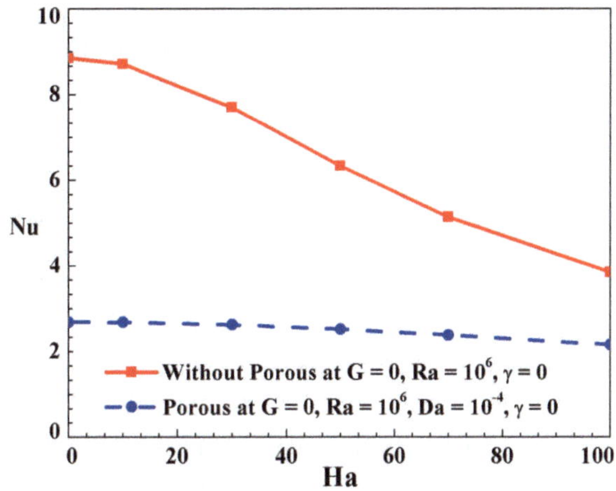

Fig. 26.12 Effect of *Ha* ($\gamma = 0°$)

26.4 Conclusions

The effect of relevant parameters of heat-generating–absorbing MHD convection in porous cavity is investigated systematically. The study reveals that both thermal and flow fields are severely affected by the magnetic field. Lower heat transfer rate is found in case of heat generation than heat absorption, irrespective of all parametric variations. At lower Ra_m due to the rejection of heat through both the sidewalls, when heat generation is higher than the heat input through the hot wall, the Nusselt number becomes very less during heat generation. Thinner boundary layers have been formed at higher *Ra* values, and the same scenarios have been observed for higher *Da* values too. Heat transfer at increased *Da* value enhances due to resulting augmented convective velocity. With the increase of Hartmann number *Ha*, *Nu* decreases irrespective of heat generation–absorption. Lorenz force increases proportionally with *Ha*, which in turn reduces heat transfer. The magnetic field angle is another important issue. However, for the present case, its effect is found marginal.

In comparison to the porous medium, change in heat transfer is marginal at lower *Ra* in absence of porous medium, whereas at increased *Ra*, heat transfer rate also increases and the increment is found higher without porous medium. At lower *Ha*, the increment in heat transfer of porous medium enclosure is found 70% higher than that of without porous medium.

References

1. Sheikholeslami, M.: Numerical simulation of magnetic nanofluid natural convection in porous media. Phy. Lett. A. **381**, 494–503 (2017)
2. Oztop, H., Bilgen, E.: Natural convection in differentially heated and partially. Int. J. Heat Fluid Flow. **27**, 466–475 (2016)

3. Sheremet, M.A., Pop, I.: Conjugate natural convection in a square porous cavity filled by a nanofluid using Buongiorno's mathematical model. Int. J. Heat Mass Transf. **79**, 137–145 (2014)
4. El-Amin, M.F.: Combined effect of internal heat generation and magnetic field. J. Magn. Magn. Mater. **270**, 130–135 (2004)
5. Ghasemi, B., Aminossadati, S.M., Raisi, A.: Magnetic field effect on natural convection in a nanofluid-filled square enclosure. Int. J. Ther. Sci. **50**, 1748–1756 (2011)
6. Sivaraj, C., Sheremet, M.A.: MHD natural convection in an inclined square porous cavity with a heat conducting solid block. J. Magn. Magn. Mater. **426**, 351–360 (2017)
7. Selimefendigil, F., Öztop, H.F.: Natural convection in a flexible sided triangular cavity with internal heat generation under the effect of inclined magnetic field. J. Magn. Magn. Mater. **417**, 327–337 (2016)
8. Bondareva, N.S., Sheremet, M.A.: Effect of inclined magnetic field on natural convection melting in a square cavity witha local heatsource. J. Magn. Magn. Mater. **419**, 476–484 (2017)
9. Mahmoudi, A., Mejri, I., Abbassi, M.A., Omri, A.: Lattice Boltzmann simulation of MHD natural convection in a nanofluid-filled cavity with linear temperature distribution. Powder. Technol. **256**, 257–271 (2014)
10. Sheikholeslami, M., Gorji-Bandpy, M., Ganji, D.D.: Numerical investigation of MHD effects on Al_2O_3–water nanofluid flow and heat transfer in a semi-annulus enclosure using LBM. Energy **60**, 501–510 (2013)
11. Sheikholeslami, M., Ganji, D.D.: Heat transfer of Cu–water nanofluid flow between parallel plates. Powder. Technol. **235**, 873–879 (2013)
12. Sheikholeslami, M., Ganji, D.D., Ashorynejad, H.R.: Investigation of squeezing unsteady nanofluid flow using ADM. Powder. Technol. **239**, 259–265 (2013)
13. Kefayati, G.H.R.: Lattice Boltzmann simulation of MHD natural convection in a nanofluid-filled cavity with sinusoidal temperature distribution. Powder. Technol. **243**, 171–183 (2013)
14. Kefayati, G.H.R.: Effect of a magnetic field on natural convection in an open cavity subjugated to water/alumina nanofluid using lattice Boltzmann method. Int. Commun. Heat. Mass. Transf. **40**, 67–77 (2013)
15. Mliki, B., Abbassi, M.A., Omri, A., Zeghmati, B.: Augmentation of natural convective heat transfer in linearly heated cavity by utilizing nanofluids in the presence of magnetic field and uniform heat generation/absorption. Powder. Technol. **284**, 312–325 (2015)
16. Malik, S., Nayak, A.K.: MHD convection and entropy generation of nanofluid in a porous enclosure with sinusoidal heating. Int. J. Heat. Mass. Transf. **111**, 329–345 (2017)
17. Mojumder, S., Rabbi, K.M., Saha, S., Hasan, M.N., Saha, S.C.: Magnetic field effect on natural convection and entropy generation in a half-moon shaped cavity with semi-circular bottom heater having different ferrofluid inside. J. Magn. Magn. Mater. **407**, 412–424 (2016)
18. Bondareva, N.S., Sheremet, M.A., Pop, I.: Magnetic field effect on the unsteady natural convection in a right-angle trapezoidal cavity filled with a nanofluid. Int. J. Numer. Methods. Heat Fluid Flow. **25**, 1924–1946 (2015)
19. Sheremet, M.A., Pop, I., Rosca, N.C.: Magnetic field effect on the unsteady natural convection in a wavy-walled cavity filled with a nanofluid: Buongiorno's mathematical model. J. Taiwan Inst. Chem. Eng. **61**, 211–222 (2016)
20. Karimipour, A., Taghipour, A., Malvandi, A.: Developing the laminar MHD forced convection flow of water/FMWNT carbon nanotubes in a microchannel imposed the uniform heat flux. J. Magn. Magn. Mater. **419**, 420–428 (2016)
21. Malvandi, A.: Film boiling of magnetic nanofluids (MNFs) over a vertical plate in presence of a uniform variable-directional magnetic field. J. Magn. Magn. Mater. **406**, 95–102 (2016)
22. Malvandi, A., Safaei, M.R., Kaffash, M.H., Ganji, D.D.: MHD mixed convection in a vertical annulus filled with Al_2O_3–water nanofluid considering nanoparticle migration. J. Magn. Magn. Mater. **382**, 296–306 (2015)

23. Rashad, A.M., Armaghani, T., Chamkha, A.J., Mansour, M.A.: Entropy generation and MHD natural convection of a nanofluid in an inclined square porous cavity: effects of a heat sink and source size and location. Chinese J. Phy. **56**, 193–211 (2018)
24. Gibanov, N.S., Sheremet, M.A., Oztop, H.A., Al-Salem, K.: MHD natural convection and entropy generation in an open cavity having different horizontal porous blocks saturated with a ferrofluid. J. Magn. Magn. Mater. **452**, 193–204 (2018)
25. Yu, P.X., Qiu, J.X., Qin, Q., Tian, Z.F.: Numerical investigation of natural convection in a rectangular cavity under different directions of uniform magnetic field. Int. J. Heat Mass Transf. **67**, 1131–1144 (2013)
26. Grosan, T., Revnic, C., Pop, I., Ingham, D.B.: Magnetic field and internal heat generation effects on the free convection in a rectangular cavity filled with a porous medium. Int. J. Heat Mass Transf. **52**, 1525–1533 (2009)
27. Revnic, C., Grosan, T., Pop, I., Ingham, D.B.: Magnetic field effect on the unsteady free convection flow in a square cavity filled with a porous medium with a constant heat generation. Int. J. Heat Mass Transf. **54**, 1734–1742 (2011)
28. Jiang, C., Feng, W., Zhong, H., Zeng, J., Zhu, Q.: Effects of a magnetic quadrupole field on thermomagnetic convection of air in a porous square enclosure. J. Magn. Magn. Mater. **357**, 53–60 (2014)
29. Nayak, A.K., Malik, S., Venkateshwarlu, K., Jena, P.K.: Magneto-convection and its effect on partially active thermal zones in a porous square domain. Int. J. Heat Mass Transf. **95**, 913–926 (2016)
30. Nkurikiyimfura, I., Wang, Y., Pan, Z.: Heat transfer enhancement by magnetic nanofluids—a review, Renew. Sustain Energy Rev. **21**, 548–561 (2013)
31. Sarkar, S., Ganguly, S., Biswas, G.: Buoyancy driven convection of nanofluids in an infinitely long channel under the effect of a magnetic field. Int. J. Heat Mass Transf. **71**, 328–340 (2014)
32. Chamkha, A.: Hydromagnetic combined convection flow in a vertical lid-driven cavity enclosure with internal heat generation or absorption. Numer. Heat Transf. Part A **41**, 529–546 (2002)
33. Sheikholeslami, M., Hayat, T., Alsaedi, A.: MHD free convection of Al_2O_3-water nanofluid considering thermal radiation: a numerical study. Int. J. Heat Mass Transf. **96**, 513–524 (2016)
34. Sheikholeslami, M., Rashidi, M., Hayat, T., Ganji, D.: Free convection of magnetic nanofluid considering mfd viscosity effect. J. Mol. Liq. **218**, 393–399 (2016)
35. Selimefendigil, F., Oztop, H.F.: MHD mixed convection and entropy generation of power law fluids in a cavity with a partial heater under the effect of a rotating cylinder. Int. J. Heat Mass Transf. **98**, 40–51 (2016)
36. Nield, D.A., Bejan, A.: Convection in Porous Media, 3rd edn. Springer, Berlin, Germany (2006)
37. Patankar, S.V.: Numerical Heat Transfer and Fluid Flow. Taylor and Francis, New York, Hemisphere (1980)
38. Datta, P., Mahapatra, P.S., Ghosh, K., Manna, N.K., Sen, S.: Heat transfer and entropy generation in a porous square enclosure in presence of an adiabatic block. Transp. Porous Media **111**, 305–329 (2016)
39. Ghasemi, B., Aminossadati, S.M., Raisi, A.: Magnetic field effect on natural convection in a nanofluid-filled square enclosure. Int. J. Ther. Sci. **50**, 1748–1756 (2011)

Chapter 27
Studies on Fluid Flow Through an Elliptical Microchannel of Different Aspect Ratios

Sudip Simlandi, Soumyanil Nayek and Raunak Joshi

Nomenclature

A	Aspect ratio ($=b/a$)
B	Coefficient defined in Eq. (27.12)
a	Major semi-axis of ellipse (m)
b	Minor semi-axis of ellipse (m)
D_h	Hydraulic diameter (m)
dA	Differential area (m^2)
F	Tangential momentum accommodation coefficient (=1)
$I[U]$	Functional defined in Eq. (27.15)
Kn	Knudsen number ($=\lambda/D_h$)
Po	Poiseuille number
$\frac{dp}{dz}$	Pressure gradient in the flow direction (N/m^3)
U	Dimensionless velocity considering no-slip condition
$\overline{U}$	Normalized dimensionless velocity considering slip
u	Velocity along the axis of the channel (m/s)
u_m	Mean velocity along the axis of the channel (m/s)
u_s	Slip velocity at the walls (m/s)
x	X-axis coordinate
X	Dimensionless coordinate ($=x/a$)
y	Y-axis coordinate
Y	Dimensionless coordinate ($=y/b$)

Greek symbols

μ	Dynamic viscosity (Pa s)
β	Slip coefficient

S. Simlandi (✉) · S. Nayek · R. Joshi
Department of Mechanical Engineering, Jadavpur University, Kolkata, India
e-mail: sudip.simlandi@gmail.com

P. Sahoo and J. P. Davim (eds.), *Advances in Materials, Mechanical and Industrial Engineering*, Lecture Notes on Multidisciplinary Industrial Engineering,
https://doi.org/10.1007/978-3-319-96968-8_27

Subscripts

m Mean
ns No-slip
s Slip

27.1 Introduction

In recent days, fluid flow within microchannels has become a key important area of research by their wide applications in computer chips, biomedical and medical, as well as chemical separations [1–5]. Components like microvalves, micropumps, microchannel heat sinks, and actuators have been miniaturized, integrated, and assembled developing a variety of microfluidic devices and systems. In this connection, the micro-electro-mechanical systems (MEMS) has become a new research area in which non-continuum behavior is significant. Microchannels are defined as channels whose characteristic dimensions are within 1 μm to 1 mm. It is observed that fluid flow through channels having characteristic dimensions above 1 mm exhibits continuum behavior [6]. The basic concepts of flow behavior such as velocity profile play an important role in design and process control for microfluidic applications [2]. The common channel shapes that are produced through microfabrication are non-circular cross sections such as elliptic, rectangular, and trapezoidal.

For gas flow through a microchannel or microtube, the mean free path (λ) of gas molecules is comparable with the characteristic length of the system. The velocity slip as well as temperature jump conditions on the solid wall need to be considered under such situation as the no-slip boundary conditions become no longer valid. If the fluid is at low pressure, it is required to incorporate the slip boundary condition. The Knudsen number (Kn), which is the ratio of molecular mean free path of gas to characteristic dimension of the duct (λ/D_h) is generally used to represent rarefaction effect of gas. The mean free path is the average distance traveled by a moving molecule between successive collisions which modify its direction of travel, or energy or other molecule properties. For example, the mean free path of some gases is shown in Table 27.1 at atmospheric conditions.

Continuum equations are applicable for $Kn < 10^{-3}$. As the Kn increases, the flow reaches the regime of slip flow. The slip-flow regime where the value of Knudsen number lies between 0.001 and 0.1, is termed as "slip flow" [7–9]. For transition flow regime Knudsen number ranges $10^{-1} < Kn < 10$, and eventually for

Table 27.1 Mean free path of some atmospheric gases at room temperature

Nitrogen	Air	Hydrogen	Helium
66 nm	68 nm	125 nm	194 nm

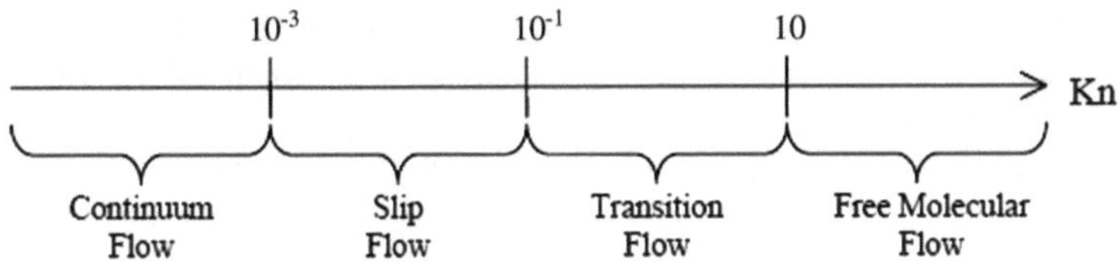

Fig. 27.1 Flow regimes classification

the free-molecular flow regime $Kn > 10$. These different four regimes are shown in Fig. 27.1. It is observed that the regime of slip flow closely behaves as that of a continuum regime. Hence, the heat transfer characteristics in the slip-flow regime can be modeled using continuum equations subjected to slip-flow boundary conditions [10]. For such flow, the Navier–Stokes equation needs to be combined with the slip-flow boundary condition [11, 12]. The basic concepts of flow characteristics such as velocity distribution play a crucial role in design and process control for applications in the areas of microfluidic [2].

Hence, understanding of the fluid flow characteristics within microchannel is very essential for determination of velocity distribution, temperature distribution, and transport properties of the flow field. In this connection, Hooman [13] presented a superposition technique to study forced convection phenomena within a microduct of arbitrary cross section for the regime of slip flow. It is shown in their work that the no-slip or no-jump results can still be used with some little modifications if one can apply an average slip velocity and temperature jump condition. On the other hand, a 2-*D* temperature field and Nusselt numbers as function of aspect ratio are also predicted by Morini [12] in case of fully developed thermal region for ducts of rectangular cross section. He considered laminar, fully developed velocity profile with constant wall temperature. Das and Tahmouresi [14] studied an analytical solution for fully developed gaseous slip flow within a microchannel of elliptic cross section. The governing equation is solved analytically for different aspect ratios in an elliptic cylindrical coordinate system. They applied integral transform technique to analyze the gaseous slip flow within the microchannel. Kuddusi [15] studied the slip flow within microchannels of rectangular cross section with constant and uniform wall temperature conditions. He considered eight possible thermal versions which are formed by different combinations of adiabatic and heated walls. Theofilis et al. [16] predicted velocity profile for fluid flow in a rectangular channel by solving the Navier–Stokes equation. They considered that the pressure gradient along the length of the channel is constant. Peng et al. [17] presented analytical results for viscous flow through a tube of equilateral triangular to irregular triangular cross sections. They determined average velocity as well as volume flow rate utilizing the velocity profile for the triangular channel. Rybiński and Mikielewicz [18] predicted Fanning friction factor and for Nusselt number analytically for laminar fully developed fluid flow within straight rectangular mini channels. They determined analytical solutions, both for velocity and temperature distributions. Chakraborty [19] considered flow problems within microchannels of various arbitrary cross sections using analytical solution methods, namely membrane vibration analogy, complex function analysis, and variational method. Kundu et al. [20] established approximate analytical techniques to predict the velocity profile considering laminar flow and no-slip boundary condition through straight rectangular channels. They adopted some exact and approximate analytical solutions for the prediction of velocity and temperature distributions. Lee and Garimella [21] studied heat transfer characteristics for laminar flow through ducts of rectangular cross section. They considered the thermal entrance region at different values of aspect ratios ranging from 1 to 10. A generalized relationship for local and average Nusselt numbers in the thermal entrance region has been presented. The local and average Nusselt numbers in the entrance region have also been determined numerically as functions of channel aspect ratio and non-dimensional axial length. Mala and Lee [2] carried out an experiment to investigate water flow characteristics in microtubes made of stainless steel and fused silica and attempted to explain the obtained results. The ranges of microtube diameter from 50 to 254 μm have been considered in their analysis. They concluded that the necessary pressure gradient to sustain the liquid flow within the microtube is more compared to that proposed in the conventional theory considering a fixed volume flow rate. The experimental result is in a rough agreement with the conventional theory when the flow rate or Reynolds number is low. Finally, a substantial deviation from the conventional theory is noticed when the Reynolds number is increased. Zhang et al. [22] analytically investigated a 2-*D* flow in microchannels considering no-slip conditions. They solved the strongly non-linear ordinary differential equation by adopting homotopy analysis method (HAM). It is concluded from the analysis that the velocity distribution becomes flatter due to the effect of slip at solid wall. They also observed that the flow velocity becomes constant at a location across the duct for any value of Knudsen number. Zhu and Liao [23] analyzed heat transfer characteristics considering laminar-forced convection for gas flow in a microchannel of arbitrary cross section. They used the orthonormal function method in the slip-flow regime considering constant heat flux along the axial direction and unsymmetric wall temperature circumferentially. They studied the heat transfer characteristics for different aspect ratio, Knudsen number, and thermal boundary conditions for microchannels of both rectangular and triangular cross sections. Their results showed that the orthonormal function method is valid for analysis of the fluid flow and heat transfer characteristics for varyingly heated microchannel of an arbitrary cross section with the incorporation of wall slip and temperature jump conditions. Dongari et al. [24] studied an analytical gas flow characteristics through a long microchannel with a second-order exact wall slip boundary condition. They determined solutions as the function of the slip coefficients and compared proposed work against experimental data. Their predictions are also valid for rarefied gas within a macrochannel. Ngoma and Erchiqui [25] numerically investigated liquid flow through a microchannel between two parallel plates with slip boundary and imposed heat flux conditions. They considered both the effect of pressure-driven and electroosmosis flow in their analysis. They obtained a system of governing differential equations using the Poisson–Boltzmann, the modified Navier–Stokes, and the energy equations. They showed that the fluid flow and heat transfer characteristics in a microchannel get affected greatly by the pressure difference, the wall slip coefficient, the heat flux, and the electrokinetic separation distance. Larrode et al. [26] investigated slip-flow heat transfer considering both the wall slip and the temperature jump boundary conditions through a circular tube. They found that both the degree of rarefaction and the surface accommodation coefficients can affect the heat transfer behavior. Tunc and Bayazitoglu [27] investigated convection heat transfer in a fully developed microchannel of rectangular cross section. They considered boundary condition of H2 type considering constant peripheral and axial heat flux conditions at the channel solid walls. The integral transform technique adopted for calculation of velocity and temperature distributions. The velocity distribution as the functions of rarefaction and channel size is shown in their work. Kuddusi [28] studied fluid flow through microchannels of rectangular cross section heated at constant and uniform heat flux. They determined the temperature profile in the microchannel using mathematical similarity between the heat conduction and convection problems. They showed that heat transfer in the microchannels is decreasing effected by the rarefaction. The microchannel has been exposed to any eight thermal versions of heat flux. The proposed results of their work agreed well with the results obtained for macrochannels in the existing literature for the no-slip-flow boundary condition.

It is found from the above literatures that the analytical solutions for fluid flow through an elliptic microchannel are less developed and the methods applied in the existing work may be complex and lengthy. Therefore, the objective of the present study is to establish a simple analytical solution for gaseous flow within an elliptical microchannel for different aspect ratio. The Navier–Stokes equation subjected to slip boundary conditions is solved analytically by variational method. It is observed from the existing literature that the Poiseuille number and slip coefficient are parameters of interest for flow within a microchannel. The variational method plays a significant role, because the momentum equation which becomes very challenging to solve using an exact analytical technique can be easily solved with the help of those techniques. It is predicted that incorporating an average slip velocity, the no-slip results with some little modifications can still be used. Predicted results for gaseous slip flow in elliptic microchannel agree well with the existing literature. Finally, the work is extended for a rectangular microchannel using the same variational formulation. Finally, the results obtained have been shown in a comparative way to easily recognize the difference in hydrodynamic phenomena between an elliptical and a rectangular cross sections of microchannels.

27.2 Review of Slip Models for Gases

In this section, the case of gas flow within microchannels which have Knudsen numbers between 0.001 and 0.1 is considered. In such case, it is already mentioned that the hydrodynamic equations can be applied but the boundary conditions need to be changed. Specifically, the gas can slip along the surface of solid wall. This condition of null velocity at the level of the stationary surface can be replaced by the following condition [29]

$$u_s = C\lambda \frac{\partial u}{\partial n}\bigg|_{\text{wall}} \tag{27.1}$$

where C is a coefficient, λ is the mean free path, and n the direction normal to the surface. The following approach suggested by Maxwell to determine the velocity slip [30]. Over a control surface, 50% of the molecules at a distance of λ/2, come from one mean free path away from the surface with tangential velocity, and the remaining 50% of the molecules get reflected from the surface. Maxwell assumed that a fraction F of the molecules get reflected diffusively at the surface and the remaining $(1 - F)$ of the molecules are reflected. Based on the assumption, Maxwell expressed the velocity slip as

$$u_s = \frac{2-F}{F}\left[Kn\left(\frac{\partial u}{\partial n}\right)_{\text{wall}} + \frac{Kn^2}{2}\left(\frac{\partial^2 u}{\partial n^2}\right)_{\text{wall}}\right] \tag{27.2}$$

In the above equation, F is called tangential momentum accommodation coefficient. The values of F equals to one for fully diffused surfaces and zero for specular surfaces are considered. n is the distance normal to the solid wall. The above model has been modified by Lam [31]. He proposed the following alternative expression of velocity slip for simplicity of analysis.

$$u_s = \frac{2-F}{F}\left[\frac{Kn}{1-bKn}\left(\frac{\partial u}{\partial n}\right)_{\text{wall}}\right] \tag{27.3}$$

where, b is a parameter which is determined based on empirical relations. The above slip model was further modefied by Beskok and Karniadakis [32] and Karniadakis and Beskok [33], for solution of gas flow at the level of microscales. Beskok and Karniadakis [34] expanded the velocity field as the function of Kn by perturbation for determination of the value b. They proposed a second-order expression of b in the regime of slip flow (where $Kn \leq 0.1$) as

$$b = \left(\frac{\partial^2 u}{\partial n^2}\right)_{\text{ns}} \bigg/ \left(2\frac{\partial u}{\partial n}\right)_{\text{ns}} \tag{27.4}$$

where subscript ns stands for no-slip and n is the distance normal to the surface. Beskok and Karniadakis [32] carried out analysis for transition and free-molecular regimes adopting the above model. On the other hand, Xue and Fan [35] proposed a model to determine the slip velocity replacing Kn by $\tanh(Kn)$ and neglecting the second-order term in Eq. (27.2). In this connection, Sreekanth [36] suggested a second-order slip model for slip velocity as

$$u_s = -C_1\lambda\left(\frac{\partial u}{\partial n}\right)_{\text{wall}} - C_2\lambda^2\left(\frac{\partial^2 u}{\partial n^2}\right)_{\text{wall}} \tag{27.5}$$

which can be rewritten as

$$u_s = -C_1 Kn\left(\frac{\partial u}{\partial n}\right)_{\text{wall}} - C_2 Kn\left(\frac{\partial^2 u}{\partial n^2}\right)_{\text{wall}} \tag{27.6}$$

The above model is analogous to the model proposed by Maxwell except two independent coefficients C_1 and C_2 unlike a single coefficient (2 − F)/F in Eq. It is seen that general agreement on the values of the slip coefficients C_1 and C_2 is absent. Though, the slip velocity model based on Eq. (27.5) is well established.

On the other hand, Hooman [13] used the first-order slip velocity as

$$u_s = \frac{F-2}{F} Kn D_h \left.\frac{\partial u}{\partial n}\right|_{\text{wall}} \tag{27.7}$$

For present study, the above Eq. (27.7) is adopted for slip velocity as the model is simple and useful.

27.3 Description of the Physical Problem

The present work considers laminar fully developed pressure-driven gas flow within a straight microchannel of elliptical cross section as presented in Fig. 27.2. The center of channel cross section is chosen as origin of the coordinates. The gas flows under steady-state condition only along the z direction. The major axis of the cross section parallel to x-axis is $2a$, and the minor axis of the cross section parallel to y-axis is $2b$. The gas properties are considered constant. The following assumptions are made in the present study.

(a) The Navier–Stokes equation subjected to slip boundary condition is used as the governing equations in the present analysis.

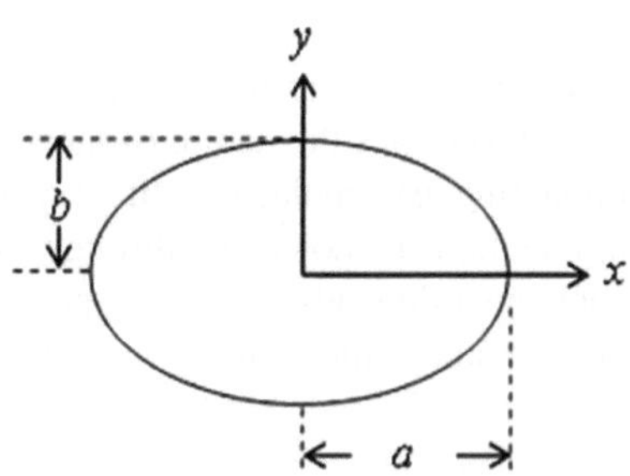

Fig. 27.2 Schematic diagram of the cross section

(b) The fluid is considered incompressible with constant properties.
(c) The two-dimensional steady-state flow condition is considered.
(d) The flow is fully developed laminar.
(e) The body forces as well as the influence of viscosity heating are disregarded.
(f) The heat transfer analysis is not carried out in the present analysis.

27.4 Mathematical Analysis

In the present work, a viscous fully developed laminar fluid flow through a straight microchannel of elliptical cross section is considered. Further, the work is extended for a rectangular cross section to draw a comparative study with the elliptical one. For a hydrodynamically developed flow through the microchannel, the momentum conservation equation along axial direction can be written as [13, 15, 19, 37].

$$\frac{\partial^2 u}{\partial x^2} + \frac{\partial^2 u}{\partial y^2} = \frac{1}{\mu}\frac{\mathrm{d}p}{\mathrm{d}z} \tag{27.8}$$

The following slip boundary conditions for the above Eq. (27.8) are used

$$u = u_s \text{ at } x = \pm a, \text{ and } y = \pm b \tag{27.9a}$$

$$\frac{\partial u}{\partial x} = 0 \text{ at } x = 0, \text{ and } \frac{\partial u}{\partial y} = 0, \text{ at } y = 0 \tag{27.9b}$$

The mean flow velocity is determined as

$$u_m = \int_0^a \int_0^b u \mathrm{d}x\mathrm{d}y \Bigg/ \int_0^a \int_0^b \mathrm{d}x\mathrm{d}y$$

The slip velocity u_s at the wall is expressed as [13, 15, 38].

$$u_s = \frac{F-2}{F} KnD_h \frac{\partial u}{\partial n}\bigg|_{\text{wall}} \tag{27.10}$$

Here, F refers to tangential momentum accommodation coefficient. The coefficient F signifies the nature of tangential momentum between the solid wall and the impinging gas molecules. It may be explained as the fraction of the tangential momentum of molecules through collisions with the solid surface. Normally, the coefficient depends on the temperature, surface roughness, and type of gas. In the present study, the analysis is performed by considering F equals to 1.

The above boundary conditions are non-homogeneous. Hence, to eliminate non-homogeneity of the boundary conditions and making the results unit free, the governing Eq. 27.8 has been non-dimensionalized incorporating the following dimensionless variables as [13].

$$U = -\frac{\mu}{a^2(\mathrm{d}p/\mathrm{d}z)}(u - u_s),\ X = \frac{x}{a},\ Y = \frac{y}{b}\ \text{and}\ A = \frac{b}{a}$$

Substituting these variables in the above Eq. 27.8, the governing momentum equation becomes

$$\frac{\partial^2 U}{\partial X^2} + \frac{1}{A^2}\frac{\partial^2 U}{\partial Y^2} = -1 \tag{27.11}$$

The non-dimensional form of the boundary conditions becomes

$$U = 0, \quad \text{at}\ X = \pm 1\ \text{and}\ Y = \pm 1$$

$$\frac{\partial U}{\partial X} = 0 \quad \text{at}\ X = 0,\ \text{and}\ \frac{\partial U}{\partial Y} = 0, \quad \text{at}\ Y = 0$$

Now, it is observed that Eq. 27.11 becomes the non-dimensional form of Navier–Stokes equation for fully developed flow subjected to no-slip boundary condition. Hence, the flow velocity U can be considered as velocity for no-slip condition and may be expressed as U_{ns}. However, as flow is taking place through a microchannel, hence flow velocity at the boundary of the microchannel is not equal to zero. Assuming slip velocity (u_s) to be constant at the boundary of the microchannel, the normalized velocity for flow through the microchannel can be expressed as [13].

$$\overline{U}(\mathrm{X}, \mathrm{Y}) = B\overline{U}_{ns} + 1 - B \tag{27.12}$$

where the term $B = \left[1 + \left(\frac{2-F}{FU_{ns,m}} Kn\left(\frac{2A}{1+A}\right)^2\right)\right]^{-1}$, the normalized no-slip velocity $\overline{U}_{ns} = U/U_{ns,m}$, and mean no-slip velocity $U_{ns,m} = \int_0^1\int_0^1 U\mathrm{d}X\mathrm{d}Y \Big/ \int_0^1\int_0^1 \mathrm{d}X\mathrm{d}Y$

Therefore, Eq. 27.12 has been solved to obtain the velocity distribution for slip flow using variational method. For flow through microchannel, the slip coefficient and Poiseuille number are important parameters of interests. The slip coefficient ($\beta = u_s/u_m$) is measured as the ratio of velocity slip at the solid wall to the mean flow velocity. The Poiseuille number is expressed as $Po = fRe$. Hence, both the above parameters have also been evaluated using the following equations [13].

$$\beta = 1 - B \tag{27.13}$$

$$Po = B\left(\frac{2}{U_{\mathrm{ns,m}}}\right)\left(\frac{2A}{1+A}\right)^2 \tag{27.14}$$

27.4.1 Solution Method

An approximate analytical solution of the governing Eq. 27.11 is given by using variational formulation. The velocity profile can be obtained by minimizing the following functional [18].

$$I[U] = \iint \left\{ \frac{1}{2}\left[\left(\frac{\partial U}{\partial X}\right)^2 + \frac{1}{A^2}\left(\frac{\partial U}{\partial Y}\right)^2\right] - U \right\} \mathrm{d}A \tag{27.15}$$

where $U(X, Y)$ is a function that satisfies the essential boundary condition $U(X, Y) = 0$ at the solid wall. The functional $I[U]$ reaches optimum value when Eq. 27.11 is satisfied. The boundary of the elliptical cross section for the present flow problem is considered as

$$\varphi(X, Y) = X^2 + Y^2 - 1 = 0 \tag{27.16}$$

Now, a trial function which is to be considered in the analysis is taken as

$$U(X, Y) = m\left(X^2 + Y^2 - 1\right) \tag{27.17}$$

The unknown coefficient m is determined when $I[U]$ is minimum with respect to m. Therefore, Eq. 27.15 can be rewritten as

$$I[U] = \iint \left\{ 2m^2\left[X^2 + \frac{1}{A^2}Y^2\right] - m\left(X^2 + Y^2 - 1\right) \right\} \mathrm{d}X\mathrm{d}Y \tag{27.18}$$

Hence, to determine the value of m, $I[U]$ should be minimized and the corresponding necessary condition is $\mathrm{d}I/\mathrm{d}m = 0$. The value of m is determined as

$$m = A^2/2\left(1 + A^2\right) \tag{27.19}$$

Now, the normalized velocity profile for no-slip condition is obtained as

$$\overline{U}_{\mathrm{ns}} = \frac{2A^2}{3\pi(1 + A^2)}\left(X^2 + Y^2 - 1\right) \tag{27.20}$$

Finally, the velocity distribution within the microchannel is obtained using Eq. 27.12 as

$$\overline{U}(\mathrm{X},\mathrm{Y}) = \frac{BA^2}{2(1+A^2)}\left(X^2+Y^2-1\right)+1-B \tag{27.21}$$

where $B = \left[1+\frac{6\pi(2-F)}{F}Kn\frac{(1+A^2)}{(1+A)^2}\right]^{-1}$

The Poiseuille number (Po) is determined as

$$Po = \frac{12\pi(1+A^2)}{(1+A)^2}\left[1+\frac{6\pi(2-F)}{F}Kn\frac{(1+A^2)}{(1+A)^2}\right]^{-1} \tag{27.22}$$

27.4.2 Solution Method for a Rectangular Microchannel

In this section, an effort is given to obtain an approximate analytical solution for a rectangular microchannel. Hence, the same governing Eq. (27.11) has been solved using the variational formulation technique.

A trial function for rectangular microchannel which is to be considered in the analysis is taken as [39]

$$U(X,Y) = m\left(1-X^2\right)\left(1-Y^2\right) \tag{27.23}$$

The value of m is determined when $I[U]$ should be minimized and the corresponding necessary condition is $\mathrm{d}I/\mathrm{d}m = 0$. Finally, the normalized velocity profile is found as

$$\overline{U}(X,Y) = \frac{9B}{4}\left(1-X^2\right)\left(1-Y^2\right)+1-B \tag{27.24}$$

where $B = \left[1+\frac{18(2-F)(1+A^2)}{5F}Kn\left(\frac{2}{1+A}\right)^2\right]^{-1}$

The Poiseuille number is expressed as

$$Po = \frac{36}{5}\left[\frac{1}{(1+A^2)}+\left\{\frac{18(2-F)}{5F}Kn\left(\frac{2}{1+A}\right)^2\right\}\right]^{-1}\left(\frac{2}{1+A}\right)^2 \tag{27.25}$$

27.5 Results and Discussion

The present work offers an analytical solution for gaseous flow within a microchannel of elliptical cross section. The Navier–Stokes equation subjected to slip boundary conditions is solved using variational technique to determine the velocity distribution, corresponding slip coefficient (β), and Poiseuille number (*Po*). The results are taken for velocity distribution based on the present analysis to validate the proposed method with the existing work. Hence, in Fig. 27.3, the velocity profile obtained is compared with published work by Das and Tahmouresi [14] at $X = 0$ and $A = 1.0$. It is seen that the present prediction agrees well with the existing literature. Therefore, the present analysis can be extended to predict velocity distribution, *Po* and β for different aspect ratio and Knudsen number.

In Fig. 27.4, velocity distribution in *y*-direction is plotted at $X = 0$ for $A = 1.0$, 0.75, 0.5, 0.25 and $Kn = 0.01$. For a particular value of aspect ratio, velocity increases toward the center of the channel and attains maximum at the center. It can also be noticed that there is a variation between the velocity distributions in all values of aspect ratio. In all the cases, the deviation is more as the flow reaches toward the center of the channel. Therefore, it is concluded that the flow velocity has a strong aspect ratio dependency.

It is already stated that the Poiseuille number (*Po*) is of particular interest while analyzing flow behavior through microchannels. The *Po* depends on the *Kn* which normally varies from 0.001 to 0.1 for gaseous flow through microchannels. Therefore, in Fig. 27.5, Poiseuille number as a function of *Kn* is presented for $A = 1.0$, 0.75, 0.5, and 0.25. The Poiseuille number decreases as the Knudsen number is increased for any value of aspect ratio. It is observed that the value of *Po* decreases sharply with an increase of *Kn* up to 0.02 for a particular aspect ratio.

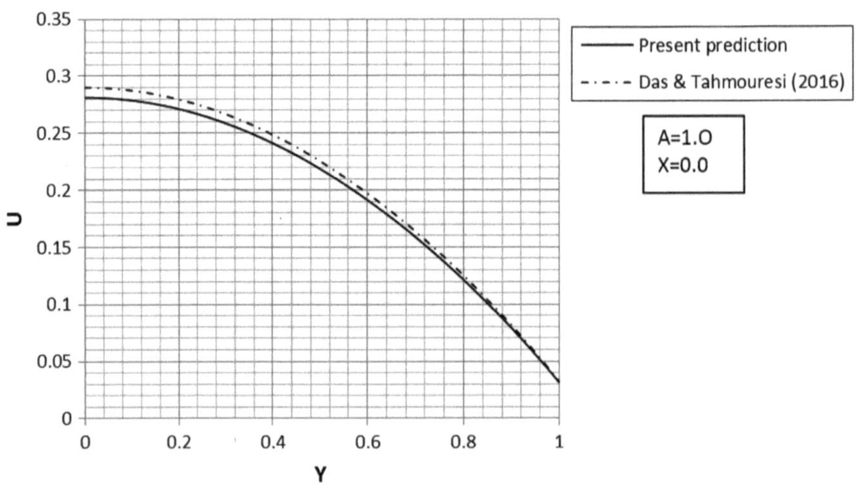

Fig. 27.3 Comparison of present prediction with published literature ($A = 1$, $X = 0.0$, $Kn = 0.01$)

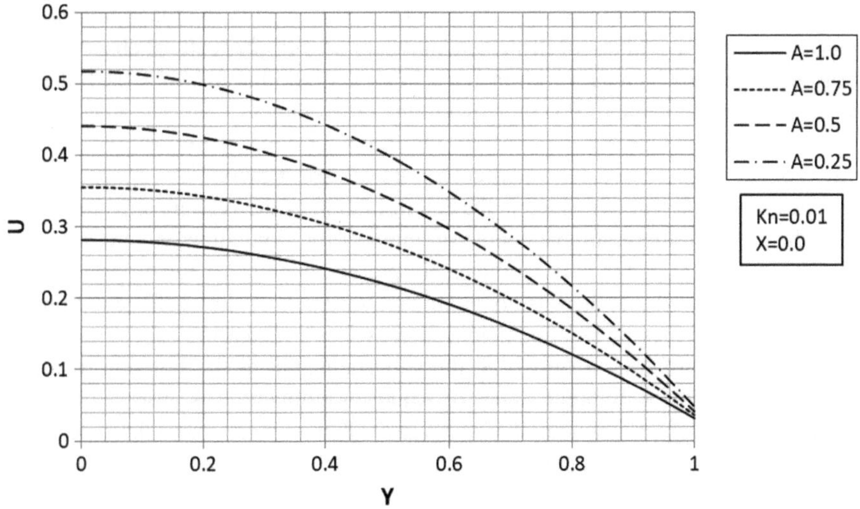

Fig. 27.4 Variation of velocity distribution for different aspect ratio (X = 0.0, Kn = 0.01, F = 1)

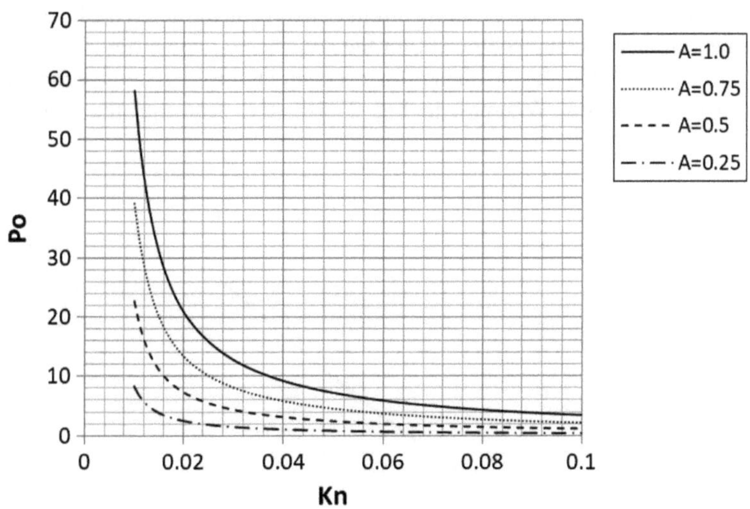

Fig. 27.5 Variation of Poiseuille number (Po) with Knudsen number (Kn) for different aspect ratio (Kn = 0.01, F = 1)

After that its rate of decrease falls and finally attains a steady value. Ebert and Sparrow [40] also established based on theoretical study that the Po reduces with increase in Kn due to the effect of rarefaction. Finally, it may be concluded that the reduction of Poiseuille number is influenced by the geometry of the cross section. The Po decreases with decreasing aspect ratio for a particular value of Kn.

The results obtained for Poiseuille number is presented as the function of aspect ratio for different values of Knudsen numbers in Fig. 27.6. It is observed that Poiseuille number increases with increase in aspect ratio but for $Kn = 0.01$, there is a steep rise in the value of Po compared to the other values of Kn. The steepness Po decreases as the value Kn is increased. On the other hand, the Po increases as the value of Kn is decreased for a particular vale of aspect ratio.

The slip coefficient (β) which is the determination of the velocity slip at the solid boundary is an important parameter of interest for fluid flow in microchannel. Hence, in Fig. 27.7, the slip coefficient is presented with Kn for different aspect ratio. It is observed that there is no significant variation of slip coefficient with Kn. However, the values of slip coefficient increase with increasing aspect ratio for a particular value of Kn. In Fig. 27.8, the slip coefficient as the function of aspect ratio is presented for different values of Knudsen numbers. It is observed that at a particular value of Knudsen number, the slip coefficient increases as the aspect ratio is increased. Whereas, at a particular value of aspect ratio, slip coefficient increases with decrease value of Knudsen number.

In Figs. 27.9 and 27.10, represent the different analytical solutions for Poiseuille number and slip coefficient with Knudsen number in a comparative way at aspect ratio $A = 0.5$, for the elliptical and rectangular cross sections, respectively. It is observed that both the figures show similar nature in the analytical solutions of Poiseuille number and slip coefficient for the considered geometries. In Fig. 27.9, it is seen that the value Po is higher for rectangular cross section compared the elliptical cross section. Whereas, in Fig. 27.10, it is seen that the value of slip

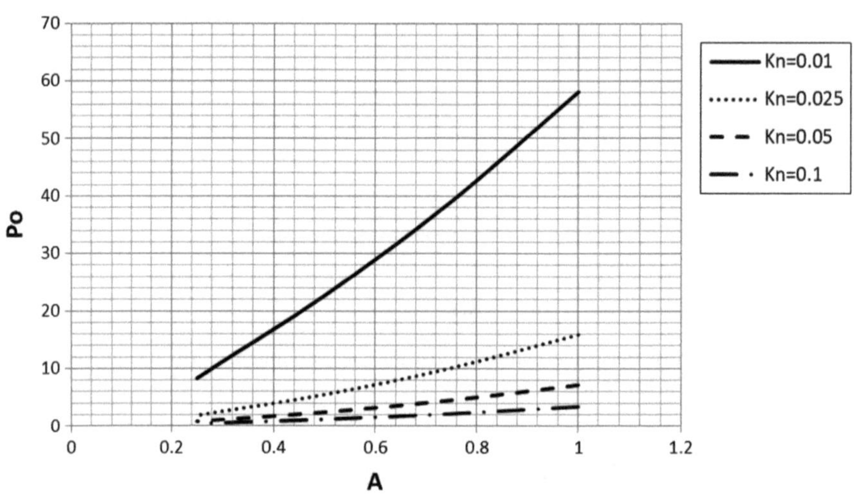

Fig. 27.6 Variation of Poiseuille number (Po) with aspect ratio for different Knudsen number (Kn)

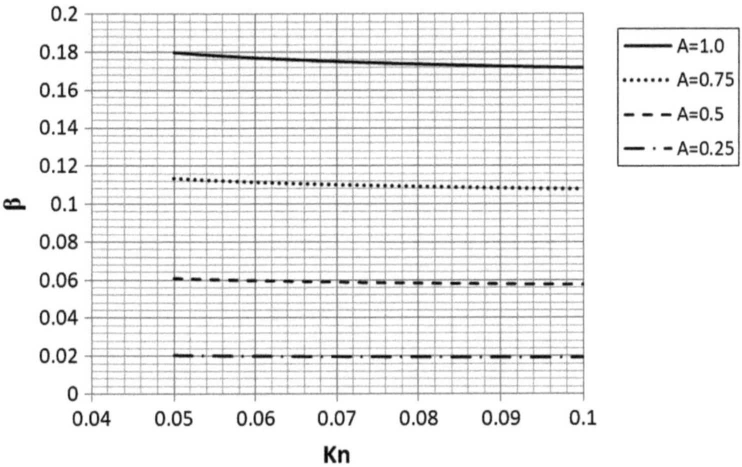

Fig. 27.7 Variation of slip coefficient (β) with Knudsen number (Kn) for different aspect ratio ($Kn = 0.01$, $F = 1$)

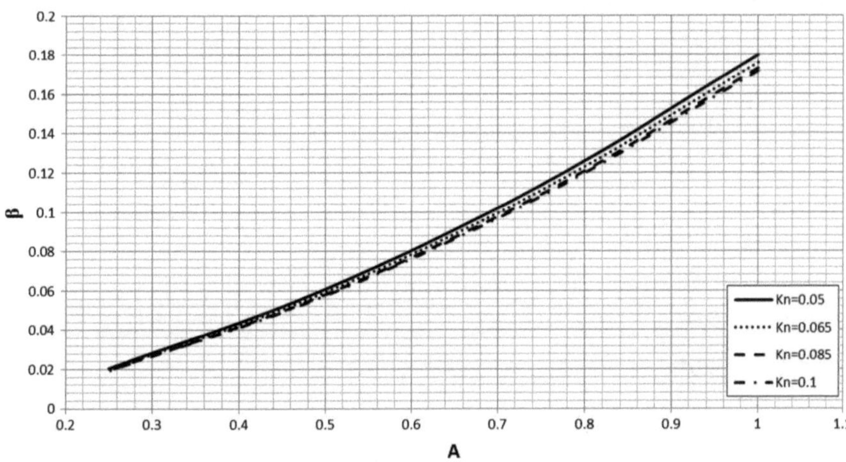

Fig. 27.8 Variation of slip coefficient (β) with aspect ratio (A) for different values of Knudsen numbers (Kn)

coefficient for the elliptical cross section is higher compared to a rectangular cross section. It is observed that the slip coefficients for both the elliptic and rectangular cross sections have same nature of variations. The solutions for both the ducts have been presented in a comparative way, and finally the reliability of the present proposed study has been proven.

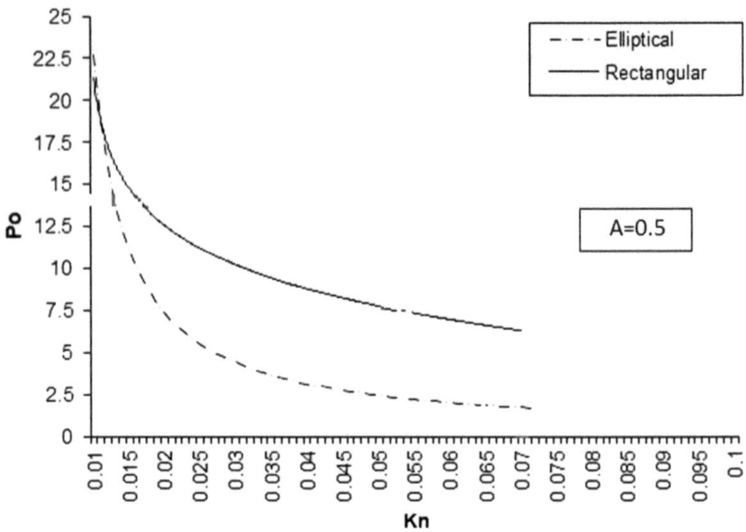

Fig. 27.9 Comparison of Poiseuille number (*Po*) with Knudsen numbers for elliptical and rectangular microchannels (*Kn*)

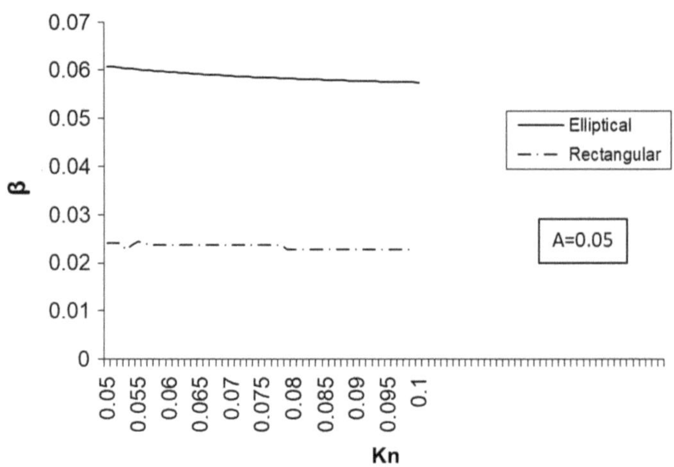

Fig. 27.10 Comparison of slip factor (β) with Knudsen numbers (*Kn*) for elliptical and rectangular microchannels

27.6 Conclusions

The present work analyzes gaseous slip flow within a microchannel of elliptical cross section for different aspect ratio. An analytical solution for velocity distribution is determined using variational method. The Navier–Stokes equation subjected to slip boundary conditions at the solid walls is solved with the variational method. Proposed result for velocity profile is compared with an existing literature and shows good agreement. The velocity distribution thus found is used to determine Poiseuille number and slip coefficient. The proposed results show strong aspect ratio dependency. It is observed that the value of Poiseuille number reduces with increasing Knudsen number, whereas slip coefficient shows no dependency on the Knudsen number. The work is extended for a rectangular microchannel using the same variational formulation. Finally, the proposed results have been shown in a comparative way to easily understand the difference in hydrodynamic phenomena between microchannels of elliptical and rectangular cross sections.

References

1. Duan, Z.P., Muzychka, Y.S.: Slip flow in non-circular microchannels. MicrofluidNanofluid **3**(4), 473–484 (2007)
2. Mala, G.H.M., Li, D.: Flow characteristics of water in microtubes. Int. J. Heat Fluid Flow **20**, 142–148 (1999)
3. Weng, C.I., Li, W.L., Hwang, C.C.: Gaseous flow in microtubes at arbitrary Knudsen numbers. Nanotechnology **10**, 373–379 (1999)
4. Cercignani, C., Lampis, M., Lorenzani, S.: Variational approach to gas flows in microchannels. Phys. Fluids **16**, 3426–3437 (2004)
5. Chen, C.H.: Slip-flow heat transfer in a microchannel with viscous dissipation. Heat Mass Transf. **42**(9), 853–860 (2006)
6. Yu, S., Ameel, T.A.: Slip flow heat transfer rectangular microchannels. Int. J. Heat Mass Transf. **44**, 4225–4234 (2001)
7. Agrawal, A., Agrawal, A.: Three-dimensional simulation of gas flow in different aspect ratio microducts. Phys. Fluids **18**, 103604 (2006)
8. Colin, S.: Rarefaction and compressibility effects on steady and transient gas flows in microchannels. Microfluid. Nanofluid. **1**, 268–279 (2005)
9. Barber, R.W., Emerson, D.R.: Challenges in modeling gas-phase flow in microchannels: from slip to transition. Heat Transf. Eng. **27**, 3–12 (2006)
10. Satapathy, A.K.: Slip flow heat transfer in an infinite microtube with axial conduction. Int. J. Therm. Sci. **49**, 153–160 (2010)
11. Rostami, A.A., Mujumdar, A.S., Saniei, N.: Flow and heat transfer forgas flowing in microchannels: a review. Int. J. Heat Mass Transf. **38**(4–5), 359–367 (2002)
12. Morini, G.L.: Analytical determination of the temperature distribution and Nusselt numbers in rectangular ducts with constant axial heat flux. Int. J Heat Mass Transf. **43**, 741–755 (2000)
13. Hooman, K.: A superposition approach to study slip-flow forced convection in straight microchannels of uniform but arbitrary cross-section. Int. J. Heat Mass Transf. **51**, 3753–3762 (2008)
14. Das, S.K., Tahmouresi, F.: Analytical solution of fully developed gaseous slip flow in elliptical microchannel. Int. J. Appl. Math. Mech. **3**(3), pp. I-15 (2016)
15. Kuddusi, L.: Prediction of temperature distribution and Nusselt number in rectangular microchannels at wall slip condition for all versions of constant wall temperature. Int. J. Therm. Sci. **46**, 998–1010 (2007)
16. Theofilis, V., Duck, P.W., Owen, J.: Viscous linear stability analysis of rectangular duct and cavity flows. J. Fluid Mech. **505**, 249–286 (2004)

17. Peng, J., Mingzhe, D., Liming, D., Jun, Y.: Slow viscous flow through arbitrary triangular tubes and its application in modelling porous media flows. J. Transp. Porous Media **74**, 153–167 (2008)
18. Rybiński, W., Mikielewicz, J.: Analytical solutions of heat transfer for laminar flow in rectangular channels. Arch. Thermodyn. **35**(4), 29–42 (2014)
19. Chakraborty, G.: A note on methods for analysis of flow through microchannels. Int. J. Heat Mass Transf. **51**, 4583–458 (2008)
20. Kundu, B., Simlandi, S., Das, P.K.: Analytical techniques for analysis of fully developed laminar flow through rectangular channels. Int. J. Heat Mass Transf. **47**(10), 1289–1299 (2011)
21. Lee, P.S., Garimella, S.V.: Thermally developing flow and heat transfer in rectangular microchannels of different aspect ratios. Int. J. Heat Mass Transf. **49**, 3060–3067 (2006)
22. Zhang, T.T., Jia, L., Wang, Z.C., Li, X.: The application of homotopy analysis method for 2-dimensional steady slip flow in microchannels. Phys. Lett. A **372**, 3223–3227 (2008)
23. Zhu, X., Liao, Q.: Heat transfer for laminar slip flow in a microchannel of arbitrary cross section with complex thermal boundary conditions. Appl. Therm. Eng. **26**, 1246–1256 (2006)
24. Dongari, N., Agrawal, A., Agrawal, A.: Analytical solution of gaseous slip flow in long microchannels. Int. J. Heat Mass Transf. **50**, 3411–3421 (2007)
25. Ngoma, G.D., Erchiqui, F.: Heat flux and slip effects on liquid flow in a microchannel. Int. J. Therm. Sci. **46**, 1076–1083 (2007)
26. Larrode, F.E., Housiadasb, C., Drossinosa, Y.: Slip flow heat transfer in circular tubes. Int. J. Heat Mass Transf. **43**, 2669–2680 (2000)
27. Tunc, G., Bayazitoglu, Y.: Heta transfer in rectangular microchannels. Int. J. Heat Mass Transf. **45**, 765–773 (2002)
28. Kuddusi, L., Cetegen, E.: Prediction of temperature distribution and Nusselt number in rectangular microchannels at wall slip condition for all versions of constant heat flux. Int. J. Heat Fluid Flow **28**, 777–786 (2007)
29. Tabelling, P.: Introduction to Microfluidics. Oxford University Press (2006)
30. Maxwell, J.C.: On stresses in rarefied gases arising from inequalities of temperature philos. Trans. Royal Soc. **170**, 231–256 (1879)
31. See Professor A. Beskok's website: www.cfm.brown.edu/people/beskok/lam.slip.html
32. Beskok, A., Karniadakis, G.E.: A model for flows in channels, pipes, and ducts at micro and nano scales. Microsc. Thermophys. Eng. **3**, 43–77 (1999)
33. Karniadakis, G.E., Beskok, A.: Microflows-Fundamentals and Simulations. Springer, New York (2002)
34. Cercignani, C., Daneri, A.: Flow of a rarefied gas between two parallel plates. J. Appl. Phys. **34**, 3509–3513 (1963)
35. Xue, H., Fan, Q.: A high order modification on the analytic solution of 2-D microchannel gaseous flow. In: Proceedings of ASME, Fluids Engineering Division Summer Meeting, Boston, June (2000)
36. Sreekanth, A.K.: Slip flow through long circular tubes. In: Trilling, L., Wachman, H.Y. (eds.) Proceedings of the Sixth International Symposium on Rarefied Gas Dynamics, Academic Press, pp. 667–680 (1969)
37. Bahrami, M., Yovanovich, M.M., Culham, J.R.: Pressure drop of fully developed, laminar flow in microchannel of arbitrary cross-section. ASME J. Fluids Eng. **128**(5), 1036–1044 (2006)
38. Xue, H., Fan, Q., Shu, C.: Prediction of microchannel flows usingndirect simulation Monte Carlo. Probab. Eng. Mech. **15**, 213–219 (2000)
39. Arpaci VS (1980) Conduction Heat Transfer, Wesley Publishing Company
40. Ebert, S.W., Sparrow, E.M.: Slip flow in rectangular and annular ducts. J. Basic Eng. **87**, 1018–1024 (1965)

Chapter 28
Stresses and Deformation in Rotating Disk During Over-Speed

Rajesh Kumar and Rajeev Jain

Nomenclatures

E	Young's modulus of elasticity of disk material
$f_1(\lambda), f_2(\lambda)$	Function depending on disk material properties
h	Disk thickness at radial location r (non-dimensional form $\bar{h} = h/h_0$) $\bar{h}' = \mathrm{d}\bar{h}/\mathrm{d}\bar{r}$
h_0	Disk thickness at the bore
H	Profile parameter of hyperbola disk
H_m	$H_m = \eta\sigma_0/E$
K_1, K_2, K_3, K_4	Integration constant
p	Profile parameter of hyperbola disk
r	Radial location from axis of rotation (non-dimensional form $\bar{r} = r/r_2$)
r_1	Disk bore radius (non-dimensional form $\bar{r}_1 = r_1/r_2$)
r_2	Disk rim radius (non-dimensional form $\bar{r}_2 = r_2/r_2$)
r_p	Elastic plastic interface radius (non-dimensional form $\bar{r}_\mathrm{p} = r_\mathrm{p}/r_2$)
u	Radial displacement at r (non-dimensional form $\bar{u} = uE/r_2\sigma_0$)
u^e	Elastic displacement (non-dimensional $\bar{u}^\mathrm{e} = u^\mathrm{e}E/r_2\sigma_0$)
u^p	Plastic displacement (non-dimensional $\bar{u}^\mathrm{p} = u^\mathrm{e}E/r_2\sigma_0$)
$\varepsilon_\theta, \varepsilon_\mathrm{r}$	Tangential and radial strain (non-dimensional form $\bar{\varepsilon}_\theta = \varepsilon_\theta E/\sigma_0$, $\bar{\varepsilon}_\mathrm{r} = \varepsilon_\mathrm{r}E/\sigma_0$)
ε_eq	Equivalent plastic strain (non-dimensional form $\bar{\varepsilon}_\mathrm{eq} = \varepsilon_\mathrm{eq}E/\sigma_0$)
$\varepsilon_\theta^\mathrm{p}, \varepsilon_\mathrm{r}^\mathrm{p}, \varepsilon_\mathrm{z}^\mathrm{p}$	Plastic tangential, radial and axial strain (non-dimensional form $\bar{\varepsilon}_\theta^\mathrm{p} = \varepsilon_\theta^\mathrm{p}E/\sigma_0, \bar{\varepsilon}_\mathrm{r}^\mathrm{p} = \varepsilon_\mathrm{r}^\mathrm{p}E/\sigma_0, \bar{\varepsilon}_\mathrm{z}^\mathrm{p} = \varepsilon_\mathrm{z}^\mathrm{p}E/\sigma_0$)
η	Hardening parameter
λ	Constant depending on disk material properties
ϑ	Poisson's ratio
ρ	Density
σ_0	Initial yield stress

R. Kumar (✉) · R. Jain
Gas Turbine Research Establishment, Banglore 560093, India
e-mail: rajeshmbsi@gmail.com

P. Sahoo and J. P. Davim (eds.), *Advances in Materials, Mechanical and Industrial Engineering*, Lecture Notes on Multidisciplinary Industrial Engineering,
https://doi.org/10.1007/978-3-319-96968-8_28

σ_y	Yield stress (non-dimensional form $\bar{\sigma}_y = \sigma_y/\sigma_0$)
σ_θ, σ_r	Tangential and radial stress (normalized form $\bar{\sigma}_\theta = \sigma_\theta/\sigma_0$, $\bar{\sigma}_r = \sigma_r/\sigma_0$)
$\sigma_\theta^e, \sigma_r^e$	Tangential and radial stresses in elastic regime
$\sigma_\theta^p, \sigma_r^p$	Tangential and radial stresses in plastic regime
ω	Angular velocity in radians per second (non-dimensional form $\Omega = \sqrt{\rho\omega^2 r_2^2/\sigma_0}$)

28.1 Introduction

Theoretical research in rotating disks has always remained an interesting topic due to its wide range of engineering applications in rotating machineries such as aero-engine compressor and turbine disks, high-speed gears, flywheels. Closed form solution for axisymmetric elastic disk is available in many textbooks [1, 2].

As per design philosophy for aero-engine disks, since safety is utmost priority, stresses should remain within yield stress for all practical purposes. With the increasing demand of high thrust to low weight ratio, engine disks are being designed for higher rotating speeds. Anisotropic disk with functionally graded material can be one of the viable solutions to meet this stringent requirement. Closed-form solution is available [3] for such disks to predict stresses and displacements. Later, Jain et al. [4] developed a methodology to optimize such type of disks by tailoring the material properties. Isotropic material disks are operating nearer to material strength of the disk. Also, in certain circumstances such as failure of control mechanism or decoupling of shaft, disk may overshoot its speed and operate beyond its design limit and undergo plastic deformation. As per certification standards mention in MIL-E-5007E [5] standard, designer must specify an acceptable value of permanent growth after disk over-speed (115% of redline speed). Although magnitude of acceptable permanent growth is not mentioned in any international standard, as a rule of thumb engine manufacturers are keeping it closer to 0.1% of the disk diameter at a particular disk location.

Stresses produced in an elastic-plastic annular disk for constant angular velocity is available in standard textbooks by Chakrabarty [6] and Johnson and Mellor [7]. In most of these works, Tresca and von-Mises yield criteria are used to analyze elastic-plastic rotating disks. The application of Tresca's yield criterion for linearly hardening material results in linear differential equations permitting the analytical treatment of the problem. Extensive work has been done to obtain exact solution for linearly strain hardening material to evaluate elastic-plastic stresses, strains and radial growths in disks using Tresca's yield criteria and its associate flow rule. Gamer [8] first solved analytical solution for uniform thickness solid rotating disk. Eraslan and Argeso and Ma et al. [9, 10] studies concluded that stresses in variable thickness disk are much lower than constant thickness disks operating at the same angular velocity. Hence, for a better utilization of the material, it is appropriate to allow variation in the thickness of the disk. Güven [11] extended the work of Gamer [8] and obtain solution for hyperbolic annular disk of variable thickness and variable density. Closed-form solution is also derived using displacement function approach in terms of hypergeometric functions for convex and concave solid disks [12, 13].

Elastic-plastic analysis of rotating solid disks using von-Mises yield criterion requires numerical treatment due to the nonlinearities inherent in this criterion and in the flow rule associated with it. Rees [14] studied elastic-plastic deformation of rotating uniform thickness solid and annular disks using Tresca and von-Mises yield criteria for a perfectly plastic material. For uniform thickness disk, von-Mises solution can be simplified to an ordinary differential equation using conventional elliptic substitution. You et al. [15] and You and Zhang [16] applied a polynomial stress-plastic strain relation to obtain an approximate analytical solution for rotating solid disks of uniform thickness with nonlinear strain hardening material properties. Later, You et al. [17] extended obtain stresses and displacement in elastic-plastic rotating disk using Runge-Kutta's algorithm. Recently, Bhowmick et al. [18] used variational method for elastic-plastic disk to investigate growth of elastic-plastic front.

Although prediction of permanent growth in rotating disk is very crucial aspect for engine designer from safety point of view, there is hardly any published work in this area. Wilterdink [19] did extensive experiments to obtain permanent growth in uniform thickness rotating disk for thermally gradient disk. With the increase in computational facility and fast algorithms of numerical techniques, finite element method [20] can be used for the estimation of permanent growth. Finite element is numerical technique and shows very large discrepancy in results if mesh refinement is not properly done in the area of high stress gradients. Analysis shows that results are not very accurate near bore and rim which are the two most important regions for experimental comparison. Recently, Ayyappan et al. [21] obtained permanent growth in annular disk using horizontal spinning facility and compared their experimental result with finite element using $\pm 3\sigma$ stress–strain curve.

In this paper, a closed-form solution has been developed to predict permanent growth in elastic-plastic rotating disk using Tresca's yield criteria for linearly strain hardening material. Results of analytical solution for stresses and residual stresses have been compared with commercially available finite element software ABAQUS [22] for uniform thickness annular disk. Thereafter, permanent growth obtained analytically and through FEM techniques have been compared with experimental test results for annular steel disk for 10,000 to 13,000 rpm rotating speed.

28.2 Mathematical Formulation

Following Timoshenko and Goodier [1], the equilibrium equation of the rotating disk for constant angular velocity in non-dimensional form is given as

$$\bar{h}\bar{\sigma}_\theta = (\bar{h}\bar{r}\bar{\sigma}_{\rm r})' + \Omega^2\bar{h}\bar{r}^2, \tag{28.1}$$

where $\bar{h} = h/h_0$ is non-dimensional thickness, h_0 is the thickness of disk at the bore; $\bar{r} = r/r_2$ is dimensionless radial location, r_2 is the radius of the disk at rim; $\Omega = \sqrt{\rho\omega^2 r_2^2/\sigma_0}$ is the non-dimensional angular speed, ρ is the density of disk material; ω is the angular speed in radians per second, σ_0 is the initial yield stress; $\bar{\sigma}_\theta = \sigma_\theta/\sigma_0$ and $\bar{\sigma}_{\rm r} = \sigma_{\rm r}/\sigma_0$ are non-dimensional tangential and radial stresses, respectively.

The strain displacement relation for small strains in axisymmetric disk is given as

$$\begin{bmatrix} \bar{\varepsilon}_\theta \\ \bar{\varepsilon}_{\rm r} \end{bmatrix} = \begin{bmatrix} \bar{u}/\bar{r} \\ {\rm d}\bar{u}/{\rm d}\bar{r} \end{bmatrix}, \tag{28.2}$$

where $\bar{u} = uE/r_2\sigma_0$ is non-dimensional displacement, u is the radial displacement, E is the Young's modulus of elasticity; $\bar{\varepsilon}_\theta = \varepsilon_\theta E/\sigma_0$ and $\bar{\varepsilon}_{\rm r} = \varepsilon_{\rm r} E/\sigma_0$ are normalized strain in tangential and radial direction, respectively.

Stress–strain relations for elastic-plastic disk in non-dimensionalized form is

$$\begin{bmatrix} \bar{\varepsilon}_\theta \\ \bar{\varepsilon}_{\rm r} \end{bmatrix} = \begin{bmatrix} 1 & -\vartheta \\ -\vartheta & 1 \end{bmatrix} \begin{bmatrix} \bar{\sigma}_\theta \\ \bar{\sigma}_{\rm r} \end{bmatrix} + \begin{bmatrix} \bar{\varepsilon}_\theta^{\rm p} \\ \bar{\varepsilon}_{\rm r}^{\rm p} \end{bmatrix}, \tag{28.3}$$

where ϑ is Poisson's ratio; $\bar{\varepsilon}_\theta^{\rm p}$ and $\bar{\varepsilon}_{\rm r}^{\rm p}$ are normalized plastic component of strain in radial and tangential direction, respectively.

If it is assumed that material of rotating disk is linearly strain hardening, then yield stress can be written as

$$\bar{\sigma}_{\rm y} = \left(1 + H_m\bar{\varepsilon}_{\rm eq}\right), \tag{28.4}$$

where $\bar{\sigma}_{\rm y}$ is the normalized initial yield stress $\bar{\sigma}_{\rm y} = \sigma_{\rm y}/\sigma_0$; $H_m = \eta\sigma_0/E$ is a constant, η is the strain hardening parameter; $\bar{\varepsilon}_{\rm eq} = E\varepsilon_{\rm eq}/\sigma_0$ is the normalized equivalent plastic strain, $\varepsilon_{\rm eq}$ is the equivalent strain.

According to the flow rule associated with Tresca's yield criteria

$$\bar{\varepsilon}_{\rm eq} = \bar{\varepsilon}_\theta^{\rm p},\ \bar{\varepsilon}_{\rm r}^{\rm p} = 0,\ \bar{\varepsilon}_\theta^{\rm p} = -\bar{\varepsilon}_{\rm z}^{\rm p}, \tag{28.5}$$

Using (28.2) strain compatibility relation is given as

$$(\bar{r}\bar{\varepsilon}_\theta)' = \bar{\varepsilon}_{\rm r}. \tag{28.6}$$

Substituting $\bar{\varepsilon}_\theta^{\rm p}$ from (28.4) in stress–strain relation (28.3) and using strain compatibility Eq. (28.6) we get a differential equation which relate tangential stress with radial stress. Substituting tangential stress from equilibrium Eq. (28.1) we get

$$\bar{r}^2\bar{\sigma}_r'' + \left(3 + \frac{\bar{h}'}{\bar{h}}\bar{r}\right)\bar{r}\bar{\sigma}_r' + \left[\frac{\bar{h}''}{\bar{h}}\bar{r}^2 - \frac{\bar{h}'^2}{\bar{h}^2}\bar{r}^2 + (2+\vartheta\lambda)\frac{\bar{h}'}{\bar{h}}\bar{r} + (1-\lambda)\right]\bar{\sigma}_r = -(3+\vartheta\lambda)\Omega^2\bar{r}^2 + \lambda/H_m, \quad (28.7)$$

where $\lambda = (1+E/\sigma_0\eta)^{-1}$ is a constant depending on the material properties of the disk.

28.2.1 Elasto-Plastic Solution in Hyperbola Disk

Equation (28.7) can be converted into equi-dimensional equation for hyperbola disk profile. In hyperbola disk, thickness of disk is the function of radial location and varies like $h = Hr^{\mathrm{p}}$, where H and p are arbitrary constants of hyperbola disk profile. In non-dimensional form thickness variation is $\bar{h} = \bar{r}^{\mathrm{p}}$, substituting $\bar{h}'\bar{r}/\bar{h} = p$ and $\bar{h}''\bar{r}^2/\bar{h} = p(p-1)$ in (28.7) we get

$$\bar{r}^2\bar{\sigma}_r'' + (3+p)\bar{r}\bar{\sigma}_r' + [p(1+\lambda\vartheta) + (1-\lambda)]\bar{\sigma}_r = -(3+\vartheta\lambda)\Omega^2\bar{r}^2 + (1-\lambda). \quad (28.8)$$

Equation (28.8) is a Cauchy's Euler equation and its solution is given by

$$\bar{\sigma}_r^{\mathrm{p}} = K_1\bar{r}^{\alpha} + K_2\bar{r}^{\beta} - f_1(\lambda)\Omega^2\bar{r}^2 + f_2(\lambda), \quad (28.9)$$

where $\alpha, \beta = 0.5\left(-(p+2) \pm \sqrt{p^2 + 4\lambda(1-p\vartheta)}\right)$ are constants; $f_1(\lambda) = (3+\vartheta\lambda)/(3+\vartheta\lambda)p + (9-\lambda)$ and $f_2(\lambda) = (1-\lambda)/(1+\vartheta\lambda)p + (1-\lambda)$ are constants depending on disk material properties and hyperbolic disk profile parameters.

Substituting non-dimensional radial stress $(\bar{\sigma}_r)$ from (28.9) in equilibrium Eq. (28.1), tangential stress $(\bar{\sigma}_\theta)$ is given as

$$\bar{\sigma}_\theta^{\mathrm{p}} = \alpha_0 K_1\bar{r}^{\alpha} + \beta_0 K_2\bar{r}^{\beta} + (1-(3+p)f_1(\lambda))\Omega^2\bar{r}^2 + (1+p)f_2(\lambda), \quad (28.10)$$

where $\alpha_0 = \alpha + 1 + p$ and $\beta_0 = \beta + 1 + p$. Radial displacement in a disk can be determined by substituting (28.9) and (28.10) in stress–strain relation (28.3) in combination with strain displacement relation (28.2) we get

$$\bar{u}^{\mathrm{p}} = \bar{r}\left[\begin{array}{r}\left(\frac{\alpha_0}{\lambda} - \vartheta\right)K_1\bar{r}^{\alpha} + \left(\frac{\beta_0}{\lambda} - \vartheta\right)K_2\bar{r}^{\beta} + \left(\frac{\vartheta^2\lambda - 1}{3+\vartheta\lambda}\right)f_1(\lambda)\Omega^2\bar{r}^2 \\ + [1-(1+p)\vartheta]\, f_2(\lambda)\end{array}\right]. \quad (28.11)$$

28.2.2 Elastic Solution

In elasto-plastic disk, stresses are plastic till $\bar{r} = \bar{r}_p$ and thereafter for $\bar{r} > \bar{r}_p$ stresses are elastic. It is interesting to note that elastic stresses and radial displacement can be obtained from plastic solution developed in previous section by substituting $\lambda = 1$ in (28.9), (28.10) and (28.11). This is due to hardening parameter approaches Young's modulus of elasticity and yield stress approach to infinity which reduces λ to unity. Approaches to elastic equations have two more additional constants K_3, K_4 given as

$$\bar{\sigma}_r^e = K_3\bar{r}^{\alpha_1} + K_4\bar{r}^{\beta_1} - f_1(1)\Omega^2\bar{r}^2, \tag{28.12}$$

where $\alpha_1, \beta_1 = 0.5\left(-(p+2) \pm \sqrt{p^2 + 4(1 - p\vartheta)}\right)$

$$\bar{\sigma}_\theta^e = \alpha_{01}K_3\bar{r}^{\alpha_1} + \beta_{01}K_4\bar{r}^{\beta_1} + (1 - (3+p)f_1(1))\Omega^2\bar{r}^2 \tag{28.13}$$

where $\alpha_{01} = \alpha_1 + 1 + p$ and $\beta_{01} = \beta_1 + 1 + p$. Substituting (28.12) and (28.13) in (28.3) and using (28.2) we get

$$\bar{u}^e = \bar{r}\left[(\alpha_{01} - \vartheta)K_3\bar{r}^{\alpha_1} + (\beta_{01} - \vartheta)K_4\bar{r}^{\beta_1} + \left(\frac{\vartheta^2 - 1}{3 + \vartheta}\right)f_1(1)\Omega^2\bar{r}^2\right] \tag{28.14}$$

28.2.3 Elastic-Plastic Stresses

The general expressions for the stresses and displacement in elastic and plastic regions contains five unknown constants K_1, K_2, K_3, K_4 and the interface radius $\bar{r}_p$. These five constants can be determined from five non-redundant conditions available, i.e., $\bar{\sigma}_r, \bar{\sigma}_\theta, \bar{u}$ are individually continuous at interface radius $\bar{r}_p$, and $\bar{\sigma}_r$ vanishes at the inner and outer boundaries, i.e., at bore $\bar{r} = \bar{r}_1$ and rim $\bar{r} = \bar{r}_2$ of the disk. Mathematically it can be expressed as

$$\begin{bmatrix} \bar{\sigma}_r^p(\bar{r}_1) \\ \bar{\sigma}_\theta^p(\bar{r}_p) \\ \bar{\sigma}_\theta^e(\bar{r}_p) \\ \bar{\sigma}_r^e(\bar{r}_2) \end{bmatrix} = \begin{bmatrix} 0 \\ 1 \\ 1 \\ 0 \end{bmatrix} \tag{28.15}$$

These equations are linear in K_1, K_2, K_3, K_4 and can be determined by solving simultaneously in terms of $\bar{r}_p$. Applying (28.15) boundary conditions in (28.9), (28.10), (28.12) and (28.13) one can write in matrix form as

$$\begin{bmatrix} \bar{r}_1^{\alpha} & \bar{r}_1^{\beta} & 0 & 0 \\ \alpha_0 \bar{r}_\mathrm{p}^{\alpha} & \beta_0 \bar{r}_\mathrm{p}^{\beta} & 0 & 0 \\ 0 & 0 & \alpha_{01} \bar{r}_\mathrm{p}^{\alpha 1} & \beta_{01} \bar{r}_\mathrm{p}^{\beta 1} \\ 0 & 0 & \bar{r}_2^{\alpha 1} & \bar{r}_2^{\beta 1} \end{bmatrix} \begin{bmatrix} K_1 \\ K_2 \\ K_3 \\ K_4 \end{bmatrix} = \begin{bmatrix} C_1 \\ C_2 \\ C_3 \\ C_4 \end{bmatrix} \tag{28.16}$$

where C_1, C_2, C_3, C_4 are

$$\begin{bmatrix} C_1 \\ C_2 \\ C_3 \\ C_4 \end{bmatrix} = \begin{bmatrix} f_1(\lambda)\Omega^2 \bar{r}^2 - f_2(\lambda) \\ 1 - (1 - (3+p)f_1(\lambda))\Omega^2 \bar{r}^2 - (1+p)f_2(\lambda) \\ 1 - (1 - (3+p)f_1(1))\Omega^2 \bar{r}^2 \\ f_1(1)\Omega^2 \bar{r}^2 \end{bmatrix} \tag{28.17}$$

Integration constants K_1, K_2, K_3, K_4 can be determined in terms of $\bar{r}_\mathrm{p}$ by solving (28.16) simultaneously.

Since, radial stress and displacement is continuous at the interface radius $(\bar{r}_\mathrm{p})$; therefore, a relation exist which relates interface radius $\bar{r}_\mathrm{p}$ with Ω such that $\Phi(r_\mathrm{p}, \Omega) = 0$.

28.2.4 Residual Stresses and Displacement

Residual stresses and displacements have been determined analytically using J_2 deformation theory in infinite [23] and finite sheet [24] by applying pressure load using Budiansky's criterion. This work is useful to enhance fatigue life of the engine and automobile components. Callioğlu et al. [25] mentioned a procedure to obtained elastic plastic residual stresses in orthotropic disk. On similar concept, residual stresses in elastic-plastic disk are obtained by superimposing plastic stresses on elastic stresses. Mathematically it is given as

$$\begin{bmatrix} \bar{\sigma}_\theta \\ \bar{\sigma}_\mathrm{r} \end{bmatrix}_\mathrm{res} = \begin{bmatrix} (\bar{\sigma}_\theta^\mathrm{p} - \bar{\sigma}_\theta^\mathrm{e}) \\ (\bar{\sigma}_\mathrm{r}^\mathrm{p} - \bar{\sigma}_\mathrm{r}^\mathrm{e}) \end{bmatrix} \tag{28.18}$$

Residual permanent growth is given as

$$\bar{u}_\mathrm{p} = \bar{u}^\mathrm{p} - \bar{u}^\mathrm{e} + (\bar{\sigma}_\theta - \vartheta \bar{\sigma}_\mathrm{r})_\mathrm{res} \tag{28.19}$$

28.3 Experimental Analysis for Permanent Growth

Experimentally permanent disk growth has been obtained using test spin facility which consists of 485 kW variable speed DC motor as a prime mover whose speed can be varied from 0 to 1500 rpm using a thyristor controller with an accuracy of ±0.01% of maximum speed. The output shaft of the DC motor is coupled with step-up gear box to increase the speed of test disk up to nearly 20,000 rpm. The test setup is equipped with proper instruments to measure vibration, temperature, lubricating oil flow rate for monitoring health of the test facility in real time. Schematic diagram of this cyclic spin test facility is shown in Fig. 28.1.

28.3.1 Specimen Disks

The specimen disks are manufactured from steel forging of ASTM A-105 specifications. Figure 28.2 shows disk configuration arrived after analyzing various configurations using finite element method (FEM) for 260 and 550 MPa disk yield and tensile strength, respectively. Numerical analysis shows that disk with bore diameter 50 mm and rim diameter 500 show appreciably disk growth at the bore and rim locations, if subjected to centrifugal load pertaining to rotating speed of 10,000–15,000 rpm without disk burst based on Robinson [26] criterion.

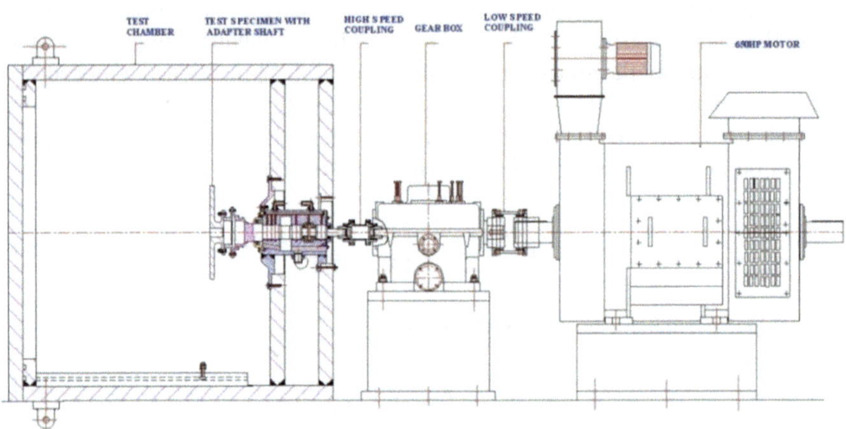

Fig. 28.1 Schematic diagram of cyclic spin test facility

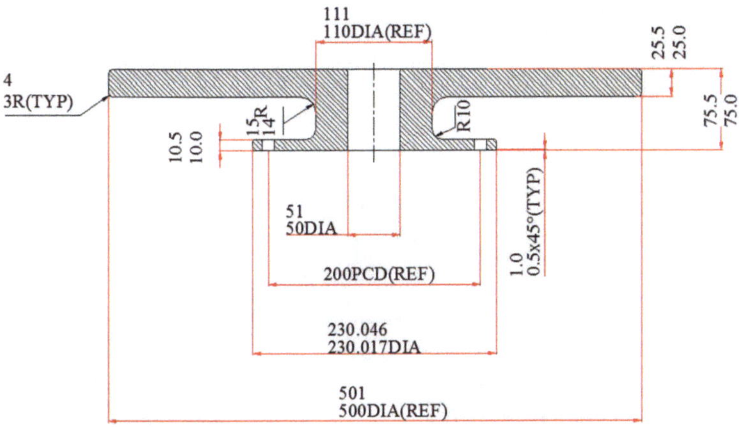

Fig. 28.2 Disk dimension for investigation

28.3.2 *Test Procedure*

Test is carried out as per detailed procedure layout in [21]. Each disk with it's attachments is balanced as per ISO 1940 Grade 2.5 and assembled in the spin rig for various rotating speeds. Residual unbalance is maintained below 1 gmm. Balanced disk assembly is assembled in test facility between lower and upper half of the bearing housing with the support of journal and thrust bearings.

After assembly, speed is increased gradually to 10,000 rpm and kept at this speed for the duration of 5 min before reducing to zero as per over-speed test cycles specified in MIL-STD 5007E [5] standard. Thereafter, disk is disassembled from the test facility and subjected to dimensional inspection in 3D coordinate measurement machine which has accuracy of 20 μm. Permanent growth at the bore and at the rim is evaluated on the basis of posttest and pretest reference dimensions. This procedure is repeated for each disk for 11,000, 11,500, 12,000, 12,500 and 13,000 rpm.

28.4 Finite Element Analysis

Commercially available finite element code ABAQUS has been used to determine permanent growth in rotating disk after applying centrifugal load. Finite element model of 30 sector disk assembly with flange, shaft and bolt shown in Fig. 28.3 consist of 174,271 nodes and 152151 C3D8R eight-noded elements.

This model is subjected to centrifugal rotation about *Z*-axis in cylindrical coordinate system. Rigid body motion is avoided by restraining the node in axial direction. Centrifugal load is simulated by applying constant angular speed in axial direction.

Fig. 28.3 Finite element model of disk assembly

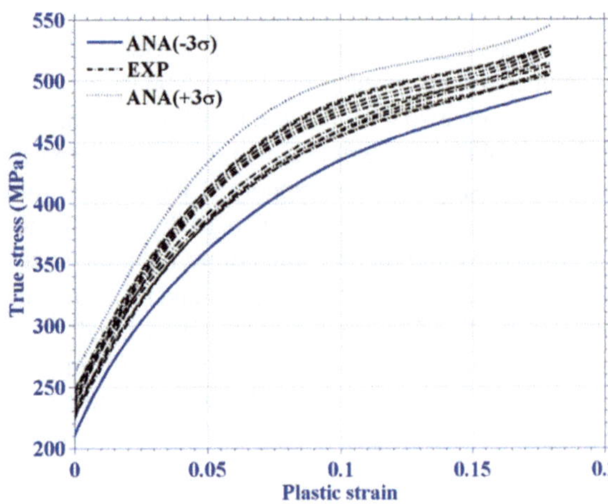

Fig. 28.4 True stress plastic strain curves along with $\pm 3\sigma$ stress strain curves

28.4.1 Material Property

Twenty-five tensile test specimens of ASTM standard (E8M) have been extracted from one of the forged disk of same melting batch at different radial locations in radial and tangential directions. These specimens are machined to 50 mm gauge length and tested in universal test machine (UTM). True-stress vs Plastic-strain variation for ten specimens is shown in Fig. 28.4. This figure indicates that although trend in the variation of True-stress vs Plastic-strain curve is almost identical among all the test specimens, there is significant variation (about 10%) between different specimens which can show variation in permanent growth in the disk. This figure also indicates upper and lower bound of these curves which have been obtained using statistical techniques. Maximum permanent growth is obtained on selecting lower bound (-3σ) True-stress vs Plastic-strain curve. Standard physical properties required for the analysis such as density (7860 kg/m^3), Young's modulus (200 GPa) and Poisson's ratio (0.3) for this material of disk material (steel A-105) is taken from standard handbook.

28.5 Results and Discussions

In this section, lower bound (-3σ) stress–strain curve has been taken to present stresses and growth in non-dimensional form. Stresses and residual stresses in tangential and radial directions using closed-form solution developed in this paper have been compared with FEM results using ABAQUS solver. Permanent growth obtained experimentally for four identical disks have been compared with closed form and FEM solutions for 10,000–13,000 rpm in the interval of 500 rpm.

28.5.1 Analytical and FEM Comparison

Closed-form analytical results based on Tresca's yield criterion have been first compared with results obtained using commercial available finite element software ABAQUS based on von-Mises yield criterion.

28.5.1.1 Stresses

Figures 28.5, 28.6, 28.7, 28.8, 28.9 and 28.10 show variation of tangential $(\bar{\sigma}_\theta)$ and radial stresses $(\bar{\sigma}_\mathrm{r})$ in elastic-plastic disk at 10,000, 11,000, 11,500, 12,000, 12,500 and 13,000 rpm obtained analytically have been compared with finite element results. These figures indicate that trend in the variation of tangential and radial stresses matches well at all rotating speed considered in this investigation. These figures further reveal that elastic plastic interface radius $\left(\bar{r}_\mathrm{p}\right)$ is more in closed-form result based on Tresca's yield criterion in comparison with FEM result which is based on von-Mises yield criterion. In the closed-form solution, the disk becomes completely plastic at nearly 11,200 rpm while in FEM it is at 11,750 rpm. Due to this, there is a significant difference in tangential stress between closed-form solution and FEM results near disk rim as seen in Figs. 28.6 and 28.7. In analysis disk attains fully plastic stage at 11,500 rpm. It is interesting to note that till 11,500 rpm maximum radial stress obtain using FEM is more than closed-form solution and thereafter close form solution result shows higher value in comparison with FEM.

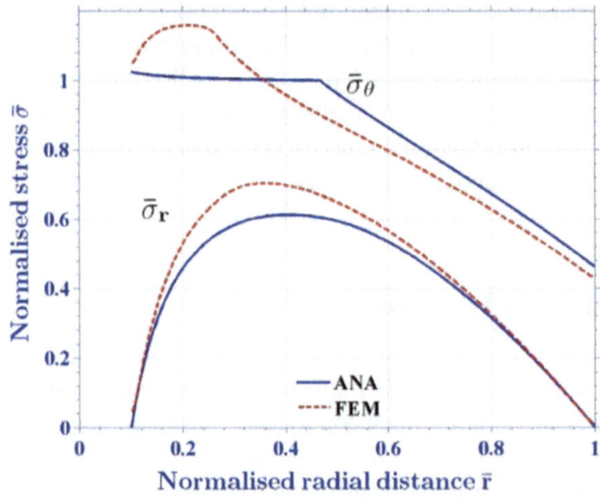

Fig. 28.5 Variation of tangential and radial stress in rotating disk at 10,000 rpm

Fig. 28.6 Variation of tangential and radial stress in rotating disk at 11,000 rpm

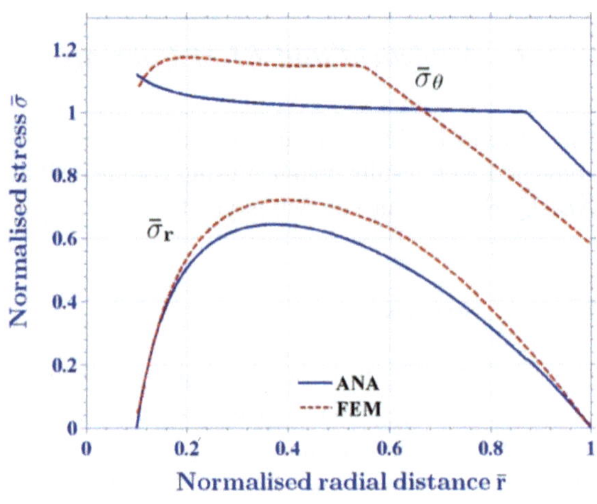

28.5.1.2 Residual Stresses

Residual stresses obtained analytically in tangential $(\bar{\sigma}_{\theta r})$ and radial directions $(\bar{\sigma}_{rr})$ have been compared with finite element method for all the rotating speeds considered in this investigation. Figures 28.11, 28.12, 28.13, 28.14, 28.15 and 28.16 show that trend in the variation of residual stresses compares well for all the rotating speeds and tangential residual stress at the bore increases with increase in speed.

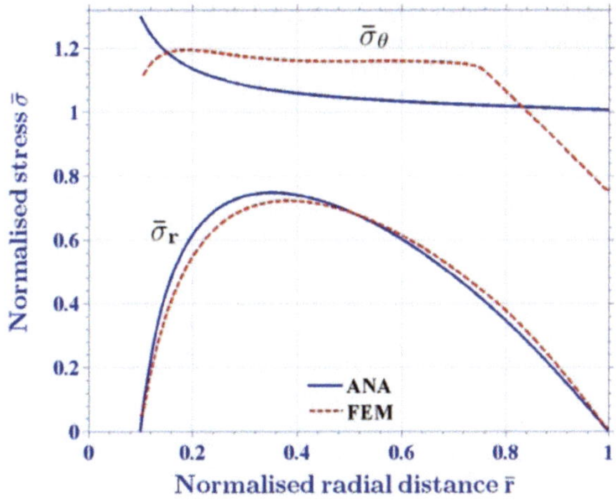

Fig. 28.7 Variation of tangential and radial stress in rotating disk at 11,500 rpm

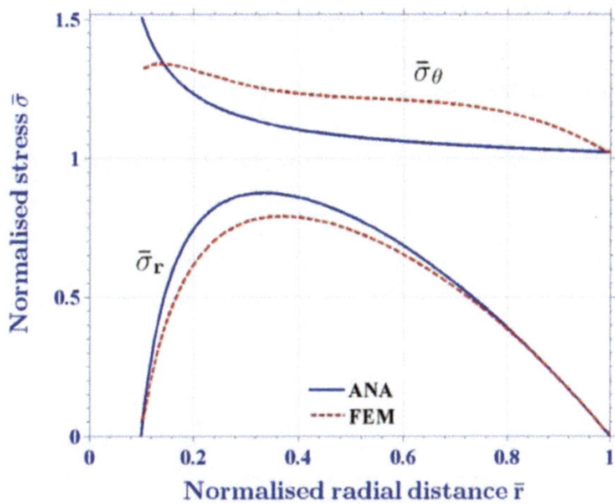

Fig. 28.8 Variation of tangential and radial stress in rotating disk at 12,000 rpm

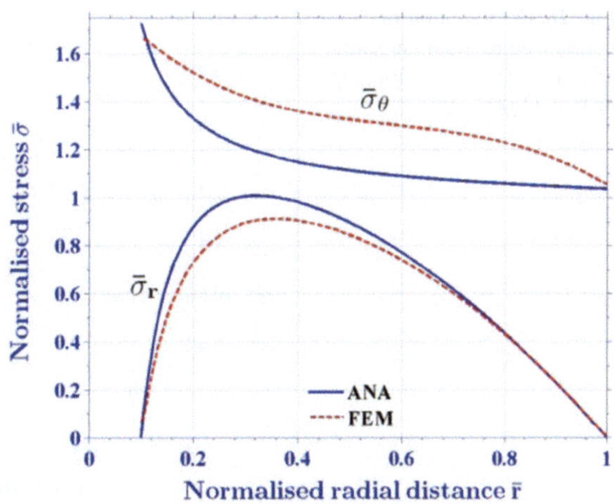

Fig. 28.9 Variation of tangential and radial stress in rotating disk at 12,500 rpm

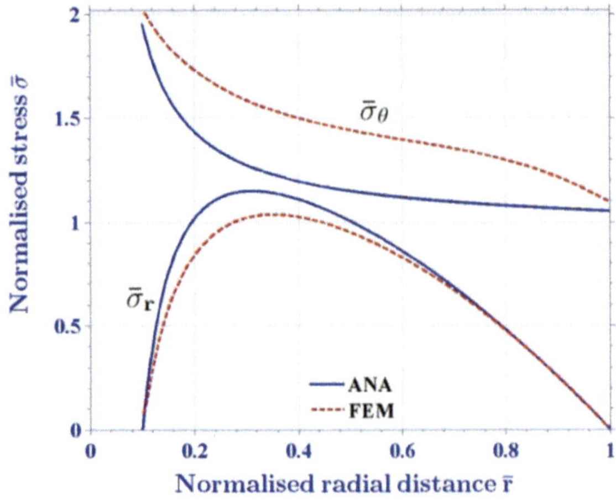

Fig. 28.10 Variation of tangential and radial stress in rotating disk at 13,000 rpm

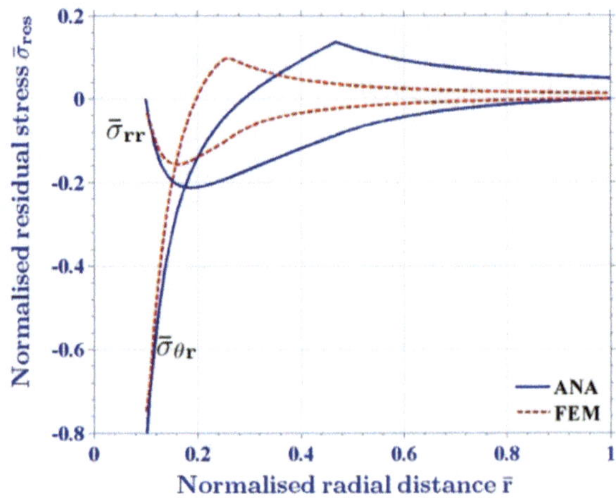

Fig. 28.11 Variation of normalized residual tangential and radial stresses in rotating disk at 10,000 rpm

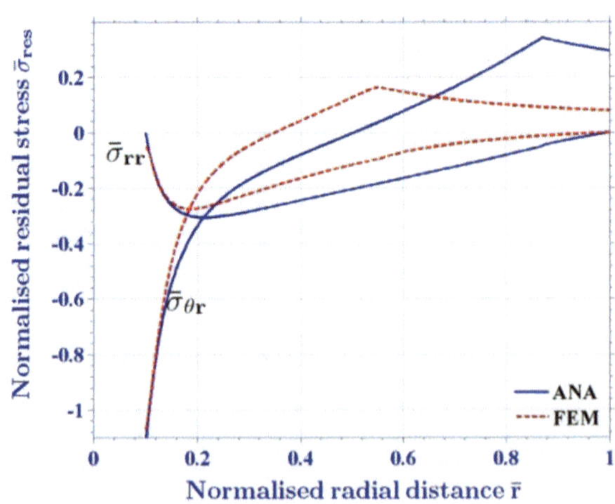

Fig. 28.12 Variation of normalized residual tangential and radial stresses in rotating disk at 11,000 rpm

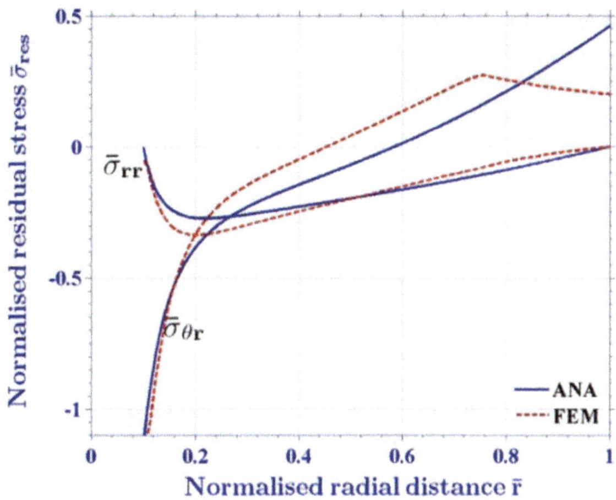

Fig. 28.13 Variation of normalized residual tangential and radial stresses in rotating disk at 11,500 rpm

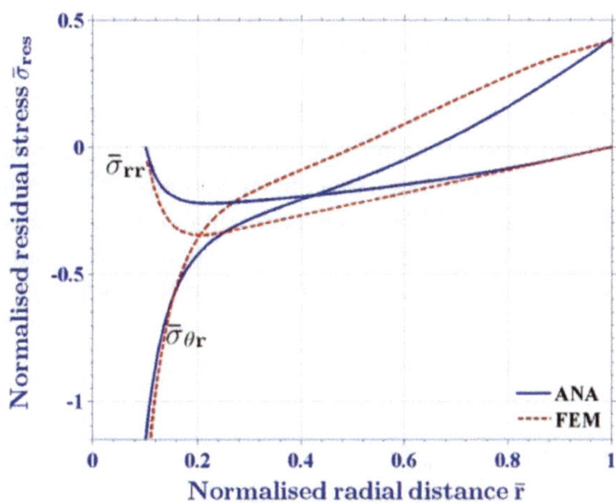

Fig. 28.14 Variation of normalized residual tangential and radial stresses in rotating disk at 12,000 rpm

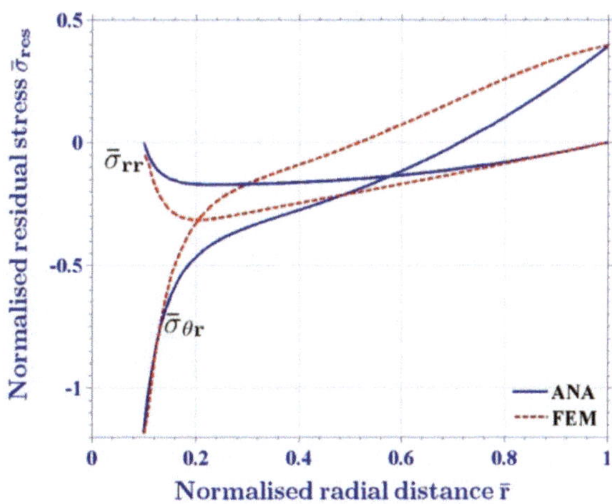

Fig. 28.15 Variation of normalized residual tangential and radial stresses in rotating disk at 12,500 rpm

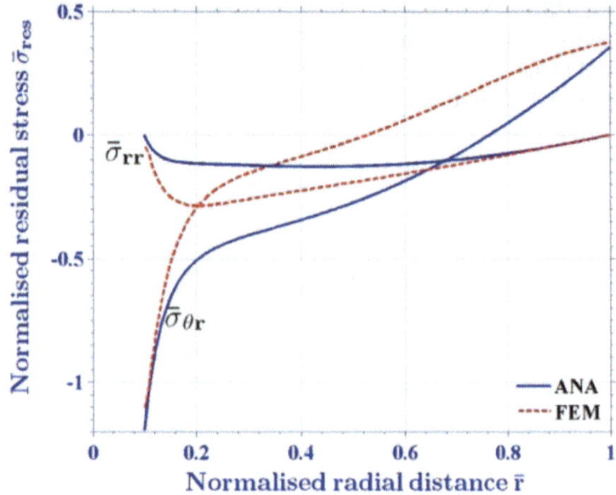

Fig. 28.16 Variation of normalized residual tangential and radial stresses in rotating disk at 12,500 rpm

28.5.1.3 Permanent Growth

Figures 28.17 and 28.18 show permanent growth in rotating disk obtained using closed-form solution, FEM techniques and experimental results at the bore and rim of the disk. Although trend in the variation of permanent growth is similar in closed form solution, FEM and experimental results, there is difference in growth values at higher rotating speeds due to lower bound stress–strain curve taken for closed-form solution and FEM simulation. These figures further reveal that permanent growth increases appreciably after 11,500 rpm ($\Omega = 1.7$), the speed at which entire disk attains fully plastic stage.

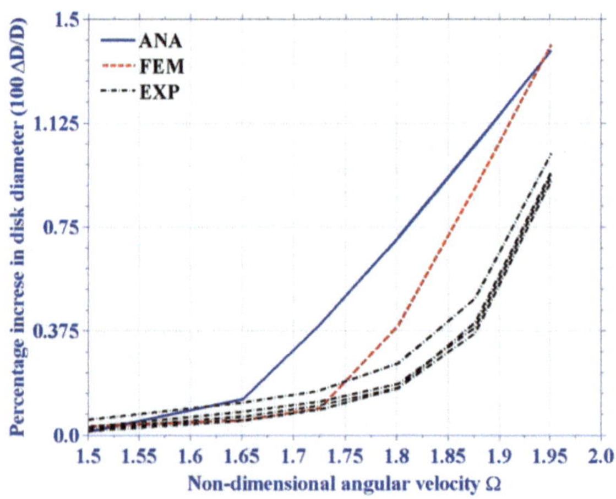

Fig. 28.17 Percentage increase in disk diameter at the bore of the disk for 10,000–13,000 disk speed

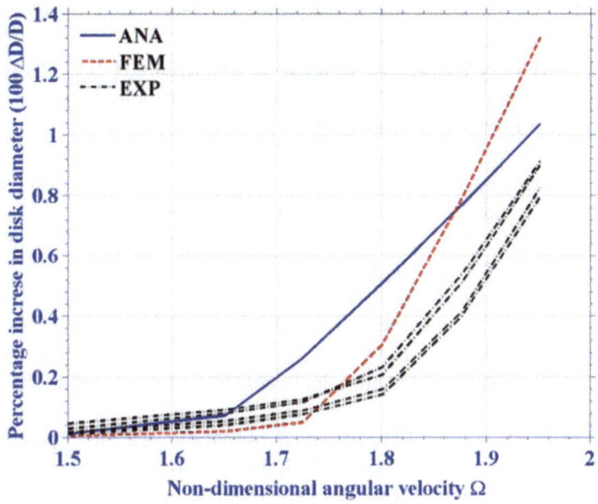

Fig. 28.18 Percentage increase in disk diameter at the rim of the disk for 10,000–13,000 disk speed

28.6 Conclusions

A closed-form solution has been developed using Tresca's yield criterion to obtain permanent growth in disk after removing centrifugal load due to rotation of the disk. Stresses and residual stresses have been compared with FEM results using ABAQUS software. The elastic plastic interface radius $(\bar{r}_p)$ obtained from closed form result based on Tresca's yield criterion is found to be more as compare to FEM result which is based on von-Mises yield criterion.

The closed-form solution, FEM as well as spin test results show that the permanent deformation for uniform thickness disk is higher at the disk bore as compared to that of the rim for all the speeds. The trend of permanent deformation between closed form and FEM matches well. The small difference in the results is due to different yield criteria adopted by both these methods. Experimental results show that there is variation in permanent growth for all the four disks tested in the spin rig and growth value is lesser than analytical and FEM as expected as these predictions are carried out for lower bound stress–strain curve.

Acknowledgements Authors are thankful to Director, GTRE Dr. C. P. Ramanarayan, Outstanding Scientist for allowing this paper to publish in international referred journal.

References

1. Timoshenko, S., Goodier, J.N.: Theory of Elasticity. McGraw-Hill, New York (1970)
2. Stodola, A.: Dampf und Gasturbinen, 6th edn. Julius Springer, Berlin (1924)
3. Reddy, T.Y., Srinath, H.: Elastic stresses in a rotating anisotropic annular disk of variable thickness and variable density. Int. J. Mech. Sci. **16**, 85–89 (1974)

4. Jain, R., Ramachandra, K., Simha, K.R.Y.: Rotating anisotropic disk of uniform strength. Int. J. Mech. Sci. **41**, 639–648 (1999)
5. Military Handbook. Department of Defence, USA: Mil-HDBK-5H (1998)
6. Chakrabarty, J.: Theory of Plasticity. McGraw-Hill, New York (1987)
7. Johnson, W., Mellor, P.B.: Engineering Plasticity. Ellis Horwood, Chichester, UK (1983)
8. Gamer, U.: Tresca's yield condition and the rotating disk. J. Appl. Mech. **50**:676–678 (1983)
9. Eraslan, A.N., Argeso, H.: Limit angular velocities of variable thickness rotating disks. Int. J. Solids Struct. **39**, 3109–3130 (2002)
10. Ma, G., Hao, H., Miyamoto, Y.: Limiting angular velocity disc with unified yield criterion. Int. J. Mech. Sci. **43**, 1137–1153 (2001)
11. Güven, U.: Elastic-plastic stresses in a rotating annular disk of variable thickness and variable density. Int. J. Mech. Sci. **34**, 133–138 (1992)
12. Eraslan, A.N.: Elastic–plastic deformations of rotating variable thickness annular disks with free, pressurized and radially constrained boundary conditions. Int. J. Mech. Sci. **45**, 643–667 (2003)
13. Eraslan, A.N.: Elastoplastic deformations of rotating parabolic solid disks using Tresca's yield criterion. Eur. J. Mech. A Solids **22**, 861–874 (2003)
14. Rees, D.W.A.: The Mechanics of Solids and Structures, 1st edn. McGraw-Hill, New York (1990)
15. You, L.H., Long, S.Y., Zhang, J.J.: Perturbation solution of rotating solid disks with non-linear strain hardening. Mech. Res. Commun. **24**, 649–658 (1997)
16. You, L.H., Zhang, J.J.: Elastic-plastic stresses in a rotating solid disk. Int. J. Mech. Sci. **41**, 269–282 (1999)
17. You, L.H., Tang, Y.Y., Zhang, J.J., Zhen, C.Y.: Numerical analysis of elastic-plastic rotating disks with arbitrary variable thickness and density. Int. J. Solids Struct. **37**, 7809–7820 (2000)
18. Bhowmick, S., Misra, D., Nath, K.: Variational formulation based analysis on growth of yield front in high speed rotating solid disks. Int. J. Eng. Sci. Technol. **2**, 200–219 (2010)
19. Wilterdink, P.I., Holms, A.G., Manson, S.S.: A theoretical and experimental investigation of the influence of temperature gradient on the deformation and burst speeds of rotating disks. Lewis Flight Propulsion Laboratory Cleveland, Ohio, Technical Note 2803 (1952)
20. Zienkiewicz, O.C.: The Finite Element Method in Engineering Science. McGraw-Hill, London (1971)
21. Ayyappan, C., Rajesh, K., Ramesh, P., Jain, R.: Experimental and numerical studies to predict residual growth in an aero-engine compressor disk after over-speed. In: Procedia Engineering, 6th International Conference on Creep Fatigue and Creep-Fatigue Interaction, vol. 55, pp. 625–30 (2013)
22. Karlsson, H.: ABAQUS/Standard User's Manual, vol. I & II, Version 5.4. SorensenInc. Pawtucket, Rhode Island, USA (1994)
23. Hsu, Y.C., Forman, R.G.: Elastic-plastic analysis of an infinite sheet having a circular hole under pressure. ASME J. Appl. Mech. **42**, 347–352 (1975)
24. Wanlin, G.: Elastic-plastic analysis of a finite sheet with a cold-worked hole. Eng. Fract. Mech. **46**, 465–472 (1993)
25. Callioğlu, H., Topcu, M., Tarakcilar, A.R.: Elastic-plastic stress analysis of an orthotropic rotating disc. Int. J. Mech. Sci. **48**, 985–990 (2006)
26. Robinson, E.L., Schenectady, N.Y.: Bursting tests of steam turbine disk wheels. Trans. ASME **66**, 373–386 (1944)

Chapter 29
Flow and Heat Transfer Characteristics of Surface-Mounted Cylinder in Presence of Rectangular Winglet Pair

Hemant Naik and Shaligram Tiwari

Nomenclature

A	Area of the heated surface
C_{p}	Coefficient of pressure
D	Cylinder diameter
f	Friction factor
H	Cylinder height
h	Convective heat transfer coefficient
h_{VG}	Height of vortex generator
JF	Thermal performance factor
k	Turbulent kinetic energy
L_1	Length of computational domain
L_2	Width of computational domain
L_3	Height of computationaldomain
l_{VG}	Length of vortex generator
Nu	Nusselt number
P	Non-dimensional pressure
Re	Reynolds number
Se	Secondary flow intensity
T	Temperature
U_∞	Free stream velocity
U_j	Cartesian velocity component in X_j-coordinate direction
X_j	Non-dimensionalized Cartesian space coordinates in X, Y, Z direction
X, Y, Z	Non-dimensionalized Cartesian space coordinates

Greek symbols

α_{t}	Turbulent dynamic thermal diffusivity
β	Angle of attack of vortex generator

H. Naik · S. Tiwari (✉)
Department of Mechanical Engineering, Indian Institute of Technology Madras, Chennai 600036, India
e-mail: shaligt@iitm.ac.in

P. Sahoo and J. P. Davim (eds.), *Advances in Materials, Mechanical and Industrial Engineering*, Lecture Notes on Multidisciplinary Industrial Engineering,
https://doi.org/10.1007/978-3-319-96968-8_29

ΔX Streamwise center distance between tube and winglet
ΔY Spanwise center distance between tube and winglet
ε Dissipation rate
λ Thermal conductivity
ν Kinematic viscosity of fluid
ν_t Turbulent kinematic viscosity of fluid
Ω Volume of the computational domain
ω^n Vorticity component normal to a cross section
ρ Density of fluid
σ_k Turbulent Prandtl numbers for k
σ_ε Turbulent Prandtl numbers for ε
θ Non-dimensional temperature
y^+ Wall y-plus

Subscripts

b Bulk-mean value
∞ Inlet
local Local value
n Normal direction
o Absence of vortex generator
w Wall

Abbreviations

AR Aspect ratio
CFD Common flow down
CFU Common flow up
DWP Delta winglet pair
FFR Friction factor ratio
HTR Heat transfer ratio
RWP Rectangular winglet pair
SFIR Secondary flow intensity ratio
VG Vortex generator

29.1 Introduction

Flow and heat transfer past surface-mounted obstacles are investigated by many researchers both experimentally and numerically. Heat and flow behaviors around surface-mounted obstacles get much attention not only because of academic attractiveness but also due to various real-life problems associated with structural aspects and energy conservation. A significant amount of research on flow field and heat transfer characteristics for flow past obstacles has been carried out in past for variety of practical applications such as cooling of electronic equipment, chimneys, cooling towers, building aerodynamics, automotive design and heat exchanger systems. It becomes important to understand the mechanism by which a surface-mounted obstacle influences the flow structure and enhances heat transfer. Various topological flow fields develop around the obstacles which can affect the heat transfer significantly. Understanding the flow phenomenon and its effect on heat transfer is much complex due to its three-dimensionality.

Investigations on flow and thermal fields from finite height surface-mounted obstacles (circular, square, rectangular and other shapes cylinders) have been carried out by various researchers in past. Flow and heat transfer around obstacles of different shapes like circular cylinder, cone, square bar have been carried out by Ghisalberti and Kondjoyan [1]. They reported that aspect ratio (AR, height to diameter ratio of cylinder) influences the enhancement in heat transfer. Different types of vortex patterns from flow past finite height circular cylinder mounted on a horizontal plate have been reported in the literature which affects the behavior of heat transfer significantly. For example, horseshoe vortex develops in front of cylinder and near the junction of cylinder-bottom wall, Karman vortex shedding takes place from the sides of the cylinder and trailing vortices generate above the free-end of cylinder. Heat transfer and flow field characteristics of mounting surface obstacles as finite height circular cylinders have been investigated by Kawamura et al. [2]. They reported that heat transfer characteristics depend on AR, boundary layer thickness and Reynolds number (*Re*). They also reported that dominance of Karman vortex shedding and trailing vortices depend on critical AR which strongly influences heat transfer characteristics. Giordano et al. [3] reported that AR of the cylinder and *Re* are the parameters that influence both heat transfer and flow field characteristics. Tsutsui et al. [4] and Tsutsui and Kawahara [5] investigated flow past circular cylinder and reported that the horseshoe vortex plays a dominant role in augmentation of heat transfer. For low values of AR, higher heat transfer enhancement is reported just below the leading edge in downstream of cylinder. Three-dimensional flow topology for flow past finite height circular cylinder for different AR and *Re* has been studied numerically by Naik and Tiwari [6, 7]. They reported that horseshoe vortices enhance fluid mixing and disrupt the boundary layer growth thereby augmentation in heat transfer takes place. Sahin et al. [8] and Rostamy et al. [9] reported that vortices generated from and around the cylinder surface influence the heat transfer characteristics. They also concluded that contribution toward heat transfer enhancement is more due to horseshoe vortex than vortices from the side surface of a cylinder. A brief review of flow field and heat transfer characteristics of flow past surface-mounted circular cylinder has been reported by Sumner [10].

Development of thermal boundary layer around obstacles always decreases heat transfer rate due to no-slip boundary condition. Interruption in the development of thermal boundary layer causes enhancement in heat transfer which gives rise to better mixing of fluid. This can be achieved by introducing secondary flow in the flow path [11, 12]. For generation of secondary flow, mounting vortex generator (VG) on plate surface by punching or stamping is one of the effective methods. Presence of VGs generates longitudinal vortices in the main flow direction such as in streamwise direction which enhances fluid mixing by disrupting the growth of thermal boundary layer and thereby augmentation in heat transfer is achieved [13–15]. Vortex generators are generally mounted on a plate surface at an angle of attack (β) with respect to the streamwise direction. Most commonly used geometric form of VG is winglet type VG in which delta winglet and rectangular winglet are much popular. Numerical investigations on enhancement in heat transfer have been performed by Tian et al. [16] for a pair of winglet type vortex generators having both common flow up (CFU) and common flow down (CFD) configurations. They considered the two commonly known types of winglet pair such as delta winglet pair (DWP) and rectangular winglet pair (RWP). They reported that surface-mounted RWP gives higher heat transfer enhancement than DWP. For surface-mounted RWP, they found that thermal performance of CFD configuration is better than that of CFU configuration. Experimental as well as numerical investigations on heat transfer and flow field characteristics due to various geometric parameters of VGs have been carried out by various researchers Wu and Tao [17] investigated the flow and heat transfer analysis of rectangular winglet type VG by punching it out from the fin surface. They studied the effect of location of VG, i.e., streamwise distance between inlet of channel and leading edge of VG and transverse spacing between winglet pair on flow characteristics and heat transfer enhancement. They reported that augmentation in heat transfer gets reduced when transverse spacing between wings of VGs is reduced. Also, heat transfer of the channel is found to decrease when the location of VG is shifted away from the inlet of the channel. Numerical investigations on flow and heat transfer by surface-mounted RWPs have been performed by Naik and Tiwari [18]. They have investigated the effect of different geometric parameters of RWP such as angle of attack of VG (β), length of RWP and spacing between RWP on flow and heat transfer characteristics. They also considered the effect of different values of *Re*. They found that augmentation in heat transfer increases with increase in β and it attains its maximum at $\beta = 35°$, further increase in β resulting into decrease in heat transfer. Enhancement in heat transfer is reported for the case of increasing length of RWP as well as increasing *Re*.

It can be concluded from above literature that horseshoe vortices generated around surface-mounted circular cylinder and longitudinal vortices generated from surface-mounted VGs play important role on heat transfer enhancement. Thorough literature review on VGs and finite height cylinders have been done and it has been observed that the study on effect of combination of surface-mounted finite height single cylinder and RWP on heat and flow characteristics is rather limited in the literature. Location of RWP corresponding to cylinder center is expected to influence the thermal performance of the cylinder and RWP system. Hence, present work aims to investigate the effect of different locations of RWP with respect to cylinder center for the combination of both cylinder and RWP on heat and flow characteristics. Streamlines plots and temperature contours are used for the investigation of flow features and heat transfer characteristics. Nusselt number and friction factor are used for the estimation of heat transfer enhancement and pressure loss. A non-dimensional number known as secondary flow intensity is used to estimate the secondary flow generated by the VGs and cylinders. Variation of secondary flow intensity is used to examine the role of secondary flow on heat transfer enhancement. Thermo-hydraulic performance is also estimated by thermal performance factor which shows the combined effect of flow and heat transfer.

29.2 Problem Formulation

The problem under consideration is fluid flow behavior and heat transfer characteristics past surface-mounted circular cylinder of a finite height in the presence of rectangular winglet pair (RWP) placed at different locations relative to center of the cylinder. Figure 29.1a depicts the schematic of three-dimensional computational domain used in the present study. Circular cylinder having diameter '*D*' and height '*H*' is mounted on a horizontal plate at a distance of 10*D* from inlet. Aspect ratio of the cylinder, i.e., the height to diameter ratio has been kept fixed as 2.0. Dimensions of the computational domain such as length, width and height are fixed and taken as 30*D*, 10*D* and 7*D*, respectively. Rectangular winglet pair (RWP) as VG with CFD configuration is considered for the present study. The length and height of VG are kept fixed as 1*D* and 0.5*D*, respectively. The angle of attack (β) has been kept fixed as 35°. Location of RWP from center of cylinder is shown in Fig. 29.1b. Effect of different possible streamwise (ΔX) and spanwise (ΔY) locations of RWP corresponding to center of cylinder such as $(\Delta X, \Delta Y) = (-2, \pm 0.5)$, $(-1, \pm 0.5)$, $(-0.5, \pm 0.5)$, $(1, \pm 0.5)$, $(2, \pm 0.5)$, $(-2, \pm 1)$, $(-1, \pm 1)$, $(-0.5, \pm 1)$, $(0, \pm 1)$, $(0.5, \pm 1)$, $(1, \pm 1)$, $(2, \pm 1)$ and $(-2, \pm 1.5)$, $(-1, \pm 1.5)$, $(-0.5, \pm 1.5)$, $(0, \pm 1.5)$, $(0.5, \pm 1.5)$, $(1, \pm 1.5)$, $(2, \pm 1.5)$ on heat and fluid flow behavior have been investigated. Computations are carried out for the flow Reynolds number (*Re*) ranging from 1000 to 4000 for all considered locations of RWP.

29.3 Numerical Methodology

29.3.1 Governing Equations and Boundary Conditions

In the present study, air has been used as working fluid and the assumptions employed for the flow of air are three-dimensional, incompressible, steady and turbulent with constant thermo-physical properties. Viscous dissipation effects and radiation effects are considered to be negligible. The non-dimensional form of governing equations in Cartesian coordinates including mass, momentum and energy equations are as follows

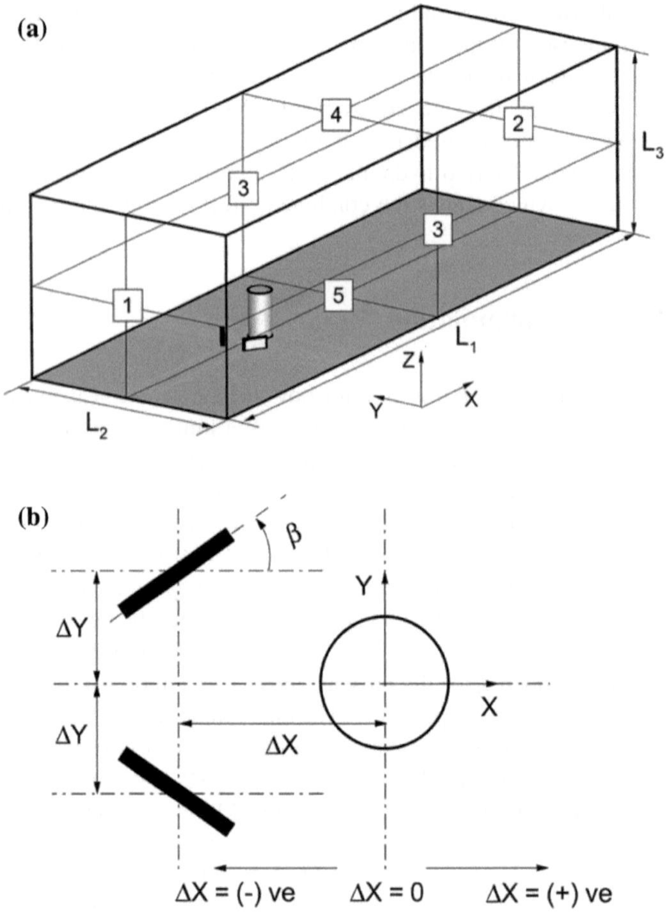

Fig. 29.1 **a** Computational domain, **b** location of rectangular winglet pair

$$\frac{\partial U_j}{\partial X_j} = 0 \tag{29.1}$$

$$U_j \frac{\partial U_i}{\partial X_j} = -\frac{\partial (P + 2k/3)}{\partial X_i} + \frac{1}{Re}\frac{\partial}{\partial X_j}\left[(1+\nu_t)\left(\frac{\partial U_i}{\partial X_j} + \frac{\partial U_j}{\partial X_i}\right)\right] \tag{29.2}$$

$$U_j \frac{\partial \theta}{\partial X_j} = \frac{1}{Re\,\Pr}\frac{\partial}{\partial X_j}\left[(1+\alpha_t)\frac{\partial \theta}{\partial X_j}\right] \tag{29.3}$$

Here the non-dimensional Cartesian component of velocity in non-dimensional X_j-space coordinate direction is denoted as U_j. For non-dimensionalization of space coordinates, considered characteristic length scale is the diameter of cylinder (D), and hence, the non-dimensional form of space coordinates are $X = \frac{x}{D}$, $Y = \frac{y}{D}$ and

$Z = \frac{z}{D}$. The free stream velocity (U_∞) has been chosen as the characteristic velocity for non-dimensionalization of velocities. The non-dimensional form of pressure terms can be written as $P = \frac{p}{\rho U_\infty^2}$, where p is the pressure and ρ is density of the fluid. The non-dimensional form of temperature can be written as $\theta = (T - T_\infty)/(T_w - T_\infty)$, here the temperature of the fluid at free stream is represented by T_∞ and the constant wall temperature is denoted by T_w which is the temperature at the surfaces of cylinder, RWP and bottom plate.

The flow Reynolds number (Re) based on cylinder diameter, uniform velocity at the inlet, i.e., free stream velocity and the kinematic viscosity of the fluid (ν) is evaluated by

$$Re = U_\infty D/\nu \tag{29.4}$$

To obtained the values of turbulent kinetic energy (k) and its dissipation rate (ε), the two equations Realizable $k - \varepsilon$ turbulence model has been chosen and corresponding transport equations are expressed as follows

$$U_j \frac{\partial k}{\partial X_j} = \frac{1}{Re} \frac{\partial}{\partial X_j} \left[\left(\frac{\nu_t}{\sigma_k} \right) \frac{\partial k}{\partial X_j} \right] + G_k - \varepsilon \tag{29.5}$$

and

$$U_j \frac{\partial \varepsilon}{\partial X_j} = \frac{1}{Re} \frac{\partial}{\partial X_j} \left[\left(\frac{\nu_t}{\sigma_\varepsilon} \right) \frac{\partial \varepsilon}{\partial X_j} \right] + C_1 S \varepsilon - C_{2\varepsilon} \frac{\varepsilon^2}{k + \sqrt{\nu \varepsilon}} \tag{29.6}$$

Shih et al. [19] are referred for the values of constants and definitions of the terms ν_t, α_t, G_k, C_1, S, $C_{2\varepsilon}$, σ_k and σ_ε.

Boundary conditions are described at the different faces of the computational domain, and these different faces have been numbered to identify the boundaries in Fig. 29.1a. The boundary conditions employed at different boundaries such as inlet, outlet, walls of the computational domain and surfaces of cylinder and RWP are as summarized below.

- Uniform velocity in streamwise direction with constant temperature is taken at inlet of domain (face-1).

$$(u = U_\infty, v = w = 0, T = T_\infty)$$

  At the inlet of the computational domain, the correlation $I = 0.16 Re^{-1/8}$ available in Min et al. [20] is used to calculate the turbulent intensity (I). The computed turbulent intensity lies between 5 and 10% for the considered velocity range.
- Pressure outlet condition has been imposed at the exit of the domain (face-2).

$$(p = p_\infty)$$

- No-slip, impermeable and constant temperature boundary conditions have been considered at bottom wall (face-5) and surfaces of cylinder and RWP.

$$(u = v = w = 0, T = T_{\infty})$$

- Free-slip and impermeable boundaries are considered at side boundaries (faces-3).

$$\left(v = 0, \frac{\partial u}{\partial y} = \frac{\partial w}{\partial y} = 0, \frac{\partial p}{\partial y} = 0, \frac{\partial T}{\partial y} = 0\right)$$

- Free-slip and impermeable boundaries are prescribed at top boundaries (faces-3).

$$\left(w = 0, \frac{\partial u}{\partial z} = \frac{\partial v}{\partial z} = 0, \frac{\partial p}{\partial z} = 0, \frac{\partial T}{\partial z} = 0\right)$$

29.3.2 Parameter Definition

In order to estimate the heat transfer and fluid flow characteristics, the expressions of the relevant parameters are defined as follows.

The overall Nusselt number (*Nu*) is used to estimate the heat transfer performance and can be expressed as

$$Nu = \frac{hD}{\lambda} \tag{29.7}$$

Here h is the overall convective heat transfer coefficient and is calculated in terms of bulk mean temperature of fluid $(T_{\mathrm{b}}(x))$, temperature gradient normal to the surface $(\partial T/\partial z)_{\mathrm{n}}$ and differential area of the heated surface $(\mathrm{d}A)$.

$$h = \frac{\int \left(\frac{-\lambda(\partial T/\partial z)_{\mathrm{n}}}{T_{\mathrm{w}} - T_{\mathrm{b}}(x)}\right) \mathrm{d}A}{\int \mathrm{d}A} \tag{29.8}$$

where λ is the thermal conductivity of fluid and n is normal direction with respect to the surface. At a given location, the bulk mean temperature of fluid can be expressed as

$$T_{\mathrm{b}}(x) = \frac{\int u(x,y,z)T(x,y,z)\mathrm{d}z}{\int u(x,y,z)\mathrm{d}z} \tag{29.9}$$

The friction factor (*f*) is obtained by calculating pressure drop (Δp) in the domain and can defined as

$$f = \frac{\Delta p}{1/2\rho U_{\infty}^{2}} \times \frac{D}{L_{1}} \tag{29.10}$$

Quantification of secondary flows generated due to the presence of obstacles such as cylinder and RWPs in the flow path can also be used to estimate the performance of heat transfer. Secondary flow is calculated with the help of secondary flow intensity (Se) which is absolute vorticity flux based a non-dimensional number [21]. The volume-averaged *Se* is defined in terms of vorticity component normal to a cross section (ω^{n}) and the differential volume of the considered domain ($\mathrm{d}\Omega$) and expressed as

$$\mathrm{Se} = \frac{\rho D^{2}}{\mu}\left(\frac{\int_{\Omega}|\omega^{\mathrm{n}}|\mathrm{d}\Omega}{\int_{\Omega}\mathrm{d}\Omega}\right) \tag{29.11}$$

Secondary flow intensity ratio (SFIR) is used to compare the secondary flow generated due to surface-mounted cylinder in the presence of RWP (Se) to the secondary flow generated in absence of RWP ($\mathrm{Se_o}$).

The thermal enhanced factor (JF) is used to estimate combined effect of flow and heat transfer and is calculated as

$$\mathrm{JF} = \frac{(\mathrm{HTR})}{(\mathrm{FFR})^{1/3}} \tag{29.12}$$

where heat transfer ratio (HTR) and friction factor ratio (FFR) are the ratio of overall surface-averaged *Nu* and friction factor (*f*) in the presence of RWP to overall surface-averaged Nusselt number (Nu_{o}) and friction factor (f_{o}) in the absence of RWP, respectively.

29.3.3 Grid Mesh and Grid Independence

Figure 29.2a presents the schematic of the grid mesh around the cylinder with RWP mounted on the bottom plate. Figure 29.2b shows the top view of grid mesh around cylinder and RWP by keeping other features hidden. A non-uniform hexahedral structured grid has been generated with the help of a commercial software widely known as ANSYS ICEM CFD 17.2. A grid mesh strategy has been followed to capture the fluid flow and heat transfer characteristics appropriately. Refined O-type

(a)

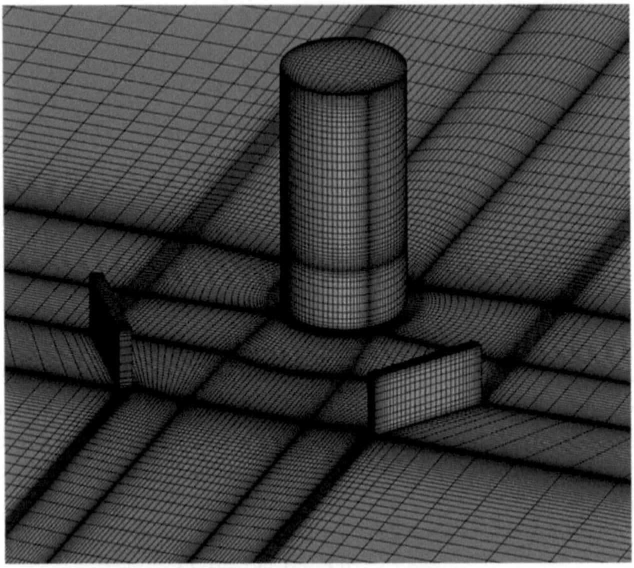

(b)

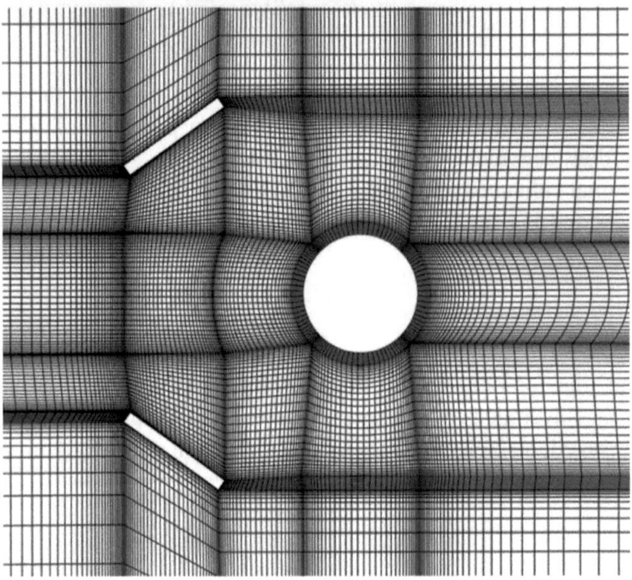

Fig. 29.2 Grid mesh around. **a** Cylinder and RWP, **b** top view of grid

non-uniform grid structure is created close to the cylinder and RWP while non-uniform hexahedral structured grid has been generated everywhere else in the domain.

Rigorous grid independence study has been carried out for better accuracy of the computations and to understand the effect of grid sizes on different parameters as presented in subsection 'Parameter Definition'. Initially grid independence has been studied for different number of grid nodes around the circumference of cylinder and along the length of RWP. Further grid independence has been studied for the upstream and downstream of the cylinder along the length of the domain. Grid independence study for a fixed location of RWP $(\Delta X, \Delta Y)$ equal to (−2, ±1), (0, ±1) and (2, ±1) has been carried out by considering four different grid sizes. Figure 29.3 depicts the variation of overall surface-averaged Nusselt number (*Nu*) with change in grid size from 0.5 million to 6.0 million for different streamwise location of RWP (ΔX) at fixed spanwise location of RWP $(\Delta Y) = \pm 1$ and fixed value of *Re* = 3000. It has been observed that for all three ΔX locations, the value of *Nu* increases with grid refinement, i.e., increase in the number of grid nodes from 0.5 million to 2.36 million. Further increase in the number of grid nodes, *Nu* does not change significantly. Percentage change in the value of *Nu* for the number of grid nodes higher than 2.36 million are found to be less than 0.29, 0.15 and 0.23% for $\Delta X = -2$, 0 and 2, respectively. Therefore, the flow and heat transfer characteristics of the cases considered for the present study have been investigated by performing all the computations having 2.36 million grid nodes.

29.3.4 Numerical Technique

Finite volume-based commercial software ANSYS Fluent 17.2 is employed to discretize and solve the governing equations. Two equations Realizable $k - \varepsilon$ turbulence model associated with enhanced wall treatment is used to improve the accuracy of computations near heated surfaces under turbulent flow conditions. To capture wall effect in turbulence, *y*-plus (y^+) equivalent to 1 has been considered for near wall modeling. To resolve the coupling between pressure and velocity on a Cartesian grid, Semi-Implicit Method for Pressure Linked Equation (SIMPLE) algorithm has been used in the present study. For discretization of diffusive terms

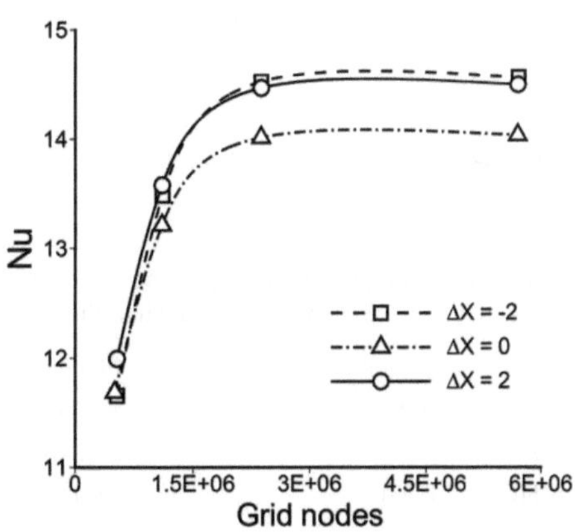

Fig. 29.3 Variation of *Nu* with change in number of grid nodes for different values of ΔX at $\Delta Y = \pm 1$ and *Re* = 3000

present in the governing equations, second-order central difference scheme has been employed and convective terms present in the governing equations have been discretized by using second-order upwind scheme. Absolute convergence criteria considered for mass, momentum and energy equations are taken as 10^{-6} while for k and ε equations it is set as 10^{-5}.

29.3.5 Validation of Computations

Computations have been carried to validate the suitability of numerical technique and the results are compared against those reported in the literature as shown in Fig. 29.4 and Table 29.1. For finite height circular cylinder at fixed value of AR and *Re* equal to 2.0 and 32,000, respectively, the results obtained from the present numerical work and the experimental work reported by Kawamura et al. [2] have been compared. Circumferential variation of coefficient of pressure (C_p) at mid-height of cylinder obtained from the present work is plotted, and comparison has been carried out with the results of Kawamura et al. [2]. The values of C_p obtained from the present computations show close match with the C_p values reported by the experimental work Kawamura et al. [2].

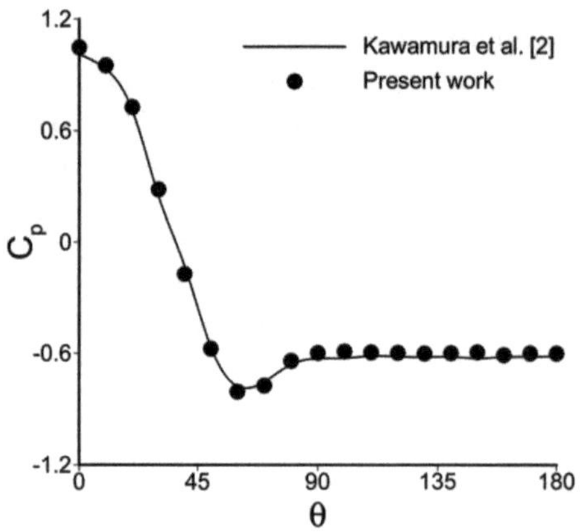

Fig. 29.4 Comparison of circumferential variation of C_p from present computations and Kawamura et al. [2]

Table 29.1 Comparison of overall *Nu*, Nu_o and HTR from present computations using different forms of $k-\varepsilon$ turbulence model against those of Tiggelbeck et al. [14]

	Tiggelbeck et al. [14]	Standard $k-\varepsilon$	Realizable $k-\varepsilon$	RNG $k-\varepsilon$
Nu_o	16.6	18.257	17.549	18.544
Nu	24.26	27.043	25.972	27.827
HTR	1.461	1.481	1.480	1.501

Further, the validation has been strengthen by validating the flow conditions in the presence of surface-mounted RWP. Similar, geometry and boundary conditions have been considered for the computation as reported in the experimental work of Tiggelbeck et al. [14]. In their experiment, RWPs are punched on the channel wall. The considered angle of attack of RWP and flow *Re* (based on hydraulic diameter of the channel which is equal to the twice of the height of RWP) for comparison are chosen from the Tiggelbeck et al. [14] and are equal to 30° and 4600, respectively. Here three different turbulence models such as Standard $k-\varepsilon$, RNG $k-\varepsilon$ and Realizable $k-\varepsilon$ turbulence model are considered for the computations. Comparison of the results obtained from these $k-\varepsilon$ turbulence models and from Tiggelbeck et al. [14] are shown in Table 29.1. The difference between the results from present computations and experimental data is found to be lowest for Realizable $k-\varepsilon$ turbulence model as compared to that from other two turbulence models. The percentage differences between the present computed results obtained by the use of Realizable $k-\varepsilon$ turbulence model and those obtained from the experimental work of Tiggelbeck et al. [14] are 7.1% for Nu_o (overall *Nu* in the absence of RWP), 5.7% for *Nu* (overall *Nu* in the presence of RWP) and 1.3% for HTR. Hence, for all the computations, Realizable $k-\varepsilon$ turbulence model has been implemented in the present study.

29.4 Results and Discussion

29.4.1 Flow and Temperature Field Analysis

Interesting flow features generated due to various locations of RWP for cylinder mounted on plate and respective effects on heat transfer characteristics have been illustrated with the help of streamline plots and non-dimensional temperature contours, corresponding to the time-averaged field. In general, different types of vortices are generated from the surfaces of cylinder and RWP. Generation of horseshoe vortices takes place in upstream of cylinder near the junction of cylinder and bottom plate and also in upstream pf RWP near the leading edges of RWP and bottom plate junction [7, 18]. From the free-end of the cylinder, generation of trailing vortices has been observed while from the side surface of the cylinder vortex shedding takes place which is widely known as Karman vortex shedding [2]. From different edges of RWP such as from leading, trailing and top edges of RWP, longitudinal and transverse vortices are generated [18]. In longitudinal vortices, vortices rotate around an axis parallel to the streamwise flow direction, which enhance the mixing of fluid and disturb the growth of boundary layers. Hence, convective heat transfer gets enhanced due to longitudinal vortices [15, 18]. Similarly, horseshoe vortices also enhance the fluid mixing near bottom plate surface and destabilize the growth of the boundary layer and thereby enhancement in heat transfer occur due to horseshoe vortices [4]. In general, transverse vortices

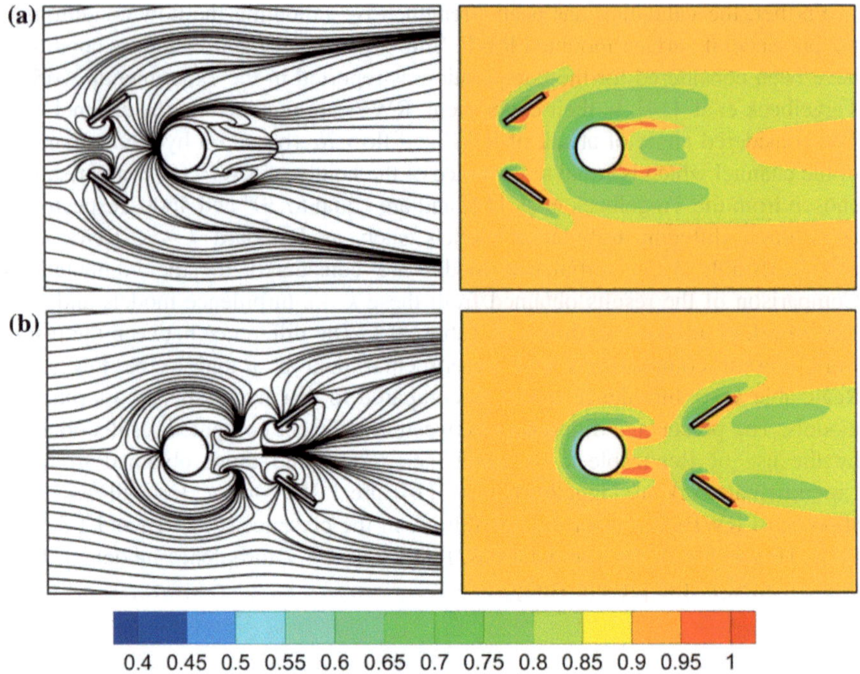

Fig. 29.5 Streamlines and temperature contours close to bottom plate ($Z = 0.1$) for locations. **a** $\Delta X = -2$. **b** $\Delta X = 2$ at $\Delta Y = \pm 0.5$ at $Re = 3000$

and other vortices from cylinder surfaces are less efficient in causing heat transfer enhancement. Figure 29.5 depicts the streamlines plots and non-dimensional temperature contours corresponding to the time-averaged field close to the bottom plate ($Z = 0.1$) for different streamwise locations of RWP, i.e., ΔX locations of RWP at spanwise location of RWP $\Delta Y = \pm 0.5$ and $Re = 3000$. From streamline plots in Fig. 29.5, horseshoe vortices can be seen in front of RWP (Fig. 29.5a) while horseshoe vortices appear in front of cylinder (Fig. 29.5b). From temperature contour of Fig. 29.5, appearance of low-temperature zone in upstream and near the front face of RWP and cylinder can be seen which is due to horseshoe vortices and this shows enhancement in heat transfer. Recirculation region is observed just behind the cylinder and RWP in downstream locations. Near the rear face of the cylinder and RWP, higher temperature zone is observed due to the presence of recirculation regions which shows the poorer transport of fluid there, and hence, heat transfer reduces. In downstream of cylinder, size of recirculation region is found to be higher in case of $\Delta X = -2$ than $\Delta X = -2$. Reason for this may be that for $\Delta X = 2$, RWP suppresses the development of recirculation region.

At fixed spanwise location of RWP $\Delta Y = \pm 1$ and fixed value of $Re = 3000$, Fig. 29.6 presents streamlines and temperature contours close to bottom plate ($Z = 0.1$) for different ΔX locations of RWP. Size of recirculation region behind the

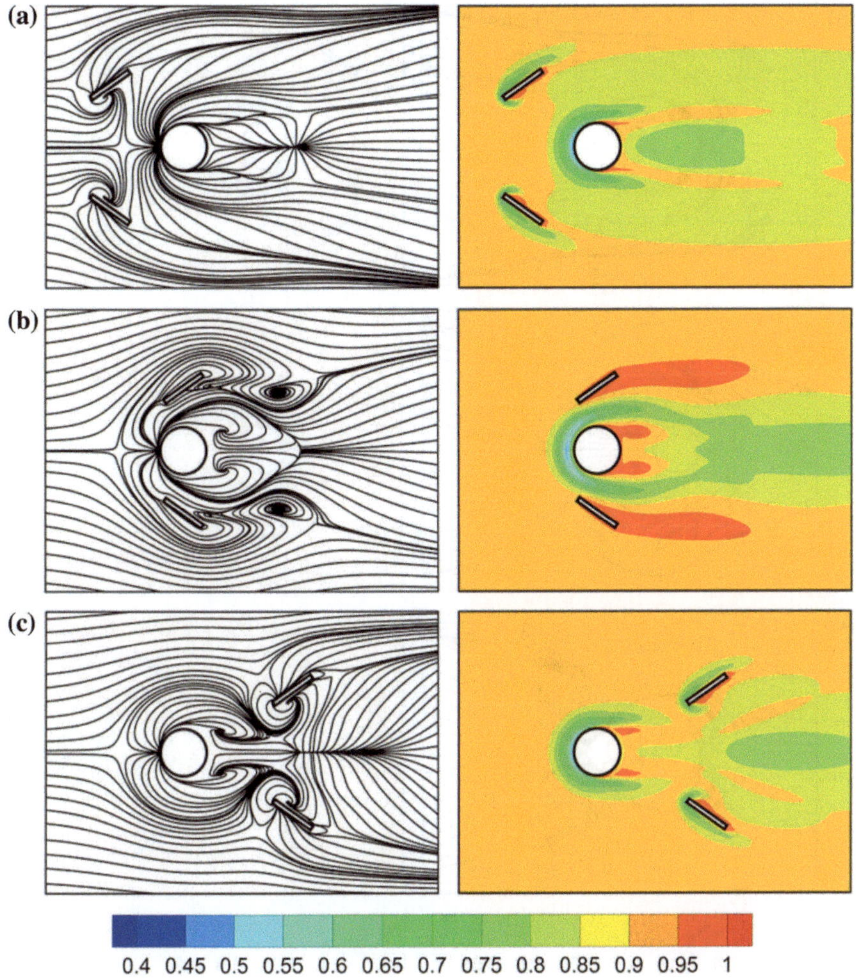

Fig. 29.6 Streamlines and temperature contours close to bottom plate ($Z = 0.1$) for locations. **a** $\Delta X = -2$. **b** $\Delta X = 0$. **c** $\Delta X = 2$ at $\Delta Y = \pm 1$ at *Re* = 3000

cylinder is found to be higher for $\Delta X = 0$ followed by $\Delta X = -2$ and $\Delta X = 2$. Correspondingly, the size of high-temperature zone behind the cylinder can also be seen to follow same order as of recirculation. For $\Delta X = -2$ and 2, effect of horseshoe vortices can be seen for both cylinder and RWP. Similarly, low-temperature zone can be observed in front of RWP and cylinder for $\Delta X = -2$ and 2. On the other hand, for the location of RWP $\Delta X = 0$, generation of horseshoe vortices is observed only due to the cylinder which eliminates the low-temperature zone in front of RWP.

For different streamwise locations of RWP, streamlines and temperature contours close to the bottom plate ($Z = 0.1$) at the location of RWP $\Delta Y = \pm 1.5$ and

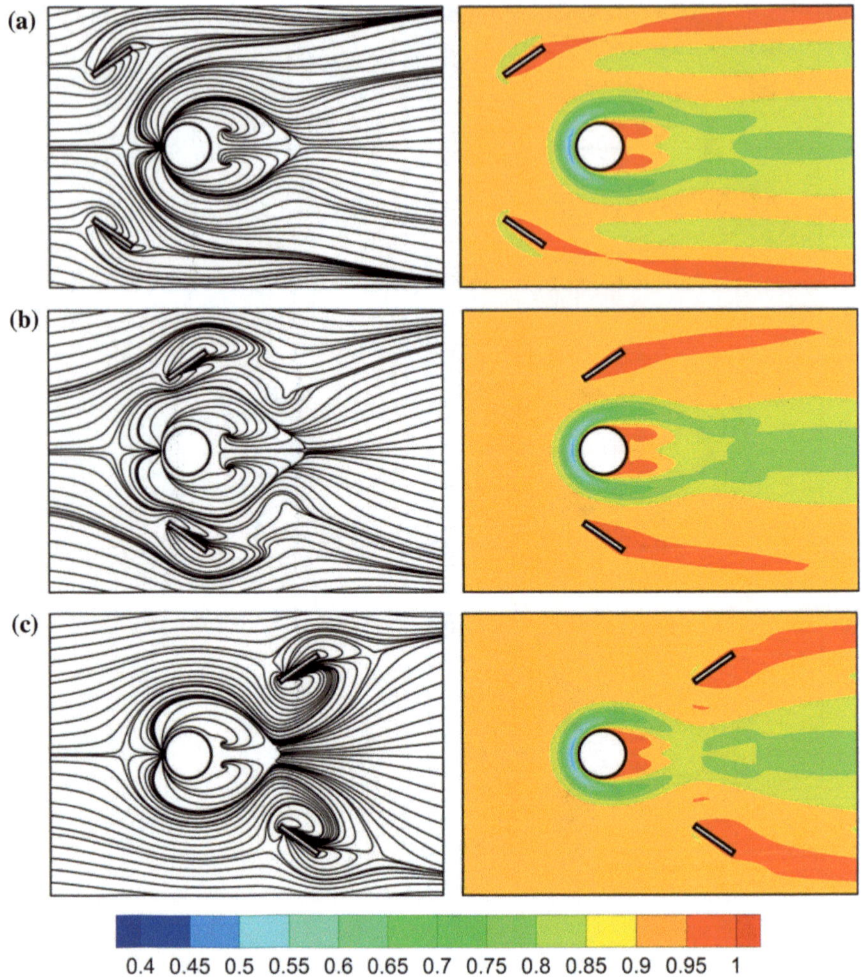

Fig. 29.7 Streamlines and temperature contours close to bottom plate ($Z = 0.1$) for locations. **a** $\Delta X = -2$. **b** $\Delta X = 0$. **c** $\Delta X = 2$ at $\Delta Y = \pm 1.5$ at $Re = 3000$

$Re = 3000$ are presented in Fig. 29.7. Due to large ΔY for all ΔX, horseshoe vortices appear for both cylinder and RWP. Correspondingly, low-temperature zone can also be seen in front of cylinder and RWP. For $\Delta X = -2$, effect of horseshoe vortices due to RWP is relatively stronger than those for other two ΔX values. Size of recirculation region behind the cylinder is higher for $\Delta X = 0$ and lower for $\Delta X = 2$. Consequently, size of high-temperature zone is higher for $\Delta X = 0$ and lower for $\Delta X = 2$. In general, for $\Delta X = -2$ and 2, with increase in the value of ΔY, i.e., the vortex generators of RWP shift away from each other in spanwise direction, horseshoe vortices due to both cylinder and RWP become independent of each other. Effect of horseshoe vortices due to influence of both is visible. For lower ΔY, such that $\Delta Y = \pm 0.5$, horseshoe vortex generation takes place either only from RWP or only from cylinder.

29.4.2 Heat Transfer and Flow Loss Analysis

Figure 29.8 shows the variation of surface-averaged *Nu* for different streamwise as well as spanwise locations of RWP at a fixed value of *Re* equal to 3000. Observations made for the different locations of RWP on heat transfer characteristics show that spanwise location of RWP plays crucial role on thermal performance. For each fixed ΔX location, $\Delta Y = \pm 1.5$ shows higher value of *Nu* while it is lower for $\Delta Y = \pm 0.5$. Difference in the values of *Nu* for $\Delta Y = \pm 0.5$ and ± 1 are less than those for $\Delta Y = \pm 1.5$. For all the ΔY locations, it has been observed that the value of *Nu* increases when the location of RWP has been shifted away from the cylinder. For upstream located RWPs, increase in the value of *Nu* is higher than for downstream located RWPs. Variation of friction factor (f) for different streamwise as well as spanwise locations of RWP at a fixed value of *Re* equal to 3000 is presented in Fig. 29.9. For all the ΔY locations, as the location of RWP shifts away from the cylinder, the value of f is found to decrease. For upstream located RWP, the RWP location $\Delta X = -2$, $\Delta Y = \pm 1.5$ is associated with higher value of f and lower for $\Delta Y = \pm 0.5$. On the other hand, for downstream located RWP, the RWP location $\Delta X = 2$, $\Delta Y = \pm 1$ shows higher value of f while the lower value is for $\Delta Y = \pm 0.5$.

Heat transfer ratio (HTR) shows the enhancement of heat transfer of the present computational model having heated cylinder, RWP and bottom plate as compared to computational model in the absence of RWP. Figure 29.10a–c depicts the variation of HTR with change in *Re* for different spanwise locations of RWP at $\Delta X = -2$, 2 and 0, respectively. With increase in *Re*, HTR is found to increase for

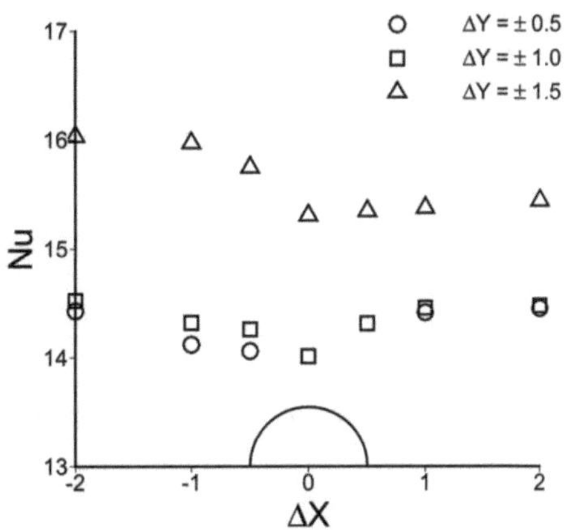

Fig. 29.8 Variation of overall surface-averaged *Nu* for different locations of RWP

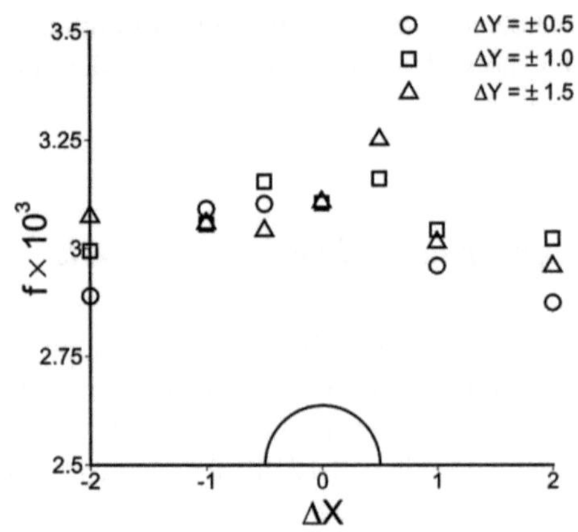

Fig. 29.9 Variation of *f* for different locations of RWP

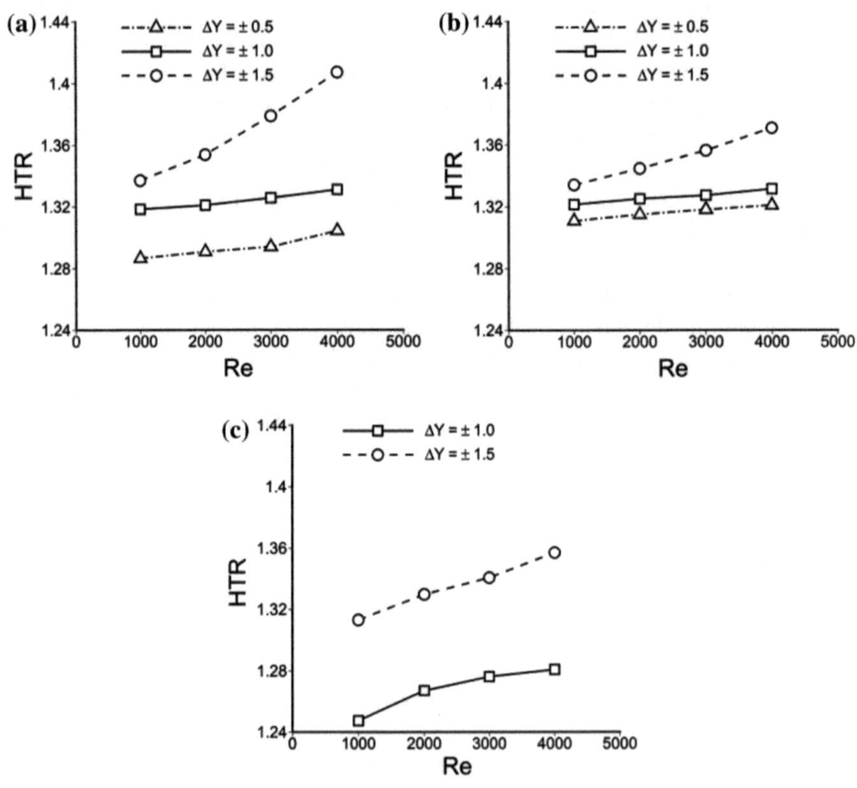

Fig. 29.10 Variation of HTR with change in *Re* for different ΔY locations at **a** $\Delta X = -2$. **b** $\Delta X = 2$. **c** $\Delta X = 0$

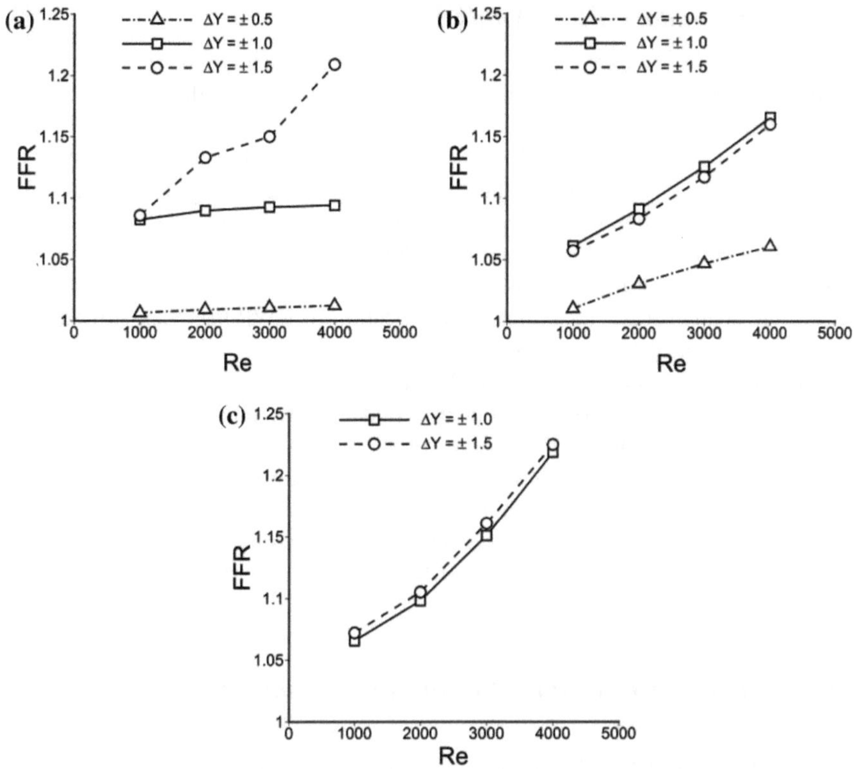

Fig. 29.11 Variation of FFR with change in *Re* for different ΔY locations at **a** $\Delta X = -2$. **b** $\Delta X = 2$. **c** $\Delta X = 0$

all the ΔX locations. For each fixed *Re* and ΔX, $\Delta Y = \pm 1.5$ shows higher HTR value followed by $\Delta Y = \pm 1$ and ± 0.5. For $\Delta Y = \pm 1$ and ± 1.5, upstream located RWP such as $\Delta X = -2$, shows higher HTR value than other two ΔX locations. For $\Delta Y = \pm 0.5$, $\Delta X = -2$ has lower HTR value for all values of *Re*. It has been observed that the enhancement in heat transfer is maximum for upstream located RWP at $\Delta X = -2$ and $\Delta Y = \pm 1.5$ at *Re* = 4000, which is equal to 40.72%. For downstream located RWP at $\Delta X = 2$ and $\Delta Y = \pm 1.5$, observed higher heat transfer enhancement is 37.14% for *Re* = 4000.

Increase in pressure loss can be estimated by friction factor ratio (FFR), which is the ratio of value of *f* in the presence of RWP to the value in the absence of RWP. At different streamwise locations of RWP, i.e. $\Delta X = -2$, 2 and 0, for different spanwise locations of RWP, variations of FFR with change in *Re* are presented in Fig. 29.11a–c, respectively. Value of FFR increases with increase in *Re* for all the values of ΔX and ΔY. It has been observed that location $\Delta Y = \pm 1.5$ shows higher value of FFR for all the ΔX location except for $\Delta X = -2$. At $\Delta X = -2$, FFR is higher for $\Delta Y = \pm 1$. Location $\Delta Y = \pm 0.5$ gives lower FFR value for all the considered cases.

29.4.3 Secondary Flow Analysis

Analysis of secondary flow in the flow field gives better understanding of heat transfer performances. Estimation of augmentation in heat transfer can also be done by quantifying the secondary flows generated by obstacles such that RWP and cylinder. Figure 29.12 illustrates the variation of volume-averaged secondary flow intensity (Se) for different streamwise and spanwise locations of RWP at a fixed value of *Re* equal to 3000. For each fixed ΔX location of RWP, the value of Se is found to be higher for the spanwise location $\Delta Y = \pm 1.5$ and lower for $\Delta Y = \pm 0.5$. For all the spanwise ΔY locations, as the location of RWP has been shifted away from the cylinder, the value of Se increases.

Secondary flow intensity ratio (SFIR) has been used to compare the secondary flow generated due to surface-mounted cylinder in the presence of RWP to the secondary flow generated in the absence of RWP. For different spanwise locations of RWP, Figs. 29.13a–c, present the variation of SFIR with change in *Re* at $\Delta X = -2$, 2 and 0, respectively. With increase in the value of *Re*, the value of SFIR increases for all the ΔX locations of RWP. It is observed that for each fixed *Re* and ΔX, SFIR is higher for $\Delta Y = \pm 1.5$ and lower for $\Delta Y = \pm 0.5$.

29.4.4 Thermo-Hydraulic Performance Analysis

The thermal performance factor (JF) which is the ratio of HTR to 1/3rd power of FFR is used to analyze the thermo-hydraulic performance such that the combined effect of heat and fluid flow. Figure 29.14 depicts the variation of JF for different streamwise and spanwise locations of RWP at a fixed value of *Re* equal to 3000. In upstream as well as downstream locations of cylinder, increase in the value of JF is observed for all the values of ΔY, as ΔX location of RWP has shifted away from the

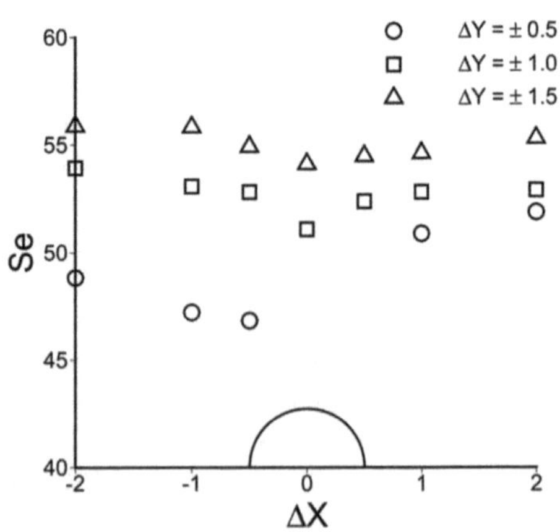

Fig. 29.12 Variation of volume-averaged Se for different locations of RWP

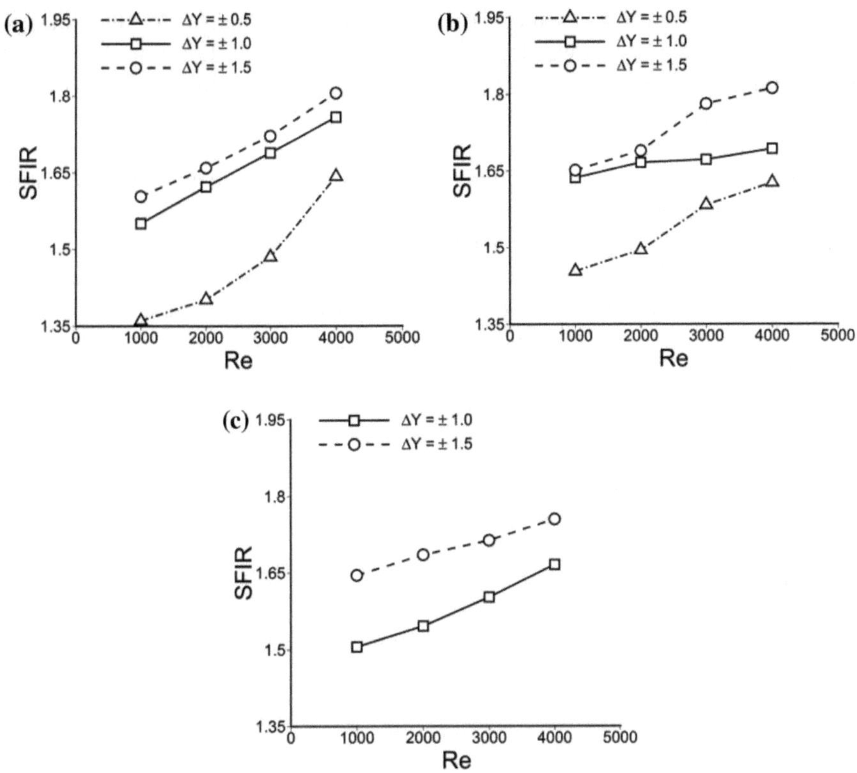

Fig. 29.13 Variation of SIFR with change in *Re* for different ΔY locations at **a** $\Delta X = -2$. **b** $\Delta X = 2$. **c** $\Delta X = 0$

Fig. 29.14 Variation of JF for different locations of RWP

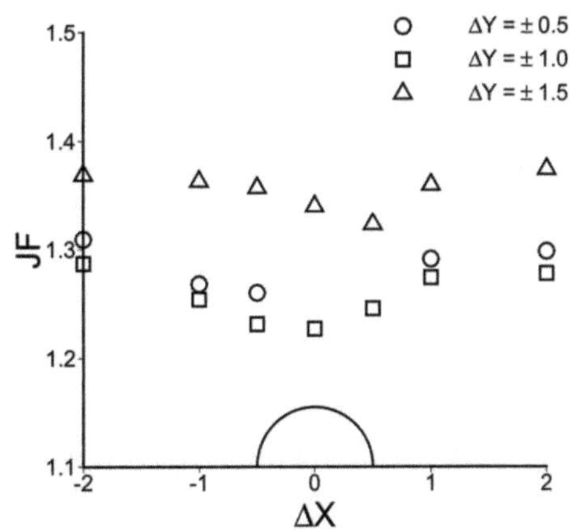

cylinder. For fixed ΔX location, the value of JF is higher for the location $\Delta Y = \pm 1.5$, while it in lower for the location $\Delta Y = \pm 1.0$. Location $\Delta Y = \pm 0.5$ shows higher JF than $\Delta Y = \pm 1.0$ which is primarily due to $\Delta Y = \pm 1.0$ being associated with higher pressure loss than $\Delta Y = \pm 0.5$ because the difference in heat transfer between the two cases is quite small.

For different values of ΔY location of RWP, Figs. 29.15a–c present the variation of thermal performance factor with change in Reat streamwise locations of RWP $\Delta X = -2$, 2 and 0, respectively. With increase in the value of *Re*, for all ΔX and ΔY locations of RWP considered it has been observed that the value of JF decreases. Moreover, higher value of JF is observed for the spanwise location $\Delta Y = \pm 1.5$ followed by $\Delta Y = \pm 0.5$ and $\Delta Y = \pm 1.0$ for each fixed value of *Re* and streamwise location ΔX.

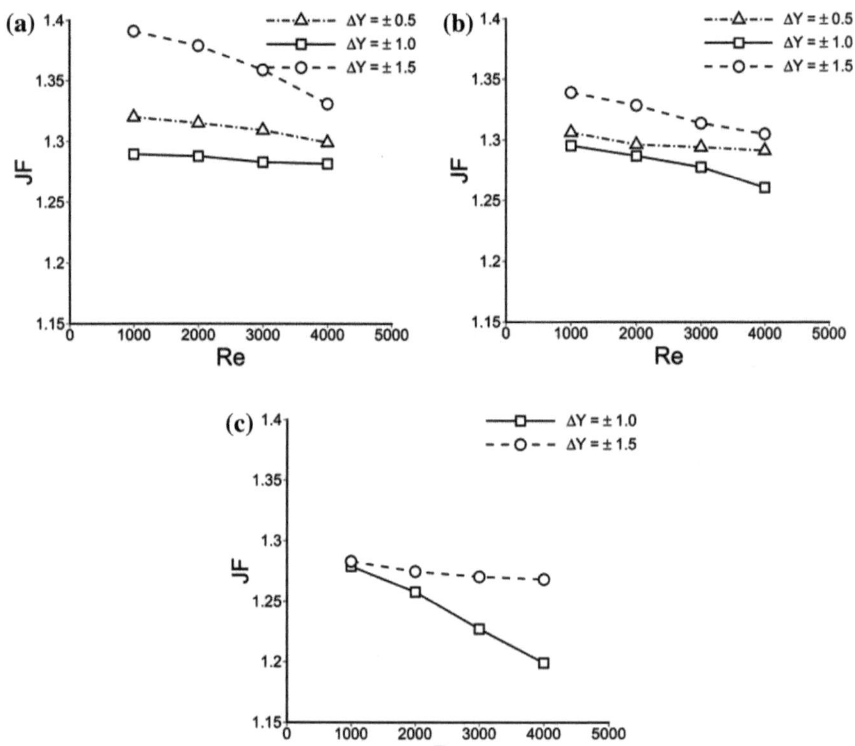

Fig. 29.15 Variation of JF with change in *Re* for different ΔY locations at **a** $\Delta X = -2$. **b** $\Delta X = 2$. **c** $\Delta X = 0$

29.5 Conclusion

Combined effect of finite height circular cylinder and rectangular winglet pair (RWP) mounted on horizontal plate has been studied using three-dimensional computations. Effect of different locations of RWP corresponding to the cylinder center on heat and fluid flow characteristics has been considered in the present study. Behavior of flow field and the characteristics of heat transfer around the finite height cylinder and RWP have been investigated with the help of streamlines plots and temperature contours, respectively. Streamline plots illustrate how generation of longitudinal and horseshoe vortices depends upon location of RWP. Similarly, temperature fields show strong dependence on locations of RWP due to disruption of thermal boundary layer. For RWP located away from cylinder both in upstream and downstream, as VGs of RWP shift away from each other in spanwise direction, horseshoe vortices due to both cylinder and RWP become characteristically independent of each other. For lower value of ΔY, horseshoe vortices are generated only from either cylinder or RWP. Horseshoe and longitudinal vortices promote fluid mixing that improves convective heat transfer. Recirculation regions observed behind the cylinder and RWP are associated with poor transport of fluids and giving rise to relatively higher temperature zone. Estimation of heat transfer augmentation has been carried out by using overall Nusselt number (Nu) and is found to be higher for the spanwise location of RWP, $\Delta Y = \pm 1.5$ at each fixed ΔX location of RWP. For each fixed ΔX location, overall Nu is found to be higher for $\Delta Y = \pm 1.5$. Enhancement in heat transfer is maximum and equal to 40.72% for the upstream located RWP at $\Delta X = -2$ and $\Delta Y = \pm 1.5$ for Re = 4000. All the performance parameters such as HTR, FFR and SFIR show an increase trend with increase in the value of Re, except the thermal performance factor which decreases with increase in Re.

References

1. Ghisalberti, L., Kondjoyan, A.: Convective heat transfer coefficients between air flow and a short cylinder, Effect of air velocity and turbulence, Effect of body shape, dimensions and position in the flow. J. Food Eng. **42**, 33–44 (1999)
2. Kawamura, T., Hiwada, M., Hibino, T., Mabuchi, T., Kumada, M.: Heat transfer from a finite circular cylinder on the flat plate. Bull. JSME **27**, 2430–2439 (1984)
3. Giordano, R., Ianiro, A., Astarita, T., Carlomagno, G.M.: Flow field and heat transfer on the base surface of a finite circular cylinder in crossflow. Appl. Therm. Eng. **49**, 79–88 (2012)
4. Tsutsui, T., Igarashi, T., Nakamura, H.: Fluid flow and heat transfer around a cylindrical protuberance mounted on a flat plate boundary layer. JSME Int. J. **43**, 279–287 (2000)
5. Tsutsui, T., Kawahara, M.: Heat transfer around a cylindrical protuberance mounted in a plane turbulent boundary layer. ASME J. Heat Transf. **128**, 153–161 (2006)
6. Naik, H., Tiwari, S.: Heat transfer and fluid flow characteristics from finite height circular cylinder mounted on horizontal plate. Procedia Eng. **127**, 71–78 (2015)
7. Naik, H., Tiwari, S.: Three-dimensional flow characteristics near a circular cylinder mounted on horizontal plate at low Reynolds number. Prog. Comput. Fluid Dyn. **17**, 102–113 (2017)

8. Sahin, B., Ozturk, N.A., Gurlek, C.: Horseshoe vortex studies in the passage of a model plate-fin-and-tube heat exchanger. Int. J. Heat Fluid Flow **29**, 340–351 (2008)
9. Rostamy, N., Sumner, D., Bergstrom, D.J., Bugg, J.D.: Local flow field of a surface-mounted finite circular cylinder. J. Fluids Struct. **34**, 105–122 (2012)
10. Sumner, D.: Flow above the free end of a surface-mounted finite-height circular cylinder: a review. J. Fluids Struct. **43**, 41–63 (2013)
11. Schubauer, G.B., Spangenberg, W.G.: Forced mixing in boundary layers. J. Fluid Mech. **8**, 10–32 (1960)
12. Jacobi, A.M., Shah, R.K.: Heat transfer surface enhancement through the use of longitudinal vortices: a review of recent progress. Exp. Thermal Fluid Sci. **11**, 295–309 (1995)
13. Fiebig, M., Kallweit, P., Mitra, N.K., Tigglebeck, S.: Heat transfer enhancement and drag by longitudinal vortex generators in channel flow. Exp. Thermal Fluid Sci. **4**, 103–114 (1991)
14. Tiggelbeck, S., Mitra, N.K., Fiebig, M.: Comparison of wing-type vortex generators for heat transfer enhancement in channel flows. J. Heat Transf. **116**, 880–885 (1994)
15. Fiebig, M.: Vortex generators for compact heat exchangers. J. Enhanced Heat Transf. **2**, 43–61 (1995)
16. Tian, L.T., He, Y.L., Lei, Y.G., Tao, W.Q.: Numerical study of fluid flow and heat transfer in a flat-plate channel with longitudinal vortex generators by applying field synergy principle analysis. Int. Commun. Heat Mass Transf. **36**, 111–120 (2009)
17. Wu, J.M., Tao, W.Q.: Numerical study on laminar convection heat transfer in a channel with longitudinal vortex generator part B: parametric study of major influence factors. Int. J. Heat Mass Transf. **51**, 3683–3692 (2008)
18. Naik, H., Tiwari, S.: Effect of rectangular winglet pair in common flow down configuration on heat transfer from an isothermally heated plate. Heat Transf. Eng. (2017). https://doi.org/10.1080/01457632.2017.1388946
19. Shih, T.H., Liou, W.W., Shabbir, A., Yang, Z., Zhu, J.: A new eddy viscosity model for high Reynolds number turbulent flows. Comput. Fluids **24**, 227–238 (1995)
20. Min, C., Qi, C., Wang, E., Tian, L., Qin, Y.: Numerical investigation of turbulent flow and heat transfer in a channel with novel longitudinal vortex generators. Int. J. Heat Mass Transf. **55**, 7268–7277 (2012)
21. Song, K.W., Wang, L.B.: The effectiveness of secondary flow produced by vortex generators mounted on both surfaces of the fin to enhance heat transfer in a flat tube bank fin heat exchanger. J. Heat Transf. **135**, 041902-1–041902-11 (2013)

Chapter 30
GA Optimization of Cooling Rate of a Heated MS Plate in a Laboratory-Scale ROT

Abhyuday Aditya, Prahar Sarkar and Pranibesh Mandal

30.1 Introduction

Run-out tables (ROTs) have been extensively used in industries in order to provide uniform cooling to the hot-rolled steel billet. They allow greater control over the cooling rate, which is important for producing various grades of steel with different metallurgical properties. ROTs have significantly eased up the process of steel production as compared to the tedious traditional methods. Studies on performance analysis on ROTs have also been done since long. Kumar et al. [1] have performed simulation-based studies on run-out tables for hot strip mills. Phaniraj et al. [2] have mentioned about the relevant importance of controlled run-out tables in hot rolling techniques. Although the material used was low carbon steel, the process is quite relevant for production-grade steel across all industries. They allow the use of improvised techniques such as ultra-fast cooling (UFC) under spray impingement along with precisely controlled cooling rates for large-scale production of steel having required grade variations and refined metallurgical properties. Ultra-fast cooling is that domain which is concerned with the use of extremely high flow rates of coolants spread over a uniformly distributed surface area in motion to rapidly cool down material which will result in influencing both the metallurgical and mechanical properties of the material, generally hot-rolled steel in industry. Bhattacharya et al. [3] have successfully used evaporation cooling technique by using spray-based coolant to achieve ultra-fast cooling in run-out tables. Mohapatra et al. [4] have used the air-atomized spray technique with a salt solution coolant on heated steel plate to achieve ultra-fast cooling. The study of Aamir et al. [5] shows the importance of ultra-fast cooling in dissipating high heat fluxes by critically controlling the coolant droplet diameter by the method of estimation from the actual observed cooling rate being achieved from the experimental study. Ultra-fast

A. Aditya (✉) · P. Sarkar · P. Mandal
Department of Mechanical Engineering, Jadavpur University, Kolkata, India
e-mail: abhyuday.aditya@gmail.com

P. Sahoo and J. P. Davim (eds.), *Advances in Materials, Mechanical and Industrial Engineering*, Lecture Notes on Multidisciplinary Industrial Engineering,
https://doi.org/10.1007/978-3-319-96968-8_30

cooling has significant effect on microstructure and is very much required for specific steel production as studied by Zhou et al. [6] on the properties of high strength steel for shipbuilding.

Specific gradation of steel and their particular uses are quite important in today's industry; hence, controlled cooling is a sought-after process as it allows for production of the steel to the exact grade required, as can be seen from the study of Longhai and Qiong [7] as they analysed quenching and controlled cooling of a moderately thick plate by analysing the various factors that actually seemed to influence the procedure. Mishra et al. [8] have given a detailed study of the cooling parameters and how they affect the heat transfer rate during spray impingement cooling of steel plate. Heat flux dissipation by spray cooling has been shown by Visaria and Mudawar [9] for thermal management of electronic devices and is equally applicable in run-out table applications.

ROTs used in industries in the past were rather bulky and consumed a lot of the available space. Use of age-old technology in various parts of the ROT resulted in the production process becoming rather time-consuming and tedious. None the less, ROTs were found to be quite useful and hence it was absolutely necessary to improve and modernize the technology used in them. Experimentation in laboratory-scale setup of industrial ROTs hence became a necessity. This allowed for work in experimentation, simulation and theoretical research on ROTs. Various aspects of the cooling process could now be individually studied and improved upon simultaneously with the others. Modern industrial ROTs having relatively smaller dimensions, higher cooling rate and huge production capacity are the result of prolonged experimentation on laboratory-scale ROTs. Cooling rate optimization is a very important task both in industries and in research laboratories. The cooling performance of laboratory-scale ROT depends upon a number of parameters which individually and collectively influence the concerned cooling rates to a significant extent.

The cooling performance of the laboratory-scale ROT used in the current study depends on different combinations of cooling parameters such as initial temperature at which cooling starts (T_{i}), the air flow rate (Q), the nozzle bank distance (d) and the velocity (v) of the billet under cooling bay, allowing for a range of variation in the cooling rate which results in achieving different steel grades. The effect of change in initial temperature (T_{i}) on the cooling rate of a static mild steel plate on a laboratory-scale run-out table, with constant air flow rate (Q) and average nozzle bank distance (d), has been shown by Sarkar et al. [10]. The work was concerned with the examination of the cooling characteristics of a heated mild steel plate on a laboratory-scale ROT setup by varying the air flow rate (Q) and the nozzle bank distance (d) individually while keeping the other cooling parameters at suitable constant values. However, large-scale production requirements make it essential to optimize the multiple cooling parameters system for maximum cooling rate, thereby improving cost efficiency and considerably reducing production time. This is where a bio-inspired optimization process can become a useful choice.

Bio-inspired optimizers have been successfully used in past for performance enhancement of real-time engineering systems. A binary-coded genetic algorithm

(GA) optimization process has been utilized by Mandal et al. [11] successfully to optimize a single-variable control parameter. Singh et al. [12] have found GA to be a fruitful modern optimization technique in engineering developments. The paper mentions the wide-scale use of several optimization techniques which are rather useful in certain domains. Mandal et al. [13] have used a real-coded genetic algorithm to estimate the parameters for coupling a model free, fuzzy controller with a feedforward controller. The optimizer was used to identify the best-suited parameter for a fuzzy controller structure for a real-time electrohydraulic actuation system. GA makes use of several evolutionary study techniques upon sets of sample data clustered in cycles. These samples are obtained from objective function evaluation of parameters which govern the entire optimization problem. Use of multiple parameters in GA optimization techniques has shown to provide results leading to better optimization of the system, as studied by Eksin [14] in a nonlinear electric circuit model. Chen et al. [15] have used the GA optimization technique in designing a fuzzy sliding mode controller capable of rapidly and efficiently controlling complex and nonlinear systems. The major difficulty faced by them was to select the most appropriate initial values for the parameter vector. This was solved by the use of GA optimization technique on the parameter vector to generate an optimum range of initial values for the controller. Herreros et al. [16] have applied genetic algorithms for PID-type controller design. Multiple parameters are required to be entertained in such type of controllers, and significance of genetic algorithms in optimizing these multi-objective parameters is quite profound. Sadegheih [17] has used a general-purpose schedule optimizer for manufacturing shop scheduling using genetic algorithms and simulated annealing. Here, optimization has been used to find the best arrangement of molecules in order to minimize the energy of the system. If the cooling process is fast, the solid will not attain the ground state, but a locally optimal structure, and hence, optimization becomes quite significant. Zhai et al. [18] have used an improvised genetic algorithm to optimize the water consumption of the secondary cooling zone based on the heat transfer model of the off-line bloom caster. GA optimizes the existing cooling system in steel casting to produce steel with better quality. Geometric shapes of traditional plate-fin heat sinks were optimized by Abdelsalam et al. [19] using a genetic algorithm optimizer, and the improved effectiveness of the fins greatly affected the overall power and raw material consumption for the same. The plate-fin heat sinks were optimized in a forced cross-flow using a multi-objective genetic algorithm (MOGA) combined with CFD simulations, by varying the fin angles. The main objective was to improve the heat dissipation rate by modifying geometric parameters. The optimized fins reported an effective increase in a range of 11.2–18.1%. It is quite evident from these works that genetic algorithm (GA) optimizers were quite successful in optimizing parameters related to cooling rates of various industrial and laboratory setups.

As far as the requirement of optimized cooling in ROT is concerned, GA can be the most useful process for achieving the same. Hence, the binary-coded genetic algorithm optimization technique used by Mandal et al. [11] has been adopted here for parameter optimization of the present laboratory-based ROT in order to enhance the cooling rate. A three-parameter optimization process, using genetic algorithm optimizer, has been carried out with the air flow rate (Q), the upper nozzle bank distance (du) and the lower nozzle bank distance (dl) being the three parameters. The study can be said to be first of its kind in terms of using bio-inspired optimization technique for parameter optimization of ROT in order to maximize the cooling rate.

30.2 System Description

The laboratory-scale ROT setup, used by Sarkar et al. [10], is used for the present set of experiments. The laboratory-scale ROT setup consists of a furnace, two nozzle banks, air supply line to the nozzles, fixtures to hold the specimen on the cooling bay consisting of roller fitted rails and mechanical handling zone in between the cooling bay and the furnace.

As shown in Fig. 30.1a, the furnace is chamber type and has 18 numbers of Kanthal Globar SD Silicon Carbide heating elements lining on the two inner side walls of the chamber. Maximum heating rate of the furnace is 15 °C/min. The control panel of the furnace is installed separately, as shown in Fig. 30.1b. There are voltage and current indicators on the control panel. There is a safety controller to limit the maximum achievable temperature. It can be set on different threshold temperature values. The maximum achievable temperature in the furnace is 1200 °C. The operator can control the heating process with the help of an automatic PID-type programmable temperature indicating controller. The furnace door operation is vertical lifting type and manually controlled by a handle and chain drive system. A limit switch is provided to ensure switching off of the heaters once the door is opened. The limit switch stays in contact with the furnace door when the furnace is in operation and is in a pushed down position. As the door is lifted, the switch returns back to its original position and the furnace heating system is switched off. The furnace working chamber has dimensions of 800 mm in length, 800 mm in width and 300 mm in height. The heating procedure inside the furnace mainly takes place by convection and radiation. Joule's heating heats up the lined heating elements on the inner side of the walls. Heat radiates from these heating elements and significantly raises the inside furnace temperature. This heat is then transferred to the test sample inside the furnace by convection. Uniform heating is ensured inside the furnace, thus eliminating experimental errors based on inconsistent heating. Asbestos-lined gloves are used while handling the test sample to and from the furnace in order to prevent any mishaps and injuries to the handler.

The nozzle bank cooling bay setup cools the mild steel plate, while a set of thermocouples records the temperatures at various positions. The nozzle bank setup is shown in Fig. 30.1c. Two nozzle banks, each consisting of twenty nozzle units, are on both sides of the cooling bay so that both the upward and downward surfaces of the specimen plate can be sprayed upon at the same time. There are two types of nozzles: round cross-sectioned full-cone hydraulic nozzles and rectangular

Fig. 30.1 **a** Closed chamber type furnace, **b** furnace control panel, **c** the nozzle bank setup, **d** the air compressor, **e** the compressed air flow circuit with digital flowmeter, **f** the test MS plate with eight thermocouples

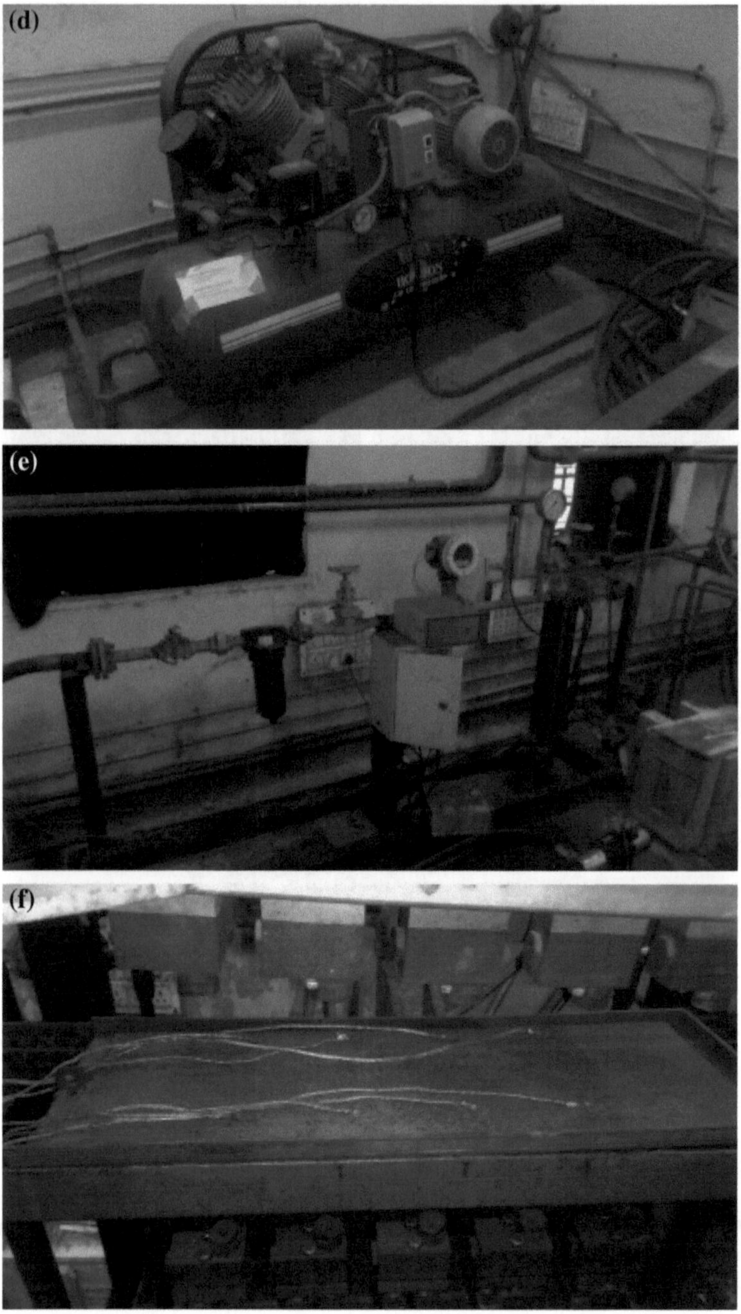

Fig. 30.1 (continued)

cross-sectioned air mist nozzles. There are 2 rows of hydraulic nozzles, 5 in each row on both sides of the plate which makes it a total of 20 of them. Similarly, there are 2 rows of air mist nozzles, 5 in each row on both sides of the plate which makes it a total of 20 of them too. Air mist and hydraulic nozzles are arranged in alternative rows. For any of the air-atomizing or hydraulic nozzles, four nozzles on the four end positions of the two rows are connected to a common header and the rest of the six nozzles are connected to another header for each nozzle bank. Headers of the air-atomizing nozzles have square cross section, and headers of the hydraulic nozzles have circular cross section. Water is supplied to the headers on both the sides of the cooling bay from a common pump. Pressurized air from a compressor comes to the square header leading to the mist spraying nozzles. Supplied air and water, as at high pressure, get thoroughly mixed in the square header and come out as mist through the nozzles. If the water supply line to the mist spraying nozzles is closed, then those nozzles can function as air spraying nozzles and cooling can be achieved using air only. For the present work, only air cooling has been used. Heights of the nozzle banks can be adjusted using simple screw and nut mechanism. So, the distance of the nozzle banks from the specimen plate can be varied to have different cooling rates. There is a clearance of 45 mm between the nozzle bank and the specimen plate on any side when the nozzle bank is set at the zero mark on the scale attached with the setup. On that scale, the distances of the nozzle banks can be varied from 0 to 140 mm.

Air is first pressurized in a centrifugal compressor driven by a three-phase induction motor, having a maximum working pressure of 12 kgf/cm^2, and then supplied to the nozzle banks through a simple circuit that is connected to the upstream compressor line through a ball valve (BV). The circuit contains an air filter with manually operated regulator, a vortex flow meter (FM) and a pressure gauge associated with a needle valve to control the pressure setting. There is a 1″ globe valve (GV) to control the flow of air in the line. There is a pilot-operated diaphragm-type solenoid valve (SV), which is normally closed to regulate the flow of air. Before entering the headers of the mist spraying nozzles, the air supply line gets divided into two branches. One branch delivers air to the headers connecting 8 nozzles and the other branch to the headers connecting 12 nozzles. Each branch is provided with separate ¾″ GVs. Figure 30.1d, e shows the air compressor used in the experiment and the air flow circuit, respectively.

Water supply comes from an overhead tank and goes into a simplex basket filter first through a BV. Then, a vertical in-line multistage centrifugal pump pressurizes the water and delivers it through a pressure gauge, fitted with a needle valve. The line is divided into a bypass line and main line after that. The bypass line contains a normally open pilot-operated diaphragm-type SV and a 1″ GV to regulate and control the water flow. The main line gets divided into two branches after another filtration. Each of the lines contains a normally closed SV and a flow meter. The line which has larger diameter goes to the mist spraying nozzles, and the other line goes to the water spraying nozzles. Both lines are provided with a GV and a pressure gauge each. Each line again divides into two lines to supply water to the headers connecting 8 nozzles and 12 nozzles separately. Globe valves are provided to control flow in every line to the headers. In the present work, the water supply line is not used as only air cooling has been studied.

As shown in Fig. 30.1f, a mild steel (MS) plate of 600 mm length, 200 mm width and 7 mm thickness has been used for the experiment as sample test plate. Mild steel is readily available in the market and is a perfect laboratory-scale representative of industrial steel for experimentation purpose. A plate thickness of 7 mm has been conveniently chosen as this is a rational choice to ensure that both heating and cooling procedures take place uniformly and not result in different locations on the plate having different temperatures at a time. The test plate is replaced with a similar plate after prolonged experimentation so as to ensure that the change in material property due to repeated heating and cooling, although in minute quantity, does not affect the actual experimentation.

The system to record the acquired data consists of eight nos. of K-type thermocouples attached along the upper surface of the MS plate, along the length of the plate and at equal distance from the centre line of the plate and from each other. The thermocouple positions though systematic are taken at random so as to cover maximum positions of the MS plate and to determine its temperature in real time. The average of the eight temperature readings has been taken as the average temperature. The temperature is recorded in real time using a NI 9211 Input Module in an NI cRIO that acts as the interface device between the hardware and a host PC. NI LabVIEW is used as the software platform in the PC for obtaining the experimental data.

30.3 Experimental Procedure

As evident from the system description, there are several interdependent components present in the system, all of which are significantly sensitive to external disturbances. For, e.g. the flow rate in the air circuit is extremely sensitive to any fluctuations in the compressor speed. Even the thermocouples are to be handled with care while taking readings, as any slight movement may cause variations in the reading.

The initialization of the setup starts with switching on the air compressor and adjusting the air flow rate through the nozzle banks by using the globe valve attached in the air circuit. The flow rate recorded by the flow meter varies over a range, and the criteria for stabilization of the flow rate are that the range should be as small as possible and should be maintained throughout the reading.

Then, the nozzle banks are adjusted to the desired distance from the plate using a hand-wheel attached to each of the nozzle banks. The final distance from the plate is checked manually, and a 30-cm measuring scale is used for that purpose. Care is taken to account for any parallax errors. The stabilization of the air flow rate takes a significant amount of time. The plate has been kept static however for the whole experimental procedure, and no water flow or air–water mixture jet is used.

The heating furnace in the setup can be heated up to a maximum temperature of 1200 °C. Keeping this in mind, as well as in an attempt to finish each reading within a reasonable period of time, all readings in this study have been taken starting at an average T_i of 600–605 °C and cooled to an average plate temperature close to 99 °C.

The temperature data are acquired in real time on a computer, by using an NI cRIO to transmit the thermocouple data into the computer. On the computer, NI LabVIEW is used as the software platform to obtain this data. Real-time temperature–time variations are plotted in LabVIEW, and the data are written into a Microsoft Excel file simultaneously. Since data from eight different thermocouples are simultaneously available on LabVIEW, an average temperature variation has also been incorporated into the readings which helps in monitoring the average plate temperature at the most crucial periods, i.e. while starting the reading from an approximate average plate temperature of 600 °C and ending it at an average plate temperature close to 100 °C.

The data obtained in the Excel file are in the form of discrete points of time, starting from zero, with the corresponding thermocouple temperatures and the average temperature in adjacent columns. The tool used for analysis of the recorded data is MATLAB, which has been used for plotting of all the data, as well as for the optimization process.

In a temperature–time graph, if initial temperature of two (or more) curves is the same, then to check the curve which is being cooled at a higher rate, one has to check which of the curves has the lowest temperature at any given time. This concept can be extended to study long cooling processes, such as the ones in this study, by calculating and comparing the areas under each curve, up to a fixed time value. The lower the area, the faster the cooling, given that all curves start at the same initial temperature.

Since the data are obtained in the form of discrete points, the areas under the curves were approximated by calculating area of a rectangle with base equal to the difference in adjacent time values (along the abscissa), and the height as the average of the two temperature values (along the ordinate) corresponding to the aforementioned time values, and then summing up all of these areas. The total number of data points was different for different curves owing to the variation in cooling behaviour; thus, total areas under different curves were calculated up to 50,000 data points (up to 750 s) only.

The following MATLAB code is used to calculate the approximate areas under the curves (A), and n was equal to 50,000:

```
A=0;
for i=1:n-1
ym(i)=0.5*(y(i)+y(i+1));
xm(i)=x(i+1)-x(i);
   a(i)=xm(i)*ym(i);
A=A+a(i);
end
```

As the objective of the present study is to optimize the parameter values, namely the air flow rate '*Q*', the upper nozzle bank distance 'du' and the lower nozzle bank distance 'dl' for maximization of cooling rate, *A* can be used as the optimization objective function. However, the GA used in this study is capable of only maximizing the fitness function, whereas our requirement is that of minimization of *A* to obtain the highest cooling rate of the ROT. Thus, the fitness function (*f*) is defined as,

$$f = 1/A \tag{30.1}$$

The ranges of the parameters have been taken as per the available setup specification. The range for the parameter *Q* is taken as 4.0–7.0 m^3/h. The upper and lower limits for each of the nozzle bank distance parameters du and dl are taken to be 50 and 180 mm, respectively. As far as the optimization procedure is concerned, a number of samples for each generation are required for the process. For the present case, the number of samples has been taken to be 20. So, for each generation of the optimization process, the experiment should be done with 20 nos. of different sets of parameter values in order to find out corresponding values of *A* and subsequent values of *f* using (30.1). The optimization procedure is detailed in the following subsection.

30.4 Optimization Technique

The binary-coded GA used by Mandal et al. [11] has been adopted here to be used as the optimization process. GA is a metaheuristic optimization technique inspired by the process of natural selection that belongs to the larger class of evolutionary algorithms. Genetic algorithms are commonly used to generate high-quality solutions to optimization and search problems by relying on bio-inspired operators such as mutation, crossover and selection. It is an iterative process which starts from a population of randomly generated individuals, known as 'generation' in each iteration. An objective function is used to specify the relation between parameters to optimize the system, and a value of this function is assigned to every individual, known as 'fitness function'. Once the genetic representation of the optimization problem and the fitness function are defined, a GA proceeds to initialize a population of solutions and then to improve it through repetitive application. Each individual's fitness is compared with the rest, and the more fitter individuals are selected from the current population. The rest of the individuals have their fitness improved, and together, a new generation is formed at the start of the next iteration. Termination of the process takes place either after a fixed number of generations or if a certain fitness level is achieved. Here, the system with three parameters is optimized for a fitness function *f* given by (30.1) as already mentioned earlier.

The procedural steps for a GA optimization are:

(I) Initialization: The nature of the optimization problem is studied to determine the initial population size, generally consisting of infinite number of possible solutions. It may be generated randomly to include the entire range of possible solutions or within an estimated range where the solution is likely to be present. As is already mentioned in the previous section, the sample number is assigned to be 20. The same numbers of samples are randomly generated from various combinations of the three available parameters. This marks the formation of the first generation of samples. Then, the experiment is run for each set of parameters and the corresponding value of *f* is calculated. A 10-bit binary coding system is used here. The ranges of the parameters have already been mentioned in the previous subsection.

(II) Genetic Operations: The main optimization process is comprised of three steps, selection, crossover and mutation. After initialization, the next step of GA optimization is selection. In this step, the parameter sets having higher fitness values are selected and carried over to the next generation along with the elimination of the sets with lower fitness values. Next, in the crossover process, the optimizer modifies some of the sets with best-found fitness values in a generation and randomly generates new sets of samples which are closer to the best-found fitness values. Here, a two-point 2-bit crossover is used with a crossover probability of 0.7 [11]. Afterwards, in the mutation process, the optimizer generates random samples resembling the genetic mutation process. The mutation probability used for the current study is fixed at 0.4 [11]. All these together form the sample set for the new generation. The experiment and optimization process thereafter continue in a cyclic way till it reaches the termination conditions.

(III) Termination: The above procedure is continued until a termination criterion has been reached. Predetermined optimal solution, fixed number of generations, limit of resource pool, etc., are some of the generally used termination conditions. Here, a simple termination criterion has been used. If the highest fitness function value does not get modified for two consecutive generations, the optimization process is going to be stopped. Also, the maximum number of generations has been fixed at 20. As far as rigorous experimental procedures are concerned, the choice can be taken to be very much rational.

30.5 Results and Discussion

The values of fitness functions, *f*, as obtained from the GA predictions, along with the corresponding values of *Q*, du and dl in each generation are tabulated in Tables 30.1, 30.2, 30.3, 30.4 and 30.5.

As is observed in Table 30.1, a random set of fitness values are generated from experiment with the maximum fitness value of 4.5445E−6 being highlighted in

Table 30.1 Parameter data and fitness values for first generation of GA optimization

Q (m³/h)	du (mm)	dl (mm)	f (°C^{-1} s^{-1})
4.2	156	108	1.3725e−6
4.7	111	78	2.4410e−6
4.1	168	166	1.3228e−6
6.6	170	174	3.8445e−6
4.2	163	174	1.2967e−6
4.4	122	87	3.4812e−6
5.3	118	61	2.7812e−6
4.1	102	135	2.4316e−6
4.6	85	176	1.6741e−6
5.1	98	142	3.2165e−6
4.0	55	79	1.2162e−6
4.5	93	118	3.6514e−6
6.0	148	150	3.1668e−6
4.5	121	133	2.8215e−6
4.9	111	65	1.3265e−6
4.0	165	175	8.4473e−7
5.6	95	68	3.2456e−6
5.3	**114**	**175**	**4.5445e−6**
4.3	126	158	3.1117e−6
3.9	132	176	1.0741e−6

Table 30.2 Parameter data and fitness values for second generation of GA optimization

Q (m³/h)	du (mm)	dl (mm)	f (°C^{-1} s^{-1})
5.3	114	175	4.5445e−6
6.2	113	174	4.5921e−6
4.5	93	118	3.6514e−6
6.0	53	117	4.9556e−6
6.2	114	57	4.9311e−6
5.3	118	61	2.7812e−6
4.3	126	158	3.1117e−6
5.3	114	175	4.5445e−6
6.9	175	116	4.6542e−6
5.0	55	54	4.8971e−6
5.3	114	175	4.5445e−6
6.7	**55**	**117**	**4.9708e−6**
5.5	91	102	3.4597e−6
5.3	174	173	4.4841e−6
5.3	114	175	4.5445e−6
4.8	116	115	4.5968e−6
6.9	115	55	4.9557e−6
5.3	114	175	4.5445e−6
6.2	174	113	4.5827e−6
4.5	93	118	3.6514e−6

Table 30.3 Parameter data and fitness values for third generation of GA optimization

Q (m^3/h)	du (mm)	dl (mm)	f (°C^{-1} s^{-1})
6.9	115	55	4.9557e−6
5.0	55	55	4.8971e−6
6.0	55	117	4.9556e−6
6.9	115	55	4.9557e−6
6.2	114	57	4.9311e−6
6.7	55	117	4.9708e−6
6.0	55	117	4.9556e−6
6.8	175	55	4.7905e−6
6.7	55	117	4.9708e−6
5.3	114	175	4.5445e−6
6.0	55	117	4.9556e−6
5.0	55	55	4.8971e−6
7.0	115	117	4.8681e−6
6.1	**56**	**55**	**4.9719e−6**
6.9	115	55	4.9557e−6
6.7	55	117	4.9708e−6
4.8	116	115	4.5968e−6
6.2	114	57	4.9311e−6
6.7	55	117	4.9708e−6
6.2	114	57	4.9311e−6

bold. This being only the initialization set though, the maximum value can be taken to be only a random one with no optimization steps still being involved in finding the same. The set with maximum fitness value from the first generation, as given above, is carried over to the second generation, as observed in Table 30.2. In fact, this set is repeated multiple times in the second generation. The ones with lower fitness values are found to be eliminated in the second generation. Furthermore, the maximum fitness value gets modified towards betterment on this generation as highlighted in bold text in Table 30.2. The optimization process continues in the third generation, where further improvement can be observed. Again, the maximum fitness value is seen to have increased to 4.9719E−6, while the fittest set of generation 2 has been repeated multiple times in the third generation. All these values are shown in Table 30.3. Proceeding to the fourth generation, as seen in Table 30.4, a lot of repetition of the higher values of f from the previous generation is observed, and again, a new maximum value of f = 5.1211E−6 is obtained, highlighted in bold text. In the fifth generation, as per Table 30.5, the fitness value of f = 5.1211E−6 is repeated several times among the 20 samples while the other fitness values are close to it but never exceed it. The closest values are also just repetitions of the previous higher values of f. All the occurrences of f = 5.1211E−6 in the fifth generation are highlighted in bold text in Table 30.5. This brings us to the termination as per the termination criteria as the maximum fitness value is not modified in the fifth generation. The parameter set with the maximum fitness value of this generation can be

Table 30.4 Parameter data and fitness values for fourth generation of GA optimization

Q (m^3/h)	du (mm)	dl (mm)	f (°C^{-1} s^{-1})
6.7	55	117	4.9708e−6
6.0	55	117	4.9556e−6
6.7	55	117	4.9708e−6
7.0	**53**	**55**	**5.1211e−6**
6.1	56	55	4.9719e−6
6.9	115	55	4.9557e−6
6.7	55	117	4.9708e−6
6.1	56	55	4.9719e−6
6.9	115	55	4.9557e−6
6.1	56	55	4.9719e−6
6.9	115	55	4.9557e−6
6.7	55	117	4.9708e−6
6.9	115	55	4.9557e−6
6.7	55	117	4.9708e−6
6.1	56	55	4.9719e−6
6.9	115	55	4.9557e−6
6.1	56	55	4.9719e−6
6.7	55	117	4.9708e−6
6.1	56	55	4.9719e−6
6.7	55	117	4.9708e−6

taken to be the optimized set for achieving highest possible cooling rate of the laboratory-scale ROT.

A simple statistical analysis of the generated fitness values provides valuable insight into how the GA is improving upon the previously generated values in each subsequent generation. Shown in Fig. 30.2 is a plot of the 20 fitness values from each generation, arranged according to the generation they occur in. Two simple parameters, the mean and the standard deviation, are calculated for the fitness values of each generation which become the basis of measuring the improvement in the fitness values in each generation.

Observing Fig. 30.2 carefully, the fitness values of the first generation appear very scattered. The thing worth observing is that the standard deviation is of the same order as the mean, which basically shows the high randomness and variation in the fitness values of the first generation. This is obvious as the first generation is only a randomly chosen initialization with optimization steps yet to be involved.

In the plot of the second generation, even by visual inspection the vast improvement in the fitness values can be observed. The data appear closer together compared to the first generation. The improvement is more accurately reflected by the increase in the mean and decrease in the standard deviation values. Both these trends point towards the elimination of low fitness values of the first generation, as well as an increase in the frequency of the higher fitness values. The maximum

Table 30.5 Parameter data and fitness values for fifth generation of GA optimization

Q (m^3/h)	du (mm)	dl (mm)	f (°C^{-1} s^{-1})
6.1	56	55	4.9719e−6
6.7	55	117	4.9708e−6
7.0	**53**	**55**	**5.1211e−6**
6.1	56	55	4.9719e−6
7.0	**53**	**55**	**5.1211e−6**
6.7	55	117	4.9708e−6
6.1	56	55	4.9719e−6
6.1	56	55	4.9719e−6
7.0	**53**	**55**	**5.1211e−6**
7.0	**53**	**55**	**5.1211e−6**
6.7	55	117	4.9708e−6
7.0	**53**	**55**	**5.1211e−6**
6.7	55	117	4.9708e−6
6.9	115	55	4.9557e−6
7.0	**53**	**55**	**5.1211e−6**
6.7	55	117	4.9708e−6
6.1	56	55	4.9719e−6
7.0	**53**	**55**	**5.1211e−6**
6.9	115	55	4.9557e−6
7.0	**53**	**55**	**5.1211e−6**

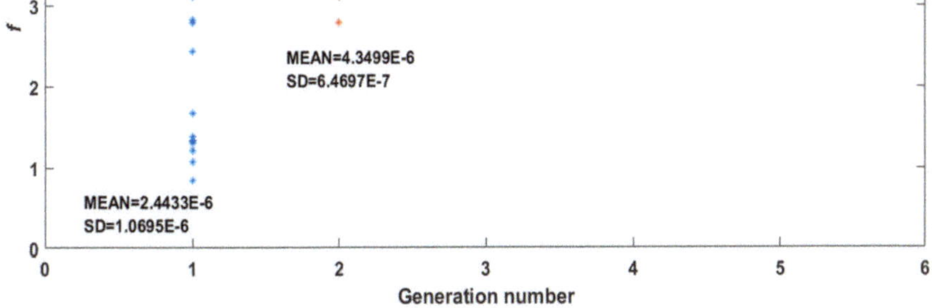

Fig. 30.2 Statistical analysis of generation-wise GA predictions

fitness value has also increased from the previous generation, thus indicating a significant overall improvement.

The fitness values of the third generation also see an improvement, albeit smaller compared to the previous improvement. The mean of the fitness values in the third generation is 4.8988E−6, while the standard deviation is 1.2022E−7. The most significant improvement is observed in the standard deviation, suggesting that the higher fitness values from the previous generation are being readily promoted. Again, the maximum fitness value also increases slightly, thus contributing to an overall improvement from the previous generation.

Coming to the fourth generation, a slight improvement in both the mean and the standard deviation is observed. The mean of the fitness values is 4.9725E−6, and the standard deviation is 3.5806E−8. While the mean increases very slightly over the previous generation, the standard deviation shows very significant improvement, which can be attributed to the repetition of higher values of fitness function from the previous generation.

In the fifth generation, while a slight improvement is observed in the mean, which increases to 5.0297E−6, the standard deviation also increases to 7.5505E−8, indicating more scatter in the data. This trend, while appearing counter-intuitive, can be attributed to the randomness of the optimization process. This increase in the standard deviation is, however, much smaller than the improvement in standard deviation from the third generation to the fourth generation. Also, since we observe several repetitions of the previous best-fitted parameter set in this generation, it can be said that the optimization process has worked well.

Figure 30.3 shows the experimental curves of the actual system. The temporal temperature variation curve for the MS plate cooled under the air jet has been plotted for the parameter sets having best fitness values in each generations of the GA optimization process. As is evident from the plot that marked improvement of the actual cooling rate has been achieved. So, not only the statistical data, but also

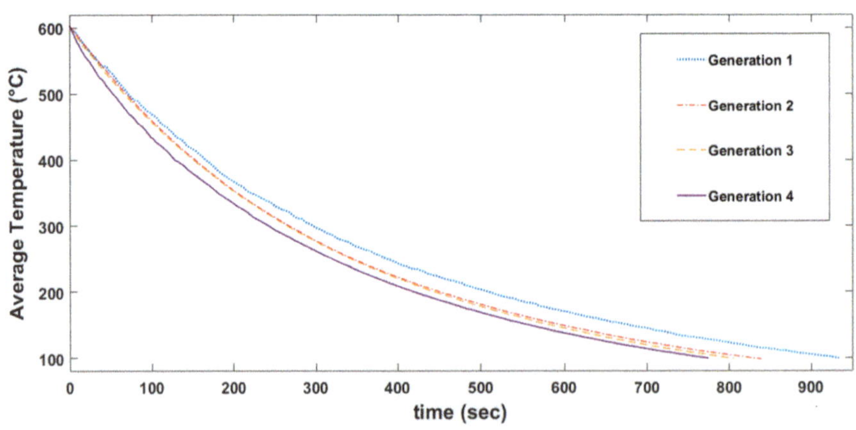

Fig. 30.3 Comparison of best GA predictions in each generation by experimental cooling curves

the actual experimental curves of the actual system show significant improvement in cooling rate, thereby proving the applicability of GA optimizer for the case.

On closer inspection of the GA prediction, the GA seems to suggest that either decreasing the nozzle bank distance or increasing the air flow rate from the nozzle banks, or both simultaneously, improves the performance of the ROT, i.e. increases the cooling rate of the MS plate.

These predictions are validated by the experimental plots in Fig. 30.3. On drawing an imaginary line parallel to the ordinate (temperature axis), passing through 750 s along the abscissa (time axis), one can clearly observe that the area under the curve is minimum for the generation 4 curve, which represents the best prediction of the GA. This in turn justifies the selection of 750 s as the limit for calculating the areas under different cooling curves and calculating the fitness function f as inverse of that area from (30.1).

Another rough indicator of the improvement in cooling rates across generations is observing the time taken by each curve to reach the lower limit of 100 °C. If the cooling is faster, time taken will be less. This is especially helpful in differentiating between generation 2 and generation 3 curves, with generation 3 curve taking lesser time in reaching 100 °C; thus, it cools at a higher rate compared to generation 2. This is consistent with the GA process, which incurs an improvement in cooling rate (thus decrease in cooling time) from generation 2 to generation 3. The generation 4 curve takes the least time to reach 100 °C, as evident from the plot.

30.6 Conclusion

An optimization-based experimental study of a laboratory-scale ROT has been performed with an objective of maximizing the cooling rate while varying the upper and lower nozzle bank distances from the heated plate along with the air flow rate. The optimization has provided suitable improvement in system performance as is evident from both statistical and experimental plots of results.

References

1. Kumar, R.K., Sinha, S.K., Lahiri, A.K.: Real time simulator for the runout table of hot strip mills. In: Industry Applications Conference, 199, Thirty-First IAS Annual Meeting, IAS'96, Conference Record of the 1996 IEEE, vol. 4, pp. 197–2618 (1996)
2. Phaniraj, M., Shamasundar, S., Lahiri, A.K.: Relevance of ROT control for hot rolled low carbon steels. Steel Res. Int. **72**(5–6), 221–224 (2001)
3. Bhattacharya, P., Samanta, A.N., Chakraborty, S.: Spray evaporative cooling to achieve ultra-fast cooling in runout table. Int. J. Therm. Sci. **48**(9), 1741–1747 (2009)
4. Mohapatra, S.S., Ravikumar, S.V., Jha, J.M., Singh, A.K., Bhattacharya, C., Pal, S.K., Chakraborty, S.: Ultra-fast cooling of hot steel plate by air atomized spray with salt solution. Heat Mass Transf. **50**(5), 587–601 (2014)

5. Aamir, M., Qiang, L., Xun, Z., Hong, W., Zubair, M.: Ultra fast spray cooling and critical droplet diameter estimation from cooling rate. J. Power Energy Eng. **2**(04), 259–270 (2014)
6. Zhou, C., Ye, Q., Yan, L.: Effect of ultra-fast cooling on microstructure and properties of high strength steel for shipbuilding. In: HSLA Steels 2015, Microalloying 2015 & Offshore Engineering Steels 2015, Springer, Cham, pp. 1179–1185 (2016)
7. Longhai, L., Qiong, Z.: Analysis on influence factors of the cooling efficiency of moderate thick plates in the quenching with controlled cooling. In: Second International Conference on Computer Modeling and Simulation 2010, vol. 2, pp. 390–393 (2010)
8. Mishra, P.C., Nayak, S.K., Chaini, R., Ghosh, D.P., Samantaray, B.B.: Effect of controlling parameters on heat transfer during spray impingement cooling of steel plate. Am. J. Eng. Res. (AJER) **2**(9), 8–16 (2013)
9. Visaria, M., Mudawar, I.: Application of two-phase spray cooling for thermal management of electronic devices. IEEE Trans. Compon. Packag. Technol. **32**(4), 784–793 (2009)
10. Sarkar, P., Aditya, A., Ghosh, A., Mandal, P.: Experimental observation of cooling characteristics of hot steel plate in an ROT. J. Basic Appl. Eng. Res. **4**(3), 296–298. p-ISSN: 2350-0077; e-ISSN: 2350-0255 (2017)
11. Mandal, P., Sarkar, B.K., Saha, R., Mookherjee, S., Acharyya, S.K., Sanyal, D.: GA-optimized fuzzy-feedforward-bias control of motion by a rugged electrohydraulic system. IEEE/ASME Trans. Mechatron. **20**(4), 1734–1742 (2015)
12. Singh, J., Chutani, S.: A survey of modern optimization techniques for reinforced concrete structural design. Int. J. Eng. Sci. Invention Res. Dev. **2**(1), 55–62 (2015)
13. Mandal, P., Sarkar, B.K., Saha, R., Mookherjee, S., Sanyal, D.: Designing an optimized model-free controller for improved motion tracking by rugged electrohydraulic system. Proc. Inst. Mech. Eng. Part I J. Syst. Control Eng. **230**(5), 385–396 (2016)
14. Eksin, C.: Genetic algorithms for multi-objective optimization in dynamic systems. In: Proceedings of the 26th International System Dynamics Conference (2008)
15. Chen, P., Chen, C., Chiang, W.: GA-based modified adaptive fuzzy sliding mode controller for nonlinear systems. Expert Syst. Appl. **36**(3), 5872–5879 (2009)
16. Herreros, A., Baeyens, E., Perán, J.R.: Design of PID-type controllers using multi-objective genetic algorithms. ISA Trans. **41**(4), 457–472 (2002)
17. Sadegheih, A.: Scheduling problem using genetic algorithm, simulated annealing and the effects of parameter values on GA performance. Appl. Math. Model. **30**(2), 147–154 (2006)
18. Zhai, Y.Y., Li, Y., Ma, B.Y., Yan, C., Jiang, Z.Y.: The optimisation of the secondary cooling water distribution with improved genetic algorithm in continuous casting of steels. Mater. Res. Innovations **19**(sup1), 1–26 (2015)
19. Abdelsalam, Y.O., Alimohammadi, S., Pelletier, Q., Persoons, T.: A multi-objective genetic algorithm optimisation of plate-fin heatsinks. In: 2017 23rd International Workshop on IEEE Thermal Investigations of ICs and Systems (THERMINIC), pp. 1–6 (2017)

Chapter 31
Experimental and Numerical Study of Velocity Profile of Air over an Aerofoil in a Free Wind Stream in Wind Tunnel

Pringale Kumar Das, Sombuddha Bagchi, Soham Mondal and Pranibesh Mandal

31.1 Introduction

A substantial amount of research has been performed on aerofoils [1–3] for years. It is used in the development of aircraft [4], windmills [5], wind turbines and such machineries operating in flow of air. On the suction side of a first-stage turbine vane, a study through experimentation was conducted by Ethridge et al. [1] for the investigation of the film-cooling performance. Dassen et al. [3] performed measurements in wind tunnel. It showed the self-noise of a series of flat plates and aerofoils with the teeth at the trailing edge of varying planforms and orientations in order to analyse for the case of more realistic flows and geometries the previously reported noise reducing potential of serrated trailing edges. Selig et al. [4] investigated a natural-laminar-flow aerofoil, the NLF (1)-0115, which had been designed for general-aviation aircraft at the NASA Langley Research Center. Majola [5] discussed the performance data of the Savonius rotor and established the design criteria. Angle-of-attack and shape are the predominant factors affecting the lift on an aerofoil [6]. Unlike cambered aerofoils generating lift at zero angle-of-attack, most aerofoils require a positive angle-of-attack for generation of lift. The circulation of air in the surroundings of the aerofoil creates curved streamlines due to which on one side lower pressure is formed and higher pressure on the other accompanied by a velocity difference, via Bernoulli's principle, so the flow field about the aerofoil resulting from this creates higher average velocity on one surface than on the other [7, 8]. Englar [8] found from experimental observations that 2D and 3D aerofoils employing nominal blowing demonstrated lift gain double to triple that of the conventional flapped aerofoil. The analysis for negative angles-of-attack is seldom performed, but the drag and lift coefficient for such positions are

P. K. Das (✉) · S. Bagchi · S. Mondal · P. Mandal
Department of Mechanical Engineering, Jadavpur University, Kolkata, India
e-mail: rikd37@gmail.com

P. Sahoo and J. P. Davim (eds.), *Advances in Materials, Mechanical and Industrial Engineering*, Lecture Notes on Multidisciplinary Industrial Engineering,
https://doi.org/10.1007/978-3-319-96968-8_31

important factors as far as tail lift generation of an airplane is concerned [9]. Maybury et al. [9] measured the generation of lift by a bird tail on mounted, frozen European starlings (*Sturnus vulgaris*) taking into consideration the variation with tail spread. The experiment was conducted in a wind tunnel at a typical airspeed with particular tail angle-of-attack and body for validation of predictions from current aerodynamic theories modelling tail lift. Boundary layer being yet another important concept has also been studied for aerofoils and refers to the film of fluid in the close proximity of a bounding surface where the effects of viscosity are indicative [10]. Sturm et al. [10] introduced for the detection of reattachment on aerofoils, the boundary layer after experiencing separation (flow separation, stall), a sensor concept. Results obtained from the measurements clearly show for all sensor positions the separation points (zero flow) and even negative flow values (back flow). Determination of the boundary layer is essential in studies concerning aerodynamics due to the fact that it helps in further determination of the pressure and velocity distribution, thereby helping to find out the effect of some object placed in wind relative to it. As far as experimental aerodynamic studies on aerofoils are concerned, wind tunnels have been used for a long time to create the required environment [11–14]. Suitably instrumented models of the object for study are placed inside wind tunnels for experimentation on aerodynamic forces, pressure distribution and other aerodynamic characteristics [15–19]. McCroskey et al. investigated over a wide range of nominally 2D flow conditions with Mach numbers and Reynolds numbers ranging up to 0.30 and 4×10^6, respectively, the static and dynamic characteristics of a fixed-wing supercritical aerofoil along with seven helicopter sections [13]. The experimental results have shown an important disparity between aerofoils, which usually gets concealed by disparity in wind tunnels, particularly in steady cases. All tested aerofoils show considerable improvement in performance over the conventional NACA0012 profile. Generally, it is seen that the parameters of the unsteady motion turn out to be more decisive than the shape of the aerofoil in determination of dynamic stall air loads. Roberts [15] described a semi-empirical theory for the characterization and estimation of the augmentation and bursting of laminar separation bubbles. This theory used a correlation obtained experimentally in relating the length of the laminar free shear layer to the free stream turbulence and a modification of Horton's method in order to calculate the properties and reattachment position of the turbulent shear layer. It was used for the prediction of the advancement of laminar separation bubbles for NACA aerofoils. Timmer and Van Rooij [16] gave an overview of the design and wind tunnel test results of the wind turbine dedicated aerofoils developed by Delft University of Technology. In order to determine the lift, drag and pitching moment coefficients for various configurations, measurements of surface pressure distributions and wake profiles were obtained experimentally for a NACA0012 aerofoil by Li et al. [18]. There was an increase in the maximum lift coefficient from 1.37 to 1.74 with a drag increment at low-to-moderate lift coefficient on addition of a Gurney flap. The measurements of boundary layer profile were also done by the use of a rake of total pressure probes at the 90% chord location on the suction side. The effective height of a Gurney flap is about 2% of chord length, providing highest lift-to-drag ratio amongst the configurations invested when compared with the immaculate NACA0012 aerofoil. In this case, the device remains within the boundary layer. Albeit the experimental study of aerofoils employed in aircraft [20] is predominant, the ones employed in automobiles or race cars have also been studied [21]. Alford et al. [20] performed experiments in a wind tunnel at supersonic and transonic speeds to determine aerodynamic characteristics of variable-sweep wing aircraft—configuration. Jasinski and Selig [21] performed a study in a low-speed wind tunnel through experimentation in determining the flow field effects and performance of two-element open-wheel race car front wing configurations. Data presentation included surface pressure data, downstream flow measurements and balance force measurements by the use of a seven-hole probe. From the results obtained, it can be concluded that these elementary factors have a significant effect on behaviour of the downstream flow field and wing performance in the design of race car front wings.

In recent times, the advancement in computational fluid dynamics (CFD) modelling has reduced the demand for wind tunnel testing [22]. Patel et al. [22] analysed for NACA6412, NACA4412 and NACA0012 aerofoil sections the lift and drag forces in a wind turbine blade. PYTHON (x, y) 2.7.9.0 programming was used to carry out lift and drag force analysis for different parameters. Moreover, instruments and experimental set-ups require involvement of huge costs and physically performing an experiment or making a product, thereby having its own limitations. The same reason explains the reason why is it not always possible to check every intricate detail or reach very critical locations. Thus, the present computing power in public domain, simulation has come up as one of the predominant steps in performing any experiment or in the development of any new product [23]. However, CFD results are still not completely reliable and wind tunnels are used to verify CFD predictions because of the accuracy of results obtained [24]. XFOIL code and Transition SST *k-omega* model were used by Gunnel et al. [24] to predict the aerodynamic performance at low Reynolds number. The results were compared, and CFD results and XFOIL code result were observed to be compatible with each other until stall angle.

It has been observed that the results from simulation differ considerably from studies conducted through experimentation owing to the fact that too many assumptions are made in computational studies [25]. McCroskey [25] obtained for more than 40 wind tunnels on a single, well-known two-dimensional configuration a large body of experimental results which was critically investigated and correlated. An assessment was done to figure out the possible sources of inaccuracy for each facility, and identification of suspected data has been done. From the aggregate of data, the representative properties of the NACA0012 aerofoil can be identified over wide ranges of angles-of-attack, Mach number and Reynolds number with reasonable confidence. Hence, assessment and validation of current or subsequent results for wind tunnel and evaluation of advanced computational fluid dynamics code can be done by the use of this synthesized information. However, the fundamental characteristics of the data remain the same making the experimental data essential in getting practical and real-life perspective of velocity profiles of air flowing over an aerofoil to validate the simulated data to determine the degree of accuracy of such results [26–28]. This is the reason behind conducting simultaneous experimental [29] and computational studies [30] to be performed to find velocity profile and various related results. Rubel et al. [26] did a numerical and experimental investigation of NACA0015 aerofoil at different AoAs for a different velocity of air by determining the forces at every two degrees from 0° to 180° in a low-speed wind tunnel. Panigrahi and Mishra [30] used CFD simulations to study the selection procedure of a convenient aerofoil blade profile for the fan blades for enhancement in energy efficiency of axial flow mine ventilation fans which is one of the key aspects for designing mine ventilation fans. Similar comparative studies have been performed in the field of motorsports [31]. Wordley and Saunders [31] used numerical methods in developing multi-element wing profiles in agreement with FSAE rules that generates high-negative lift coefficients. In demonstrating their performance, a range of full-scale testing data done in wind tunnel was presented for these designs, both in confinement (free stream) and on the car.

Despite the fact that a substantial amount of the literature is presented in analysis of aerofoils, the study of conventional aerofoil is predominant amongst them [32–34]. A NACA0015 aerofoil was studied by Troolin et al. [32] for $Re_c = 2.0 \times 10^5$ with and without a Gurney flap in a wind tunnel for examining the flow structure of the wake evolving through time-resolved PIV and correlating this framework with measurements of the lift coefficient done by time-averaged method. The second shedding mode was responsible for a significant portion for the overall increase in lift. This was inferred from comparing the flow around 'filled' and 'open' flap configurations. McCroskey et al. [34] studied erratic boundary layer separation and dynamic stall at moderately low Reynolds numbers in incompressible flow. Three different types of stall were produced along with vortex-shedding phenomenon being predominant feature of each by variation in the leading-edge geometry of a NACA0012 aerofoil. In most cases, dynamic stall was found not to emerge from bursting of a laminar separation bubble but with a breakdown of the turbulent boundary layer, including the leading-edge stall on the basic NACA0012 profile. An in-depth analysis of unconventional aerofoils is very scarce. Unconventional aerofoils can be the answer to our need for obtaining a more aerodynamically efficient aerofoil, and hence, a detailed analysis is necessary [35–37]. Amitay et al. [36] investigated through a series tests conducted in a wind tunnel the controlling of flow separation on a non-conventional aerofoil, symmetric in nature by the use of synthetic (zero net mass flux) jet actuators. Control effectiveness is distinctly different for high and low actuation frequencies, influenced by the discrepancy in the characteristic timescale of the actuation. Ranzenbach and Barlow [37] studied a critical aspect concerning the performance of Formula One or Indy race car front wing through idealization of a negatively cambered two-dimensional aerofoil operating in ground effect and determination of the flow field at various heights. The elementary objective of the work is elucidating the phenomenon of reduction in force for the definitive case of an inverted NACA4412 aerofoil proceeding above ground in still air at high Reynolds number. The auxiliary objective was comparing and contrasting the flow field about this aerofoil in road conditions and when operated in a wind tunnel environment, i.e. when there is no relative movement between the aerofoil and the ground.

The present study attempts to determine velocity profile over an unconventional aerofoil, NACA3119 [38] placed in free wind stream over five separate angles-of-attack inside a wind tunnel which has been validated with the data obtained from numerical simulation. The aerofoil studied here is not amongst the standard aerofoils used till date, and more specifically the negative angle-of-attacks have been taken into considerations. As it can be seen from the nomenclature of the aerofoil, the maximum camber of the aerofoil as a percentage of the chord length is 3%, considerably higher than the unconventional aerofoils analysed commonly. This may lead to modification of properties in a manner which may have utilities in various fields where aerofoils are frequently employed. This is the main motivation behind the study.

31.2 Experimental Set-up and Procedure

For the experimental part, the aerofoil is schematically shown in Fig. 31.1 and the set-up is presented in Figs. 31.2 and 31.3. Figure 31.2 shows the model of the subsonic closed-circuit wind tunnel with the lower test section of dimension 4, 1 and 1 m length, height and width, respectively, and the upper test section of dimension 3, 4 and 4 m in length, height and breadth, respectively. Figure 31.3 portrays the aerofoil section as placed inside the actual wind tunnel. For the present study, we have used the lower test section. A wooden asymmetric aerofoil has been used with 21.9-cm chord length, the leading-edge radius and maximum thickness being 2.1 and 4.2 cm, respectively. The maximum camber is 0.7 cm occurring at 10% of chord length, at the point A shown in Fig. 31.1. As per the NACA four-digit nomenclature [38], the cambered aerofoil used can be said to be a NACA3119 which is not amongst the standard ones. The initial digit of the NACA nomenclature denotes the maximum camber of the aerofoil as a percentage of the

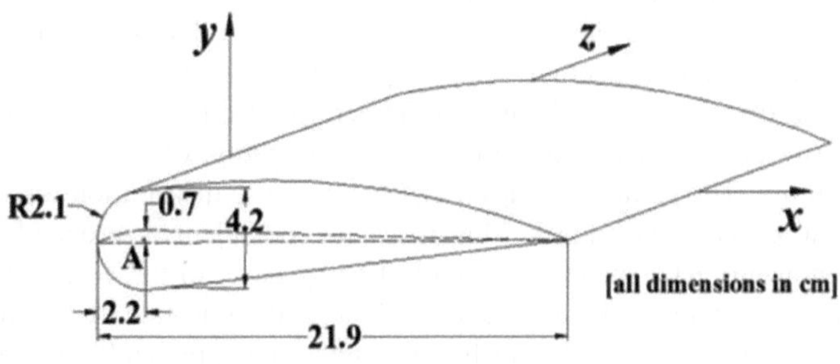

Fig. 31.1 Aerofoil geometry

Fig. 31.2 Model of the subsonic closed-circuit wind tunnel (Jadavpur University)

chord length. In the present study, it is found to be 3%, which is considerably higher than the cambered aerofoils analysed experimentally till date. In general, high camber aerofoil increases the maximum lift coefficient and minimizes the stalling speed of aircraft. Greater camber normally generates greater lift at slower speed. Aircraft having wing design on the basis of cambered aerofoils tend to have a lower value of stalling speed in comparison with similar aircraft having wing design on the basis of symmetric aerofoils. The camber is there to extend the range of AoAs at which the wing can operate. Camber of an aerofoil is varied to increase its efficiency. While using a cambered aerofoil, the chances of producing a much higher amount of lift are more as laminar flow can be maintained by air which in turn changes the static pressures resulting in creation of a pressure difference between the flow fields which are above and below the wing. Drag will also be reduced as a uniform motion is maintained by the airflow around the smooth curves of the wing. The resultant of this phenomenon would be a less turbulent flow around the trailing edge of the wing, and hence the wing can be pitched upwards higher without hampering the ability of the aerofoil to create lift. All these aspects call for wind tunnel experimentation with such an aerofoil section in order to find out different aerodynamic coefficients. As a primary work, this paper focuses on the determination of velocity profile around the aerofoil section placed in free stream with various AoAs. That in turn will allow determination of various aerodynamic factors in future. The camber can itself cause a pressure distribution for generation of lift even at zero angle-of-attack. This means lift coefficient at zero angle-of-attack should be nonzero. The span of the aerofoil being close to the width of the wind tunnel, the data obtained should be averaged along the mid-line of the structure so as to minimize the end effects as much as possible. The upper and lower surfaces of the aerofoil have been marked for grid points at which the experimental data are to be taken. Two wooden cylindrical beams help the aerofoil to be held firmly in the longitudinal direction. The coordinate system has also been shown in Fig. 31.4 with the x-axis being along the chord and z-axis being along the span, and the height from the aerofoil surface is denoted as y-axis. The thickness of the aerofoil is considered to be negligible in order to avoid the small difference in the y-direction for the upper and lower surfaces. The aerofoil can be rotated about the cylindrical beam at angles corresponding to multiples of 10° in order to achieve different AoAs. The free air stream is started over the aerofoil after placing it longitudinally in the wind tunnel with a free stream velocity v_∞. The device used for the flow of wind within the wind tunnel is an axial flow fan placed inside the tunnel, which is run by a Siemens make variable speed motor. An ABB make variable frequency drive is used to change the speed of the motor. Frequency drive range has been 0–50 Hz. Corresponding v_∞ variation in the wind tunnel can be achieved in the range of 0–32 m/s. Wind velocity is measured using a Pitot tube, Spectrum make, attached to digital differential manometer. The digital manometer is Kimo make (Model No. CP300-HP), capable of measuring from 0/+10 to −10,000/+10,000 Pa, as shown in Fig. 31.4. The Pitot tube has been attached to a robotic arm having three degrees of freedom which helps in moving the Pitot tube to various points as shown in Fig. 31.3. It is raised and lowered in the y-direction with the help of a stepper motor whose actions can be controlled from a control panel placed outside the wind tunnel (Fig. 31.4). The x- and z-axes movement of the same, however, has been provided manually. For the present study, the aerofoil has been set at five

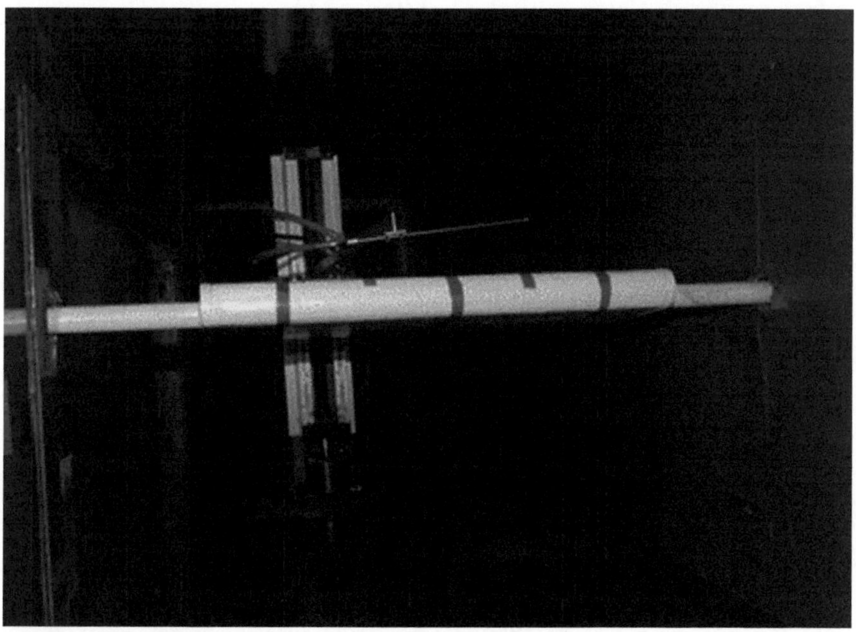

Fig. 31.3 Test section of the aerofoil along with Pitot tube mounted on the robotic arm

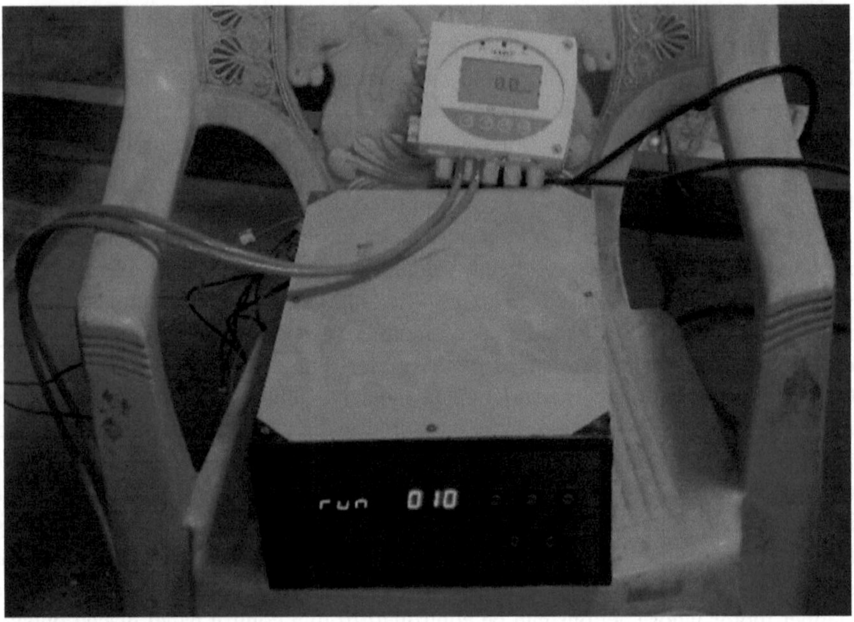

Fig. 31.4 Robotic arm controller with digital manometer on top of it

different angles-of-attack of −20°, −10°, 0°, 10° and 20°, while the free stream wind velocity has been held constant at 20 m/s for all the cases. Readings have been taken for three coordinates of z-axis and have been averaged about $z = 0$ which is the mid-point of the aerofoil in order to take care of any end effects if available. A total of ten grid points on the x-axis for each z-coordinate have been taken. At each point, y-axis velocity variation data have been taken at intervals of 10 mm starting from the surface to a height where free stream is found to prevail.

31.3 Numerical Methodology

A rectangular flow domain (4 m × 1 m) is used to simulate a wind tunnel, and an aerofoil section is placed within it. The aerofoil used is shown in Fig. 31.1 with dimensions exactly matching the one used for experimental observations detailed in the previous subsection.

The well-known continuity, momentum and Bernoulli's equations have been used for performing the simulation which are given, respectively, by

$$\frac{\partial \rho}{\partial t} + \nabla \cdot \left(\rho \vec{V}\right) = 0 \tag{31.1}$$

$$\frac{\partial(\rho\vec{V})}{\partial t} + \nabla\cdot\left(\rho\vec{V}\vec{V}\right) = -\nabla p + \nabla\cdot\vec{\tau} + \rho\vec{g} \tag{31.2}$$

where

p static pressure,
$\vec{\tau}$ stress tensor,
$\rho\vec{g}$ body force (gravitational),
$\vec{F}$ body force (external) such as that arising from interaction with the discrete phase.

The stress tensor is given by

$$\vec{\tau} = \mu\left[\left(\nabla\vec{V} + \nabla\vec{V}^{\mathrm{T}}\right) - \frac{2}{3}\left(\vec{V}I\right)\right] \tag{31.3}$$

A finite volume-based CFD code has been used to perform the required numerical simulations. A pressure-based solver is chosen as the numerical scheme, and a steady analysis has been performed. A viscous and turbulent standard *k*-ε two-equation model has been used with standard wall functions and curvature corrections. The model constants are $C_{\mathrm{mu}} = 0.09$; $C_{2-\varepsilon} = 1.92$; $C_{1-\varepsilon} = 1.44$; TKE Prandtl no. = 1 and TDR Prandtl no. = 1.3. The solution method is based on the SIMPLE scheme which provides better and faster convergence results at steady turbulent models. A least-squares cell-based gradient has been used, and the momentum, turbulent kinetic energy and turbulence dissipation rate have been solved using the second-order upwind scheme for more accurate results. The inlet conditions of the model were set to velocity inlet with a free stream velocity of 20 m/s. The turbulent intensity was limited to 5% and the turbulent viscosity ratio set to 10. The upper and lower walls of the wind tunnel geometry were set to no-slip wall condition. The outlet conditions of the model were set to outflow to simulate the closed-circuit nature of the wind tunnel. The standard initialization has been done from inlet, and the solution was converged within 1048 iterations with the total calculation duration of 20 min for all the cases.

'All Triangles' based unstructured grid has been used to form the mesh as shown in Fig. 31.5. The maximum size and face size of element were limited to 0.05 m. The disfeaturing tolerance was reduced to 1.e−005 m. For more accurate and better calculations, inflation was performed around the aerofoil geometry. It had a final layer height of 1.e−05 m and maximum number of layers 50 with a growth rate of 1.5. The total number of nodes and elements was 240,915 and 245,296, respectively.

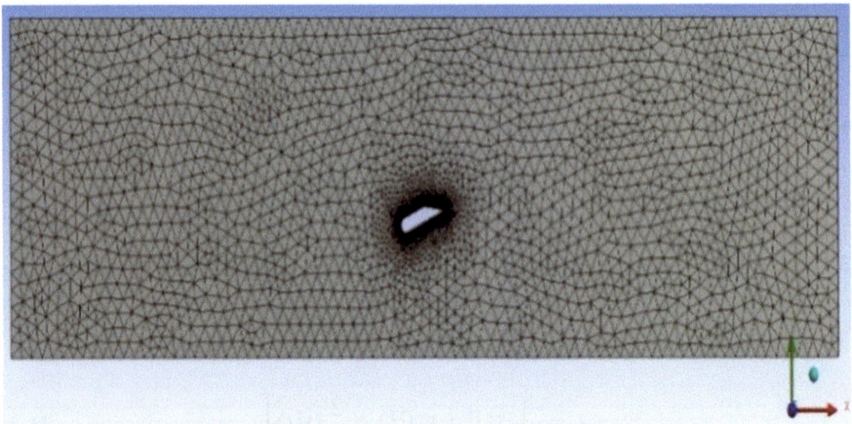

Fig. 31.5 Mesh details for numerical study

31.4 Results and Discussions

The velocity is observed at various points in the Cartesian coordinate system consisting of the aerofoil in the *XZ* plane and the vertical coordinate being the *Y*-axis using the manometer. The experimental and numerical results are plotted and compared. The velocity profile along the *Y*-axis is calculated from the experimental data. For five numbers of different angles-of-attack, i.e. 0°, 10°, −10°, 20° and −20°, the experimental velocity profiles and velocity profile obtained in numerical simulation have been shown in Figs. 31.6, 31.7, 31.8, 31.9, 31.10, 31.11, 31.12, 31.13, 31.14 and 31.15, while Figs. 31.16, 31.17, 31.18, 31.19, 31.20, 31.21, 31.22, 31.23, 31.24 and 31.25 show corresponding velocity variations with height from the aerofoil surface both for upper and lower surfaces. For zero or positive

Fig. 31.6 Contours of velocity from experimental data for 0° angle-of-attack

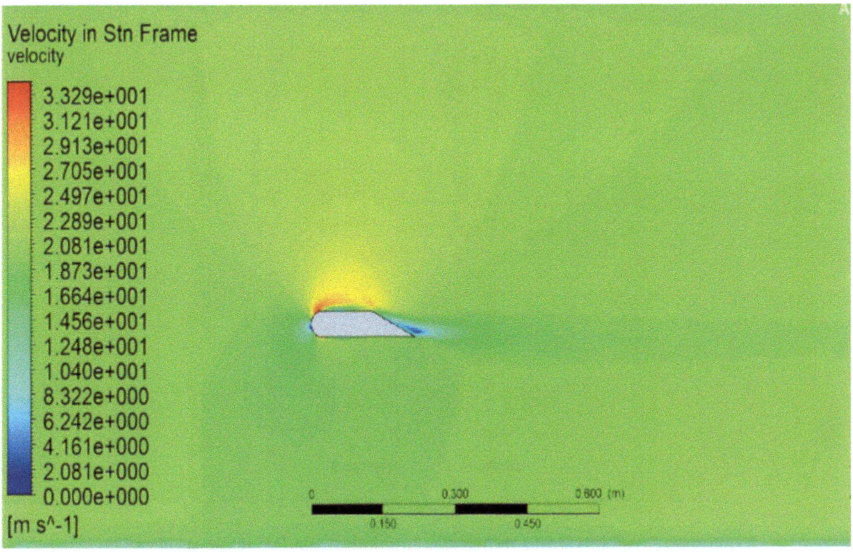

Fig. 31.7 Contours of velocity from numerical simulation for 0° angle-of-attack

Fig. 31.8 Contours of velocity from experimental data for 10° angle-of-attack

angles-of-attack, separation of boundary layer can be observed at the upper surfaces near trailing edge, which is obvious from the reversal of flow velocity. It is observed from Figs. 31.16 and 31.18, that with increase in positive angle-of-attack, the position of separation moves forward, i.e. towards the leading edge. While the separation of streamline from the upper surface for zero angle-of-attack occurs at the ninth channel only as is evident from reversal of velocity for $x = 9$ in Fig. 31.16, the same phenomenon starts at the seventh channel, denoted by $x = 7$ for 10° angle-of-attack in Fig. 31.18. However, in case of very high positive

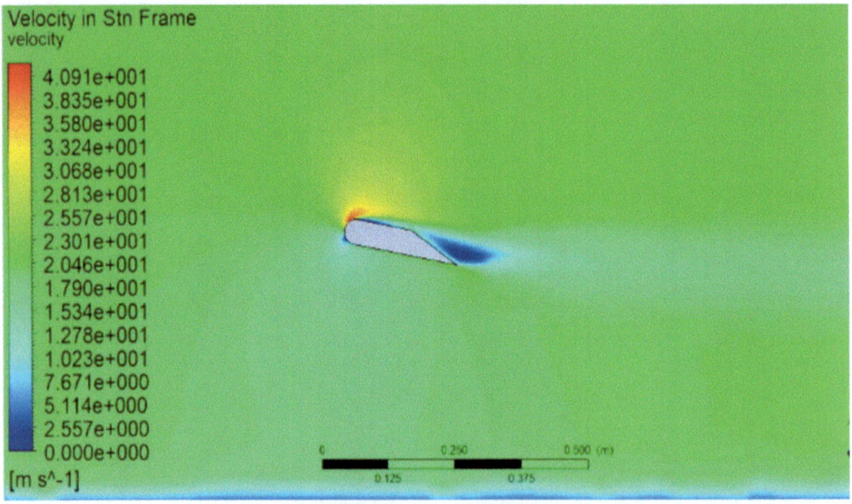

Fig. 31.9 Contours of velocity from numerical simulation for 10° angle-of-attack

Fig. 31.10 Contours of velocity from experimental data for −10° angle-of-attack

angle-of-attack of 20° in Fig. 31.22, a small reverse velocity occurs very near to the leading edge at $x = 2$, which may be taken to be the first point of separation. Afterwards, no reverse velocity can be observed. This can be attributed to high turbulence at the region causing measurement of velocity extremely difficult. So after the separation very near to the leading edge, the current measurements could not actually reflect the actual phenomenon due to high turbulence. For zero or positive angle-of-attack, the lower surface velocity variations have shown no separation as evident from Figs. 31.17, 31.19 and 31.23, which is normal as far as aerodynamic behaviour is concerned. As far as negative angles-of-attack are concerned, Figs. 31.21 and 31.25 clearly show separation of streamline near trailing edge from the lower surface. For both the cases, the flow reversal started from the

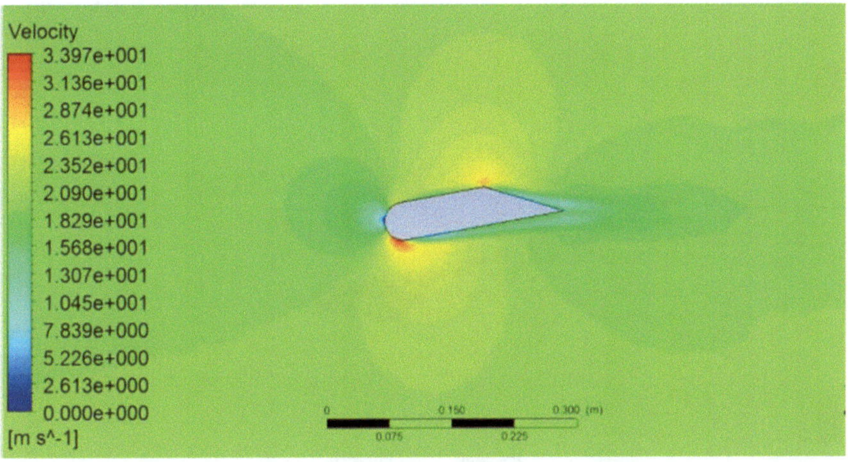

Fig. 31.11 Contours of velocity from numerical simulation for −10° angle-of-attack

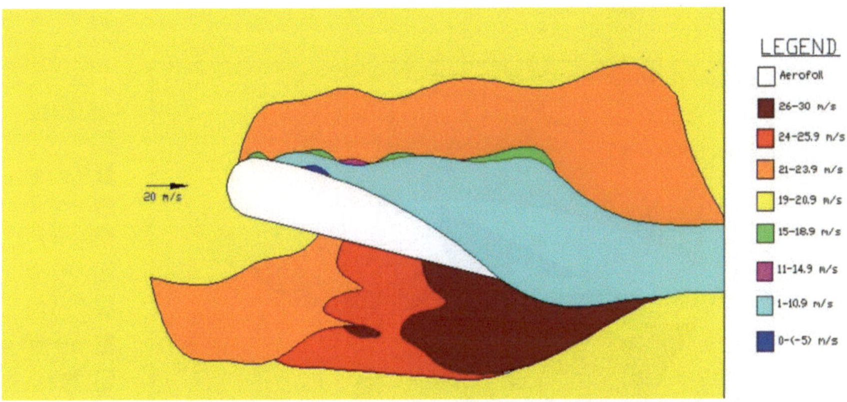

Fig. 31.12 Contours of velocity from experimental data for 20° angle-of-attack

fourth channel, but for the latter case in Fig. 31.25, near the trailing edge for $x = 9$, the flow reversal could not be observed which may be caused due to high turbulence as is already discussed for high positive angle-of-attack. One interesting aspect is that for high positive angle-of-attack of 20°, the measurement process becomes very much unreliable. But for negative angles-of-attack up to −20°, the experiments show more or less good results matching with theoretical knowledge. Only at the last channel very near to the trailing edge, the experiment could not show the flow reversal. So, the same magnitudes of positive and negative angles-of-attack behave separately as far as upper and lower surface velocity measurements are concerned. This can be attributed to the asymmetric nature of the aerofoil, having positive camber. Again, as far as the experimental and numerical

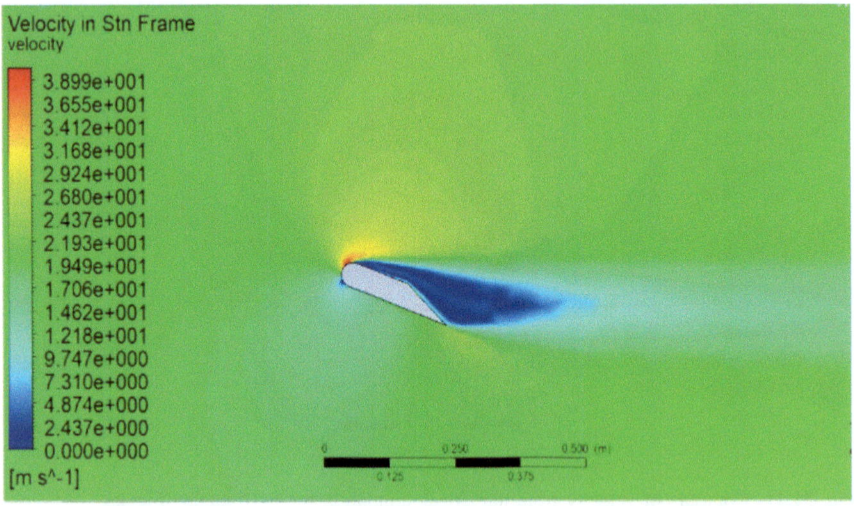

Fig. 31.13 Contours of velocity from numerical simulation for 20° angle-of-attack

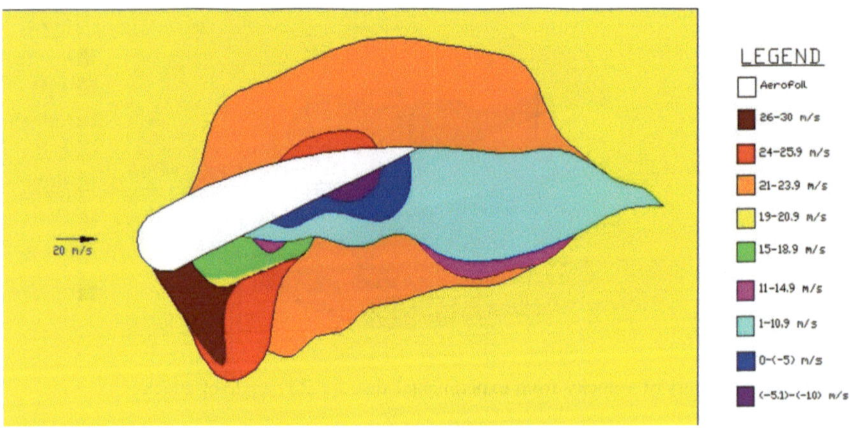

Fig. 31.14 Contours of velocity from experimental data for −20° angle-of-attack

velocity contours are concerned, for each case, the results obtained from the experiments performed at the closed-circuit wind tunnel and those from numerical simulations show excellent convergence. Some discrepancies which are visible from the velocity contours can be ascribed to the working conditions and experimental errors. Moreover, the experimental velocity contours shown in Figs. 31.6, 31.8, 31.10, 31.12 and 31.14 build the platform for more detailed analysis of aerodynamic coefficient for the aerofoil. It is obvious from Fig. 31.6 that due to lower velocity near the lower surface, higher pressure will prevail which will cause

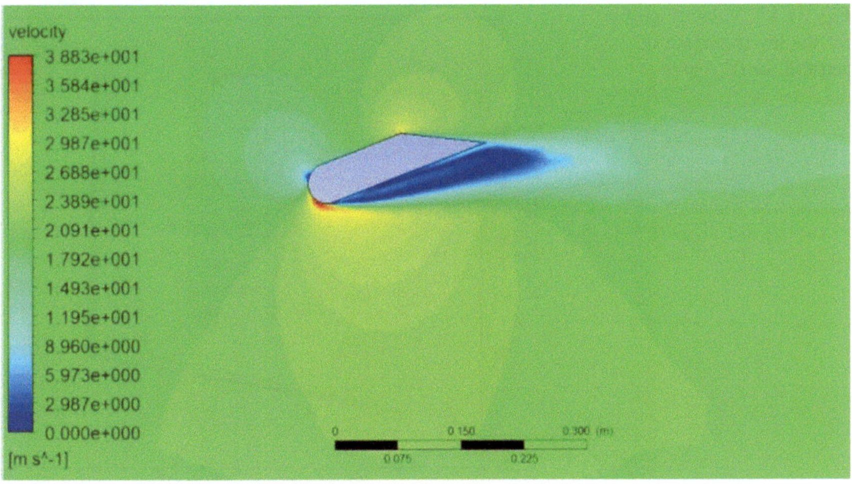

Fig. 31.15 Contours of velocity from numerical simulation for −20° angle-of-attack

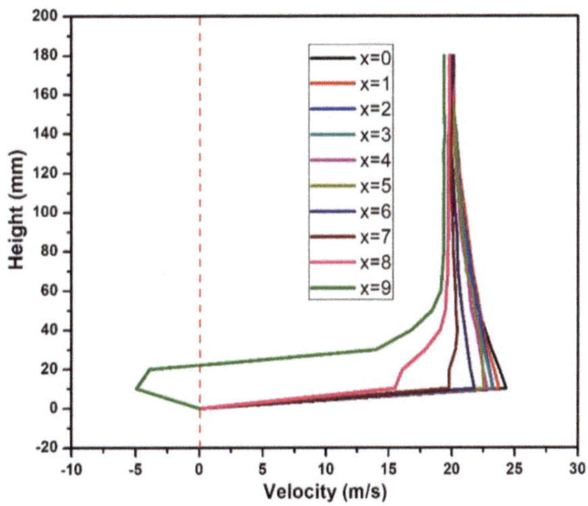

Fig. 31.16 Velocity profile for the upper surface of aerofoil for 0° AoA

a net upward force contributing to lift. In Fig. 31.8, as the angle-of-attack increases, the difference between the velocities adjacent to upper and lower surfaces gets increased, thereby resulting in higher lift force. Obviously, the observation conforms to known aerodynamic theories of higher lift for higher AoA. However, for higher angle-of-attack of 20° in Fig. 31.12, the difference in velocities near both the surfaces is not much prominent. As is already mentioned, due to high turbulence, the reliability of velocity measurement can be questionable for this case. Also, the phenomenon can be attributed to stalling, where high turbulence causes reduction in flow energy, thereby decreasing lift generation. For the negative angles-of-attack as

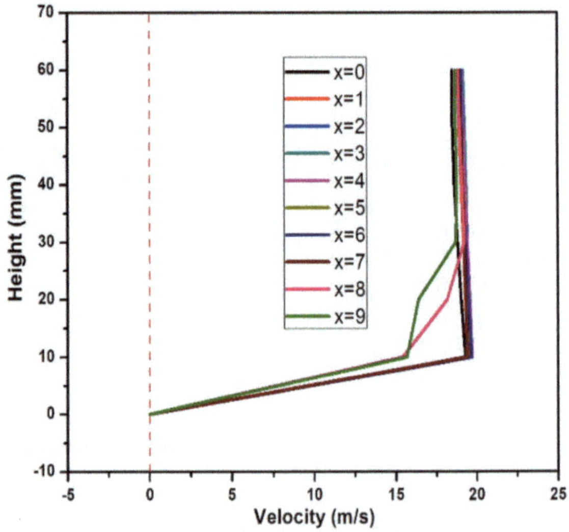

Fig. 31.17 Velocity profile for the lower surface of aerofoil for 0° AoA

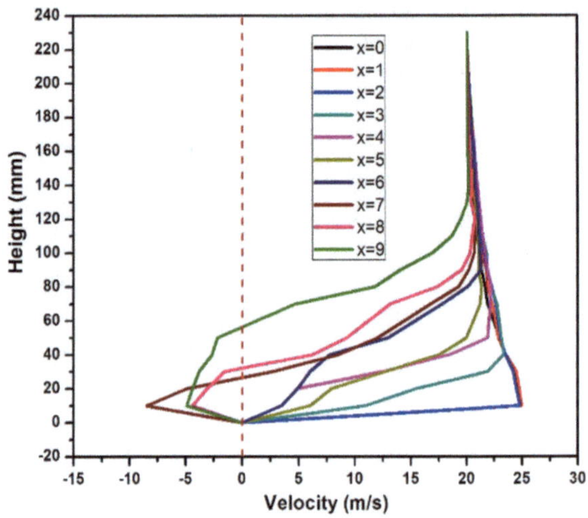

Fig. 31.18 Velocity profile for the upper surface of aerofoil for 10° AoA

depicted in Figs. 31.10 and 31.14, the velocity profile shows higher velocity near the lower surface, thereby resulting in lower pressure and corresponding downward lift generation. Also, the phenomenon is more prominent with increase in magnitude of AoA, thereby causing higher downward lift generation for the latter case. With much sophisticated instruments like load cells, the actual lift and drag forces can be found experimentally, which in turn may provide the lift and drag coefficient. However, the force determination study is kept as a future scope.

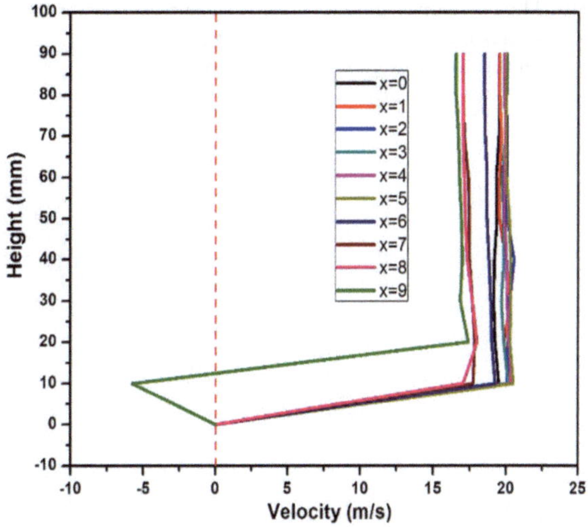

Fig. 31.19 Velocity profile for the lower surface of aerofoil for 10° AoA

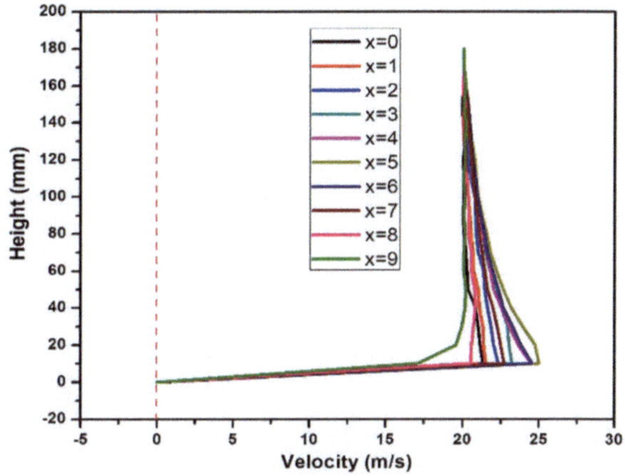

Fig. 31.20 Velocity profile for the upper surface of aerofoil for −10° AoA

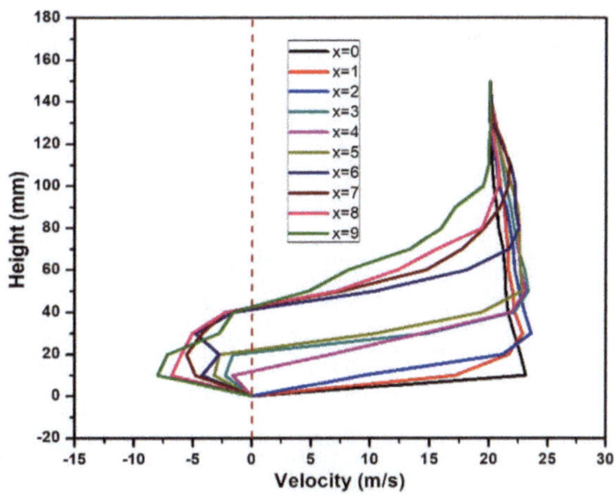

Fig. 31.21 Velocity profile for the lower surface of aerofoil for −10° AoA

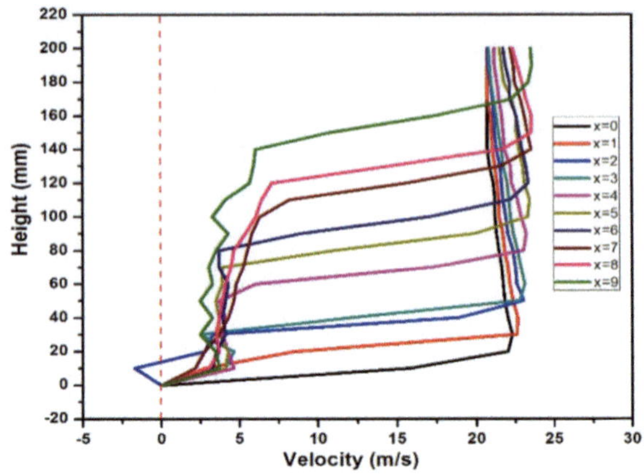

Fig. 31.22 Velocity profile for the upper surface of aerofoil for 20° AoA

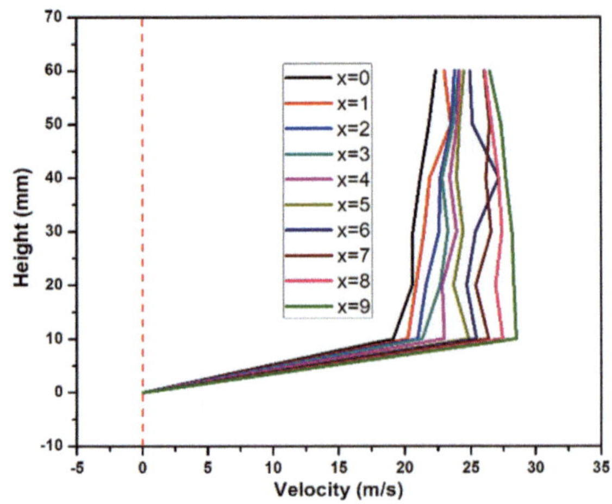

Fig. 31.23 Velocity profile for the lower surface of aerofoil for 20° AoA

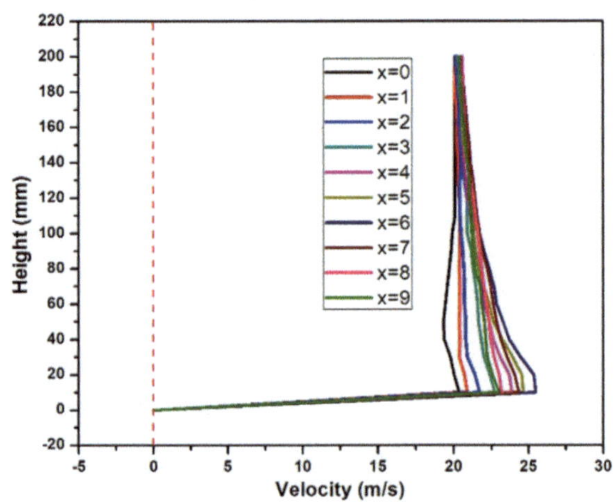

Fig. 31.24 Velocity profile for the upper surface of aerofoil for −20° AoA

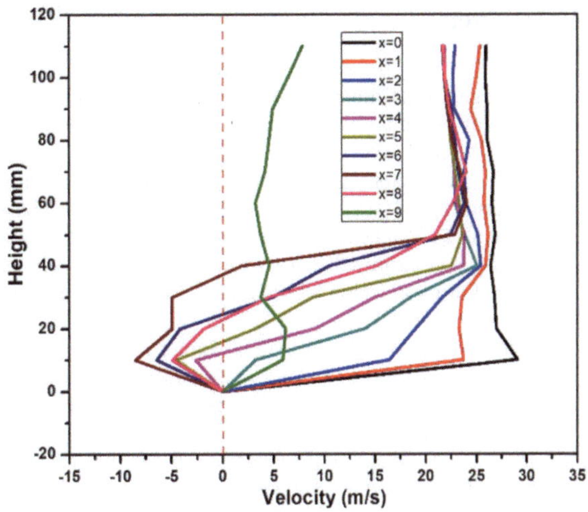

Fig. 31.25 Velocity profile for the lower surface of aerofoil for −20° AoA

31.5 Conclusion

The experimental and numerical results are plotted and compared for five different angles-of-attack, i.e. 0°, 10°, −10°, 20° and −20°. The velocity profile along the *Y*-axis is calculated from the experimental data and observed. The findings obtained from the experiments performed at the closed-circuit wind tunnel and those from numerical simulations show admirable convergence, as far as the experimental and numerical velocity contours are concerned. Boundary layer separation is observed near the trailing edge for zero or positive angles-of-attack due to reversal of flow velocity. The position of separation moves towards the leading edge with increase in angle-of-attack. No separation point is observed for extremely high positive angles-of-attack due to high turbulence in the region which makes the measurement of velocity an extremely gruelling task. The streamline separates near trailing edge from the lower surface with the separation advancing with rise in negative angle-of-attack. Upward lift force is generated for nonnegative angles-of-attack. The same magnitudes of positive and negative angles-of-attack behave differently for upper and lower surfaces, as far as velocity variations are concerned, due to the asymmetric nature of the aerofoil, having positive camber. The velocity contours found experimentally primarily allow one to predict the direction lift force generation and also establish the platform of further determination of aerodynamic force coefficient and other aerodynamic variables for the unconventional aerofoil section.

References

1. Ethridge, M.I., Cutbirth, J.M., Bogard, D.G.: Scaling of performance for varying density ratio coolants on an airfoil with strong curvature and pressure gradient effects. In: ASME Turbo Expo 2000: Power for Land, Sea, and Air, vol. 3, p. V003T01A047 (2000)
2. Addy, H., Broeren, A., Zoeckler, J., Lee, S.: A wind tunnel study of icing effects on a business jet airfoil. In: 41st Aerospace Sciences Meeting and Exhibit, p. 727 (2003)
3. Dassen, T., Parchen, R., Bruggeman, J., Hagg, F.: Results of a wind tunnel study on the reduction of airfoil self-noise by the application of serrated blade trailing edges. In: Procedure of the European Union Wind Energy Conference and Exhibition, No. NLR TP 96350, pp. 5–8 (1996)
4. Selig, M.S., Maughmer, M.D., Somers, D.M.: Natural-laminar-flow airfoil for general-aviation applications. J. Aircr. **32**(4), 710–715 (1995)
5. Mojola, O.O.: Aerodynamic design of the Savonius windmill rotor. J. Wind Eng. Ind. Aerodyn. **21**(2), 223–231 (1985)
6. Jacobs, E.N., Ward, K.E., Pinkerton, R.M.: The characteristics of 78 related airfoil sections from tests in the variable-density wind tunnel. NACA Technical Report 460, No. NACA-TR-460, PB-177874, pp. 3–58 (1933)
7. Liebeck, R.H.: Design of subsonic airfoils for high lift. J. Aircr. **15**(9), 547–561 (1978)
8. Englar, R.L.L.: Circulation control for high lift and drag generation on STOL aircraft. J. Aircr. **12**(5), 457–463 (1975)
9. Maybury, W.J., Rayner, J.M.V., Couldrick, L.B.: Lift generation by the avian tail. Proc. Roy. Soc. Lond. B Biol. Sci. **268**(1475), 1443–1448 (2001)
10. Sturm, H., Dumstorff, G., Busche, P., Westermann, D., Lang, W.: Boundary layer separation and reattachment detection on airfoils by thermal flow sensors. Sensors **12**(11), 14292–14306 (2012)
11. Devinant, P., Laverne, T., Hureau, J.: Experimental study of wind-turbine airfoil aerodynamics in high turbulence. J. Wind Eng. Ind. Aerodyn. **90**(6), 689–707 (2002)
12. Selig, M.S., McGranahan, B.D.: Wind tunnel aerodynamic tests of six airfoils for use on small wind turbines. J. Sol. Energy Eng. **126**(4), 986–1001 (2004)
13. McCroskey, W.J., McAlister, K.W., Carr, L.W., Pucci, S.L.: An Experimental Study of Dynamic Stall on Advanced Airfoil Sections, vol. 1. Summary of the Experiment. National Aeronautics and Space Administration Moffett Field Ca Ames Research Center, No. NASA-A-8924-VOL-1, pp. 1–93 (1982)
14. Das, P.K., Mondal, S., Bagchi, S., Mandal, P.: Experimental study of velocity profile of air over an aerofoil in a free wind stream in wind tunnel. In: Proceedings of First International Conference on Mechanical Engineering (INCOM 2018), Department of Mechanical Engineering, pp. 329–332 (2018)
15. Roberts, W.B.: Calculation of laminar separation bubbles and their effect on airfoil performance. AIAA J. **18**(1), 25–31 (1980)
16. Timmer, W.A., Van Rooij, R.P.J.O.M.: Summary of the Delft University wind turbine dedicated airfoils. J. Sol. Energy Eng. **125**(4), 488–496 (2003)
17. Blackwell Jr., J.A.: Preliminary study of effects of Reynolds number and boundary-layer transition location on shock-induced separation. Langley Research Center Langley Station, Hampton, National Aeronautics and Space Administration, Washington, D.C., Nasa Technical Note NASA TN D-5003, pp. 1–25 (1969)
18. Li, Y., Wang, J., Zhang, P.: Effects of Gurney flaps on a NACA0012 airfoil. Flow Turbul. Combust. **68**(1), 27 (2002)
19. Migliore, P., Oerlemans, S.: Wind tunnel aeroacoustic tests of six airfoils for use on small wind turbines. J. Sol. Energy Eng. **126**(4), 974–985 (2004)
20. Alford Jr., W.J., Henderson, W.P., Luoma, A.A.: Wind-tunnel studies at subsonic and transonic speeds of a multiple-mission variable-wing-sweep airplane configuration. Technical Memorandum X-206, pp. 1–73 (1959)

21. Jasinski, W.J., Selig, M.S.: Experimental study of open-wheel race-car front wings. SAE Technical Paper Series, Motorsports Engineering Conference Proceedings, vol. 1, No. 983042 (1998)
22. Patel, T.B., Patel, S.T., Patel, D.T., Bhensdadiya, M.: An analysis of lift and drag forces of NACA airfoils using python. Int. J. Appl. Innovation Eng. Manage. **4**(4), 198–205 (2015)
23. Patel, K.S., Patel, S.B., Patel, U.B., Ahuja, A.P.: CFD analysis of an aerofoil. Int. J. Eng. Res. **3**(3), 154–158 (2014)
24. Gunel, O., Koc, E., Yavuz, T.: Comparison of CFD and Xfoil airfoil analyses for low Reynolds number. Int. J. Energy Appl. Technol. **3**(2), 83–86 (2016)
25. McCroskey W. J.: A critical assessment of wind tunnel results for the NACA 0012 airfoil. Technical Memo National Aeronautics and Space Administration Moffett Field Ca Ames Research Center, No. ADA193182 (1987)
26. Rubel, R.I., Uddin, K., Islam, Z., Rokunuzzaman, M.D.: Numerical and experimental investigation of aerodynamics characteristics of NACA 0015 aerofoil. Int. J. Eng. Technol. **2** (4), 132–141 (2016)
27. Conner, M.D., Tang, D.M., Dowell, E.H., Virgin, L.N.: Non-linear behaviour of a typical airfoil section with control surface freeplay: a numerical and experimental study. J. Fluids Struct. **11**(1), 89–109 (1997)
28. Cummings, R.M., Morton, S.A., Siegel, S.G.: Numerical prediction and wind tunnel experiment for a pitching unmanned combat air vehicle. Aerosp. Sci. Technol. **12**(5), 355–364 (2008)
29. Murphy, J.T., Hu, H.: An experimental study of a bio-inspired corrugated airfoil for micro air vehicle applications. Exp. Fluids **49**(2), 531–546 (2010)
30. Panigrahi, D.C., Mishra, D.P.: CFD simulations for the selection of an appropriate blade profile for improving energy efficiency in axial flow mine ventilation fans. J. Sustain. Min. **13** (1), 15–21 (2014)
31. Wordley, S., Saunders, J.: Aerodynamics for formula SAE: a numerical, wind tunnel and on-track study. SAE Technical Paper Series, Vehicle Aerodynamics, No. 2006-01-0808 (2006)
32. Troolin, D.R., Longmire, E.K., Lai, W.T.: Time resolved PIV analysis of flow over a NACA 0015 airfoil with Gurney flap. Exp. Fluids **41**(2), 241–254 (2006)
33. Sheldahl, R.E., Klimas, P.C.: Aerodynamic Characteristics of Seven Symmetrical Airfoil Sections Through 180-Degree Angle-of-Attack for Use in Aerodynamic Analysis of Vertical Axis Wind Turbines. United States Department of Energy, Sandia National Labs, Albuquerque, NM (USA), No. SAND-80-2114, pp. 3–111 (1981)
34. McCroskey, W.J., Carr, L.W., McAlister, K.W.: Dynamic stall experiments on oscillating airfoils. AIAA J. **14**(1), 57–63 (1976)
35. Mohamed, M.H.: Performance investigation of H-rotor Darrieus turbine with new airfoil shapes. Energy AIAA J. **47**(1), 522–530 (2012)
36. Amitay, M., Smith, D.R., Kibens, V., Parekh, D.E., Glezer, A.: Aerodynamic flow control over an unconventional airfoil using synthetic jet actuators. AIAA J. **39**(3), 361–370 (2001)
37. Ranzenbach, R., Barlow, J.: Cambered airfoil in ground effect-an experimental and computational study. SAE Technical Paper Series, Vehicle Aerodynamics: Wind Tunnels, CFD, Aeroacoustics and Ground Transportation systems, No. 960909 (1996)
38. Anderson, J.D.: Fundamentals of Aerodynamics, 3rd edn. McGraw-Hill Series in Aeronautical And Aerospace Engineering (2001)

Chapter 32
Numerical Simulation of Vortex Shedding from a Cylindrical Bluff-Body Flame Stabilizer

Sombuddha Bagchi, Sourav Sarkar, Uddalok Sen, Achintya Mukhopadhyay and Swarnendu Sen

32.1 Introduction

In recent times, NOx emission control has become a primary concern of aerospace and automobile industries due to the stringency of emission control rules and regulations. High-temperature combustion is identified as the primary source of NOx formation. So, often lean combustion is preferred over-rich combustion to reduce NOx emission. However, lean combustion is susceptible to be unstable and leads to large-amplitude oscillations, unacceptable noise, and even structural damage of the combustion systems [1, 2]. Bluff-body flame stabilizers are sometimes installed in practical combustors like ramjet and turbojet afterburner. The mixing shear layer formed at the downstream of the bluff body helps to stabilize the flame. However, complex flow patterns behind the wake are observed and it is identified as one of the classical problems of fluid mechanics as it has several interesting flow features. Asymmetrical Von Karman vortex shedding pattern at certain values of Reynolds number is one of the most interesting features in flow past a cylinder situation. This shedding pattern is intensively studied as it is of prime importance in bluff-body stabilized combustors. Delhaye et al. [3] simulated the structure and dynamics of a spatially growing reactive mixing layer. A vortex in an incompressible flow was defined in terms of the eigenvalues of the symmetric tensor and anti-symmetric parts of the velocity gradient tensor which arrests the pressure minimum in a plane at right angles to the vortex axis in high-speed flows

S. Bagchi (✉) · S. Sarkar · A. Mukhopadhyay · S. Sen
Department of Mechanical Engineering, Jadavpur University, Kolkata, India
e-mail: sombuddha.bagchi@gmail.com

U. Sen
Department of Mechanical Engineering, University of Illinois at Chicago, Chicago, USA

P. Sahoo and J. P. Davim (eds.), *Advances in Materials, Mechanical and Industrial Engineering*, Lecture Notes on Multidisciplinary Industrial Engineering,
https://doi.org/10.1007/978-3-319-96968-8_32

by Jeong and Hussain [4]. Poszdiech and Grundmann [5] used the spectral element method for the numerical calculation of fundamental quantities of the 2D flow past a circular cylinder at Reynolds number between 5 and 250. Wang et al. [6] studied the proper orthogonal decomposition (POD) analysis of a finite-length cylinder, which is wall mounted, near wake, and found the relation between the orientation of the vortex shedding and the coefficients of the POD mode. The flow past a circular cylinder centrally placed inside a channel was studied by Singha and Sinhamahapatra [7] using an unstructured collocated grid-based finite volume method based on the primitive variable formulation. Williamson [8] reviewed the dynamics of vortex shedding in a cylinder wake. The decreased channel height resulted in a wake of shorter length and the lateral boundaries present delayed the shift to an unsteady detached wake flow from a steady attached wake. Zhu et al. [9] used a hybrid U-RANS/PDF method to study the vortex shedding behind a bluff-body (triangular) flame stabilizer. Flame stabilization for laminar flow has been a prime focus of importance by researchers in recent times. Raghavan et al. [10] have done a considerable amount of work in flame stabilisation. Vortices generated by flame holding devices in a combustor and interactions between the flame and vortex in separated CH_4–air cross-flow flames anchored behind three bluff bodies, namely an isosceles triangular cylinder, a square cylinder, and a half V-gutter. In the reactive flow, the mixing phenomenon was explained using species velocities' streamlines of CH_4 and O_2. From the POD and discrete Fourier transform (DFT) of OH mass fraction data, two separate unstable layers were visible. The interactions between factors influencing fluid dynamics and chemical kinetics causing several fluctuating layers in the hydroxyl ion concentrations were illuminated. Many experimental and theoretical studies on laminar boundary layer non-premixed flames performed over a flat plate have been described [11, 12]. A gaseous fuel was injected through a horizontal porous plate which was kept parallel to the direction of air stream by Hirano and Kanno [13] and significant acceleration of the gas stream around the flame zone was observed. Similar work was performed by Hirano and Kinoshita [14] on flame stabilization over liquid fuel pool. Gopalakrishnan and Raghavan [15] carried out numerical simulations on laminar boundary layer flame, which was stabilized by a backward facing step placed upstream of the fuel injection port. The effects of the step and its position were analysed. The formation of a recirculation zone behind the step resulted in increased stability limits. The step height played a less important role as compared to the step location in affecting the stability characteristic of flames. Peters [16] gave an outline of flame calculations with reduced mechanisms. Ueda et al. [17] elucidated the effects of gravity on aerodynamic structure of laminar boundary layer non-premixed flame which was stabilized over a flat plate along with the local pressure distribution. Ramachandra and Raghunandan [18] experimented by varying the direction of injected fuel with respect to the direction of the gravitational field and brought to attention the absence of velocity overshoot with fuel injection in the downward direction. Stability diagrams were drawn out in detail to give a description of the stability limits and stabilization regimes by Rohmat et al. [19] in term of air and fuel flow velocities for different cases in the presence and absence of impediments of varying dimensions. Chen and T'ien [20] numerically analysed laminar non-premixed flames, using a 2D model along with a single-step chemical reaction. Wang et al. [21] used computational techniques to study laminar non-premixed flames over a plane wall in a two-dimensional field using single-step reaction chemistry. Along the flame sheet, a sequence of counter-rotating vortices was observed, which, when powerful enough, crumpled the flame. Shijin et al. [22] presented a comprehensive numerical study of laminar cross-flow diffusion CH_4–air flames in the presence of a square cylinder and analysed cell Damköhler number, local equivalence ratio, velocity of the species, and net consumption rate of methane. It displayed the variations in the locations where the flame was anchored along with the analysis of unsteadiness in the case of the separated flame. An extensive investigation of the boundary layer reacting flow has been performed by Lavid and Berlad [23] over a flat plate, where the supply of fuel is perpendicular to the air stream. Experimental studies on flame stabilization using a diaphragm were described [24]. In the past, many attempts have been made to study the mutual interactions between the vortices and the flame zone in the flow field. Marble [25] used infinitely fast chemical kinetics to perform analytical calculations. A detailed report of the physics of a flame sheet roll-up in the occurrence of a single vortex was given. Along the boundary between a fuel and oxidizer, a horizontal non-premixed flame was assumed to exist. After the introduction of a vortex at the origin, analysis of the local flame structure and kinematics of flame extinction were performed. Karagozian and Marble [26] analysed non-premixed flames in laminar flow in the incidence of a vortex in a single plane and a stretched line vortex. The numerical study by Alain and Candel [27] on diffusion flame–vortex interaction using infinitely fast chemistry assumption validated the results by Marble [25]. Ashurst and Mcmurtry [28] conducted a similar study considering finite rate chemistry and dynamically incompressible flow for a premixed flame. They discovered the coupling between vortex flow field and chemical heat release. Lewis et al. [29] studied the mutual interactions between a laminar co-flowing jet flame and periodic vortical motion experimentally. A loudspeaker was used to acoustically excite the fuel stream to generate periodic vortices. Optical planar imaging techniques were used to analyse the flame structure infused in the vortical flow. The interaction of non-premixed flame and vortex shedding was studied by Macaraeg et al. [30] using numerical investigations and asymptotic analysis. Rolon et al. [31] studied the interaction of a vortex ring with a counter-flow non-premixed flame in turbulent flow experimentally. The interaction of the flame with strong vortices causes flame extinction and a consequent blowout. In contrast, the interaction with a weaker vortex caused the flame to recover. A bluff body was used by Fan et al. [32] in a micro-combustor to prolong the blow-off limit. Altay et al. [33] studied the combustion dynamics of propane–H_2 mixtures in an atmospheric pressure, lean, premixed backward-facing step combustor and varied inlet temperature, the equivalence ratio, and fuel composition systematically for determination of the stability map of the combustor. Barlow et al. [34] measured temperature, the major reacting species, OH, and NO in steady laminar opposed-flow partially premixed flames of methane and air, using the non-intrusive techniques of Raman scattering and laser-induced fluorescence. Mondal et al. [35] experimentally studied a bluff-body stabilized laboratory-scale pulse combustor to investigate the effects of different parameters on combustion instability. Sen et al. [36] used an unstructured grid finite volume method to study the transient, 2D laminar flow past a circular cylinder with cross- and co-counter injection of methane. The fluctuations of Strouhal number were observed to study the effect of injection on vortex shedding. The two injection arrangements were observed to affect the vortex shedding downstream of the bluff body in distinctly dissimilar manners. Investigation of the effect of lateral walls on the flow dynamics was performed. The conception of aperiodic wake structures was deferred by the presence of lateral walls. The walls promoted better mixing in the wake downstream of the cylinder. The proper orthogonal decomposition analysis was performed to detect the dominant modes and their respective entropy contents. Fujii et al. [37] have compared non-reacting and reacting flows around a bluff-body flame stabilizer for homogeneous propane/air mixture. Bagchi et al. [38] investigated the laminar flow past a slotted circular cylinder with cross injection of methane which was compared between two different concentrations of methane. The contrast between a reacting and non-reacting flow was also studied in detail [39]. Recently, researchers have been focusing on more complex aspects of combustion. A detailed study of vortex ring and its interaction with a flame has been analysed. Bharadwaj et al. [40] studied the characteristics of non-premixed flame–wall interaction in a reacting vortex ring configuration. Safta and Madnia [41] performed direct numerical simulations of non-premixed methane flame–vortex ring interactions. Renard et al. [42] experimentally studied the interactions of counter-flow flame and a vortex ring. Hermanns et al. [43] conducted a numerical investigation of a counter-flow non-premixed flame in a flow field and its interaction with a laminar vortex ring. A computational study was reported by Mishra et al. [44] of the interaction of convective line vortex with a laminar diffusion flame. The increased area of the interface and diffusion across it caused the mixing phenomena. Similar trends were reported for a non-reacting flow [45]. A detailed experimental and numerical investigation of a toroidal vortex and its interaction with a stagnation premixed flame was carried out by Thiesset et al. [46] with the aim of quantifying the ability of such a vortex to stretch the flame. Stöhr et al. [47] experimentally studied the interaction of a helical precessing vortex core (PVC) with turbulent swirl flames in a gas turbine model combustor.

In our present work, transient flow past a circular cylinder has been studied. The bluff body, i.e. the cylinder, has two slots which are perpendicular to the flow direction, both of them 180° apart. This arrangement is actually referred to as a cross-flow arrangement. Methane is injected through the two slots at a particular velocity. Different comparisons have been drawn for different changes in parameters. The injection velocity of methane from the two slots has been varied along with simultaneous changes in the mass fraction of methane. A qualitative study has been performed to analyse the different characteristics of combustion. The motivation of our current study is not to restrain the vortex shedding, but to study the phenomenon of mixing and have a stable combustion so that the flame does not wash out and is sustained; i.e., the flame is stabilized by the wake of the bluff body. A substantial amount of literature in this field is present although none of those are studied in laminar zone have attempted to stabilize the flame using the current configuration. The simulations were carried out at a free stream Reynolds number of 100. Fast Fourier transforms have been done to determine the frequency of vortex shedding for all the cases.

32.2 Problem Geometry

A rectangular unconfined flow domain (200 mm × 200 mm) is considered and the slotted cylinder is placed centrally within the flow domain. The maximum diameter of the cylinder is 6 mm and two diametrically opposite slots which are placed at right angles to the direction of the flow 180° apart. Schematic of geometry is presented in Fig. 32.1. In this configuration, the direction of injection velocity of methane is along the positive and negative *y*-direction and it is varied from $\varepsilon = 1$ to $\varepsilon = 4$ where ε denotes the ratio of the injection velocity to that of the free stream velocity. Simultaneously, the mass fraction of methane is varied from $w = 1$–0.5 at steps of 0.25, where w denotes the mass fraction of methane, with dilution being performed by nitrogen. The free stream Reynolds number of 100 is considered for

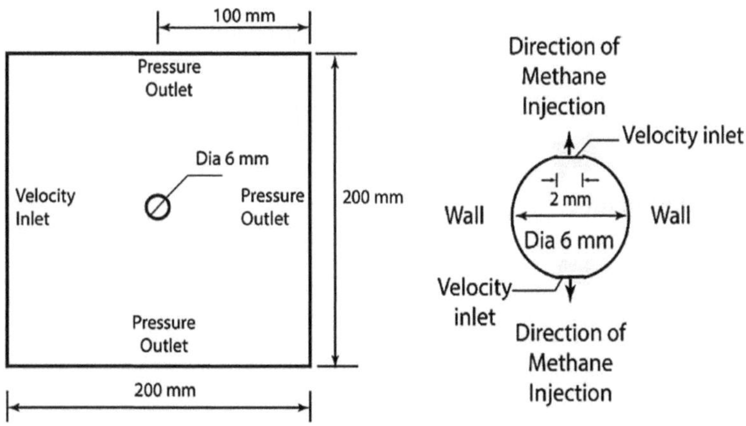

Fig. 32.1 Problem geometry and boundary conditions

the present study. A velocity inlet condition is specified at the inlet, pressure-outlet condition at the outlet and a specified shear of zero magnitude at the two walls.

32.3 Governing Equations

For incompressible, laminar, 2D transient flow, the following governing equations have been used:

(i) **Continuity**:

$$\frac{\partial \rho}{\partial t} + \nabla \cdot (\rho \vec{V}) = 0 \tag{32.1}$$

(ii) **Momentum**:

$$\frac{\partial(\rho \vec{V})}{\partial t} + \nabla \cdot (\rho \vec{V}\vec{V}) = -\nabla p + \nabla \cdot \vec{\tau} + \rho \vec{g} \tag{32.2}$$

where $\overrightarrow{\tau}$ is the stress tensor, p is the static pressure, $\rho \vec{g}$ is the gravitational body force, and $\vec{F}$ is any other external body force, for example, which arises from interaction with the discrete phase.

The stress tensor is given by

$$\vec{\tau} = \mu\left[\left(\nabla \vec{V} + \nabla \vec{V}^{\mathrm{T}}\right) - \frac{2}{3}\left(\vec{V} I\right)\right]$$

(iii) **Lift coefficient**:

The lift coefficient is defined by the following relation

$$C_{\mathrm{L}} = \frac{F_{\mathrm{L}}}{\frac{1}{2}\rho U_{\infty}^{2} D} \tag{32.3}$$

(iv) **Strouhal number**:

$$\mathrm{St} = \frac{fD}{U_{\infty}} \tag{32.4}$$

(v) **Energy Equation**

$$\frac{\partial(\rho E)}{\partial t} + \nabla\cdot\left[\vec{V}(\rho E + p)\right] = \nabla\cdot\left[(k\nabla T) - \sum_j h_j J_j + \nabla\cdot\left(\bar{\bar{\tau}}\vec{V}\right)\right] + S_h \tag{32.5}$$

The energy is given by

$$E = h - \frac{p}{\rho} + \frac{V^2}{2}$$

where

$$h = \sum_j Y_j h_j + \frac{p}{\rho}$$

$$h_j = \int_{T_{\mathrm{ref}}}^{T} C_{p,j}\mathrm{d}T$$

The reference temperature was taken to be 298.15 K.

(vi) **Species Transport Equation**

$$\frac{\partial(\rho Y_i)}{\partial t} + \nabla\cdot\left(\rho\vec{V}Y_i\right) = -\nabla\cdot J_i + R_i + S_i \tag{32.6}$$

where R_i is a reaction source term and S_i represents other source terms, which is generated from the discrete phase. Generally, $N-1$ species equations are solved for N species. Nitrogen is taken to be the Nth species.

J_i is the diffusion flux of species i

$$J_i = -\rho D_{i,m}\nabla Y_i - D_{T,i}\frac{\nabla T}{T} \tag{32.7}$$

The laminar finite rate model was used to calculate source term R_i. As gas flows are laminar in this study, laminar finite rate model is chosen. For a reversible reaction, the molar rate of generation of a species i in reaction r is given by the expression

$$R_i = \Gamma(v_i'' - v_i')\left[k_{f,r}\prod_{j=1}^{N} C_{j,r}^{v'_{j,r}} - k_{b,r}\prod_{j=1}^{N} C_{j,r}^{v''_{j,r}}\right] \tag{32.8}$$

The rate constants are calculated as

$$k_{f,r} = A_r T^{\beta_r} e^{\frac{E_r}{RT}}$$

$$k_{b,r} = k_{f,r}/k_r$$

where k_r is the equilibrium constant of the *r*th reaction. A reduced reaction mechanism with 16 species and 46 reactions was used to model chemistry of the combustion phenomena [48]. The thermodynamic and transport database files in FLUENT provide all other data required for calculation of the rate constants.

32.4 Numerical Method and Validation

The finite volume-based CFD code has been used to perform the required numerical simulations. The laminar viscous model is used as the Reynolds number of the flow is 100. A least square cell-based scheme is employed for gradient calculations. SIMPLE scheme is used for pressure–velocity coupling. The pressure-based solver is chosen as the numerical scheme, and second-order implicit transient solutions are performed. Second-order upwind schemes are used for spatial discretization of momentum, species, and energy.

A triangle-based unstructured grid is used for meshing. For better results and finer meshing around the central region, inflation is carried out with least element size of 10^{-7} m having 80 layers and a growth rate of 1.5. Grid independence study along with time independence is carried out to select the optimum mesh for non-reacting flow (Table 32.1) and time step (Table 32.2) for simulation. It is

Table 32.1 Strouhal number values at different grid sizes for non-reacting flow

Mesh no.	Nodes	Elements	Strouhal no.
M1	14,328	17,253	0.1711
M2	17,828	22,570	0.1631
M3	18,024	22,361	0.1658
M4	23,029	26,721	0.1651

Table 32.2 Strouhal number values at different time steps

Time step size (s)	Strouhal no.
0.1	0.0018
0.01	0.1211
0.001	0.1631
0.0001	0.1652

Table 32.3 Validation study with the present literature

Authors	Strouhal number
Braza et al. [49]	0.160
Liu et al. [50]	0.165
Park et al. [51]	0.165
Meneghini et al. [52]	0.165
Shi et al. [53]	0.1640
Ding et al. [54]	0.166
Posdziech and Grundmann [5]	0.1644
Mittal [55]	0.1644
Rajani et al. [56]	0.1569
Li et al. [57]	0.164
Harichandan and Roy [58]	0.161
Sen et al. [36]	0.1634
Present study	0.1631

Table 32.4 Mesh independence study for reacting flow

Mesh	Nodes	Elements	Strouhal number
M3	24,602	30,995	0.183
M2	17,828	22,570	0.183
M1	11,546	13,236	0.140

carried out by replacing the bluff body by a regular cylinder of the same diameter. The Reynolds number of the free stream is 100.

Mesh M2 and a time step size of 0.001 are considered to be optimum for simulation to reduce the computational time without any considerable loss in accuracy.

The Strouhal number obtained from the temporally fluctuating lift coefficient at the cylinder surface has been observed and compared with the present literature to validate the present model for the non-reacting flow (Table 32.3).

Mesh independence study has also been performed for reacting flow. Mesh M2 (Table 32.4) has been selected as the optimum mesh considering both accuracy and computational time.

Hence, our present model has been successfully validated and optimised for analysing of the non-reacting and reacting flow.

32.5 Results and Discussion

The velocity contours of the non-reacting and reacting flow for different cases have been considered and are presented in Figs. 32.2a, 32.3, 32.4, 32.5, 32.6, 32.7, 32.8, 32.9, 32.10, 32.11, 32.12, and 32.13. As there is no significant change in momentum and other parameters due to 0.25 mass fraction decrease in methane from $w = 1$, the results of non-reacting flow for $w = 0.75$ have been omitted from this present study. For $\varepsilon > 1$, the velocity contours are observed to oscillate at the frequency of the time-dependent lift coefficient in Figs. 32.4a, 32.5, 32.6, 32.7, 32.8, 32.9, 32.10, 32.11, 32.12, and 32.13. With increase in ε for non-reacting flow, a larger zone of stagnation has been observed downstream. For reacting flow, a more compressed Von Karman vortex street has been detected for rise in ε.

The vorticity contours of the non-reacting and reacting flow have been delineated and compared in Figs. 32.14a, 32.15, 32.16, 32.17, 32.18, 32.19, 32.20, 32.21, 32.22, 32.23, 32.24, and 32.25. Von Karman vortex street is observed for

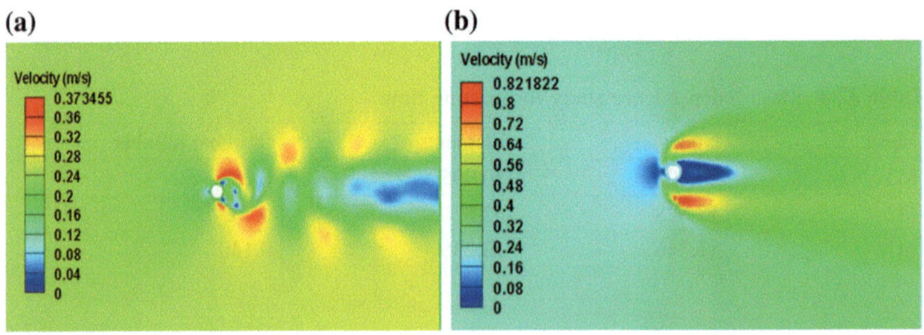

Fig. 32.2 **a** Velocity at $\varepsilon = 1$ and $w = 1$ for non-reacting flow. **b** Velocity at $\varepsilon = 1$ and $w = 1$ for reacting flow

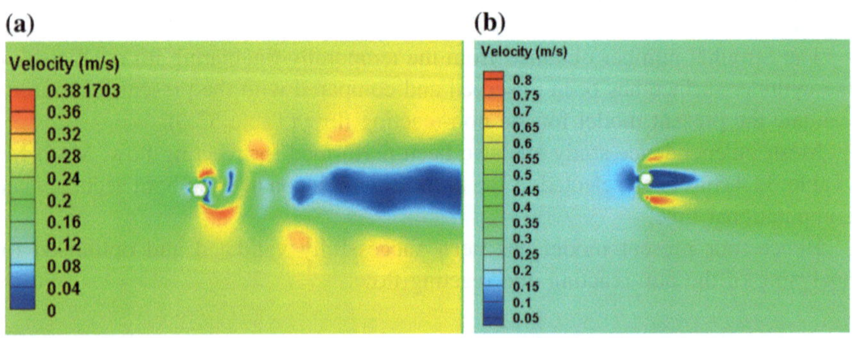

Fig. 32.3 **a** Velocity at $\varepsilon = 1$ and $w = 0.5$ for non-reacting flow. **b** Velocity at $\varepsilon = 1$ and $w = 0.5$ for reacting flow

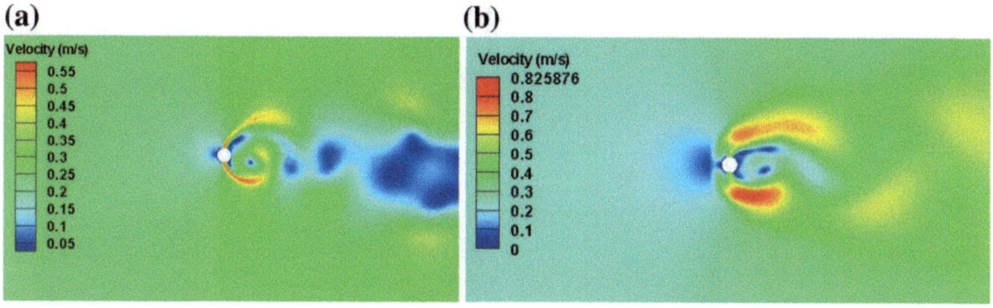

Fig. 32.4 **a** Velocity at ε = 2 and w = 1 for non-reacting flow. **b** Velocity at ε = 2 and w = 1 for reacting flow

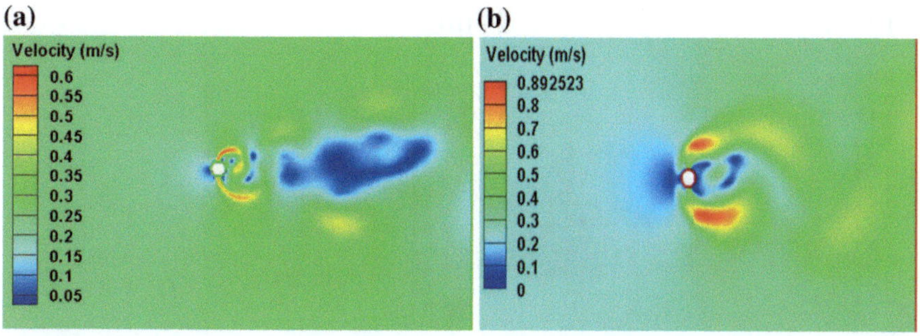

Fig. 32.5 **a** Velocity at ε = 2 and w = 0.5 for non-reacting flow. **b** Velocity at ε = 2 and w = 0.5 for reacting flow

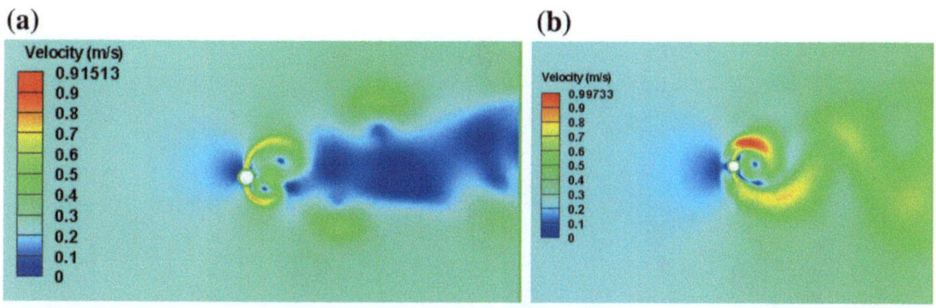

Fig. 32.6 **a** Velocity at ε = 3 and w = 1 for non-reacting flow. **b** Velocity at ε = 3 and w = 1 for reacting flow

reacting flow at all injection velocities greater than 1 (Figs. 32.16b, 32.17b, 32.18b, 32.19b, 32.20b, 32.21b, 32.23, 32.24, and 32.25). The swirling vortices are separated for all non-reacting flows (Figs. 32.14a, 32.15a, 32.16a, 32.17a, 32.18a, 32.19a, 32.20a, and 32.21a), whereas for reacting flows they are attached.

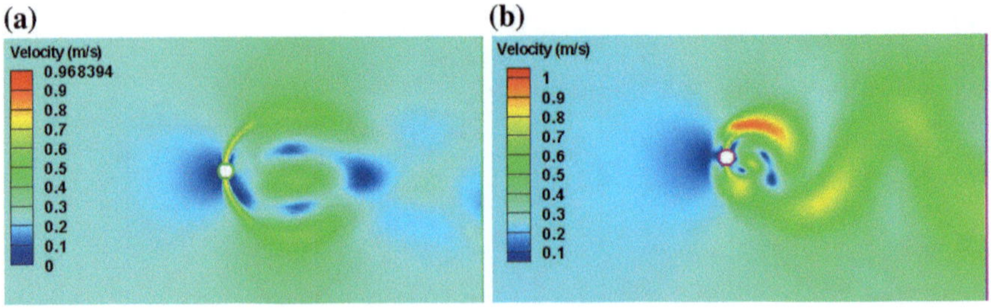

Fig. 32.7 **a** Velocity at ε = 3 and w = 0.5 for non-reacting flow. **b** Velocity at ε = 3 and w = 0.5 for reacting flow

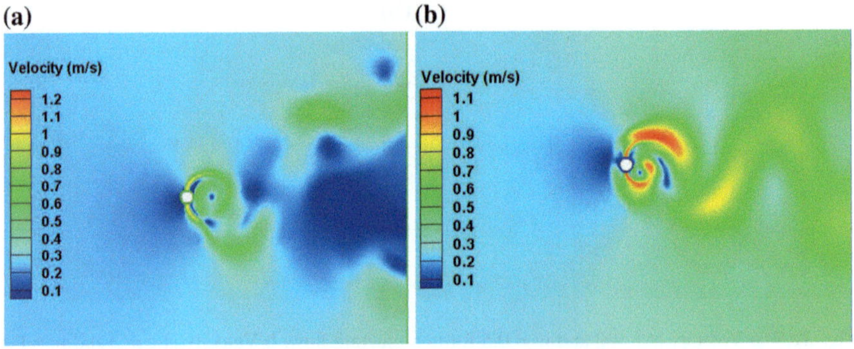

Fig. 32.8 **a** Velocity at ε = 4 and w = 1 for non-reacting flow. **b** Velocity at ε = 4 and w = 1 for reacting flow

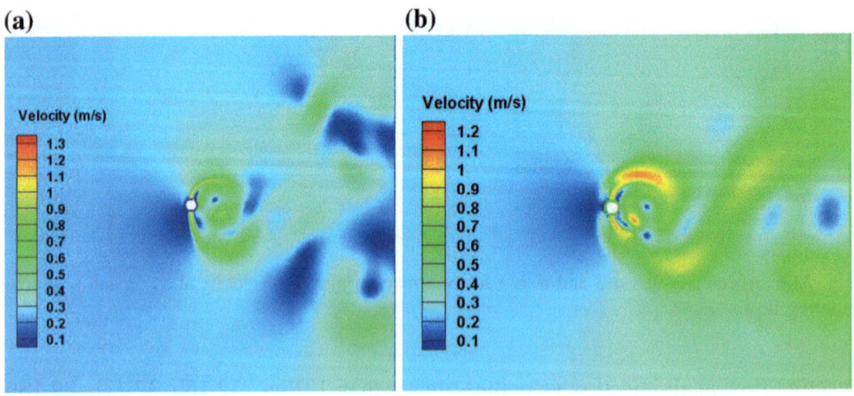

Fig. 32.9 **a** Velocity at ε = 4 and w = 0.5 for non-reacting flow. **b** Velocity at ε = 4 and w = 0.5 for reacting flow

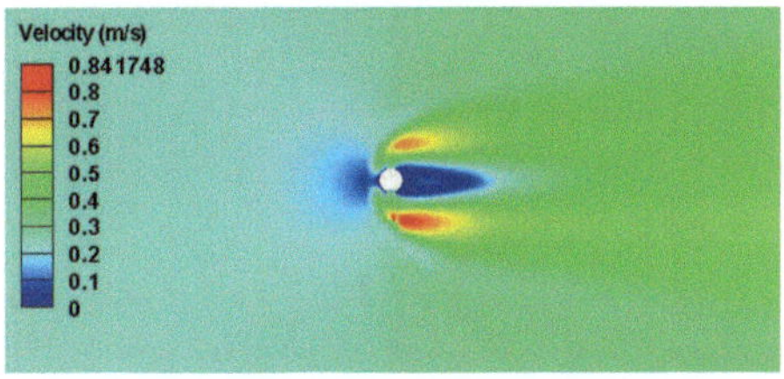

Fig. 32.10 Velocity at ε = 1 and w = 0.75 for reacting flow

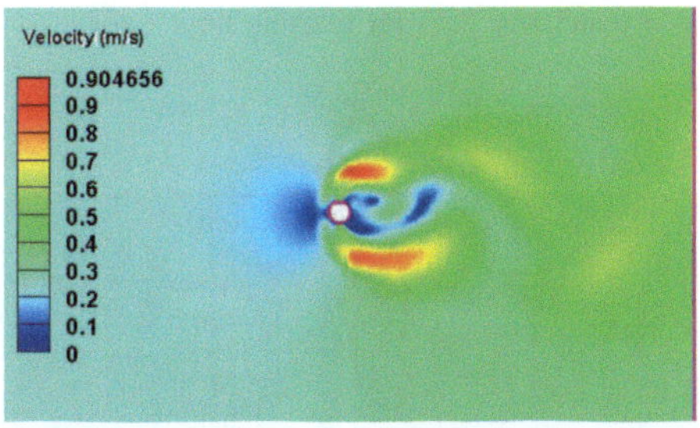

Fig. 32.11 Velocity at ε = 2 and w = 0.75 for reacting flow

Fig. 32.12 Velocity at ε = 3 and w = 0.75 for reacting flow

No recognisable vortex shedding phenomenon is observed at ε = 1 reacting flow (Figs. 32.14b, 32.15b, and 32.22). This can be attributed to the fact that the combustion suppresses the vortex shedding occurring downstream of the slotted cylinder. No Von Karman vortex street has been observed for non-reacting flow.

The methane mass fraction contours for the reacting and non-reacting flow have been outlined and compared below in Figs. 32.26a, 32.27, 32.28, 32.29, 32.30,

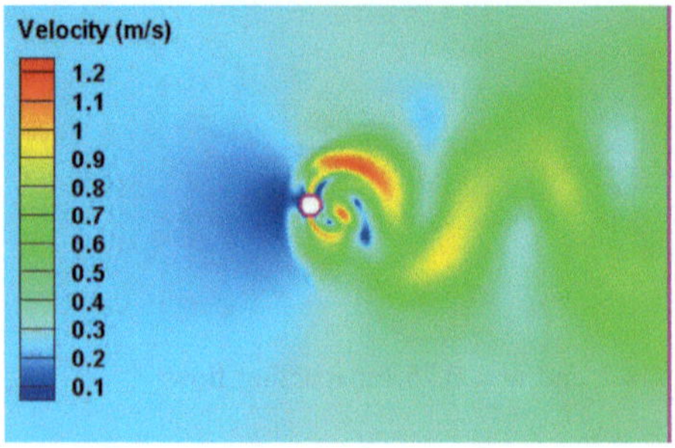

Fig. 32.13 Velocity at ε = 4 and w = 0.75 for reacting flow

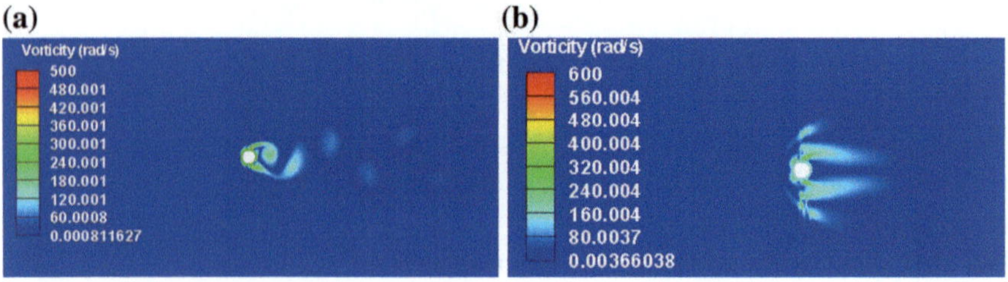

Fig. 32.14 **a** Vorticity at ε = 1 and w = 1 for non-reacting flow. **b** Vorticity at ε = 1 and w = 1 for reacting flow

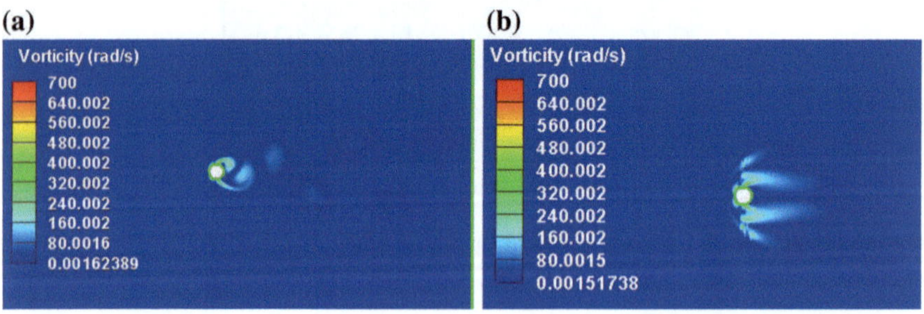

Fig. 32.15 **a** Vorticity at ε = 1 and w = 0.5 for non-reacting flow. **b** Vorticity at ε = 1 and w = 0.5 for reacting flow

32.31, 32.32, 32.33, 32.34, 32.35, 32.36, and 32.37. The methane contours are observed to oscillate at the frequency of vortex shedding. The mixing shear layer downstream of the slotted cylinder has facilitated the mixing of the methane being injected from the ports.

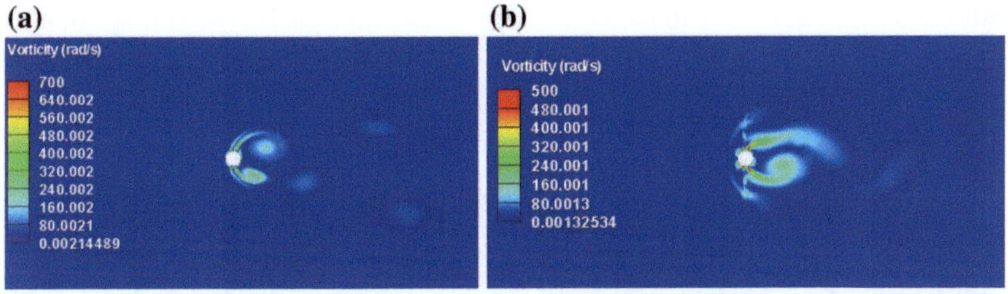

Fig. 32.16 **a** Vorticity at ε = 2 and $w = 1$ for non-reacting flow. **b** Vorticity at ε = 2 and $w = 1$ for reacting flow

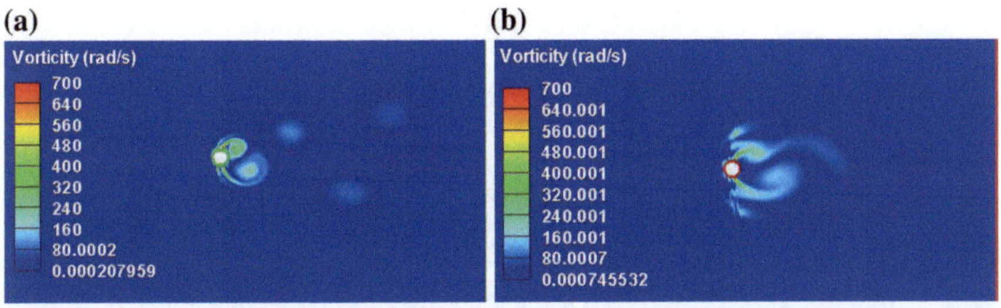

Fig. 32.17 **a** Vorticity at ε = 2 and $w = 0.5$ for non-reacting flow. **b** Vorticity at ε = 2 and $w = 0.5$ for reacting flow

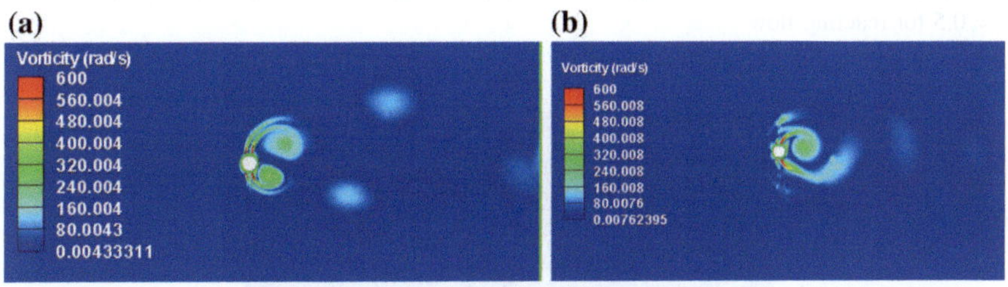

Fig. 32.18 **a** Vorticity at ε = 3 and $w = 1$ for non-reacting flow. **b** Vorticity at ε = 3 and $w = 1$ for reacting flow

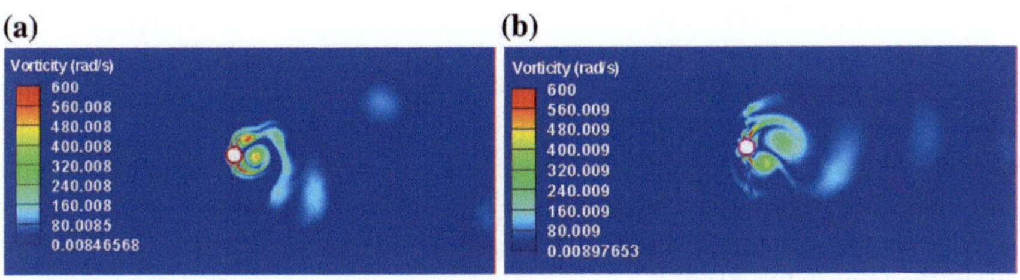

Fig. 32.19 **a** Vorticity at ε = 3 and $w = 0.5$ for non-reacting flow. **b** Vorticity at ε = 3 and $w = 0.5$ for reacting flow

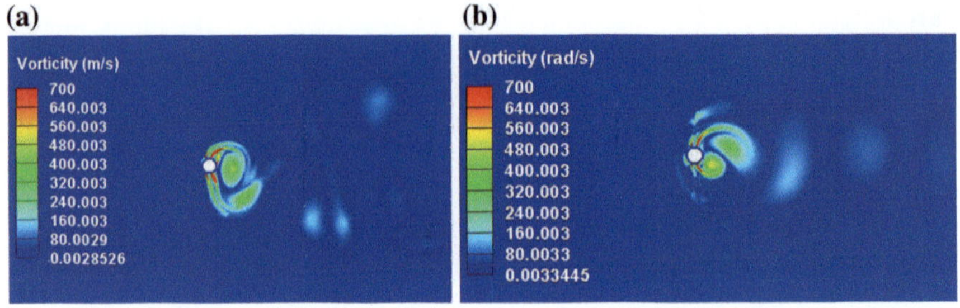

Fig. 32.20 **a** Vorticity at ε = 4 and w = 1 for non-reacting flow. **b** Vorticity at ε = 4 and w = 1 for reacting flow

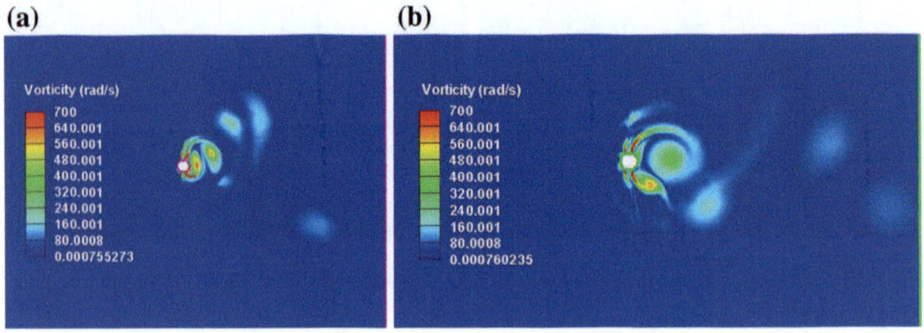

Fig. 32.21 **a** Vorticity at ε = 4 and w = 0.5 for non-reacting flow. **b** Vorticity at ε = 4 and w = 0.5 for reacting flow

Fig. 32.22 Vorticity at ε = 1 and w = 0.75 for reacting flow

The temperature contours for the reacting flow are observed and demarcated below in Figs. 32.38, 32.39, 32.40, 32.41, 32.42, 32.43, 32.44, 32.45, 32.46, 32.47, 32.48, and 32.49. The temperature contours are observed to oscillate along with the vortex shedding. There is a non-reacting region just downstream of the slotted cylinder and the reacting test region is present at the outline of the zone where the vortex shedding takes place.

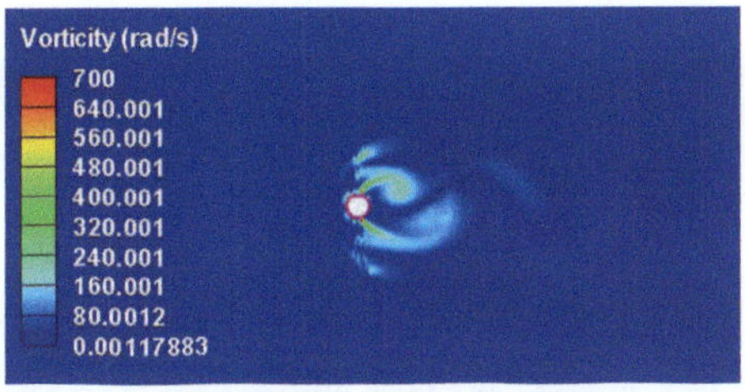

Fig. 32.23 Vorticity at ε = 2 and $w = 0.75$ for reacting flow

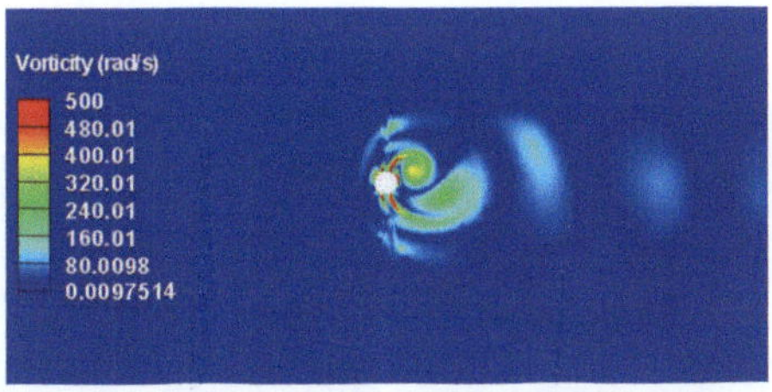

Fig. 32.24 Vorticity at ε = 3 and $w = 0.75$ for reacting flow

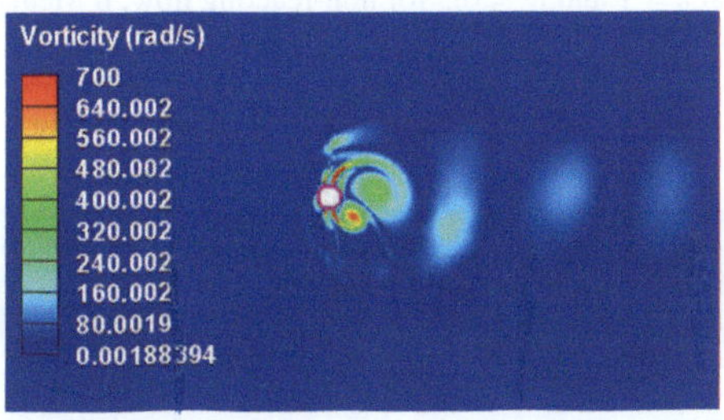

Fig. 32.25 Vorticity at ε = 4 and $w = 0.75$ for reacting flow

The hydroxyl ion contours for reacting flow which represents where the actual combustion has taken place has been presented below for the different cases in Figs. 32.50, 32.51, 32.52, 32.53, 32.54, 32.55, 32.56, 32.57, 32.58, 32.59, 32.60 and 32.61. The presence of hydroxyl ion has mainly been observed at the outline of zone where the vortex shedding phenomenon takes place and it oscillates at the frequency of vortex shedding. For all of the cases considered, the OH^- ion concentration is observed to decline with decrease in the mass fraction of methane (Figs. 32.50, 32.51, 32.52, 32.53, 32.54, 32.55, 32.56, 32.57 and 32.58), except for

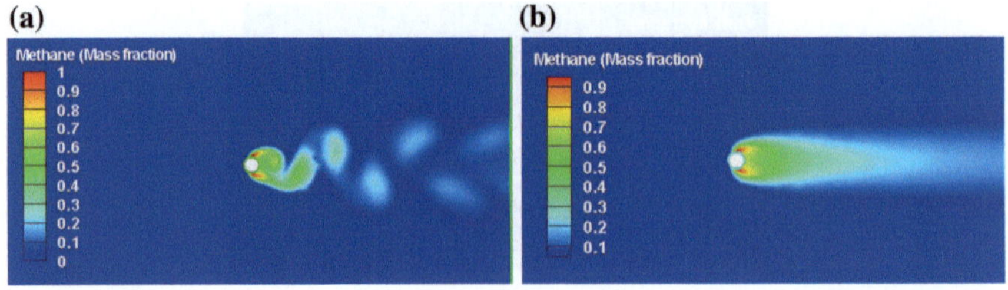

Fig. 32.26 **a** Methane at $\varepsilon = 1$ and $w = 1$ for non-reacting flow. **b** Methane at $\varepsilon = 1$ and $w = 1$ for reacting flow

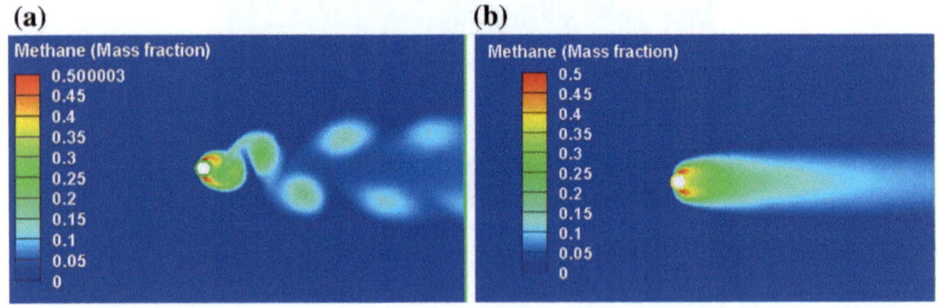

Fig. 32.27 **a** Methane at $\varepsilon = 1$ and $w = 0.5$ for non-reacting flow. **b** Methane at $\varepsilon = 1$ and $w = 0.5$ for reacting flow

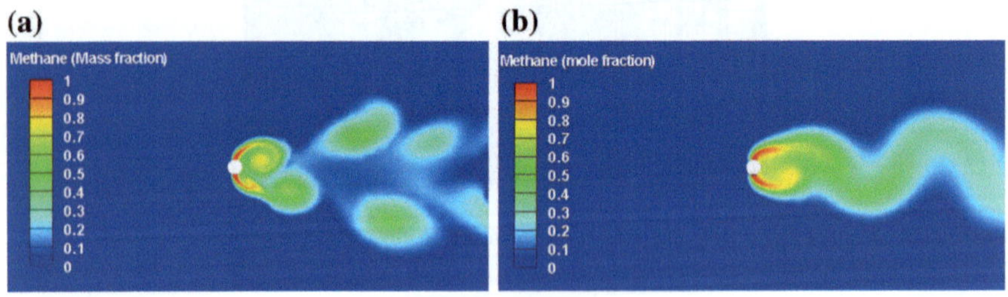

Fig. 32.28 **a** Methane at $\varepsilon = 2$ and $w = 1$ for non-reacting flow. **b** Methane at $\varepsilon = 2$ and $w = 1$ for reacting flow

$\varepsilon = 4$, where an opposite phenomenon has been observed, i.e. increase in concentration of OH^- with decrease in methane mass fraction (Figs. 32.59, 32.60 and 32.61). The occurrence can be attributed to the high injection velocity, which results in better mixing for leaner fuel compositions. From the OH^- contours, it has been observed that no combustion occurs at any ε for $w = 0.25$ (not documented in this present study).

The fast Fourier transform plots of the temporally fluctuating lift coefficients for

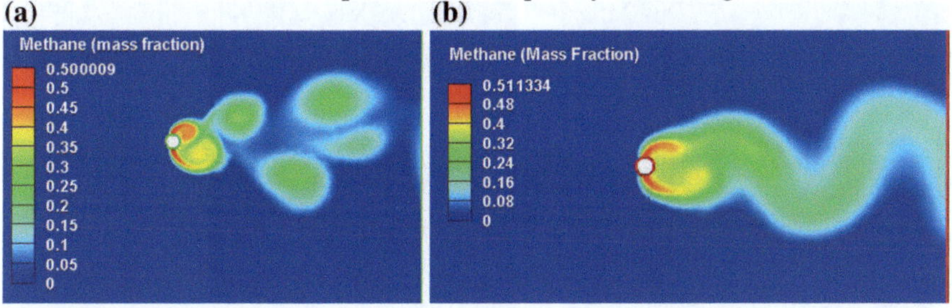

Fig. 32.29 **a** Methane at ε = 2 and w = 0.5 for non-reacting flow. **b** Methane at ε = 2 and w = 0.5 for reacting flow

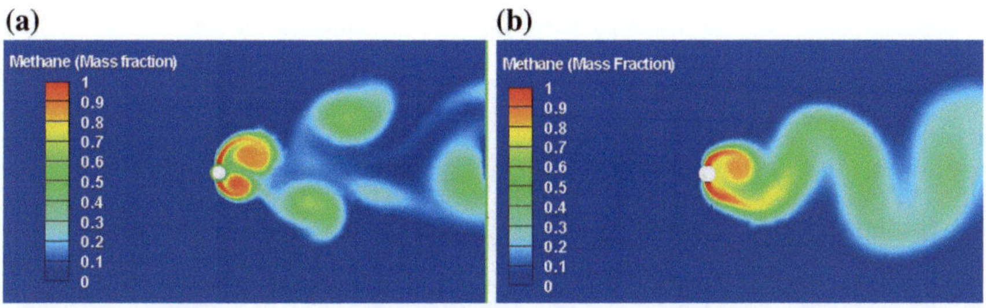

Fig. 32.30 **a** Methane at ε = 3 and w = 1 for non-reacting flow. **b** Methane at ε = 3 and w = 1 for reacting flow

both reacting and non-reacting flow have been performed using MATLAB and

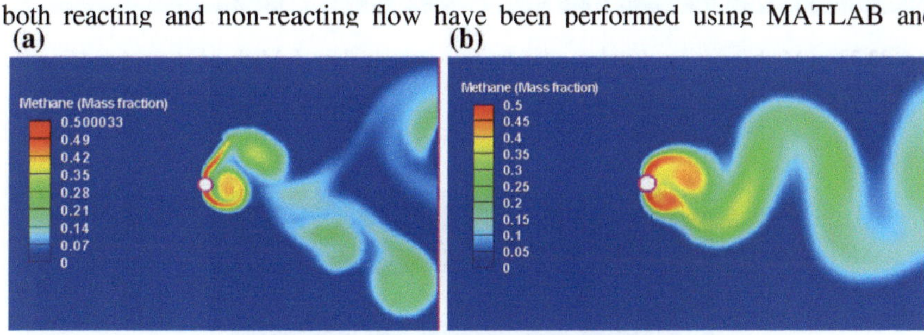

Fig. 32.31 **a** Methane at ε = 3 and w = 0.5 for non-reacting flow. **b** Methane at ε = 3 and w = 0.5 for reacting flow

compared below in Figs. 32.62a, 32.63, 32.64, 32.65, 32.66, 32.67, 32.68, 32.69, 32.70, 32.71, 32.72 and 32.73. It has been observed that for most of the cases, the amplitude of frequency of lift coefficient is substantially greater for the non-reacting flow compared to their corresponding reacting flow. This particular observation can

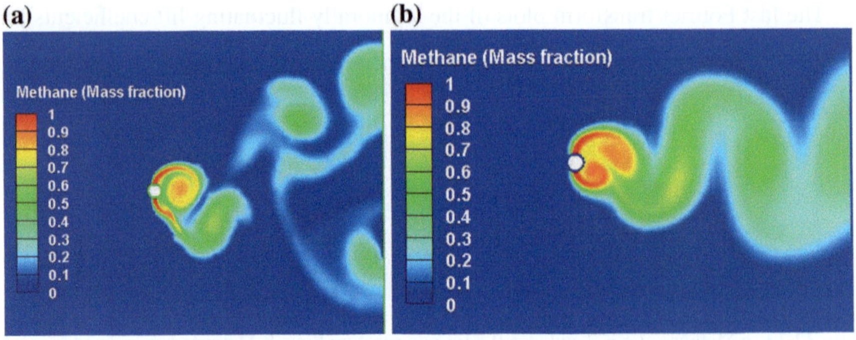

Fig. 32.32 **a** Methane at $\varepsilon = 4$ and $w = 1$ for non-reacting flow. **b** Methane at $\varepsilon = 4$ and $w = 1$ for reacting flow

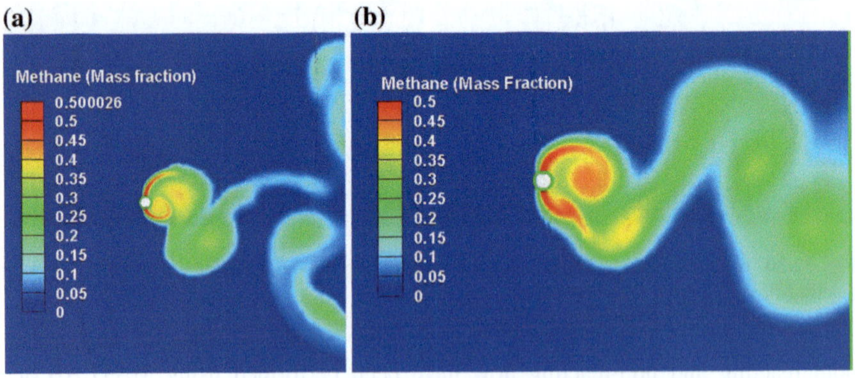

Fig. 32.33 **a** Methane at $\varepsilon = 4$ and $w = 0.5$ for non-reacting flow. **b** Methane at $\varepsilon = 4$ and $w = 0.5$ for reacting flow

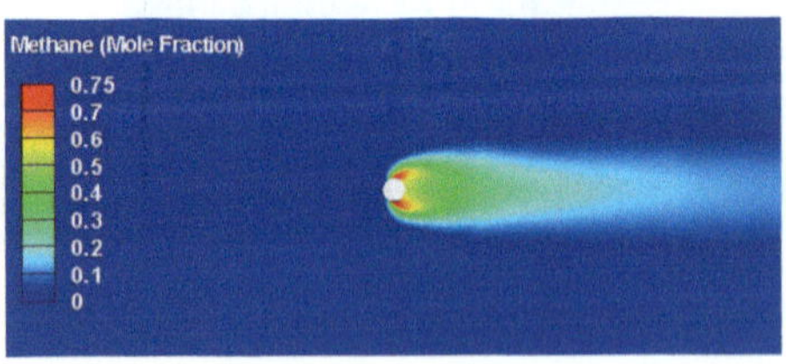

Fig. 32.34 Methane at $\varepsilon = 1$ and $w = 0.75$ for reacting flow

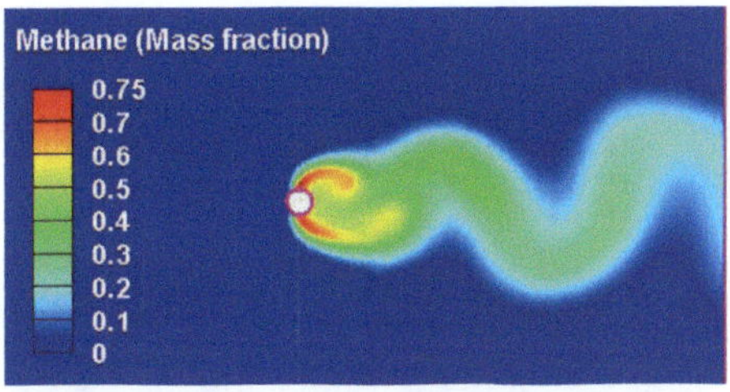

Fig. 32.35 Methane at ε = 2 and w = 0.75 for reacting flow

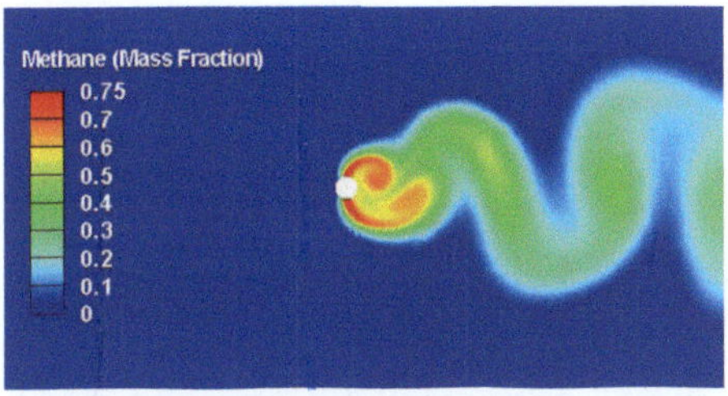

Fig. 32.36 Methane at ε = 3 and w = 0.75 for reacting flow

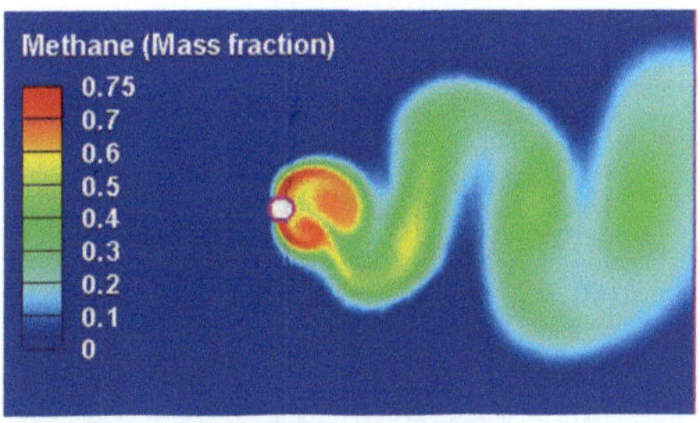

Fig. 32.37 Methane at ε = 4 and w = 0.75 for reacting flow

be due to the expansion of the gases after combustion, which causes decrease in pressure around the bluff body, which in turn minimises the lift coefficient. However, for ε = 3 and w = 0.5 a high peak has been observed for the reacting flow which is considerably higher than its corresponding non-reacting flow as shown in Fig. 32.67a, b. Similarly, for ε = 4 and w = 0.5, the frequency amplitude of

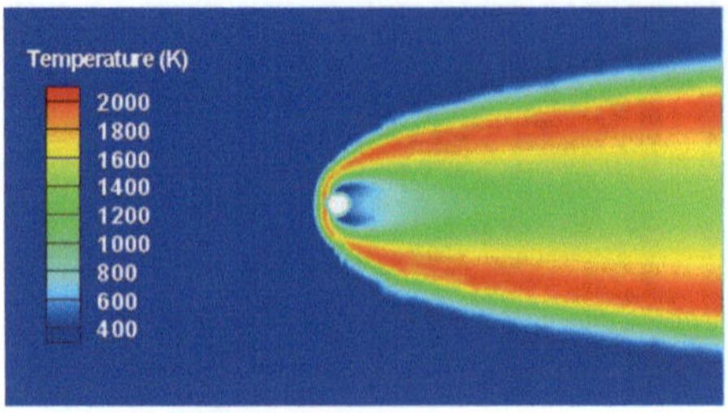

Fig. 32.38 Temperature at ε = 1 and $w = 1$

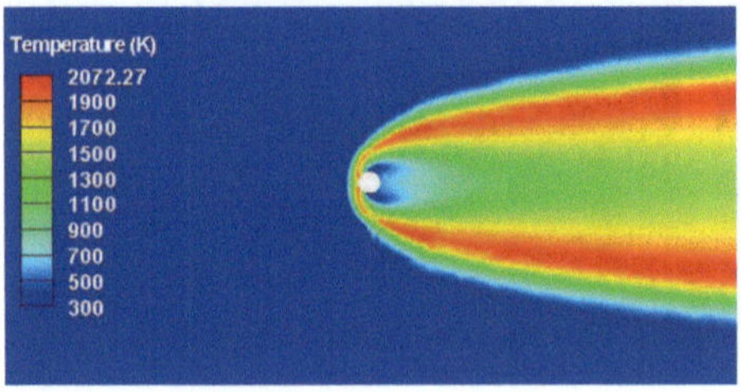

Fig. 32.39 Temperature at ε = 1 and $w = 0.75$

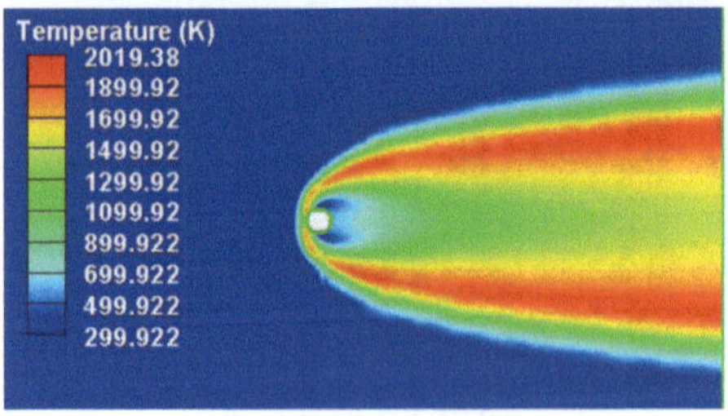

Fig. 32.40 Temperature at ε = 1 and $w = 0.5$

reacting flow is observed to be of same order as that of its corresponding non-reacting flow as shown in Fig. 32.69a, b. The frequency amplitude of majority of the cases of reacting flow is observed to be negligible. For reacting flow occurring at ε = 1 (Figs. 32.62b and 32.63b), i.e. injection velocity equal to free stream velocity, no recognisable sharp peak of the frequency of oscillating lift

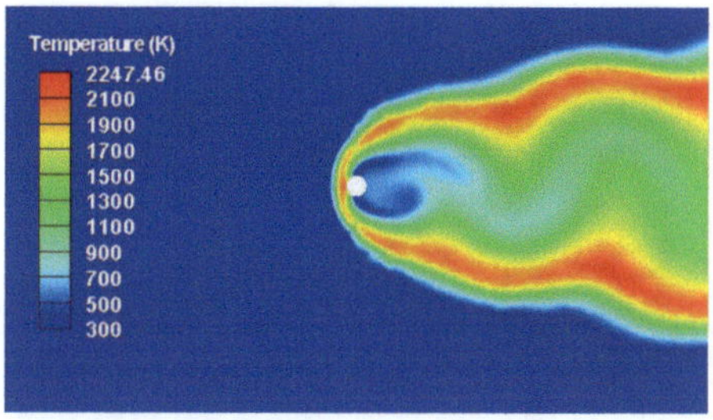

Fig. 32.41 Temperature at ε = 2 and w = 1

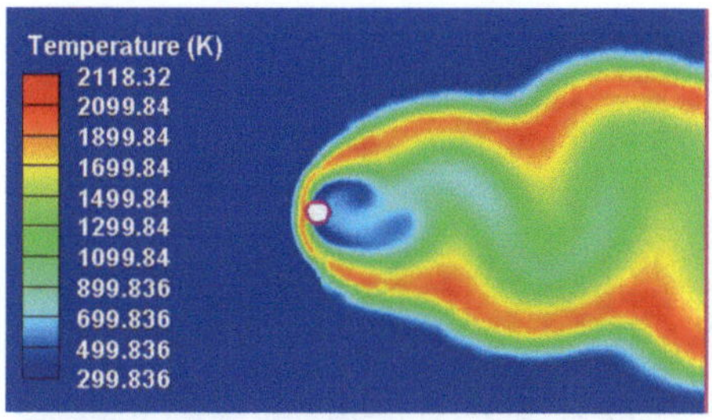

Fig. 32.42 Temperature at ε = 2 and w = 0.75

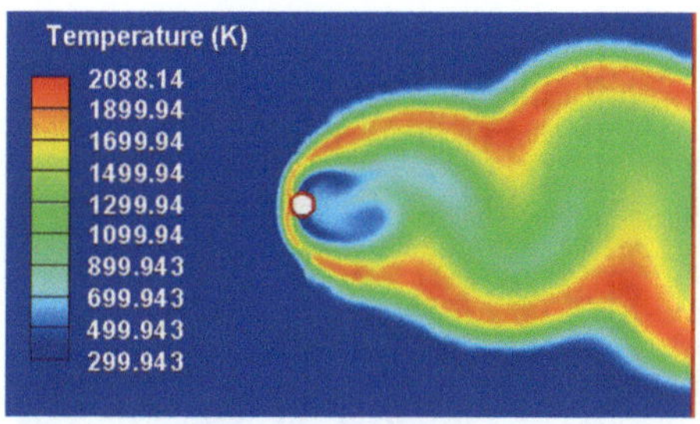

Fig. 32.43 Temperature at ε = 2 and w = 0.5

coefficient has been observed, thereby indicating negligible and highly suppressed vortex shedding phenomena.

Since for ε = 3 and w = 0.5, the peak of amplitude of the reacting flow from the fast Fourier transform plots has been observed to be substantially higher from its corresponding non-reacting flow, and the transient variations of the vorticity

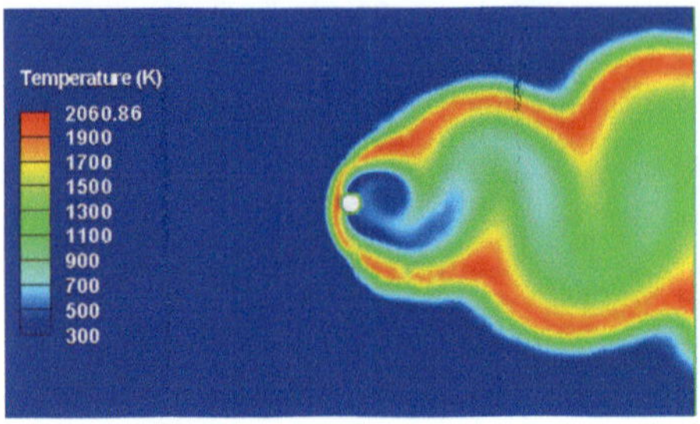

Fig. 32.44 Temperature at ε = 3 and w = 1

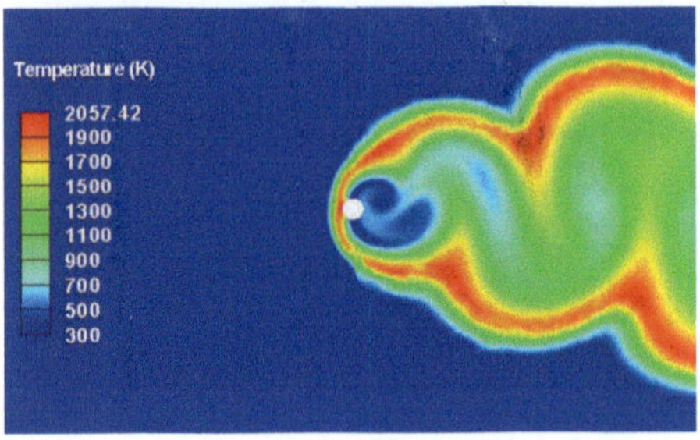

Fig. 32.45 Temperature at ε = 3 and w = 0.75

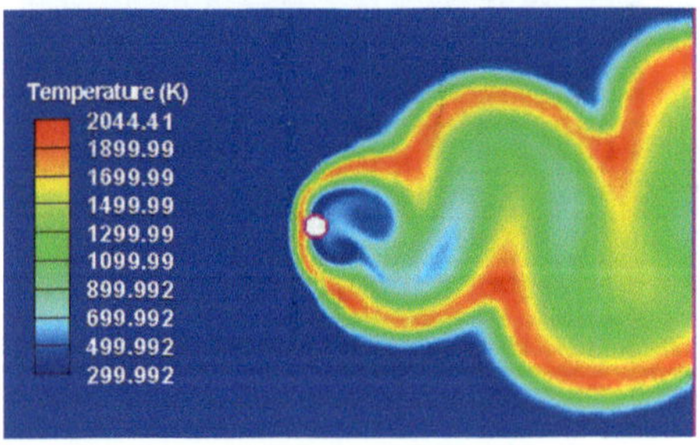

Fig. 32.46 Temperature at ε = 3 and w = 0.5

contours have been scrutinised with minute variations in time as shown in Figs. 32.74 and 32.75.

For the non-reacting flow, the vortex shedding is observed to be in 2P mode, i.e. 2 pairs of counter-rotating vortices shed per cycle, while for the reacting flow, the vortex shedding is in normal 2S mode, i.e. two single vortices coming off each cycle.

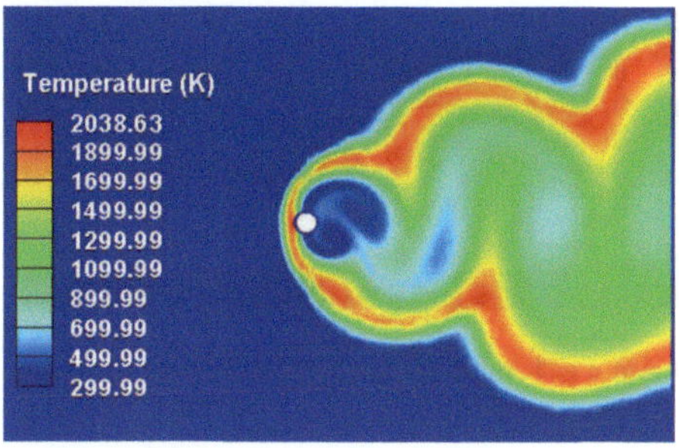

Fig. 32.47 Temperature at $\varepsilon = 4$ and $w = 1$

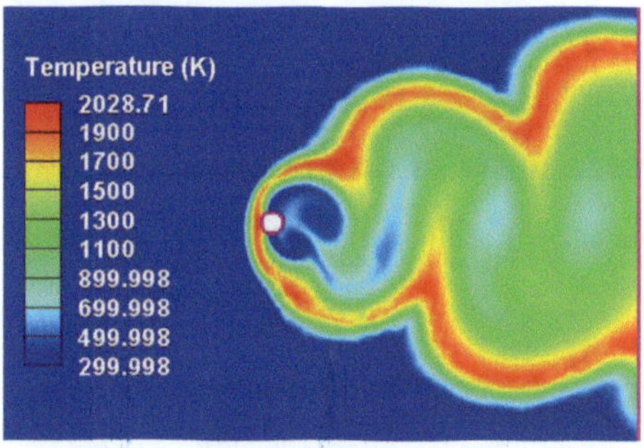

Fig. 32.48 Temperature at $\varepsilon = 4$ and $w = 0.75$

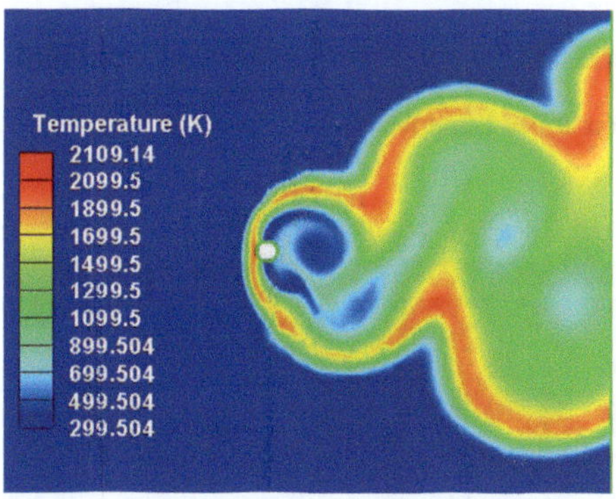

Fig. 32.49 Temperature at $\varepsilon = 4$ and $w = 0.5$

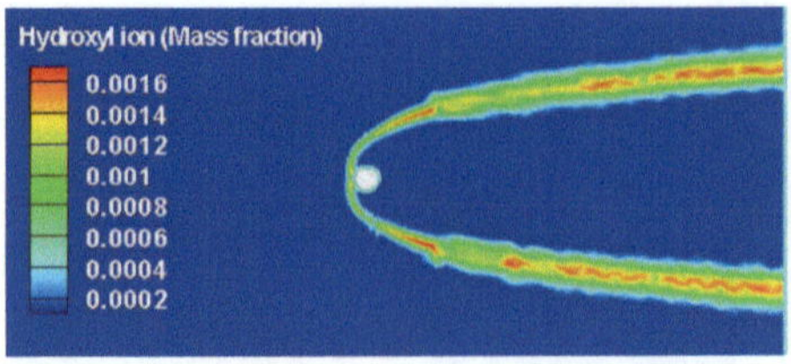

Fig. 32.50 Hydroxyl ion at ε = 1 and w = 1

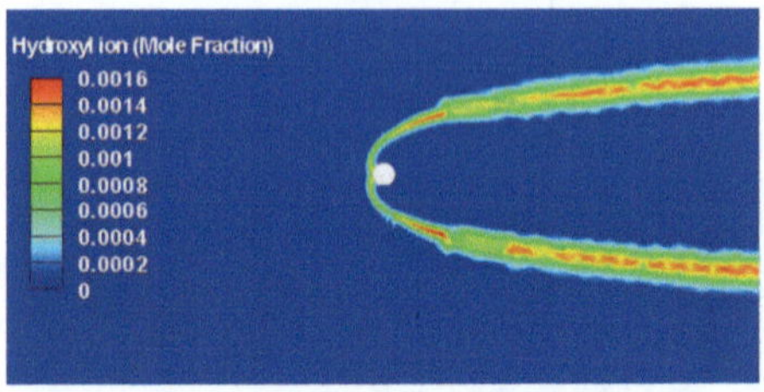

Fig. 32.51 Hydroxyl ion at ε = 1 and w = 0.75

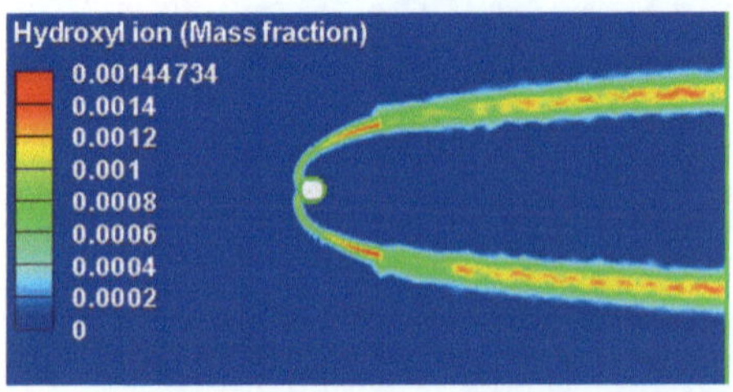

Fig. 32.52 Hydroxyl ion at ε = 1 and w = 0.5

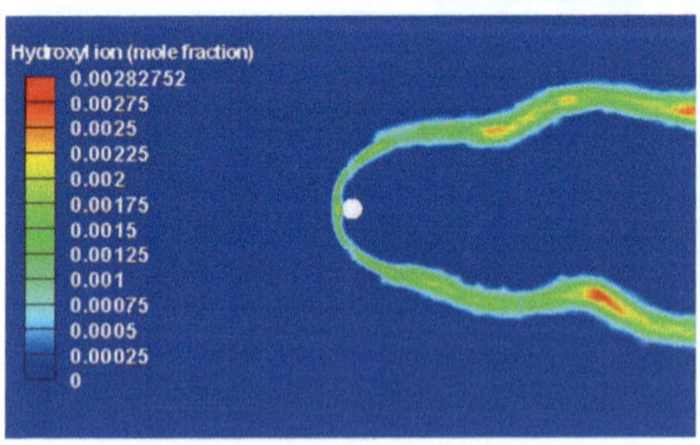

Fig. 32.53 Hydroxyl ion at ε = 2 and w = 1

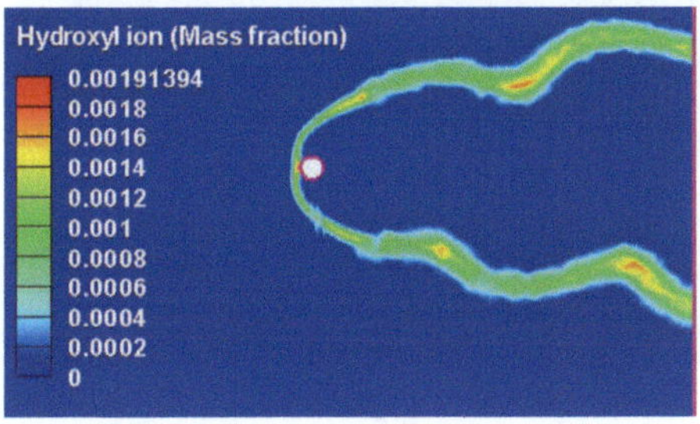

Fig. 32.54 Hydroxyl ion at ε = 2 and w = 0.75

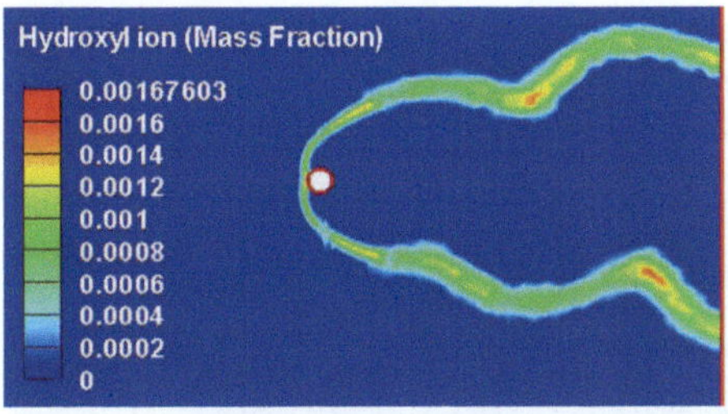

Fig. 32.55 Hydroxyl ion at ε = 2 and w = 0.5

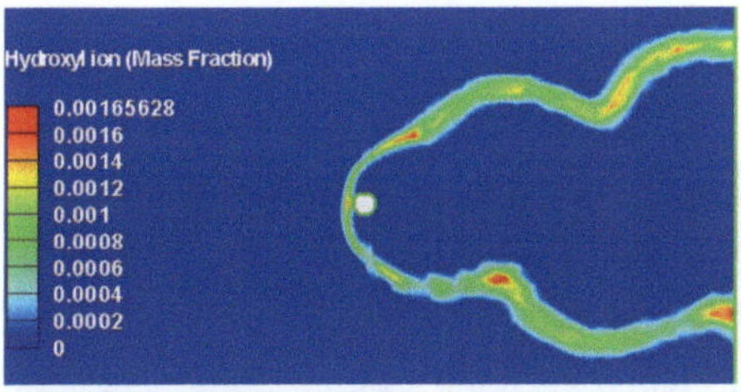

Fig. 32.56 Hydroxyl ion at ε = 3 and w = 1

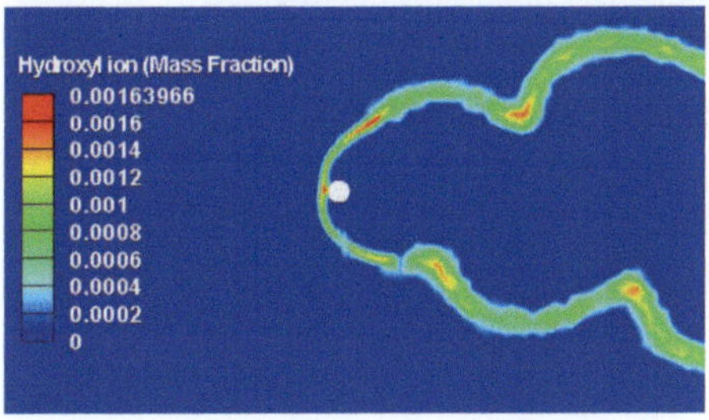

Fig. 32.57 Hydroxyl ion at ε = 3 and w = 0.75

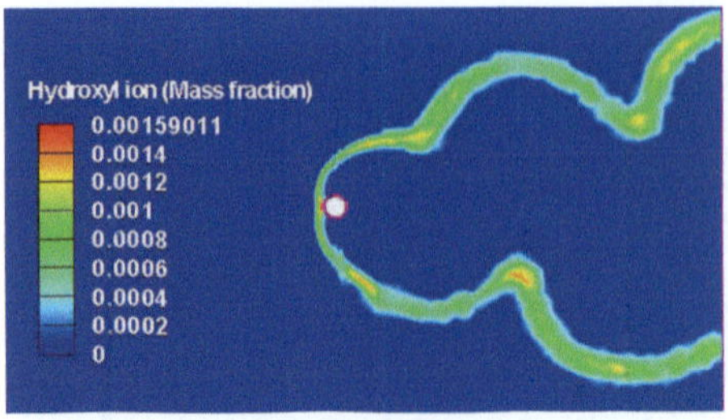

Fig. 32.58 Hydroxyl ion at ε = 3 and $w = 0.5$

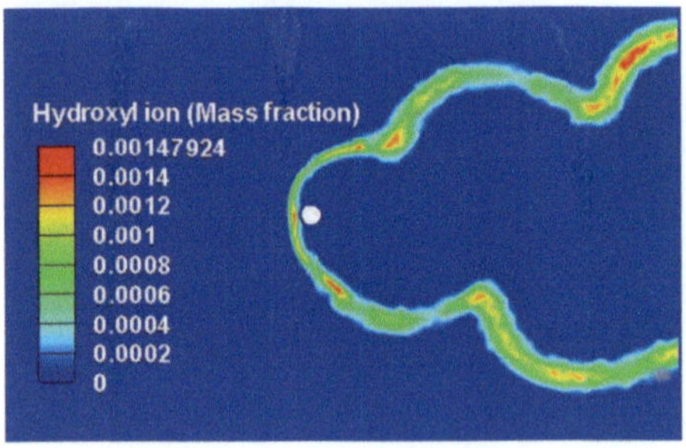

Fig. 32.59 Hydroxyl ion at ε = 4 and $w = 1$

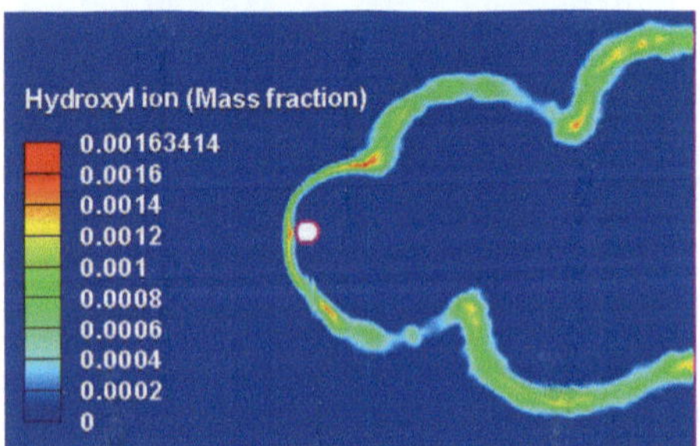

Fig. 32.60 Hydroxyl ion at ε = 4 and $w = 0.75$

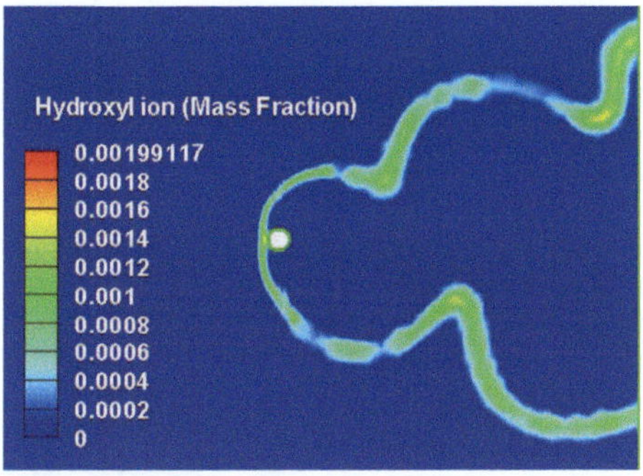

Fig. 32.61 Hydroxyl ion at ε = 4 and w = 0.5

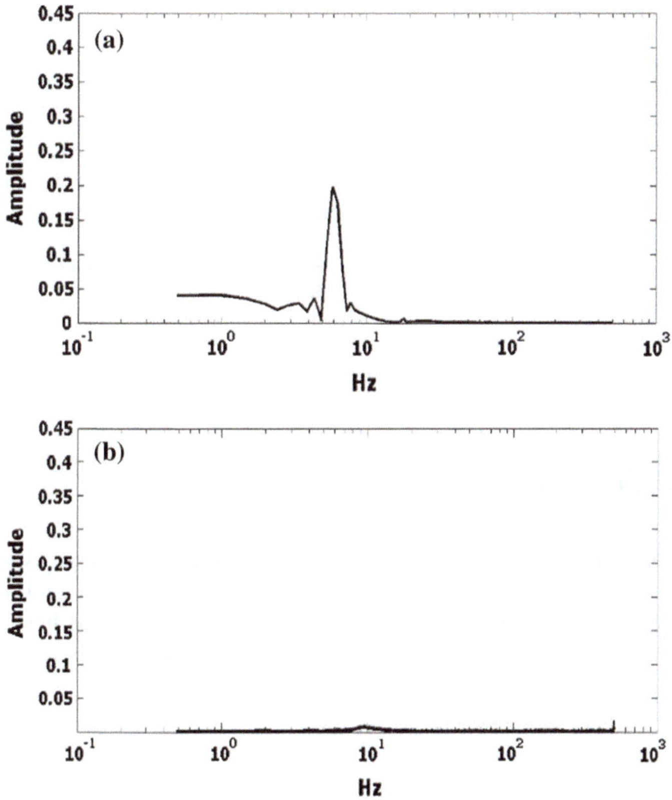

Fig. 32.62 **a** FFT plot for ε = 1 and w = 1 for non-reacting flow. **b** FFT plot at ε = 1 and w = 1 for reacting flow

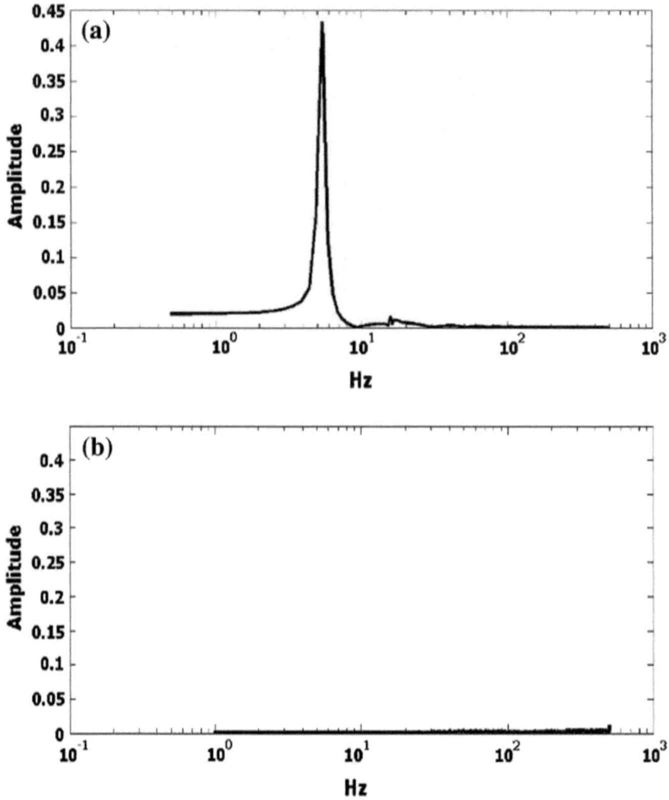

Fig. 32.63 **a** FFT plot for $\varepsilon = 1$ and $w = 0.5$ for non-reacting flow. **b** FFT plot at $\varepsilon = 1$ and $w = 0.5$ for reacting flow

32.6 Conclusion

The combustion and fluid mechanics for flow around a slotted cylinder with cross-flow methane injection have been studied in detail. Different parameters have been altered and its effect on combustion and fluid mechanics has been observed. For reacting flow, the flame is anchored right in front of the cylinder for all injection

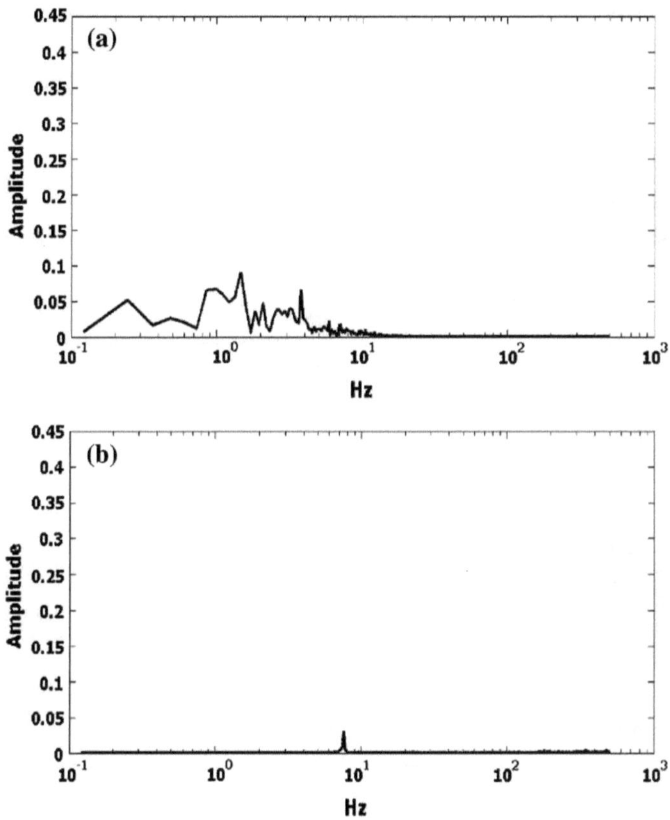

Fig. 32.64 **a** FFT plot for ε = 2 and *w* = 1 for non-reacting flow. **b** FFT plot at ε = 2 and *w* = 1 for reacting flow

velocity and methane mass fraction variations except for *w* = 0.25. The different contours of velocity, temperature, methane mass fraction, and OH^- mass fraction oscillate at the frequency of vortex shedding. The contours become more compacted in case of reacting flow with increase in injection velocity.

The swirling vortices from the Von Karman vortex street downstream of the cylinder are separated in case of the non-reacting flow due to wake instability.

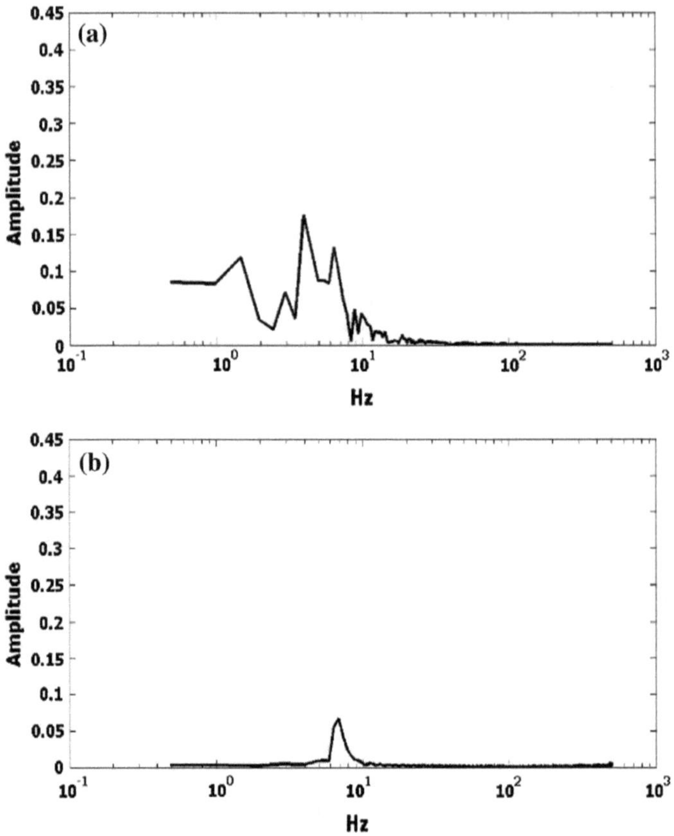

Fig. 32.65 **a** FFT plot for $\varepsilon = 2$ and $w = 0.5$ for non-reacting flow. **b** FFT plot at $\varepsilon = 2$ and $w = 0.5$ for reacting flow

There is a significantly better mixing after the combustion has taken place which is evident from the attached eddies constantly shed from either side of the bluff body.

From the power spectral density plots for reacting flow occurring at $\varepsilon = 1$, i.e. injection velocity equal to free stream velocity, no sharp peak of the frequency of oscillating lift coefficient is extant, thereby indicating negligible and highly

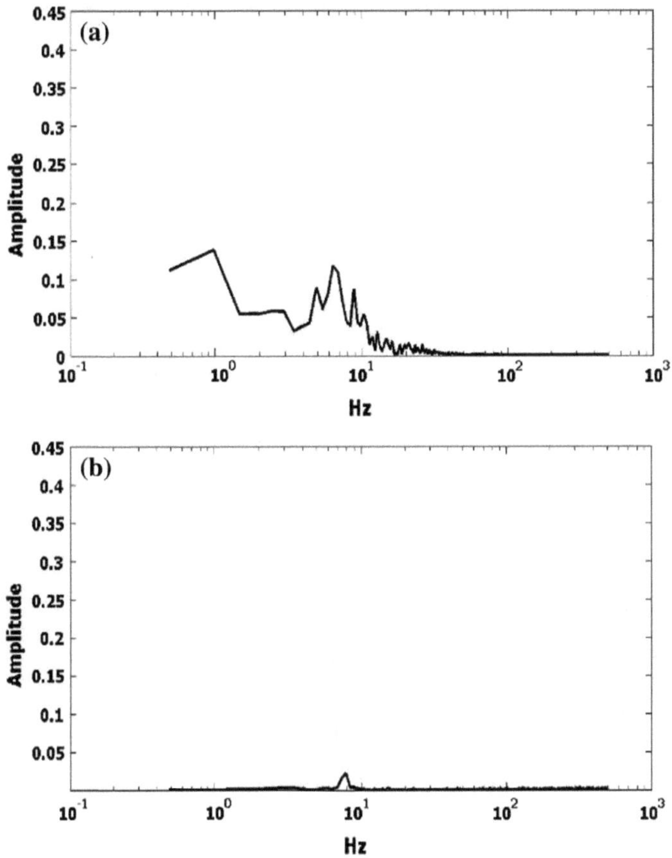

Fig. 32.66 **a** FFT plot for ε = 3 and w = 1 for non-reacting flow. **b** FFT plot at ε = 3 and w = 1 for reacting flow

suppressed vortex shedding phenomena. For non-reacting flow, there is sharp peak of amplitude of lift coefficient at ε = 1; however with increase in injection velocity, there are multiple peaks with no unique dominant frequency of vortex shedding. There is a change in vortex shedding phenomena in case of ε = 3 and w = 0.5 from

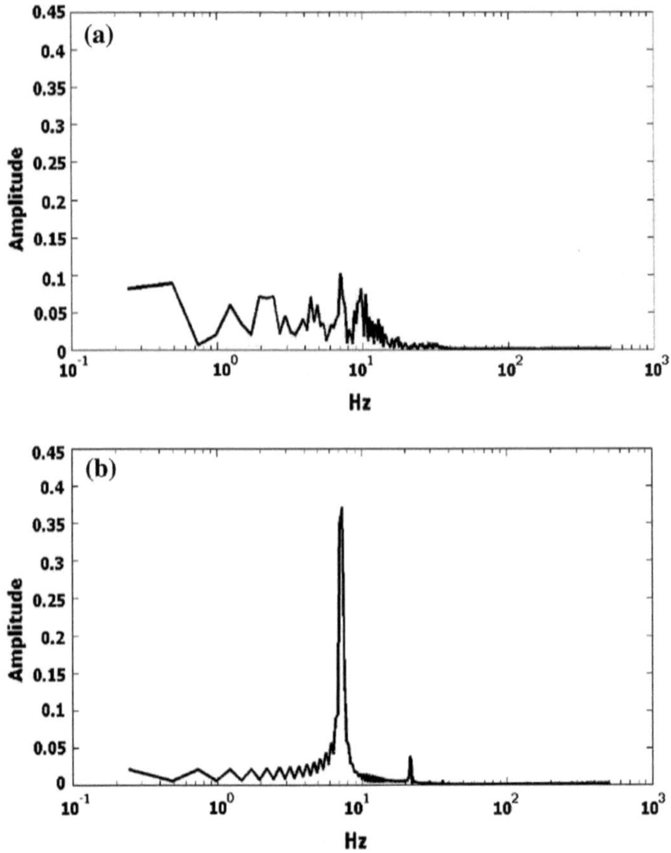

Fig. 32.67 **a** FFT plot for $\varepsilon = 3$ and $w = 0.5$ for non-reacting flow. **b** FFT plot at $\varepsilon = 3$ and $w = 0.5$ for reacting flow

non-reacting to reacting flow, as the 2P mode of vortex shedding transforms to 2S mode. Moreover, there is marked increase in the amplitude of lift coefficient. For the reacting flow, there is a decline in amplitude with increase in methane mass fraction from $w = 0.5$ to $w = 1$.

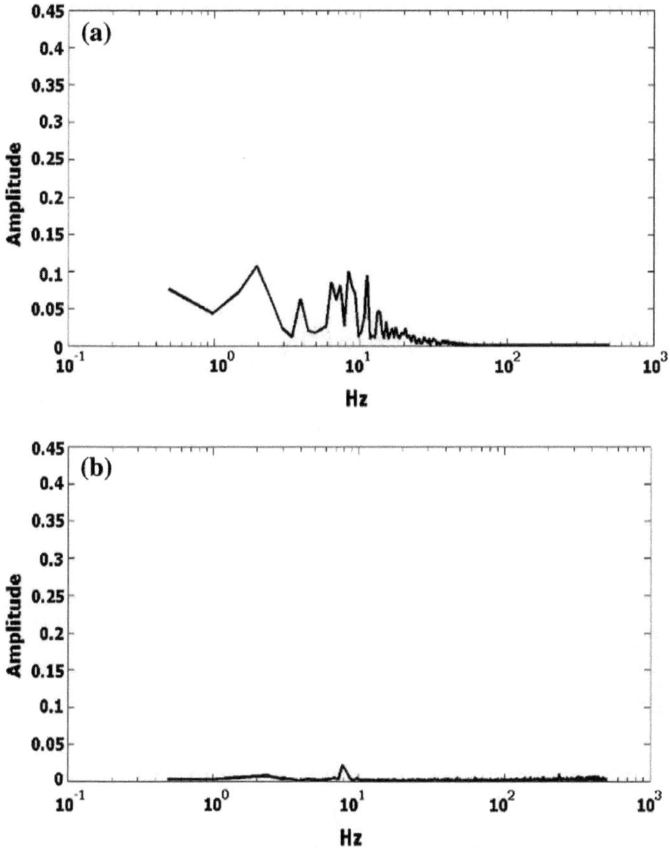

Fig. 32.68 **a** FFT plot for ε = 4 and *w* = 1 for non-reacting flow. **b** FFT plot at ε = 4 and *w* = 1 for reacting flow

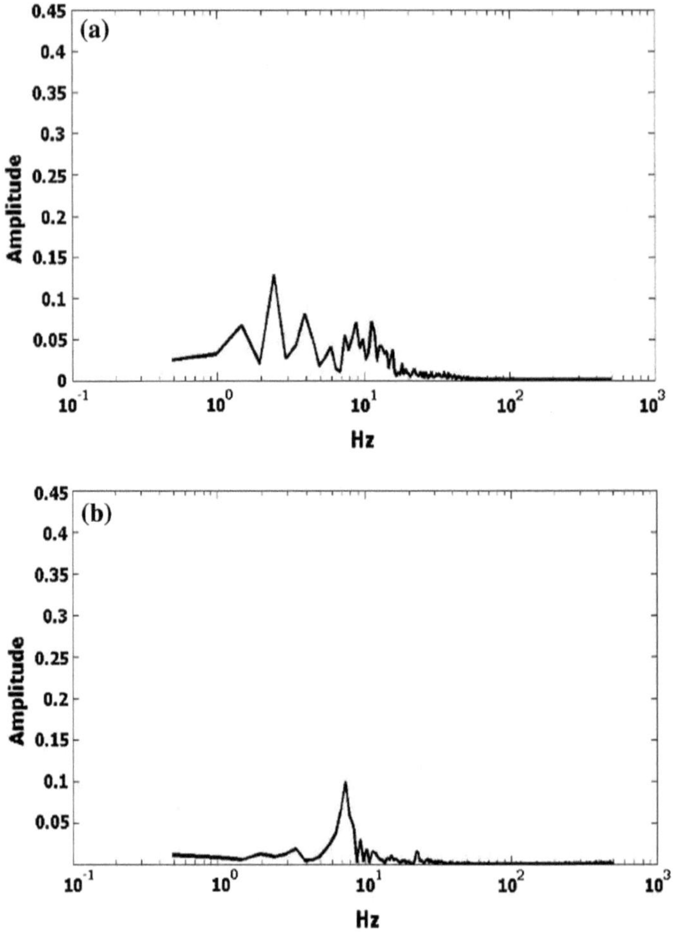

Fig. 32.69 **a** FFT plot for $\varepsilon = 4$ and $w = 0.5$ for non-reacting flow. **b** FFT plot at $\varepsilon = 4$ and $w = 0.5$ for reacting flow

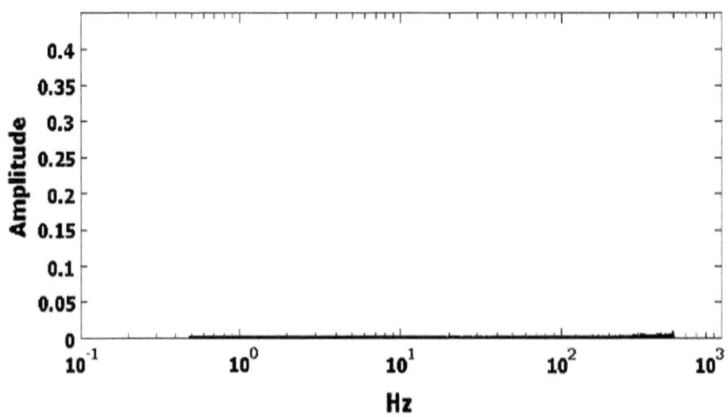

Fig. 32.70 FFT plot for $\varepsilon = 1$ and $w = 0.75$ for reacting flow

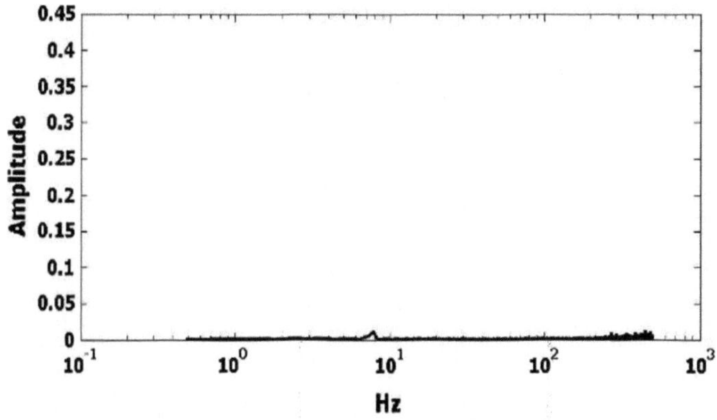

Fig. 32.71 FFT plot at ε = 2 and w = 0.75 for reacting flow

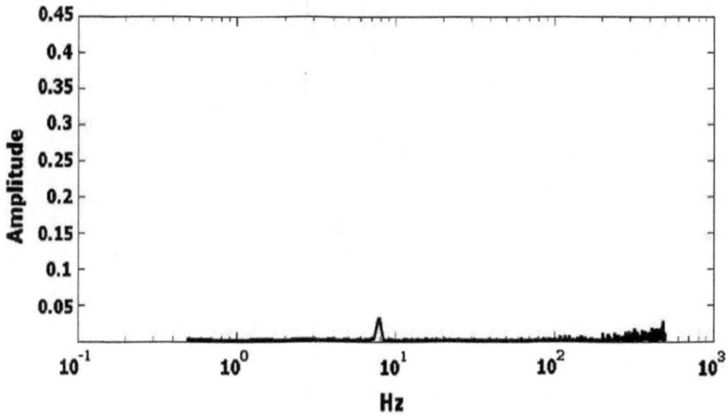

Fig. 32.72 FFT plot for ε = 3 and w = 0.75 for reacting flow

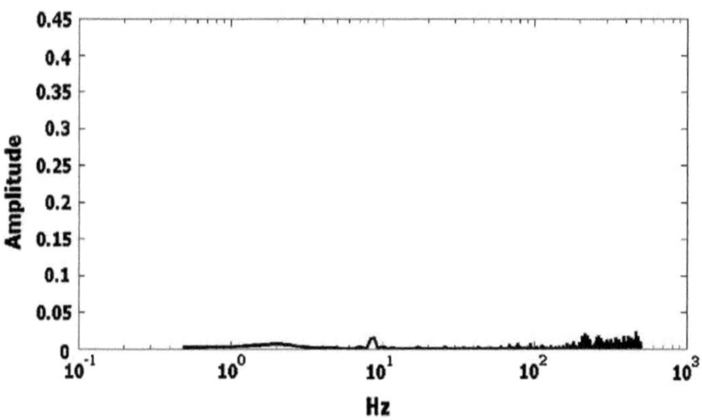

Fig. 32.73 FFT plot at ε = 4 and w = 0.75 for reacting flow

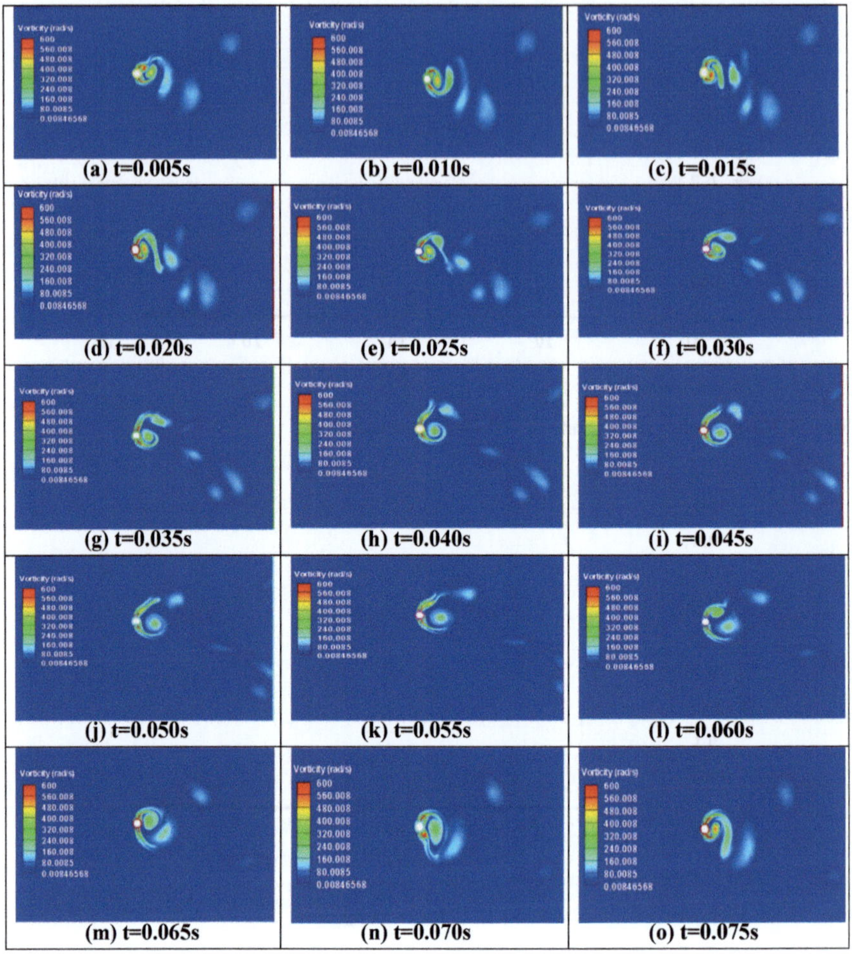

Fig. 32.74 Transient variations of vorticity contours for non-reacting flow at $\varepsilon = 3$ and $w = 0.5$

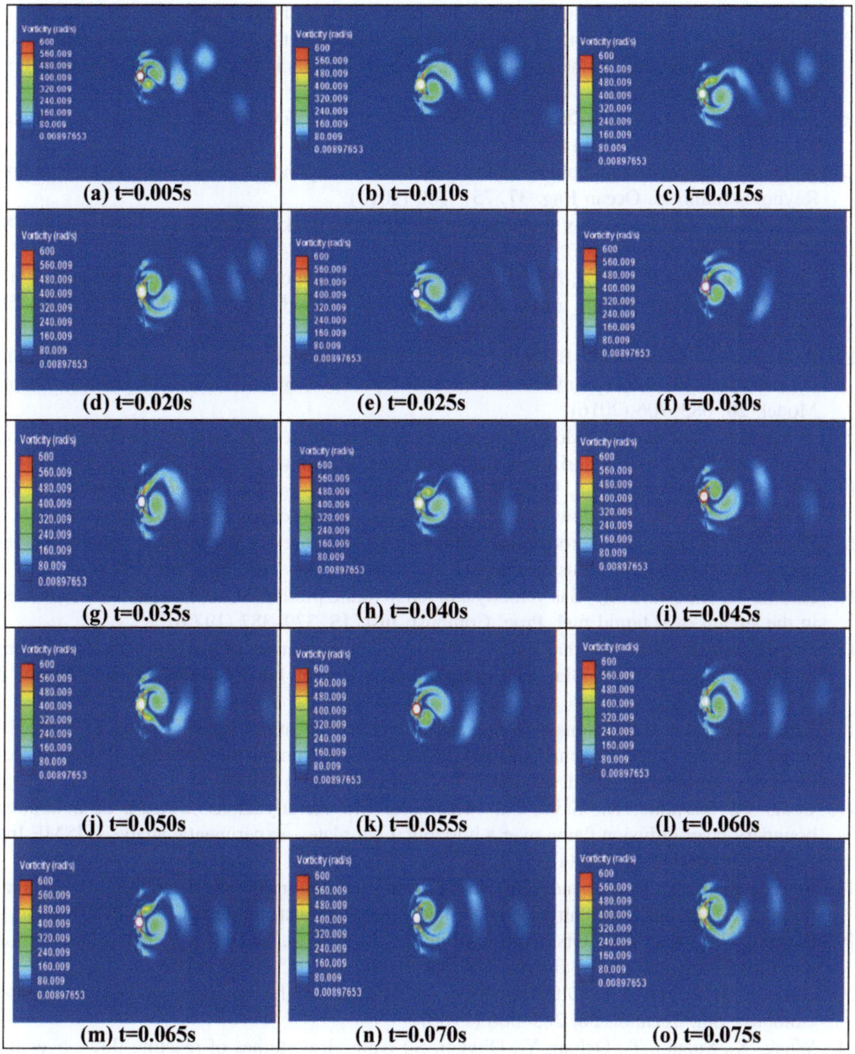

Fig. 32.75 Transient variations of vorticity contours for reacting flow at $\varepsilon = 3$ and $w = 0.5$

References

1. Renard, P.H., Thévenin, D., Rolon, J.C.: Dynamics of flame/vortex interactions. Prog. Energy Combust. Sci. **26**(3), 225–282 (2000)
2. Candel, S.: Combustion dynamics and control: progress and challenges. Proc. Combust. Inst. **29**, 1–28 (2002)
3. Delhaye, B., Veynante, D., Candel, S.M., Minh, H.H.: Simulation and modeling of reactive shear layers. Theor. Comput. Fluid Dynam. **6**, 67–87 (1994)

4. Jeong, J., Hussain, F.: On the identification of a vortex. J. Fluid Mech. **285**, 69–94 (1995)
5. Poszdiech, O., Grundmann, R.: A systematic approach to the numerical calculation of fundamental quantities of the two-dimensional flow over a circular cylinder. J. Fluids Struct. **23**, 479–499 (2007)
6. Wang, H.F., Cao, H.L., Zhou, Y.: POD analysis of a finite-length cylinder near wake. Exp. Fluids **55**, 1790 (2014)
7. Singha, S., Singhamahapatra, K.P.: Flow past a circular cylinder between parallel walls at low Reynolds numbers. Ocean Eng. **37**, 757–769 (2010)
8. Williamson, C.H.K.: Vortex dynamics in the cylinder wake. Annu. Rev. Fluid Mech. **28**, 477–539 (1996)
9. Zhu, M.M., Zhao, P.H., Ge, H.W., Chen, Y.L.: Simulation of vortex shedding behind a bluff body flame stabilizer using a hybrid U-RANS/PDF method. Acta. Mech. Sin. **28**, 348–358 (2012)
10. Raghavan, V., Shijin, P.K., Babu, V.: Numerical investigation of flame–vortex interactions in laminar cross-flow non-premixed flames in the presence of bluff bodies. Combust. Theor. Model. **20**, 683–706 (2016)
11. Torero, J.L., Bonneau, L., Most, J.M., Joulain, P.: The effect of gravity on a laminar diffusion flame established over a horizontal flat plate. Proc. Combust. Inst. **25**, 1701–1709 (1994)
12. Ha, J.S., Shim, S.H., Shin, H.D.: Boundary layer diffusion flame over a flat plate in the presence and absence of flow separation. Combust. Sci. Technol. **75**, 241–260 (1991)
13. Hirano, T., Kanno, Y.: Aerodynamic and thermal structures of the laminar boundary layer over a flat plate with a diffusion flame. Proc. Combust. Inst. **14**, 391–398 (1973)
14. Hirano, T., Kinoshita, M.: Gas velocity and temperature profiles of a diffusion flame stabilized in the stream over liquid fuel. Proc. Combust. Inst. **15**, 379–387 (1975)
15. Gopalakrishnan, E.D., Raghavan, V.: Numerical investigation of laminar diffusion flames established on a horizontal flat plate in a parallel air stream. Int. J. Spray Combust. Dynam. **3** (2), 161–190 (2011)
16. Peters, N.: Flame calculations with reduced mechanisms—an outline. In: Peters, N., Rogg, B. (eds.) Reduced Kinetic Mechanisms for Applications in Combustion Systems. Lecture Notes in Physics Monographs, vol. 15, pp. 3–14. Springer, Berlin (1993)
17. Ueda, T., Ooshima, A., Saito, N., Mizomoto, M.: Aerodynamic structure of a laminar boundary layer diffusion flame over a horizontal flat plate—experimental analysis. JSME Int. J. Ser. **234-II**(4), 527–532 (1991)
18. Ramachandra, A., Raghunandan, B.N.: Buoyancy effects on the characteristics of a laminar boundary layer diffusion flame in a confined flow. Combust. Flame **58**, 191–196 (1984)
19. Rohmat, T.A., Katoh, H., Obara, T., Yoshihashi, T., Ohyagi, S.: Diffusion flame stabilized on a porous plate in a parallel airstream. AIAA J. **36**(11), 1945–1952 (1998)
20. Chen, C.-H., T'ien, J.S.: Diffusion flame stabilization at the leading edge of a fuel plate. Combust. Sci. Technol. **50**, 283–306 (1986)
21. Wang, X., Suzuki, T., Ochiai, Y., Ohyagi, S.: Numerical studies of reacting flows over flat walls with fuel injection: Part 1—velocity anomaly and hydrodynamic instability. JSME Int J. Ser. B **41**(1), 19–27 (1998)
22. Shijin, P.K., Sundaram, S.S., Raghavan, V.: Numerical investigation of laminar cross-flow non-premixed flames in the presence of a bluff-body. Combust. Theor. Model. **18**(6), 692–710 (2014)
23. Lavid, M., Berlad, A.L.: Gravitational effects on chemically reacting laminar boundary layer flows over a horizontal flat plate. Proc. Combust. Inst. **16**, 1557–1568 (1976)
24. Raghunandan, B.N., Yogesh, G.P.: Recirculating flow over a burning surface—flame structure and heat transfer augmentation. Proc. Combust. Inst. **22**, 1501–1507 (1988)
25. Marble, F.E.: Growth of a diffusion flame in the field of a vortex. Recent advances in the aerospace sciences. In: Casci, C., Bruno, C. (eds.) Honor of Luigi Crocco on His 75th Birthday, Springer, Boston, MA, pp. 395–413 (1985)
26. Karagozian, A.R., Marble, F.E.: Study of a diffusion flame in a stretched vortex. Combust. Sci. Technol. **45**, 65–84 (1986)

27. Alain, M., Candel S.M.: A numerical analysis of a diffusion flame–vortex interaction. Combust. Sci. Technol. **60**, 79–96 (1988)
28. Ashurst, W.T., Mcmurtry, P.A.: Flame generation of vorticity: vortex dipoles from monopoles. Combust. Sci. Technol. **66**, 17–37 (1989)
29. Lewis, G.S., Cantwell, B.J., Vandsburger, U., Bowman, C.T.: An investigation of the structure of a laminar non-premixed flame in an unsteady vortical flow. Symp. (Int.) Combust. **22**, 515–522 (1989)
30. Macaraeg, M.G., Jackson, T.L., Hussaini, M.Y.: Ignition and structure of a laminar diffusion flame in the field of a vortex. Combust. Sci. Technol. **87**, 363–387 (1993)
31. Rolon, J.C., Aguerre, F., Candel, S.: Experiments on the interaction between a vortex and a strained diffusion flame. Combust. Flame **100**, 422–442 (1995)
32. Fan, A., Wan, J., Maruta, K., Yao, H., Liu, W.: Interactions between heat transfer, flow field and flame stabilization in a micro-combustor with a bluff body. Int. J. Heat Mass Trans. **66**, 72–79 (2013)
33. Altay, H.M., Speth, R.L., Hudgins, D.E., Ghoniem, A.F.: Flame–vortex interaction driven combustion dynamics in a backward-facing step combustor. Combust. Flame **156**(5), 1111–1125 (2009)
34. Barlow, R.S., Karpetis, A., Frank, J.H., Chen, J.Y.: Scalar profiles and NO formation in laminar opposed-flow partially premixed methane/air flames. Combust. Flame **127**, 2102–2118 (2001)
35. Mondal, S., Mukhopadhyay, A., Sen, S.: Dynamic characterization of a laboratory scale pulse combustor. Combust. Sci. Tech. **186**(2), 139–152 (2014)
36. Sen, U., Mukhopadhyay, A., Sen, S.: Effects of fluid injection on dynamics of flow past a circular cylinder. Eur. J. Mech. B/Fluids **61**, 187–199 (2017)
37. Fujii, S., Eguchi, K.: A Comparison of non-reacting and reacting flows around a bluff-body flame stabilizer. J. Fluids Eng. **103**(2), 328–334 (1981)
38. Bagchi, S., Sarkar, S., Sen, U., Mukhopadhyay, A., Sen, S.: Numerical investigation of vortex shedding from a bluff body stabilised flame with cross injection. Proc. Sustain. Energy Environ. Challenges **23**, 250–254 (2018)
39. Bagchi, S., Sarkar, S., Mukhopadhyay, A., Sen, S.: Numerical simulation of vortex shedding from a cylindrical bluff body flame stabilizer. In: Proceedings of International Conference of Mechanical Engineering, pp. 678–682 (2018)
40. Bharadwaj, N., Safta, C., Madnia, C.K.: Flame-wall interaction for a non-premixed flame propelled by a vortex ring. Combust. Theor. Model. **11**(1), 1–19 (2007)
41. Safta, C., Madnia, C.K.: Characteristics of methane diffusion flame in a reacting vortex ring. Combust. Theor. Model. **8**(3), 449–474 (2004)
42. Renard, P.H., Rolon, J.C., Thévenin, D., Candel, S.: Investigations of heat release, extinction, and time evaluation of the flame surface, for a non-premixed flame interacting with a vortex. Combust. Flame **117**, 189–205 (1999)
43. Hermanns, M., Vera, M., Liñán, A.: On the dynamics of flame edges in diffusion-flame/vortex interactions. Combust. Flame **149**, 32–48 (2007)
44. Mishra, S., Santhosh, R., Basu, S.: Scalar transport in diffusion flame wrapped up by an air and fuel side vortex. Int. Comm. Heat Mass Trans. **47**, 32–40 (2013)
45. Cetegen, B.M., Mohamad, N.: Experiments on liquid mixing and reaction in a vortex. J. Fluid Mech. **249**, 391–414 (1993)
46. Thiesset, F., Maurice, G., Halter, F., Mazellier, N., Chauveau, C., Gökalp, I.: Flame-vortex interaction: effect of residence time and formulation of a new efficiency function. Proc. Combust. Inst. **36**(2), 1843–1851 (2017)
47. Stöhr, M., Boxx, I., Carter, C.D., Meier, W.: Experimental study of vortex-flame interaction in a gas turbine model combustor. Combust. Flame **159**(8), 2636–2649 (2012)
48. Smooke, M.D., Puri, I.K., Seshadri, K.: A comparison between numerical calculations and experimental measurements of the structure of a counter flow diffusion flame burning diluted methane in diluted air. In: 21st Symposium (International) on Combustion. The Combustion Institute, pp. 1783–1792 (1986)

49. Braza, M., Chassaing, P., Ha, M.H.: Numerical study and physical analysis of the pressure and velocity fields in the near wake of a circular cylinder. J. Fluid Mech. **165**, 79–130 (1986)
50. Liu, C., Zheng, X., Sung, C.H.: Preconditioned multigrid methods for unsteady incompressible flows. J. Comput. Phys. **139**, 35–57 (1998)
51. Park, J., Kwon, K., Choi, H.: Numerical solutions of flow past a circular cylinder at Reynolds number up to 160. KSME Int. J. **12**(6), 1200–1205 (1998)
52. Meneghini, J.R., Saltara, F., Siqueira, C.L.R., Ferrari Jr., J.A.: Numerical simulation off low interference between two circular cylinders in tandem and side-by-side arrangements. J. Fluids Struct. **15**, 327–350 (2001)
53. Shi, J.-M., Gerlach, D., Breuer, M., Biswas, G., Durst, F.: Heating effect on steady and unsteady horizontal laminar flow of air past a circular cylinder. Phys. Fluids **16**(12), 4331–4345 (2004)
54. Ding, H., Shu, C., Yeo, K.S., Xu, D.: Numerical simulation of flows around two circular cylinders by mesh-free least square-based finite difference methods. Int. J. Numer. Methods Fluids **53**, 305–332 (2007)
55. Mittal, S.: Instability of the separated shear layer in flow past a cylinder: forced excitation. Int. J. Numer. Methods Fluids **56**, 687–702 (2008)
56. Rajani, B.N., Kandasamy, A., Majumdar, S.: Numerical simulation of laminar flow past a circular cylinder. Appl. Math. Model. **33**, 1228–1247 (2009)
57. Li, Y., Zhang, R., Shock, R., Chen, H.: Prediction of vortex shedding from a circular cylinder using a volumetric Lattice-Boltzmann boundary approach. Eur. Phys. J. **171**, 91–97 (2009)
58. Harichandan, A.B., Roy, A.: Numerical investigation of low Reynolds number flow past two and three circular cylinders using unstructured grid CFR scheme. Int. J. Heat Fluid Flow **31**, 154–171 (2010)

Chapter 33
Entrepreneurial Culture-Driven Improvement of Technical Facets for Product Quality and Customer Satisfaction

Sudeshna Roy, Nipu Modak and Pranab K. Dan

33.1 Introduction

Globalization leads the firms towards an international competitive environment. This environment causes them to adopt new and innovative strategies for industrial sustainability [1]. New product development (NPD) is one of those activities essential for developing innovative products to achieve success in global competition. The successful development of new product is one of those vital strategies which are critically challenging for management [2]. Development of new products as per customer requirements within the estimated budget can provide competitive advantage to the company. Researchers over the past few decades identified the factors which are critical to success for NPD which are famously known as critical success factors [3]. Previous studies recognized technology [4–6], research and development [7, 8], market analysis [9, 10], managerial support [11, 12] as widely used success factors for NPD success. The interactions of these success factors are equally important to develop a comprehensive idea about their impacts on NPD activities. Sadeghi et al. [13], Akgün et al. [14] and Sun and Wing [15] have highlighted several common success factors which are critical for NPD success.

Innovation and entrepreneurial culture (EC) have become an inseparable entity for industrial growth in a highly competitive market scenario [16]. Entrepreneurial culture of a firm is the propensity to welcome innovative and novel ideas to develop new products. It is the most promising factor of the firms to achieve success in the competitive market environment. Risk-taking tendency of the firm helps to accept

S. Roy (✉) · N. Modak
Mechanical Engineering Department, Jadavpur University, Kolkata, India
e-mail: sudeshnaroy689@gmail.com

P. K. Dan
Rajendra Mishra School of Engineering Entrepreneurship, IIT Kharagpur, Khragpur, India

P. Sahoo and J. P. Davim (eds.), *Advances in Materials, Mechanical and Industrial Engineering*, Lecture Notes on Multidisciplinary Industrial Engineering,
https://doi.org/10.1007/978-3-319-96968-8_33

the unidentified substances with proper strategy and tact [17]. Entrepreneurial culture enhances knowledge-based capital, such as human capital, social capital and organizational capital, and helps in improving the NPD activities. This culture is also related to organizational factors (OFs), system integration (SI) and financial resources (FRs) which are considered as crucial variables to control NPD activities. Organizational factors can be defined by the determinants like organizational structure, organizational culture, firm life cycle, size of organization, up-to-datedness [13]. Entrepreneurial culture of the firm encourages these organizational factors and reconstructs them as required for the sake of NPD. Similarly, entrepreneurial culture offers accumulation of resources of various functional groups for efficient development of innovative products. This incorporation can be addressed as system integration. This integration refers to assimilation of ideas, knowledge, internal and external capabilities [18]. Product advantage is inseparably associated with product meaningfulness along with innovativeness and superiority of products. Availability of resources ensures the attainment of quality products with competitive advantage [19]. Financial resources are one of the essential components which must be specially treated for NPD success. Entrepreneurial culture promotes optimal usage of financial resources for NPD success. The constant change in the market changes the technology itself. Escalation of entrepreneurial culture encourages organizational factors, system integration and financial resources which also offer technical upgradation of the firm. Technical improvement has become necessary to sustain in the global competition. Understanding of new leading-edge technology is highly recommended for attaining technical improvements and making them feasible for the implementation [5]. The ultimate aim of the firm is achieving higher profit through successful NPD. The rapid changes in customers' preferences make volatile business environment which reduces the chance of success [20]. Development of quality products escalates operational effectiveness of a firm's NPD activities that can persuade the customers' expectations [21]. A comprehensive framework considering entrepreneurial culture along with the entrepreneurial culture-driven associated variables for better NPD performance practically remains untouched. The drivers and their antecedents for both input and output variables are rarely been addressed in a single model. The scarcity of higher-order model depicting the intermediate factor controlling the NPD success is clearly identified. Considering these serious research gaps, this study enquires a few vital questions which need to be answered. These research questions are: (1) what is the role of entrepreneurial culture for technical improvements to attain NPD success? (2) What are the intermediate factors driven by entrepreneurial culture to enhance technical improvements and how they impacted on? (3) What are the antecedents of entrepreneurial culture and related constructs including technical improvements? (4) What are the key performance outcomes that measure NPD success? The responses of these questions able to draw significant inferences which help the organization to understand the necessity of these constructs and make them developing required support system and providing suitable environment to implement these practices.

The objective of this study is to develop a higher-order model to realize the impact of entrepreneurial culture on technical improvements cascaded through intermediate variables such as organizational factors, system integration and financial resources to achieve NPD success. The contribution of this research, at first, is the identification of entrepreneurial culture-driven factors as well as their antecedents to draw their impact on technical improvements. These antecedents of entrepreneurial culture and its associated factors are collated from the available literatures and also from primary responses obtained from manufacturing experts in the form of opinion and views. Second, recognition of the role of technical improvements is critically linked with entrepreneurial culture for NPD success. The manifests of technical improvements are also been identified. The third contribution is in associating the product quality and customer satisfaction as performance outcome to quantify the NPD success and thereby portraying cascading interaction mechanism of entrepreneurial culture to NPS success. Finally, this empirical research draws responses from small- and medium-sized enterprises (SMEs) of Indian manufacturing industries for analysing the developed linkage framework which contributes an additional novelty.

This paper is clustered as follows: the very next section is theoretical background; it provides a brief description of the background to develop this conceptual framework of the entrepreneurial culture and associated constructs along with the technical improvements and success measures such as product quality and customer satisfaction. It also incorporates the path model showing the linkage of constructs to achieve NPD success. Research methodology, the third section, includes research objectives, methodology of data analysis, data collection methods and profiles of Indian manufacturing companies involved in this empirical research. Results and discussions section comprises of analysis of measurement validity, results and discussion of measurement model, results and discussion of structural model for hypotheses testing and formation of SEM model using IBM SPSS Amos 21.0 software packages. Lastly, conclusion section explains theoretical contribution of this study along with the managerial implications depicting their opinion accompanied by the limitations of this research.

33.2 Theoretical Background

Attainment of industrial success requires technological expansion for achieving manufacturing advancement to provide world-class system [22]. This advancement can be achieved not only through modification of existing products but it requires the introduction of new products to fulfil customer demands. Identification of factors critical for NPD success must be identified to emphasize their implementation in practical field to make the success feasible. Among various success factors, entrepreneurial culture is one of the most discussed variants which have become mostly essential to promote new ideas inevitable for sustaining in the global competition. Assimilation of all constituents related to entrepreneurial culture contributes to NPD success through exploring technical advancements for developing quality products as per customers' requirements. A brief theoretical background helps in developing hypotheses relating to the constructs and their effects on NPD success for better performance.

33.2.1 Hypotheses Development

This study focuses on entrepreneurial culture for technical advancement which leads towards the development of quality products as well as achievement of customer satisfaction. To make the study comprehensive, the factors related to entrepreneurial culture are identified and their effect on technical advancement is acknowledged. Control of entrepreneurial culture on technical advancement cascaded through its associated variables for NPD success is discussed in this section which helps in hypotheses development for elucidating their standalone relationship for NPD success.

33.2.1.1 Impact of Entrepreneurial Culture on Technical Improvements

Entrepreneurial culture is a practice to salute the novel ideas succeeded by innovation. Innovative product development shapes the fresh ideas into new products which can be proved as a milestone in the history of the company. But it is also associated by the uncertainty of failure to develop the targeted product as per the customers' requirements. This risk-taking capability of a firm directly impacts on firm's growth rate which can be achieved through entrepreneurial orientation to accept the unpredictable substances strategically and tactfully [17, 23]. This culture of developing something innovative is the resultant of continuous support and effort of top managers for influencing risk-taking behaviour of the firm in developing new products essential for growth. The culture acquaintances manufacturing capabilities to the market need to enrich the organizational performance [24]. This adds the potentiality of the firm to generate new knowledge by adding tangible values [25]. With due support of top management, entrepreneurial culture enriches the organizational factors like organizational structure, organizational culture, up-to-datedness which in turn encourages NPD success of the firm [13]. This practice helps in developing organizational vision accompanied by mission, strategy, learning, knowledge sharing, technological competitiveness to explore innovate ideas and escalate innovation capability of the firm [26–28]. Identification of constituents of organizational factors motivated by entrepreneurial culture is essential for technical improvements for NPD success. Entrepreneurial culture also involves in system integration which is the incorporation of all the resources of various functional groups within the organization to innovative product development. The product development requires various kinds of resources, and depending upon the resources, the integration category differs. Resources are segmented in tangible and intangible categories, among which intangible ones are emphasized as they are often being neglected. Among intangible assets, knowledge is one of the very essential ones for NPD. Both internal and external knowledge is welcomed to achieve competitive advantage to sustain in the global market [29, 30]. Integration of internal capabilities such as various functional groups working together and the external relations like supplier and customer involvement are equally imperative for NPD success [31]. NPD process possesses various stages such as idea generation, screening, implementation, and from the very early stage, the processes should be performed concurrently. These phases must be changed concurrently as times passes [32]. In practical field, there are various types of integration required for successful completion of NPD. Integration of cross-functional teams, integration of R&D activities, integration in system level and production level [33], and integration in the manufacturing systems [34] are the common practices each firm should be aware of. Though essentiality of entrepreneurial culture is unavoidable, the financial support is one of the vital constructs for achieving technical developments. Financial uncertainties can hamper the NPD activities irrespective of availability of all other resources [19]. Entrepreneurial culture develops the practice of utilizing financial resources for technical improvements to develop new products without compromising the quality to achieve customer satisfaction. Along with financial resources, escalation of organizational factors and integration of various systems associated with NPD activities motivate strategic orientation of the firm ensuring technological developments through correlating time, costs, quality, people and organization for better innovative product development activity. From the above discussions, it can be stated that the entrepreneurial culture drives three variables such as organizational factors, system integration and financial resources. The combined effect of these three factors enriches technical improvements which in turn cascaded from entrepreneurial culture. From this, the inferences can be drawn:

Hypothesis 1a (H1a) An effort of entrepreneurial culture escalates the organizational factors.

Hypothesis 1b (H1b) A better entrepreneurial culture enriches system integration among several functions of the firm.

Hypothesis 1c (H1c) Entrepreneurial culture encourages proper appliance of financial resources for NPD success.

Hypothesis 2 (H2) Organizational factors positively encourage the technical improvements and help in NPD.

Hypothesis 3 (H3) An effort of system integration escalates the technical improvements of the firm.

Hypothesis 4 (H4) Availability of adequate financial resources helps in technical developments essential for the betterment of NPD.

33.2.1.2 Impact of Technical Improvements on NPD Success

Technology advancement helps in developing innovative products by utilizing updated technologies leading towards firm's success providing competitive advantages [5]. Technical improvements provide additional high-tech features to firms which help them to develop products with updated distinctiveness essential to achieve success. Technological uncertainty is often treated as a control variable of NPD which signifies that the advanced technological support increases the probability of firm's success by effective NPD [35]. As previously discussed, entrepreneurial culture and its associated variables such as organizational factors, system integration and financial resources motivate technical facets such as usage of computer-aided design, computer-aided manufacturing, group technology, lean manufacturing, flexible manufacturing which in turn escalate NPD success of the firm. There are various types of outcome performances to measure the NPD success. Mainly, qualitative and quantitative criteria are used to measure the outcome of the firm. Quantitative criteria cover all the parameters which can be measured in numbers such as return on investments, profitability, cost reduction, number of customers and many more. In case of qualitative criteria, these cannot be directly measured in numbers like customer satisfaction, customer loyalty, ability to change along with the changing environment, product quality and organizational reputation [36]. This study considers product quality and customer satisfaction as the success measure to quantify the NPD success. This discussion helps to draw the hypotheses such as:

Hypothesis 5a (H5a) Technical facets improve in developing high-quality new products facilitating the updated technologies to achieve NPD success.

Hypothesis 5b (H5b) Technical facets provide better customer satisfaction by developing new products as per their requirements.

This theoretical background assists in comprehensive framework development by using the above-discussed constructs exploring the set hypotheses that examine the effect of driver (i.e. entrepreneurial culture) and outcome (i.e. product quality and customer satisfaction) cascaded through organizational factors, system

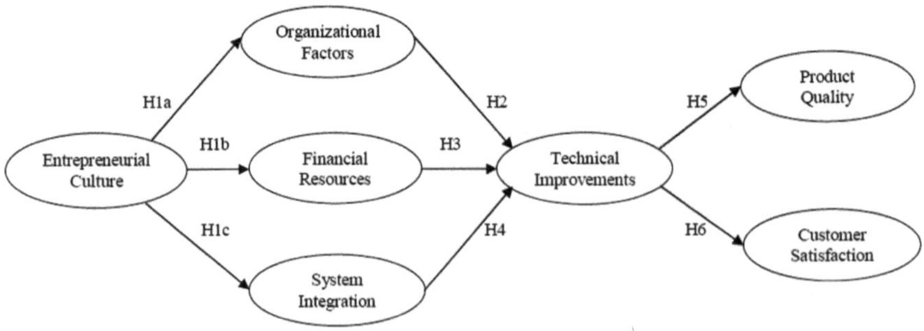

Fig. 33.1 Path model of constructs depicting the effect of entrepreneurial culture on technical improvements cascaded through its associated variables for NPD success

integration and financial resources. Figure 33.1 depicts the path model entrepreneurial culture and its associated variables to control technical improvements essential for NPD success.

33.3 Research Methodology

33.3.1 Research Objectives

Identifying the importance of NPD for firm's success and sustainability, this study investigates the firm's entrepreneurial culture for technical improvements to achieve NPD success by providing quality products and achieving customer satisfaction. The effect of entrepreneurial culture is cascaded through organizational factors, system integration and financial resources to positively influence the technical improvements within NPD framework in Indian scenario. As per our knowledge, this effort is one of the few approaches that investigate the interrelationship of entrepreneurial culture on organizational factors, system integration and financial resources which in turn escalate technical improvements for NPD success quantified by the quality of the newly developed products and customer satisfaction.

33.3.2 Methodology

Structural equation modelling (SEM) is a methodology for representing, estimating and testing a theoretical network of (mostly) linear relations between variables [37]. These variables are latent constructs and manifest variables among which latent constructs cannot be measured directly, but be quantified by measuring the manifest variables or the measured variables. SEM is a statistical approach for analysing hypotheses depicting the relations among latent constructs on the basis of the available researches and primary data collected from industrial experts [38]. This approach is used here to build the relationship among entrepreneurial culture and technical improvements linked through organizational factors, system integration and financial resources for NPD success. This empirical study accumulates primary data from NPD personnel and managers of Indian manufacturing industries for realizing the practical impression. Cronbach's alpha (α) reliability testing has been performed for checking the reliability and internal consistency of the collected data.

33.3.3 Data Collection

In this research, data have been collected from 76 manufacturing experts of Indian companies. A semi-structured questionnaire has been developed which is

Table 33.1 Composition of samples participated in survey

Firm-level characteristics	Frequency
Industry	-
Fabrication	13
Electrical equipment	10
Industrial valves	10
Textile machineries	9
Firefighting equipment	9
Hydraulics and pneumatic	6
Burners and heaters	6
Material handling equipment	6
Cell and battery	3
R&D sectors	2
Air ventilators	2
Total number of respondents	76
Firm size (number of employees) <25 26–100 >100	- 36 18 22
Annual turnover (in crore) >1 1–5 <5	- 27 38 11

segmented in importance and implementation of the manifest variables of the latent constructs to realize their impact in practical scenario. Mostly, directors, vice presidents and managers of design and development departments are considered as right persons to be questioned as their valuable knowledge and work experience are helpful and highly dependable for future analysis and implications. Data accumulation has been performed maximum by visiting and direct interviewing, few over telephonic interviews and e-mails sharing. The profiles of companies participated in this survey work have been enlisted in Table 33.1.

33.4 Results and Discussions

33.4.1 Analysis of Measurement Validity

A thorough data survey has been performed in Indian manufacturing industries as per the questionnaire. The constructs are associated with their respective manifests which are again segmented in two sections as importance to adopt the manifests for developing the practice of the related latent constructs and another is a degree of implementation of the manifests in the practical scenario for emerging the entrepreneurial culture. A 7-point Likert scale is used to quantify the degree of importance and implementation of the indicators as per the experts' opinion. In case of importance, 1 denotes strongly disagree and 7 denotes strongly agree about the importance of the indicators, whereas in case of implementation, 1 denotes very low and 7 denotes very high degree of implementation and same in the case of the NPD success. Respondents are given full freedom to contribute their very own opinion about the manifests they have faced to measure the latent constructs in the practical situation but they are not provided in the given list. This practice adds an extra novelty to this research by accumulating compactness. Composite reliability (CR), average variance extracted (AVE) and Cronbach's alpha reliability testing have been performed to test the reliability of the survey data by using IBM SPSS Amos 21.0 software. This portrays the reliability as well as internal consistency of the survey data. The CR values greater than 0.5 are considered as highly reliable, whereas its values in between 0.3 and 0.5 are moderate. For AVE, values greater than 0.5 indicate the reliability. In case of α, values should be either greater or equal to 0.8, is treated as reliable one and can be further used for analysis purpose [39, 40]. The standardized regression weights (SRWs) of the indicators of latent constructs along with reliability indices (CR, AVE and α) have been listed in Table 33.2. The interrelationship of EC with technical improvements for developing high-quality new products for customer satisfaction is structured model using IBM SPSS Amos 21.0 software packages.

33.4.2 Measurement Model

Confirmatory factor analysis has been performed for estimating the model fit. Estimation of path values for analysing the model fitness is the most vital stage of SEM analysis. Besides, evaluation of model fitness can also be performed by using multiple methods such as chi-square including degree of freedom [41], goodness of fit (GFI), adjusted goodness-of-fit (AGFI) and root mean square of error approximation (RMSEA) [42, 43]. Performing the fitness tests, it can be depicted that the model fitting values are within the desired range as $\chi^2 = 408.159$, degrees of freedom = 246, χ^2/degrees of freedom = 1.659, RMSEA = 0.050, GFI = 0.904, AGFI = 0.892 [44]. The SRWs of the manifest variables range from 0.46 to 0.97 as listed in Table 33.2. Values of reliability indices which are CR and AVE range from 0.69 to 0.84 and 0.49 to 0.63 mostly greater than 0.5 as standard estimated values and values of α range from 0.724 to 0.892 mostly greater than 0.8, representing the satisfactory data reliability for future analysis.

The development of measurement model performs confirmatory factor analysis to validate the occurrence of the manifest variables on their respective latent construct. Manifest variables are identified from the available literatures and also from experts' opinion from manufacturing sectors of India. They have pointed out the degree of importance and implementation of the manifests with respect to the latent variables which helps in analysing their impact in measurement. In case of entrepreneurial culture, the measured variables that have been identified which help in

Table 33.2 Constructs and indicators including reliability indices and path estimates

Constructs and manifests including reliability indices of constructs of both success factors and success measures	Standardized regression weights (SRWs)
1. Entrepreneurial culture (EC) [CR = 0.69; AVE = 0.49; α = 0.724]	–
m1: Risk-taking capability	0.80
m2: New product development culture	0.89
m3: Technological leadership for R&D and innovation	0.93
m4: Development of many new lines of products	0.72
m5: Initiative actions	0.77
m6: First to market	0.48
m7: Highly competitive approach	0.78
m8: Productivity for high-risk projects	0.70
m9: Bold and wide-ranging acts	0.69
m10: Exploration of opportunities	0.76
2. Organizational factors (OFs) [CR = 0.84; AVE = 0.63; α = 0.892]	
m11: Responsiveness to change	0.94
m12: Availability of qualified human resources	0.63
m13: Availability of production resources	0.76
m14: In-house development of technology	0.96
m15: Involvement of project leaders	0.92
m16: Focus on core competency areas	0.46
3. System integration (SI) [CR = 0.72; AVE = 0.50; α = 0.793]	–
m17: Availability of formal department	0.47
m18: Possession of experienced engineers	0.53
m19: Availability of middle management to handle system integration	0.78
m20: Necessity of a product manager	0.65
4. Financial resources (FRs) [CR = 0.70; AVE = 0.50; α = 0.799]	–
m21: Available funds for product development	0.82
m22: Funds for R&D	0.73
m23: Funds for marketing	0.77
5. Technical improvements (TIs) [CR = 0.73; AVE = 0.51; α = 0.801]	–
m24: Investment to upgrade technological infrastructure	0.88
m25: Technology forecasting	0.63
m26: Usage of CAD/CAM	0.84
m27: Group technology	0.82
m28: Lean manufacturing	0.90
m29: Flexible manufacturing	0.82

(continued)

Table 33.2 (continued)

Constructs and manifests including reliability indices of constructs of both success factors and success measures	Standardized regression weights (SRWs)
m30: Design for manufacturability	0.78
m31: Enterprise resource planning	0.85
m32: Number of machine tools	0.92
m33: Trained manpower	0.85
6. Product quality (PQ) [CR = 0.76; AVE = 0.57; α = 0.858]	–
m34: Meet quality guidelines	0.93
m35: Achieved product performance	0.76
m36: Achievement of design goal	0.67
7. Customer satisfaction (CS) [CR = 0.81; AVE = 0.60; α = 0.889]	–
m37: Customer delight: cost aspects	0.81
m38: Customer delight: quality aspects	0.97
m39: Customer delight: specification and features aspects	0.90
m40: Customer delight: after-sales services aspects	0.80

quantifying the construct are risk-taking capability to achieve something that is uncertain (m1), new product development culture (m2), technological leadership for R&D and innovation (m3), development of many new lines of products (m4), initiative actions to which competitors then respond (m5), first to market (m6), i.e. the introduction of new products/services/techniques/technologies, highly competitive approach (m7), productivity for high-risk projects (m8), bold and wide-ranging acts to achieve firm's objectives (m9) and exploration of opportunities (m10). The measurement variables for organizational factors are responsiveness to change (m11), availability of qualified human resources (m12), availability of production resources (m13), in-house development of technology related to the product to be developed (m14), involvement of project leaders (m15) and focus on core competency areas (m16). System integration includes the measurement of availability of formal department that manages the system integration (m17), possession of experienced engineers (m18), availability of middle management to handle system integration (m19) and necessity of a product manager for system integration. In case of financial resources, available funds for product development (m21), funds for R&D (m22) and funds for marketing (m23) help to quantify the construct. The technical improvements are measured by investment to upgrade technological infrastructure (m24), technology forecasting (m25), usage of CAD/CAM (m26), degree of implementation of group technology/cellular manufacturing (m27), implementation of lean manufacturing practices (m28), presence of flexible manufacturing system (m29), design for manufacturability (m30), enterprise resource planning (m31), number of machine tools (m32) and trained manpower (m33). The performance outcomes in this study are product quality and customer satisfaction. Product quality is been measured by the achievement of quality guidelines as per requirements (m34), attainment of product specifications as per customer requirement (m35) and achievement of design specifications (m36). In case of customer satisfaction, the identified measures are customer delight: cost aspects (m37), customer delight: quality aspects (m38), customer delight: specification and features aspects (m39) and customer delight: after-sales services aspects (m40). The standardized regression weights (SRWs) have been calculated by using SEM approach which links values of predictor variables to the dependent variables. The SRWs with positive values are considered for further analysis to develop the structural model. Table 33.2 enlists the manifest variables of latent constructs along with their SRWs which represent positive values ranging from 0.46 to 0.97.

33.4.3 Structural Model

Analysis of measurement model is followed by the structural model formation. In case of structural model, also the fitness tests are performed showing the appropriate model-to-data fit as $\chi^2 = 440.136$, degrees of freedom = 256, χ^2/degrees of freedom = 1.719, RMSEA = 0.048, GFI = 0.912, AGFI = 0.903 [44]. The structural model represents the hypothesized relationships between entrepreneurial culture and technical improvements which in turn control the NPD success which can be measured by manifests of product quality and customer satisfaction. The structural model is developed by using IBM SPSS Amos 21.0 software packages which is shown in Fig. 33.2. Path estimates between latent constructs range from 0.41 to 0.92 representing that the assumed hypotheses are supported, and their values are listed in Table 33.3.

Structural model depicts the interrelationship among latent constructs by analysing the hypotheses among them constructed from the available theoretical background. It calculates the correlation among the constructs by using several statistical analyses. The path values among these constructs are calculated by multiple regression analysis using SEM approach. The obtained results from the structural model are listed in Table 33.3 which clearly mentions the path values among the latent constituents. These values interpret the interrelationship among the two related constructs as described. It shows that entrepreneurial culture relates to few variables which are treated as its associated variables such as organizational factors, system integration and financial resources with positive path values of 0.92, 0.65 and 0.79. The direction of the arrowheads is also from entrepreneurial culture to the related factors, as mentioned. So, this depicts that entrepreneurial culture escalates these associated variables for future developments. Similarly, these associated variables, i.e. organizational factors, system integration and financial resources, motivate technical improvements of the firm representing through the path values 0.73, 0.41 and 0.63, respectively. This technical improvement enriches the product quality and customer satisfaction as well represented through the path

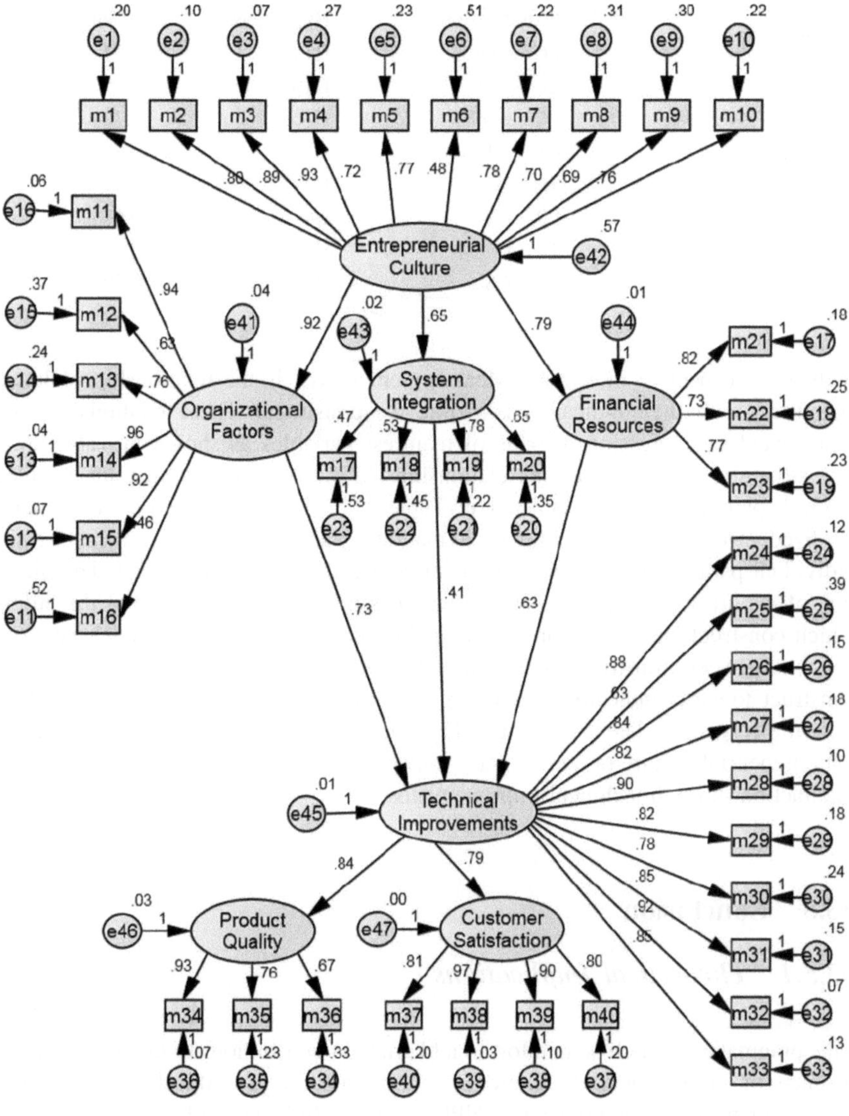

Fig. 33.2 Structural equation modelling (SEM) model interrelating latent constructs and their measures after execution

values 0.84 and 0.79, respectively. The product quality and customer satisfaction are treated as the outcome performance corresponding to NPD success of the firm as the ultimate objective of any firm is to develop high-quality new products to achieve customer satisfaction.

Figure 33.2 symbolizes the comprehensive framework consisting of both measurement model and structural model merging all the latent constructs along

Table 33.3 Statistics of path estimates

Path description	Hypotheses	Estimates
EC → OF	H1a	0.92
EC → SI	H1b	0.65
EC → FR	H1c	0.79
OF → T	H2	0.73
SI → T	H3	0.41
FR → T	H4	0.63
T → Q	H5a	0.84
T → CS	H5b	0.79

with their respective indicators. Measurement model is the relationship of latent constructs with their indicators to define the constructs. These relationships are represented by regression weights of manifest variables as shown in Table 33.2. The structural model is the interrelationships of the constructs which are depicted through path estimates as enlisted in Table 33.3. The figure also contains standardized errors associated with each manifest variable. Similarly, the error for individual path coefficient for each latent constructs is also mentioned. The direction of the arrows linking the latent constructs brings high relevance as this infers which construct has impact on the other. In case of measurement model, i.e. connection between constructs and their indicators, the direction of arrows is from construct towards indicators describing the connection of the manifest variables with the respective latent construct. This model ultimately depicts the impact of entrepreneurial culture on performance outcome, i.e. product quality and customer satisfaction through technical improvements.

33.5 Conclusion

33.5.1 Theoretical Implications

This pragmatic research is an effort to add crucial contributions to the theory on the essence of entrepreneurial culture in an organization for developing technical improvements which in turn leads to NPD success in Indian manufacturing industries. First, entrepreneurial culture and its associated variables along with their indicators relative to technical improvements are recognized. Second, the impact of these constructs on technical improvements is verified incorporating the identification of indicators of technical improvements as well. Third, the effect of technical improvements on NPD success which in turn is quantified by product quality and customer satisfaction is framed for better realization of these constructs in practical field.

33.5.2 Managerial Implications

The inferences of this study portray the vitality of NPD success to achieve industrial sustainability through achieving competitive advantages through developing high-quality new products for attaining customers' expectations. Entrepreneurial culture plays a major role to achieve this target to make NPD activities a common practice. The results contribute significant managerial implications highly valuable for future developments.

First, entrepreneurial culture of the firm needs to be a common practice for realizing the requirement of continuous NPD activities along with the regular product development process. This culture develops the risk-taking capability in the urge of developing something new and innovative. The entrepreneurial culture is also associated with few variables which are highly related to technical improvements. Organizational factors, system integration and financial resources are those associated variables driven by entrepreneurial culture as recognized from the available literatures linking to technical improvements for NPD success. The indicators of entrepreneurial culture have also been identified to quantify the construct. The study shows risk-taking capability, culture of developing new products in the firm, providing technological leadership to continue research activities and developing new products for innovation, new line product development, initiative actions, development of products those are first to market, competitive approach, productivity for high-risk projects, bold and wide-ranging actions to achieve objectives and exploration of new opportunities are antecedents to define the entrepreneurial culture of a firm. These practices help to generate the practice of this culture in the firm to sustain in the competition. Entrepreneurial culture also relates to organizational factors which can be improved through developing the culture of responsiveness to change, hiring qualified human resources, accumulation of production resources, betterment of in-house development technologies that are required for development of new products, dedicated involvement of project leaders to ensure the success of the project and urge to enrich the core competency of the firm. System integration is another most vital constituent of entrepreneurial culture which can be upgraded through ensuring the availability of formal department for system integration, possessing experienced and skilled engineers, availability of middle management to handle integration activities and hiring an experienced product manager. In case of financial resources, availability of funds for product development, R&D and marketing help to escalate technical improvements through a well-planned entrepreneurial culture.

Second, the practice of entrepreneurial culture and organizational factors, system integration and financial resources driven by entrepreneurial culture provide a concrete support for technical facets which can be attained by investing to upgrade the technological infrastructure with newly developed machines, technology forecasting to predict the future trend, using computer-aided systems like CAD/CAM, implementing group technology, adopting lean manufacturing practices, using flexible manufacturing system for manufacturing, applying design for manufacturability in the firm, adopting customized enterprise resource planning for accurate planning, using various machine tools and appointing trained manpower.

Ultimately, the aim of all organizations is to attain competitive advantages to sustain in the global competition. Development of new products with high quality for achieving customer satisfaction is the only way to retain in the volatile market environment. Product quality is one of the vital performance outcomes which can be achieved by attaining quality guidelines as required by the customers, achieving product specifications as per customer requirement and also achieving design specifications as well. Attaining customer satisfaction is one of the toughest jobs as the customer requirements are changing day by day which cannot be traced easily. Customer satisfaction is predicted by customers' delight in several aspects such as cost, quality, specification and features and after-sales services.

The structural model framed in this study has clearly expressed the impact of entrepreneurial culture and its associated variables (organizational factors, system integration and financial resources) on technical improvements of the firm which helps in achieving NPD success by producing high-quality new products for customer satisfaction. This practice ensures the success as well as the survival of the firm by providing better performance.

33.5.3 Limitations

The limitation is one of the inseparable parts of the research, and this study is not an exception. The empirical research model gathers responses from a large number of samples to analyse and deliver results. Accumulation of data from a huge number of industry experts is quite a difficult task to be performed. Convincing the experts to share their company's information for the sake of the better is a very complicated assignment. Moreover, the reliability of the obtained data is a matter of concern to use them for the future analysis. Identification and elimination of inconsistent responses are critical task before using them for analysis to draw the final results.

Acknowledgements The research work was substantially supported by a grant from the Department of Science and Technology (DST) of Government of India as a DST INSPIRE Fellowship to carry out the doctoral research of the corresponding author. The authors are also thankful to the industry personnel and experts from manufacturing industries from India who shared their views and opinions.

Appendix

- Interview Protocol with description of latent variables

1. *Entrepreneurial Culture (EC)*
    - Risk-taking capability (m1): Risk-taking capability which enhances the probability of profit making

- New product development culture (m2): Emphasis on development of new and innovative products
- Technological leadership for R&D and innovation (m3): Emphasis on R&D, technological leadership and innovations
- Development of many new lines of products (m4): Development of many new lines of products or services to increase profitability
- Initiative actions (m5): Initiative actions to which competitors then respond
- First to market (m6): First to introduce new products/services, techniques, technologies
- Highly competitive approach (m7): Adoption of very competitive, "undo-the-competitors" posture
- Productivity for high-risk projects (m8): Strong productivity for high-risk projects (with chances of very high returns)
- Bold and wide-ranging acts (m9): Bold and wide-ranging acts are necessary to achieve the firm's objectives
- Exploration of opportunities (m10): Adoption of bold, aggressive posture to maximize the probability of exploiting opportunities.

2. *Organizational Factors (OFs)*

- Responsiveness to change (m11): Quick responses to the changes occur during new product development.
- Availability of qualified human resources (m12): Available qualified human resources successfully perform new product development activities.
- Availability of production resources (m13): Available production resources for new product development.
- In-house development of technology (m14): In-house development of technology related to the product to be developed.
- Involvement of project leaders (m15): Involvement of project leaders in different activities at working level.
- Focus on core competency areas (m16): Focus on core competency areas within the organization.

3. *System Integration (SI)*

- Availability of formal department (m17): Availability of formal department that manages the system integration
- Possession of experienced engineers (m18): Possession of experienced engineers with product knowledge in the system integration department who help to integrate several teams working on separate modules
- Availability of middle management to handle system integration (m19): Availability of middle management, with operational skills and understanding of corporate strategy, to handle system integration activities
- Necessity of a product manager (m20): Necessity of a product manager for system integration.

4. *Financial Resources (FRs)*

- Available funds for product development (m21): Availability of funds to be spent on new product development activities
- Funds for R&D (m22): Allotment of funds for continuous research activities and development of innovative products based on the findings of research work
- Funds for marketing (m23): Funds spend on marketing activities to get promote the newly developed products for increasing profit.

5. *Technical Improvements (TIs)*

- Investment to upgrade technological infrastructure (m24): Percentage of funds invested for upgrading of technological infrastructure to make the development process high-tech
- Technology forecasting (m25): Implementation of technology forecasting to trace the difficulties
- Usage of CAD/CAM (m26): Usage of CAD/CAM and various state-of-the-art technologies and newly launched technologies
- Group technology (m27): Degree of implementation of group technology/ cellular manufacturing
- Lean manufacturing (m28): Implementation of lean manufacturing practices
- Flexible manufacturing (m29): Presence of flexible manufacturing system
- Design for manufacturability (m30): Degree of adoption of Design for Manufacturability and Assembly (DFMA)
- Enterprise resource planning (m31): Use of enterprise systems (ERP) for product data management
- Number of machine tools (m32): Number of machine tools used in the company for new product development process
- Trained manpower (m33): Availability of trained manpower to handle the new product development activities skilfully and efficiently.

6. *Product Quality (PQ)*

- Meet quality guidelines (m34): Achieve the quality guidelines as per requirements
- Achieved product performance goal (m35): Attainment of product specifications as per customer requirement
- Achievement of design goal (m36): Achievement of design specifications.

7. *Customer Satisfaction (CS)*

- Customer delight: cost aspects (m37): Achievement to develop the new products within the cost as stated by customers
- Customer delight: quality aspects (m38): Attainment of quality of new products as per customer requirements
- Customer delight: specification and features aspects (m39): Successful development of new products as per specification and features mentioned by customers

- Customer delight: after-sales services aspects (m40): High-end after-sales service facilities that company can provide after selling their products.

References

1. Buyukozkan, G., Arsenyan, J.: Collaborative product development: a literature overview. Prod. Plan. Control **23**, 47–66 (2012)
2. Tzokas, N., Hultink, E.J., Hart, S.: Navigating the new product development process. Ind. Mark. Manage. **33**, 619–626 (2004)
3. Ernst, H.: Success factors of new product development: a review of the empirical literature. Int. J. Manage. Rev. **4**, 1–40 (2002)
4. Kobeda, E., Isaacs, P., Pymento, L.: Critical success factors for electronic manufacturing services. Presented at Pan Pacific Microelectronics Symposium (Pan Pacific) (2016)
5. Mendes, G.H.D.S., Ganga, G.M.D.: Predicting success in product development: the application of principal component analysis to categorical data and binomial logistic regression. J. Technol. Manag. Innov. **8**, 83–97 (2013)
6. Haverila, M.J.: Product–firm compatibility in new product development in technology companies. J. High Technol. Manag. Res. **23**, 130–141 (2012)
7. Medeiros, J.F.D., Ribeiro, J.L.D., Cortimiglia, M.N.: Success factors for environmentally sustainable product innovation: a systematic literature review. J. Clean. Prod. **30**, 1e11 (2014)
8. Yang, E., Ma, G., Chu, J.: The impact of financial constraints on firm R&D investments: empirical evidence from China. Int. J. Technol. Manage. **65**, 172–188 (2014)
9. Graner, M.: Are methods the key to product development success? An empirical analysis of method application in new product development. In: Impact of Design Research on Industrial Practice, pp. 23–43, Springer International Publishing, (2016)
10. Shinno, H., Hashizume, H.: Structured method for identifying success factors in new product development of machine tools. CIRP Ann. Manuf. Technol. **51**, 281–284 (2002)
11. Globocnik, D., Salomo, S.: Do formal management practices impact the emergence of bootlegging behavior? J. Prod. Innov. Manag. **32**, 505–521 (2015)
12. Felekoglu, B., Moultrie, J.: Top management involvement in new product development: a review and synthesis. J. Prod. Innov. Manag. **31**, 159–175 (2014)
13. Sadeghi, A., Azar, A., Rad, R.S.: Developing a fuzzy group AHP model for prioritizing the factors affecting success of high-tech SME's in Iran: a case study. Procedia-Soc. Behav. Sci. **62**, 957–961 (2012)
14. Akgün, A.E., Byrne, J.C., LynnGS, Keskin H.: Team stressors, management support, and project and process outcomes in new product development projects. Technovation **27**, 628–639 (2007)
15. Sun, H., Wing, W.C.: Critical success factors for new product development in the Hong Kong toy industry. Technovation **25**, 293–303 (2005)
16. Matsuno, K., Zhu, Z., Rice, M.P.: Innovation process and outcomes for large Japanese firms: roles of entrepreneurial proclivity and customer equity. J. Prod. Innov. Manag. **31**, 1106–1124 (2014)
17. Miller, D., Friesen, P.H.: Archetypes of strategy formulation. Manage. Sci. **24**, 921–933 (1978)
18. Sun, H., Wong, S.Y., Zhao, Y., Yam, R.: A systematic model for assessing innovation competence of Hong Kong/China manufacturing companies: a case study. J. Eng. Tech. Manage. **29**, 546–565 (2012)
19. Healy, B., O'Dwyer, M., Ledwith, A.: An exploration of product advantage and its antecedents in SMEs. J. Small Bus. Enterp. Dev. (2017)
20. Awwad, A., Akroush, D.M.N.: New product development performance success measures: an exploratory research. EuroMed J. Bus. **11**, 2–29 (2016)

21. Acur, N., Kandemir, D., Boer, H.: Strategic alignment and new product development: drivers and performance effects. J. Prod. Innov. Manag. **29**, 304–318 (2012)
22. Eid, R.: Factors affecting the success of world class manufacturing implementation in less developed countries: the case of Egypt. J. Manuf. Technol. Manag. **20**, 989–1008 (2009)
23. Soininen, J., Martikainen, M., Puumalainen, K., Kyläheiko, K.: Entrepreneurial orientation: growth and profitability of Finnish small-and medium-sized enterprises. Int. J. Prod. Econ. **140**, 614–621 (2012)
24. Chavez, R., Yu, W., Jacobs, M.A., Feng, M.: Manufacturing capability and organizational performance: the role of entrepreneurial orientation. Int. J. Prod. Econ. **184**, 33–46 (2017)
25. Kickul, J.R., Griffiths, M.D., Jayaram, J., Wagner, S.M.: Operations management, entrepreneurship, and value creation: emerging opportunities in a cross-disciplinary context. J. Oper. Manage. **29**, 78–85 (2011)
26. Martín-Rojas, R., García-Morales, V.J., García-Sánchez, E.: The influence on corporate entrepreneurship of technological variables. Ind. Manag. Data Syst. **111**, 984–1005 (2011)
27. Lin, H.F.: Knowledge sharing and firm innovation capability: an empirical study. Int. J. Manpower **28**, 315–332 (2007)
28. Nellore, R., Balachandra, R.: Factors influencing success in integrated product development (IPD) projects. In: IEEE Transactions on Engineering Management, **48**, pp. 164–174 (2001)
29. Gu, Q., Jiang, W., Wang, G.G.: Effects of external and internal sources on innovation performance in Chinese high-tech SMEs: a resource-based perspective. J. Eng. Tech. Manage. **40**, 76–86 (2016)
30. Leitner, K.H.: The effect of intellectual capital on product innovativeness in SMEs. Int. J. Technol. Manage. **53**, 1–18 (2011)
31. Johnson, W.H.A., Filippini, R.: Integration capabilities as mediator of product development practices–performance. J. Eng. Tech. Manage. **30**, 95–111 (2013)
32. Nihtilä, J.: R&D–Production integration in the early phases of new product development projects. J. Eng. Tech. Manage. **16**, 55–81 (1999)
33. O'Sullivan, A.: Dispersed collaboration in a multi-firm, multi-team product-development project. J. Eng. Tech. Manage. **20**, 93–116 (2003)
34. Lindstrom, V., Winroth, M.: Aligning manufacturing strategy and levels of automation: a case study. J. Eng. Tech. Manage. **27**, 148–159 (2010)
35. Chiang, Y.H., Shih, H.A.: Knowledge-oriented human resource configurations, the new product development learning process, and perceived new product performance. Int. J. Hum. Resour. Manage. **22**, 3202–3221 (2011)
36. Limpibunterng, T., Johri, L.M.: Complementary role of organizational learning capability in new service development (NSD) process. Learn. Organ. **16**, 326–348 (2009)
37. Rigdon, E.E.: Structural equation modelling. In: Marcoulides, G.A. (ed.) Modern Methods for Business Research, pp. 251–294. Lawrence Erlbaum Associates Publishers, Mahwah, NJ (1998)
38. Hoyle, R.H.: The structural equation modelling approach: basic concepts and fundamental issues. In: Hoyle, R.H. (ed.) Structural Equation Modelling: Concepts, Issues, and Applications, pp. 1–15. Sage Publications Inc., Thousand Oaks, CA (1995)
39. Ong, C.S., Lai, J.Y., Wang, Y.S.: Factors affecting engineers' acceptance of asynchronous e-learning systems in high-tech companies. Inf. Manag. **41**, 795–804 (2004)
40. Nunnally, J.C.: Psychometric Theory. McGraw Hill, New York (1978)
41. Hu, L., Bentler, P.M.: Fit indices in covariance structure modeling: sensitivity to under parameterized model misspecification. Psychol. Methods **3**, 424–453 (1998)
42. Bentler, P.M.: Comparative fit indexes in structural models. Psychol. Bull. **107**(2), 238 (1990)
43. Chen, H.C.: The impact of children's physical fitness on peer relations and self-esteem in school settings. Child Indic. Res. pp. 1–16 (2015)
44. Hair, J.F., Anderson, R.E., Tatham, R.L., Black, W.C.: Multivariate Data Analysis with Readings. Prentice Hall, Englewood Cliffs, NJ (1995)

Chapter 34
Managerial Support in R&D Operations and Practices for Realizing Technological developments

Sudeshna Roy, Nipu Modak and Pranab K. Dan

34.1 Introduction

Successful development of new products has become one of the most compelling strategies for top management for sustaining in the competition [1, 2]. This competition is becoming more challenging due to globalization insisting companies to adopt innovative practices for industrial sustainability [3]. Products are no longer been bounded to geographical limitations. Globalization makes new product development (NPD) as an influential mechanism for strengthening the position of the organization in the competitive market. It has become a vital determinant of the endurance of the organization to attain competitive advantage accordingly [4]. Availability of several constituents, critical for NPD success, is famously known as critical success factors (CSFs) [5, 6], which help to develop new products as per customers' demand, e.g., Cooper and Kleinschmidt [7]. The identification of various CSFs and the interactions among these constructs has been highlighted by various researchers in the previous studies [8–11].

The objective of any organization is the success and survival in the competitive market environment. This can be achieved through involving in continuous developments by opting resources available for product development. Research and development (R&D) is a vital activity every firm should be involved not only for innovative product development but also for modifying the available products as well [12–14]. Previously, organizations were unaware of R&D activities and its requirements for achieving organizational success. But the globalization makes them realize the need of developing new products for sustaining in the competition.

S. Roy (✉)
Mechanical Engineering Department, Jadavpur University, Kolkata, India
e-mail: sudeshnaroy689@gmail.com

N. Modak · P. K. Dan
Rajendra Mishra School of Engineering Entrepreneurship, IIT Kharagpur, Kharagpur, India

P. Sahoo and J. P. Davim (eds.), *Advances in Materials, Mechanical and Industrial Engineering*, Lecture Notes on Multidisciplinary Industrial Engineering,
https://doi.org/10.1007/978-3-319-96968-8_34

This can be achieved by R&D operations and practices. Developing R&D operations and practices in the organization requires huge investment to achieve something that is unpredictable. But available researches portrayed the profitability obtained by investing for R&D which able to provide high market share through innovative product development [2].

Top management support is the motivation of top managers in NPD activities along with their active participation. Managerial support not only encourages attaining something new but it also motivates in taking risks associated with the innovative actions to achieve uncertain goals. The product development process is segmented in different stages from idea generation to commercialization [11]. Senior management's participation and their encouraging behavior to complete the development activities effectively are essential throughout the idea generation to commercialization which is the key of NPD success. It helps to sustain in the competitive market environment by enriching NPD capacity of the organization through developing urge for innovations [15, 16]. Managerial support helps in implementing various state-of-the-art technologies with proper learning through knowledge sharing and utilizing the intellectual capital as well. Adoption of modern technologies is closely associated with the NPD performance for developing new products by thorough R&D activities by providing the updated technological support required for the development [17]. Technological advancement is the replication of R&D as R&D operations and practices add modern characteristics to the available technologies and these modified technology resources can be able to enrich the R&D activities by providing the updated features as well [18]. Again, learning is a practice of skill development through sharing ideas, knowledge, and information among the NPD teams which are helpful for NPD activities. The cultural barriers among the various teams can be overcome by this practice which helps in developing products as per their best ideas and knowledge [19, 20]. It is an approach of restoring all the previous data and information for future application purpose [21]. Like technology and learning, intellectual capital is the constituent escalating R&D operations and practices for innovative product development. Intellectual capital is the intangible asset to acquire competitive advantage for NPD [22]. Intellectual capital comprises three types of capitals such as human capital, structural capital, and relational capital [23]. Among these three capitals, relational capital is mainly linked with R&D activities [24]. As the identification of the success factors is critical to NPD success, the success measures for obtaining the performance outcome of the organization are also been recognized. These performance outcomes are different in different cases as applied by the researchers [25]. To identify the role of R&D activities for NPD success, the recognition of technological developments as a success measure for quantifying the NPD success is essential to be taken care of [26]. Researches concerning the R&D activities to develop new products with better technological developments have been performed previously. But, the comprehensive approach considering the R&D practice along with the role of top managers and other associated factors are remained unnoticed. Consideration of managerial support and the other associated factors for R&D

practice along with their antecedents is rarely been addressed in the previous studies. A single model depicting these constructs along with the technological development as performance measure for quantifying NPD success is clearly been recognized. These vital research gaps create a platform where the answers of some vital questions needed to be responded for the sake of the success of the organization. These research questions are: (1) What is the impact of R&D activities for achieving NPD success? (2) What is the role of top management support to develop the R&D practice in the organization? (3) What are the intermediate factors influenced by managerial support which in turn escalate the R&D activities? (4) What are the major performance outcomes that used to quantify NPD success of the organization? The responses of these questions are able to realize the role of R&D for NPD success by drawing the valuable inferences which help in realizing the essentiality of above-mentioned constructs for innovative product development practices.

The primary objective of this study is the formation of a higher-order model realizing the responsibility of top management to motivate the associated factors for developing a positive environment for R&D practice which in turn ensures the NPD success. First, this research identifies the importance of managerial support and motivation to encourage the factors such as technology resources, learning, and intellectual capital for R&D practice. It also recognizes the indicators of these constituents necessary to quantify these factors required for model development. Second, the indicators of R&D operations and practices have recognized for realizing the impact of R&D on NPD success. The third contribution is identification of performance outcome which is technological developments for attaining NPD success. Ultimately, a gap concerning the role of R&D activities has been recognized in small-and medium-sized enterprises (SMEs) in India. This empirical study accumulates responses from experts of SMEs of Indian manufacturing industries for developing the comprehensive model. The implications are drawn analyzing the responses from manufacturing experts as well as the available researches.

This paper is segmented in the following sections: After introduction of the study, the very next section is the theoretical background which portrays a description of the theories on the basis of which the conceptual framework is developed. It helps in developing the research hypotheses which need to develop the structural model depicting the role of managerial support to escalate the technological resources, learning, and intellectual capital which in turn implement R&D practices to achieve NPD success through technological developments. This section also contains the path model developed from the hypotheses made from theoretical background. The next section is the research methodology which includes research objectives, methodology used for data analysis purpose along with the data collection methods, and profiles of the Indian manufacturing companies participated in this research. Next, the results and discussion section involves the analysis of measurement validity, results and discussion of measurement model as well as structural model for hypotheses testing, and developing the SEM model by using IBM SPSS AMOS 21.0 software packages. Finally, the conclusion section concludes by inferring the theoretical contribution along with the managerial implications about the role of top management and its associated constructs to escalate R&D practices for NPD success.

34.2 Theoretical Background

Technological development is the key performance outcome to provide world-class system which helps in achieving industrial success and sustainability [27]. R&D activities are the one of the most vital practices to attain the technological developments which helps in modifying the existing products as well as developing the innovative products as per customer requirements. There are number of factors associated with the R&D operations and practices which needed to be identified and analyzed to realize their effect on R&D for NPD success. Top management support is the most vital and essential constituent of any development activities as because without help and support from senior management, the development of new products is impossible task to be performed. In practical scenario, top managers optimize the utilization of technology resources, adoption of learning practice for knowledge sharing, and exploitation of intellectual capital for R&D practices. The R&D practice again motivates the NPD success through technological developments of the organization achieved by adopting the technological developments. On the basis of the theoretical background, the hypotheses interrelating the constructs for better NPD performance have been developed which are further analyzed for final analysis.

34.2.1 Hypotheses Development

This study identifies the NPD success as a vital concern required for industrial sustainability. Development of new products requires innovative actions achieved through research activities along with the development of results of research findings. Successful implementation of R&D activities is supported by top managers by providing technology resources for adopting learning practices for utilizing intellectual capital for NPD success. The performance outcome of NPD success is measured by technological developments attained by the newly developed products as a fruitful result of R&D activities.

34.2.1.1 Impact of Top Management Support on R&D Operations and Practices

Role of top management support to control NPD activities is one of the vital concerns which is analyzed in the previous empirical researches [28–30]. These available researches mostly highlight certain particular areas on which the impact of managerial support is analyzed. Top management plays an active role by assessing and providing guidance for utilizing organization's capabilities to ensure the future growth. They develop strategic planning of the organization along with taking the decisions about which new products the organizations are going to develop, the investment amount and area for profitability, and so forth. The research agenda of the organization is another most vital decision top management bodies take for NPD success [31]. Efficient managerial support and motivation escalates R&D activities of the organization cascaded through few imperative constituents. Technology is one of those factors to which R&D is closely associated for innovation [17, 32]. Identification of technology resources along with their controllability on R&D enhances the firm success [9]. Technology is often considered as the control variable signifying the impact of technological support for R&D operations and practices [33]. Like technology, learning is another factor requires managerial support to escalate R&D culture of the organization. Top management approves the fund adequate for technology researches and their developments and reforms the work process to optimally utilize the technology opportunities in the organization. Managerial bodies expedite technology transfer throughout the organization which helps in identical circulation of technological facilities among all teams within the organization [34]. Sometimes, managers arrange sponsors for funding the technology developments, but the knowledge of sponsors about that technology is essential for ease of investment by knowing the importance of the development for that project [35]. Prior technology experience of the senior management and the team is also essential for successful application of technology resources for R&D operations and practices [36]. Learning is the result of strong collaborative culture which provides the interfaces of customers with the developed products. It develops the overall information about the customers' expectations essential for NPD success [37]. Learning helps to triumph over the cultural barriers among the different teams which is essential for R&D [38]. It escalates the knowledge-oriented human resource for improving R&D activities for NPD [33]. Impact of top management support on organizational learning for better innovation performance is a well-discussed topic by previous researchers [39]. Top managers encourage knowledge-sharing practice among the same team as well as different teams by providing all types of necessary resources on appreciating the beneficial impression of knowledge sharing on NPD [40]. Top managers with expert knowledge in multi-project encourages organizational learning process to build superior organizational structure for R&D operations and practices [41]. Moreover, the enthusiasm to contribute and accumulate knowledge to and from all classes develops the practice of global knowledge integration beneficial for innovation [16, 40, 42]. The intangible assets of the organization known as intellectual capital cover learning, knowledge, abilities, and strategies of the firm to achieve the objectives [23]. Knowledge sharing considers the process of communicating about the personal intellectual capital, and accumulation of knowledge concerns the encouraging the others to share their knowledge and ideas. This practice improves the innovation capability of the organization [43]. There are mainly three types of intellectual capitals such as human capital, structural capital, and relational capital. Among these three, relational capital is the most important succeeded by human capital and structural capital [24]. For enrichment of R&D activities, structural capital is mostly effective concerning the investments in R&D and the number of R&D experts. The mutual impact of these three constituents escalates R&D activities of the organization which is previously cascaded from top management support. From this discussion, the inferences can be drawn:

Hypothesis 1a (H1a) Efficient managerial support helps in proper utilization of technology available for developing R&D operations and practices in the organization.

Hypothesis 1b (H1b) Top management support escalates the learning capabilities of the organization for successful implementation of R&D operations and practices.

Hypothesis 1c (H1c) Support and motivation from senior management helps to optimally utilize the available intellectual capital of the organization for successful adoption of R&D operations and practices.

Hypothesis 2 (H2) Technology positively enriches the R&D operations and practices of the organization, essential for NPD success.

Hypothesis 3 (H3) Learning activities positively escalate the R&D operations and practices of the organization opt for NPD success.

Hypothesis 4 (H4) Availability of plenty of intellectual capital enriches the R&D operations and practices for NPD success.

34.2.1.2 Impact of R&D Operations and Practices on NPD Success

Innovative product development brings fortune to the organization through continuous research activities and development of research findings for successful NPD. Investing for R&D activities grows the culture developing creative ideas, own designs and additional features which helps in structuring unique products. R&D is the approach of developing the new products with such specifications that cannot be substituted by their competitors [11]. Investments in R&D operations and practices mainly concern the investment for betterment of organizational infrastructure for better NPD activities, appointment of skilled human resources, and sustainable product development [32, 38]. Though R&D practice provides the NPD success as well as the competitive advantage to the organization, it also associated with high risks and uncertainties as the development of innovative ideas is always not feasible [44]. Marketing department identifies the customer requirements, and based on that, the new products are developed. Synchronization among the various teams is required for the development of the unique products as expected by the consumers. This practice improves the manufacturability of the organization for innovation [45]. R&D teams require qualified experts with huge NPD experiences for development of new products as per requirement [46]. This discussion draws the statement:

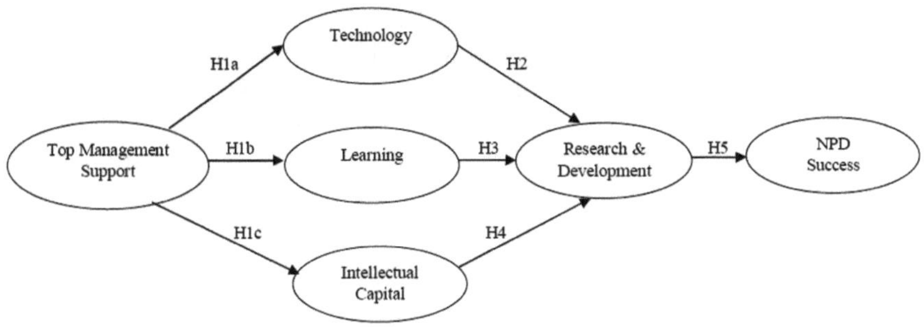

Fig. 34.1 Path model of constructs depicting the effect of entrepreneurial culture on technical improvements cascaded through its associated variables for NPD success

Hypothesis 5 (H5) R&D operations and practices help in developing new products with technological developments which facilitated the NPD success of the organization.

The theoretical background develops the framework considering the effect of drivers (i.e., managerial support) on the R&D operations and practices to achieve better performance outcome (i.e., NPD success in turn the technological developments) cascaded through technology, learning, and intellectual capital. From the hypotheses developed from the discussed theoretical background, a comprehensive path model has been configured as shown in Fig. 34.1.

34.3 Research Methodology

34.3.1 Research Objectives

This study explores the role of top management support for developing efficient R&D activities in the organization cascaded through technology, learning, and intellectual capital for achieving NPD success which in turn be measured by technological developments. The development of the structural model accumulates the primary data from Indian manufacturing industries.

34.3.2 Methodology

Structural equation modeling (SEM) is an approach used here for developing the interrelationship among the factors, critical for the NPD and correlates them with organizational success. The comprehensive statistical approach is used for testing the hypotheses among latent and manifest variables and depicts their relationships as well [47]. The representation of the structural model, estimation of path values, and finally the testing of the developed model are performed by using SEM analysis [48]. In this study, the interrelationship model has been developed by using top management support as a success factor which enriches R&D practices of the firm cascaded through technology, learning, and intellectual capital. Finally, R&D positively influences technological developments of the newly developed products for attaining NPD success.

34.3.3 Data Collection

This research accumulates the primary data from 76 manufacturing companies of various corners of India. This empirical approach performs development of a semi-structure questionnaire clustered into two sections, such as importance of the manifests to measure the latent constructs and the implementation of those antecedents in the practical field for successful execution of the respective constituents. The targeted respondents are mainly the Directors, Presidents, Vice-Presidents, and Managers of design and development departments concerned about the R&D activities. Data have been collected mostly by direct interviewing method. Very few responses have also been acquired through telephonic interviews and e-mail sharing. The profiles of the companies actively participated in this primary survey have been listed in Table 34.1.

Table 34.1 Composition of samples participated in survey

Firm-level characteristics	Frequencies
Industry	–
Fabrication	13
Electrical equipment	10
Industrial valves	10
Textile machineries	9
Fire-fighting equipment	9
Hydraulics and pneumatic	6
Burners and heaters	6
Material handling equipment	6
Cell and battery	3
R&D sectors	2
Air ventilators	2
Total number of respondents	76
Firm size (number of employees)	-
<25	36
26-100	18
>100	22
Annual turnover (in crore)	-
>1	27
1-5	38
<5	11

34.4 Results and Discussion

34.4.1 Analysis of Measurement Validity

A semi-structure questionnaire has been developed for summing up the opinion and suggestions from experts of Indian manufacturing industries. Each latent constructs are associated with their respective manifest variables to quantify the latent variables. Each manifest again be segmented in two sections, such as importance of the manifests to measure the latent construct and implementation of that manifest in the practical field. There is another section associated with each latent construct which is obstacle to implement those factors in the company. A 7-point Likert scale has been used to quantify the latent variables. In case of importance, 1 denotes strongly disagree and 7 denotes strongly agree. For implementation purpose, 1 represents very low and 7 stands for very high degree of implementation. Respondents are also given an open space to show their views about the obstacles they have been faced during the development process for implementing the constructs like top management support and R&D practices. CR, AVE, and Cronbach's alpha reliability testing have been performed for testing the reliability of the collected data using IBM SPSS 21.0 software. The CR values above 0.5 are considered as highly reliable, whereas CR values in between 0.3 and 0.5 are also moderately reliable. Again in case of AVE, values greater than 0.5 are trustworthy for analysis. α values greater than 0.8 are highly reliable for data analysis though values near 0.7 can also be acceptable [49, 50]. The standardized regression weights (SRWs) have been calculated and listed in Table 34.2 along with the values for reliability indices, such as CR, AVE, and α. Finally, the structural model depicting the interrelationships among the latent variable as shown in Fig. 34.1 is developed by using IBM SPSS AMOS 21.0 software.

34.4.2 Measurement Model Results and Discussion

The estimation of model fit has been performed by using confirmatory factor analysis. The estimation of model fitness is the vital part of the analysis. This can be achieved through calculating the model fitness. The model fitness can be analyzed by multiple fitness tests such as chi-square including degree of freedom [51], goodness-of-fit (GFI), adjusted-goodness-of-fit (AGFI) and root mean square of error approximation (RMSEA) [52, 53]. The results of the fitness tests show that the values are within the desired range—$\chi 2$ = 398.056, degrees of freedom = 216, $\chi 2$/degrees of freedom = 1.843, RMSEA = 0.051, GFI = 0.912, and AGFI = 0.905 [54]. Though the values of GFI and AGFI are less than 0.90, their values are quite satisfactory. In Table 34.2, the SRWs of the manifests have been enlisted which show that their values range from 0.41 to 0.99 which depicts the feasibility of the estimated relationships among the latent and manifest variables. Reliability indices

Table 34.2 Constructs and indicators including reliability indices and path estimates

Constructs and manifests including reliability indices of constructs of both success factors and success measures	Standardized regression weights (SRWs)
1. Top management support (TMS)	–
[CR = 0.55; AVE = 0.43; α = 0.693]	
m1: Support and motivation from top management	0.49
m2: Commitment throughout development	0.74
m3: Frequency of annual meeting	0.49
m4: Delegation of top management	0.71
m5: Leadership by example	0.78
m6: Willingness of taking risk of NPD	0.51
m7: Support for entrepreneurial culture	0.80
m8: Planning and objective	0.69
2. Technology (T)	–
[CR = 0.67; AVE = 0.49; α = 0.759]	
m9: Investment to upgrade technological infrastructure	0.69
m10: Technology forecasting	0.48
m11: Usage of CAD/CAM	0.73
m12: Group technology	0.82
m13: Lean manufacturing	0.68
m14: Flexible manufacturing	0.42
m15: Design for manufacturability	0.45
m16: Enterprise resource planning	0.56
3. Learning (L)	–
[CR = 0.74; AVE = 0.56; α = 0.836]	
m17: Debriefing of all NPD experiences of NPD team members	0.42
m18: Adoption of well-defined process	0.96
m19: Adoption of NPD manuals to assist managerial decision making	0.95
m20: Collective review to assess the progress and performance	0.77
m21: Trend of attending in-house training program	0.88
m22: Providing on-the-job training to individuals	0.69
m23: Reporting about progress of NPD projects	0.73
m24: Maintenance of database containing factual information	0.83
m25: Maintenance of contact list of potential persons	0.86
m26: Development of guidelines to assist managerial decision making	0.98
m27: Frequent updating of NPD management guidelines or manuals	0.90
m28: Managers' participation in committees to expand knowledge	0.78
m29: Managers' attendance in meetings and seminars	0.64
m30: Managers' informal sharing and exchange information	0.89

(continued)

Table 34.2 (continued)

Constructs and manifests including reliability indices of constructs of both success factors and success measures	Standardized regression weights (SRWs)
4. Intellectual capital (IC)	–
[CR = 0.70; AVE = 0.52; α = 0.804]	
m31: Enrichment of human capital	0.52
m32: Enhancement of process capital	0.89
m33: Improvement of innovation capital	0.87
m34: Relational capital	0.44
5. Research & Development (R&D)	–
[CR = 0.69; AVE = 0.50; α = 0.792]	
m35: Number of R&D persons	0.64
m36: Experience (years) of the R&D team members	0.84
m37: Qualification of the R&D team members	0.75
m38: Investment in R&D infrastructure	0.84
m39: R&D management vision and direction	0.70
m40: Number of patents	0.93
m41: R&D-oriented culture	0.62
m42: Investment in cleaner technology research	0.41
m43: Adoption of methods for sustainable product development	0.56
m44: Scope and facility	0.99
6. Technological developments (TD)	–
[CR = 0.61; AVE = 0.48; α = 0.726]	
m35: Technological breakthrough	0.67
m36: Beating competition technologically	0.91
m37: Expanding product family	0.89
m38: Rate of product failure	0.85
m39: Frequency of product launching	0.56
m40: Reduction of risks	0.60

such as CR and AVE are mostly greater than 0.5 and range from 0.55 to 0.74 and 0.43 to 0.56, respectively. The α range from 0.693 to 0.792 which is not greater than 0.80 can be considered as reliable. These indices show the reliability of the collected data which can be used for further analysis.

The measurement model of SEM analysis validates the occurrence of manifest variables which correspond to their latent constructs. The manifest variables to quantify the latent constructs as mentioned above are identified from the available literatures as well as the experts' opinion. This study covers the importance of manifests in the respective organizations to measure their respective latent constructs, and at the same time, the degree of implementation of these measures in the practical field is also considered. First, the role of top management support to motivate technology, learning, and intellectual capital is identified. Top management support is a practice which is measured by motivation from senior management to develop new products (m1), commitment of senior managers throughout the development process (m2), frequency of annual meeting (m3), delegation of top managers for developing new products successfully (m4), efficient leadership provided by the senior managers (m5), willingness of taking risk of NPD (m6), support for developing the entrepreneurial culture in the organization to welcome new challenges by taking the risks (m7), and setting of well-defined plan according to the objective of the organization (m8). The loading values of top management support range from 0.49 to 0.80 showing the incidence of manifest variables to the construct, i.e., top management support. Now, top management support influences technology for controlling R&D activities. Investment to upgrade technological infrastructure (m9), practice of technology forecasting (m10), application of computer-aided design and computer-aided manufacturing (m11), adoption of group technology (m12), application of lean manufacturing (m13), adoption of flexible manufacturing systems (m14), design for manufacturability (m15), and adoption of customized enterprise resource planning measure the technology to successfully motivate the R&D practices in the organization. The values of measures show range from 0.42 to 0.82 which are positive. This depicts the fact of occurrence of the above-identified measures to quantify technology construct. Learning is the another variable influenced by managerial support which is measured by debriefing of all NPD experiences of NPD team members (m17), adoption of well-defined process to guide the formation of any NPD project (m18), preparation of NPD manuals to assist managerial decision making (m19), collective review to assess the progress and performance of NPD projects (m20), trend of attending in-house training program (m21), opportunities for on-the-job training to individuals who are relatively new to managing NPD activities (m22), reporting about progress of NPD projects (m23), maintenance of database containing factual information on each of its NPD projects (m24), maintenance of contact list of potential persons (m25), development of checklists or guidelines to assist managerial decision making and actions while managing NPD activities (m26), frequent updating of NPD management guidelines or manuals (m27), managers' participation in committees to expand knowledge (m28), their attendance in meetings, seminars, or retreats to exchange NPD-related information and experiences (m29), and informal sharing and exchange information among them (m30). The values of factor loadings of these mentioned measures are positive and range from 0.42 to 0.98 showing the contribution of these manifests to measure the learning. Another construct to escalate the R&D activities is intellectual capital. Enrichment of human capital (m31), process capital (m32), innovation capital (m33), and relational capital (m34) develops the intellectual capital of the organization valued from 0.44 to 0.89 to enrich the R&D activities essential for NPD. R&D practice is escalated by managerial support which is cascaded from the above-mentioned variables. Number of R&D persons (m35), their experiences (m36), their qualification (m37), investment in R&D infrastructure (m38), vision and direction of R&D management (m39), number of patents (m40), R&D-oriented culture in the organization encouraging the innovative ideas (m41), investment in cleaner technology research (m42), adoption of methods for sustainable product development (m43), and available scopes and facilities in the organization to develop the R&D culture (m4) are used to quantify the R&D culture in the practical field of product development. The occurrence of these measures is ranging from 0.41 to 0.99. The outcome performance of NPD success is measured by technological developments which is again quantified by technological breakthrough (m45), attaining the success through achieving strong technological characteristics to beat the competitors (m46), expansion of product family (m47), rate of product failure (m48), frequency of product launching (m49), and reduction of risks (m50). The regression weights of outcome performance range from 0.56 to 0.91 showing the positive values and can be considered for further analysis. These SRWs are calculated by SEM using IBM SPSS AMOS 21.0 that has been listed in Table 34.2.

34.4.3 Structural Model Results and Discussion

The next step of formation of measurement model is the development of structural model. In case of structural model, also the previously mentioned fitness tests have been performed. The values of the parameters show the appropriate model-to-data fit as $\chi^2 = 425.901$, degrees of freedom = 235, χ^2/degrees of freedom = 1.812, RMSEA = 0.047, GFI = 0.921, and AGFI = 0.915 [54]. This structural model defines the role of top management support to control the R&D activities which in turn enhance the technological improvements of the company and the NPD success as well. IBM SPSS AMOS 21.0 software has been used to develop the structural model shown in Fig. 34.2. The value of path estimates among the constructs range from 0.41 to 0.98. These depict that the assumed hypotheses are supported which states that top management support motivates R&D activities linked through technology, learning and intellectual capital which positively influence NPD success of the firm measured by technological developments as enlisted in Table 34.3.

Interrelationships among the latent constructs are depicted by the structural model developed by using SEM approach by IBM SPSS AMOS 21.0. The interrelationships have been established by testing set hypotheses depicting the relation between two constructs. The hypotheses among the constructs have been set from the available theoretical background. The correlation among the constructs has been analyzed by number of statistical analyses. Multiple regression analysis calculates the path values among the constructs, and the values are listed in Table 34.3. These values show the interrelationship of the constructs such as if the value is positive that portrays the significance of that path between two constructs that means the constructs are interrelated, whereas negative value shows the insignificance. In this developed structural model, top management support relates to technology, learning, and intellectual capital by 0.98, 0.56, and 0.69, respectively. These constructs again are connected with R&D activities showing the path values as 0.81, 0.78, and 0.41, respectively. The path value between R&D practice and NPD success which

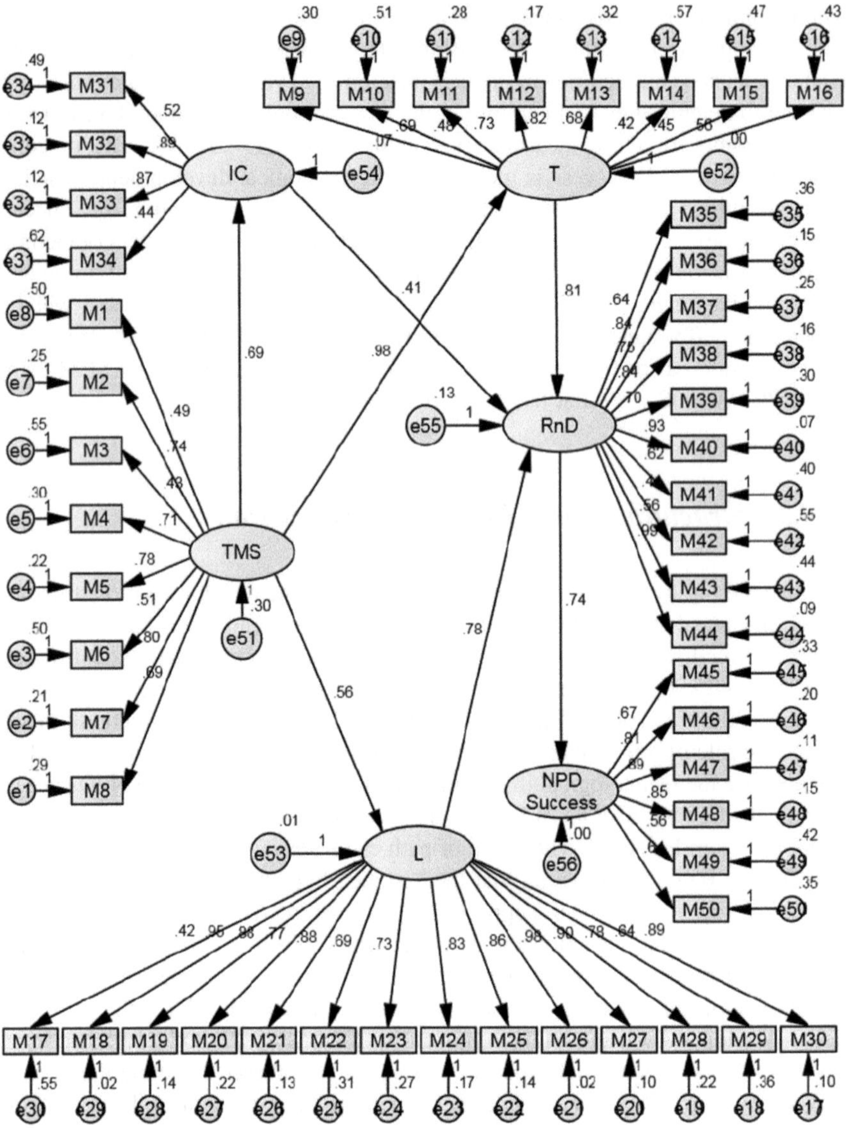

Fig. 34.2 Structural equation modeling (SEM) model interrelating latent constructs and their measures after execution

in turn be measured by technological developments is estimated by the value as 0.74. All the estimated values are positive and quite high which depicts that the assumed hypotheses are correct. Finally, the ultimate objective of the firm to achieve NPD success through technological developments by R&D culture is achieved.

Table 34.3 Statistics of path estimates

Path description	Hypotheses	Estimates
TMS → T	H1a	0.98
TMS → L	H1b	0.56
TMS → IC	H1c	0.69
T → R&D	H2	0.81
L → R&D	H3	0.78
IC → R&D	H4	0.41
R&D → TD	H5	0.74

Figure 34.2 represents a comprehensive framework of latent constructs along with their manifest variables combining both measurement model and structural model together. This model also includes the manifest variables along with their latent constructs. The measurement model is the measurement of constructs by quantifying their values which are shown by linking among them. The connection of latent constructs with their measures is calculated by regression weights as mentioned in Table 34.2. The structural model is the representation of the inter-relationship among the latent constructs measured by the path estimates as listed in Table 34.3. The figure consists of standardized error terms along with each manifest variable showing the difference of the actual value and the desired value. Similarly, the errors corresponding to the each latent constructs are also calculated. The direction of arrows in between the latent constructs is highly significant as this shows that which construct has the impact on which construct. For example, the direction of arrow in between top management support and technology is from top management support toward technology which means top management support escalates technology to achieve NPD success. In case of measurement model, the direction of arrow is from construct to the indicators. This means that these measures are used to quantify that particular construct. Finally, the model depicts the role of managerial support for successful adoption of R&D activities through technology, learning, and intellectual capital for achieving NPD success.

34.5 Conclusions

34.5.1 Theoretical Implications

This empirical research is a novel approach to add the quintessence of R&D in the organization for NPD success by providing technological developments in Indian manufacturing industries. First, antecedents of R&D operations and practices for technological developments are identified. Second, the impact of top management and its related factors to improve R&D culture is recognized. The indicators of these constructs are documented to quantify these constituents for developing the detailed framework. Third, the outcome performance of the organization is measured by technological developments providing the platform to achieve competitive advantages making the organization superior from its competitors which in turn intensifies NPD success in challenging market environment.

34.5.2 Managerial Implications

The interpretations of this study designate the essentiality of NPD practice for attaining success and survival in the global competition. The consumer behavior is the most uncertain thing that always should be kept in track to achieve the target. R&D operations and practices are the most dynamic practice to fulfill the customer expectations which help to enrich NPD performance in continuous basis. The findings of this study draw substantial managerial implications extremely valuable for future developments.

First, R&D activities of the firm are a very crucial practice every organization should be aware of to perform NPD on regular basis. The requirement of this culture for all size of organizations is already an established phenomenon. Despite of its vitality, R&D is an overlooked practice in SMEs due to dearth of innovative mentalities developed from the chance of failure. This study illustrates the way of developing the R&D practice with the thorough support and motivation from top management. The commitment of higher authority throughout the development process makes this approach a successful one. Moreover, frequent annual meetings, delegation of top management, and strong leadership make the R&D activities feasible in practical scenario. Moreover, the urge of taking risks of NPD avoiding the chances of failure builds the R&D practice in the organization. This risk-taking capability is the result of entrepreneurial culture which can be obtained through proper planning and objectives of senior management. There are certain constituents desperately vital for R&D practice development which requires constant support from management bodies. Technology is one of them which needs to be taken care of for continuous implementing the creative ideas developed from research activities. Investment of funds for upgrading technological infrastructure helps in providing updated technologies for product development process. Implementation of technology forecasting makes the organization aware of the difficulties in future days. Moreover, usage of computer-aided designs, manufacturing various state-of-the-art technologies, group technology, lean manufacturing, flexible manufacturing, design for manufacturability, and enterprise resource planning help to achieve high-tech facilities in the organization for successful product development process. Not only the available technologies, but also R&D requires proper learning for fruitful adoption of these technologies in the practical field. Learning is another practice essential for R&D which includes debriefing of all NPD experiences of NPD ream members, both prior and current which helps in framing the future developments. It develops culture of adoption of well-defined process which helps in guiding the configuration of any NPD project. Learning can be more structured through adoption of NPD manuals to assist managerial decision making while forming or managing NPD activities. The collective review to assess the progress and performance of NPD projects along with reporting of progress provides the learning activities a concrete shape. Moreover, trend of attending externally conducted training programs and practical training to individuals who are new in managing NPD activities develops skills of NPD team members. Maintenance of database containing factual information on each of its NPD projects such as data and purpose of the NPD project, names of NPD team managers, maintenance of contact list of persons (insiders/outsiders) who have potential to assist NPD activities helps in continuing the R&D operations and practices with due from the senior management. Development of checklists or guidelines to assist managerial decision making and actions while managing NPD activities along with the frequent updating of these guidelines directs the organization to develop the products in a well-defined manner. As the role of managerial support is inevitable to build the learning culture a successful practice, managers' participation in committees to expand knowledge, their attendance in meetings, seminars, or retreats to exchange NPD-related information and experiences, also the informal sharing and exchange of NPD-related information, and know-how with peers or colleagues make NPD activities a successful process. The intellectual capital is an asset of the organization which boosts up the R&D practice to a different level ensuring the success of the organization. Enrichment of human capitals such as competence, experience, knowledge, skills, attitude, commitment, and wisdom of employees and managers, enhancement of process capital by workflow, operation processes, specific methods, business development plans, information technology systems, and cooperative culture, improvement of innovation capital by intellectual properties such as patents, copyrights, trademarks, and know-how, augmentation of level of relational capital which includes all value of stakeholders, customers, and supplier relations enrich the R&D practice by providing the intellectual capital which makes the difference in performance from the competitors.

Second, the R&D operations and practices get a motion from appropriate managerial support and motivation escalated through constituents like technology, learning, and intellectual capital improving the quality of R&D for successful development of new products in an incessant manner. Availability of R&D department and concerned team is the first and foremost criterion for structuring the creative ideas to attain the NPD target. This practice requires sufficient number of R&D persons having experiences in NPD teams. The qualification of R&D team members is also a matter of concern for selecting the R&D team members. This team should have a well-defined vision and direction crucial for success. The number of patents, scope and facility, and R&D-oriented culture makes the development quality more enriched through continuous enhancement of NPD performance by research activities. The most vital thing above all of these manifests is investments in R&D infrastructure, cleaner technology research, and adoption of methods for sustainable product development which make the research activities feasible to shape the innovative ideas.

Finally, the primary objective of the firm is the achievement of NPD success to attain the success and survival of the organization. R&D is an activity providing competitive advantage to the firm to acquire better performance helping to develop products innovative as well as superior than their competitors. This study highlights the technological developments of the developed products which have been considered as the performance outcome of NPD success. The R&D operations and practices boost the technological breakthrough which helps in beating the competitors technologically. It also expands the product family providing the varieties of products. R&D practice increases the frequency of product launching producing creative ideas. It also reduces the rate of product failure and also lowers the risk of development as well.

The structural model developed in this study shows the importance of R&D operations and practices for successful NPD escalated from managerial support to attain required constituents of R&D activities such as technology, learning, and intellectual capital which helps in developing new products with high technological aspects. This practice helps to achieve the success as well as endure in the competitive market environment overcoming the obstacles.

34.5.3 Limitations

Limitation is a part of every research which cannot be avoided. This research also contains certain limitations which provide the scope of further improvement. The major limitation of this research is the large number of responses required for analysis. Collection of data by convincing the industrial experts is quite a tough and laborious task to be performed. Experts are sometimes not interested to share their company's information which makes the job more difficult. Another limitation is the question of reliability of the obtained data. Recognition of untrustworthy data and removal of those data are the required tasks to be done before using them for the future analysis.

Acknowledgements The research work was substantially supported by a grant from the Department of Science and Technology (DST) of Government of India as a DST INSPIRE Fellowship to carry out the doctoral research of the corresponding author. The authors are also thankful to the industry personnel and experts from manufacturing industries from India who shared their views and opinions.

Appendix I

- Interview protocol with description of latent variables

1. *Top Management Support (TMS):*

    - Support and motivation from top management (m1): Strong support and continuous motivation from senior management to develop new products,

- Commitment throughout development (m2): Commitment of senior managers throughout the development process to get the job done perfectly as required,
- Frequency of annual meeting (m3): Frequent arrangement of meetings to continue the development process uninterruptedly,
- Delegation of top management (m4): Delegation of top managers for developing new products successfully
- Leadership by example (m5): Efficient leadership provided by the senior managers,
- Willingness of taking risk of NPD (m6): Risk-taking attitude of senior management for NPD,
- Support for entrepreneurial culture (m7): Support for developing the entrepreneurial culture in the organization to welcome new challenges by taking the risks,
- Planning and objective (m8): Setting the well-defined plan according to the objective of the organization.

2. *Technology (T):*

- Investment to upgrade technological infrastructure (m9): Percentage of funds invested for upgrading of technological infrastructure to make the development process high-tech,
- Technology forecasting (m10): Implementation of technology forecasting to trace the difficulties,
- Usage of CAD/CAM (m11): Usage of CAD/CAM and various state-of-the-art technologies and newly launched technologies,
- Group technology (m12): Degree of implementation of Group technology/ Cellular manufacturing,
- Lean manufacturing (m13): Implementation of lean manufacturing practices,
- Flexible manufacturing (m14): Presence of Flexible manufacturing system,
- Design for manufacturability (m15): Degree of adoption of Design for manufacturability and assembly (DFMA),
- Enterprise resource planning (m16): Use of enterprise systems (ERP) for product data management.

3. *Learning (L):*

- Debriefing of all NPD experiences of NPD team members (m17): Tradition of debriefing of prior and/or current NPD experiences of NPD team members in your company,
- Adoption of well-defined process (m18): Adoption of well-defined process to guide the formation of any NPD project,
- Adoption of NPD manuals to assist managerial decision making (m19): Adoption of NPD manuals to assist managerial decision making and/or actions while forming or managing NPD activities,

- Collective review to assess the progress and performance (m20): Collective review to assess the progress and performance of NPD projects of your company,
- Trend of attending in-house training program (m21): Managers' attendance in meetings, seminars, or retreats to exchange NPD-related information and experiences,
- Providing on-the-job training to individuals (m22): Opportunities for on-the-job training to individuals who are relatively new to managing NPD activities,
- Reporting about progress of NPD projects (m23): Reporting about progress and performance of NPD projects,
- Maintenance of database containing factual information (m24): Maintenance of database containing factual information on each of its NPD projects (such as data and purpose of the NPD project, names of NPD team managers),
- Maintenance of contact list of potential persons (m25): Maintenance of contact list of persons (insiders/outsiders) who have potential to assist NPD activities,
- Development of guidelines to assist managerial decision making (m26): Development of checklists or guidelines to assist managerial decision making and actions while managing NPD activities,
- Frequent updating of NPD management guidelines or manuals (m27): Frequent updating of NPD management guidelines or manuals,
- Managers' participation in committees to expand knowledge (m28): Managers' participation in forums such as committees or task forces to expand knowledge,
- Managers' attendance in meetings and seminars (m29): Managers' attendance in meetings, seminars, or retreats to exchange NPD-related information and experiences,
- Managers' informal sharing and exchange information (m30): Managers' informal sharing and exchange of NPD-related information and know-how with peers or colleagues.

4. *Intellectual Capital (IC):*

   - Enrichment of human capital: Enrichment of company's human capital by employees' and managers' competence, experience, knowledge, skills, attitude, commitment, and wisdom,
   - Enhancement of process capital: Enhancement of company's process capital by workflow, operation processes, specific methods, business development plans, information technology systems, and cooperative culture,
   - Improvement of innovation capital: Improvement of company's innovation capital by intellectual properties such as patents, copyrights, trademarks, and know-how,
   - Relational capital: Level of relational capital which includes all value of stakeholders, customers, and supplier relations.

5. *Research & Development (R&D):*

- Number of R&D persons (m35): Number of R&D persons associated with the R&D activities,
- Experience of the R&D team members (m36): Experience (number of years) of R&D team members having skills in framing innovative ideas to a concrete shape which helps in the new product development,
- Qualification of the R&D team members (m37): Qualification of the R&D team members which affects the continuous flow of the product development,
- Investment in R&D infrastructure (m38): Amount of funds invested to develop the R&D infrastructure,
- R&D management vision and direction (m39): Concrete vision and direction of R&D management to develop new products,
- Number of patents (m40): Number of patents of the organization,
- R&D-oriented culture (m41): R&D-oriented culture in the organization encouraging the innovative ideas,
- Investment in cleaner technology research (m42): Investment to adopt the cleaner technology in the organization,
- Adoption of methods for sustainable product development (m43): Various types of methods are adopted for sustainable product development,
- Scope and facility (m44): Available scopes and facilities in the organization to develop the R&D culture.

6. *Technological developments (TD):*

- Technological breakthrough (m45): Achievement of technological breakthrough,
- Beating competition technologically (m46): Attainment of the success though achieving strong technological characteristics to beat the competitors,
- Expanding product family (m47): Expansion of product family through developing various kinds of products by R&D activities,
- Rate of product failure (m48): Fail to develop the products as per requirements,
- Frequency of product launching (m49): Frequent launch of the new products,
- Reduction of risks (m50): Reduction the risks associated with NPD.

REFERENCES

1. Tzokas, N., Hultink, E.J., Hart, S.: Navigating the new product development process. Ind. Mark. Manage. **33**, 619–626 (2004)
2. Lau, A.K.W.: Critical success factors in managing modular production design: six company case studies in Hong Kong, China, and Singapore. J. Eng. Technol. Manage. **28**, 168–183 (2011)

3. Buyukozkan, G., Arsenyan, J.: Collaborative product development: a literature overview. Prod. Plann. Control **23**, 47–66 (2012)
4. Sarja, J.: Explanatory definitions of the technology push success factors. J. Technol. Manage. Innov. **10**(1), 204–214 (2015)
5. Ernst, H.: Success factors of new product development: a review of the empirical literature. Int. J. Manage. Rev. **4**, 1–40 (2002)
6. Bhuiyan, N.: A framework for successful new product development. Int. J. Ind. Eng. Manage. **4**, 746–770 (2011)
7. Cooper, R.G., Kleinschmidt, E.J.: Success factors in product innovation. Ind. Mark. Manage. **16**, 215–223 (1987)
8. Sadeghi, A., Azar, A., Rad, R.S.: Developing a fuzzy group AHP model for prioritizing the factors affecting success of high-tech SME's in Iran: a case study. Procedia-Soc. Behav. Sci. **62**, 957–961 (2012)
9. Tsai, C.C.: A research on selecting criteria for new green product development project: taking Taiwan consumer electronics products as an example. J. Clean. Prod. **25**, 106–115 (2012)
10. Akgün, A.E., Byrne, J.C., Lynn, G.S., Keskin, H.: Team stressors, management support, and project and process outcomes in new product development projects. Technovation **27**, 628–639 (2007)
11. Sun, H., Wing, W.C.: Critical success factors for new product development in the Hong Kong toy industry. Technovation **25**, 293–303 (2005)
12. Ernst, H., Hoyer, W.D.H., Rübsaamen, C.: Sales, marketing, and research-and-development cooperation across new product development stages: implications for Success. J. Mark. **74**(5), 80–92 (2010)
13. Kawazoe, M., Abetti, P.A.: Transition of strategy, marketing, R&D and new product development policies after mergers and acquisitions: a case study of superpower inc. under US, Dutch and Japanese ownership. Int. J. Technol. Manage. **66**, 32–56 (2014)
14. Nicholas, J., Ledwith, A., Aloini, D., Martini, A., Nosella, A.: Searching for radical new product ideas: exploratory and confirmatory factor analysis for construct validation. Int. J. Technol. Manage. **68**, 70–98 (2015)
15. Felekoglu, B., Moultrie, J.: Top management involvement in new product development: a review and synthesis. J. Prod. Innov. Manag. **31**, 159–175 (2014)
16. Rahab, C., Sulistyandari, S., Sudjono, S.: The development of innovation capability of small enterprises through knowledge sharing process: an empirical study of Indonesian creative industry. Int. J. Bus. Soc. Sci. **2**, 112–123 (2011)
17. Mendes, G.H.D.S., Ganga, G.M.D.: Predicting success in product development: the application of principal component analysis to categorical data and binomial logistic regression. J. Technol. Manage. Innov. **8**, 83–97 (2013)
18. Samli, A.C., Weber, J.A.E.: A theory of successful product breakthrough management: learning from success. J. Prod. Brand Manage. **9**, 35–55 (2000)
19. Eder, P.: Expert inquiry on innovation options for cleaner production in the chemical industry. J. Clean. Prod. **11**, pp. 347e364 (2003)
20. Roy, M.J., The´rin, F.: Knowledge acquisition and environmental commitment in SMEs. Corp. Soc. Responsib. Environ. Manag. **15**, 249–259 (2008)
21. Brown, S.L., Eisenhardt, K.M.: Product development: past research, present findings, and future directions. Acad. Manag. Rev. **20**, 343–378 (1995)
22. Zhou, A.Z., Fink, D.: The intellectual capital web: a systematic linking of intellectual capital and knowledge management. J. Intellect. Capital **4**, 34–48 (2003)
23. Hsu, Y.H., Fang, W.: Intellectual capital and new product development performance: The mediating role of organizational learning capability. Technol. Forecast. Soc. Chang. **76**, 664–677 (2009)
24. Chen, Y.S., Lin, M.J.J., Chang, C.H.: The influence of intellectual capital on new product development performance—the manufacturing companies of Taiwan as an example. Total Qual. Manag. **17**, 1323–1339 (2006)

25. Huang, X., Soutar, G.N., Brown, A.: Measuring new product success: an empirical investigation of Australian SMEs. Ind. Mark. Manage. **33**, 117–123 (2004)
26. Hart, S.: Dimensions of success in new product development: an exploratory investigation. J. Mark. Manage. **9**, 23–41 (1993)
27. Eid, R.: Factors affecting the success of world class manufacturing implementation in less developed countries: the case of Egypt. J. Manuf. Technol. Manage. **20**, 989–1008 (2009)
28. Gomes, J., Weerd-Nederhof, P., Pearson, A., Fissher, O.: Senior management support in new product development process. Creativity Innov. Manage. **10**, 234–242 (2001)
29. Rodríguez, N.G., Pérez, M.J.S., Gutiérrez, J.A.T.: Can a good organizational climate compensate for a lack of top management commitment to new product development? J. Bus. Res. **61**, 118–131 (2008)
30. Swink, M.: Technological innovativeness as a moderator of new product design integration and top management support. J. Prod. Innov. Manag. **17**, 208–220 (2000)
31. Halman, J.I.M., Hofer, A., vanVuuren, W.: Platform-driven development of product families: linking theory with practice. J. Prod. Innov. Manag. **20**, 149–162 (2003)
32. Haverila, M.J.: Product–firm compatibility in new product development in technology companies. J. High Technol. Manage. Res. **23**, 130–141 (2012)
33. Chiang, Y.H., Shih, H.A.: Knowledge-oriented human resource configurations, the new product development learning process, and perceived new product performance. Int. J. Hum. Res. Manage. **22**, 3202–3221 (2011)
34. Martín-Rojas, R., García-Morales, V.J., García-Sánchez, E.: The influence on corporate entrepreneurship of technological variables. Ind. Manage. & Data Syst. **111**, 984–1005 (2011)
35. Reilly, R.R., Chen, J.Y., Lynn, G.S.: Power and empowerment: the role of top management support and team empowerment in new product development. In: PICMET'03 Conference on Technology Management for Reshaping the World Proceedings,pp. 282–89 (2003)
36. Bao, G.M., Yang, J.:Dynamic competences and technological innovation in Chinese enterprises. In: Engineering Management Conference Proceedings,IEEE International, 1, pp. 454–458 (2004)
37. Johnson, W.H.A., Filippini, R.: Integration capabilities as mediator of product development practices–performance. J. Eng. Tech. Manage. **30**, 95–111 (2013)
38. Medeiros, J.F.D., Ribeiro, J.L.D., Cortimiglia, M.N.: Success factors for environmentally sustainable product innovation: a systematic literature review. J. Clean. Prod. **30**, 1e11 (2014)
39. Bolívar-Ramos, M.T., García-Morales, V.J., Mihi-Ramírez, A.: Influence of technological distinctive competencies and organizational learning on organizational innovation to improve organizational performance. Econ. Manage. **16**, 670–675 (2011)
40. Lin, H.F.: Knowledge sharing and firm innovation capability: An empirical study. International Journal of Manpower **28**, 315–332 (2007)
41. Unger, B.N., Kock, A., Gemünden, H.G., Jonas, D.: Enforcing strategic fit of project portfolios by project termination: An empirical study on senior management involvement. Int. J. Project Manage. **30**, 675–685 (2012)
42. Kleinschmidt, E.J., De Brentani, U., Salomo, S.: Performance of global new product development programs: A resource-based view. J. Prod. Innov. Manag. **24**, 419–441 (2007)
43. Jantunen, A.: Knowledge processing capability and innovative performance: an empirical study. European J. Innovation Manage. **8**, 336–349 (2005)
44. Yang, E., Ma, G., Chu, J.: The impact of financial constraints on firm R&D investments: empirical evidence from China. Int. J. Technol. Manage. **65**, 172–188 (2014)
45. Kim, B., Kim, J.: Structural factors of NPD (new product development) team for manufacturability. Int. J. Project Manage. **27**, 690–702 (2009)
46. Chang, P.L., Chen, K.L.: 'The influence of input factors on new leading product development projects in Taiwan. Int. J. Project Manage. **22**, 415–423 (2004)
47. Hoyle, R.H.: The structural equation modelling approach: basic concepts and fundamental issues. In: Hoyle, R.H. (ed.) Structural equation modelling: Concepts, Issues, and Applications, pp. 1–15. Sage Publications Inc., Thousand Oaks, CA (1995)

48. Rigdon, E.E.: Structural Equation Modeling. In: Marcoulides, G.A. (ed.) Modern Methods for Business Research, pp. 251–294. Lawrence Erlbaum Associates Publishers, Mahwah, NJ (1998)
49. Ong, C.S., Lai, J.Y., Wang, Y.S.: Factors affecting engineers' acceptance of asynchronous e-learning systems in high-tech companies. Inf. Manag. **41**, 795–804 (2004)
50. Nunnally, J.C.: Psychometric Theory. McGraw Hill, New York (1978)
51. Hu, L., Bentler, P.M.: Fit indices in covariance structure modeling: sensitivity to under parameterized model misspecification. Psychol. Methods **3**, 424–453 (1998)
52. Bentler, P.M.: Comparative fit indexes in structural models. Psychol. Bull. **107**(2), 238 (1990)
53. Chen, H.C.: The impact of children's physical fitness on peer relations and self-esteem in school settings. Child Indic. Res. pp. 1–16 (2015)
54. Hair, J.F., Anderson, R.E., Tatham, R.L., Black, W.C.: Multivariate Data Analysis with Readings. Prentice Hall, Englewood Cliffs, NJ (1995)